NASA SP-4005

ASTRONAUTICS AND AERONAUTICS, 1964

Chronology on Science, Technology, and Policy

Prepared by the NASA Historical Staff,
Office of Policy Planning

Scientific and Technical Information Division   1 9 6 5
NATIONAL AERONAUTICS AND SPACE ADMINISTRATION
Washington, D.C.

Foreword

The year 1964, the seventh of the space age, was a year of achievement, especially in space science. It was also a building year in which many components and techniques basic to manned flight capability were proved out. Because of the foundations laid in 1964, we can proceed with more sureness toward the long-duration manned missions, culminating with the moon landing, scheduled for the coming years.

The dramatic flight of RANGER VII leads the list of accomplishments. Before crashing into the moon, the satellite snapped more than 4,000 photographs of its surface. Many of the pictures were of such high resolution that they amounted to a thousand-fold improvement over the best taken by earth-mounted telescopes. MARINER IV is well started on its long journey to Mars. Early in its flight, it was ordered to perform an intricate midcourse maneuver. The spacecraft followed its instructions to the letter; and barring accident it will pass within 5,400 miles of its objective as planned.

NIMBUS I, the first of the second generation of meteorological satellites, functioned long enough to demonstrate that its TV cameras and infrared scanners could provide day and night weather coverage far superior to that supplied by its predecessors.

The data produced by these and the numerous other investigations of space noted in this chronology were sent to more than 2,300 scientists and engineers in more than a hundred university laboratories and space centers for study and analysis. These researchers are thus obtaining a broader and deeper scientific and technical base for their future activities. Scientists and engineers not in space-related work benefit from the creation of a facility that permits storage and retrieval of data resulting from space experiments.

The buildup of the national capability in advanced research and technology in the area of aeronautics and astronautics has been more rapid than in any previous period.

As this chronology shows, many of the basic components and techniques required for manned flight were proved out during 1964. The seventh successful test launch of the Saturn I booster paved the way to the next phase of the Apollo project. More particularly, it assured that this booster's more powerful successors, the Saturn I and the Saturn V, can be relied upon to discharge their appointed roles in the lunar landing program.

Centaur passed the half-way mark on its way to operational status. Along with the other test activities, the Centaur flights proved that the use of liquid hydrogen as an engine fuel had been mastered.

The unmanned Gemini capsule testing was completed except for a final ballistic shot to determine the ability of the heat shield to withstand the temperatures generated by high-speed reentry. The way was thus opened to proceed in 1965 with manned flights, to practice rendezvous and docking, and to undertake long-duration orbital missions. Such missions will provide essential information on the effect of weightlessness and other aspects of the space environment on the human organism. Additionally, the checkout of many of the components, stages, and subsystems that will go into the Saturns, the Apollo capsule, and the LEM was finished.

The construction of new ground facilities for housing, handling, and launching the giant boosters to come was well advanced. Progress was made in expanding Cape Kennedy to meet the future needs of the program.

During 1964, good headway was made in focusing the resources needed for the execution of the Apollo program and other manned missions. Some 1,600 prime contractors and thousands of subcontractors went forward with the design, production, and test of a variety of equipment made for ground testing. NASA's comprehensive system of ground-testing was largely completed. In this system, many aspects of the space environment are simulated on the ground, giving us a capability to subject the equipment to conditions that will be encountered in flight and thus to uncover difficulties and find solutions for them before flight testing begins.

This has a double significance. In the coming months, equipment will be coming off the production line in greatly increased volume, imposing a heavy requirement for testing which we will be able to meet. Secondly, the problems that the simulated environmental testing uncovers, the need for alterations or improvements that are revealed, can be quickly fed back to the production lines so that what has been learned can be promptly applied to equipment in the process of fabrication.

As a result, when the flight-test stage is reached we have greater assurance that the flight equipment delivered by the contractors will perform according to specifications and that we can execute the missions that have been scheduled. Our experience with ground simulation of the space environment has also supplied information which has enabled us to upgrade equipment so that it will outperform its design specifications. Because of this system of ground testing, the thrust of some of the new boosters has been substantially increased.

Improved management techniques which enhance the effectiveness of the all-up system wherein the complete unit is tested at once have been put into effect. Configuration control has been tightened to ensure that systems and subsystems are compatible and that design requirements are met.

These and other significant improvements were made in NASA's overall management system for the production and flight of very large and very complex rockets and spacecraft. In many ways this system is now further advanced than any previously developed on an equivalent scale.

The sum of this chronology is an encouraging increase in our space capabilities and experience. The developments cited herein are proof that the space effort is moving forward on a broad front at a steadily accelerating pace. These facts give assurance that the National requirement of preeminence in space will be met.

ROBERT C. SEAMANS, JR.,
Associate Administrator.

Contents

	PAGE
FOREWORD	III
Associate Administrator Robert C. Seamans, Jr.	
PREFACE	VII
JANUARY	1
FEBRUARY	40
MARCH	90
APRIL	124
MAY	162
JUNE	199
JULY	234
AUGUST	271
SEPTEMBER	301
OCTOBER	335
NOVEMBER	373
DECEMBER	403
APPENDIX A: Satellites, Space Probes, and Manned Space Flights, 1964	445
APPENDIX B: A Chronology of Major NASA Launchings, 1964	481
APPENDIX C: Abbreviations of References	489
INDEX	491

Preface

This chronology is a first step in the historical process of documenting the dynamic and complex undertaking of space exploration and exploitation. It was prepared from open public sources to provide a ready reference for current use as well as for future analysts and historians. This chronicle of the seventh year of the Space Age also will assist other scholars, students, and writers. As the years pass perhaps its value may actually increase.

Previous volumes in this series may be of interest to new readers. They are:

* *Aeronautics and Astronautics, 1915–1960* (Washington: NASA, 1961, available from Supt. of Documents, GPO, $1.75).
* *Aeronautical and Astronautical Events of 1961* (published by the House Committee on Science and Astronautics, June 1963, available from Supt. of Documents, GPO, $1.50).
* *Astronautics and Aeronautics, 1963* (Washington: NASA, 1964, SP-4004, available from Supt. of Documents, GPO, $2.00).

In the preparation of *Astronautics and Aeronautics, 1964*, Dr. Frank W. Anderson, Assistant NASA Historian, carried major responsibility. Mrs. Helen T. Wells drafted the bulk of the text and also prepared the list of abbreviations. In the last two months of the year, the Science and Technology Division of the Library of Congress took over responsibility for the text. Miss Nancy Ebert (ATSP) prepared the index, while Miss Sara Corbett was responsible for preparation of the copy, ably assisted by Dema Nappier (AFEE–1). Lloyd Robbins and Creston Whiting (ATSS–T) continued to provide timely translations of Russian materials. Center historians and historical monitors were helpful in validating material as well as providing items. Without the comments of busy scientific and technical personnel, this volume would have been much the poorer.

Appendix A, "Satellites, Space Probes, and Manned Space Flights —1964," was prepared by Dr. Anderson, as was Appendix B, "Major NASA Launchings, 1964." As historical documentation and analysis are dynamic processes, comments and criticism are welcomed at any time.

EUGENE M. EMME
NASA Historian (APPH)
Office of Policy Planning

January 1964

January 1: NASA position paper on DOD's Manned Orbiting Laboratory, embodying eight principles of NASA policy toward the MOL, was made public. The paper stated:

"The DOD Manned Orbital Laboratory is a single military project (a specific goal; not a broad program) within the overall National Space Program. . . .

"The MOL should not be construed as the national space station, a separate program currently under joint study by NASA and the DOD. Such a space station project would be considered under the terms of the Webb/McNamara agreement on manned orbital research and development systems larger than Gemini and Apollo. The MOL is, rather, a specific experimental test bed utilizing NASA's Gemini project and the Titan III for certain potential military space applications not within the scope of NASA's activities. . . ." (Text; Simons, *Wash. Post,* 1/2/64, 1; *A&A,* 2/64, 90; *AF* Mag., 74)

- Head designer of Soviet spacecraft, name not identified, was interviewed in *Izvestia.* His predictions for the next five years: ". . . permanently operating orbital stations consisting of sputniks having various purposes such as providing service in investigating the Earth, the Sun, forecasting weather, monitoring the radiation conditions in space, etc.; . . . an all-purpose radio and TV system for transmitting through space, with the aid of the so-called 'hanging' ('stationary') satellites; . . . ways of using space . . . for mail-freight deliveries, requiring high speed, and finally passenger service; . . . permanent orbiting stations [where] regular shifts of scientific workers of the most diversified specialties will live and work. . . . In general, during this period flights into near space and return to Earth will become very commonplace. I think that in the next five years, many comrades will sojourn in space."

  Regarding "direct investigations in the region of the Moon and on its surface," the designer said: "In all likelihood, in the imminent years such studies can be accomplished only with the aid of small automatic devices transmitting the data to Earth by radio." (*Izvestia,* 1/1/64, 1, ATSS-T Trans.)

- Two mathematicians at Lawrence Radiation Laboratory of Univ. of Calif. achieved first accurate description of the orbit of Mars, Univ. of Calif. announced. The mathematicians, Joseph L. Brady and Edna Vienop, calculated the planet's orbit with aid of computers used to develop nuclear weapons. Brady said calculation left Mercury as the solar system's only planet without accurate description of its orbit. (AP, *Wash. Post,* 1/2/64, A3)

- Beginning of cable circuit connecting New York and Offenbach, West Germany, announced by Dr. Robert M. White, Chief of U.S. Weather Bureau, for exchange of meteorological information between North America and Europe. Inauguration of cable circuit, expected to yield improved reception of weather data because of freedom from radio propagation occurring on radioteletypewriter circuit, represented major achievement in global exchange of weather information. International meteorological community now was planning for future exchanges using high-speed computers and communications satellites. (Commerce Dept. Release WB 64-1)

January 1: Fiftieth anniversary of first scheduled airline flight, from St. Petersburg to Tampa, Fla., by Benoist flying boat. Plane made two round-trip flights daily, six days per week across Tampa Bay. 18-mi. flights took 2 hrs. by steamer, 6 hrs. by car and 12 hrs. by rail. During 50th year (1963), U.S. scheduled airlines carried 71,418,000 revenue passengers 50,361,300,000 mi. (*Aerospace,* Summer 1964)

January 2: NASA Deputy Administrator Dr. Hugh L. Dryden, appearing on CBS radio program "Capitol Cloakroom," said that before U.S. and U.S.S.R. could cooperate in a manned space flight effort, the U.S.S.R. would have to radically change its secrecy policy in space exploration. Some changes might be required in U.S. space policies also. "To conduct cooperative manned flights," Dr. Dryden said, "I think you'd have to . . . start from the beginning with a joint team to consider all the problems together." Regarding U.S.-U.S.S.R. agreements reached on weather satellites, communications satellites, and magnetic field studies, Dr. Dryden commented that "implementation of the agreements has been proceeding rather slowly. . . . We hope they'll give us some clear indication as to whether they are ready to exchange cloud pictures in 1964, or not."

Regarding Project Apollo timetable, Dr. Dryden said that if Congress supported the proposed budget for the project it should be possible to land a man on the moon by 1970. However, two year's "slack" built into the schedule to allow for contingencies was lost by the FY 1964 budget reduction. (UPI, *N.Y. Her. Trib.,* 1/3/64; Wilson, *Minneapolis Morning Trib.,* 1/3/64)

• President Lyndon B. Johnson said in article for *The Journal of the National Education Association:*

"The nation that has the schools has the future. No less than our rocket thrusts into outer space, there is urgently needed a national education thrust of massive proportions that will help us master the problems of our age.

". . . figuratively and physically we are building not only a pathway to the stars but also—hopefully—bridges of understanding here on earth.

"It is here on earth that our task begins. And for that task, the education attainments of our fellow Americans are, in some categories, far less than they should be. . . ." (AP, *NYT,* 1/3/64, 21)

• Rep. George P. Miller (D.-Calif.), Chairman of House Committee on Science and Astronautics, announced agreement with National Academy of Sciences that the academy would serve as adviser to the Congress. In recent years NAS, created a century ago by act of Congress, has established active working relationship with Executive Branch. Today's agreement was to establish similar relationship with the Congress. Agreement was result of recommendation by Subcommittee on Science, Research and Development, chaired by Rep. Emilio Q. Daddario (D.-Conn.), which has been investigating relationship of science and Government. (Finney, *NYT*, 1/3/64, 9; Simons, *Wash. Post*, 1/3/64)

January 2: AFSC established Hq. National Range Division (NRD), Provisional, at Patrick AFB, Fla., with Maj. Gen. Leighton I. Davis (USAF), as commander. Replacing General Davis as AFMTC commander was Brig. Gen. Harry J. Sands, Jr., vice commander of AFMTC. At the same time, AFSC established Air Force Space Test Center, Provisional (AFSTC), at Vandenberg AFB, with Brig. Gen. Jewell C. Maxwell, commander of 659th Aerospace Test Wing (SSD) as commander. To become fully operational on July 1, Hq. NRD would coordinate activities of AMR (managed by AFMTC commander) and of that portion of PMR transferred from USN to USAF (managed by AFSTC commander). (AFSC Release 41–5–1; *SBD*, 1/7/64, 26; *M&R*, 1/13/64, 10)

- DOD announced additional $30,414,000 had been obligated to Martin Marietta Corp., Denver, on cost-plus-incentive-fee contract for design, development, and fabrication of Titan III space booster. (DOD Release 3–64)
- According to survey by the Foundation Library Center, U.S. philanthropic foundations have assets totaling $14.5 billion, increase of $3 billion since 1960. Foundations make grants at rate of $779 million per year, increase of more than $150 million since 1960. Education receives most in foundation grants, $315 million; sciences receive fifth largest amount, $86 million. (Hechinger, *NYT*, 1/3/64, 18)

January 3: RELAY I communications satellite continued transmitting despite an onboard timer set to turn it off after one year. Electrolytic solution was supposed to have eaten through main power lead to switch off transmission power. NASA and RCA experts speculated erosion was slowed because environmental temperatures were cooler than anticipated. Launched into orbit Dec. 13, 1962, RELAY I has been used for 2,000 communications experiments with more than 290 hours of transmission time, most of which was wideband transmission or TV. (NASA Release 64–1)

- AEC announced preliminary evaluation of mockup space reactor re-entry flight (RFD–1), conducted May 22, 1963, with a Scout booster, indicated that "space reactors can be designed to break apart and disintegrate when they re-enter the earth's atmosphere." During re-entry, telemetry recorded disassembly events and measured heating rates at critical points on the reactor "until blackout of radiofrequency occurred—just before disassembly of the reactor. The data on the temperature rises and the times of the disassembly events transmitted by the telemetry system were essentially as predicted." Test results were still being evaluated by Sandia Corp. and Atomics International in an effort to substantiate the conclusions inferred from optical tracking of re-entry. (AEC Release G–1)
- NASA Manned Spacecraft Center announced selection of RCA to build and install ultrahigh-vacuum chamber and associated equipment at MSC, the fixed-price contract totaling $245,000. (MSC Release 64–41)
- Seven noctilucent clouds visible in southwest U.S. in past six months probably were caused by PMR rocket launchings, according to Dr. Aden B. Meinel and Carolyn P. Meinel in *Science* magazine. The Meinels, astronomers at Univ. of Arizona's Stewart Observatory (Tucson), described cloud appearing Nov. 2 at mean altitude of 35 mi. By measuring speed and direction of cloud's drift, Meinels traced it back

to vicinity of PMR, just as they had traced 44-mi. altitude cloud of June 15. Dr. Meinel and colleagues had described the clouds in *Science* last September, tentatively linking the June 15 cloud to Scout launching at PMR. Now, despite fact that no firing was announced for Nov. 2, Meinels concluded that "the coincidence of the clouds of June 15 and 2 November 1963 with a drift vector indicating origin from the Pacific Missile Range would appear to remove the probability of a fortuitous occurrence that could be argued from one cloud alone." (*Science*, 1/3/64; *CSM*, 1/6/64)

January 4: 1,100 students at 131 colleges and universities would participate in NASA graduate training program during 1964–65, NASA announced. Students in space-related predoctoral fields would be selected by the participating universities. Each student would receive $2,400 stipend for twelve months, plus dependents' allowance up to $1,000. Program is administered by Grants and Research Contracts Div. of NASA Office of Space Science and Applications. (NASA Release 64-2)

January 5: Evidence obtained from instruments carried aloft by balloons supports the theory that supernovae are a source of electrons in cosmic rays reaching vicinity of earth, scientists reported in *Physical Review Letters*. Experiments revealed primary cosmic rays in earth's upper atmosphere contain average of three times as many electrons as positrons. The high-altitude balloon experiments, conducted last summer near Ft. Churchill, Manitoba, by physicists of Univ. of Chicago and Argonne National Laboratory, also demonstrated for first time the existence of positrons in primary cosmic rays. (*Wash. Post*, 1/6/64; Schmeck, *NYT*, 1/6/64, 47)

- Dr. Richard Head, formerly Associate Director in NASA Office of Plans and Program Evaluation, was appointed Deputy Director of Policy Planning Division, NASA Office of Technology Utilization and Policy Planning. Dr. Thomas P. Murphy, Assistant to Assistant Administrator for Technology Utilization and Policy Planning, was appointed Acting Director of Policy Planning Div. pending selection of permanent Director. (NASA Announcement 64-3)

January 6: Supersonic Transport Evaluation Group, formed under overall management of FAA, convened in Washington to begin its task of evaluating airframe and engine manufacturers' initial design proposals for SST. Its 210 members representing FAA, NASA, CAB, USAF, USN, and Dept. of Commerce, the group would complete evaluation of proposals by early March. Concurrently, 10 airlines would conduct independent evaluations of the proposals, and on March 25–26 joint Government-airline discussions would formally review the results of their respective evaluations. By May 1, FAA Administrator Najeeb E. Halaby would announce which direction the U.S. SST effort would take. (FAA Release 64-2)

- Dr. Edward C. Welsh, Executive Secretary and Acting Chairman, National Aeronautics and Space Council, said in *Space Business Daily* interview that major U.S. space accomplishments during 1963 were: decision to develop Manned Orbiting Laboratory (MOL), progress of Saturn launch vehicle, orbiting of Syncom communications satellites, L. Gordon Cooper's 22-orbit Project Mercury flight, the successful Atlas-Centaur flight test, and the military achievements in space. Dr. Welsh pointed out that in 1963 "70% of all U.S. orbital shots" were made by the military. (*SBD*, 1/6/64, 18–19)

January 6: NASA MSC announced Gemini test vehicle for training astronauts in recovery procedures had arrived at MSC, and water tests with the astronauts and recovery forces would be conducted in nearby Galveston Bay. (MSC Release 64-2)

- Lt. Col. Peter Romo, staff meteorologist of AFMTC, said his office was using photographs from TIROS VIII in its weather predictions for missile launchings from Cape Kennedy. APT (Automatic Picture Transmission) ground-receiving station was installed in mobile trailer at AFMTC for receiving the TIROS VIII photographs. (*M&R*, 1/6/64, 16)
- USAF F-104 squadron completed first nonstop tactical fighter deployment from U.S. West Coast to Europe, flying from George AFB, Calif., 6,150 mi. to Moron Air Base, Spain, in 10 hrs. 20 min. The 18 aircraft were refueled in flight six times. (DOD Release 11-64)

January 7: NASA Manned Spacecraft Center issued request for proposals for study of deep space laser acquisition and tracking techniques in manned space missions. Specifications called for high priority development of laser tracking techniques between earth and a manned deep space vehicle or between a satellite and two manned deep space vehicles. System must be capable of handling two-way telemetry and voice communications, as well as spacecraft-to-ground TV. 22 companies were invited to bid on the six- to nine-month study. (MSC Release 64-3)

- NASA Manned Spacecraft Center announced that tests of flashing beacon were being conducted with a T-33 aircraft for evaluation of similar beacons for use in Gemini rendezvous operations. Beacon is a white sphere, weighs 10 lbs., has flashing power of 12 candleseconds. (MSC Release 64-1)
- First power tool built specifically for use in space was demonstrated by its co-producers, Martin Co. and Black & Decker Manufacturing Co., who developed it for USAF. Similar in shape and size to home power-drill, the battery-powered "electric minimum reaction space tool" has 99.97% less reactive torque than ordinary utility drills, a feature made possible by its freely revolving casing rotating in opposite direction from revolution of the motor. Tool has been tested in simulated 100-mi. altitude space environment. (Hines, *Wash. Eve. Star*, 1/8/64; *SBD*, 1/8/64, 37)
- NASA submitted to Bureau of the Budget its proposed reports on three patent bills: H.R. 4482, to prescribe national policy encouraging invention by providing for administration of property rights to inventions resulting from research financed by public funds; S. 1290, to establish uniform national policy for property rights to inventions resulting from publicly financed work; and S. 1444, to amend National Aeronautics and Space Act of 1958 to waive U.S. proprietary rights to inventions. (NASA LAR II/2)
- Senate Committee on Commerce held hearings on S. 1278, to provide that the National Bureau of Standards study practicability of U.S. adoption of the metric system. (NASA LAR II/1)
- John Stack, vice president and director of engineering at Republic Aviation, predicted that early in the 1970's a traveler flying westward in supersonic transport will be able to keep up with the sun: "Whether it is a Mach 2 plane or a Mach 3 or better, you could leave London at 8 A.M., arrive in New York at 8 A.M., refuel and be in Los Angeles before 8 A.M., refuel again and be in Hawaii at 8 A.M." (UPI, *NYT*, 1/8/64, 13)

January 7: In keynote address to 10th National Symposium on Reliability and Quality Control, Lt. Gen. Howell M. Estes, Jr. (USAF), AFSC Vice Commander, reviewed the reasons for reliability in missile and space systems and specific measures taken by AFSC to improve reliability and quality control. Discussing space systems he said: ". . . Payload weight will always be a major factor in the systems effectiveness equation for space systems. In unmanned systems, redundancy can never be acceptable but only reluctantly tolerated. Even in manned space systems, any provisions for in-flight maintenance or replacement of components will reduce payload weight which might otherwise be available for more productive use. Our greatest challenge thus lies in space. . . ." (Text, AFSC Release 41–R–4)

- USN launched Polaris A–3 missile from land site at Cape Kennedy in successful flight test down AMR. This was 14th success in 23 land launches of the A–3. (M&R, 1/13/64, 10)

First Week in January: NASA lifted hiring freeze on about six manned space flight projects, including Apollo work at North American Aviation, Inc. (*L.A. Times, Miami Herald*, 1/10/64; NAA *S&ID Skywriter*, 1/10/64, 1)

- North American Aviation, Inc., began testing prototype Apollo environmental control system, developed by Garrett Corp.'s AiResearch Div. Unmanned tests would continue for about six months, with manned tests scheduled to begin June 1. (*Av.Wk.*, 12/25/63, 23)

January 8: X–15 No. 1, piloted by Capt. Joe H. Engle (USAF), reached 139,500-ft. altitude and 3,716 mph (mach 5.36) in test near Edwards AFB, Calif. Flight plan called for Capt. Engle to shut off X–15's Stability Augmentation System (SAS), then the Alternate Stability Augmentation System (ASAS), to evaluate basic stability of the aircraft during re-entry. At peak of flight the X–15's altimeter, air speed, and rate-of-climb instruments failed, but Capt. Engle followed flight plan and landed the X–15 safely on Rogers Dry Lake with assistance from a ground control station. (FRC Release; AP, Balt. *Sun*, 1/9/64; UPI, *N.Y. Her. Trib.*, 1/9/64)

- In his State of the Union message to Congress, President Lyndon B. Johnson said:

    ". . . our ultimate goal is a world without war, a world made safe for diversity, in which all men, goods and ideas can freely move across every border and boundary.

    "We must advance toward this goal in 1964 in at least ten different ways, not as partisans but as patriots. . . .

    "Fourth, we must assure our pre-eminence in the peaceful exploration of outer space, focusing on an expedition to the moon in this decade—in cooperation with other powers if possible, alone if necessary." (Text, *NYT*, 1/9/64, 16)

- Second rocket in India's space program, a Nike-Apache, was launched from Thumba Range on southern tip of Indian peninsula, releasing sodium vapor clouds used to study high-altitude conditions. The Nike-Apache's peak altitude was estimated at 106 mi. The Physical Research Laboratory of Ahmedabad provided the instrumented payload for the NASA rocket. (NASA Rpt. SRL; Reuters, *NYT*, 1/9/64, 15)

- Dr. Edward C. Welsh, NASC Executive Secretary and Acting Chairman, spoke at 10th National Symposium on Reliability and Quality Control on misconceptions about national space policy:

"First, the lunar project should not be confused with our total space effort. Of course, it is a complex, expensive, and difficult task, but it's just a part of our young and growing national space program. Its major significance is that it sets a goal for an orderly development of competences—all of which will be useful for a wide range of space activities in the future.

"Second, the national security aspects of space are not being neglected. The National Aeronautics and Space Act requires that they receive attention and they are receiving it. Careless thinking prompts critics to assert that practically nothing is being done by the military in the space field. For example, about two-thirds of all our space flights in the past two years have been launched by the Department of Defense. And just as careless thinking assumes that a larger space budget for NASA than for Defense means a neglect of the latter. While one need not be satisfied with the rate of progress in either the civilian or non-civilian aspects of space, nevertheless it is worth noting that national security objectives are being pursued by both NASA and the Defense Department. . . . The fact that we do not plan space programs for aggression does not mean that we have a national security space gap or that we will allow one to develop.

"Third, there is another misconception to the effect that the national space program is employing an unduly large portion of our engineering and scientific manpower. Added to this misconception is the parallel view that the space program is creating a shortage of such talents which otherwise would be applied in our civilian economy. Actually, profits in private enterprise are at an all-time high and, in general, higher in the non-space and non-defense companies than in those whose business is primarily government-financed for space and defense. If those very profitable companies do not hire scientists and engineers to engage in research and development, it is not because they can't get them as their profit position gives them remarkable bargaining power. Rather, failure to employ such scarce talents in large numbers is the result of deliberate decisions on the part of private entrepreneurs and not because there are no such persons available. . . . Furthermore, government-financed research and development—particularly in such challenging fields as space—is having a salutary effect in stimulating an increase in the number of highly trained scientists and engineers.

"Fourth, there has also been some criticism and confusion on the subject of international cooperation in the space program. We have formed a useful base for peaceful and constructive relationships with some 60 other countries in space activities. The benefits are mutual. Some of our projects could just not have been conducted successfully without the tracking facilities, sites, and other cooperative efforts of countries around the world. In no case do we impair our national security through such cooperation, and we do not plan to do so. . . .

"Fifth, there are those who argue that it would be better to slow down—or even stop—our space effort in order to divert the resources to education, slum clearance, medical research, and other such purposes. As much as we need to do more in those important areas, it is unrealistic to believe that the resources are transferable—even if it were desirable to attempt the shift. . . . What is more, there is no assurance that the Congress would vote more funds for education or public housing just because it voted less for space. . . ." (Text)

January 8: Edward Z. Gray, Director of Advanced Studies in NASA Office of Manned Space Flight, said in *Space Business Daily* interview that when NASA builds a manned space station it would be "more sophisticated" than the DOD Manned Orbiting Laboratory (MOL). NASA now has 13 studies dealing with concepts leading toward development of an orbiting manned space station; when these studies are completed, NASA would be able to appraise "the space station requirements and what will be the best method of pursuing the development of a manned orbiting space station system." (*SBD*, 1/8/64, 34)

- DOD announced W. S. Bellows Construction Corp. and Peter Kiewit and Sons, Inc., were awarded $4,211,377 contract for construction of Mission Simulation and Training Facility and the MSC Center Support Facility at NASA Manned Spacecraft Center, Houston, Tex. Contract was awarded by Army Corps of Engineers. (DOD Release 21-64)
- Mating assembly of Saturn V rocket sections would be accomplished with aid of air cushions, *Marshall Star* reported. General Motors' Defense Research Laboratories was developing for MSFC a special tool—basically 20 "air pads"—for lifting rocket section and its wheeled carrier. Air pads would lift the huge sections about 1½ in. from the floor. Tool designers were also working on mechanical mover for the air-cushion device so that one man would be able to move the stages forward and laterally. (*Marshall Star*, 1/8/64, 1-2)
- Secretary of the Air Force Eugene M. Zuckert, addressing Harvard Business School of Washington, presented an appraisal of U.S. aerospace power in 1964:

    "My appraisal for 1964 indicates continuing increase in airpower. All of the ATLAS and TITAN missile squadrons were operational by the end of 1963. Our total U.S. strategic power, the backbone and starting point of deterrence, will more than double in 1964 when new MINUTEMAN strategic missile wings become operational. . . .

    "Today's appraisals of airpower must also cover the space front.

    "The Air Force space program will have a Manned Orbital Laboratory as its central project. The switch from Dynasoar to MOL reflects a change in emphasis toward on-orbit experiment rather than re-entry experiments. . . ." (Text, DOD Release 18-64)
- CAB Chairman Alan S. Boyd told House Committee on Interstate and Foreign Commerce's Public Health and Safety Subcommittee hearings on recent jet airliner crashes that investigation showed air turbulence no more hazardous for jets than for other airliners. (Witkin, *NYT*, 1/9/64, 52)
- Washington *Evening Star* editorialized about recent discovery that "Mars is covered by great yellow clouds of gaseous compounds of nitrogen and oxygen, or smog" and about scientists' conclusion that the clouds preclude any chance of man's finding life on Mars:

    "Quite the contrary. This clearly indicates that Martian life exists and the Martians are one, hopefully two, generations ahead of us in smogging up the atmosphere. As for landing there, all we need are spacemen from Los Angeles." (Wash. *Eve. Star*, 1/8/64)

January 9: USN announced a solar-powered, 120-lb. satellite was launched pickaback last Sept. 28 along with a previously announced nuclear-powered satellite. On board the solar-powered satellite were six transistors and several radiation detectors, testing means of protecting the transistors from radiation damage. (AP, *NYT*, 1/10/64, 86)

January 9: Sen. Barry Goldwater (R.–Ariz.) stated in Portsmouth, N.H., press conference: "I don't feel safe at all about our missiles, to tell you the truth. I wish the Defense Department could tell the American people how undependable the missiles in our silos actually are." Secretary of Defense Robert McNamara announced that Senator Goldwater's charge was "completely misleading, politically irresponsible, and damaging to the national security. There is no information, classified or otherwise, to support the false implications that our long-range missiles cannot be depended upon to accomplish their mission. . . ." (*NYT,* 1/10/64, 1, 14)

- Following Senator Goldwater's charge that U.S. missiles are undependable, unnamed official sources at Cape Kennedy pointed out that newest U.S. ICBM's—Minutemen and Titan II's—can be fired in less than a minute's notice, have been successful in nearly 7 out of 10 test firings, partially successful in 2 out of 10, and completely unsuccessful in only 1 out of 10 firings. (UPI, *NYT,* 1/10/64, 14)
- NASA Administrator James E. Webb told University Club in New York that NASA's last 25 attempts to place satellites in orbit were all successful. Of the 25, only one payload failed to function. Mr. Webb pointed out that in 1959, NASA's first full year of operation, the record showed 6 successes and 13 failures. By 1961, rocket reliability "had improved to the point that we had 54 successful space flights with five successes for each failure." (*NYT,* 1/10/64, 85)

January 10: Static-test firing of Titan II rocket for first Project Gemini flight was postponed at Cape Kennedy until Jan. 14 because of possible malfunction of first-stage propellant valve during countdown. (UPI, *Wash. Post,* 1/11/64; AP, Wash. *Eve. Star,* 1/11/64)

- Preliminary design for re-entry spacecraft for manned Mars missions during 1971–75 period was submitted to NASA Manned Spacecraft Center after six-month study by Lockheed Missiles and Space Co., MSC announced. The earth re-entry module would travel to Mars and back attached to 4- to 6-man Mars spacecraft, then would be detached and used for last eight hours. It is designed to withstand re-entry heating of about 35,000° F. at speeds up to 44,000 mph, protecting crew with ablative heat shield about 3½ in. thick. (MSC Release 64–6; MSC *Roundup,* 1/22/64, 1, 3)
- USAF Office of Aerospace Research announced Dr. Sanford Freedman of Tufts Univ. and Dr. Richard Held, Prof. of Experimental Psychology at MIT, had been conducting experiments of human performance in space and had expressed less optimism toward man's adaptability in space than previously had been felt. The experimenters studied human performance under conditions of reduced and abnormal visual and audial inputs related to restriction of body movements. They found that the subject was able to adapt only when he was allowed free movement of his body or appropriate limb. Besides supporting the view that accuracy of performance was degraded under prolonged exposure to conditions not permitting a correlated motor-sensory feedback, the experimenters concluded that stimulus changes accompanying body movements were an important source of order in human adaptation. This order was dependent upon relative constancy of environmental factors, such as gravity. Where one or more of these factors was absent—such as in space—body coordinations depending upon their regularizing influence may in time be lost. (OAR Release 1–64–1)

January 10: Contradictory speculation among scientists about nature of red spots observed recently on the moon, William J. Perkinson of Baltimore *Sun* said, should force NASA "to step up its badly lagging Ranger and Surveyor programs for unmanned exploration of the moon, and also force NASA to reconsider the standards that at this time bar any trained scientist from becoming an astronaut." (Balt. *Sun*, 1/10/64)

- More than 20 Martlet rockets were launched from Barbados site during first year of the high-altitude research project conducted by Canada's McGill Univ., *New York Times* reported. Fired from 16-in. naval cannon, Martlet has motor that ignites once the rocket is in flight, sending rocket to altitudes between 60 and 100 mi. It lands about 100 mi. southeast of Barbados in the Atlantic. 200 shots, financed by U.S., were planned for 1964. (AP, *NYT*, 1/11/64)
- DOD announced award of $15,599,775 contract to Boeing Co. for Project Hibex (Hi-G Booster Experiment), an ARPA research experiment investigating high performance missile boosters. (DOD Release 38–64)

January 11: Philco Corp. had begun cable and pneumatic tube installation in the Mission Control Center under construction at NASA Manned Spacecraft Center, MSC announced. Philco was prime contractor for flight control equipment for the Project Gemini/Apollo facility, which was scheduled for occupancy in mid-1964. (MSC Release 64–8)

- USAF launched Thor-Agena D booster from PMR with unidentified satellite payload. It was later disclosed that five satellites were orbited with the single booster. (HHN–48)
- James J. Haggerty, Jr., predicted that assignment of Manned Orbiting Laboratory to DOD was "an ominous harbinger of a reversal in trend, an indication that the military services may play a more prominent role in future space exploration at NASA's expense. . . .

    "MOL is a starting point. Whether you label it development platform, satellite or laboratory, it is clearly intended as a beginning for space station technology. It is also clearly the intent of this Administration that, at least in the initial stages, space station development shall be under military rather than civil, cognizance. . . ." (*A–N–AF J&R*, 1/11/64, 10)

January 12: NASA Nike-Apache sounding rocket with sodium-vapor payload was launched to 106-mi. altitude from Thumba Range, Trivandrum, India, in joint effort of NASA Goddard Space Flight Center and Physical Research Laboratory, Ahmedabad, India. Rocket performed as predicted, sodium ejected as predicted, and temperature and turbulence measurements were made with ground-based cameras and instrumentation. (NASA Rpt. SRL)

- Three USAF test pilots entered Apollo command module at Martin Co., Baltimore, for simulated 7-day lunar landing mission. (AP, *NYT*, 1/13/64, 9)
- USN evacuated staff and dependents (25 persons) at NASA tracking station in Zanzibar, where military coup overthrew the government of the one-month-old nation. (GSFC Historian)
- Cosmonaut Valentina Nikolayeva-Tereshkova, wife of Cosmonaut Andrian G. Nikolayev, was an expectant mother, Moscow sources reported. (UPI, *NYT*, 1/13/64, 5)

January 13: Management changes in propulsion programs in NASA's Office of Advanced Research and Technology were effective this date. John L. Sloop, head of OART Div. of Chemical Propulsion and Power Generation, was appointed to new position of Assistant Associate Administrator for Advanced Research and Technology, assisting Dr. Raymond L. Bisplinghoff. Propulsion division was redesignated Div. of Chemical Propulsion, with Adelbert O. Tischler of NASA Office of Manned Space Flight as Director. At the same time, management of M–1 engine project was transferred from OMSF to Div. of Chemical Propulsion, OART. (NASA Release 64–4)

- In White House Ceremony, President Johnson awarded 1963 National Medal of Science to the recipients: Drs. Jerome B. Wiesner, Cornelius B. van Neil, Luis W. Alvarez, Vannevar Bush, John R. Pierce, and Norbert Wiener. (Simons, *Wash. Post*, 1/14/64)
- USN launched three Polaris A–2 missiles in first operational tests of new steam ejection system, the launchings conducted from U.S.S. *Nathan Hale* submerged off Cape Kennedy. Second of the three firings was successful, the first failing when second-stage motor did not ignite and the third veering off course after 15 sec. of flight (destroyed by range safety officer). (*M&R*, 1/20/64, 11)
- DOD announced Army Materiel Command award of $1,170,751 contract to Western Electric Co. for work in support of Project Sleighride, re-entry vehicle vulnerability test program. (DOD Release 36–64)

January 14: Nike-Apache sounding rocket was launched from Wallops Island, Va., with Rice Univ. experiment "Sammy I" to measure and study causes of auroras and airglow. Vehicle carried 68-lb. payload, designed and built by professors and students in Rice Space Science Dept., to peak altitude of 56 mi. before impacting 74 mi. downrange in the Atlantic Ocean. No recovery operation was conducted; telemetry data during flight would be analyzed by the Space Science Dept. This was first of four similar experiments to be conducted under NASA grant; remaining three flights would take place at Ft. Churchill, Canada. (Wallops Release 64–3)

- NASA and Canada's Defence Research Board announced reaching agreement for joint multi-satellite ionosphere-monitoring research program, designated Isis (International Satellites for Ionospheric Studies). Agreement called for orbiting of a second Alouette and Isis A, B, and C satellites at intervals between 1965 and 1970. The satellites would continue and expand experiments of ALOUETTE I topside sounder, and data obtained from the satellites would be made freely available to the scientific community. Under terms of the agreement, DRB would design, construct, and test the satellites, provide basic onboard experiment, and operate at least one ground station to supply data on spacecraft operation. NASA would provide four launch vehicles, conduct launch operations, and provide and launch up to five sounding rockets to test components, subsystems, and experiments for the satellites. NASA also would launch an Explorer satellite in conjunction with the Alouette to obtain coordinated direct measurements. (NASA Release 64–6)
- Static test-firing of Titan II at Cape Kennedy postponed for a week, this second postponement due to failure of gas generator starter cartridge. (*M&R*, 1/20/64, 12)

January 14: Senate received President Johnson's nomination of Dr. Donald F. Hornig to be Director of Office of Science and Technology, succeeding Dr. Jerome B. Wiesner. (*CR,* 1/14/64, 371)

- In keeping with NASA reorganization, subcommittees of the House Committee on Science and Astronautics were revised as follows: Manned Space Flight; Space Sciences and Applications; Advanced Research and Tracking; and Science, Research, and Development. (NASA LAR III/6)
- Maj. Gen. Ben I. Funk, Commander of AF Space Systems Div., said in speech before Tennessee Valley Post of American Ordnance Association, Redstone Arsenal, Ala.:

    "Space can be nothing less than a total national undertaking.

    "Space has to do with our future, with our country's future, and with the progress of civilization. It also has to do with our personal and our national security. . . .

    "I personally believe that the working partnership between NASA and the Department of Defense will grow increasingly close in the months and years just ahead, and that joint efforts in ventures such as Gemini, and the lunar expedition, will serve to identify this country's slate of space activities as national programs. . . ."

    Discussing role of the military in space, he said: "As a national policy, we have declared that space will be reserved for peaceful purposes. This, I submit, serves to put the military into space—not restrict space to purely non-military activities. . . .

    "In describing the military mission, I like to say that the Defense Department's job is to insure that space is used for us—not against us." (*Marshall Star,* 1/15/64, 1–2)

- Engineers Joint Council would request this session of Congress for charter to establish National Academy of Engineering, EJC president Eric A. Walker said in Washington press conference. NAE would be patterned after National Academy of Sciences, which was working with EJC in setting up NAE, but the two bodies would have no organizational ties. (*M&R,* 1/20/64, 10)
- Boeing Co. scientists announced invention of lightweight telescope with more than 100-mi. range for use by astronaut in satellite interceptor in spotting another orbiting spacecraft. (*SBD,* 1/15/64, 74)

January 15: President Johnson sent to Congress his proposed $5.304 billion authorization for NASA in FY 1965, $600 million less than NASA earlier had requested. The proposed legislation (H.R. 9641) was introduced in the House by Rep. George P. Miller (D.-Calif.) and referred to Committee on Science and Astronautics. Of this sum, $641,000,000 would be for administrative operations; $281,000,000 for construction of facilities; and $4,382,000,000 for research and development. Of R&D amount, $3,011,900,000 would be for manned space flight—$2,677,500,000 for Project Apollo; $308,400,000 for Project Gemini; and $26,000,000 for advanced missions. At press conference on the budget, NASA made public its plans to request $141 million supplemental appropriation for FY 1964, to restore some of the budget cut by Congress. The NASA position, as expressed by Associate Administrator Dr. Robert C. Seamans, Jr.: "We have now been cut back to a point where we'll have to admit we cannot carry out the [lunar landing] program in this decade without the full support of Congress for the supplemental and for our 1965 request." Of supplemental request, $31 million would be for Apollo spacecraft development and the rest for Saturn launch vehicle. (NASA LAR III/7; Finney, *NYT*, 1/16/64, 1; *WSJ*, 1/16/64; Dodd, *Chic. Trib.*, 1/16/64)

January 15: Nike-Apache launched from Wallops Island, Va., in flight to measure light intensity and particle fluxes from 150,000 ft. to apogee. Good data were received throughout the flight, but experiment was only partially successful because Nike-Apache reached only 56.1-mi. altitude. (NASA Rpt. SRL)

- USAF launched Titan II rocket from AMR on successful 5,000-mi. flight to test its ability to function both as an ICBM and as booster for Project Gemini manned spacecraft. The Titan II carried Gemini malfunction detection system and damping device to counteract vibrations. (*M&R*, 1/20/64, 11)

- Senator Clinton P. Anderson (D.-N. Mex.), Chairman of Senate Committee on Aeronautical and Space Sciences, addressed National Capital Section of AIAA and questioned whether hazard of intense solar radiation in the 1968-70 period might postpone Project Apollo lunar flight. He pointed out that 1968-70 is period of maximum solar activity in the 11-year solar cycle, and radiation intensity might require "very substantial shielding for the Apollo capsule." This raised question of whether heavy shielding imposes too great a weight penalty on Apollo. Citing recent scientific recommendation that there was "relatively low probability" of Apollo crew's encountering solar flare, he questioned whether maintaining the schedule would be worth such a gamble. (Finney, *NYT*, 1/16/64, 1; *SBD*, 1/16/64, 80-81)

- Army Corps of Engineers awarded $2,295,220 contract for Saturn S-II stage test-site development at NASA Mississippi Test Operations to C. H. Leavell and Co. and Peter Kiewit and Sons, Inc., DOD announced. (DOD Release 50-64)

- Bendix Products Aerospace Div. was awarded $99,973 NASA study contract for shock absorbing device to permit soft landing on moon, NASA Manned Spacecraft Center announced. Bendix would test various sizes and shapes of crushable aluminum honeycomb structures to determine performance characteristics under lunar environmental conditions. (MSC Release 64-9)

- Three airframe manufacturers and three engine manufacturers submitted initial design proposals for the U.S. supersonic transport, FAA Deputy Administrator for Supersonic Transport Development Gordon M. Bain announced. The airframe manufacturers: Lockheed Aircraft, Boeing Co., and North American Aviation; engine builders: Curtiss-Wright, General Electric Co., and Pratt & Whitney Div. of United Aircraft. (FAA Release 64-4)

- Rep. Harold M. Ryan (D.-Mich.) said on House floor that if NASA's proposed electronics research center were located in Michigan it "would provide a stimulus to the heartland of America, and bring into geographical balance the national capability in the field closely related with the electronic industry.

   "It is important that the resources of our part of the country be taken into consideration and the fact recognized that they are not presently being used to their fullest capacity. . . ." (*CR*, 1/15/64, 434-35)

January 15: France's preparations for hydrogen bomb development in the South Pacific were reaching unexpectedly large proportions, *New York Times* reported. Gaston Palewski, French Minister for Scientific Research, had recently said in Papeete that "France has engaged herself in Polynesia for a very big task." Test site and military installations were being constructed at Mururoa in the Tuamotu Islands, 720 mi. southeast of Tahiti. French sources indicated France would become a thermonuclear power by late 1967 or early 1968. (*NYT*, 1/16/64, 4)

January 15–16: Series of four sodium-vapor experiments was launched on Nike-Apache sounding rockets from NASA Wallops Station in sunset-to-sunrise comparative measurements of winds and turbulence in the upper atmosphere. First two rockets were launched at sunset, third at midnight, and fourth at sunrise. First flight was unsuccessful because Nike-Apache reached only 28.4-mi. altitude instead of desired 118 mi. The remaining three rockets fired successfully, their 70-lb. payloads ejecting sodium vapor trails from about 25-mi. to as high as 123-mi. altitudes. Pink and reddish clouds were visible for several hundred miles. Photographic observations of the clouds would be analyzed by Geophysics Corp. of America, which provided payloads for the GSFC-managed project. (NASA Release 64–7; Wallops Release 64–5; NASA Rpts. SRL)

Mid-January: Work on test site for Project Apollo at White Sands Missile Range was halted by strike of operating engineers. (SBD, 1/15/64, 72)

January 16: NASA announced selection of Air Products and Chemicals, Inc., to supply liquid hydrogen to NASA Mississippi Test Operations and Marshall Space Flight Center for Saturn launch vehicle test program. Contract called for supply of up to 30 tons of liquid hydrogen per day by no later than mid-1965, would be renewable annually through December 1970. (NASA Release 64–10)

- X–15 No. 3 flown by NASA pilot Milton O. Thompson to 3,170 mph (mach 4.80) and 70,000-ft. altitude in test to investigate heat-transfer rates with sharp-leading-edged vertical tail. The sharp leading edge on top portion of aircraft's vertical tail is $1/100$-inch radius, as compared to previous blunt edge of $1/2$-inch radius. Change was designed to reduce complexity of air flow around aircraft's tail surfaces; this would cause difference in heat-transfer rates. During flight Thompson performed several control maneuvers to evaluate aircraft stability. (FRC Release; FRC X–15 Flight Log)

- The lowest altitude at which demarcation line between air space and outer space should be assigned is 100 km. (60 mi.), according to Norman Sissenwine, USAF Cambridge Research Laboratories scientist and co-chairman of U.S. Committee on the Extension of the Standard Atmosphere. Sissenwine's definition had been requested by Hyman, Hayman, and Harris law firm, which was attempting legislative interpretation for the International Astronautical Federation and the International Institute of Space Law. Sissenwine acknowledged that ultimately the line must be drawn arbitrarily, explained considerations favoring 100 km.: "Aerodynamic forces on most of the ballistic re-entry, lifting re-entry, and boost-glide orbital aerospace vehicles can generally be neglected at altitudes above 100 km. (This may vary from 80 to 120 km, depending upon the vehicle concept, but I believe 100 km is a good nominal altitude to use.) . . ." He pointed out that atmosphere's basic composition begins to change near 100 km. "Specifically it has been determined that disassociation of molecular oxygen into atomic oxygen should be depicted at an altitude of 90 km in the Standard Atmosphere. . . ." Another consideration was that relationships of temperature and pressure fields to wind fields are generally constant to nearly 100 km. This is also true for speed of sound, coefficient of viscosity, kinematic viscosity, and thermal conductivity. Finally, he pointed out that meteors entering the atmosphere generally disintegrate below 120-km. altitude, and most meteor trails formed in atmospheric entry are observed below 100 km. (OAR Release 1-64-4)

January 16: U.S. Army Corps of Engineers awarded $63,366,378 contract for NASA to joint bidders Paul Hardeman, Inc., Morrison-Knudsen Co., Inc., and the Perini Corp., for construction of Saturn V-Apollo assembly building at Cape Kennedy. (*NYT*, 1/17/64, 32; *WSJ*, 1/17/64)

- Sen. John Stennis (D.-Miss.), Chairman of Senate Armed Services' Preparedness Investigating Subcommittee, said he had directed his staff "to up-date and review the current information and data relating to the reliability and dependability of the Nation's long range ballistic missiles as soon as possible." (AP, *Wash. Post*, 1/17/64; Raymond, *NYT*, 1/17/64, 13)

- In interview with *Space Business Daily*, Rep. George P. Miller (D.-Calif.) cited two factors he believed would improve NASA's relations with House Committee on Science and Astronautics this session. First, committee members now have more knowledge of NASA's problems and needs, having visited NASA installations and learned first-hand about NASA's activities. Last session, he pointed out, 10 of committee's 31 members were new to space legislation. Second, NASA was perfecting its procedures of informing Congress, improving the information flow and clarifying its justifications. (*SBD*, 1/16/64, 80)

- 34 scientists fled East Germany to West Germany in 1963, German Ministry for Refugees reported. Since Jan. 1, 1958, the Ministry added, 1,710 scientists had emigrated from East to West Germany. (UPI, *NYT*, 1/16/64, 2)

- USAF fired two Minuteman ICBM's, one from AMR and one from PMR. AMR flight was test of devices on warhead design to confuse enemy defenses; PMR flight was routine training launch by SAC crew. (*M&R*, 1/27/64, 10)

January 17: Four-stage Javelin (Argo D-4) rocket vehicle was launched at 12:01 a.m. EST from Wallops Station, Va., carrying 120-lb. instrumented payload to 587.22-mi. altitude. Payload experiments: magnetic mass spectrometer to identify ions in upper atmosphere and measure their distribution; Lyman-alpha radiation experiment to determine spatial variation of atomic hydrogen in atmosphere; and experiment to measure altitude distribution and intensity of night airglow from atomic oxygen. Impact occurred 727 mi. downrange in the Atlantic Ocean. Data were telemetered to ground stations and no recovery operation was attempted. Experiment was designed and built by E. O. Hulburt Center for Space Research at U.S. Naval Research Laboratory and flown in cooperation with NASA Goddard Space Flight Center. (Wallops Release 64-6)

- In Columbus, Ohio, Lt. Col. John H. Glenn, Jr., first American astronaut to orbit the earth, announced that he would be a candidate for the Ohio

Democratic nomination for U.S. Senator in the May 5 primary. He added that he was retiring from the U.S. Marine Corps after 22 years of service. After explaining his party affiliation, he said Congress "is an area in which I have had a life-long interest. To serve in a body whose actions help mold the destiny of America and the free world is certainly a challenge and a high calling."

Regarding his departure from the space program, Glenn said: "It could very well be wishful thinking on my part to train for another six or seven years for flights for which I might be too old.

"I have chosen this course because I feel that it provides the best opportunity to make use of the experience I have gained in 22 years of Government service. Today, well over half our national budget goes toward the support of our military forces, our space program and other research and development activities. These are areas with which I am well acquainted. . . ." (AP, Parks, *Wash. Post,* 1/18/64, A1; Finney, *NYT,* 1/17/64, 17)

January 17: James C. Elms, Deputy Director of NASA Manned Spacecraft Center, resigned to accept a position with industry, Director Dr. Robert R. Gilruth announced. Resignation would be effective in February. (MSC Release 64–13)

- *New York Times* editorialized that delay of the 1969 target date for Project Apollo "may increase the chance that the project of landing a man on the moon will become an international effort, mobilizing all the resources of world technology. That certainly would be most appropriate for man's first landing in the cosmos beyond this tiny planet." (*NYT,* 1/17/64, 42)
- MSC reported that Astronaut Alan B. Shepard had had a small benign tumor removed from his thyroid gland by surgery. (AP, *Wash. Post,* 1/17/64, A2)

January 18: At ceremonies dedicating the new National Geographic Society headquarters in Washington, D.C., President Lyndon B. Johnson called for international sharing of knowledge: "The last four centuries of human experience have been centuries of exploration, discovery and advancement of the frontiers of man's knowledge . . .

"America, as we know it—freedom, as we know it—could well not exist tomorrow for our children and their children if we should lose from our national life that confidence in the future and that enthusiasm for exploration which has brought us to this high moment of history and hope.

"All the seas have been sailed—all the continents explored.

"The highest mountains have been scaled—the darkest jungles penetrated.

"We have reached into the realms of space—and outward toward the domain of the stars.

"This generation of Americans is challenged to live a life of high adventure. If we are to keep our trust of freedom, we must—in these last four decades of this century—undertake explorations in many realms which dwarf all those of the last four centuries. . . .

"Scientific exploration and research know no national boundaries. Human knowledge is never the captive of international blocs.

"Common sense dictates that all nations lend their learning to all other nations. This is a loan in which the science of all nations is the beneficiary and the good of all mankind is advanced. The more we share with each other, the less we misunderstand each other . . .

"This Nation is committed now to the most intensive effort ever made by any peoples to advance the frontier of human knowledge.

"We shall remain committed . . .

"The United States shall welcome any who wish to join with us in seeking to serve the common good of mankind. But if others are not willing—or if they are not able—to join with us our own endeavors will not slacken . . ." (Text, *Wash. Post*, 1/19/64, A6)

January 18: President Johnson sent letter to Soviet Premier Khrushchev, replying to Khrushchev's Dec. 31 letter which declared "the use of force for the solution of territorial disputes is not in the interest of any people or any country." President Johnson said: "Let us emphasize . . . our agreement on the importance your letter places on preserving and strengthening peace, and on the need to accompany efforts for disarmament with new efforts to remove the causes of friction and to improve the world's machinery for peacefully settling disputes. In this spirit, let us both present new proposals to the Geneva Disarmament Conference, in pursuit of the objectives we have previously identified: To prevent the spread of nuclear weapons; to end the production of fissionable material for weapons; to transfer large amounts of fissionable materials to peaceful purposes; to ban all nuclear weapons tests; to place limitations on nuclear weapons systems; to reduce the risk of war by accident or design; to move toward general disarmament. . . ." (Text, *CR*, 1/22/64, 812–13)

January 19: George M. Low, NASA Deputy Associate Administrator for Manned Space Flight, was named Deputy Director of NASA Manned Spacecraft Center, replacing James C. Elms. Low would assume his new duties in February, would continue present position until May 1. Low joined NACA Lewis Research Center in 1949, became Chief of Manned Space Flight Programs upon formation of NASA in October 1958. (NASA Release 64–13)

• By this date U.S. still did not know whether U.S.S.R. would participate in tracking Echo II balloon satellite, scheduled for launching Jan. 23. Earlier this month NASA sent cable to Soviet Academy of Sciences outlining launch schedule and inquiring about Soviet tracking plans, agreed on in principle last summer. NASA received cable in reply, acknowledging receipt of U.S. message but providing no details of Soviet plans. NASA planned to launch the satellite as scheduled, hoped to hear further word from U.S.S.R. (Finney, *NYT*, 1/19/64)

• USAF launched Thor-Agena D booster from Vandenberg AFB with unidentified satellite. It was later disclosed that two satellites were placed in orbit. (*M&R*, 1/27/64, 10; HHN–48)

January 20: NASA Deputy Administrator Hugh L. Dryden announced U.S.S.R. had notified U.S. that it would participate in space communications experiments using Echo II passive communications satellite which NASA would orbit in the coming week. Soviet officials planned to begin optical and radio experiments with the 135-ft.-diameter balloon satellite a week after it was orbited. Soviets would photograph Echo II and track it with telescopes as it passed over U.S.S.R., bounce radio signals off the balloon between U.S.S.R. and England. U.S.–U.S.S.R. cooperative experiments were based on 1963 agreement reached by

Dr. Dryden and Soviet Academician A. A. Blagonravov. (Simons, *Wash. Post*, 1/21/64)

January 20: In his Economic Report to the Congress, President Lyndon B. Johnson reviewed progress toward economic goals and presented his programs for "achieving our national economic potential." On transportation and technology, he said that the ". . . Federal Government provides major support for the research and development which underlie our striking technological advances. In the past much of our research and development has been connected with national defense. Now, as military outlays level off, we face:

"• A challenge to apply the nation's growing scientific and engineering resources to new socially profitable uses;

"• An opportunity to accelerate the technological progress of our civilian industries.

"The Federal Government should join with private business and our universities in speeding the development and spread of new technology. I have directed the Department of Commerce to explore new ways to accomplish this." (Text, *NYT*, 1/21/64, 16)

- Sen. Clinton P. Anderson (D.-N. Mex.), Chairman of Senate Committee on Aeronautical and Space Sciences, amplified his concern for solar radiation danger in Project Apollo manned flights in 1969–70, saying it would be a "sound" decision to delay the Apollo flight into the 1970's to increase safety. "We don't know all there is to know about these flares." He repeated that adding shielding to Apollo spacecraft could impose weight penalty, added that NASA was "intensifying" its study of solar radiation. (*SBD*, 1/21/64, 104)

- Bill to authorize appropriations to NASA for FY 1965 (S. 2446) was introduced in the Senate by Sen. Clinton P. Anderson, Chairman of Senate Committee on Aeronautical and Space Sciences, and referred to the Committee on Aeronautical and Space Sciences. (NASA LAR III/9)

- Bill to establish a Commission on Federal Expenditure Policy (S. 2441) was introduced in the Senate by Sen. Jacob K. Javits (R.-N.Y.) on behalf of minority of Joint Economic Committee, and was referred to the Committee on Government Operations. The proposed Commission would be composed of 16 members from the Executive Branch, Congress, and private life, would review Government expenditure policy and make specific studies and policy recommendations. Bill was also introduced in the House, by Rep. Thomas B. Curtis (R.-Mo.). (NASA LAR III/10; *CR*, 1/20/64, 592–96)

- NASA Deputy Associate Administrator for Manned Space Flight Operations Walter C. Williams said in *Space Business Daily* interview that "we are over the hump financially on the Gemini program and that the Apollo target of mid-1969 is still realistic with proper funding and if an inordinant amount of problems doesn't appear." He added that Project Gemini was on schedule if the December 1964 flight was successful and if no new "pacing headaches" arose. (*SBD*, 1/20/64, 98)

- AFSC Ballistic Systems Div. awarded $1,218,500 obligation on previous contract to Aerojet-General Corp. for Titan II propulsion systems and related spare parts; and $2,578,000 obligation to AC Spark Plug Div. of General Motors for production of gyros and accelerometers for Titan II. (DOD Release 67–64)

January 20: USN successfully launched Polaris A-3 missile from surface ship off Cape Kennedy on 2,000-mi. flight down the Atlantic Missile Range. (AP, *NYT*, 1/21/64, 4)
- USAF planned four tests of the six-vehicle Asset re-entry program during the coming year. The first test, in September 1963, was very successful. The second test shot, in March, was planned to attain re-entry speed of 18,000 fps; later tests would attain maximum series velocity of 19,500 fps. First five vehicles would have zirconium-rod nose caps and sixth vehicle a coated tungsten nose cap. (*M&R*, 1/20/64, 17; USAF Astronautics Div., DCS/R&D)
- Program to encourage development and use of "crash locator beacons" was announced by FAA, which was inquiring whether industry can produce crash locator beacons at reasonably low prices and whether individual aircraft owners would buy or rent them if available. Director of FAA Flight Standards Service George S. Moore said tests conducted by FAA "clearly indicated that such a beacon can successfully radiate energy to permit suitably equipped search aircraft to identify and 'home-in' on the transmitted signal." (FAA Release 64–5)
- FAA announced Australia's Qantas Empire Airways had requested six supersonic transport aircraft, bringing total of reservations to 51 held by 8 airlines. (FAA Release 64–6)
- With nuclear explosives new sea-level canal could be built across Isthmus of Panama for $500 million or less, according to Lawrence Radiation Laboratory's Gerald W. Johnson, Director of AEC Project Plowshare. Technical and economic feasibility of such a project was discussed in letter from Johnson to Sen. John O. Pastore (D.-R.I.), Chairman of Joint Congressional Committee on Atomic Energy. Use of nuclear explosives to dig new Panama canal had been under study since 1958, when Panama Canal Co. requested Lawrence Radiation Laboratory for preliminary feasibility study of the project. (Finney, *NYT*, 1/21/64, 10)
- Retiring science editor of *New York Times* William L. Laurence was honored at dinner in New York by wide range of scientific societies. Laurence was science journalist for 34 years, winner of two Pulitzer Prizes. (*NYT*, 1/21/64, 1)

January 21: RELAY II active communications satellite launched into orbit with Thor-Delta launch vehicle from AMR (4,606-mi. apogee; 1,298-mi. perigee; 195-min. period; 46.3° inclination to the equator). The 184-lb. satellite would continue communications tests of RELAY I, still orbiting the earth, and provide evaluation of improvements in the new comsat. RELAY II completed its first communications test during its first orbit, receiving radio signals and TV test pattern from Mojave, Calif., ground station and beaming them back to earth. RELAY II was initially in excellent mutual visibility position between North America and Europe for TV transmission between the two hemispheres. Mutual visibility with Japan, about 40 min. at maximum, would decrease gradually to nothing by mid-July and continue with no mutual visibility again until mid-November. RELAY II was equipped for transmitting one-way wideband communications (one-way TV, 300 one-way voice channels or high-speed data) or two-way narrow-band communications (12 two-way telephone conversations or teletype, photo-facsimile and data). Unlike its predecesor, RELAY II had no automatic cutoff device;

it had negative-on-positive (n-on-p) solar cells for greater resistance to radiation than RELAY I's p-on-n solar cells. RELAY II was 22nd straight launch success of Thor-Delta launch vehicle. (GSFC Historian: *Goddard News*, 1/27/64, 1; AP, *NYT*, 1/22/64, 17)

January 21: Both stages of first Titan II launch vehicle for Project Gemini were successfully static-fired for 30 seconds at Launch Complex 19, Cape Kennedy, producing combined total thrust of more than half a million pounds. Purpose of test was to evaluate overall Gemini launch vehicle system performance—fueling, countdown, ignition and cutoff commands, guidance control, and telemetry; and to verify engine performance by thrust generation and engine gimbaling. The stages were mounted side-by-side on separate mounts; electrical signal cut off first-stage engine after 30 sec., then started second-stage engine, which was cut off by radio signal from ground computer just as in actual flight. Charles W. Matthews, Gemini Project Manager, MSC, said static-firing test met all prelaunch requirements. (MSC Release 64–9; *M&R*, 1/27/64, 10)

- President Johnson sent $97.9 billion FY 1965 budget to Congress. Of space research and technology, he said:

"Our plan to place a man on the moon in this decade remains unchanged. It is an ambitious and important goal. In addition to providing great scientific benefits, it will demonstrate that our capability in space is second to no other nation's. However, it is clear that no matter how brilliant our scientists and engineers, how farsighted our planners and managers or how frugal our administrators and contracting personnel, we cannot reach this goal without sufficient funds. There is no second-class ticket to space.

"Appropriations enacted for 1964 for the National Aeronautics and Space Administration were $600 million below the amount requested. As a result, major development programs leading to the manned lunar landing have fallen behind schedule. Careful replanning of the entire program, including a reduction in the number of test flights, will offset some of this delay. Even so, more funds are needed in 1964, and I am therefore recommending a supplemental appropriation of $141 million for this year.

"For 1965, I am requesting appropriations of $5.3 billion, $63 million above the 1964 amount, including the proposed supplemental appropriation. The 1964 and 1965 recommendations represent the minimum amount needed to achieve our goals in space. The estimated increase of $590 million in expenditures in 1965 is due principally to payments required by commitments made in 1964 and earlier years. With the leveling off of appropriations, annual outlays should remain relatively stable in subsequent years.

"In addition to the manned space flight program, though related to it, funds are included to support unmanned space flights for lunar exploration and supporting research and development. Funds are also included for scientific satellites, planetary probes, and experiments with meteorological and communications satellites.

"The programs of the National Aeronautics and Space Administration are designed to maintain American supremacy in space and to demonstrate this supremacy by achieving a manned landing on the moon in this decade. . . .

"The level of activity in support of the 3-man Apollo lunar spacecraft development program will reach its peak during 1965 with 13 flight spacecraft in production by the end of the year. The Saturn V rocket program will be proceeding rapidly along its development cycle in preparation for launching the Apollo spacecraft, and will account for about $850 million of the estimated 1965 expenditures. . . ." (Text, *NYT*, 1/22/64, 19–21)

January 21: The Federal budget for FY 1965 estimated expenditures for R&D in FY 1965 at $15.3 billion, only a three per cent increase over FY 1964. For past decade Federal R&D spending averaged 20% increase each year over the preceding year; a decade ago, Government spent $3.3 billion on R&D. (Finney, *NYT*, 1/22/64, 18)

- In the President's proposed budget for FY 1965, DOD expenditures were estimated at $51.2 billion, of which $5.5 was for research and development (including military space expenditure). Requested budgets for operation of ranges: $231 million for AMR, $159 million for PMR, and $93 million for WSMR. (*NYT*, 1/22/64, 21, 22; *M&R*, 2/3/64, 9)

- 17-nation Disarmament Conference resumed in Geneva. In his letter of Jan. 19 to the Conference, President Lyndon B. Johnson said: "There is only one item on the agenda of this conference—it is the leading item on the agenda of mankind—and that one item is peace.

  "Already this conference has led to more concrete and effective results than any disarmament conference in modern history. Your efforts and deliberation laid the ground work for the nuclear test ban treaty—for the communications link between Washington and Moscow—and for the U.N. General Assembly action against nuclear weapons in space. . . ."

  President Johnson then proposed five major types of potential agreement: to stop spread of nuclear weapons to nations not now possessing them; to accept observation posts on both sides; to stop all production of fissionable material; to ban threat or use of force in changing boundaries or controlling territories; and to freeze numbers and kinds of strategic nuclear-carrying vehicles (aircraft and missiles). (Text, *CR*, 1/22/64, 812; "Statement to American People," *CR*, 122/64, 813)

- Vice president of Soviet Academy of Sciences Mikhail A. Lavrentev's disclosure that U.S.S.R. had been forced to abandon attempt to drill through the earth's crust was reported by Howard Simons in *Washington Post*. Soviet drillers reached depth of about 20,000 ft. near Caspian Sea when economic and technical difficulties forced them to give up the project. Lavrentev said Soviet scientists now were "seeking more effective methods of drilling and shoring up the hole." Project was akin to U.S. Project Mohole, directed by National Science Foundation, which would attempt to drill through sea floor to penetrate earth's crust and sample the mantle beneath. (Simons, *Wash. Post*, 1/21/64)

- Dr. Abe Silverstein, Director of NASA Lewis Research Center, was presented 1964 Sylvanus Albert Reed Award at AIAA dinner in New York. Dr. Silverstein was cited for "major contributions toward the development of propulsion systems for aircraft and spacecraft and for outstanding leadership in the nation's programs of scientific satellites and manned space flight." Also presented were AIAA's Research Award to Henry M. Shuey and Space Science Award to Herbert Friedman. (LRC Release 64–6; "AIAA Honors & Awards")

January 21: Rep. Edward J. Gurney (R.-Fla.) included in *Congressional Record* a resolution by Florida Library and Historical Commission appealing to the President, the Florida Governor, and the U.S. Board of Geographic Names "to carefully consider the possibility of retaining the name, Cape Canaveral, as the geographic designation only of the easternmost tip of the land formation known as Cape Canaveral (that tip upon which no portion of the missile complex is located) thereby detracting none whatever from the widespread desire to honor our late President John F. Kennedy." (*CR*, 1/21/64, A214)

- Testimonial dinner for Dr. Jerome B. Wiesner, resigning to return to MIT as Dean of Science, held in Washington. For three years Dr. Wiesner had been Director of the Office of Science and Technology, Chairman of the Federal Council of Science and Technology, Chairman of the President's Science Advisory Committee, Special Assistant to the President, and Acting Director for Telecommunications Management. (Miller, *Wash. Daily News*, 1/22/64)

January 22: First intercontinental communications tests conducted with RELAY II communications satellite were successful; NASA officials described the satellite's performance as "excellent." The tests were conducted between ground stations at Nutley, N.J., and Raisting, West Germany, consisted of voice transmissions and radio signals. (AP, *NYT*, 1/23/64, 7)

- NASA announced selection of Eclipse-Pioneer Div. of Bendix Corp. for negotiation of cost-plus-incentive-fee contract for stabilization platforms for Saturn IB and Saturn V launch vehicle guidance systems. The ST-124-M platform systems were modifications of Bendix ST-124 systems used in Saturn I launch vehicle, the modifications resulting in simpler system and reduced weight and cost. (NASA Release 64-17)

- Cape Canaveral Missile Test Annex was officially redesignated Cape Kennedy Air Force Station. (Special Order GA-7, issued by USAF CofS Curtis E. LeMay, in KHN-1)

- Evidence that the electron is smallest indivisible unit and that the proton lacks a core, as had been postulated, was reported at annual meeting of American Physical Society in New York. Experiments were made with electron accelerator built by Harvard Univ. and MIT and were described by Dr. Roy Weinstein of Northeastern Univ. (Boston), Dr. Richard Wilson of Harvard, and Dr. Louis S. Osborne of MIT. (Sullivan, *NYT*, 1/23/64, 20)

- Dr. Edward Teller, nuclear physicist, testified before House Select Committee on Government Research on the Federal Government's role in research. Dr. Teller criticized DOD's secrecy in scientific developments, saying practice had been to keep them secret unless it could be proved without a doubt that making them public would be in the national interest. "I think the burden of proof should be on the other side," Dr. Teller stated. "I am pretty well convinced that the Russians have all our secrets and I am even afraid they have the secrets we are going to discover in the next two years. Our industry and our citizens don't have those secrets." (NASA LAR III/11; AP, *NYT*, 1/23/64, 9)

January 23: Nike-Cajun sounding rocket was fired from Wallops Island with 82-lb. payload containing grenades, detectors, and associated equipment, the grenades ejecting and exploding at altitudes ranging from 24- to 59-mi. altitudes. This flight was followed 20 min. later by Nike-Cajun sounding rocket launched from Fort Churchill, Canada, with identical payload, a less successful flight because second-stage motor failed to ignite. Grenades were ejected and detonated but at altitudes too low to obtain useful data. The close-order firings were first two in current NASA series to gather comparative data on winds and atmospheric temperatures up to 55-mi. altitude at widely separated locations on near-simultaneous basis. (Wallops Release 64–7)

January 23: Rep. Thomas M. Pelly (R.-Wash.) introduced a concurrent resolution to change the goal of the manned lunar landing to before 1975 instead of before 1970 (H. Con. Res. 257). "The change of policy . . . is to allow our scientists greater flexibility in meeting the challenging technological problems of the Apollo project. Especially, I have in mind the desirability of eliminating the pressures of meeting time schedules, of minimizing perils such as solar radiation, and frankly I would stretch out the fantastic cost of the manned space exploration program over a longer period of time. . . .

"My new resolution is to invite the opinions of qualified experts and knowledgeable persons on this subject. . . . Congress should provide for an overall reappraisal of such a costly manned flight activity which accounts for two-thirds of the $5 billion space budget. Certainly, I cannot believe there is any valid reason why the Space Administration and those engaged in the manned space research venture should not be given more latitude in the scheduling of the program. Any slippage in time schedules due to reasons of safety and increasing assurance of success should not require any apology or excuse. Nor should the program be speeded for prestige at the cost of added risk. Congress by adopting my resolution would be setting a policy of proceeding at an orderly and normal speed and of avoiding undue risks in establishing any deadline. Finally this greater latitude should allow for tighter budgeting and contribute to reductions in the level of annual appropriations." (*CR*, 1/23/64, 932)

- New evidence of possible explosions in Milky Way galactic core, reported by Soviet astronomer Dr. Iosif S. Shklovsky, was discussed by British astronomer Dr. Fred Hoyle at American Physical Society meeting in New York. Dr. Shklovsky discovered that spurlike structures above and below the galaxy are connected in a single figure-8 configuration, its center at the galactic core. It is possible that the loops, which are strong sources of radio emissions, are magnetically trapped particles which may have been forced from the core by a past eruption. (Sullivan, *NYT*, 1/24/64, 48)

- Rep. John W. Davis, (D.-Ga.) said DOD had turned over the large solid-propellant rocket motor project to NASA, the decision made because no definite military mission was assigned for the technology. In FY 1965 NASA will cover cost of the large-motor R&D; in FY 1964 USAF carried the cost. NASA officials said they planned to request $13 million for the project, with about half that going to contractors Thiokol Chemical Corp. and Aerojet-General. (*Atlanta Const.*, 1/24/64)

- NASA Administrator James E. Webb stated Boston still appeared to be the most suitable site for NASA electronics research center. NASA's Electronics Site Selection Committee had reviewed presentations by 50 groups representing 19 geographical areas, in accordance with P.L. 88–113 (NASA Authorization Act for FY 1964). Mr. Webb still had not officially notified Congress of NASA's site selection. (*SBD*, 1/24/64, 129)

January 23: NASA Manned Spacecraft Center awarded $9.2 million increase to contract with North American Aviation, Inc., calling for construction at Downey, Calif., of seven new buildings and modifications to existing structures to permit expansion of NAA's research and development. NAA is prime contractor for Apollo command and service module under contract signed with MSC Aug. 14, 1963. (AP, *Tulsa Daily World*, 1/24/64)

- USAF launched Titan II ICBM from underground silo at Vandenberg AFB in successful flight 5,000-mi. down Pacific Missile Range. (*M&R*, 2/3/64, 10)
- Paper on Government's relationship with universities, by Delphis C. Goldberg, staff member of Subcommittee on Intergovernmental Relations of House Committee on Government Operations, inserted in *Congressional Record:*

  "There is much to be said for disbursing a substantial portion of research project funds through universities as a partial substitute for the selection of individual grantees at the Federal level. However, I would doubt the political feasibility and wisdom of allocating very large sums to institutions to be expended at their discretion. Instead, I would propose that Federal agencies set standards for university project review committees and require approval of the grant management policies and procedures of an institution as a condition to its eligibility for distributing Federal research funds. . . .

  "As the funds available for science and for each of its major components become more scarce relative to demand, Federal agencies will be increasingly confronted with the necessity of making hard choices in the allocation of money between research fields as well as between competing projects within each specialized field. The multi-disciplinary committee, consequently, will likely become a valuable device in helping to determine priorities for Federal science expenditures as the pressure for efficient resource allocation grows. The university project review committee, which included representation from a variety of disciplines and specialties, might serve in this capacity at the local level. . . ." (Text, *CR*, 1/23/64, A286–87)
- Dr. Romney H. Lowry, former head of Boeing Co. bioastronautics organization, was named Chief of Research and Education Div. in FAA Office of Aviation Medicine by Dr. M. S. White, FAA Federal Air Surgeon. (FAA Release T64–1)

January 23–24: More than 100 representatives of European electronics and aeronautical firms met in Brussels where they discussed feasibility of development project for manned shuttle service between earth and orbiting space stations (U.S. and Soviet). Recent cancellation of U.S. Dyna Soar project spurred the discussions. German delegation was headed by Dr. Eugen Saenger. (*M&R*, 2/3/64, 9)

January 24: Paul L. Styles, Chief of NASA Marshall Space Flight Center's Industrial Relations Office, was named Labor Director of NASA. In the new position, which he assumed in addition to his present one, Styles would advise NASA management officials at Hq. and all field installations on all aspects of labor-management relations, also would serve as NASA representative with Dept. of Labor, NLRB, President's Missile Sites Labor

Commission, and Federal Mediation and Conciliation Service. (NASA Release 64–14)

January 24: Development of astronaut techniques for egress from Gemini spacecraft began this week at NASA Manned Spacecraft Center, MSC announced. Boilerplate model of Gemini spacecraft floats in water-filled tank; two subjects in the closed spacecraft practice egress and the manner, procedure, and time required are recorded for study. (MSC Release 64–19)

- NASA Manned Spacecraft Center announced seven floors of the nine-floor Project Management building had been accepted for occupancy. The structure was expected to be ready for tenancy Feb. 20. To date, 13 MSC facilities at the Clear Lake site had been certified as ready for occupancy. (MSC Release 64–18)
- NASA Lewis Research Center employees Robert Steinberg and William B. Schwab won $700 NASA incentive award for invention of method and device for rapidly mapping neutron flux of each section of nuclear research reactor. Other LRC awards: John W. Macomber won $500 for invention of nuclear reactor control rod assembly with improved driving mechanism; Joseph M. Savino and Chester D. Lanzo, $200 award for developing simulated fuel assembly for nuclear reactor; and John C. Sturman, $100 award for transistorized logic circuit to handle data accumulated by satellites and transmit it back to earth. (LRC Release 64–8)
- Pierre Chatenet, president of Euratom, said in Paris that nuclear energy developments and advances in space technology called for revision of NATO. "The Atlantic alliance originally was drafted in a certain context which has now changed. For one thing, access to space presents us with a new battlefield." He also declared that six-nation European Economic Community planned $7 billion nuclear energy development program over the next 15 years, to include construction of "80 to 90 power stations for the peaceful use of nuclear energy." (O'Toole, *NYT*, 1/26/64, 17)

January 25: NASA's ECHO II passive communications satellite was placed in near-polar orbit by Thor-Agena B launch vehicle launched from Vandenberg AFB, Calif. The second-stage Agena fired shortly after booster separation, coasted, then reignited over South Africa, injecting into orbit the canister containing the laminated mylar plastic and aluminum balloon. Canister opened and released the 535-lb. balloon-satellite which inflated to 135-ft. diameter. Initial orbital data: apogee, 816 mi.; perigee, 642 mi.; period, 109 min.; inclination to equator, 81.5°. (NASA Release 64–11; AP, *Wash. Post*, 1/26/64; Becker, *NYT*, 1/26/64, 1)

- NASA Nike-Apache sounding rocket with Univ. of New Hampshire instrumented payload was launched from Thumba Range, Trivandrum, India, to 102-mi. altitude. Payload instrumentation included densitometer and baroswitch, to determine altitude and intensity of electric current systems in the ionosphere near the magnetic equator. Performance of both rocket and instrumentation was considered good; telemetry signal was received for 395 sec., and good data were expected. (NASA Rpt. SRL)
- Liquid-hydrogen-powered rocket engines for Saturn I's S–IV stage exploded during preparations for static-firing at Douglas Aircraft Co.'s Sacramento, Calif., facility. Explosion, which demolished the test stand, was evaluated by NASA to determine if forthcoming Saturn SA–5

flight test at AMR should be delayed; NASA officials decided the explosion was not sufficient reason to interrupt the SA-5 launch preparations. (AP, Wash. *Eve. Star*, 1/25/64; AP, *Wash. Post*, 1/26/64)

January 25: Britain had expressed interest in joining U.S. and European countries in financing and use of global communications satellite system, John Finney reported in *New York Times*. After months of favoring cables for international communications in the immediate future, British Post Office expressed approval of participation in global satellite system and urged similar decisions by European nations, at meeting of satellite committee of European Conference of Postal and Telecommunications Administration in Karlsruhe, Germany. (Finney, *NYT*, 1/26/64, 1)

- DOD announced plan for aiding economic readjustment of contractors performing work on major projects that are canceled, including Dyna Soar and Typhon projects. Plan called for continuing work with a continuing value to the Government, to identify additional work requiring these skills, and to expedite placement of contracts which would normally be awarded to these contractors. Military departments must submit detailed plans in accordance with this policy as soon as possible and no later than 45 days after decision to terminate a major project. (*SBD*, 1/27/64, 135; *M&R*, 2/3/64, 13)

- *Army-Navy-Air Force Journal and Register* Space Editor James J. Haggerty, Jr., said: "The major reservation about MOL is the fact that it will employ existing or in-development hardware . . . rather than advances in technology. . . .

    "With the Dyna-Soar experience in mind, it could be a big minus from the standpoint of developmental continuity in later years. Dyna-Soar was killed for a number of reasons, but one of them was the fact that it was not a very advanced spacecraft for its day (1965-66 flights). Suppose that, along about 1966 when MOL is in 'tin-bending' status but still a year or two away from flight status, someone takes a long look at it and says 'What have we got?'

    "By that time, NASA will have accumulated close to 1,000 hours of manned space flight experience and will have tested a number of new and advanced systems; it will have conducted the all-important rendezvous tests with the Gemini spacecraft; the earth-orbiting version of Apollo, which can stay aloft for two weeks as a space laboratory, will be in or near flight status. In that frame of reference, the short-duration, non-rendezvousing MOL could go the way of Dyna-Soar, offering 'insufficient technological advancement to justify the cost of completing the project.' . . ." (*A-N-AF J&R*, 1/25/64, 21, 30)

January 26: Potential of the gas-core nuclear rocket was outlined at recent press conference by National Aeronautics and Space Council staff-member Maxwell W. Hunter, Jr., and Welsey Kuhrt and George McLafferty of United Aircraft Corp., which is performing research on gas-core nuclear rocket under NASA contract. Maximum specific impulse for gas-core nuclear rocket would be between 2,000 and 3,000 (contrasted with 400 for the best chemical propellants and 1,000 for solid-core nuclear rocket). United Aircraft calculated useful payload sent to moon and back by gas-core nuclear rocket would be 10-30% of take-off weight, compared to a fraction of 1% for chemical rockets. Power of gas-core rocket would be so great there would be no need for rocket staging; single-stage rocket could take off from earth, land on moon, take off, return to earth, and land safely. Such a rocket vehicle would be reusable. (Witkin, *NYT*, 1/26/64, 32)

January 26: DOD was reconsidering its tentative plans for a military communications satellite system, instead would rely on satellites of Communications Satellite Corp., John Finney reported in *New York Times*. Partly in response to Congressional pressure, Finney reported, DOD was proposing to delay development of ground stations and defer or prolong the completion of its military comsats. Final decision would be dependent upon outcome of DOD negotiations with ComSatCorp, still underway. (Finney, *NYT*, 1/27/64)

January 27: Soviet Academy of Sciences announced Soviet ground stations were tracking ECHO II passive communications satellite, in first joint U.S.–U.S.S.R. space experiment. Academician A. A. Blagonravov said Soviet ground stations observed the 135-ft. balloon as it inflated in space and three observatories succeeded in photographing the satellite. Observation stations reported the satellite appeared very bright, varying from magnitude–1 to zero. "Observers in Arkhangelsk report that two additional objects are moving in front of and slightly faster than the satellite proper, and a third object is moving behind it. The results of the first observations are being forwarded to the U.S. The optical stations will continue tracking the ECHO II. The radio communications experiment is slated to begin somewhat later." (*NYT*, 1/27/64; *N.Y. Her. Trib.*, 1/27/64; *Izvestia*, 1/28/64, 6, ATSS–T Trans.)

- NASA Hq. received telegrams from Soviet Academy of Sciences containing first data from Soviet optical observations of ECHO II passive communications satellite. The data were obtained Jan. 25 from Soviet optical observation stations at Tashkent, Ashklabad, Erevan, Vologda, Archangel, Dushanbe, and Vorkuta, were being analyzed along with observations made at NASA stations around the world. Soviet Academy also indicated photographs of ECHO II were obtained at five stations. (NASA Release 64–21)

- President Johnson transmitted to Congress his annual space report, *United States Aeronautics and Space Activities, 1963* (H. Doc. 207). In his accompanying message the President said:

  ". . . Our space program, in both its civilian and military aspects, is peaceful in purpose and practice. Moreover, it combines such objective with a policy of international cooperation based upon a mutuality of participation and benefits as well as the wide dissemination of knowledge.

  "Space progress is essential if this Nation is to lead in technology and in the furthering of world peace. Such progress requires the use of substantial resources, which must be employed efficiently and effectively in order that we obtain the maximum benefits with a minimum of waste. . . ." (Text, *CR*, 1/27/64, 1045)

- Launching of Saturn I SA–5 rocket from AMR was postponed at T−100 min. because blind flange in liquid oxygen line to S–I stage had not been removed after recent pressurization check. Launch was rescheduled for Jan. 29. (*Wash. Post*, 1/28/64; *Marshall Star*, 1/29/64, 1)

- Performance of its two nuclear-test detection satellites has been excellent in all respects, DOD announced. The two satellites, launched more than two months ago for ARPA by USAF, were performing almost perfectly, and their reliability has been excellent. DOD anticipated the satellites would provide data on background radiation and other measurements in space for useful lifetime of more than six months. Success of this initial detection effort "has moved the research program forward by more than one year," DOD said. (DOD Release 30–64)

January 27: Secretary of Defense Robert S. McNamara testified in military posture hearings before House Committee on Armed Services, emphasizing that U.S. missile and bomber force was neither too large nor too small for today's "damage limiting" military strategy. He pointed out that missiles are more reliable than bombers, being mechanically reliable, capable of surviving enemy attack, and able to penetrate enemy defenses in war-time, while the bombers are vulnerable to enemy attack and less likely to reach target enemy defenses. (Fryklund, Wash. *Eve. Star*, 1/28/64)

- NASA and Rocketdyne announced Saturn IB launch vehicle would use cluster of eight uprated H–1 engines for manned earth-orbital missions, each engine developing 200,000 lbs. of thrust for total of 1,600,000 lbs. (*M&R*, 1/27/64, 9; *Marshall Star*, 1/29/64, 1)

- FAA released summary statements on the supersonic transport by the six companies that submitted initial design proposals—three airframe companies (Boeing, Lockheed, and North American) and three engine manufacturing companies (Curtiss-Wright, General Electric, and Pratt & Whitney Div. of United Aircraft). Boeing's proposed SST was based on basic arrow-wing shape, designed to carry 150 passengers, and cruise at mach 2.7. Lockheed's proposed SST was based on advanced "double delta" wing shape, designed to carry 218 passengers, and cruise at mach 3. North American's proposed SST was based on modified delta shape wing, was designed to carry up to 187 passengers, cruise at mach 2.65. Curtiss-Wright's summary said its proposed turbojet engine would produce thrust about twice that of current transport jet engines. GE's proposal offered choice between advanced turbojet or turbofan engine types suitable for cruising speeds at mach 2.5 to mach 3. Pratt & Whitney offered two configurations of turbofan engine for aircraft flying up to mach 3. (FAA Release 64–8)

- DOD announced $6,261,000 contract had been awarded by Army Corps of Engineers to Blount Construction Co. for construction of NASA Central Instrumentation Facility at Merritt Island, Fla. (DOD Release 84–64)

- Revival of George H. Darwin's 1879 theory that moon was formed by breaking off the earth, by two independent scientists, was reported in *Newsweek*. Geologist Donald U. Wise believes that when earth was molten mass, heavy material flowed into core, causing instability in its rotation and irregular elongation of its form. Then centrifugal forces pulled smaller knob away, sending it into orbit to become the moon. Fragments breaking away at same time fell on moon and caused pock-marked surface. Related theory offered by A. G. W. Cameron, Canadian physicist at NASA's Goddard Institute for Space Studies, was that earth lost at least half its mass, "enough material for ten moons," in large chunks, some of which clumped together to form the moon while others shot out to become asteroids. (*Newsweek*, 1/27/64)

January 27: First flight tests of new 16-in. rocket motor in Project Harp (High-Altitude Research Project) were expected in May at Barbados Is., B.W.I., *Missiles and Rockets* reported. Static tests on the motor were now underway. The rocket motor would be used to boost gun-launched payloads in the McGill University-U.S. Army project. The May series would include some 30 firings of the rocket vehicle and new versions of the Martlet-series missile. Project sources indicated use of multiple staging with new rocket motor could make possible the orbiting of 50–70 lb. payloads into 60–100 mi. earth orbit. (*M&R*, 1/27/64, 22)

- Article advising small business firms on obtaining space contracts and subcontracts, by former counsel to House Science and Astronautics Committee Spencer Beresford, introduced into *Congressional Record* by Rep. Joe L. Evins (D.-Tenn.):
"There are a few simple rules to follow. They can be listed quickly, but are often ignored.
"An even bigger stumbling block is attitude. Paradoxically, many businessmen take a political rather than a businesslike view of the Government, even when doing business with it. What needs to be stressed is that they would do better to treat the Government like a prospective customer—and a demanding one, at that . . . ." (*CR*, 1/27/64, A304)

- DOD announced appointment of new Chairman and four new members of Defense Science Board. New Chairman was Dr. Frederick Seitz, President of National Academy of Sciences, succeeding Dr. Clifford C. Furnas, President of New York State Univ. at Buffalo, whose term expired Dec. 31, 1963. (DOD Release 82–64)

- Total face value of Titan III program increased only $\frac{1}{3}$ of 1% during first 10 months of the program, according to Maj. Gen. Ben I. Funk (AFSSD) in *Missiles and Rockets*. When all present contracts are definitized, increase will amount to about 1.2% of original values. (*M&R*, 1/27/64, 9)

January 27–30: About 400 scientists and engineers representing 34 private industries and eight Government agencies met at NASA Marshall Space Flight Center, Huntsville, Ala., for Symposium of Manned Interplanetary Missions Studies Performed by Industry for NASA in 1963/64. (*Marshall Star*, 2/5/64, 1)

January 28: 100th flight of the X–15 rocket aircraft was conducted near Edwards AFB, Calif., with Maj. Robert Rushworth (USAF) as pilot. X–15 No. 1 reached 3,682 mph (mach 5.4) and 107,000-ft. altitude in its "fully successful mission." The X–15 was launched over Delamar Lake, Nev., its engine burning 76 sec. and the flight lasting 11 min.

X–15 program was initiated in December 1954 with signing of formal memorandum of understanding by NACA, USAF, and USN. Contracts were awarded to North American Aviation for construction of three X–15 aircraft and to Thiokol Chemical Corp.'s Reaction Motors Div. for manufacture of rocket engines. X–15 was designed to fly at mach 6 and 250,000-ft. altitude, objectives often exceeded in flight. First powered flight of X–15 was made Sept. 17, 1959, by NAA test pilot A. Scott Crossfield. X–15 research program has provided information on aerodynamics, structures, flight control, and physiological aspects of high-speed, high-altitude flight. Follow-on program, begun in 1962, utilizes X–15 as test-bed for scientific experiments beyond earth's atmosphere. (TWX, FRC to NASA Hq.; FRC Release 2–64)

January 28: In *Komsomolskaya Pravda,* Soviet Academician A. A. Blagonravov was interviewed by Tass correspondent regarding the launching of ECHO II: ". . . in time, international space cooperation will acquire decisive significance. Each country will contribute its best efforts to the peaceful conquest of space. The knowledge obtained will accrue to the benefit of all mankind. Certain long-range communication problems will be clarified by the joint investigations of the U.S. and U.S.S.R. scientists. There are other equally fruitful points of contact. The coordination of the program of space experiments envisages the use of satellites for weather service on a world-wide scale. Also the combined use of sputniks for compiling charts of the Earth's magnetic field will be a significant contribution to world science. These projects are part of the Soviet-American collaboration in the interests of mankind." (Tass, *Komsomolskaya Pravda,* 1/28/64, 4, ATSS–T Trans.)

- Early telemetry and radar reports indicated ECHO II appeared to be losing its spherical shape and high reflectivity, NASA spokesman announced. Telemetry showed ECHO II pressure "has decreased more rapidly than anticipated." However, "optical observations tell us we have a beautiful balloon." Analysis of TV photographs taken by camera mounted on Agena stage showed sphere had inflated properly; but inflating gas apparently leaked out after two hours instead of lasting for 20 hours as planned. However, quality of radio signals between Rome, N.Y., and Columbus, Ohio, and between Dallas, Tex., and Stump Neck, Md. was reported to be good. Spokesman indicated telemetry reports of pressure decrease could be false due to possible error in the observations. (*Wash. Post,* 1/28/64; *Chic. Trib.,* 1/28/64; Finney, *NYT,* 1/29/64, 12)

- NASA conducted three grenade experiments with Nike-Cajun sounding rockets launched from Wallops Island, Va.; Fort Churchill, Canada; and Ascension Island in the South Atlantic, part of current series to gather comparative data on high-altitude winds and atmospheric temperatures at widely separated locations. At Wallops Island, Nike-Cajun lifted off at 11:11 p.m. EST, followed by Nike-Cajun at Fort Churchill at 11:17 p.m. EST, and Nike-Cajun at Ascension Island, 11:18 p.m. EST. Grenades were ejected and detonated at altitudes ranging from 23 to 56 miles. The series of experiments was under direction of NASA Goddard Space Flight Center. (Wallops Release 64–11)

- Aerobee 300A vehicle launched from NASA Wallops Station carried 95-lb. instrumented payload to 192-mi. altitude in experiment to measure simultaneously electron and neutral particle temperatures in the upper atmosphere. Secondary objectives were to obtain data on ion and neutral particle densities. After $10\frac{1}{2}$-min. flight, payload impacted 156 mi. downrange in the Atlantic Ocean. Experiment was fourth in series of ionosphere studies, joint project of Univ. of Michigan and NASA Goddard Space Flight Center. (Wallops Release 64–10)

- NASA Aerobee 150 sounding rocket launched from White Sands, N. Mex., sent instrumented payload to 117-mi. altitude in successful test of the sounding rocket's attitude control system with the improved inertial reference system. Payload was recovered immediately after impact and found to be in excellent condition. (NASA Rpt. SRL)

- George M. Low, MSC Deputy Director and NASA Deputy Associate Administrator for Manned Space Flight, said in *Space Business Daily* interview that crew "aboard the Apollo spacecraft will be more than adequately shielded from solar flare radiation." Low cited National Academy of Sciences' opinion that solar radiation dosages to Apollo astronauts would be five to 10 times less than maximum tolerable dosage, and "the astronauts would be safe on their trip to the Moon if it is made during its presently scheduled time." He pointed out that special solar-radiation advisory committee (see Dec. 9, 1963) did not take into account the NAS information and the GSFC tables which show that solar flares during 1969–70 would not be as dangerous as heretofore thought. (*SBD*, 1/28/64, 144)

January 28: NASA and India's Dept. of Atomic Energy announced they were jointly conducting series of four sounding rocket launchings from Thumba Equatorial Rocket Launching Station to investigate equatorial electrojet, electric current flowing along earth's magnetic equator at about 60-mi. altitude. Thumba range is located near southwestern tip of India, on the magnetic equator. In the four electrojet experiments, India's Dept. of Atomic Energy provides launching site and supporting facilities; personnel for sounding rocket launch operations, acquisition of data and data analysis; and supporting ground magnetometer and meteorological observations. NASA provides Nike-Apache rockets, loans the launcher, telemetry equipment, and supplementary magnetometers. Indian launch and telemetry personnel were trained by NASA. Rocket magnetometer was built by Univ. of New Hampshire through NASA contract. (NASA Release 64–19)

• Spain and U.S. announced agreement to construct and operate space tracking and data acquisition station about 30 mi. west of Madrid. Cooperating agencies were NASA for the U.S. and Instituto Nacional de Tecnica Aeroespacial on behalf of Spain's National Space Research Commission. To be used primarily in NASA's Deep Space Network, the $1.5 million installation would include 85-ft.-diameter parabolic antenna and equipment for transmitting, receiving, recording, data handling, and communications with spacecraft. (GSFC Historian)

• Army Corps of Engineers awarded $5,154,551 contract to Robert E. McKee, General Contractor, Inc., for construction of NASA's Lunar Excursion Module Test Facilities at White Sands Missile Range. (DOD Release 89–64)

• Discussing U.S. supersonic transport program, Lockheed Aircraft Corp. vice president Dudley E. Browne told American Bankers Association's conference in Philadelphia that Lockheed had proposed "increased use of Government capability to underwrite both risk and financing, balancing this with sufficient risk assumption and financing by the contractor to justify returning him a reasonable profit. . . .

"Whatever final plan is adopted, the banks will be called upon to play a significant part.

"When they are, I think they will find themselves participating in a program whose magnitude they have never encountered before, a total program cost that may run into something like $8 or $10 billion, may involve loan limits for aircraft of as much as 20 years in place of the present 10 or 12, and may require supplemental participation of other financing organizations like insurance companies. . . ." (Clark, *NYT*, 2/2/64, 90)

January 28: USAF fired third and last interim Minuteman missile, a Wing II version with Wing VI second stage, from Cape Kennedy. This was 30th success in 45 Minuteman attempts from AMR. (*M&R*, 2/3/64, 13)

- Distinguished Flying Crosses were presented to 14 members of U.S. Army and USAF for record mass free-fall parachute jump Dec. 16, 1963, as part of Operation Halo. (DOD Release 87–64)

January 29: NASA launched Saturn I SA–5 from AMR in first successful flight test of both stages of the rocket. The 1.5-million-lb.-thrust S–I stage performed as planned, and the 90,000-lb.-thrust upper stage (S–IV) powered by six liquid-hydrogen RL–10 engines separated and burned properly for eight minutes, attaining orbital speed. The orbited body (spent S–IV stage, instrument unit, payload adapter, Jupiter nose cone, and 11,500 lbs. of sand ballast) weighed 37,700 lbs., nearly 20,000 lbs. of which was payload. As the first of the Block II Saturn I's, SA–5 was first flight test of both Saturn stages and only the second flight test of a liquid-hydrogen powered rocket stage (see Centaur AC–2, Nov. 27, 1963). The five Block II Saturn I's differ from Block I predecessors in that they have live upper stages, S–I propellant tanks are extended to provide 100,000 lbs. usable propellant, and tail fins are added for stability. During the flight eight onboard motion picture cameras photographed various operations of the rocket and a TV camera provided real-time photographs of separation and ignition of S–IV stage. Shortly after S–I burnout, the motion picture cameras were ejected, impacting downrange where pararescue men recovered seven of them. This test of the world's largest known rocket proved flight capability of Saturn I's liquid-hydrogen, clustered-engine upper stage and demonstrated the vehicle's capability to orbit 20,000-lb. payload. (*Marshall Star*, 2/5/64, 1; *NYT*, 1/30/64, 1; *M&R*, 2/3/64, 17–18)

- Following successful launching of Saturn I two-stage SA–5, President Johnson issued statement saying U.S. had now "proved we have the capability of putting great payloads into space. . . .

  "We have come a long way from the 31 pounds of EXPLORER I on January 31, 1958, to the some 37,000 pounds which has just been placed into orbit by Saturn I.

  "We have demonstrated not only enormous boost capability, but we have proved the effectiveness and the practical use of liquid hydrogen as a rocket fuel for space flight. . . ." (Knighton, Balt. *Sun*, 1/30/64)

- At Cape Kennedy press conference following Saturn I SA–5 launch, NASA Associate Administrator Dr. Robert C. Seamans stated, "There is little question that it took us ahead of the Russians in our capability." He noted that payload equivalent of SA–5's orbiting body was between 19,000 and 20,000 lbs., far exceeding the 14,292 lbs. reported for heaviest Soviet satellites. (*M&R*, 2/3/64, 18)

- RELAY II communications satellite transmitted portions of Soviet-American championship hockey game of the 1964 Winter Olympics at Innsbruck, Austria, to the U.S., the first public demonstration of RELAY II. Transmission of the event was made via French ground station at Pleumeur-Bodou and U.S. ground station at Andover, Me. By this time numerous communications tests of the satellite had been made, and NASA scientists said that all communications experiments were of excellent quality. RELAY II also was sending back information on radiation collected by onboard instrumentation. (GSFC Historian; NASA Release 64–24)

January 29: Dr. Robert R. Gilruth, Director of NASA Manned Spacecraft Center, said in speech at Macalester College, St. Paul, Minn.: "Although we have not yet equalled the Soviets in the number of hours for manned flight, I am convinced we have gained full dollar value for our Mercury program. The total project cost was less than one percent of our military budget for one year. The project was successfully completed with Gordon Cooper's flight of 22 orbits and 34 hours length in May 1963. Needless to say I am very proud of this team effort and the short time that passed from the day we were told to get on with the project until it was finished. . . ." (Text)

- NASA spokesman said ECHO II balloon satellite was still operating with "great success" despite preliminary radar reports that it was deflating and losing its reflectivity. The 135-ft. balloon was tested in transmission between Rome, N.Y., and Ohio State University at Columbus, Ohio. (UPI, *NYT*, 1/30/64, 12)

- Senator Clinton P. Anderson (D.–N. Mex.), Chairman of Senate Committee on Aeronautical and Space Sciences, noted on Senate floor that 100th flight of X–15 was made yesterday. "During the past 4½ years the program has achieved a very commendable record, providing a wealth of research data for the aeronautical and space program of the United States. The past accomplishments of the X–15 program reflect a superb job of management, flight-test operation, and cooperation between the National Aeronautics and Space Administration, the Air Force, and the Navy. . . .

  "The X–15 program not only has contributed to the increased confidence of designers of current high-performance aircraft but, in doing so, has focused attention on a few areas which require additional research. The planned program will be directed toward these areas. This program is providing information for the next logical step in the NASA's flight research program, the hypersonic.

  "In addition . . ., the X–15 program is going to produce information of value, I believe, to the development of supersonic commercial airplanes. This data will include flight controls, materials and surface temperatures.

  "The lesson from the X–15 is that cooperation among agencies having an interest in a project pays off." (*CR*, 1/29/64, 1358–59)

- NASA Goddard Space Flight Center announced award of $1,356,000 contract to Blaw-Knox Co. for design and fabrication of three 85-ft.-diameter dish-shaped antennas to be used in manned space flight tracking network for Project Apollo lunar mission. (GSFC Release G–4–64)

- NASA announced $300,000 contract award to Wayne State University to establish program to accelerate industrial application of aerospace-related technology. Organized under Wayne State's Center for Applications of Science and Technology, the program would receive and would process technical information through NASA; evaluate applicability of the information to prospective industrial users; and train applications engineers to match needs of specific industries with specific innovations or new knowledge. (NASA Release 64–23)

- Former President Dwight D. Eisenhower, in press conference in Detroit, remarked that the U.S. since 1955 had obtained "a great deal of dependability" in ballistic missiles. " . . . I do have confidence that we do have just as good missiles as there are in the world." (Jones, *NYT*, 1/30/64, 1)

January 29: Maj. Gen. Don R. Ostrander, Commander of USAF Office of Aerospace Research, addressed National Space Club in Washington:

"I am sometimes asked how military space research differs from space research conducted by civilian institutions or other government agencies. The answer, of course, is that there is no difference, except in the ultimate purpose for which it is conducted. As a matter of fact, there is very close liaison in this field between my Command, the National Aeronautics and Space Administration, the Office of Naval Research, and the scientific community. We make every effort to keep them informed about the work we are doing or sponsoring, and we are constantly exchanging information with them about our findings on the characteristics of space. Our job, however, is specifically to support the Air Force mission in space. . . .

"I know that you are all familiar with the potential radiation hazards to man in space, particularly with the deadly proton showers that accompany certain solar flares. We have made great strides in predicting the onset of these showers at our Sacramento Peak Observatory in New Mexico.

"We can predict safe periods with increasing accuracy—better than most weather predictions here on earth—but there is still a large gray area in which we do not know whether conditions will be safe or unsafe. It is possible that we could narrow this gray area significantly with that which penetrates the atmosphere and can be picked up at the observatory.

"If we had this knowledge, it would extend the time we could operate in space with absolute assurance of safety. It would also give us advanced early warning that would enable us if necessary to get our astronauts out of danger.

"Even in the present state of the art, our Sacramento Peak predictions have been crucial to the safety of the astronauts participating in the Mercury program. We have been able to give NASA absolute assurance of safe conditions, insofar as proton showers are concerned, when the astronauts have gone into orbit; and warnings of potential danger, at other times. . . ." (Text)

- U.S.S.R. opened first section of its 21-acre radiotelescope 55 mi. south of Moscow, Tass announced. The section, consisting of east-west portion of the cross-shaped antenna, was being used to measure with great accuracy radio waves originating in the outer solar corona, part of an IQSY project. (*NYT*, 1/30/64, 12)

January 30: NASA launched RANGER VI lunar probe with Atlas-Agena B launch vehicle from AMR in lunar landing mission to photograph the moon's surface. The Atlas booster fired properly and, after stage separation, the Agena B stage ignited and attained orbital speed. After coasting in parking orbit of about 110-mi. altitude, the Agena B reignited, sending RANGER VI through the "launch window" on a course which would pass within 600 mi. of the moon. A midcourse correction maneuver was planned so that the spacecraft would impact the lunar surface. The only scientific instruments onboard the 804-lb. spacecraft were six TV cameras, designed to provide more than 3,000 photographs of the lunar surface during the last 10 min. of the 66-hr. flight. This was the first Ranger launch since October 1962, when extensive review of the Ranger project was undertaken because of previous Ranger failures. For RANGER VI, changes were made in subsystems to improve reliability and provide redundancy in some areas; also, sterilization requirements for Ranger spacecraft were relaxed, since excessive heat was suspected of causing some previous malfunctions. (NASA Release 64–16; Witkin, *NYT*, 1/31/64, 1; LRC Release 64–10; Miles, *Wash. Post*, 1/31/64, 1)

January 30: U.S.S.R. announced orbiting of two satellites, ELECTRON I and ELECTRON II, with a single launch vehicle. Soviet news agency Tass said the satellites were studying "the internal and external radiation belts of the earth and physical phenomena connected with them." ELECTRON I was orbiting at 7,100-km. (4,412-mi.) apogee; 406-km. (252-mi.) perigee; 2-hr., 49-min. period; and 61° inclination; ELECTRON II was orbiting at 68,200-km. (42,379-mi.) apogee; 460-km. (286-mi.) perigee; 22-hr., 40-min. period; and 61° inclination. (Tass; *Wash. Post*, 1/31/64, A14)

- At American Meteorological Society meeting, UCLA campus, Lewis D. Kaplan presented paper on current ideas about atmosphere of Mars—surface pressure is considerably lower than previously estimated and CO_2 concentration is considerably higher—and discussed significance of these findings for Martian-atmosphere entry spacecraft. (AMS Program, 717)

- NASA submitted to the Congress its report on H.R. 6651, to provide for study on establishment of a National Space Museum. NASA position was that since the National Air Museum of the Smithsonian Institution had undertaken to preserve and display space objects, enactment of H.R. 6651 was considered unnecessary. (NASA LAR III/15)

- NASA's equal opportunity employment activities were commended on Senate floor by Sen. Jacob K. Javits (R.–N.Y.). Senator Javits introduced in the *Congressional Record* letters from NASA indicating specific measures taken in broad-based equal opportunity program, especially in Huntsville, Ala., and Wallops Station, Va. (*CR*, 1/30/64, 1328–29)

- President Johnson appointed Dr. Herbert F. York, Chancellor of Univ. of Calif. at San Diego and La Jolla, to the President's Science Advisory Committee succeeding Dr. George B. Kistiakowsky. (AP, *NYT*, 1/31/64, 11)

- Interviewed in Houston, where he toured NASA Manned Spacecraft Center, NASC Executive Secretary Dr. Edward C. Welsh commented that U.S.S.R. leads U.S. in "total manned [space] flying hours, aggregate weight lifted into space, and some phases of biomedicine.

 "But we are ahead in space navigation, communications, weather satellites, and boosters.

 "I doubt that they now can equal the thrust of Saturn I, although I think they eventually will do so." (*Houston Chron.*, 1/31/64; *Houston Post*, 1/31/64)

- Dr. Edward C. Welsh, Executive Secretary of National Aeronautics and Space Council, addressed Texas City Chamber of Commerce, discussing these summarized points:

 "1. The space program is a sound investment to improve the economic health of the country and is not a wasteful expenditure of funds so long as the program is handled on a business-like basis.

 "2. The space program is being carried out as a partnership with private enterprise and requires cooperation with communities throughout the country.

"3. The space program does not and should not drain resources from other important projects, such as schools and housing; in fact, it stimulates both.

"4. The space program is essential to our national security—through both its NASA activities and its Department of Defense activities.

"5. The lunar program is a key feature in the space program, as it provides a clear objective and it requires the orderly development of a wide range of competences, all of which will have value for future space performance.

"6. International cooperation in our space program can be mutually beneficial and will not impair our national security or inhibit the development of our own space capabilities.

"7. The space program stimulates the supply of trained scientists and engineers, but does not require them in such quantities as to interfere with the constructive employment of the vast majority of them in other lines of endeavor." (Text)

January 30: NASA Administrator James E. Webb addressed World Affairs Council of Northern California, in San Francisco, discussed national and international aspects of the U.S. space program:

"The competition is tough. The U.S.S.R. has the ability to concentrate its efforts without public debate and without the regard for individual freedom of choice or living standards which enjoy high priority in our own country. But it does not follow that our own space effort should be similarly concentrated. The implications of space are broad, broader than we can now know, and they do not end on the moon. The necessary technologies are enormously demanding and therefore enormously promising. And, whatever we do in space, we are critically dependent upon a thorough knowledge of the space environment. Moreover, the Soviet program, which has appeared to many of us to be so narrow and directed almost exclusively to prestigious exploits, more recently gives signs of broadening to include considerable basic research as well as practical investigations, for example, through the use of weather satellites in which we pioneered.

"Therefore, we must not be trapped into a narrow competition—focusing our energies on one mission or even on manned flight alone. Our preparations must be broad enough to give us a flexible base suitable for a continuity of effort and ability to move in a direction or directions perhaps still unknown.

"The projects and programs, national and international, in which we are engaged have unprecedented scope and potential. It is this scope and this potential which make space activity a valuable instrument of national and international policy. It is a many faceted tool to be used for the economic and social advance of this and cooperating nations as well as for the preservation of our technological and political leadership in the world. This tool and its importance should be fully understood and appreciated. At the same time, as Lord Hailsham has said, we should remember that international cooperation is not a substitute for national excellence.

"Europe has not been alone in recognizing these factors. In the complex of UN organizations, great interest has been evidenced among nations great and small in the problems and benefits of space.

"Here again, space has served to bring East, West, and neutrals together in a code of legal principles to govern space activity, which was unanimously adopted in the last session of the General Assembly. In the associated forum of the International Telecommunications Union agreements were reached on the assignment of radio frequencies for space research and space communications. Without these arrangements the future of communications satellite systems would be heavily clouded. And in another UN agency, the World Meteorological Organization, a start has been made in the evaluation and planning of requirements for a world weather satellite system which also promises real economic and human benefits.

"Thus it is clear that our national effort in space research and exploration is becoming a powerful force in developing greater international cooperation—cooperation which is not limited to the nations of the Free World, but which spans the Iron Curtain, as well. It is in this context, as a force which joins all mankind against a common enemy, the hostile environment of space, rather than against each other, that the space effort may bring the most significant and enduring rewards." (Text)

January 30: In its annual report to the Congress, AEC reported that "substantial increases" had been made in explosive power of thermonuclear warheads for Polaris, Minuteman, and Titan long-range missiles. AEC also said progress had been made in developing nuclear weapons that produce less radioactive fallout. The advances were made in 1962 test series of atmospheric explosions in the Pacific. (Finney, *NYT*, 1/31/64, 1)

January 31: Midcourse maneuver for RANGER VI lunar spacecraft was executed from NASA–JPL Tracking Station, Goldstone, Calif., 16 hrs. after the spacecraft was launched from AMR. The radio signal activated an onboard motor which fired for 69 sec., aiming the spacecraft on collision course with the moon. Target for the lunar spacecraft: Sea of Tranquility. (Witkin, *NYT*, 2/2/64, 1; *Houston Chron.*, 1/31/64)

• Sixth birthday of EXPLORER I, first U.S. satellite, which was launched into orbit with a Jupiter C vehicle by JPL–ABMA team. EXPLORER I has traveled around the earth 26,315 times covering some 775 million miles. With initial orbit of 1,573-mi. apogee, 224-mi. perigee, and 114.8-min. period, EXPLORER I now orbits at 1,011.6-mi. apogee, 212.5-mi. perigee, and 104.6-min. period. It was originally expected to re-enter atmosphere and disintegrate after five years, was now expected to last until 1966. Its smaller transmitter relayed information for 105 days; the larger transmitter, 31 days. Onboard instrumentation discovered radiation belt around the earth, named after Dr. James A. Van Allen who provided the experiment. (*Marshall Star*, 1/29/64, 4; "NASA Chronicle of Earth Sat. & Space Probes, 1957–60")

• In letter to Speaker of the House and President Pro Tempore of the Senate, NASA Administrator James E. Webb indicated NASA electronics research center should be located in the Greater Boston area. "It is our carefully considered opinion that to meet the specific requirements of the NASA program in respect to time and in respect to existing resources, the center should be located in the Greater Boston area." In accordance with P.L. 88–113, NASA had reviewed some 50 locations as possible sites for the center. (NASA LAR III/18; UPI, *Houston Chron.*, 2/1/64)

January 31: Successful testing of design of radar antenna 100 times more sensitive than world's largest radiotelescope (Arecibo, P.R.) was announced by USAF Office of Aerospace Research. The multiple antenna was designed by Dr. Allan Schell of OAR Cambridge Research Laboratories, as space surveillance and tracking radar system and for use as a radiotelescope. One-year evaluation of test section verified predicted performance characteristics of full-size antenna. The multiple antenna would consist of some 5,000 flat metal plates each 20-ft.-by-20-ft. arranged in four elliptical areas around 1,000-ft. tower. Studies were continuing to determine optimum size from standpoint of economy and performance, tower stability, and antenna control system. (OAR Release 1-64-2)

- Sen. Warren G. Magnuson (D.-Wash.) was quoted as saying: "In view of the tremendous success of the Saturn flight, perhaps further savings in the cost of the space program might be possible between now and the first Saturn V flight by leap-frogging some of the intermediate steps." (*SBD*, 1/31/64, 169)

- Rep. J. Edward Roush (D.-Ind.) said in the House: "On Wednesday we saw the successful launching of the Saturn rocket carrying into orbit around the earth 18 tons. This included the largest payload ever placed in orbit by man and exceeds the largest Russian payload by 11 tons. For several years now this Nation has been plagued with the uncomfortable knowledge that the Russians were ahead of us in the space effort. This has been based primarily on the fact that they had a greater lift capability. The United States is now ahead in lift capability and in every other phase of the space program. . . ." (*CR*, 1/31/64, 1443)

- RELAY I communications satellite continued operating normally and the one-year onboard timer had not cut off as of this date. (GSFC Historian)

- Communications Satellite Corp. submitted its annual report to the President and the Congress on its activities February 1–December 31, 1963. ComSatCorp reported it had made plans for development and establishment of commercial communications satellite system with global capability, calling for launching of one or more synchronous-orbit satellites over the Atlantic Ocean in 1965 on an experimental-operational basis and initial launching of satellite for the comsat system in 1966. (ComSatCorp Rpt., 1/31/64)

- DOD announced award of $7,112,913 contract to Franchi Construction Co. for construction of NASA launch operations center headquarters building at Merritt Island, Fla. Contract was awarded by Army Corps of Engineers. (DOD Release 96-64)

During January: President Johnson sent letters to all major space/defense contractors urging them to achieve significant reductions in expenditures. In separate letters from NASA Administrator James E. Webb, contractors were told of three areas offering greatest potential for significant savings: purchasing at lowest price compatible with reliability requirements, placing greater emphasis on incentives, and reducing operating costs. (NAA *S&ID Skywriter*, 1/10/64, 1)

- NASA began installation of aerospace ground equipment at Cape Kennedy's Launch Complex 19 for Project Gemini, work beginning ahead of previously expected date in effort to expedite first Gemini flight. (*M&R*, 1/13/64, 9)

During January: Snap–10A space nuclear power system completed a record 90-day continuous test operation, in test conducted for AEC by Atomics International. Purpose of the test, conducted with mockup of the Snap–10A, was to check operation of various components as a system under conditions closely simulating space environment. (*SBD*, 1/27/64, 136)

- NASA awarded Martin Co.'s Baltimore Div. a $1,499,111 contract for Part II study of post-Saturn launch vehicles, a one-year study to include overall vehicle definition and preliminary design. (*Marshall Star*, 1/15/64, 1; *M&R*, 1/13/64, 9)
- Ryan Aeronautical Corp. was awarded contract from RCA to design landing radar for the Project Apollo Lunar Excursion Module. RCA, responsible for major LEM subsystems, would integrate landing radar with LEM's overall sensor system. (*M&R*, 1/13/64, 10–11)
- Twelfth session of Mexico-U.S. Commission for Space Observations held in Mexico City to consider report on construction at Guaymas tracking and data acquisition station to modify it for Project Gemini. Guaymas station was one of five being built or updated for Gemini tracking—Carnarvon, Australia; Bermuda; Canary Islands; and Cape Kennedy. (*SBD*, 1/13/64, 59)
- W. V. Pangborn and Co., Inc., and Lowery Electric, Inc., submitted joint bid of $1,443,289.20 for NASA–USAF contract to build frequency control and calibration facility at Cape Kennedy's Merritt Island. Facility would become part of AMR tracking network, would be aimed at supporting Titan III and Saturn V. (*M&R*, 1/20/64, 12)
- NASA released to industry its first technology utilization report, entitled "An Improved Precision Height Gage." NASA Ames Research Center developed the report. (NASA SP–5001)
- Seth W. Booth, Group Director, Re-entry Feasibility Group, San Bernardino Operations of the Aerospace Corp., was presented AFSC's Outstanding Achievement Award by Maj. Gen. W. Austin Davis, Commander of AFSC Ballistic Systems Div. Booth was cited for his exceptional technical direction to BSD contractors for various re-entry programs from 1956–1963. (*A–N–AF J&R*, 1/18/64, 10)
- 120-in. solid-propellant rocket motor being developed by United Technology Center for USAF was static-fired for nearly two minutes, a "completely successful" test. (*SBD*, 1/30/64, 163)
- Reviewing the worth of scientific results gained in the first years of U.S. space exploration, *Fortune* magazine said: ". . . new knowledge is a dukedom whose great wealth and resources cannot even begin to be estimated or exhausted. Already the new knowledge acquired in space exceeds by far the value of funds so far spent. For knowledge, more than guns and butter, is the true power of modern states." (NASA FY 1965 Auth. Hearings, Part I, 3)

February 1964

February 1: President Johnson stated in press conference:

"This past week the United States has demonstrated in at least eight different situations this Nation's determination to insure both peace and freedom in the widest possible areas.

"Progress toward these ends is frequently slow and rarely dramatic, but it should be viewed in the perspective of history and not headlines. . . ."

The President listed the first seven situations, then

"And finally, we have witnessed and the whole world has witnessed with pleasure the remarkable success of our Saturn rocket, the most powerful rocket thrust known to man. This rocket, I am happy to say, was first recommended by our committee [Senate Preparedness Subcommittee, chaired by Johnson] in 1958." (Transcript, *Wash. Post,* 2/2/64, A7)

- Editorial in *Christian Science Monitor* said:

  "It is one thing to question the wisdom of spending so much so quickly to put an American hopefully on the moon in this decade, a time limit that the economy drive in Washington may remove. But it would be less than wisdom to cut back sharply on space funding under the false assumption that the United States is pulling ahead of the Soviets. It is still coming up from behind." (*CSM,* 2/1/64)

- Roscoe Drummond, in his *Washington Post* column, said success of Saturn I SA-5 ". . . means that the major advantage which the Soviets have had so long and which has yielded them their principal space spectaculars—the long Russian lead in rocket power—is now a thing of the past. . . .

  "There should be no minimizing what the Soviets have already done. . . .

  "But the United States accomplishments have been more diverse. We have put almost four times as many spacecraft into orbit as the Soviets. We have made far more progress in the application of space technology in communications, weather, and navigation. We both have compiled a large volume of scientific data about outer space and neither is significantly ahead.

  "As Dr. Edward C. Welsh, Executive Secretary of the National Aeronautics and Space Council, puts it, 'the space race continues and it will be a long one.'" (Drummond, *Wash. Post,* 2/1/64, in *CR,* 2/1/64, A441)

- Nuclear "breeder" reactor using thorium fuel, pioneered by Vice Adm. Hyman G. Rickover, was under consideration by AEC and Calif. Dept. of Water Resources for a nuclear power plant, according to recent letter from AEC Chairman Dr. Glenn T. Seaborg to Sen. John O. Pastore, Chairman of Joint Congressional Committee on Atomic Energy. Built around the "seed-and-blanket" design developed by the Rickover team, the thorium-fueled reactor appeared to have "unusual potential for generating economic nuclear power and for breeding." If successful, Dr. Seaborg said, the concept would "represent a major advance in reactor technology and an important milestone in our national objective of conserving nuclear fuel resources." (Finney, *NYT*, 2/2/64, 1, 30)

February 1: At Washington press conference, National Bureau of Standards and team of visiting Soviet metrologists announced preliminary agreement for extended cooperation in measurement and standards data exchange. Agreement provided for exchange of (1) data on observations of standard time and frequency signals from radio stations, (2) calibration information to compare measurements of electrical and electronic quantities, and (3) publications of National Bureau of Standards and Soviet Committee on Standards, Measures, and Measuring Instruments. (*M&R*, 2/10/64, 10–11)

- Special AFSC management office for the Manned Orbiting Laboratory (MOL) program was established in AFSC Hq., headed by Col. R. K. Jacobson, Assistant Deputy Commander for Space for MOL. (AFSC Release 41-R-16)

- According to USAF authorities in Berlin, preliminary examination showed that the T-39 jet trainer which crashed in East Germany last Jan. 28, killing the three Americans aboard, was shot down by Soviet aircraft. Denying charges made by East Germany, USAF spokesman said the aircraft carried no arms or photographic equipment, was incapable of reconnaissance. (*Wash. Post*, 2/2/64, A16)

- Soviet Chief Air Marshal Konstantin A. Vershinin said in *Red Star* that Soviet air force can destroy any enemy land, sea, or air target. Soviet air power now is based on supersonic, rocket-carrying planes capable of operating in all weather conditions. (UPI, *Wash. Post*, 2/2/64, A16)

February 2: RANGER VI crashed onto the moon on target in the Sea of Tranquility, but its onboard camera system failed to return TV photographs of the lunar surface. Indications were that the two TV systems responded to Central Computer and Sequencer (CC&S) command to warm up, but the cameras did not complete the warm-up to picture taking. RANGER VI's cameras (two wide-angle and four narrow-angle) were to have made more than 3,000 photographs during its last 10 min. The photographs were to have provided scientific data on lunar topography and supported the Surveyor unmanned soft-landing spacecraft and Apollo manned lunar landing program.

Dr. William Pickering, Director of JPL which is responsible for Ranger project management as well as its tracking and communications through Deep Space Network, said of RANGER VI:

"It was a sad finish to something that was going so good.

"Our accuracy was a very significant achievement. The failure of the camera system isn't going to discourage us. I'm quite confident in our ability, and we will succeed in exploring the moon on a future shot." NASA planned three more Ranger TV flights this year. (Witkin, *NYT*, 2/3/64; UPI, *Wash. Post*, 2/3/64; AP, *Wash. Post*, 2/3/64)

- Sir Bernard Lovell, Director of Jodrell Bank Experimental Station which tracked RANGER VI on its course to the moon, said of the spacecraft's accurate impact of the moon: "I think this represents a very great achievement for American rocket engineers.

"This achievement is underlined by the fact that it is now four and a half years since either the Russians or the Americans have made a direct hit of this nature on the lunar surface, despite repeated attempts."

Kenneth W. Gatland, Director of the British Interplanetary Society, referred to the failure of RANGER VI's cameras to take pictures: "I should say this is one of the most bitter disappointments of the space age so far in having come so near to success in a vital reconnaissance." He speculated that the cameras may have failed to work because the temperature caused a relay to stick or a component to fail. (AP, *NYT*, 2/3/64)

February 2: Tass announced ELECTRON I and ELECTRON II scientific satellites were continuing their elliptical orbits of the earth. By 6:00 p.m. Moscow time Feb. 1, ELECTRON I had passed repeatedly through the inner radation belt in its 19 orbits and ELECTRON II had moved four times through the outer radiation belt in its two orbits. Tass said the following studies were being made with aid of onboard equipment: outer and inner radiation belts; charged particles having low energy; concentrations of electrons and positive ions; magnetic fields and radiation belts of the earth; nuclear component of cosmic radiation; shortwave solar radiation; propagation of radio waves; radio-radiation of galaxies; and densities of meteoritic material. Tass said the studies, being accomplished under a unified program in various areas of outer space, were being conducted for the first time. These studies were important for conducting the program of the IQSY and for permitting collection of data to assure radiation safety of manned space flights. Tass said all onboard equipment was functioning normally and ground stations were receiving scientific data from the satellites. (Tass, *Krasnaya Zvezda*, 2/2/64, 1, ATSS-T Trans.)

- 300-ft.-high helium-filled balloon with 550-lb. instrumented payload was recovered 30 mi. west of Williamsport, N.C., after seven-hour flight from Palestine, Tex., where it was launched by the National Center for Atmospheric Research. Scientists from Case Institute, which sponsored the launching, said the cosmic-ray-studying instruments performed perfectly during the flight. (*Houston Chron.*, 2/3/64)

February 3: NASA established RANGER VI Review Board, composed of members from NASA Hq., Langley Research Center, and Goddard Space Flight Center, and chaired by Deputy Associate Administrator for Industry Affairs Earl D. Hilburn, to review program effect of RANGER VI failure. The Board would review results of failure analyses and corrective procedures being prepared by JPL, then make recommendations regarding remainder of Ranger flight program.

At press conference, NASA Associate Administrator Dr. Robert C. Seamans, Jr., explained that the single aim of the Board was to determine when Ranger 7 would be launched. Until the mystery of RANGER VI's failure was solved, NASA was delaying launch of Ranger 7, which also would have mission of photographing lunar surface. (NASA Release 64-26; Simons, *Wash. Post*, 2/4/64; Finney, *NYT*, 2/4/64)

- Nike-Cajun sounding rocket was launched from Wallops Island, Va., in coordination with Nike-Apache launch at Ascension Island, third set in current NASA series to obtain data on high-altitude winds and atmospheric temperatures. A correlated launch from Ft. Churchill, Canada, was postponed due to sound interference problems. The Wallops experiment was timed to coincide with the annual east-to-west "warming trend", was to measure its effects on the upper atmosphere. The experiments were designed to obtain measurements at altitudes ranging up to 55 mi. The Wallops experiment used grenade payload; for the Ascension experiment, the grenade payload was replaced with pitot-static probe to measure densities, pressures, and temperatures. The series was being conducted for NASA Goddard Space Flight Center. (Wallops Release 64–12; NASA Rpt. SRL)

February 3: NASA and U.S. Dept. of Commerce announced agreement to develop National Operational Meteorological Satellite System (NOMSS). Signed by NASA Administrator James E. Webb and Secretary of Commerce Luther H. Hodges, agreement provided for NOMSS which would be based on NASA-developed Tiros technology and would become operational in 1965. Compared with earlier plans, this joint program was expected to save Dept. of Commerce some $125,000,000 over five-year period. Once NASA orbited the satellites, Weather Bureau would operate and control the system and analyze, process, and distribute the meteorological data gathered by the satellites. Under reimbursable order from Weather Bureau, NASA would design, procure, test, launch, and track the weather satellites. (NASA Release 64–25; Commerce Dept. Release G 64–20; Text, NMI 2–3–30)

- 14 men selected last October for astronaut training reported to NASA Manned Spacecraft Center to begin training for the Gemini and Apollo space flight crew pool. (MSC Release 64–24)
- Dr. Wernher von Braun, Director of NASA Marshall Space Flight Center, said at Washington press luncheon that the two astronauts in Project Apollo's Lunar Excursion Module would stand up during the descent to the moon. He described support gear—like elasticized parachute harness—which would cushion the landing for the crewmen. Elimination of chairs and rearrangement of LEM's gear permitted weight reduction of 1,000 lbs., he said. (Halsell, *Houston Post,* 2/4/64)
- FAA began sonic-boom study program at Oklahoma City, Okla., generating sonic booms at the levels predicted for supersonic transport operations to determine public acceptability of these levels. Data acquired from the 26-week program would be used in planning and design of U.S. supersonic transport. FAA conducted the program with cooperation of USAF and NASA; USAF provided F–104 aircraft to generate the booms and NASA researchers provided ground instrumentation to measure booms and boom effect. (FAA Release 64–3)
- Interviewed in *Boston Globe,* Dr. Albert J. Kelley, Director of Electronics and Control Section of NASA Office of Advanced Research and Technology, said NASA planned a gradual buildup of personnel for the proposed electronics research center to a maximum of 2,100 in 1969. Year-by-year buildup: 1964, 50; 1965, 250; 1966, 550; 1967, 1,000; 1968, 1,600; and 1969, 2,100. Nucleus of the center's staff would be electronics task force currently operating in Washington. (White, *Boston Globe,* 2/3/64)
- NASA submitted report to Congress on patent legislation: to House Committee on the Judiciary on H.R. 4482, to provide administration in the public interest of property rights to inventions resulting from R&D financed wholly or partially by public funds; to Senate Committee on the Judiciary on S. 1290, to establish uniform national policy for proprietary rights to inventions made through expenditure of public funds;

and to Senate Committee on Aeronautical and Space Sciences on S. 1444, to amend National Aeronautics and Space Act of 1958 to waive proprietary interests of the U.S. in inventions. (NASA LAR III/19)

February 3: Secretary of the Air Force Eugene M. Zuckert testified before House Committee on Armed Services:

"In the field of military applications of space, our views as to the future remain unchanged. We believe that we must vigorously exploit the most likely avenues of interest, though we are not yet able to be definitive enough to describe man's military space role adequately to project specific weapon systems. . . .

"The Manned Orbiting Laboratory (MOL) program, which replaced the Dyna-Soar, is a research program aimed at giving man the opportunity to operate in space so that we may determine whether and when the manned space vehicle will be militarily significant. The MOL program will provide much of the supporting knowledge for indicated future manned space systems. The timing of the program is based upon a technical decision that our primary need is to know what functions man can perform in orbit, before exploring, as completely as Dyna-Soar would have, the problems and techniques of controlled reentry. This latter will now be studied on a more limited developmental basis." (Text, *AF Info. Pol. Ltr.*, 2/64)

- Maj. Gen. T. C. Bedwell, Jr., Cdr. of USAF Aerospace Medical Div. at Brooks AFB, reported that four airmen spent 30 days in pure oxygen atmosphere with no apparent ill effects. The volunteers were confined in special chamber at Brooks AFB for six weeks. During the first two weeks they were in mixed-gas atmosphere while they were acclimating to the surroundings and practicing space-related techniques; then the chamber's atmosphere was converted to 100 percent oxygen. With no initial harmful effects, the airmen would continue under medical surveillance for a long period. "Longer tests of this type are planned," General Bedwell said. "We need many statistics, not just a few samplings, to draw any definite conclusions." (AP, *Houston Post*, 2/4/64)

- General Curtis E. LeMay, Air Force Chief of Staff, testified before House Committee on Armed Services:

"The Air Force program of research and development seeks to assure that this Nation is first in the development of military capabilities in space. We have some cause for optimism in this area. Our Titan III program is proceeding satisfactorily. In addition, we are undertaking the development of a medium altitude communications satellite program. . . .

"The next major step toward the achievement of future space capabilities is the Manned Orbiting Laboratory (MOL) which was approved for development under Air Force management. The MOL will provide a means to fulfill the compelling requirement to acquire information essential to determining accurately the threat from space, the usefulness and the capabilities of man-in-space, and the unique advantages which may accrue from military space operations. It will also serve as a platform to support testing of equipment and procedures in the environment in which they will be used.

"Although the Dyna-Soar program has been terminated, a need still exists for development of a maneuverable aerospacecraft capable of controlled reentry and precision recovery, ferrying missions to and from a space laboratory, transfer of men and equipment in space, and a wide range of other roles. . . ." (Text, *AF Info. Pol. Ltr.*, 2/64)

February 3: Rocketdyne engine reliability of 100% in 135 orbital or deep space vehicle launchings over the last three years was reported in *Space Business Daily.* (*SBD*, 2/3/64, 175)

February 3–5: At aerospace medicine conference held at Brooks AFB, NASA Langley Research Center assistant director, Clinton E. Brown, said NASA was proposing a large space telescope in earth orbit capable of orbiting with manned orbiting research laboratory. "A preliminary step towards a very large scale telescope might consist of an enlarged orbiting astronomical laboratory (OAO) with a 100 to 120 in. aperture, but specifically designed to make use of recoverable film and the adjustment and operational advantages of a manned space control center. . . ." Brown said analysis made by Dr. Lyman Spitzer, Jr., Princeton Univ. astronomer, showed that "the moon is not the place for such an observatory but rather a manned earth orbiting system is to be desired." (*Av. Wk.*, 2/24/64, 71)

- At space medicine conference, Brooks AFB, Lt. Gen. James Ferguson, USAF DC/S for R&D, said space rendezvous maneuvers of MOL with a winged ferry vehicle were under consideration: "We visualize such a ferry craft to be a lifting body with some maneuverability in the atmosphere for precision landing. Undoubtedly it would be able to carry more than two men, as well as be a logistics carrier." (*M&R*, 2/10/64, 35)

February 4: U.S. Committee for International Quiet Sun Year (IQSY) and USN announced satellite to monitor solar x-ray emission, instrumented by Naval Research Laboratory scientists, was launched recently to initiate continuous "watch on the sun" during IQSY (1964–65). First signals from the satellite showed the sun now is close to its minimum activity, reported NRL scientists Robert W. Kreplin, Talbot A. Chubb, and Herbert Friedman. Any solar activity that occurs this year should be isolated and relatively uncomplicated, permitting the satellite to record individual storm centers, in contrast to IGY which was marked by succession of overlapping storms. Observatories in 14 countries joined U.S. in preparing to utilize continuous data transmissions from the satellite.

At press conference, the NRL scientists said that astronauts on lunar flight would be under no more radiation danger during maximum period of solar cycle than during minimum period. Friedman explained: "All available evidence indicates that solar minimum is as bad as solar maximum for superflare eruption." To date, superflares occur at random and cannot be predicted. (NAS–NRC Release; USN Background Statement; Simons, *Wash. Post*, 2/5/64)

- House Committee on Science and Astronautics began hearings on NASA authorization for FY 1965. NASA Administrator James E. Webb testified:

"In the tight budget situation faced by the President for Fiscal Year 1965 it was necessary to stress with him and the Bureau of the Budget that unless the full $5.3 billion authorization requested for FY 1965, *and* the supplemental appropriation of $141 million requested for FY 1964 are approved, the manned space flight program will encounter further delays. It will then not be possible to achieve the national goal of exploring the moon with men within this decade.

"There may be some inclination to assume that the adjustment of our program to the reductions imposed for FY 1964 indicates that this will likewise be possible in the event of reductions in the request which is before you at this time. I cannot emphasize too strongly that this is not the case.

"In adjusting our program to compensate for the reductions made in the appropriation for FY 1964, we have already sacrificed the margins and early target dates which were needed, and which in reality are still needed, in our effort to achieve the goals which have been set for us. In fact, Mr. Chairman, we are not able to maintain a reasonably balanced program, and accommodate the entire reduction, with any strong assurance that we can meet the 1970 lunar target date. The best we can say is that we have a 'fighting chance.'

"To sum up our situation in a sentence, *if we do not receive the funds which the President has requested, there is nothing left to sacrifice except the National goal itself.*

". . . whether it is agreed or not that international considerations and national security factors require that we hold to the present pace of the space program, the fact is that prudence and economy will be served. Even if economy alone were to be the guiding consideration in the evaluation of the NASA request for FY 1965, the cost of establishing and maintaining superiority in space will be less if we maintain the pace, the momentum, which the supplemental appropriation and this authorization request will provide.

". . . Despite the difficulties which have been experienced, we can still put two American explorers on the moon in this decade, and we can do it for less than $20 billion. . . . [But] if the program is further curtailed, if the momentum is lost, if the Apollo program is stretched into the next decade, the cost will *not* be under $20 billion, *it will be several billions more*. The ultimate cost will increase for each year in which achievement of our goal is delayed.

". . . [According to a NASA study], the cost of the lunar exploration would increase by approximately $1 billion for each year that the landing is delayed. A three year delay would cost $3 billion, a six year delay $6 billion, with no corresponding improvement in the benefits obtained. . . ." (Testimony: *CR.* 2/5/64, A545–48)

February 4: NASA Deputy Administrator Dr. Hugh L. Dryden testified before House Committee on Science and Astronautics:

"Because of the initiative and the daring of the Wright Brothers, this nation gave man the capacity for powered flight, freeing him forever from the bonds which for thousands of years of human existence had confined his activities to land and sea. The United States became the first country in the world to possess a military airplane when, in 1908, the Army Signal Corps contracted for a Wright biplane.

"Yet, prior to World War I this nation was still so pre-occupied with conventional weapons systems that it totally neglected the development of aeronautics—the force which was to dominate warfare for the next quarter century. In 1914, the United States possessed fewer military aircraft and of inferior types than the six leading aeronautical nations (including Mexico). The United States in 1914 was the only major nation in the world not to possess an aeronautical laboratory with an up-to-date wind tunnel. By November 1918, not one aircraft of American design and manufacture had entered combat operations during World War I.

"In the 30's we were so pre-occupied with refinement of conventional piston-driven aeronautical systems that we made little progress in jet propulsion. Meanwhile, the Germans set out to build a bigger and better NACA and to a large extent, they did, developing jet propelled military aircraft and 5½-ton V-2 rockets, which almost spelled disaster in World War II.

"In the late 40's, despite the fact that Robert Goddard had demonstrated the feasibility of a liquid-fueled rocket engine in this country in 1926, and despite the memory of V-2's raining on London during the blitz, we were so pre-occupied with mating jet carriers to our exclusively held atomic bomb capability that we neglected missilry while other nations forged ahead. And finally, in the 50's, our A-bomb advantage gone, we were so pre-occupied with the development of our ballistic missile program that we neglected a clear opportunity to become first in space.

"Today, the nation faces, we all face, this question: Have we learned enough from the often bitter and always costly experience of the last half century not only to carry out with determination this effort to meet the requirements of the present in space research and exploration, but to exercise the vision which is demanded if we are not, once again, to find ourselves lagging in the next phase of this most challenging effort? . . .

"The present gap in manned flight activity is a direct consequence of a postponement of the decision to proceed beyond Project Mercury from September 1960, until May 1961, when President Kennedy recommended the present manned lunar landing project as a national goal.

"The decisions which confront us today are those which will determine whether this kind of history will repeat itself a few years hence and whether we will once again experience a bitter awakening to the fact that others have seized the initiative in the more advanced space missions of the future. . . ." (Testimony; *CR*, 2/5/64, A548–50)

February 4: Nike-Cajun sounding rockets with grenade experiments were launched at Wallops Island, Va., and Ft. Churchill, Canada, in current NASA series of comparative upper atmosphere studies. The experiments were designed to obtain wind and temperature data and to measure the effects of the "warming trend" at altitudes ranging up to 55 mi. Nike-Cajun launched from Wallops reached 73.8-mi. altitude, with all 12 of its grenades exploding; Nike-Cajun from Ft. Churchill reached 78-mi. altitude, with 11 grenades functioning normally and the twelfth ejecting but not exploding. (Wallops Release 64–14; NASA Rpt. SRL)

- Dr. Joseph F. Shea, manager of Apollo Spacecraft Office at NASA Manned Spacecraft Center, said in Milwaukee speech that the most probable date for U.S. manned lunar landing was late in 1968 or early in 1969. He discussed goals of Ranger TV probes, Surveyor lunar lander, and Lunar Orbiter probes, which would provide information about the moon preparatory to manned lunar landing. If these should fail, he said, NASA would plan manned Apollo reconnaissance flights to orbit the moon before the landing flight. "It might set our landing back six months to a year, but if the other programs don't work, we aren't dead." (Pease, *Milwaukee Journal*, 2/4/64)

February 4: Scientists at Aerospace Medical Div., Brooks AFB, said in paper presented by Maj. William B. Clark at space medicine conference that, with proper reconnaissance training, astronauts in space could easily spot missile bases, encampments, troop movements, and "unsuspected targets of opportunity" from outposts 100-mi. high. The scientists' report was based on studies of visual reports of six U.S. astronauts who made space flights and relied heavily on accounts by Astronaut Leroy Gordon Cooper (Maj., USAF). (AP, *N.Y. Her. Trib.*, 2/5/64; AP, *Phil. Eve. Bull.* 2/6/64)

- Rep. James C. Cleveland (R.–N.H.), member of House Select Committee on Government Research, wrote in *Washington World* magazine:

  "Fourteen to fifteen billion dollars of the taxpayer's money is going into R&D this year. The figure was only $2 billion 10 years ago.

  "More than 30 different departments, bureaus, and agencies of the Federal Government conduct research or have it conducted for them, often in the same fields. . . .

  "A breakdown of Federal research and development spending shows that five organizations account for 90 percent of it. These are the Department of Defense, the National Aeronautics and Space Administration, the Department of Health, Education, and Welfare, the Atomic Energy Commission, and the National Science Foundation. About 70 percent of these funds support military research and development. About 10 percent of the total funds is channeled to colleges and universities. . . ." (*CR*, 2/4/64, A563)

- NASA scientist John M. Eggleston told space medicine conference at Brooks AFB that astronauts on lunar flight would face less danger from space radiation than was once thought. "If the Apollo spacecraft as it is now designed were exposed to the largest solar flare seen in the latest solar cycle, the astronaut would be exposed to only one tenth of the critical dose designated for space flight.

  "An exposure to the critical dose would probably make an astronaut sick within a few days. . . . But it would not prevent him from finishing the mission and returning to lead a normal life." (AP, *Houston Post*, 2/5/64)

- In Washington *Evening Star* article, William Hines reported Georgetown Univ. astronomer Dr. Carl C. Kiess' conviction that life cannot exist on Mars. Dr. Kiess' spectrographic observations of the planet convinced him no oxygen or water vapor existed there. Nitrogen compounds have been detected, and Dr. Kiess believed Martian atmosphere is 97–99% pure nitrogen gas. He believed density of air at Martian surface is greater than is generally believed, meaning that nitrogen tetroxide could exist as gas, liquid, and solid—gaseous above 70° F, solid below 14° F, and liquid in between. In laboratory experiments Dr. Kiess had demonstrated that when various nitrogen compounds are subjected to temperature changes they turn color to duplicate "seasonal" color changes observed on Mars, even creating artificial "snow storm" and simulating waxing and waning of Martian "polar caps." Dr. Kiess' view was minority opinion, refuted by majority of scientists. Dr. Lewis Kaplan of JPL said in interview that Dr. Kiess' conclusions were based on spectrographic analyses which had been refuted by better spectrographic analyses of Vancouver, B.C., scientists. (Hines, *Wash. Eve. Star*, 2/4/64)

February 4: Danish Government announced it would join international telecommunication system based on global communications satellites. (AP, *NYT*, 2/6/64, 6)
- DOD announced appointment of Thomas F. Rogers, MIT scientist, as Assistant Director (Communications and Electronics), Defense Research and Engineering. (DOD Release 101–64)
- 42 FAA flight service stations were going to be converted into remote-control facilities with expected savings of between $30,000 and $40,000 per year per station, FAA announced. (FAA Release 64–9)
- DOD and AEC announced first at-sea automatic weather station powered by nuclear energy had begun operation in the Gulf of Mexico. The 60-watt Snap-7D nuclear generator was expected to last 10 years as compared to conventional batteries requiring recharging every six months. The NOMAD (Navy Oceanographic Meteorological Automatic Device) was forerunner of worldwide network of unattended weather stations designed to provide meteorological and oceanographic data. (DOD Release 103–64)

February 5: Preliminary evaluation of Saturn I SA-5 flight Jan. 29 indicated there was "no significant deviation or malfunction," according to MSFC Saturn Flight Evaluation Group. Because insertion velocity was slightly greater than predicted, the 37,700-lb. orbiting body was in an earth orbit slightly higher than expected: 470-mi. apogee, 70-mi. higher than expected; 167-mi. perigee, 7 mi. higher than expected. Timing of all significant actions in the launch sequence varied no more than half a second from prediction. First stage cutoff occurred after slightly more than 146 sec., S–IV stage ignited at 148 sec., and insertion into orbit occurred at 639 sec. (KSC Release 8–64)

- NASA Associate Administrator Dr. Robert C. Seamans, Jr., testifying before House Committee on Science and Astronautics, discussed NASA organization, management, procurement, and programing.

"The management and programing processes . . . require continuing program support, assuming reasonable accomplishment by NASA compared with our stated goals. Little that we start in one year is completed in that year. On the average, flight projects require three or more years prior to launch, while total programs have a life of ten to fifteen years. The governmental process for allocating resources can lead to a funding pattern at odds with this inherently long-duration need. We plan to accomplish major objectives over the period of a decade, but we must tailor our actual accomplishment to those resources that we receive annually.

"A budget of $5.445 billion, in our estimation, provides the necessary increment of resources to support the technical program . . . . This is divided into $141 million, requested as our FY 1964 supplemental, and a $5.304 billion budget for FY 1965. During the preparation of this budget, we were asked by the President to make major cost reductions, in part by introducing improved management practices. This was done prior to submission of the President's budget, and the estimated cost savings are therefore reflected in our budgetary request. . . . Within these estimates, we propose to carry out this year's increment of NASA's program of aeronautical research, advanced technological development, space hardware development, and space exploration." (Testimony)

February 5: In House Science and Astronautics Committee hearings, Rep. James D. Weaver (R.–Pa.) challenged the 84% figure given by NASA Associate Administrator Dr. Robert C. Seamans, Jr., as NASA's 1963 space record—11 successes out of 13 launches. In response to Rep. Weaver's questioning, Dr. Seamans said NASA had planned 42 launches for 1963, 8 of which were backups. Rep. Weaver replied that, on this basis, NASA's 1963 space record could be interpreted as 31% successful. (Wash. *Eve. Star,* 2/6/64, 15)

- Testifying before House Committee on Science and Astronautics, NASA Associate Administrator for Space Science and Applications Dr. Homer E. Newell discussed objectives, progress, and future plans of NASA's programs in the area of space science (space environment, sun-earth relationships, geodesy, investigation of moon and planets, investigation into nature of the universe, exobiology) and applications (meteorological satellites, communications program, and advanced technological satellites). Some of the space environment results obtained during the past year:

"1. The existence of an appreciable number of protons in the outer Van Allen Belts was established.

"2. The lifetime in the belt of artificially injected electrons was determined. . . .

"3. Trapping lifetimes and rates of enhancement of electrons of varying energies in the outer radiation zone have been correlated with magnetic storms.

"4. Large low energy electron fluxes have been found on lines of magnetic force which appear to trail off to immense distances out into deep space.

"5. Contours of constant counting rate near the magnetic equator were found to draw closer to the Earth on the night side of the Earth than on the day side.

"6. Repeated observations were made of simultaneous VLF electromagnetic emission, auroral optical emission, and particle precipitation into the atmosphere.

"7. It was discovered that the Earth's magnetic tail in the antisolar direction extends at least half way to the Moon's orbit.

"8. Direct observations were made of the way in which solar plasma on the sunward side of the Earth piles up outside the boundary of the Earth's magnetic field, and in so doing appreciably compresses the geomagnetic field.

"9. Direct extended observations were made of the particle density and velocity of solar interplanetary plasma. These densities and velocities were found to have direct correlations with magnetic activity on the Earth and calcium plage activity regions on the Sun.

"10. Extended measurements were made of the interplanetary magnetic field and solar and galactic cosmic ray intensities.

"11. The true energy spectrum possessed by solar cosmic rays at their source on the Sun was deduced from the observational data. . . .

"12. First simultaneous measurements were made of electron temperature, positive ion density, and neutral atmospheric constituents in the Earth's high atmosphere.

"13. A new theory was developed to account for the types of ions found in the ionosphere, specifically the ratio of helium to hydrogen.

"14. Sounding rockets furnished important information on the ionospheric D region from 35 miles to 55 miles and the origin of the sporadic effects in the E region of the ionosphere at about 75 miles.

"15. Diurnal and seasonal variations in the ionosphere were established.

"16. Sounding rockets launched simultaneously with an overhead pass of the satellite Alouette reached within a few miles of the satellite, thereby connecting up ionospheric measurements in the lower atmosphere with those being made by the satellite." (Testimony)

February 5: Senator Wayne L. Morse (D.-Ore.) said on Senate floor that "the most recent developments in the strike of the Florida East Coast Railway by the 11 nonoperating railroad unions, which is now in its second year, point to the strong likelihood that we will have a major shutdown of the vital construction work at Cape Kennedy by tomorrow night. . . .

"The National Aeronautics and Space Administration has just announced that, contrary to the recommendations of the President's Board of Inquiry last October, it will permit the Florida East Coast, operating its trains with strikebreakers, to use the 18-mile spur newly constructed by the Federal Government to move freight into Cape Kennedy. . . .

"This NASA decision, announced just yesterday, will, I fear, lead to new labor trouble in the Cape Kennedy area, for the FEC strike still continues. . . ." (CR, 2/5/64, 2035-36)

- Ernest Brackett, Director of NASA Office of Procurement, was appointed Assistant Deputy Associate Administrator for Procurement Policy. In this position, Mr. Brackett would assist NASA Deputy Associate Administrator for Industry Affairs on procurement policy, "particularly with respect to relationships with the Congress, the Department of Defense, and industry and industrial associations." George J. Vecchietti, Deputy Director of NASA Office of Procurement, was designated Acting Director. (NASA Announcement 64-30)

- DOD announced it had undertaken comprehensive review of procurement policies and practices affecting labor-management relations at missile and space sites. NASA was cooperating in the study, which in 1964 was expected to involve the following sites: Vandenberg AFB, Warren AFB, Minot AFB, Whiteman AFB, and NASA's Mississippi Test Facility. The first on-site review was made at Cape Kennedy late last month. (DOD Release 110-64)

- Dr. Albert C. Hall, DOD Deputy Director for Space, Directorate of Defense Research & Engineering, discussed the objectives of the military space program at Institute of Electrical and Electronic Engineers' Winter Convention on Military Electronics, Los Angeles:

"Our problems . . . are twofold: First: the United States military space program must be one of action, not reaction. We must concern ourselves with military space developments—not necessarily to match or offset any Soviet development, but to sustain and augment our own versatility and strength. The successes we have already obtained should encourage us in this drive. Second, mindful that in the past we often have been optimistic in the short run and pessimistic in the long run, we must expect that some of the new discoveries and developments that appear may suggest applications and capabilities in wholly unplanned ways. Accordingly, while we pursue efforts where applications are clear we must also lay a foundation of technology, knowledge, and experience which is sufficiently broad in scope to provide for future contingencies as they materialize or are identified.

"From these considerations, then, are derived the two fundamental goals of the Defense space program:

"First, to continue with the development and exploitation of space systems and capabilities which measure up on a cost-effectiveness basis to meet clearly defined high priority national defense requirements.

"Second, to develop through a program integrated with other government agencies, such as NASA and the AEC, a broad base of technology and experience to permit the timely development and exploitation of space systems and capabilities which may be needed in the future, recognizing that lead times in certain areas, such as manned military space operations, may be 10 years or longer. . . ." (Text, DOD Release 106–64)

February 5: West German government issued denial that any West German firm was producing missiles suitable for military application. Denial was in response to Soviet diplomatic complaint that missiles capable of carrying nuclear warheads were being produced in West Germany, recalling rocket test-firings last December by Weapons and Air Equipment Co. of Hamburg. German government said investigation of the Hamburg company established that it possessed no manufacturing facilities. Two days later, according to *Missiles and Rockets,* the company halted production of rockets that it "apparently planned to export to Middle Eastern Arab nations." Company officials told *M&R* the decision was made voluntarily to prevent further embarrassment to West German government. (*NYT,* 2/6/64, 5; *M&R,* 2/17/64, 9)

- Lt. Col. Robert S. Buchanan (USAF) of Aerospace Research Pilots School, Edwards AFB, told space medicine conference at Brooks AFB that the greatest potential dangers to large scientific space station with crew of 15 or more men would be fire and collision with a meteoroid or space debris. Studies were considering possibility of separating space stations into individual compartments that could be sealed off in case of catastrophes such as fire. (AP, *Houston Post,* 2/6/64)

- NASA Langley Research Center awarded two contracts for construction on Vehicle Antenna Test Facility to aid in solving spacecraft communications problems: $1,265,505 contract to Nat Harrison Associates for construction of the building, and $280,630 contract to B. F. Goodrich Sponge Products Div. for design and installation of anechoic chambers for the facility. (LaRC Release)

- British sources revealed Britain had contracted to sell 600 guided missiles to Saudi Arabia for estimated $1.4 million. (AP, Wash. *Eve. Star,* 2/6/64, 1)

February 6: In testimony before House Committee on Science and Astronautics, NASA Associate Administrator for Manned Space Flight Dr. George E. Mueller said that "1963 was a milestone year. It was the year of transition to hardware—now we can fill the pipeline and move forward to a manned lunar landing in this decade. . . .

"The resources for the Gemini and Apollo programs are moving forward. The major developmental resources are essentially completed; the manufacturing resources are under construction and approaching our goal; the test and operational resources have been initiated; a logistics program is being formulated and our personnel build-up is progressing.

"The personnel staffing at the NASA Centers . . . is increasing from 11,858 at the end of FY 1963 to 13,711 at the end of FY 1964 and a forecast total of 14,711 at the end of FY 1965. In addition to the government team of personnel there are approximately 181,000 contractor personnel who are working on Manned Space Flight programs. With the continued buildup of construction, manufacturing, test and operations, it is anticipated that this figure will approximate 300,000 personnel at the peak of the Manned Space Flight Program. . . ."

Discussing planned activity in manned space flight program, he said: "A major decision was made in 1963 to adopt the 'all-up' testing concept as a basic approach to our flight verification test program. 'All-up' testing means, all flights will be scheduled with complete space vehicles using live stages and flight type spacecraft. . . .

"There are several advantages to the 'all-up' approach. It will permit us to land an American astronaut on the moon, and return him safely to earth, in accordance with our schedule even though we are operating this year on a reduced budget. It will permit us to capitalize on successful flights. It will also allow us to gather a very large amount of data early in the flight program and thereby provide much needed information to our design organization. It is planned in the Saturn IB and Saturn V programs to launch a complete unmanned space vehicle on the first flight in an earth orbital trajectory." (Testimony)

February 6: NASA Associate Administrator for Advanced Research and Technology Dr. Raymond L. Bisplinghoff, appearing before House Committee on Science and Astronautics, outlined NASA's program of advanced research and technology and proposed activities for FY 1965. Objectives in advanced research and technology were summarized:

"1. Basic research aimed toward an understanding of the natural laws underlying aeronautical and space technology.

"2. Engineering research aimed at the application of these laws and at codifying them in terms of engineering design principles.

"3. Experimental subsystem development, using advanced concepts and components, to produce test 'know-how' for the design of future operational systems.

"Our aim in basic research is to contribute to the fund of relevant scientific knowledge and, in so doing, to maintain the best possible contact with the scientific and engineering community. Engineering research, that is the creation of engineering design principles, is also conducted by these same institutions with industry playing a somewhat greater role than in the case of basic research . . . .

"A program of [advanced] research . . . which looks beyond today's technology is believed to be a pre-requisite for continued advancement in the performance and efficiency of aeronautical and space vehicles." (Testimony)

- John F. Kennedy Space Center, NASA, announced major reorganization. KSC Director Dr. Kurt H. Debus said objectives of reorganization were to: (1) realign Apollo program management functions in accordance with concept of NASA Associate Administrator for Manned Space Flight Dr. George E. Mueller that these functions be standardized at NASA Hq. and the three manned space flight centers; (2) separate and strengthen administrative and technical support functions; and (3) reduce number of offices reporting to Dr. Debus and delegate more authority and responsibility to Assistant Directors. Under realignment there were five Assistant Directors, with authority to act for Center Director within their areas of responsibility: Program Management, Administrative Management, Launch Vehicle Operations, Technical Support Operations, and Instrumentation. Program Management was divided into two areas, Plans and Program Support Office and the Apollo Program Management Office, which was aligned as counterpart organization with Apollo program offices in NASA Hq. and the other manned) space flight centers (Manned Spacecraft Center and Marshall Space Flight Center). Unaffected by the change were Launch Support Equipment Engineering Div., Facilities Engineering and Construction, and five staff offices. New NASA Hq. office, NASA Regional Inspections, was attached to KSC. MSC's Preflight Operations Div. was renamed MSC Florida Operations but its basic relationship to KSC was unchanged. Other NASA Center elements attached to KSC, such as GSFC Field Projects Branch, remained unchanged. (KSC Release 10-64)

February 6: *Washington Post* editorialized:

"The failure of the sixth straight Ranger spacecraft hoists aloft the question of whether the United States will accomplish its goal of putting an American on the moon by the decade's end. . . .

"A delay would not simply be a case of a stretchout in which work would continue but at a slower pace. A delay would take much of the binding and motivating tension from Apollo. It would increase costs by underusing men and facilities brought together for a tighter schedule, and it would allow projects and personnel to drift out of the main stream into the eddies of lower priority. A delay also would entail loss of prestige for NASA, for the Administration and for the Nation as well.

"No schedule, however, is sacred. If it is determined that the landing of a man on the moon by Jan. 1, 1970, is simply not feasible, or cannot be done with reasonable assurance of safety for the astronauts, then the space agency and the country will have to adjust to this reality, no matter how costly and comfortless it may be. A manned flight to the moon will be no less exciting or valuable for being somewhat delayed." (*Wash. Post,* 2/6/64, 18)

- Tass announced ELECTRON I and ELECTRON II scientific satellites were continuing to send back "valuable scientific information." ELECTRON I had made 53 orbits of the earth and ELECTRON II, 6 orbits. (AP, Wash. *Eve. Star,* 2/6/64, 6)
- Soviet Academician Anatoli A. Blagonravov announced in Moscow that U.S.S.R. would participate in three-nation experiment of bouncing radio signals off U.S. balloon satellite ECHO II beginning Feb. 22. Signals would be transmitted from Jodrell Bank, England, to ECHO II, and reflected back by radioastronomy observatory near Gorki. (*NYT,* 2/7/64)
- NASA Telephone and Data Distribution Center became first technical support element of the John F. Kennedy Space Center, NASA, to become operational on Merritt Island, KSC announced. Some buildings were being serviced by telephone service through the Center and other buildings would be phased in as required. Fully equipped to handle all types of communications, the Center participated in its first technical support operation during launch of Saturn I SA-5, when it provided closed-

circuit television service to Washington and other areas. (KSC Release 11-64)

February 6: President Johnson spoke at 18th annual dinner of Weizmann Institute of Science, held in New York:

"The Weizmann Institute . . . is an international scientific institution in every sense of the word. To its buildings come students from all over the globe. It has helped to make Israel one of the foremost scientific resources of the world. . . .

"We, like Israel, need to find cheap ways of converting salt water to fresh water. . . . This Nation has begun discussions with the representatives of Israel on cooperative research in using nuclear energy to turn salt water into fresh water. . . ." (Text, *CR*, 2/17/64, A703–04)

- Rep. E. Ross Adair (R.–Ind.) said on House floor that a fifth transatlantic communications cable (TAT) would be unnecessary: "It is expected that within the next 6 weeks the . . . [ComSatCorp] will announce its plans for financing a communications satellite system and will tell Congress and potential investors when it expects to have a system in operation. I understand that CSC expects to design its first satellites to handle traffic between the United States and Europe.

". . . based on information filed with the FCC by the International Carriers having an interest in these matters . . . I am convinced that there is no need for TAT 5. If permission is granted to lay TAT 5 it would add at least 128 channels of capacity . . ., and this would mean that there would be no economic justification for orbiting a satellite system in the immediate future . . ."

He recommended adopting Mackay Radio's proposal to change the route planned for TAT 4: ". . . this route would not only make it possible to establish intermediate relay and terminal facilities in the Azores, but also to provide for extension to Spain and Portugal and even down to the Canary Islands.

"The U.S. military might well prefer cable links to the Azores and this route could also serve the interests of the National Aeronautics and Space Administration, both in Spain and the Canary Islands, where NASA tracking stations are either in being or being constructed. . . ." (*CR*, 2/6/64, 2220)

- General Accounting Office made public report to Congress charging NASA's "unrealistic delivery dates" for first two Oso satellites had cost an unnecessary $526,000. GAO was critical of NASA for ordering backup Oso eight weeks before scheduled launching of the first satellite, thus spending $273,000 that was not needed. The GAO report also said NASA had unnecessarily limited competition for Oso construction. (AP, Wash. *Eve. Star*, 2/6/64, 2; Finney, *NYT*, 2/7/64, 10)

- Col. Daniel McKee, head of AFSC Field Office at NASA Manned Spacecraft Center, said in *Houston Post* article that USAF would participate in NASA Project Gemini with 20 experiments, beginning with the second manned Gemini flight. Some of the areas in which USAF would conduct experiments include radiation, photography, communications, and extravehicular activities. (Maloney, *Houston Post*, 2/6/64)

- FAA Administrator N. E. Halaby announced formation of Future Flight Manpower Study Board to investigate future manpower requirements for civil aviation. Board was headed by Dan A. Kimball, Board Chairman of Aerojet-General Corp. and former Secretary of the Navy. (FAA Release 64–11)

February 6: Welcoming Dr. S. Dillon Ripley II as eighth Secretary of the Smithsonian Institution, Senator Claiborne Pell (D.–R.I.) quoted Dr. Ripley in the Senate:

" 'In the realm of the humanities, in art and the history of our culture, as well as in a variety of sciences ranging from the study of the ocean, of space and the natural sciences, it is implicit that the Smithsonian should assume a commanding position in education as well as leadership in the museum world. We cannot afford to overlook any possibility of increasing and diffusing knowledge among our people, young and old, at all levels—from exhibits and museum lectures and demonstrations to postgraduate specialized research. Museums and their technical resources are important as never before in these days of crisis in education.' " (CR, 2/6/64, 2161)

- FAA awarded $1.3 million contract to Skidmore, Owings and Merrill for improvements at FAA's National Aviation Facilities Experimental Center (NAFEC) in Atlantic City, N.J. (FAA Release T 64–4)

February 7: NASA announced it would join with Norwegian Space Research Committee and the Ionospheric Research Laboratory of the Royal Technical Univ. of Denmark in three sounding rocket launches from Andoeya, Norway, during February and March. The Nike-Apache sounding rockets would carry experiments to test theories of how D-region of the ionosphere is formed. Norwegian Committee was providing launch site and associated facilities; Norwegian Committee and Danish Ionospheric Laboratory together were purchasing sounding rockets and contributing scientific payloads. NASA was furnishing certain experiments for the payload, already had assisted Danish and Norwegian scientists by providing training in rocket handling and data acquisition. (NASA Release 64–27)

- NASA issued its first exclusive patent license to Satellite Div. of Union Carbide Corp. for commercial development of a nickel alloy invented by John C. Freche, metallurgist at NASA Lewis Research Center. For as long as it develops the alloy for commercial use, Union Carbide agreed to spend at least $20,000 a year for not more than three years. After fully developing it, Union Carbide would hold exclusive marketing rights for not more than four years, at which time exclusive license would expire and product would be made fully available to the public. (NASA Release 64–30)

- Australian scientist Victor Albert Bailey speculated that moon may possess far-reaching electrostatic field, which could have been responsible for RANGER VI camera failure by interfering with switches for the cameras. (SBD, 2/7/64, 207)

- Two balloons with instrumented payloads to make upper-atmosphere measurements were released at Palestine, Tex., and recovered. Balloon carrying 550 lbs. of instruments to study cosmic rays, designed by Case Institute researchers, was recovered later near Roanoke, Ala.; balloon carrying 405 lbs. of instruments to take oxygen specimens at varying altitudes, designed by MIT, was recovered near Derry, La. (UPI, Phil. *Eve. Bull.*, 2/7/64)

- Life science researchers in North American Aviation's Space and Information Systems Div. were studying effects of weightlessness and increased gravity on boa constrictors, NAA reported. The eyes of the boas were of special interest, for they offered easily observable reactions to the conditions. Normally a slit that remains vertical, the pupil of the boa's eye tilted some 15 degrees after the snakes were subjected to centrifuge. The snakes were "grossly disoriented" and, in striking at objects, they missed widely. (NAA S&ID *Skywriter*, 2/7/64, 2)

February 7: Sen. J. Howard Edmondson (D.-Okla.), member of Senate Committee on Aeronautical and Space Sciences, and Sen. Kenneth Keating (R.-N.Y.) were appointed as advisers to U.S. representatives on U.N. Committee on the Peaceful Uses of Outer Space, replacing Senators Margaret Chase Smith and Howard Cannon. (*Tulsa Daily World*, 2/8/64)

- FAA announced Air France and British Overseas Airways Corp. had reserved six delivery positions apiece for the U.S. supersonic transport aircraft, bringing total reserved positions to 63 and number of airlines holding positions to 10. (FAA Release 64-12)

February 8: NASA Manned Spacecraft Center announced $133,358 contract was awarded to AiResearch Manufacturing Co. Div. of Garrett Corp. for design and production of extravehicular pressurization ventilation system to be used by astronauts in Project Gemini. (UPI, *San Diego Union*, 2/9/64)

- Major Robert A. Rushworth (USAF), X-15 pilot, said at Texas A&M Univ. that "we stand about half-way in the amount of work we want to do" with the X-15 aircraft. "In the next four years, we want to get another one hundred flights under our belts." Describing the X-15A-2 plane for advanced missions, he said the new aircraft would range up to 80 mi. altitude at speed of 5,450 mph, with help of hydrogen-fueled ramjet engine and 13,500 additional pounds of fuel. (UPI, Omaha *World Herald*, 2/9/64)

February 9: NASA announced FAA-NASA Memorandum of Understanding specifying NASA's role in development of U.S. supersonic transport aircraft, under overall responsibility of FAA. NASA would furnish facilities, background research, technical advice, and resources to FAA and would conduct advanced flight research. In addition, NASA would conduct necessary background research to provide supporting technology and technical assistance and make available NASA's facilities and technical resources. In flight testing phase, FAA would be responsible for direction and acceptance testing, and certification trials; NASA would be responsible for flight research with its own funds, instrumentation, and resources. (NASA Release 64-28)

- Balloon with instrumented payload to measure light given off by oxygen at high altitudes was launched from Palestine, Tex., landing later that day at Roganville, Tex. (*Houston Chron.*, 2/10/64)

- Project Amos, ARPA project to develop and operate full-scale infrared astronomical observatory, was described by Howard Simons in the *Washington Post*. To be installed on 10,000-ft.-high Mt. Haleakala on Hawaiian Island of Maui, the observatory would be used in twofold program: research in locating and identifying missile nose cones and satellites; and study of infrared characteristics of missiles and satellites as well as earth's atmosphere, the sun, stars, moon, and planets. (Simons, *Wash. Post*, 2/9/64, E3)

- Information conditioning unit that converts data for computer use much more rapidly than any before had been tested on USN satellite, according to Applied Physics Laboratory of Johns Hopkins Univ., inventor of the system. System was said to reduce to hours the weeks or months previously required for conversion of satellite data into usable scientific information. After recent launch of USN satellite with the unit onboard, scientists obtained more than 50 types of information—analyzed, checked, and printed—from the satellite within 12 hours. (UPI, *NYT*, 2/9/64, 91)

February 10: President Johnson submitted to the Congress his 1963 report on the activities and accomplishments of the U.S. program to establish and operate a commercial communications satellite system. In his transmittal message, the President said:

"The year 1963 has been a period of major accomplishment toward the objectives established by the Congress in the Communications Satellite Act. The Communications Satellite Corporation has been organized, established, has employed a competent staff, and is implementing plans for a commercial communications satellite system. All agencies of Government concerned have contributed wholeheartedly to the furtherance of the objectives of the act. . . ." (*CR*, 2/10/64, 2714)

- USAF Athena missile misfired in its first overland test flight from Green River, Utah, and pieces of the vehicle and instrumented payload landed in sparsely populated southwest Colorado. No person or property was damaged, and USAF officials apologized immediately to Colorado Governor John A. Love. The 8-ton, 50-ft. missile was supposed to have made 450-mi. flight from Green River to White Sands Missile Range, N. Mex. USAF sources later indicated that the missile's nosecone was released when pyrotechnic charges were actuated by a short circuit in second-stage junction box. As result of damage to forward wires, signals for first-stage jettison were lost and second stage ignited with first stage still attached. Pressure forced off the stage fairing as well as second-stage fins, but the vehicle continued to travel—20° off course. (AP, *Balt. Sun.*, 2/11/64; UPI, *NYT*, 2/11/64, 26; *M&R*, 2/24/64, 11; *Av. Wk.*, 3/2/64, 36)

- 120 delegates from 17 European nations met with U.S. and Canadian delegations in Rome for exploratory talks on European participation in an international communications satellite system proposed by the U.S. (*NYT*, 2/11/64)

- Examining the failure of RANGER VI as well as previous Ranger lunar probes, William Coughlin pointed out in *Missiles and Rockets* editorial that Jet Propulsion Laboratory, Ranger project manager, is "the only NASA facility where NASA owns all the land, owns all the equipment, pays all the bills and allows a non-profit corporation to run the facility. . . .

  "An academic environment is neither comparable nor conducive to the kind of hard-driving industrial atmosphere required to make complex space hardware function in a highly reliable manner . . . we think Congress should reopen the whole question of the JPL-Cal Tech relationship." (*M&R*, 2/10/64, 50)

- NASA announced selection of Trans World Airlines for negotiation of contract to provide base support services at John F. Kennedy Space Center, NASA. Under this contract TWA would provide general maintenance, utilities operation, and supply operations on a cost-plus-incentive-fee basis; and janitorial, fire protection and prevention, security, and medical services on fixed-price basis. (NASA Release 64-32; KSC Release 13-64)

February 10: Dr. Eugen Saenger, eminent European space scientist, died at 58 in West Berlin. Dr. Saenger was best known for his World War II study "Rocket Propulsion of Long Range Bombers" and his post-war theoretical investigations of photon-powered rockets. He was one of the first to investigate use of liquid hydrogen as a fuel and of metal powders mixed with rocket fuels. Formerly head of Institute of Jet Propulsion Physics in Stuttgart, since last May he held chair of space flight technology at Berlin Technical Univ. Of him, Dr. Walter Dornberger said: "He didn't see the goal of human endeavor as travel just in our planetary system. He wanted to see travel to the outer galaxies. He made very extensive studies in this field, which one day—maybe 40, 50 years from now—may be looked at very seriously." (*NYT*, 2/11/64, 39)

- Rep. Glen P. Lipscomb (R.–Calif.) discussed on the House floor a document titled "Common Action for the Control of Conflict: An Approach to the Problem of International Tension and Arms Control," prepared by Vincent P. Rock of Institute for Defense Analysis under contract to U.S. Arms Control and Disarmament Agency. Rep. Lipscomb said the document "reportedly has been must reading for administration officials. . . . [It] becomes an important document because, since its release [July 1963], administration policies seem to have coincided to a high degree with its recommendations. Although it purports to be nothing more than its author's opinions, it has turned out to be a handy advance guidebook to administration actions. . . ." Citing examples of Rock report's recommendations, Rep. Lipscomb said report recommended U.S. seek Soviet cooperation in future space efforts and in scientific areas; he noted NASA–U.S.S.R. space cooperation agreement announced Aug. 16, 1963, and President Kennedy's proposal for U.S.–U.S.S.R. negotiations for lunar exploration cooperation presented Sept. 20, 1963. (*CR*, 2/10/64, 2720)
- NASA announced it would negotiate one-year cost-plus-incentive-fee contract with General Electric Co., Computer Dept., for continuation of engineering services at NASA Marshall Space Flight Center's computation laboratory. (NASA Release 64–31)
- Florida East Coast Railway moved a freight train into Merritt Island with non-union workers, as NASA had formally authorized on Feb. 8. Striking railroad workers established picket lines at all entrances to Merritt Island and Cape Kennedy. Army Corps of Engineers said apparently all 4,540 construction workers refused to cross the picket lines, halting construction on more than 30 projects. (UPI, *Houston Chronicle*, 2/10/64; AP, *NYT*, 2/11/64, 42)
- Thomas L. McGrath, director of R&D Dept. of Greater Boston Chamber of Commerce, said in *Christian Science Monitor* interview that NASA's electronic research center would be a stimulus to Boston area economy. When the center is completed in 1969, he said, a "conservative" estimate would be $250 million contribution to income of the area. "The center will pull in a lot of capable people, and it's impossible to say what they'll add to the income of the area. Also one good research discovery coming out of the center might be worth millions to industry." (Russell, *CSM*, 2/10/64)
- Soviet cosmonaut Valentina Nikolayeva-Tereshkova left London for Moscow, ending tour of Britain during which she was presented to Queen Elizabeth. (AP, *Wash. Post*, 2/11/64)

February 10: Air-India had reserved three delivery positions for U.S. supersonic transport aircraft, FAA announced. (FAA Release 64–13)

February 11: NASA Administrator James E. Webb announced at Hq. press conference that U.S.–U.S.S.R. communications experiments with ECHO II balloon satellite would begin after Feb. 21.

In response to question, Mr. Webb said failure of attempts to acquire lunar photographs had not adversely affected Apollo schedule: ". . . up to this time we have not had to make any changes in the program due to this, nor do I think we will."

Asked whether 1969 date for manned lunar landing was a "deadline" or a "planning date", Mr. Webb said:

"The view President Kennedy had . . . was that he was setting a deadline by which he wanted this nation to get the job done, and he was providing resources for us to plan the flight in 1967, or early 1968, giving some leeway for intractable technical problems, for unknown and unforeseeable things, and also to avoid having to put into parallel development systems that would be more expensive if done that way. . . .

"Now, the 1967, early 1968 target date that we were setting was to mobilize the effort, to press forward with it, but not to do so in such a way as to run this cost up from $20 billion to something much higher.

"So we were thinking of it as pretty much of a deadline that it was in the interests of this nation to meet.

"Now, with the cuts last year of $600 million and the reorganization of the effort to all-up systems testing, the cancellation of a number of Saturn flights, things of this kind, I think you've got to think of it somewhat differently. I think the target date for the flights, which is a tight date but still attainable, is mighty close to the January 1, 1970 period.

". . . I believe we can do it within the latter six months of 1969 if we proceed vigorously under the plan we have presented to the Congress. But again we don't have any leeway. . . . We have lost that, and we've got no more than what I'd call a fighting chance to do this by January 1, 1970."

Mr. Webb discussed NASA relationship with Cal Tech's Jet Propulsion Laboratory, manager of NASA's lunar and planetary exploration programs. The current three-year contract "expires next June and we have now for some months been examining together with the California Institute of Technology ways and means to make a better relationship from their standpoint and from NASA's standpoint." Mr. Webb said NASA was seeking a reorganization of JPL to achieve a "hardheaded type of industrial management." (Finney, *NYT*, 2/12/64; Loory, *N.Y. Her. Trib.*, 2/12/64; Simons, *Wash. Post*, 2/12/64; *M&R*, 2/17/64, 16)

• NASA announced that failure of RANGER VI spacecraft to take pictures of the moon may have been caused by unscheduled turn-on of cruise telemetry from the TV system at about the same time as Atlas booster separation. For unexplained reason, telemetry turned on for about one minute and relayed measurements in the TV system, including battery voltages and temperatures. It was possible that TV system could have turned on at the same time; if so, severe electrical arcing would have occurred and damaged the system. However, TV telemetry turned on 17 min. after separation from Agena stage as scheduled, and nothing at that time indicated any abnormalities in the TV system.

Tracking data showed that RANGER VI impacted the moon within 20 mi. of its target and within one third of a second of the predicted arrival time. (NASA Release 64–33)

February 11: Spokesman for European conference on satellite communications, meeting in Rome with U.S. and Canadian delegations, said conference had reached agreement on principal points of the U.S. proposal to establish an international comsat system. "They are now discussing the details, upon which the accord isn't so extensive." (*NYT*, 2/12/64, 13)

- Communications Satellite Corp. announced six companies had responded to its request for proposals for designing international commercial communications satellite. AT&T and RCA submitted a joint proposal; Thompson-Ramo-Wooldridge, Inc., submitted proposal with ITT as principal subcontractor; Hughes Aircraft Co. and Philco Corp. submitted individual proposals. ComSatCorp was evaluating the proposals. (*WSJ*, 2/12/64)
- Army Corps of Engineers awarded two NASA contracts for work to be performed at NASA Mississippi Test Facility: $3,801,146 contract to Chaney and James Construction Co., Inc., for construction of booster storage building, RP–1 dock, and booster transfer dock; and $2,401,542 contract to Farrell Construction Co. for construction of main canal. (DOD Release 128–64)
- Annual report of NORAD indicated U.S. orbited 60 payloads in 38 launches during 1963, compared with 17 by U.S.S.R. (AP, *CSM*, 2/11/64)
- Construction workers at Cape Kennedy refused for a second day to cross picket lines of Railroad Telegraphers Union, and some of the other 9,000 union members at Cape Kennedy honored the picket lines. Later, at request of National Labor Relations Board, U.S. District Court issued temporary injunction ordering withdrawal of the picket lines against Florida East Coast Railway. (UPI, *Wash. Post*, 2/12/64; AP, *Wash. Eve. Star*, 2/12/64; UPI, *Wash. Post*, 2/13/64)
- None of the supersonic transport bidders was willing to assume the prescribed share of cost of the joint Government—industry-funded SST program, John P. MacKenzie reported in *Washington Post*. Under plan presented by President Kennedy, SST would cost $1 billion, with Government paying 75% of the cost and manufacturers paying remaining 25%. MacKenzie said none of the six bidding manufacturers had met the 25% participation challenge, and "some manufacturers said the plane could not be developed for $1 billion." Officials were reported to be studying independent report by Eugene R. Black, former president of the World Bank, which unnamed sources said recommended Government finance 90% of the development cost. Report also was said to have proposed transferring SST program from FAA to an independent agency, slowing pace of the SST development and avoiding international "race" with U.K.-France Concorde development. (MacKenzie, *Wash. Post*, 2/11/64; *N.Y. Her. Trib.*, 2/13/64; *Newsweek*, 2/17/64)
- Experimental hot-electron propulsion system had completed 1,000 hrs. of testing at RCA David Sarnoff Research Center without degradation of components or performance, RCA announced. (*NYT*, 2/11/64, 53; *M&R*, 2/17/64, 21)

February 11: Comparison of amount of earth's heat resulting from radioactive breakup of elements uranium, thorium, and potassium was discussed in *Science.* Half of all potassium heat was released in about first quarter of earth's lifetime, while heat from uranium and thorium was released at a more even rate. (Sci. Serv., *NYT*, 2/11/64, 28)

- USAF successfully launched 30th Minuteman ICBM from Vandenberg AFB, a routine training launch by SAC crew. (*M&R*, 2/17/64, 13)

February 12: Continuing the current series of sounding rocket experiments to obtain comparative characteristics of the upper atmosphere at widely separated locations, NASA launched three Nike-Cajun sounding rockets at Wallops Island, Fort Churchill, and Ascension Island. The rocket payloads ejected 12 grenades which detonated at intervals up to 55-mi. altitude. (Wallops Release 64–16)

- NASA Nike-Apache sounding rocket was launched from Wallops Island, Va., in test of hinged-fin assembly with modified wedge designed to produce about 5 rps at Apache burnout and to obtain temperatures in flight at various points on the fin panel. Rocket reached 90-mi. altitude; performance of rocket and instrumentation was considered good, with exception of temperature probes, which were lost shortly after lift-off. (NASA Rpt. SRL)

- Soviet news agency Tass said ELECTRON I and ELECTRON II scientific satellites continued to function successfully. By 6:00 p.m. Moscow time Feb. 11, ELECTRON I had completed 104 orbits and ELECTRON II more than 13 orbits. (Tass, *Krasnaya Zvezda*, 2/12/64, 1, ATSS–T Trans.)

- Dr. S. Fred Singer, Director of U.S. Weather Bureau's National Weather Satellite Center, said that the modified Tiros meteorological satellite system, using presently available equipment, would be able to: provide daily global coverage of weather; use vertical cameras for more complete pictures; aid in prediction of time and location of thundershowers; determine if clouds are storm cover; aid in studying hydrology and geology, mapping earth's surface; aid airlines in tracking jet stream, especially on transatlantic flights. (*SBD*, 2/13/64, 239)

- Catalyst for use with monopropellant hydrazine rocket engines has been developed by Shell Oil Co.'s General Chemistry Dept. working with Jet Propulsion Laboratory in NASA-sponsored project, NASA announced. Development of the catalyst permits use of hydrazine for spacecraft guidance and control rockets. NASA scientists reported that hydrazine compounds produced by the catalytic reaction were 50% more energetic than monopropellant hydrogen-peroxide now used. (NASA Release 64–29)

- NASA Lewis Research Center issued requests for proposals for initial construction of 500-ft. drop tower for producing zero gravity environments. The facility would be largest of its kind in the U.S., would be able to provide weightlessness for periods up to 10 sec. (LRC Release 64–12)

- Army Corps of Engineers awarded $3,207,166 NASA contract to Peter Kiewit Sons Co. and C. H. Leavell Co. for construction of two Saturn test stand substructures at NASA Mississippi Test Facility. (DOD Release 132–64)

- Construction workers at Cape Kennedy returned to work after Federal judge temporarily halted picketing by the Railroad Telegraphers Union against the Florida East Coast Railway. (UPI, *Wash. Post*, 2/12/64)

February 12: Dynamite bomb planted along Florida East Coast Railway tracks near Titusville was disarmed by four boys seconds before freight train passed by, the bomb providing authorities with clues to similar bombing incidents involving FEC property. The attempted sabotage occurred only two miles north of Cape Kennedy, where striking unions' picket lines had halted construction for two days. (UPI, *N.Y. Her. Trib.*, 2/14/64)

- Trans-American Aeronautical Corp. of Washington had reserved a delivery position for U.S. supersonic transport aircraft, FAA announced. Trans-American was expected to make the aircraft available to a Latin American flag carrier through leasing arrangement. (FAA Release 64-14)
- FAA said during first week of sonic boom tests over Oklahoma City 60 damage claims were reported. Of the 60, only two persons actually signed affidavits, FAA said. (UPI, *Wash. Daily News*, 2/13/64)
- USAF successfully launched Atlas F ICBM from Vandenberg AFB in routine training mission. (*M&R*, 2/17/64, 13)
- USAF successfully launched Minuteman ICBM from Cape Kennedy, first in series of tests of Minuteman's penetration aids. (*M&R*, 2/17/64, 13)
- Fifth Women's Space Symposium was held in Los Angeles, with Dr. Orr E. Reynolds, Director of NASA Bioscience Programs, as principal speaker. (S. Thomas)

February 13: At conclusion of Rome meeting of European Conference on Satellite Communications with representatives from U.S. and Canada, head of U.S. delegation Abram Chayes said "the most significant result of the conference is that a number of European countries as well as Canada have now expressed serious interest in joining as copartners" in the international comsat system. Discussions would continue, and another meeting was scheduled for mid-March in Washington. (Wollemborg, *Wash. Post*, 2/14/64)

- NASA–AEC Space Nuclear Propulsion Office (SNPO) announced that the Los Alamos Scientific Laboratory had successfully completed a major cold-flow test series on the Kiwi B–4D–CF reactor. The experiments were designed to verify design changes incorporated in the basic Kiwi B-4 reactor to prevent flow-induced vibrations under nonpower producing conditions. No vibrations were observed. (SNPO)
- NASA employees Christopher C. Kraft, Jr., and Dr. John W. Townsend, Jr., were among the 10 outstanding young men in Government service awarded the 1963 Arthur S. Flemming Awards. Kraft was Assistant Director for Flight Operations, NASA Manned Spacecraft Center; Townsend, Assistant Director of Space Science and Satellite Applications, NASA Goddard Space Flight Center. (NASA Notice)
- Technology from $3 million program to eliminate vibrations in Project Gemini's Titan II launch vehicle should be applied to other large boosters, according to NASA Chief of Launch Vehicle Systems Integration Jerome B. Hammack. (Webb, UPI, *Houston Press*, 2/14/64; *SBD*, 2/13/64, 240)
- $1,224,793 NASA contract for construction of JPL Explosive Safe Facility at Cape Kennedy was awarded to Fryd Construction Corp. by Army Corps of Engineers. (DOD Release 134–64)
- Sen. Stuart Symington (D.–Mo.) speaking before the Annual Meeting of the Cape Girardeau, Mo., Chamber of Commerce, said:

"Our policies are peaceful in nature, and our space program is peaceful in purpose and content; but it is also a sound investment for our future national security, economic as well as military. . . .

"Landings on the moon will add to our knowledge of the origin of the universe, and also contribute to our knowledge of the broad spectrum of science and technology in such areas as meteorology and communications . . .

"But the lunar project is not the whole space program; and the military significance of the program cannot be judged by examining that project alone; nor can anyone today estimate the great wealth and resources that will flow from this new knowledge. Knowledge is, and will be, the basic power of modern states.

"Moreover, the Defense Department is now deeply engaged in the space effort . . .

"The true defense value of the lunar project rests with the development of our over-all space capabilities; and no one can say today how this stockpile of knowledge will be used in the defense of this Nation or how important it will be. . . ." (Text)

February 13: Dr Willy Ley predicted at Texas A&M Univ. that U.S.S.R. would attempt manned orbital flight to the moon during 1964. He did not believe such a flight would "significantly advance the time of an actual landing . . . [because] that would take great strides in new rocket power." He believed U.S. had a "50–50 chance of being the first to get there." (Maloney, *Houston Post*, 2/14/64)

• New York Gov. Nelson Rockefeller, campaigning in New Hampshire Republican primary for presidential nomination, said there was a "very serious question" whether U.S. should abandon Project Apollo manned lunar project and rely instead on instrumented spacecraft. About 95% of information obtainable by men on the moon "for a half an hour" could be gathered by instruments, he said. He said the decision for Apollo came after the abortive Bay of Pigs invasion, when the Administration decided the U.S. needed something "dramatic." Rockefeller proposed reconsidering the space plan advocated by President Eisenhower, calling for a slower, less expensive, "scientific" program. (NYT, 2/14/64, 10; AP, *Wash. Post*, 2/14/64)

• Atlas missile caught fire and exploded at silo site east of Roswell, N. Mex., demolishing the $1.5 million missile and the $10 million silo. USAF reported no personal injuries. (UPI, *Wash. Post*, 2/14/64)

February 14: NASA postponed indefinitely the launch of Ranger 7 lunar probe. New launch date would depend on further study of failure of RANGER VI's TV camera system before impacting the moon Feb. 2. NASA–JPL investigators still were concentrating on unscheduled turnon of telemetry from TV system which occurred at about same time as separation of Atlas booster engines. (NASA Release 64–38)

• NASA–DOD agreement that instrumentation ships required to support NASA and DOD space programs would be pooled and operated by DOD on behalf of both agencies was announced. By end of 1967 pool would have 20 operational tracking ships. NASA would participate with DOD in all phases of the program, including systems engineering and development of the plans and specifications. Navy Dept. would have central responsibility for design, construction, and modification of the ships. Military Sea Transportation Service (MSTS) would be responsible for ship operation, and DOD Range Agencies would be responsible for scheduling and ship instrumentation operation. Senate and House space committees released $83 million of FY 1964 funds, which had been authorized subject to NASA–DOD agreement. The funds would be used to acquire three ships from Maritime Reserve Fleet and to modify two ships for Apollo already in the joint tracking system. (NASA Release 64–34; DOD Release 137–64; Wash. *Eve. Star*, 2/14/64)

February 14: Discussing MOL's growth potential in testimony before House Armed Services Committee, Subcommittee on R&D, Director of DOD Research and Engineering Dr. Harold Brown said: "I want to make a distinction between its ultimate capability and what is in the program. . . . What is in the program is not a space station in that sense. It is an experimental laboratory. But . . . this could grow to be a space station, if and after . . . we conclude . . . that a man can have a substantial military purpose [in space]." Dr. Brown said DOD should "move more aggressively into the manned space flight arena in order to explore more fully man's utility for the performance of military space missions and to develop . . . his full capabilities in any area where potential exists." (*SBD*, 2/17/64, 260; *M&R*, 2/24/64, 15)

- NASA selected Genisco, Inc., to develop, fabricate, and assemble motion generator for research into guidance and psycho-physiological problems of manned space flight. The fixed-price contract was expected to be valued between $6 million and $6.5 million. Genisco had been selected last July for contract negotiations for the motion generator, major component of advanced space navigation simulator at NASA Ames Research Center; however, "during the negotiations a better understanding of the design problems was reached and it was determined that numerous changes were desirable." NASA then directed Ames to request revised proposals from the five bidders. When the companies submitted new proposals in January, Genisco again received highest technical rating and proposed the lowest fixed fee. (NASA Release 64–35)

- George Low, recently appointed Deputy Director of NASA Manned Spacecraft Center, said in Houston that failure of RANGER VI lunar TV probe was "a grave disappointment." Despite Ranger failure, target date for manned lunar landing was still 1969, he said, but "we certainly can't guarantee there won't be technical difficulties that won't let us do it. . . . If we have to use the Apollo as an unmanned reconnaissance craft our small margin [of time] will be eaten up." (*Houston Post*, 2/15/64)

- NASA selected Honeywell Aeronautical Div. for negotiation of contract for three suborbital spacecraft to be used in Project Scanner horizon definition experiment. Valued at about $1,000,000, the contract called for early 1965 delivery of two Scanner payloads plus one backup. Scanners would be flown on ballistic trajectories from Wallops Island, Va. (NASA Release 64–36; LaRC Release)

- President's Missile Sites Labor Commission adopted statement calling on Railroad Telegraphers Union to refrain from picketing at Cape Kennedy and Merritt Island while the Commission considered the dispute between the union and the Florida East Coast Railway. Statement also called for the union and the company to maintain "full and continuous operations at the sites" while the Commission had the dispute under consideration. (UPI, *NYT*, 2/15/64)

February 14: Patent for design of space propulsion system using jet of electrically charged liquid particles had been awarded to Washington inventor Dominique M. P. Gignoux, president of Cosmic, Inc. The 500,000 volts of electricity to charge and accelerate the particles would be produced by onboard nuclear reactor. (Jones, *NYT*, 2/15/64)

February 15: Rebuilt X-15 No. 2 was rolled out of North American Aviation, Inc., facility at Los Angeles. Damaged in landing at Mud Lake, Nev., in Nov. 1962, the X-15 had been modified to increase its top speed by nearly one third. The rebuilt rocket research aircraft would be capable of short bursts of flight at mach 8—more than 5,300 mph. It would resume test flights later this year. (*N.Y. Her. Trib.*, 2/16/64; AP, *Wash. Post*, 2/16/64; FRC *X-Press*, 2/20/64, 1)

- USAF launched Thor-Agena D booster from Vandenberg AFB with unidentified satellite. (*M&R*, 2/24/64, 11)
- Two-man crew and backup crew for first manned Gemini space flight would be selected in May or June, according to Astronaut Donald K. Slayton at NASA Manned Spacecraft Center. Flight was expected in late 1964. (*Houston Chron.*, 2/16/64)
- Dr. Wernher von Braun, Director of NASA Marshall Space Flight Center, said in press conference preceding address at Texas A&M Univ. that a geologist would be in crew of second Apollo lunar landing mission. Geologists probably would be sought when NASA recruited astronaut trainees for subsequent Apollo lunar flights. In his address, Dr. von Braun advised that U.S. must protect itself from any hostile force that attempts to "use space as an unchallenged avenue of aggression against us." (*Houston Post*, 2/15/64; *Houston Chron.*, 2/16/64)
- Rep. Olin Teague (D.-Tex.), Chairman of Subcommittee on Manned Space Flight, House Committee on Science and Astronautics, visited NASA Manned Spacecraft Center with members of the subcommittee. Rep. Teague said at MSC that the manned space flight budget for FY 1965 "is a tight budget and we have a good program. No major cuts are foreseen." (*Houston Chron.*, 2/16/64; *Houston Post*, 2/16/64)
- Question of how U.S. would use or dispose of 54 Titan I launch silos was raised in *Christian Science Monitor*. $1.5 billion had been spent on digging and equipping the silos. First Titan I site became operational in April 1962. Now that Titan II had become operational augmented by Minuteman missiles, Secretary of Defense Robert S. McNamara had declared Titan I would be eliminated after June 30, 1966. (Fleming, *CSM*, 2/15/64)
- Formation of six F-105 supersonic fighter-bombers flown by USAF pilots from Guam to New Zealand set unofficial nonstop distance record. The jets completed the flight of 4,422 mi. in 7 hrs., 45 min. (AP, *Wash. Sun. Star*, 2/16/64)

February 16: USAF's big Project Forecast, according to Richard Witkin in *New York Times*, anticipated eventual need for four new manned aircraft: Advanced Manned Precision Strike System (AMPSS), intercontinental-range attack aircraft; transport capable of 10,000–12,000 mi. range; Vtol attack-reconnaissance craft; and limited-range Vtol transport. Begun almost a year ago at request of Secretary of the Air Force Eugene M. Zuckert and Chief of Staff General Curtis E. LeMay, Project Forecast was said to be virtually completed. (Witkin, *NYT*, 2/16/64, 1)

February 17: NASA submitted to Senate Committee on Foreign Relations comments on S. Con. Res. 65, "Favoring agreements with other nations for the joint exploration and use of space and to place a man on the moon." NASA recommended: deleting two paragraphs to avoid "the unwarranted impression that substantial savings might be realized in the early stages of a cooperative arrangement between the United States and the Soviet Union in connection with manned lunar exploration"; deleting reference to U.N., "since other existing international mechanisms appear better suited to bring to bear the technical competence and the responsibility associated with national requirements necessary to contribute to cooperative projects of this type"; and, "the bipartisan Congressional support given to the civilian space program should be continued on the basis of the national benefits accruing from vigorous implementation of this program." Subject to these comments, NASA had no objection to adoption of the resolution. (NASA LAR III/27)
- According to breakdown of FY 1965 budget of NASA's Office of Advanced Research and Technology, NASA would initiate preliminary design study of advanced fluorine-hydrogen (flox) rocket engine in FY 1965. The new engine would be based on results of experimental fluorine program with RL–10 engine. (*M&R,* 2/17/64, 18)
- Congressional Joint Atomic Energy Committee had sent letters to NASA Administrator James E. Webb, Air Force Secretary Eugene Zuckert, AEC Chairman Glenn Seaborg, and Budget Director Kermit Gordon, *Missiles and Rockets* reported, indicating it would authorize funds to AEC for flight-testing Snap reactors. DOD had announced it had "no stated requirement" for the nuclear auxiliary power systems, and BOB had refused AEC request for $15 million to flight-test Snap 10A, cut $7 million from Snap 50 development, and cut $5.5 million from systems improvement. Committee action would provide funds in excess of AEC's request. (*M&R,* 2/17/64, 14)
- House Committee on Science and Astronautics was planning to hold further hearings on the NASA proposed electronics research center, although Committee Chairman George P. Miller (D.–Calif.) had described NASA Administrator Webb's report to the Congress as "very comprehensive." (*M&R,* 2/17/64, 17)
- USAF launched Titan II ICBM from Vandenberg AFB on 5,000-mi. flight down PMR. (UPI, *Houston Chron.,* 2/18/64)
- Rep. Carl Elliott (D.–Ala.), Chairman of House Select Committee on Government Research, submitted to the House the committee's first progress report. At its preliminary information-gathering hearings, the committee heard testimony by 75 persons representing Federal departments and agencies, private research organizations, institutions of higher learning, scientific and professional associations, and private industry. Rep. Elliott said: ". . . while we hope to be able to lay sound guidelines for the future, we shall also strive to insure that the incentives for engaging in research and development are strengthened and safeguarded rather than strangled by excessive controls and redtape. . . .

    "Our only bias is one in favor of wisely conceived and soundly administered research and development programs designed to strengthen our country and promote the general welfare of our people." (*CR,* 2/17/64, 2854–55)

February 17: Fuel cells producing power to operate all electronic equipment in Apollo spacecraft would also produce enough drinking water as a by-product to supply the three-man crew during 14-day journey to the moon and back, MSC announced. (MSC Release 64–32)

- New gravity gradient stabilization system utilizing earth's gravitational field to orient the satellite and earth's magnetic field to assist in damping was developed by General Electric and recently tested onboard a USN satellite, orbited Jan. 11. Within three days the system stabilized the satellite to accuracy of ±5 degrees. (*M&R*, 2/17/64, 10; *Av. Wk.*, 2/24/64, 57)
- USN fired Polaris A–2 missile from submerged submarine off the Florida coast in successful test. (UPI, *Houston Chron.*, 2/18/64)

February 18: Senate Foreign Relations Committee voted to recommend Senate ratification of radio communications treaty, drawn up at 70-nation conference in Geneva last fall. Considered a step toward establishment of global communications satellite network, treaty revised radio regulations, provided for international allocation of radio frequencies for communications, navigational, and meteorological satellites. (AP, *NYT*, 2/19/64, 15)

- NASA announced two industry teams had presented to NASA and other agencies their feasibility studies for nonmilitary navigation satellite system. General Electric Co. and Univac Div. of Sperry Rand Corp. made one study under $156,000 NASA contract, while Westinghouse Electric Corp. studied another approach under $225,000 contract. The proposed satellite network would provide worldwide, all-weather coverage for ships and aircraft. (NASA Release 64–37)
- Dr. Albert C. Hall, Deputy Director for Space, Defense Research and Engineering, said in address before National Space Club in Washington that the Manned Orbiting Laboratory would be the first phase of DOD's National Orbiting Space Station and would provide means of carrying out experiments to prove or disprove the military role of man in space. MOL would carry two astronauts for 30 days, also would carry experiments of nonmilitary nature, primarily for NASA. Discussing nonmilitary unmanned space program, Dr. Hall disclosed that the Army had secretly launched SECOR last month and that it was performing "exceptionally well." (*SBD*, 2/19/64, 274; Simons, *Wash. Post*, 2/19/64)
- NASA Associate Administrator for Manned Space Flight Dr. George E. Mueller said in hearings before Subcommittee on Manned Space Flight, House Committee on Science and Astronautics, that for Project Gemini "the year 1963 was, quite frankly, one of development headaches. It was the year of development and qualification testing of Gemini's many equipments. The program is now largely over the development hump, however, and all Gemini spacecraft and launch vehicle subsystems will be fully qualified in 1964. Significant in 1963 also were a number of hardware deliveries. . . .

"In 1964, three vitally important flights will be made—the third one manned and orbital. The year will be characterized by production, test, and delivery of spacecraft and launch vehicles. . . ."

Questioned about return to earth and landing in Project Gemini, Dr. Mueller and George Low said that plans now called for the two-man Gemini spacecraft to land on water rather than dry land. NASA had been investigating both procedures, and paraglider for land recoveries was still in development stage. Paraglider would be "available" at about the time of the last two Gemini flights but "not necessarily scheduled." Dr. Mueller added that water landing also would be "characteristic" for manned Apollo flights. (Testimony; Finney, *NYT*, 2/19/64, 19; *M&R*, 2/24/64, 21)

February 18: NASA Associate Administrator for Space Science and Applications Dr. Homer E. Newell discussed Lunar Orbiter in testimony before Subcommittee on Space Science and Applications, House Committee on Science and Astronautics: "In 1963, the Lunar Orbiter was defined and a contractor selected. This program will move into high gear this year and will require $49.3 million in Fiscal Year 1965 to meet its targeted first launch in 1966. This will be a tight schedule but because of the importance of this mission in support of Apollo, every effort will be made to meet that date. . . ."

Questioned by Subcommittee Chairman Rep. Joseph E. Karth (D.-Minn.), Dr. Newell said NASA was thinking about adding five Lunar Orbiters to the five already approved. The proposed increase was because the Lunar Orbiter "will be a most important spacecraft not only for the support of the manned lunar landing effort, but for continuing research on the moon, on its geodetic properties, and experiments on radioactivity of the lunar surface."

Of the recently orbited ECHO II passive communications satellite, Dr. Newell said: "The ECHO II has shown some scintillation in the radio wavelengths which might indicate that it has some defect in shape—perhaps a tear, maybe a dent, or something of that sort. However, it shows no such scintillation on the optical observations, which is puzzling. It remains, however, a completely adequate reflector so that whatever distortion in shape may have occurred has not interfered with its usefulness." (Transcript)

• Rep. Carl Vinson (D.-Ga.), Chairman of House Armed Services Committee, commented on Sen. Barry Goldwater's charges of missile unreliability:

". . . I, the Armed Services Committee, the Secretary of Defense, and the Secretary and Chief of Staff of the Air Force are all wholly persuaded that our missiles—and I am particularly referring to our intercontinental ballistic missiles—are fully capable of doing the job for which they were intended. . . ."

Rep. Vinson was appearing before the House Rules Committee to set forth provisions of proposed $19.6 billion military R&D authorization bill, already approved by Armed Services Committee. (This bill, in a procedure initiated this year, covered separately the R&D portion of overall $50.9 billion proposed defense authorization.) One provision of bill would authorize $52 million for development of new manned bomber, instead of $5 million requested by the Administration for studies. Rep. Vinson emphasized that his Committee favored the USAF proposal, explaining, "Solid testimony was given to the Committee by General LeMay, clearly indicating the urgency of starting out on a follow-on bomber and a new interceptor." (*NYT*, 2/19/64, 4)

• Attempt by Assistant Labor Secretary James J. Reynolds to arbitrate 13-month-old dispute between 11 striking unions and Florida East Coast Railway was unsuccessful. Unions' representative George W. Leighty refused to agree to arbitration, and Reynolds termed talks with FEC

Board Chairman Ed Ball "completely futile." Reynolds recommended to Labor Secretary Wirtz that FEC be barred from using Government-owned tracks at Cape Kennedy pending settlement. (AP, Balt. *Sun*, 2/19/64; *M&R*, 2/24/64, 11)

February 18: Rep. Roland V. Libonati (D.–Ill.) presented to the House a group report on AFSC, SAC, TAC, and Air Force Academy, after Congressional delegation visited various USAF installations. Listing highlights of AFSC during 1963, report said:

"Since March 1959, more than 460 space probes have been conducted from Air Proving Ground Center's Elgin AFB, Fla., aerospace launch facility on Santa Rosa Island. The probes included at least 87 in the Project Firefly program. The project actually is the intense investigation of the earth's atmosphere and an assault on the secrets of 'near space'. . . .

"AMRL [Aerospace Medical Research Laboratories, Wright-Patterson AFB] initiated the first lunar gravity research in 1963. Using a zero-gravity parabola, aircraft tests were conducted under actual lunar—one-sixth earth gravity—conditions. . . ." (*CR*, 2/18/64, 2906–35)

• FAA announced Canadian Pacific Air Lines had reserved three delivery positions for the U.S. supersonic transport plane, bringing total reserved positions to 70 by 12 airlines. (FAA Release 64–16)

• New York State Atomic Research and Development Authority signed contract with General Electric Co.'s Aerospace and Defense Service Engineering Dept. for continued operation of Malta Test Station, which New York recently acquired from Federal Government through competitive bidding. Station has been operated by GE for Federal Government since its establishment in 1945, when facilities were installed there to duplicate German V–2 facilities at Peenemünde. New York Authority would operate the station's expanded test and research facilities in continued support of industrial contractors of Federal defense, space, and atomic energy programs. (*CR*, 2/19/64, 2011–12)

February 19: X–15 No. 3 flown by NASA pilot Milton O. Thompson in third of current series designed to study heat transfer rates with sharp-leading-edge vertical tail. Launched from B–52 near Hidden Hills, Calif., the X–15 attained maximum speed of 3,504 mph (mach 5.26) and peak altitude of 77,500 ft. This was flight 101 for the X–15, 26 for X–15 No. 3. (FRC Release; NASA X–15 Proj. Off.)

• Boilerplate Apollo spacecraft arrived at Cape Kennedy, where it would be orbited by Saturn I SA–6 in April. Three separate USAF aircraft transported the boilerplate launch escape system, command module, service module, related ground service equipment, and adapter. Within a few hours, S–1 stage and instrument unit for SA–6 arrived by barge after 13-day trip from Huntsville, Ala. (MSC Release 64–33; MSC *Roundup*, 2/19/64, 1)

• India's Dept. of Atomic Energy and NASA announced signing memorandum of understanding for Indian-U.S. cooperation in meteorological sounding rocket program. Data gathered in the program would contribute to International Indian Ocean Expedition, a continuation of research begun during IGY, concentrating on weather and atmospheric conditions related to the Indian Ocean area. Under the agreement, India Dept. of Atomic Energy would provide, transport, assemble, and launch Loki-Dart sounding rockets and payloads; operate the Thumba Equatorial

Rocket Range; and coordinate launchings with those of other countries in International Indian Ocean Expedition. NASA would lend radar equipment, train Indian personnel, and provide technical advice. (NASA Release 64–39)

February 19: NASA announced FRC had issued requests for proposals for construction of two low-speed lifting-body gliders. NASA would furnish the selected contractor with information obtained in research with M–2 concept at Ames Research Center and the HL–10 lifting-body at Langley Research Center. Both vehicles would undergo thorough testing in Ames wind tunnel, then be tested in flight at FRC. (NASA Release 64–41)

- NASA Associate Administrator for Space Science and Applications Dr. Homer E. Newell told Subcommittee on Space Science and Applications, House Committee on Science and Astronautics, that Scout space vehicle had not been performing with desired reliability. Ground tests and modifications were conducted, and when a test vehicle was fired the fixes were shown to be "adequate." Subcommittee Chairman, Rep. Joseph E. Karth (D.–Minn.), replied that "if NASA spent half as much money on solids as it did on liquid vehicles, the solid rocket program would have 100 per cent reliability."

  Discussing Surveyor lunar probe, Dr. Newell said first nine Surveyors would be reduced from 2,500 lbs. to 2,100 lbs., and it was hoped flox upper-stage Atlas-Centaur space vehicles would be available to launch the subsequent Surveyors. (*SBD*, 2/20/64, 279)

- Edmond C. Buckley, Director of NASA Office of Tracking and Data Acquisition, testified before Subcommittee on Advanced Research and Tracking, House Committee on Science and Astronautics:

  "We are requesting $274.4 million to support the Tracking and Data Acquisition program in fiscal year 1965, as compared to $307.7 million in fiscal year 1964, a decrease of eleven per cent. The reason for this decrease is that most of our new station construction requirements in all three networks, were funded in fiscal years 1963 and 1964. . . ."

  Mr. Buckley pointed out that the largest single item in OT&DA budget for FY 1965 was $106.9 million for equipping the manned space flight network for Projects Gemini and Apollo. (Testimony)

- Unnamed NASA spokesman denied Feb. 15 statement by Dr. Wernher von Braun that scientist would be included in crew for second lunar landing flight. NASA said it had not been decided whether a scientist would be aboard the flight, and that if fully qualified scientists did make lunar expeditions they probably would go "later than the second flight." (*Houston Press*, 2/19/64; *Houston Chron.*, 2/19/64)

- USN request that NASA include four of its experiments in manned Gemini flights was revealed in NASA Gemini program review. (*Av. Wk.*, 2/24/64, 23)

- Discovery of the omega-minus particle in the atomic structure was hailed as a significant step toward construction of an ordered table of elements for the atom similar in nature and importance to the periodic table of elements. Prof. Paul T. Matthews, of the Imperial College, London, and a Fellow of the Royal Society, described his discovery in the British journal *New Scientist*. Prof. Y. Ohnuki had suggested in 1960 that the clue to orderly relationship between the 50 or more known atomic particles might lie in a type of "spin" which could only be described in mathematical terms and known as "unitary transformations." Several scientists in various parts of the world, including Dr. Murray Gell-Mann in the U.S., began working on this theory. Relationships did begin to appear among particles; also it became apparent that one more subparticle was needed. Its characteristics could be deduced from its place in the atomic relationship. At this point teams of scientists went to work with the 33-billion-volt Alternating Gradient Synchrotron at Brookhaven National Laboratory and others with equipment at the Center for European Nuclear Research (CERN) at Geneva, Switzerland. Analysis of 100,000 photographs of bubble patterns made by accelerated atomic particles traveling through "boiling" liquid hydrogen disclosed two sightings of particles with the characteristics predicted by the relationship table. (Hillaby, *NYT*, 2/20/64, 1; Sullivan, *NYT*, 2/23/64, 67)

February 19: Dr. J. Allen Hynek, Director of Lindheimer Astronomical Research Center and head of astronomy dept. of Northwestern Univ., Evanston, Ill., said in an interview that remote control telescope on the moon was practical as well as desirable. Heart of lunar telescope would be image orthicon, comparable to picture tube in TV camera; signals would travel to the telescope and pictures would travel back to earth on laser beams. ". . . If you had a telescope on the moon, you could point it where you want to by remote control and the picture would go into an image orthicon to be flashed back to earth on the same laser beam that controls the moon telescope." (Hughes, *Chic. Trib.*, 2/19/64)

February 20: NASA Nike-Apache sounding rocket launched from Ft. Churchill, Canada, carried instrumented payload to 127-mi. altitude, in successful flight to obtain data on charged particles responsible for visual aurora and on those responsible for auroral absorption. Instrumentation included five charged-particle detectors, an aspect magnetometer, and an accelerometer. (NASA Rpt. SRL)

* NASA announced selection of biological specimens to be included in first biosatellite in late 1965. Effects of radiation, weightlessness, and absence of earth's rotation would be studied on the 14 specimens, including pepper plants, wheat seedlings, amoeba, frog and sea urchin eggs, bread mold, fruit flies, and embryonic beetles. The biosatellite payload would be launched into orbit by a Delta vehicle from Cape Kennedy and recovered three days later in the Pacific. (NASA Release 64–40)

* AEC issued special report on Snap (Systems for Nuclear Auxiliary Power) program, stressing need to flight-test Snap in spacecraft in order to develop its full potential: "The Snap program has reached a point in technology where its importance to the Nation must be recognized, accepted and supported, or the development and performance of ambitious space missions will be handicapped by a lack of adequate and reliable power." Flight test of Snap–10A had been canceled earlier this year when Bureau of the Budget cut Snap funds requested for FY 1965. (AP, Wash. *Eve. Star*, 2/20/64; Simons, *Wash. Post*, 2/21/64)

* House passed (336–0) military R&D authorization bill of $16,914,800,000 for FY 1965. Amendment to delete $52 million for design and development of new jet bomber and $40 million for design and development of new interceptor aircraft was defeated (121–29). The programs, recommended by Air Force Chief of Staff General Curtis E. LeMay, had not been included in Administration budget request; Defense Secretary McNamara had requested only $5 million for studies of new bomber. (Raymond, *NYT*, 2/21/64, 1)

February 20: Prime contractors of NASA Marshall Space Flight Center were assuming additional responsibilities in Saturn launch vehicles program, MSFC Director Dr. Wernher von Braun announced. Principal contractors immediately affected would be Boeing Co., Chrysler Corp., IBM Corp., and General Electric Co. Dr. von Braun said this was "a continuation of the Marshall policy to transfer research and developmental work, as well as fabrication, from 'in house' to contractor plants as soon as practical. Not only has this been our philosophy but prime contracts in most cases have spelled out the eventual taking over by industry of full responsibility for major elements and stages of the vehicles. . . ."

MSFC would retain overall systems management responsibility for the three Saturn launch vehicles and continue to "maintain a technical and managerial competence in our own organization. Without it we cannot monitor the work of our contractors to insure that we are getting the quality and reliability that a manned space program demands. . . ." (*Marshall Star*, 2/26/64, 10)

- Edward Z. Gray, NASA Director of Advanced Manned Missions, described manned lunar bases under study for post-Apollo missions, in testimony before House Committee on Science and Astronautics' Subcommittee on Manned Space Flight. "The earliest system we can see for extending the capability of Apollo would be the Apollo logistic support system (ALSS).

"It consists of the launching on the Saturn V of an Apollo spacecraft with an unmanned LEM, wherein we have replaced the ascent stage of the LEM with a cargo compartment. The Apollo spacecraft and unmanned LEM, or LEM truck are launched to the Moon. The LEM truck is disengaged from the Apollo spacecraft in lunar orbit and lands in an unmanned mode on the surface of the Moon.

"The payload capability with the present Apollo launch vehicles and spacecraft would be approximately 7,000 pounds. There would be a payload volume of about 2,000 cubic feet. This would provide a shelter with occupancy capability for two men for about 14 days on the lunar surface. An alternate payload that could be carried by the LEM truck would be a two-man roving vehicle . . ., which would have a capability to cover about 200 to 300 miles on the lunar surface in the 14 days. . . .

"Then looking beyond to the time when much greater staytime on the lunar surface would be desired, we show a lunar exploration system for Apollo (LESA) which weighs about 25,000 pounds on the surface of the Moon. These are Earth pounds.

"It would be launched with a Saturn V without an Apollo spacecraft and would be delivered in an unmanned mode directly onto the surface of the Moon. It has a volume of over 3,000 cubic feet, and could provide satisfactory shelter and life support for 3 men to stay 90 days on the surface of the Moon.

"The mode of operation would be to launch this unmanned vehicle to the surface of the Moon, and follow it up with an Apollo launch so that the LEM would land in the near vicinity of the LESA payload, probably using a homing radar. . . ."

He described the LESA concept as expandable through addition of modules; for instance, three modules could form a base for 18 men, 24 months occupancy time. (*1965 NASA Auth. Hearings*, Part 2, 597–99)

February 20: USAF announced selection of Douglas Aircraft Co., Martin Marietta Corp., and General Electric Co. for orbital space station study-contracts. The studies would aid in defining tests and experiments for USAF Manned Orbiting Laboratory (MOL). (DOD Release 148–64)

- Lt. Col. Andrian G. Nikolayev had replaced Col. Yuri A. Gagarin as head of Soviet cosmonauts, *Krasnaya Zvezda* reported. (UPI, *NYT*, 2/22/64, 10)
- Dr. Herman J. Sander, Behavioral Scientist in USAF OSR Life Sciences Directorate, was presented Air Force Exceptional Civilian Service Medal by Maj. Gen. Don R. Ostrander, OAR Commander. (OAR Release 2–64–1)
- Martin Marietta Corp. received $1,748,450 increment to USAF contract for R&D for Titan II ICBM, DOD announced. (DOD Release 149–64)
- NASA Manned Spacecraft Center began the first increment of its mass move to the new facilities at Clear Lake from leased temporary facilities in Houston and Ellington AFB. (MSC Release 64–35; *Houston Chron.*, 2/20/64)

February 21: DOD announced Army Corps of Engineers' SECOR (Sequential Collation of Range) geodetic satellite was performing "exceptionally well," confirming Dr. Albert Hall's disclosure of Feb. 18. The 40-lb. satellite was orbiting at about 600-mi. altitude in periods of about 100 min. SECOR system tests were being conducted by Army's Geodesy, Intelligence and Mapping Research and Development Agency (GIMRADA), which had contracted for the satellite from ITT Federal Laboratories. GIMRADA ground stations were measuring distances by sending and receiving signals through satellite's onboard transponder, then acquiring ranging data for use in automatic data reduction process by Army Map Service to determine station positions. AFSC/SSD had lofted the satellite into orbit for GIMRADA; NASA Goddard Space Flight Center was providing orbital data and telemetry support. (DOD Release 156–64)

- Astronaut Alan B. Shepard, Jr., said in informal press gathering in Washington that when NASA had decided on 30 trainees for its astronaut pool, it had anticipated losing 4 of the 30 men, for various reasons, before completion of Project Apollo. Commander Shepard said, "we figured on two casualties as a result of space flight," explaining that the figure was "just pulled out of the air" but was considered a reasonable guess. NASA Deputy Administrator Dr. Hugh L. Dryden emphasized "We have no idea what the casualty rate will be. We're hoping it will be zero." (Troan, *Knoxville News-Sentinel*, 2/21/64)
- 200-ft.-diameter balloon was launched by AF Cambridge Research Laboratories personnel from Holloman AFB, N. Mex., with special telescope to make spectroscopic records of the planet Venus. The telescope was carried to altitude of 87,500 ft., above all but an insignificant amount of terrestrial water vapor. (USAF OAR Release 4–64–2)
- Dr. George E. Mueller, NASA Associate Administrator for Manned Space Flight, said in speech at Engineers' Week dinner, Los Angeles:

    "At present, we estimate that more than 35,000 engineers and scientists are at work in the Gemini and Apollo programs and the studies of possible advanced manned missions. This number will reach a level of 45,000, about 2.8 percent of the national employment of scientists and engineers, when our currently approved programs reach their peak of effort next year. . . .

"In recent months, we have reviewed thoroughly the status of program technology . . ., the pace of the program for its impact on total cost . . ., and the possible effects on the program of the hazards of interplanetary space. . . .

"First, with respect to technology . . . we have found all of the knotty questions involved in advancing the state of the art are yielding to hard work. We know of no technological problems that would prevent our accomplishing the Apollo program in this decade. . . .

"[Second], . . . starting with the level of funding planned for the current fiscal year there is sound economic justification for maintaining the program on its present schedule. To stretch out the program at this time would add to the cost and unnecessarily delay this first major step in space exploration.

"The final area of examination of the program has been with respect to the hazards of interplanetary space. . . . Estimates based on present information indicate that meteoroids will not constitute a major problem in the planning of the first manned lunar exploration. . . .

"Altogether, the present evidence indicates that radiation does not present a hazard that would prevent accomplishment of manned lunar exploration in this decade. Indeed, there is no serious evidence that the radiation factor would inhibit us in selecting the time of a lunar mission. . . ." (Text)

February 21: "Astrometric" telescope of the U.S. Naval Observatory went into operation at Flagstaff, Ariz. The telescope featured world's largest known telescopic mirror, 61 inches in diameter, ground from huge block of synthetic quartz. New telescope would be used chiefly to measure distances to nearby stars that are too faint for such study by other existing telescopes, thus exploring previously unknown regions of the universe. (Sullivan, *NYT*, 2/25/64, 14)

- Sen. George McGovern (D.–S. Dak.) referred on Senate floor to article in recent *Business Week* which reported sharp drop in engineering and scientific jobs due to defense cutbacks. Employment cutbacks "tend to be concentrated in certain industries—aerospace, defense installations, and atomic energy—and in certain parts of the country—the Boston area, New York, Pennsylvania, parts of the Midwest, the Pacific Northwest, and California." (*CR*, 2/21/64, 3205)
- Public sale of stock in Communications Satellite Corp. would be offered in April, New York sources revealed. (NYTNS, *Denver Post*, 2/21/64)
- Board of Directors of U.S. Chamber of Commerce recommended that Congress cut $119 million from NASA's requested FY 1965 budget and that the $141 million FY 1964 supplemental request be denied. (*Houston Post*, 2/22/64)
- Lt. Col. John H. Glenn, Jr., former astronaut, was presented the George Washington Award for 1963 by the Freedom Foundation, in Valley Forge, Pa., ceremony. (Wash. *Eve. Star*, 2/22/64)

February 22: First Soviet-American space communications experiment, with British cooperation, was conducted successfully at 2:00 a.m. Moscow time, Tass reported. Radio signal sent from Jodrell Bank Experimental Station bounced off ECHO II balloon satellite and was received by Zimenki Observatory near Gorki. Tass said, "This marks a success of the first experiment in super-long distance international cosmic radio communications on ultra short waves." (AP, Wash. *Eve. Star*, 2/22/64)

February 22: Published reports about the SST study prepared for the White House by Eugene Black and Stanley deJ. Osborne had failed to mention the report's strong endorsement of SST development, Evert Clark reported in *New York Times*. According to Clark, letter transmitting the report, still not officially released, said development of SST was "of great economic importance to the nation. . . . Failure to do so might well leave our important airline and aircraft industries in potentially dangerous competitive situations." (Clark, *NYT*, 2/22/64, 1)

- Sir Bernard Lovell, Director of Britain's Jodrell Bank Experimental Station, advocated international controls on space experiments. In *Saturday Evening Post* article, he said: "It is now time for concern about the ethical standards which man must apply to embryonic dangers which are not obvious. We cannot afford to wait until the dangers become practically apparent, since some experiments may create an irretrievable condition. . . ."

As an example, he described the ozone layer, and continued: ". . . it happens that zone is rather easy to destroy. And if it is intentionally or accidentally removed, then the ultraviolet radiation would penetrate to earth. If this happened for a sufficiently long period, then human beings migh suffer severe sunburn and possible sterilization. Further, the temperature distribution in the atmosphere could be radically altered and, in the state of our present knowledge, no one could predict what climatic changes might occur.

"It is not difficult to estimate that a few tons of suitable contaminant, deposited in the atmosphere 25 miles above the earth, would destroy the ozone over several miles for a few hours. . . . a rocket launched from earth to land a man on the moon might burn up 2,000 tons of fuel in its passage through the earth's atmosphere . . . [and] the launching of one such rocket a week might well lead to a permanent transformation of the conditions in the high atmosphere," according to recent study by the Advanced Research Projects Agency.

". . . when one turns from the contamination as an accidental by-product to an intentional act, an ugly array of possibilities is revealed. To begin with, specially selected substances could be deposited in great quantities at precise altitudes. . . . The American [ARPA] report states that 25 tons of fluorine would be sufficient to depopulate the ionosphere of electrons and so blot out all long-distance radiotelephony. The deposition of large quantities of gases like carbon monoxide at an altitude of 25 miles might radically alter the temperature distribution and the climatic conditions by absorbing some of the sun's infrared radiation. . . .

"The human race therefore faces a critical situation in the next few decades, because there is at the moment little evidence of the moral and legal controls which must be enforced if man's continued life on earth is not to be jeopardized by the accidental or intentional results of space research. There are large areas of uncertainty, so that it is not yet possible to assess whether various other space activities may affect man more directly than in the inhibition of his study of the cosmos. It is a vital necessity of our age that the political decisions on these problems should be based on the judgment of the international scientific community, and that this judgment should govern the future space launchings of the world." (*Sat. Eve. Post*, 2/22/64, 10, 14)

February 24: House Committee on Science and Astronautics, meeting in executive session, approved the NASA Electronics Research Center report selecting the Boston area for the center. (NASA LAR III/32)
- NASA's Group Achievement Award was presented to Paul F. Bikle, Flight Research Center Director, and Brig. Gen. Irving L. Branch, AFFTC Commander, on behalf of all Government personnel connected with the X-15 program. Dr. Hugh L. Dryden, NASA Deputy Administrator, made the presentation at FRC ceremony, and said in his remarks:

    "Earlier this month the Research Airplane Committee approved nine additional test-bed experiments for the X-15 Program. With the 19 experiments approved before and the flights required to complete the original research program, these additional experiments will require X-15 flight tests until the end of 1968.

    "The X-15 flight-test program has been eminently successful during the last four years. The Research Airplane Committee, by its action this month in approving these new experiments for the program, has given concrete evidence of its faith in the future program. . . ." (FRC *X-Press*; Speech Text)
- NASA announced the nine new experiments for X-15 aircraft would be: ramjet propulsion, structures research, advanced flight data system, star tracker, horizon scanner, ionization gage, and other missions involving evaluation of advanced vehicle systems and structural materials. (NASA Release 64-42)
- NASA Associate Administrator for Manned Space Flight Dr. George E. Mueller said in *Space Business Daily* interview that the "major problem" in Project Apollo "is not technical or environmental," but, rather, "proper funding." Attainment of lunar landing goal before 1970 depended "more on funding than anything else," he said. The hiring freeze imposed on Apollo contractors had been eased on a "selected basis," with the only companies not receiving the easement being Pratt & Whitney and General Dynamics. (*SBD*, 2/24/64, 289)
- Col. Clarence J. George was sworn in as NASA Executive Secretary, newly established position in Office of the Administrator. A former executive officer for the late General George C. Marshall, Col. George had more than 28 years of military service including eight years as aide to three former Secretaries of the Army, as well as General Marshall's tours as Secretary of Defense and Secretary of State. (NASA Photo)
- Atlas engine using flox (30% fluorine, 70% liquid oxygen) oxidizer with its normal RP-1 kerosene fuel was static-fired last month by General Dynamics, Associated Press reported, and successful results of the test had prompted NASA to request FY 1965 funds to accelerate development. NASA requested $17.5 million in FY 1965 for flox development, which preliminary experiments indicated would increase Atlas payload capability from 30 to 90%. (AP, Wash. *Eve. Star*, 2/24/64)
- Analysis of NASA FY 1965 budget request revealed that NASA construction had passed its peak. FY 1964 funding for construction was $637.5 million, and FY 1965 request was for $281 million. Accentuating the trend was the fact that funding of major new facilities for Project Apollo—at Cape Kennedy, MSC, MSFC, and Mississippi Test Facility—would be completed in FY 1965. (*M&R*, 2/24/64, 18)

February 24: Sen. Mike Monroney (D.–Okla.) told American Fighter Aces Association, Washington chapter, that aviation was "short-changed" in research and development. He urged that more NASA effort go into aeronautical R&D, particularly on the supersonic transport. "NASA means National Aeronautics and Space Administration. I'm worried about what's happening to the A-for-Aeronautics." (Wash. *Eve. Star*, 2/25/64)

- Lewis Research Center announced awarding of three contracts: $3,117,387 contract for continued development of Centaur launch vehicle stage and $1,800,000 contract for Oao shroud system and Pt. Loma test program, both to General Dynamics/Astronautics; and $2,110,000 contract to Aerojet-General Corp. for continued development of M–1 rocket engine. (*WSJ*, 2/24/64)

- NASA signed contract with Reaction Motors Div. of Thiokol Chemical Corp. for research on oxygen difluoride propellants. (*SBD*, 2/26/64, 304)

- USAF planned 71 Atlas-Agena firings during 1964–66 period and NASA planned 26 Atlas-Agena and five Atlas launches, *Aviation Week* reported. Initial launch rate planned for Titan III with unmanned payloads would be less than 10 a year. (*Av. Wk.*, 2/24/64, 23)

- USAF predicted reddish spots on the moon—observed last October and November—would reappear June 4 and 5. USAF astronomers at Lowell Observatory planned to take color photographs of the recurrence as well as scan Aristarchus area of moon with sensitive photoelectric spectrograph to determine nature of the "spots." (Finney, *NYT*, 2/25/64, 14)

- *Missiles and Rockets* published interview made last summer with Dr. Eugen Saenger, who died earlier this month. Dr. Saenger, whose World War II concept of rocket sled-launched orbital bomber was later influential in design of X–20 (Dyna Soar), had said: "I am assisting a German group within Eurospace in looking into the economics of space transportation. We are considering reusable vehicles. Dyna-Soar is assumed to be the necessary first step. . . . We consider it essential to develop an aerospace plane able to lift off from Earth, go up to orbit in the form of an airplane with wings, return like the Dyna-Soar along a glide path to Earth and land like an airplane so it can be re-used." He said he had modified his chemical rocket-sled idea into steam-rocket sled and had developed such a rocket. "It develops 30 tons of thrust. We ran hundreds of tests on it. We would use it to accelerate a stage to subsonic or perhaps supersonic velocity."

 Dr. Saenger described Egyptian rocket development, which he had assisted until late 1961 when West German government asked him to stop: "I was asked in the spring of 1960 by the Egyptian government to give lectures at Cairo University and help the country by doing consulting work on meteorological sounding rockets. . . .

 "It's a rocket only for vertical launching and liquid fuel. There is no guidance. Although in principle the rocket could be launched into an inclined flight, there is no way of knowing with accuracy where it will impact. So I can't imagine how it would be used for military purposes." (*M&R*, 2/24/64, 35–38)

- Minuteman ICBM fired from Vandenberg AFB after sundown left spectacular trail across the sky seen from San Francisco to San Bernardino and caused Redding and Chico areas to call civil defense alerts. Cross winds distorted the exhaust trail and it was lighted by rays from the sun below the horizon. (AP, *NYT*, 2/27/64, 17)

February 25: USAF launched Atlas-Agena D combination from Vandenberg AFB with undisclosed payload. (*M&R*, 3/2/64, 12)

- NASA Associate Administrator for Space Science and Applications Dr. Homer E. Newell said in testimony before House Committee on Science and Astronautics' Subcommittee on Space Sciences and Applications that some of the byproducts of space research included cure for some types of stuttering, diet that can sustain a man for a month on one cubic foot of synthetic chemicals mixed with water, and "fail-safe" method of producing oxygen and food for orbiting space stations through electrolytic splitting of water into hydrogen and oxygen. (AP, *Houston Chron.* 2/26/64)

Rep. Donald Rumsfeld (R.–Ill.) noted in the House that House Committee on Science and Astronautics had approved NASA's report providing for electronics research center in the Boston area. "There is, in my mind, still considerable question as to the need for this center. Furthermore, there is considerable question in my mind as to whether, if such a center is needed, as opposed to developing additional electronics capability at existing centers, the criteria used for selecting Boston as the site were sound. . . ." (*CR*, 2/25/64, A878)

- Senate Armed Services Committee approved $17,040,140,000 military R&D authorization bill for FY 1965. Like the House, the Senate Committee voted $52 million for development of new manned bomber. But Senate Committee did not support House on $40 million for new manned interceptor aircraft. Committee made other reductions, restored several House cuts. (Raymond, *NYT*, 2/26/64, 13)

- Sen. Russell B. Long (D.–La.) said on Senate floor that AT&T, "the world's largest protected monopoly," had a "100-percent monopoly in international voice communications," and that "steps should be taken to require the divestment by AT&T of its international operations." He recommended that RCA, ITT, and Western Union be allowed to compete in international communications and that the Communications Satellite Corp. become a competitor in this area. (*CR*, 2/25/64, 3389–91)

- USAF fired Atlas E ICBM on successful 5,000-mi. flight down the Atlantic Missile Range, testing advanced nose cone design. Same day, USAF also fired Minuteman ICBM from Cape Kennedy and two Minuteman ICBM's from Vandenberg AFB, Calif., within 45 min. of each other. (UPI, *NYT*, 2/27/64, 17; *M&R*, 3/2/64, 12)

- Maurice Farman, aviation pioneer, died at 96 in Paris. With his brother Henry he made the first flight of more than a kilometer in a heavier-than-air machine, in 1908. The Farmans manufactured biplanes for France, Britain, and U.S. during World War I, in 1917 constructed the Goliath, forerunner of passenger airliner. (AP, *NYT*, 2/28/64, 29)

February 26: NASA Aerobee 150 sounding rocket with Johns Hopkins Univ. instrumented payload was launched from Ft. Churchill, Canada, to 100-mi. altitude. Rocket carried spectrophotometric instrumentation which measured absolute intensity of certain spectral features in the far ultraviolet region of the airglow during an aurora. Although rocket did not reach predicted 157-mi. altitude, the experiment provided general confirmation of the only previous experiment in that region, and with much higher resolution. (NASA Rpt. SRL)

February 26: NASA announced purchase from Douglas Aircraft Co. of eight more Delta launch vehicles, some of which would be the Thrust-Augmented Delta (TAD) capable of placing 1,000-lb. payload into low earth orbit as compared to 800 lbs. for standard Delta. (NASA Release 64-46)

- A. O. Tischler, NASA Director of Chemical Propulsion, Office of Advanced Research and Technology, discussed 1.5-million-lb.-thrust hydrogen-oxygen M-1 engine and the large solid-propulsion rocket motor before Subcommittee on Advanced Research and Tracking, House Committee on Science and Astronautics:

  "Last year this engine project [M-1], which was originally planned for completion in the late 60's, was slowed down to a minimum level of effort consistent with such a development program. It is our intent to maintain the low level of funding during the present critical years of the manned lunar landing program and then to accelerate this development for completion in the early 70's. . . .

  "The M-1 can be a team-mate of the large solid propellant motors. Potentially these two, if used in vehicle systems together, can provide the United States with increased payload capability at reduced cost during the 1970's. . . .

  ". . . the large solid motor demonstrations have heretofore been funded by the Department of Defense and have been carried out by the Air Force in compliance with NASA-defined requirements. . . . This year NASA is requesting funds for the continuation of the contracted work on the 260" diameter solid propellant motors. . . .

  "Each contractor [Thiokol Chemical Corp. and Aerojet-General Corp.] will fire a half-length solid motor during the fiscal year. . . . each half-length motor will produce three million pounds of thrust, about double the thrust of the Saturn I booster. The full-length motor, for which we would intend to contract if the half-length motor firings are successful, will produce about six million pounds of thrust." (Testimony)

- Dr. George M. Knauf, NASA Acting Director of Space Medicine, described before Subcommittee on Manned Space Flight of the House Committee on Science and Astronautics the procedures which NASA had followed for coordinating with USAF in bioastronautics-space medicine programs. Upon suggestions of this Subcommittee, NASA had "made a determined effort to coordinate our program with that of the Department of Defense so that all unwarranted duplication of effort would be eliminated." Further, to assure that the coordination would be on a continuing basis, NASA established Space Medicine Liaison Office at Aerospace Medical Div., Brooks AFB. NASA also obtained assurances from Army, USN, FAA, and U.S. Public Health Service that NASA was not duplicating work being carried out by those agencies. (Testimony)

- Lockheed-California Co. released details of its recommendations to NASA Manned Spacecraft Center on a scientific space station program. Lockheed study concluded manned station with crew of 24 could be orbiting the earth in 1968. Total cost of program—including logistics spacecraft and ground support—for five years' operation was estimated at $2.6 billion. Study recommended launching the unmanned station into orbit with a Saturn V launch vehicle, then launching manned logistics vehicle to rendezvous and dock at the station. (Maloney, *Houston Post*, 2/27/64; MSC *Roundup*, 3/4/64, 8)

February 26: MSC announced issuing requests for proposals for six-month study of how astronaut on the moon could conduct selenodetic measurements (lunar surface surveys), making maximum use of scientific equipment already planned for Apollo's Lunar Excursion Module and Command Module. Proposals for the study were due March 5. (MSC Release 64-37)

- Sonny Liston-Cassius Clay heavyweight title match was first championship fight to be telecast via communications satellite; video tape of the match was transmitted to 11 European countries shortly after 1:00 a.m. EST via NASA's RELAY II communications satellite. (NASA Release 64-44)
- Amendment to $17 billion military procurement and R&D authorization bill to cut $52 million for development of new manned bomber was offered by Senators George McGovern (D.-S. Dak.), Gaylord Nelson (D. Wis.), and William Proxmire (D.-Wis.). Senator McGovern quoted minority report by the dissenting members of the House Armed Services Committee which pointed out that "the $52 million is just the beginning of a program which will cost close to $5 billion before it is completed," said his amendment would "reduce the bomber funds to the $5 million level requested by the President and the Secretary of Defense." (CR, 2/26/64, 3602-04)
- FAA disclosed it was requesting Coordinating Research Council of New York City to study comparative safety features of the two most commonly used aircraft fuels—kerosene and JP-4. (Hudson, *NYT*, 2/26/64, 21)
- John F. Kennedy Space Center, NASA, announced 14,760 visitors to the facility during 1963. (KSC Release 16-64)
- FAA announced Irish International Airlines had reserved two delivery positions for U.S. supersonic transport aircraft. (FAA Release 64-18)
- Jack Anderson, writing in Drew Pearson's "The Washington Merry-Go-Round" column in the *Washington Post*, charged that two documents existed "revealing how Texas's lanky, laconic Congressman Albert Thomas used his influence to get the telephone contract at the Houston Space Center for Southwestern Bell." The documents, Anderson charged, were: (1) a confidential memo to NASA Administrator James E. Webb from his executive assistant recounting a telephone call from Rep. Thomas, in which Thomas recommended Southwestern Bell over the other bidder, General Telephone, stating that General Telephone had a reputation in the Houston area for poor service; and (2) a Southwestern Bell memo to employees acknowledging Rep. Thomas as "extremely helpful to our company—providing necessary information and other assistance—during the negotiations for the space lab communications contract." (*Wash. Post*, 2/26/64, D13)
- Party of scientists led by Nikolai Yerpylev, scientific secretary of Soviet Academy of Sciences, left Moscow for Cuba to "start the preliminary stage of the work" to establish Soviet satellite observation station on the island. Tass said equipment for the station already had been sent to Cuba. (UPI, *NYT*, 2/27/64, 12)
- USAF fired Titan II ICBM from Cape Kennedy on successful flight down the Atlantic Missile Range. (*M&R*, 3/2/64, 12)
- Douglas Aircraft Co. made public its company-funded evaluation of its "Icarus" concept (intercontinental aerospacecraft, range unlimited system) that could transport 132 tons of cargo or 1,200 fully-equipped troops

at speeds of 17,000 mph. Concept featured liquid-hydrogen propellant tanks that could be jettisoned after takeoff boost and recovered intact. Douglas spokesmen said Icarus could be operational in little more than a decade. (*SBD*, 2/26/64, 305)

February 27: U.S.S.R. orbited COSMOS XXV scientific earth satellite from unidentified site in Soviet Union. Initial orbital data, according to Tass: 526-km. (320.6-mi.) apogee, 272-km. (169-mi.) perigee, 92.27-min. period, and 49° angle of inclination to equator. Scientific equipment and communication systems onboard the satellite were said to be functioning normally. (Tass, *Pravda*, 2/28/64, 1, ATSS-T Trans.)

- Senate passed H.R. 9637, authorizing $17,040,140,000 FY 1965 funds for DOD procurement of aircraft, missiles, and naval vessels, and R&D. The McGovern amendment to reduce by $52 million the funds for development of new manned bomber was defeated. Bill would go to Senate-House conference committee for compromise with $16,914,800,000 House-passed bill. (NASA LAR III/35; Raymond, *NYT*, 2/28/64, 1)

- At NASA's request, DOD had declassified two of its experiments for Project Gemini, *New York Herald Tribune* reported. The previously classified experiments involved "visual definition of objects in near proximity in space" and "radiometric observations of objects in space." (Loory, *N.Y. Her. Trib.*, 2/27/64)

- In NASA Hq. ceremony, NASA Deputy Administrator Dr. Hugh L. Dryden presented cash awards to 10 inventors in recognition of their outstanding contributions to space science technology.

  Awards under the Space Act: $1,000 to Conrad Josias of JPL for invention of "bipolar logarithmic current-to-voltage transducer" which made possible the discovery of interplanetary plasma; $1,000 to James D. Acord and Howard C. Vivian of JPL for invention of "space vehicle attitude control" employed in Ranger and Mariner spacecraft; $2,000 to Robert C. Baumann and Leopold Winkler of GSFC for invention of "spin adjusting mechanism" used on OSO I, SYNCOM II, and Aerobee sounding rockets; and $4,000 to Harold R. Kaufman of LRC for invention of "ion rocket engine," representing great step forward in ion propulsion R&D—the engine will be used in first Space Electric Rocket Test flight (Sert I).

  Awards under the Incentive Awards Act: $1,000 to Pleasant T. Cole, GSFC, for invention of "system for recording and reproducing pulse code modulated data," capable of achieving significantly larger packing density on magnetic tape to reduce weight requirements; and $1,400 to William C. Morgan, Jack B. Esgar, and Richard H. Kemp of LRC for invention of "thin-walled pressure vessel" for use in testing materials subjected to cryogenic temperatures. (Hootman Memo, 2/18/64; NASA Release 64–45; LRC Release 64–16)

- Senator Mike Monroney (D.-Okla.) said in interview that aircraft manufacturers must bear the 25% share of supersonic transport development cost, rejecting the 10% formula advocated by Eugene R. Black in recently reported study. Senator Monroney said USAF should be assigned task of developing SST's engine and the airframe manufacturers should join together to build the SST body. He said the builder of the aircraft would "have the greatest property ever owned in aviation, a superplane good for at least 20 years. They've got to take some risk." (Steif, *Wash. Daily News*, 2/27/64)

February 27: Harold B. Finger, Manager of AEC–NASA Space Nuclear Propulsion Office and Director of NASA Nuclear Systems and Space Power, discussed nuclear rockets before Subcommittee on Advanced Research and Tracking, House Committee on Science and Astronautics: ". . . our major effort is in the reactor area. It includes analysis and experimental work on all of the components and subsystems of our reactors with sufficient evaluation and testing to give the greatest possible confidence in the successful operation of the graphite reactors that will form the basis for the first nuclear rocket engines to be developed in the world. We are, however, not overlooking the potential advantages provided by the metallic tungsten reactors. We are, therefore, conducting a program to evaluate the feasibility and advantages of these systems.

"Our NERVA engine and non-nuclear component technology work has made substantial progress in those difficult areas of engine technology that require extension of presently available information so that system development can eventually be undertaken rapidly and with a high degree of confidence.

"Sufficient facilities are available or are being constructed for the conduct of our presently planned program. Tests to be conducted in our reactor facilities during this year should clearly demonstrate the success of the solutions we have designed to avoid the mechanical problems that have been encountered in the past and should demonstrate the operating capabilities of graphite reactors. Our advanced concept work is fundamental in nature; it is long term; but it is being pursued because of the high, theoretical performance potential of these advanced ideas for nuclear rocketry.

"Out of this over-all nuclear rocket program, I am convinced, will come a new family of rockets that will open the way to more extensive space travel than is possible with the chemical combustion rocket systems upon which we are now completely dependent. Although new in capability, the nuclear rockets we are now developing are logical follow-ons to the chemical rocket systems and, therefore, do not involve basically new test, development, and system technology or scientific breakthroughs. These nuclear rocket developments must be actively pursued to ensure that this country is not prevented from undertaking missions involving higher velocities, energies, and payloads than are required for the early Apollo manned lunar landing mission." (Testimony)

- Harlan Cleveland, Assistant Secretary of State for International Organization Affairs, addressed the Conference of U.S. National Organizations of the U.N.:

  "The technological imperative—the impulse to build worldwide technical agencies—comes of course from the headlong pace of scientific discovery. A world technical community is in the making because international communications, international transport, and international economics demand international organizations—and because you can't deal with world health or world weather or radio frequencies and a lot of other things except on a world basis. . . .

  "Taken together, three new kinds of technology—weather satellites, communications satellites, and computer technology—now make it technically possible to work out a global weather reporting and forecasting system, a prospect too valuable to all nations to leave unexploited. The

United States is now engaged in a very large program of research and development in this field that will involve the cooperation of more than 100 other countries. This, together with what other countries are doing, will fit into an overall plan for a World Weather Watch being worked out by the World Meteorological Organization, a specialized agency of the U.N.

"In just 2 years our first Tiros satellites discovered 20 hurricanes, typhoons, and tropical storms and observed the behavior of 62 others. And world data centers to process these and other reports and issue warnings have been established in Washington and Moscow...." (Text, CR, 3/18/64, 5462–64)

February 27: USAF fired Minuteman ICBM from Cape Kennedy on successful routine 5,000-mi. flight down Atlantic Missile Range. (*M&R*, 3/9/64, 8)

- Sen. J. William Fulbright (D.–Ark.) inserted in *Congressional Record* a *Wall Street Journal* article on the "declining demand" for engineers, and said:

  "According to the *Wall Street Journal* article, the current slump in demand for engineers results primarily from uncertainty about future spending on defense and space programs. Although the defense budget for the next fiscal year calls for a reduction of only $1.3 billion, the space budget is $200 million above last year's appropriation. If this minute shift in defense spending is sufficient to bring about a severe curtailment in the demand for engineers and other skilled technicians, it makes one wonder what will happen to our economy when there is more than a token cutback in defense outlays...." (*CR*, 2/27/64, 3666–67)

- USN disclosed it was developing seagoing tracking system for its ships and submarines to "avoid those areas of the oceans covered at intervals by spy satellites. Submarines would stay submerged, surfacing only when the sky above is clear." (*Wash. Post*, 2/28/64)

- Visiting Oklahoma City, where FAA sonic-boom study was being conducted, FAA Administrator Najeeb Halaby was placed under police guard after anonymous telephone caller threatened his life. (AP, Balt. *Sun*, 2/28/64)

February 28: NASA Nike-Apache sounding rocket with instrumentation to obtain data on charged particles responsible for visual aurora and on charged particles responsible for auroral absorptions was launched from Ft. Churchill, Canada, to 127-mi. altitude. Instrumented payload included five charged-particle detectors, aspect magnetometer, and accelerometer. All flight objectives were met. (NASA Rpt. SRL)

- NASA and Australian Ministry of Supply announced installation to command flight of Syncom C synchronous-orbit communications satellite would be installed at Salisbury, Australia. Installation would have two functions: monitor the Syncom's attitude and command its apogee-motor firing, and command the satellite's control system to make any necessary adjustments to maintain synchronous position. (NASA Release 64–47)

- NASA announced Saturn contract modifications totaling $16,448,169 had been awarded to Chrysler Corp. Space Div., which was assembling two Saturn I first stages and 12 Saturn IB first stages at NASA Michoud Operations. One award was for expansion of Chrysler's Saturn booster reliability testing program at Michoud. Other two awards were for increased engineering support at Marshall Space Flight Center and John F. Kennedy Space Center, NASA. (NASA Release 64-48)

February 28: Dr. Hugh L. Dryden, NASA Deputy Administrator, said in letter to NASA Center Directors and Heads of Headquarters Program and Staff Offices: "I call upon each of you . . . to demonstrate full equality of opportunity without regard to sex, and that personal attitudes must not be allowed to operate adversely against women, or men, in recruitment, selection, training, or promotion. . . . I now ask your full cooperation in meeting these stated objectives." (NASA Hq. PB 64-14)

- USAF launched Thor-Agena D launch vehicle from Vandenberg AFB, Calif., with unidentified satellite payload. (UPI, *Wash. Post*, 2/28/64; HHN-48)
- With no FY 1965 funds authorized for Snap-10A flight test, AEC would "quietly continue to press for authority to conduct flight tests of Snap-10A, even without the cooperation of a potential customer," Jonathan Spivak predicted in the *Wall Street Journal.* Estimated cost of flight test of the 500-watt nuclear power system was $12 million. (Spivak, *WSJ*, 2/28/64)
- Thomas Jefferson Wong, 40, internationally known research scientist and head of manned Mars mission studies at NASA Ames Research Center, died after apparent heart attack. Mr. Wong had been with Ames for 13 years. (*Palo Alto Times*, 3/2/64)

February 29: ECHO II passive communications satellite was used for transmission of two radiophotos between Britain's Jodrell Bank Experimental Station at Manchester and Gorki University's radioastronomy observatory at Zimenki. A third space telegram was sent from Jodrell Bank to Zimenki via the moon, and its quality was comparable to those received via ECHO II. The U.S. satellite was termed by Tass commentator "the Friendship Sputnik." (UPI, *Wash. Post*, 3/1/64; *NYT*, 3/2/64, 9; *Krasnaya Zvezda*, 3/1/64, 4, ATSS-T Trans.)

- At White House news conference, President Johnson announced:

"The United States has successfully developed an advance experimental jet aircraft, now the A-11, which has been tested in sustained flight at more than 2,000 miles an hour and at altitudes in excess of 70,000 feet.

"The performance of the A-11 far exceeds that of any other aircraft in the world today. The development of this aircraft has been made possible by major advances in aircraft technology of great significance to both military and commercial application. Several A-11 aircraft are now being flight-tested at Edwards Air Force Base in California. . . .

"The development of supersonic commercial transport aircraft will also be greatly assisted by the lessons learned from this A-11 program. For example, one of the most important technological achievements in this project has been the mastery of the metallurgy and fabrication of titanium metal which is required for the high temperatures experienced by aircraft traveling at more than three times the speed of sound.

"Arrangements are being made to make this and other important technical developments available under appropriate safeguards to those directly engaged in the supersonic transport program.

"This program was first started in 1959. Appropriate members of the Senate and the House have been kept fully informed on the program since the day of its inception. The Lockheed Aircraft Corporation at

Burbank, Calif., is the manufacturer of the aircraft. The aircraft engine, the J-58, was designed and built by the Pratt & Whitney Division, United Aircraft Corporation. The experimental fire control and air-to-air missile system for the A-11 was developed by the Hughes Aircraft Company.

"In view of the continuing importance of these developments to our national security, the detailed performance of the A-11 will remain strictly classified and all individuals have been directed to refrain from making any further disclosure concerning this program. . . ."

The President also announced he would release on March 2 Eugene Black's supersonic transport study. (Transcript, *Wash. Post*, 3/1/64, A20)

February 29: Distribution of space and military contracts was of growing concern for the Administration, John Finney reported in *New York Times* article. "Space contracts have tended to accentuate the existing research concentration on the East and West Coasts." 90% of space contracts and 83% of defense research contracts were awarded to only 10 states, and in FY 1963 50% of space contracts and 41.4% of military R&D contracts were awarded in California. Finney pointed out that NASA was "attempting to broaden the economic and industrial impact of its rapidly expanding program. . . . Working through universities, the agency is attempting to enlist new regions in its program and to apply the technological developments of its research to the civilian economy. . . . In fact, James E. Webb, the agency head, has become a salesman . . . to enlist new regions, new universities, new industries. . . ." (Finney, *NYT*, 3/1/64, 1)

• In article on "The Politics of the Space Age" in the *Saturday Evening Post*, President Lyndon B. Johnson essayed his views on the manner in which space exploration may bring an end to hostilities among nations on earth. In this abridgement of forthcoming volume edited by Lillian Levy entitled *Space: Its Impact on Man and Society*, President Johnson said:

". . . Any discussion of the 'politics' of the space age is, necessarily, a discussion of politics in the classical sense, not the convention-and-campaign sense. If the space age has not eliminated the smoke-filled room, it has at least opened a window and given air to fundamental and sometimes obscured values.

"The fate of the free society—and the human values it upholds—is inalterably tied to what happens in outer space, as humankind's ultimate dimension. While the response of our technology is important, no less important is the response and the role of our political institutions—both to the challenge of the present and to the opportunities of the future.

"The vital role of politics in the space age has been evident since October 4, 1957, when the Soviet's Sputnik I was launched. The orbiting of that first unmanned earth satellite was a feat of science. But the worldwide impact and importance were essentially political. Sputnik I was proclaimed by the Soviets as validation of Communist prophecies about the superiority of their political system. Such conclusions have been proved premature . . .

"Why did this occur? Self-examination is imperative. In retrospect, the answers seem to be these:

"1. Despite two decades of intensive scientific advance we had failed to establish adequate relationships between the scientific community and the political community—to the blame of each and the disservice of both.

"2. Inherent in the political miscalculation toward space exploration was the influence of the unfortunate anti-intellectualism of the early 1950's. . . . The open contempt with which the Sputnik I success was greeted by some in positions of political trust betrayed a degeneration of the respect for intellect which is indispensable to the governing of a free and open society.

"3. Perhaps the greatest failure of all was devaluation of the first— and the hardest—responsibility of elected representatives of the people in our system: the responsibility to lead. . . .

"In the years since 1957 the American political system and its politicians have acquitted themselves well, as have our scientists. An orderly program of space exploration has been devised and financed without unbalancing the values of our free society. . . .

"Our most important achievement politically in the space age has been this retention of basic national character, purpose and political values in our space program. It is reflected by a primary objective of our program, clearly stated at the outset, to gather knowledge for the benefit of all nations and to join with all nations in this endeavor. . .

"Thus far the space age has been characterized as a period of competition. It is important, however, that we keep in mind the fact that the competitiveness is between political systems, not between national scientific communities. In the world of science the logical instinct is toward cooperation without regard to political boundaries. This impulse we must preserve. The real challenge of the space age is for the politician to tear down the walls between men which have been erected by his predecessors and contemporaries in the political field, rather than to raise its barriers higher into the free and peaceful vastness of space. If the potentials of the age of space are fully realized, this period will someday be known—and blessed by all people on earth—as the Golden Age of Political Science." (*Sat. Eve. Post*, 2/29/64, 22, 24)

February 29: Former astronaut John H. Glenn, Jr. (Lt. Col. USMC), hospitalized with minor concussion suffered in accidental fall, announced he would remain on active duty in USMC until April 1; after this retirement date, he would be free to campaign for Democratic senatorial nomination in Ohio. (UPI, *Wash. Post*, 3/1/64, A2)

During February: Second 30-day test of effects of 100% oxygen on men began at USAF School of Aerospace Medicine, Brooks AFB. The four subjects would remain in space cabin 42 days, with 12 days at sea-level atmosphere before change to oxygen atmosphere and pressure of about 5 psia. (*Av. Wk.*, 2/24/64, 27; *M&R*, 2/24/64, 9)

- "Universal" upper stage space vehicle powered by restartable liquid-hydrogen/liquid-fluorine engine was proposed by Douglas Aircraft Co.'s Missile and Space System Div. to NASA. Douglas proposed the concept as upper stage for Thor boosters, said it also could be used with Atlas, Titan, and Saturn first stages. Douglas said the high-energy upper stage could be ready for use in three years, would have the advantages of heavy payload capability and adaptability to either earth-orbit or space-probe missions. (*SBD*, 2/25/64, 300; *A-N-AF J&R*, 2/29/64, 14)

During February: Severely criticizing NASA's Industrial Applications Program, Arthur Kranish claimed in *Western Aerospace* article: "Years of work and millions of dollars have resulted, to date, in a handful of dreary pamphlets and a few sheafs of mimeograph paper. . . ." Companies writing for details of "NASA 'innovations' " had "an unpleasant shock awaiting them. . . . Their requests sat for months on someone's desk. Finally, a brief mimeographed description of the innovation was forwarded—along with an unsubtle letter from a leading research institution offering to perform full followup and interpretive work for the interested company—at a fee." (Kranish, *Western Aerospace*, 2/64)

- Kennedy Space Center, NASA, awarded $3,061,826 contract to Ling-Temco-Vought's Range Systems Div. for services at Merritt Island Launch Area: automatic data processing, technical information, photographic support, and printing plant operation. The one-year contract would be renewable for two more years through annual options. (*SBD*, 2/25/64, 296)

- Impact of global communications satellite systems was predicted by Arthur C. Clarke in *Astronautics and Aeronautics* magazine. During the coming decade, comsat systems would bring about "orbital post office," "orbital newspapers," and "orbital telephony." And eventually global comsat system would underscore requirement for universal language, which Clarke predicted would be met with English; would educate millions with TV visual aids; and would end barriers to free flow of information (such as censorship and jamming).

    "The inexorable force of astronomical facts will destroy the political fantasies which so long fragmented our planet. For when all major artistic productions, entertainments, political and news events (not to mention disasters) can be observed and experienced simultaneously by the whole world, the parochialism and xenophobia of the past will not be able to survive.

    ". . . another [influence], perhaps even more fundamental, may be reversal of an historic trend which has proceeded with scarcely a break for 5000 years. The traditional role of the city as a meeting-place is coming to an end; Megapolis may soon go the way of the dinosaurs it now resembles in so many respects. This century may see the beginnings of a slow but irresistible dispersion and decentralization of mankind—a physical dispersion which will take place, paradoxically enough, at the same time as cultural unification." (Clarke, *A/A*, 2/64, 45–48)

- Eastern Airlines announced its research on wind-shear turbulence, or clear air turbulence (CAT), had yielded the significant finding that CAT is preceded by changes in temperatures too slight to be read on present cockpit instruments. Eastern found that temperature change of as little as one degree Centigrade over two-min. period can indicate aircraft is approaching area of turbulence, either in clear air or in clouds. Eastern developed and is flight-testing "Cat-Spy," special instrument which records outside temperatures in fractional degrees, activates warning lights when temperature changes indicate approach to area of possible turbulence. (Eastern Release)

- Frank Gifford of U.S. Weather Bureau, Oak Ridge, Tenn., postulated that "canals" on Mars are actually sand dunes 2,500 to 3,000 mi. long. Size of the dunes would indicate surface pressure of about 30 millibars, close to the 25 millibars calculated by JPL from spectrographic measurements.

In Martian gravity and air density, wind speeds would have to be five to ten times those on earth to form dunes, and such winds have been observed in Martian dust storms. (*M&R*, 2/17/64, 21)

During February: NASA Hq. contracted with RAND Corp. for evaluation of alternate approaches to conducting Project Apollo and attaining its goal. (*SBD*, 2/14/64, 248)

- USAF Space Systems Div. reinitiated its program to develop Astronaut Maneuvering Unit (AMU) for extravehicular space activity. Earlier this year SSD had requested bids for "remote space maneuvering system" but canceled this in favor of the AMU, to be used in MOL-oriented program. AMU would consist of five subsystems—hydrogen peroxide propulsion, environment control, attitude stabilization, manual control, and telemetry and voice communications. (*SBD*, 2/17/64, 255)

- General Dynamics/Convair shipped Little Joe II booster to White Sands Missile Range, where it would be used later this year to launch Apollo boilerplate spacecraft in test of the launch escape system. (MSC *Roundup*, 2/19/64, 1)

- North American Aviation's Space and Information Systems Div. was studying the planaria to help understand human healing process under conditions of weightlessness. Planaria was capable of regeneration even under extremes of temperature, vibration, and shock. (*M&R*, 2/17/64, 21)

- AFSC awarded $100,000 study contract to Sylvania Co. for feasibility study of balloon-satellite communications network. System would relay relatively interference-free signals over microwave radio beams by bouncing them off balloons orbiting earth at about 2,000-mi. altitudes. (*SBD*, 2/24/64, 290)

- Lt. Gen. Thomas P. Gerrity, AF Deputy Chief of Staff/Systems & Logistics, described the purpose of the military cost reduction program: "Cost reduction will enhance military capability. We'll save money and thus afford better weapon systems. We are being challenged as never before to perform our mission with less money; this is a way of life we can expect in the future." (*AF Mag.*, 2/64, 21–23)

- Aerospace Industries Association of America created Aerospace Technical Council to serve as technical advisory board, technical policy planning group, and overall supervisory agent for AIA's Technical Service and the AIA technical committees. (*M&R*, 2/24/64, 49)

- Indonesian air force launched Soviet-built surface-to-air missile in connection with military alert exercises, according to official Antara news agency. (*M&R*, 2/24/64, 11)

- Tory II C series of reactors expected to lead to low-altitude supersonic vehicle nuclear propulsion system was scheduled to undergo initial tests at AEC Nevada Test Site early this year. (*M&R*, 2/17/64, 9)

- In 70-sec. firing, modified Minuteman first stage was used to test prototype gimbaling nozzle half as big as one needed for 156-in. solid motor and four times as big as any gimbaled nozzle built or tested to date. (*M&R*, 2/17/64, 21)

March 1964

March 1: Sen. Richard B. Russell (D.–Ga.), Chairman of Senate Armed Services Committee, said in CBS network interview that the U.S. had developed 11 or 12 of the A–11 2,000-mph experimental jet. He said the A–11 had been "put through all kinds of tests," and tests were continuing. The prototype was nearly ready for USAF acceptance. President Johnson had disclosed the A–11's existence at press conference the previous day. (Raymond, *NYT*, 3/2/64, 1; AP, *Wash. Post*, 3/2/64, A1)

- At Oklahoma City, where FAA was conducting sonic boom tests in preparation for future supersonic transport flights, complaints about the booms were outnumbered by supporters of the program. After one month of tests, Mayor Jack S. Wilkes said his mail was 5-to-1 in favor of testing. Some 2,100 claims for damages had been filed with FAA, but an FAA spokesman said there had "been no proof yet of damage directly attributable to the sonic booms." Testing would continue five more months. (Janson, *NYT*, 3/2/64, 12)

- Frederic A. Louden, 66, former aeronautical engineer at Navy Bureau of Aeronautics until retirement in 1957, died after long illness. Louden was former head of the Bureau's special research liaison office for NACA and a former member of NASA's Committee on Aerodynamics. (*Wash. Post*, 3/3/64)

March 2: NASA and DOD established the National Space Station Planning Sub-Panel of the NASA–DOD Aeronautics and Astronautics Coordinating Board (AACB), charged with studying and then making recommendations to AACB on the best configuration for follow-on manned space station to the MOL. Co-chairmen were Dr. Michael I. Yarymovych of NASA Advanced Manned Missions Office and Col. Kenneth Schultz, Deputy Director of Development Planning for Space, Hq. USAF. (*M&R*, 3/9/64, 13; *SBD*, 3/11/64, 383)

- Apollo spacecraft boilerplate command module and related equipment for first full-scale test flight of Apollo launch escape system arrived at White Sands Missile Range, N. Mex., where it would be flight-tested with Little Joe II booster. (MSC Release 64–39)

- NASA announced Hughes Aircraft Co. had been awarded contract to develop and build five Advanced Technological Satellites (ATS) based on Advanced Syncom design study. The 650-lb. satellites would be used to test several spacecraft techniques, particularly spacecraft stabilization and orientation in higher altitudes. (NASA Release 64–50; *Goddard News*, 3/9/64, 8)

- White House released report by Eugene R. Black and Stanley deJ. Osborne recommending revision of U.S. program to develop supersonic transport. As previously disclosed (Feb. 11, 22), the report urged Government financing be increased from 75% to 90% and the program be administered by a new, independent agency. (MacKenzie, *Wash. Post*, 3/3/64, A24)

March 2: Dr. Charles A. Berry (M.D.) was named Chief of Medical Programs for NASA Manned Spacecraft Center, MSC Director Dr. Robert R. Gilruth announced. In this new position Dr. Berry would be in charge of all medical and physiological matters in the MSC mission. Promoted to Dr. Berry's vacated position, Chief of Medical Operations, was Dr. D. Owen Coons (M.D.), formerly Deputy Chief. (MSC Release 64-40)

- AEC released report from its Advisory Committee on Reactor Safeguards concerning NASA Plumb Brook Reactor Facility. After reviewing NASA's application to convert its provisional operating license to 10-year operating license, the Committee concluded that "with careful planning and operation, the NASA PBRF organization can continue to operate this facility without undue hazard to the health and safety of the public." (AEC Release G-46)
- Sen. Stephen M. Young (D.-Ohio) spoke in Senate of NASA's selecting Boston as site for new electronics research center, charging that "there is little to indicate that thorough evaluation was given to potential locations distant from the Boston area" and pointing out that "in all respects, Cleveland eminently qualifies for this important new facility.

    ". . . Each time a new facility is proposed, we hear the same well-worn argument that it should go to an area which already has institutions with experience in this field. As a result a vicious cycle has formed, and it can almost be predicted that any new important space facility will go to Boston, Houston, or to California. . . .

    "Frankly, I am tired of the argument. I rise today to speak briefly in protest of what has been going on. . . ." (*CR*, 3/2/64, 3943-44)
- Milton B. Ames, Jr., Director of Space Vehicles in NASA Office of Advanced Research and Technology, testified before House Committee on Science and Astronautics' Subcommittee on Advanced Research and Tracking:

    ". . . space vehicle research, first of all, produces knowledge that permits the accomplishment of new missions. When the knowledge resulting from research is obtained well in advance of the planning of new missions, we may proceed with confidence that the missions are feasible. For this reason, the generation of new knowledge through research should, when possible, precede developments that are beyond the current state of the art. Only when the pressures for immediate development are extremely great, as has been the case in these first years of our national space program, are we justified in violating the principle of order in favor of expediency. When we possess adequate research data, we not only are confident of the feasibility of any new missions for which we may wish to plan, but we also . . . lay the basis for improved reliability and efficiency and for reductions in the weight and cost of the space vehicles required for these missions. The truth of these principles has been demonstrated many times in the past by costly developments which either had to be abandoned for lack of the necessary technical capability to complete them or which could only be salvaged by crash research programs.

    "We must, therefore, strive through research to develop new knowledge that will make new space missions possible, if and when national policy should dictate that any one of these missions must be accomplished. At the same time we must conduct the research required to ensure the success of on-going projects, and we must also strive through research to pave the way for increased reliability and efficiency of space vehicles of proven feasibility. . . ." (Testimony)

March 2: In interview with *Aviation Week and Space Technology*, Air Force Secretary Eugene M. Zuckert replied to question about MOL:

"I am sure that the MOL must go forward. I am in sympathy with Mr. McNamara's pressing us to try to define what we hope to do with it. This can have a lot to do with the subsystems that you plan to put in it. Also, the definition effort has to do with the cost, and with the time it is going to take us to do the job. We haven't suffered at any time in the last three years that I have been here by taking a few months to try and work out with reasonable precision what we are planning to do. I am convinced, for one, that the studies and program definition that we went through in the Titan 3 materially improved our chances of getting a fine booster at lower cost. I know, as Secretary McNamara knows, that there is a limit to the precision with which you can define what you are going to do with something you haven't got. But the effort should be made to set out the parameters. It has the merit, at least, of excluding some of the things you know you are not going to try. I'm convinced Mr. McNamara means exactly what he says—that we have to get this MOL capability, and I'm not disturbed by the efforts that are being made to attempt to define what we mean by a capability. . . ." (*Av. Wk.*, 3/2/64, 73)

- Air Force Chief of Staff General Curtis E. LeMay said in testimony before House Armed Services Committee that the Titan III launch vehicle program was "proceeding satisfactorily" and that USAF was "undertaking the development of a medium altitude communications satellite program.

". . . the next major step toward the achievement of future space capabilities is the Manned Orbiting Laboratory (MOL) which was approved for development under Air Force management. The MOL will provide a means to fulfill the compelling requirement to acquire information essential to determining accurately the threat from space, the usefulness and the capabilities of man-in-space, and the unique advantages which may accrue from military space operations. It will also serve as a platform to support testing of equipment and procedures in the environment in which they will be used." (AF Info. Pol. Ltr.)

- *New York Times* editorialized that the A–11 experimental aircraft "represents an engineering breakthrough on several fronts and a tribute to Lockheed, but it will not satisfy the need for an advanced manned interceptor, a long-range bomber or a new high-speed commercial craft." (*NYT*, 3/2/64, 26)

- Col. John C. Nickerson (USA), 47, a key figure in USA-USAF Thor-Jupiter controversy, died in automobile accident in Alamagordo, N. Mex. (UPI, *Wash. Post*, 3/3/64, A2)

March 3: Albert J. Evans, NASA Acting Director of Aeronautics, said before Subcommittee on Advanced Research and Tracking, House Committee on Science and Astronautics:

"The Aeronautical Research effort of the National Aeronautics and Space Administration has one specific goal: to provide technical information for use by the aircraft manufacturers of the United States, information which is sufficiently far advanced to permit the design and production of military and civil air vehicles which are superior to those of any other nation.

"Our aeronautical program is classified in four broad disciplinary categories . . .: Aerodynamics, Structures, Air-Breathing Propulsion, [and] Operating Problems.

"The resources utilized in accomplishing the NASA Aeronautical Research program are found primarily in the highly-competent technical staffs of the four Research Centers of the NASA. The in-house effort is supported, as required, by a certain amount of out-of-house contract research performed by universities, non-profit organizations, and private industry. The funds requested for the NASA Aeronautics program support these in-house and out-of-house research efforts. The program, however, is predominantly an in-house effort, conducted in the established framework of the theoretical and experimental research activity of the NASA Research Centers. . . ." (Testimony)

March 3: AEC announced plans to explode two experimental nuclear reactors—one a Snap–10A for spacecraft use and the other a ground-based reactor for civilian power—during forthcoming two weeks. The tests, to be made at AEC's National Reactor Testing Station, Arco, Idaho, were intended to provide "vital safety information." (AP, *NYT*, 3/4/64, 5; AP, *Wash. Post*, 3/4/64)

- Dr. Evelyn Anderson, research scientist at NASA Ames Research Center, was among the six distinguished career women to receive Federal Women's Award at banquet in Washington. (NASA Notice)
- Douglas Aircraft Co. was awarded follow-on study contract for Manned Orbital Research Laboratory by NASA Langley Research Center, to refine the NASA concept and examine the feasibility of cylindrical six-man space station using equipment providing intermittent artificial gravity. Selection of Douglas for contract negotiation was announced last December. (LARC Release)
- Sen. Gordon Allott (R.–Colo.) said on Senate floor:

    ". . . how is it that this country develops a plane which costs a vast sum . . . and now it is declared that we have fully and successfully developed an advanced experimental jet aircraft, but the very committee which has to approve the appropriation required for the development of that aircraft knows nothing about it?

    ". . . having sat on the Defense Subcommittee of the Appropriations Committee, beginning in 1959, and having attended at least 98 percent of all the meetings all the time that subcommittee has met, the first time I ever heard a word mentioned about the A–11 was when I turned on the television, over the weekend, and saw the President make the announcement. . . . Unfortunately, I have not been able to ask the distinguished chairman of the subcommittee . . . about this matter; but all the rest of my colleagues on the subcommittee have told me they know nothing about the A–11. . . ." (*CR*, 3/3/64, 4006–07)
- Sen. William Proxmire (D.–Wis.) inserted into the *Congressional Record* an editorial from the *Wall Street Journal* concerning Government participation in financing the supersonic transport:

    ". . . The question, as we see it, is not whether there will eventually be operational supersonic airlines, or even whether the Government has any business in that business. Rather, it is a question whether the Government is trying to rush the thing before its economic time, with too few charts and too little knowledge.

"A more cautious flight into the unknown could well make for speedier progress in the end—not to mention sounder investments." (*CR*, 3/3/64, 4067)

March 3: Spokesman for Maser Optics, Inc., said "laser rifle" had been delivered to Army's Frankford Arsenal. "Laser rifle," producing intense bursts of light, could be used for such tasks as igniting objects or detonating explosives. (UPI, *NYT*, 3/4/64, 23)

March 3–6: British aviation officials met with FAA representatives in Washington for technical policy discussions of civil aviation matters. Methods for exchanging information on environmental aspects of supersonic transport were among the matters discussed. (FAA Release 64–19)

March 4: NASA Administrator James E. Webb, appearing before Senate Committee on Astronautical and Space Sciences, said:

"It is apparent to most persons who are concerned with the national space effort that we have reached what you described recently, Mr. Chairman, as a 'critical mid-point' in our effort to achieve space pre-eminence for the United States. We have moved from a circumstance in which the Soviet Union held clear superiority in space to one in which, as a result of the driving effort put forth in the past five years, we have achieved a degree of parity with the Russians in our ability to penetrate and operate in the space environment.

"As a consequence of this significant progress, the nation is in a position to consider not merely what it is *able* to do in space, but also what it is *wise* to do in view of our greatly increased scientific and technical ability. We have, in short, moved from a period in which we did everything we could in space, and regretted our inability to do more, to one in which we have a very large capability and are increasingly confronted with hard decisions in the selection of the programs which we will undertake.

"This is the position in which NASA found itself in preparing the program which is the basis for the FY 1965 authorization request. We have deferred or eliminated many projects which were worthy contenders for a place in the program and which, I believe, when budgetary limitations are less restrictive, the nation will decide to undertake. The activities which are proposed for FY 1965 will, in my judgment and that of my associates, produce the greatest benefits for the country within the limitations of the national resources which the President has decided are available and can prudently be spent for the NASA program. . . ."

Mr. Webb cited mounting evidence that NASA's need for scientists and engineers was not adversely affecting other areas:

"At the beginning of this calendar year, approximately 74,000 scientists and engineers were employed in the NASA program—about 12,000 within NASA, and 62,000 under NASA contracts and grants. This amounted to approximately 4.9 percent of the 1.5 million scientists and engineers in the nation's work force.

"By next January, it is estimated that about 82,000 scientists and engineers will be working on the NASA program; about 5.2 percent of the available national supply. If the space effort were to continue at the present level of funding for the remainder of the decade, NASA's program is not expected to require more than 5.5 percent of the national supply of scientists and engineers.

"Of particular interest, however, is the fact that NASA's requirement for additional engineers has peaked at an earlier period than had been anticipated. Current statistics indicate that the number employed on NASA work increased by about 30,000 during the last calendar year, rather than 20,000 as had been anticipated. As a consequence, our requirement for the current year will be only about 8,000 instead of 18,000 as had been projected. We will thus require a much smaller share than had been anticipated of the 45,000 engineers who will complete their education this year. . . ." (Testimony)

March 4: NASA and French National Center for Space Studies (CNES) agreed to proceed with cooperative space project of launching satellite to investigate characteristics of VLF electromagnetic wave propagation in the ionosphere. The French satellite would be launched from PMR using a Scout vehicle provided by NASA. This would be Phase II of two-phase CNES–NASA program for VLF investigations, pursuant to Memorandum of Understanding dated Feb. 18, 1963. Phase I consisted of two suborbital launches of French experiments with U.S. Aerobee sounding rockets, to demonstrate the feasibility of scientific experiments to be performed by the satellite. (NASA Release 64–61)

- NASA announced signing preliminary $800,000 contract with Ball Brothers Research Corp. for construction of five additional Oso (Orbiting Solar Observatory) satellites. Ball Brothers already had built two Oso's— one of which was orbited in 1962 and the other scheduled for orbit this year—and was in the process of building a third. The five additional Oso's would study solar cycle through periods of increased activity during the late 1960's. (NASA Release 64–51)

- NASA Administrator Webb submitted a report on long-range studies in space exploration to the House Committee on Science and Astronautics' Subcommittee on Advanced Research and Tracking. Emphasizing that "all decisions on future missions must be national, not solely NASA, decisions," the report outlined "the areas of space mastery that require serious attention by the Nation," surveying all categories of space exploration—including scientific, weather, and navigational satellites, manned and unmanned flights to the moon and the planets. The report said that "development and operation of manned space stations in earth orbit will be a necessary preliminary to extension of manned operations in space," described feasibility and design studies being made for manned lunar bases. (Text, *1965 NASA Auth. Hearings*, Part 4, 2729–2804)

- NASA announced granting 28 waivers of U.S. commercial rights to inventions resulting from work performed under NASA contracts, bringing total of NASA's patent waivers to 74. Under policy "designed to encourage prompt public and industrial benefit from the scientific byproducts of its research and development activities," NASA granted the patent rights to 12 companies, one university, one research organization, and two individuals (jointly). NASA retained royalty-free license for any governmental use of the inventions. (NASA Release 64–52)

- Rep. Bob Wilson (R.–Calif.) said in the House:
    "Atlas has more scheduled launches during the 1960's than ever before for missions of the Air Force Systems Command and NASA, a demand so great that a new standardized Atlas space launch vehicle called SLV–3 or Atlas-3 has been designed. This standardized vehicle will produce

lower long-term costs while increasing flexibility and flight reliability. Development of this standardized vehicle is an excellent example of how existing space systems can be improved to satisfy new requirements. Improvement of a flight proven system such as Atlas is inherently more efficient and economical than designing new vehicles for new missions. Research and development costs are already expended and Atlas has certainly proven its reliability. . . ." (*CR*, 3/4/64, A1115)

March 4: ComSatCorp applied to FCC for authority to construct and orbit over the mid-Atlantic a synchronous-orbit communications satellite to provide commercial service on an "experimental-operational" basis between North America and Europe. The satellite was described by ComSatCorp as an "initial step in the development of a global communications system." It would provide 24-hour communications service capable of relaying TV broadcasts or up to 240 telephone calls. Plans called for the satellite to be orbited in spring of 1965 with a TAT-Delta booster supplied by NASA on a reimbursable basis. (ComSatCorp Release)

- USAF planned to launch six manned orbiting laboratories (MOL) over an 18-month period, beginning in late 1967 or early 1968, unnamed sources revealed in Washington. (UPI, Omaha *Eve. World-Herald*, 3/4/64)
- FAA announced Lufthansa-German Airlines had reserved three delivery positions for U.S. supersonic transport aircraft, bringing total of reserved positions to 75 by 14 airlines. (FAA Release 64-20)
- DOD announced establishing fundamental policies for application and conduct of Project Definition Phase (PDP). The technique would be applied to all proposed engineering or operational system development projects with development, test, and evaluation costs amounting to $25 million or more or with production costs amounting to $100 million or more. Major benefits were expected to include reduction of technical changes during development-production cycle, decreased cost, improved operational effectiveness or early cancellation of projects. (DOD Release 186-64)
- Soviet Cosmonaut Yuri Gagarin, touring Copenhagen, predicted that U.S.S.R. would conduct another manned space flight sometime this year. (AP, Wash. *Eve. Star*, 3/4/64)

March 5: 1,300-lb. instrumented payload was carried to 97,000 ft. by balloon in Coronoscope II project, sponsored jointly by ONR, NSF, and NASA. Released from National Center for Atmospheric Research near Palestine, Tex., the balloon traveled about 105 mi. eastward, then doubled back to travel another 100 mi. before dropping its instrument payload on radio command signal, 50 mi. from the launch site. Dr. Gordon Newkirk of High Altitude Observatory at Boulder, Colo., said photographs made by the instruments were expected to aid in study of effects of sun's corona upon earth's magnetic field and upper atmosphere. (*Houston Chron.*, 3/6/64; *Houston Post*, 3/6/64)

- NASA Deputy Administrator Dr. Hugh L. Dryden, appearing before Senate Committee on Aeronautical and Space Sciences, reported on NASA's activities in international cooperation during 1963:

  "Six new international satellites, to be built and financed by three other nations, were programmed as contributions to the NASA scientific satellite effort over the next six years, bringing to ten the number of such cooperative satellite projects completed, in process, or agreed. . . .

Seven foreign experiments were selected in competition with domestic proposals for flight principally on NASA observatory satellites during the next three years....

"Nineteen sounding rockets carrying scientific payloads were launched here or abroad in joint projects with nine countries, raising to forty-six the total of such launchings to date.

"New agreements were concluded with India and Pakistan under which both countries will join the UK and Australia in a year-long test of meteorological sounding rocket systems...."

After listing NASA activities in the categories of international ground-based experiments and relationships with European agencies (ELDO and ESRO), Dr. Dryden discussed implementation of the U.S.-U.S.S.R. bilateral agreement for space cooperation:

"... Correspondence has been maintained [with the Soviet Academy of Sciences], although the Soviet side has not met time schedules on most agreed action items. However, the Soviet Academy of Sciences has just conducted optical observations of the ECHO II satellite during its early orbits and we were able to begin the telecommunications experiments with ECHO II between the Jodrell Bank Observatory in Great Britain and the U.S.S.R. on February 21. While we have not yet received detailed data of the optical observations and have still to evaluate the results of the telecommunications tests, which we have received, I think the real significance is that we are taking advantage of existing programs, approved and executed on their own merits, to provide an opportunity for scientists and engineers of both countries to gain experience in working together for their mutual benefit. This is obviously a pioneer venture and it has purposefully been designed as a *coordinated* rather than a *joint* effort. There is no provision for exchange of funds between the two countries, no exchange of equipment is involved and all exchanges of data are to be on a reciprocal basis and made available to all other interested countries...." (Testimony)

March 5: In testimony before Senate Committee on Aeronautical and Space Sciences, NASA Associate Administrator Dr. Robert C. Seamans, Jr., discussed NASA's relationship with JPL:

"... [A] major contracting advance is taking place in our operating relationships with the California Institute of Technology ... [which] manages the NASA's Jet Propulsion Laboratory under contract. Originally, the Laboratory and the contract for its operation were transferred to NASA from the Army in 1959. As the functions of the Laboratory have expanded and changed, we have modified the NASA/CIT relationships to meet the current needs of the program. The present contract expires at the end of this year, and we have carefully negotiated its extension under terms more applicable to current agency requirements. A comprehensive review has been made of the operating experience gained since the inception of the NASA/CIT/JPL relationship. As a result of this study, improved operating procedures in both the technical and administrative aspects have been developed and implemented. In addition to reflecting clauses which implement the above improvements in management, the new arrangements include three major departures from earlier agreements:

"First, the earlier contracts were based on a principle of mutually requiring CIT/JPL concurrence before NASA directions would be carried out. This often resulted in lengthy 'negotiations' of assigned tasks. The new contract restricts the principle of mutuality to the broad terms of the operating relationship, but clearly gives NASA the unilateral right to issue task orders, require reports, and take other administrative actions as desired.

"Second, a separate facilities contract is also being negotiated which will provide the government a more direct control over the planning, construction, and utilization of the facilities at JPL.

"Third, a new principle is provided for determining CIT's fee for the operation of the Laboratory. The new contract contains a schedule which sets the upper and lower limits of the fee on the basis of the dollar volume of work which NASA places with CIT/JPL. The exact fee within these limits, will be determined by how well CIT/JPL performs in the management of the Laboratory." (Testimony)

March 5: First of series of background briefings on scientific and technical aspects of NASA programs was held at NASA Hq., with Dr. John E. Naugle, Director of Geophysics and Astronomy Programs, reviewing the broad scientific objectives of the space program. Dr. Naugle revealed NASA planned to ask three separate companies for independent design studies of Advanced Ogo's and Oao's (Orbiting Geophysical Observatories and Orbiting Astronomical Observatories). The Advanced Oso (Orbiting Solar Observatory) design work by contractor Republic Aviation was progressing, and Aoso should be in operation in 1969. (NASA Memo; *SBD*, 3/6/64, 360)

- In press conference, Secretary of Defense Robert S. McNamara announced reduction of B-70 development program from three aircraft to two. "This action . . . is a result of a very comprehensive review of the project we have just completed, and it reflects our concern over the continued delays in the program . . . brought about by severe technical difficulties.

". . . the first flight [originally] was scheduled for December 1962. The program is already some 18 months behind that schedule and the first aircraft has not yet been completely assembled. . . .

"We have been plagued with fuel leaks in . . . [the Number Three] airplane. . . . I am told . . . that the first plane, if it flies this spring . . . or certainly before mid-year, will probably fly with one tank not usable because of continued leak problems.

"Secondly, we have had serious problems in mating the wing to the fuselage. It is a very, very complex form of construction. We are using what amounts to a steel sandwich in the B-70 to solve the high temperature problems associated with supersonic speed, and the welding problems associated with mating this steel sandwich wing and fuselage construction have been severe, indeed. . . ." (Transcript, *NYT*, 2/6/64, 12; DOD Release 194-64)

- Senate passed compromise $16,976,620,000 military authorization bill for FY 1965 procurement and R&D. The Senate-House conferees retained $50 million for initiating development of new manned bomber, dropped $40 million approved in House bill for improved manned interceptor aircraft. (*NYT*, 2/6/64, 3)

- AFSC issued annual report of its 1963 highlights and accomplishments. In preface summarizing and evaluating significant activities, AFSC Commander General B. A. Schriever said:

"... In 1963 AFSC made major contributions to the nation's defense and space exploration programs. These included delivery of ballistic missiles at an average of one per day, completion of vital early warning and air defense systems, first flight of the transport aircraft which will make world wide airlift a reality, and delivery of the launch vehicles which made possible the orbital flights of Commander Schirra and Major Cooper. . . .

"Cancellation of the X–20 DynaSoar program in late 1963 and assignment to the Air Force of responsibility for development of a near-earth Manned Orbiting Laboratory (MOL) represent a shift in emphasis from reentry technology to on-orbit experiments. MOL, which is designed to provide a means of obtaining hard, factual data concerning man's capabilities in operational space vehicles, is among AFSC's most urgent projects. . . ." (AFSC Release 42–R–24; AFSC 1963 Ann. Rpt.)

March 6: "Boosted Arcas" sounding rocket was launched from NASA Wallops Station to 59.5-mi. altitude with sodium-lithium payload designed and built by Sweden's Uppsala Ionospheric Observatory on contract to Swedish Space Committee. The wide and reddish sodium-lithium cloud ejected in the atmosphere was visible for many miles and was photographed from several locations. The test was in preparation for series of such launchings in the auroral zone in northern Scandinavia later this year, to study causes of aurora and airglow. This first launch of Boosted Arcas from Wallops was quite successful. Small booster included in the Arcas increased its altitude from normal 225,000 ft. to 314,000 ft. (Wallops Release 64–24)

• Nike-Cajun sounding rocket was launched from NASA Wallops Station to 71-mi. altitude with payload of 12 explosive charges to provide data on atmospheric winds and temperatures and effects of the "warming trend" in the upper atmosphere. Previous measurements had been made during February "warming trend," which moved in westerly direction across the northern hemisphere. This second and unexpected warming trend appeared to be moving in easterly direction. (Wallops Release 64–25; NASA Rpt. SRL)

• NASA Administrator James E. Webb, speaking at Chamber of Commerce dinner in Oxford, N.C., clarified the cost of a U.S. manned lunar landing in this decade:

"This $20 billion outlay for manned space flight for the period 1961 through 1970 is for Projects Mercury, Gemini, and Apollo. . . .

"Of the $20 billion marked for manned space flight in the present decade, $18 billion will be expended for the launch vehicles and spacecraft, the ground facilities, the astronaut training, and the management skills and other human factors that are an integral part of our nation's capital investment. These are all things we must have—in any event—to achieve the across-the-board competence in every major area of manned space flight that security, well-being, and progress of the nation may demand.

"The portion of this total effort devoted to landing on the moon will cost only about $2 billion—that is, sending three Americans on a round-trip voyage of lunar exploration, along with everything in the way of spacecraft and special equipment that the expedition will require. . . .

"I hope that this brief and somewhat simplified explanation will put the correct price tag on the extra cost of the lunar mission. Let us all endeavor to put the emphasis where it really belongs: on the rapid growth of American competence in all areas of manned space flight particularly in the area near earth which is the prime importance for national security." (Text)

March 6: Before Senate Committee on Aeronautical and Space Sciences, NASA Assistant Administrator for Technology Utilization and Policy Planning Dr. George L. Simpson, Jr., said that, on the basis of the Aerospace Research Application Center (ARAC) pilot program, NASA "judged the first year of the Indiana University experimental program promising enough to warrant support for a second year." ARAC receives all NASA reports, sends pertinent abstracts to subscribing companies based on their "interest profiles." This "model of a regional technology transfer program" was being observed by other organizations in other regions. Another such program was underway at Wayne State Univ. and two others were beginning. (Testimony)

March 6–7: NASA astronauts made trip to the bottom of Grand Canyon and back up, in field exercise to train the astronauts as "competent observers" of rock and mineral formations. Geologists from NASA Manned Spacecraft Center and U.S. Geological Survey served as field instructors. (UPI, *Houston Press*, 3/7/64)

March 7: TIROS VIII meteorological satellite, launched December 21, 1963, was performing highly successfully in automatic transmission of cloud pictures to ground stations around the world, NASA announced. "Reports from 47 ground stations around the world, including five foreign stations, show that with one exception results have been satisfactory. TV signal strength has been good, tracking of the spacecraft based on data furnished daily by NASA has been easy for the ground stations, and picture contrast of the cloud cover photos generally has been excellent." (NASA Release 64–53)

- Beckman Instruments, Inc., of Fullerton, Calif., said it received a contract for more than $1 million from Grumman Aircraft to produce a data acquisition system to be used in development and test of LEM at White Sands Missile Range. (UPI, *Houston Chron.*, 3/7/64)
- Univ. of Miami President Henry King Stanford announced the university's School of Environmental and Planetary Sciences, to open next autumn, would be the first of its kind in the nation. With Dr. S. Fred Singer as dean, the school would be composed of the Institute of Marine Sciences and three new institutes—planetary bioscience, atmospheric science, and space physics. The institutes would offer courses leading to both masters' degrees and doctorates. (Mann, *Miami Herald*, 3/8/64)
- Patent for system of space communications using x-rays has been granted to Henry R. Chope, executive vice president of Industrial Nucleonics Corp., Columbus, Ohio. The 1½-lb. transmitter employed radioisotope as source of energy. (NYTS, Cleveland *Plain Dealer*, 3/7/64)

March 8: At meeting in Geneva, U.S. delegation to the U.N. Committee on the Peaceful Uses of Outer Space Legal Subcommittee submitted to the Subcommittee drafts of two treaties. One draft treaty dealt with aid to astronauts in distress, calling for nations to take "all possible steps to assist or rescue promptly the personnel of spacecraft" having difficulty or making emergency landings. Astronauts making emergency landings would be returned to their own country "promptly and safely," and any spaceship or its parts falling to earth would be returned to country of ownership at the country's request. The second treaty, dealing with liability for damage caused by spacecraft, would establish a "simple and expeditious procedure to provide financial protection against damage." (NYT, 3/10/64, 8)

March 8: Plans for two-acre U.S. Space Park at New York World's Fair were announced by DOD and NASA. The Space Park would feature largest exhibit of full-scale USAF and NASA rockets and spacecraft ever assembled outside Cape Kennedy. (NASA Release 64–54; DOD Release 197–64)

March 9: Subcommittee on Advanced Research and Tracking of the House Committee on Science and Astronautics voted (5–3) to not authorize the $10 million NASA requested for FY 1964 for beginning construction of electronics research center. (AP, *NYT*, 3/10/64, 26; *SBD*, 3/11/64, 385)

- House passed H.R. 9637, FY 1965 authorization for military procurement of aircraft, missiles, and naval vessels, and for research, development, test, and evaluation. Of the $16,976,620,000 total authorization, $6,363,320,000 was for R&D. (NASA LAR III/42)
- Atlas ICBM exploded in its underground silo near Roswell, N. Mex., as it was being fueled. USAF said there were no casualties. (UPI, *NYT*, 3/11/64, 14)
- USAF conducted briefing on the technologies associated with A–11 advanced experimental aircraft for officials of airlines and aviation manufacturing companies involved in U.S. supersonic transport development. (FAA Release 64–21)
- *Aviation Week and Space Technology* reported that the A–11 advanced experimental aircraft had "already flown long-range reconnaissance missions over communist territory." Government officials flatly denied the report, maintaining that the mach 3.5 aircraft had never been on any mission outside U.S. territory. (*Av. Wk.*, 3/9/64, 16; UPI, *Wash. Post*, 3/10/64)
- Federation of Italian-American Democratic Organizations of New York presented to NASA a plaque honoring the late President Kennedy and Italian explorer Amerigo Vespucci. The plaque, which commemorated Vespucci's sighting the area now called Cape Kennedy, was accepted by NASA Administrator Webb in ceremonies at NASA Hq. (NASA Announcement 64–47. It was subsequently hung in the Hq. building of Kennedy Space Center, NASA; KSC Release 61–64)
- Appointment of David S. Johnson as Director of National Weather Satellite Center, effective April 1, was announced by Chief of U.S. Weather Bureau Dr. Robert M. White. Mr. Johnson succeeded Dr. S. Fred Singer, resigning to become Dean of new School of Environmental and Planetary Sciences, Univ. of Miami. Deputy Director of the Center since its formation in 1962, Mr. Johnson had been engaged in research, development, and technical management of meteorological satellites and sounding rockets since 1958. (Commerce Release WB 64–6)

March 10: Senate Committee on Aeronautical and Space Sciences voted (10–4) to approve NASA's report on the electronics research center. (House Science and Astronautics Committee had approved the report Feb. 24.) (NASA LAR III/43)

- Soviet draft treaty for aid to astronauts in distress and for return of spacecraft was submitted to Legal Subcommittee of U.S. Committee on the Peaceful Uses of Outer Space meeting in Geneva. Draft treaty differed from U.S. draft by providing that country of origin reserve "exclusive right" to search unassisted for the downed spacecraft on the high seas. Soviet draft also specified two conditions for returning spacecraft: (1) the spacecraft must have been launched "for the purposes of peaceful exploration and use of outer space" and (2) the purpose and launching of the spacecraft must have been "officially announced." (*NYT*, 3/11/64, 5)

March 10: Search for fragments from the moon would be conducted this spring in six western Iowa counties by NASA and U.S. Dept. of Agriculture, Goddard Space Flight Center scientist Dr. John A. O'Keefe announced. Western Iowa was chosen after Dr. O'Keefe had found some "promising" specimens there on an exploratory trip last summer. Residents of select counties were requested to turn over any unusual stones found during spring plowing of loess soil areas to local USDA offices. Stones seeming most likely to have nonterrestrial origins would be forwarded to GSFC for further scientific examination and study. (GSFC Release G–6–64)

• Astronauts Neil A. Armstrong, Lt. Cdr. James A. Lovell, Jr., and Maj. Thomas P. Stafford practiced in the Gulf of Mexico developing a training technique for escaping from Gemini spacecraft, both with and without flotation collar. (MSC Release)

• Dr. George G. Manov, Director of Exhibits, NASA Office of Public Affairs, died in Washington after a heart attack. (NASA Announcement 64–55)

• Army Corps of Engineers awarded $1,860,884 NASA contract to C. H. Leavell Co. and Morrison Knudsen for construction of vibration test laboratory at NASA Manned Spacecraft Center, Houston. (DOD Release 207–64)

• Patent grant to USN space surveillance system, which detects and catalogs silent and "lost" satellites from its stations stretched across the Southern U.S., was announced. (NYTNS, Louisville *Courier-Journ.*, 3/10/64)

March 11: House Committee on Science and Astronautics, considering NASA's FY 1965 authorization bill, cut the $3.54 billion manned space flight request by $41.6 million. "Not one single penny was cut from the Gemini and Apollo programs," Rep. Olin Teague pointed out. The cuts included $3.9 million for advanced planning; 10% of $224.9 million requested for construction; and 5% of $304.9 million requested for administration and operating expenses. Of $777 million requested for space sciences, $742 million was approved by the Committee, cutting NASA's request by $4\frac{1}{2}$%. (L.A. Times, *Wash. Post*, 3/12/64; *WSJ*, 3/12/64; Hines, Wash. *Eve. Star*, 3/12/64)

• USAF announced Atlas-Agena D launch from Pt. Arguello, Calif., with undisclosed satellite payload. (UPI, *Wash. Post*, 3/12/64)

• NASA announced selection of Lear Siegler, Inc., Electronics Instrumentation Div. for negotiation of contract to install new electronic systems and modify present systems at East Test Area of Marshall Space Flight Center, Huntsville, Ala. Contract value was estimated to exceed $4,000,000. (NASA Release 64–59)

• Rep. J. Edward Roush (D.–Ind.) proposed on the House floor that NASA establish a Midwest office, similar to its North Eastern Office and Western Operations Office. ". . . The benefits of such an office to NASA would be great and certainly it would be a step in establishing a working relationship between NASA and the Midwest. It would also assist NASA in exploiting the potential of the Midwest. . . .

"I cannot stress enough the need for a more equitable geographic distribution of our defense and space activities. It is in the national interest that steps be taken to spread these activities throughout the country. My proposal for a NASA Midwest office would be a step in that direction." (*CR*, 3/11/64, 4725)

March 11: Soviet news agency Tass said Soviet scientists had concluded the moon's surface was a "loose porous substance" and the interior was solid rock. Conclusion was said to be based on study of radio waves from the moon. (AP, *Wash Post*, 3/12/64)

March 12: Scientific results from the EXPLORER XVIII satellite—Interplanetary Monitoring Platform (IMP)—was topic of discussion at GSFC symposium. The NASA satellite was orbited in November 1963, with onboard instrumentation including two magnetometers, three plasma probes, and five sets of low and medium energy particle detectors. IMP data revealed a shock wave existed 53,600 mi. above earth's sunlit surface, formed by interaction of interplanetary magnetic field and high-speed constant wind of particles from the sun. Between shock wave and upper edge of Van Allen belts, a region ranging in depth from 12,000 to 20,000 mi., there is great turbulence; moderately energetic particles constantly flow into the region and drain away. (Simons, *Wash. Post*, 3/13/64; *NYT*, 3/13/64; *Goddard News*, 3/23/64, 1–2)

- NASA announced establishment of new research program with Lockheed Missiles and Space Co. to develop advanced technology for large nuclear and chemical rockets. The program would continue certain research begun under Project Rift (Reactor-In-Flight Test), principally insulation development, cryogenic testing, structural development, application of Rift technology to Saturn stages, and documentation of pertinent technology for future use. NASA had committed an initial $500,000 letter contract to the program; cost of seven-month period was expected to be about $1.5 million. (NASA Release 64–57)

- House Committee on Science and Astronautics approved $764 million for advanced research and technology in NASA's FY 1965 budget authorization, a cut of nearly $30 million from NASA's request. The Committee restored $10 million for initiating construction of electronics research center, which the Advanced Research and Tracking Subcommittee had cut. (*Wash. Post*, 3/13/64)

- NASA and Australian Ministry for Supply announced agreement for establishment of space tracking and data acquisition station in the Orroral Valley near Canberra, Australia, for ground support of NASA's Orbiting Observatory satellites. The facility would be a sister facility to NASA stations near Fairbanks, Alaska, and Rosman, N.C., which monitor, control, and command the Orbiting Observatories, as well as receive, record and forward to the control center information transmitted from the satellite to the ground. (NASA Release 64–55)

- Army Missile Command's new R&D center at Redstone Arsenal was dedicated. The facility was named for Maj. Gen. Francis J. McMorrow, AMC commander at the time of his death last year. (*M&R*, 3/9/64, 9)

March 13: X–15 No. 3 flown by NASA pilot John B. McKay near Edwards AFB with a microphone embedded in the X–15's side to record noise of turbulent air sweeping over the surface, to help in prediction of metal fatigue. The X–15 reached maximum speed of 3,239 mph (mach 4.91) and peak altitude of 76,000 ft. This was 102nd flight of the X–15. (NASA X–15 Proj. Off.; UPI, *Wash. Post*, 3/14/64; UPI, *NYT*, 3/14/64, 39)

March 13: Fourth stage of four-stage Blue Scout rocket failed to ignite after USAF launch from Cape Kennedy. The booster was to have sent a USAF scientific probe 23,000 mi. into space to study earth's magnetic field. Preliminary estimates indicated the 33-lb. payload reached altitude of several hundred miles before it fell back into the Atlantic Ocean. (AP, Wash. *Eve. Star*, 3/13/64)

* S-I booster for Saturn I SA-9 was test-fired for 30 sec. in its initial firing at NASA Marshall Space Flight Center, Huntsville, Ala. This S-I was the last Saturn I booster built by MSFC; future stages were being built by Chrysler Corp., and would be shipped to Huntsville for static firings. Since the first Saturn I booster firing March 28, 1960, MSFC had conducted more than 50 firings on eight flight S-I's and one ground-test booster, all of which were assembled at MSFC. (*Marshall Star*, 3/18/64, 1)

* Publication of Advanced Bearing Technology (SP-38) by Lewis Research Center engineers Edmond E. Bisson and William J. Anderson, was announced by NASA. The book was described as "an important step in NASA's Technology Utilization Program aimed at making new space technology available to all branches of industry." (NASA Release 64-58; LRC Release 64-22)

* USAF fired Titan II ICBM from Vandenberg AFB more than 5,000 mi. down the Pacific Missile Range in 10th and last of series of R&D tests at PMR. (AP, *Wash. Post*, 3/14/64)

* AEC conducted low-yield underground nuclear explosion at its Nevada Test Site. (UPI, *NYT*, 3/14/64, 38)

March 14: Soviet Air Marshal Vladimir Aleksandrovich Sudets, Commander-in-Chief of Soviet antiaircraft defenses, said in *Izvestia* interview that aircraft of the A-11 type were "perfectly attainable targets" for Soviet antiaircraft weapons. (*NYT*, 3/15/64, 21)

* National Science Foundation plans for initiating its Science Development Program this spring were reported. The program would make available grants to colleges and universities "that could be given a shot in the arm that would have appreciable results and get them started toward becoming a center of excellence," as NSF Director Dr. Leland J. Haworth said. (Finney, *NYT*, 3/15/64, 39)

March 15: Aerospace Corp. confirmed rumors that Walter C. Williams, NASA Deputy Associate Administrator for Manned Space Flight and veteran of 24 years' service with NASA and NACA, had resigned to become vice president of Aerospace Corp. and general manager of its Manned Space Systems Div. Williams' resignation was effective March 16. (*Houston Chron.*, 3/15/64; *Houston Post*, 3/17/64; *NYT*, 3/16/64, 34; MSC *Roundup*, 4/1/64, 1)

March 16: 38th anniversary of first liquid-fuel rocket flight—that of Dr. Robert H. Goddard whose rocket flew 184 ft. in 2½ sec. at Auburn, Mass., in 1926. (*A&A/1915-60*, 21)

* NASA Associate Administrator for Manned Space Flight, Dr. George E. Mueller, testified before Senate Committee on Aeronautical and Space Sciences to present FY 1965 funding requirements for the manned space flight program and to discuss Project Apollo schedule considerations in detail. Dr. Mueller observed:

"The funds thus far appropriated by the Congress represent about one-third of the cost of the manned space flight program—up to and including the first manned lunar exploration. We now approach the period of maximum effort. The measure before you will bring us about half way to the moon." (Testimony)

March 16: NASA announced requesting from industry proposals for feasibility and design study of possible payloads for manned lunar exploration with an Apollo Logistic Support System (ALSS). One of the payloads was a mobile laboratory capable of supporting and transporting two Apollo astronauts for about 14 days; the other payload concept included both a stationary base and a smaller vehicle. The ALSS concept, making maximum use of Apollo-developed systems, provided for Saturn V launch vehicle to launch into earth-moon trajectory a manned Apollo command module, service module, and lunar excursion module modified for landing equipment rather than men on the moon and called the LEM Truck. Following ALSS landing, astronauts would land nearby in unmodified Apollo LEM. MSFC would manage the nine-month study. (NASA Release 64–62)

- Boeing study of post-Apollo manned lunar bases for NASA was basis for LESA concept (lunar exploration system for Apollo). LESA provided that a 25,000-lb. module would be soft-landed on the moon by a Saturn V launch vehicle. Three crewmen would follow by means of Saturn V/Apollo. The module would have 3,000 cu. ft. volume and could support three men for 90 days on the moon. Boeing's lunar base was expandable simply by adding more of the basic shelter modules, and longer visits by larger crews could be made with larger lunar bases. (*M&R*, 3/16/64, 26–27)

- Article in *Pravda* described Soviet radar observations of the planets Venus, Mercury, and Mars as "an outstanding achievement in Soviet science and technology." The value of the astronomical unit was computed, as a result of radarastronomy measurements, to be equal to 149,599,300 km., with an error of ±2,000 km. Radar fixes on planets Mercury and Mars in June 1962 and February 1963 confirmed their reliability in determining the astronomical unit. First radio communications via Venus, transmitted in November 1962, also were described. Article was signed by M. Subbotin, Director, Institute of Theoretical Astronomy, and Member of U.S.S.R. Academy of Sciences. (*Pravda*, 3/16/64, 4, ATSS–T Trans.)

- FCC announced Communications Satellite Corp. planned negotiation of $8 million contract with Hughes Aircraft Co. for development, fabrication, assembly, test, maintenance, and service for "Early Bird" communications satellite—the proposed experimental-operational comsat system. The proposed contract would be subject to FCC approval. (UPI, *Houston Chron.*, 3/17/64; Wash. *Eve. Star*, 3/17/64; *M&R*, 3/23/64, 8)

March 17: Radio command signal to SYNCOM II communications satellite pulsed onboard gas jets to increase the satellite's 6,800-mph speed. The satellite began drifting westward at rate of about 1.3 degree per day, was expected to reach vicinity of the International Date Line in mid-May. There SYNCOM II would serve as backup for a third Syncom, scheduled for synchronous orbit over the Pacific in May and transpacific communications experiments during the summer. (Finney, *NYT*, 3/19/64, 14)

March 17: Launch of NASA's S–48 ionosphere explorer satellite ("Topsi") was postponed when malfunction was discovered in Scout vehicle's wiring system during prelaunch checkout at Pt. Arguello. The Scout was returned to manufacturer Ling-Temco-Vought in Dallas for repair. "Topsi" was to have continued NASA's program to gather information on structure of the upper ionosphere pioneered by three satellites—EXPLORER VIII, ARIEL I, and ALOUETTE I—and a number of sounding rocket flights. (NASA Release 64–49; *Av. Wk.*, 3/23/64, 23)

* House Committee on Science and Astronautics approved H.R. 10456, NASA's FY 1965 authorization bill, for $5,193,810,500, which was $110 million less than the $5.304 billion NASA had requested. Of this amount, $4,327,950,000 was for R&D; $248,335,000 was for CofF; and $617,525,500 was for Administrative Operations. (H.R. 10456; NASA LAR III/48 and III/49)

* Dr. Homer E. Newell, NASA Associate Administrator for Space Science and Applications, reviewed 1963 highlights in testimony before Senate Committee on Aeronautical and Space Sciences:

  "Although direct measurements of the physical properties of the Moon await successful lunar flights, our Earth-based work has continued with some exciting results. Lowell Observatory scientists detected evidence of gas escaping from three different spots in the area of the lunar crater Aristarchus. This observation suggests that the Moon may be far from inert, and it has an important bearing on how we may conduct our lunar investigation. For example, if the Moon is seismically active, the seismograph may play an important role in the investigation of the lunar surface. . . .

  "Also, much effort was applied to clarifying the data from Earth-based or balloon observations of Mars. One important subject of controversy developed. Whereas it had been believed previously that the pressure at the surface of Mars is around 85 millibars, where 1000 millibars is roughly one atmosphere of pressure, it is now believed by many that the surface pressure on Mars may be as low as 10 or 11 millibars. This has serious implications with regard to landing a capsule on Mars. If the lower pressure is correct then retrorockets have to be used, whereas with the higher pressure atmospheric drag could be used to slow down the incoming capsule.

  "Significant new knowledge on the fundamental physical nature of the universe may have to await the launching of the first OAO. In the meantime, sounding rocket experiments continue to provide data in ultraviolet and radio astronomy. . . .

  "Although the formal development of Scout had been completed, additional work was required to improve its reliability. The primary objective of the December 1963 Scout firing was to determine the efficacy of the measures taken to correct observed deficiencies. The virtually perfect vehicle performance during this firing encourages one to look for a high success rate in future Scout firings. . . .

  "Performance of all Atlas-boosted launch vehicles would be considerably improved by introducing a mixture of fluorine and oxygen (FLOX) as the oxidizer. Careful study shows this can be done with admixtures of up to 30% fluorine without major redesign of the propulsion system. We are, therefore, planning to FLOX the Atlas. A primary objective of this program is to provide for additional payload in the Surveyor investigations. . . ." (Testimony)

March 17: Sixth anniversary of VANGUARD I satellite, launched into orbit Mar. 17, 1958. The satellite's stability of orbit had provided geodetic observations, including the determination that the earth is slightly pear-shaped. VANGUARD I was the only satellite orbited before 1959 that was still transmitting. (*A&A/1915–60*)

- Dr. Eugene Fubini, Assistant Secretary of Defense (Deputy Director, DDR&E), told Military Operations Subcommittee of the House Government Operations Committee that DOD was "very close to an agreement" with the Communications Satellite Corp. on working out a system for military satellite communications. Although decision had not been reached as to whether DOD would develop its own comsat system or lease channels from ComSatCorp satellites, Fubini said leasing channels "has proved to be more feasible than our fondest hopes." He emphasized that there would be "no compromise on security" if the commercial system were chosen. (Finney, *NYT*, 3/18/64, 15; *Av. Wk.*, 3/23/64, 19)

March 18: COSMOS XXVI scientific earth satellite was launched into orbit. Soviet news agency Tass said the satellite's initial orbital parameters were: apogee, 403 km. (250 mi.); perigee, 271 km. (168 mi.); period, 91 min.; angle of inclination to the equatorial plane, 49°. Onboard scientific equipment, radio system, and instrumentation were said to be functioning normally. (Tass, *Krasnaya Zvezda*, 3/19/64, 1, ATSS–T Trans.)

- First in series of 16 Judi-Dart meteorological sounding rockets was launched by Pakistan Space and Upper Atmosphere Research Committee (SUPARCO) under cooperative agreement with NASA, from Pakistan's Sonmiani Range near Karachi. This was first launching planned as a direct contribution to the International Indian Ocean Expedition (IIOE), a multinational effort by scientific groups to obtain and correlate information on the Indian Ocean area's oceanic and atmospheric conditions. The NASA-SUPARCO project was designed to gather data on the dynamics of air circulation at 100,000 to 200,000 ft. above West Pakistan's coast during the coming year. Pakistan purchased the Judi-Dart rockets and chaff payloads from the U.S. for assembly and launch from Sonmiani Range; NASA loaned to Pakistan for one year the radar needed for tracking the chaff—thousands of tiny metallic threads released at peak altitude—and was training Pakistani personnel in radar equipment operation. (NASA Release 64–64; Wallops Release 64–28)

- NASA Nike-Apache sounding rocket launched from Ft. Churchill, Canada, with Rice Univ. instrumented payload reached 94-mi. altitude in successful experiment to measure light intensities and particle fluxes. Rocket performed excellently and instrumentation—two photometers, two scintillators, three Geiger counters, and two magnetometers—functioned properly except for one scintillator, which failed at nose cone ejection. The Sammy II instrument package was designed and developed by Rice University's Satellite Techniques Laboratory. (NASA Rpt. SRL)

- Rep. Olin Teague (D.-Tex.), chairman of Subcommittee on NASA Oversight of the House Committee on Science and Astronautics, held special hearings "to emphasize our interest in seeing that there is progress" in advanced propulsion systems. At opening session, representatives of four industrial firms recommended development of nuclear rocket power be accelerated and amplified. Officials testifying were George H. McLafferty, United Aircraft; William C. House, Aerojet-General; Dr. Woodrow E. Johnson, Westinghouse Astronuclear Laboratory; and Dr. Theodore B. Taylor, General Dynamics. (*SBD*, 3/19/64, 433; *M&R*, 3/23/64, 12)

March 18: House passed and sent to the Senate $1.6 billion military construction authorization bill for FY 1965 (H.R. 10300). Discussing the bill on the floor, Rep. William H. Bates (R.–Mass.) observed:

"This year, research, development, testing, and engineering for our Federal Government projects will approximate $15 billion. This represents about 17 percent of our entire Federal budget and a much larger percent of those funds over which we actually have control.

"In the various military services in fiscal year 1965, we will spend approximately $7.6 billion on research, development, testing and engineering. In the bill before you today about $90 million is set aside for research and development facilities. . . ." (NASA LAR III/50; *CR*, 3/18/64, 5356)

• Dr. Raymond L. Bisplinghoff, NASA Associate Administrator for Advanced Research and Technology, testified before Senate Committee on Aeronautical and Space Sciences. He mentioned that NASA's Advanced Research and Technology program encompassed more than 2,000 separate tasks and projects, and highlighted a few areas for illustration. ". . . Considerable effort is concentrated on commercial supersonic transports, a subject of NASA research for many years. Various aircraft configurations embodying widely differing design features have resulted from this research. All of these designs appear promising, particularly with regard to supersonic performance. In September 1963, at the Langley Research Center, the results of preliminary design studies and of related NASA in-house research programs were presented to all interested government agencies and major aircraft manufacturers. In general, it was concluded that a supersonic transport for airline use is technically feasibly. It was stressed, however, that additional research will be required in order to raise the level of technology to the point where a supersonic transport can be designed which is economically attractive to the airlines. . . ." (Testimony)

• In testimony before Senate Committee on Aeronautical and Space Sciences, Director of NASA Office of Tracking and Data Acquisition Edmond C. Buckley said:

"We assure a thorough coordination between NASA and DOD in the planning, implementation, and use of ground instrumentation facilities through several mechanisms and procedures. . . .

"I feel that the planning within NASA and the cooperation between NASA and DOD is such that both NASA and DOD are achieving an optimum utilization of the existing ground facilities and of the currently planned augmentations to these facilities. I can assure you that this type of coordination will continue.

"I would like to emphasize one point, however. The planning for the support of a mission involves so many trade-offs and compromises in launch vehicles, trajectories, weight, complexity, and cost that the fundamental responsibility must lie in the hands of the agency responsible for the program. This is not only true in the case of NASA projects, but it is equally true that DOD has to have full responsibility for the planning and trade-offs on such projects as Discoverer, Dyna-Soar, and the upcoming Manned Orbiting Laboratory. These projects are so complex and the things we are trying to do are so new that divided responsibility would most seriously jeopardize the possibility of mission success. . . ." (Testimony)

March 18: Dr. Eugene G. Fubini, Assistant Secretary of Defense (Deputy Director, DDR&E), disclosed in testimony before House Committee on Government Operations, Subcommittee on Military Operations, that DOD planned to use one of the Syncom communications satellites for experimental communications, probably between Hawaii and the Philippines. (Finney, *NYT*, 3/19/64, 14)

- Dr. William A. Mrazek, Director of MSFC Propulsion and Vehicle Engineering Laboratory since 1956, was appointed chief engineer of Saturn launch vehicle project in MSFC Industrial Operations. (*Marshall Star*, 3/18/64, 1)
- Eleanor Pressly, head of Vehicle Section, Spacecraft Integration, Sounding Rocket Div. at NASA Goddard Space Flight Center and winner of a 1963 Federal Women's Award, was guest speaker at Mrs. Lyndon B. Johnson's White House luncheon for "women-doers." (*Wash. Post*, 3/17/64)
- Dr. Kurt H. Debus, Director of John F. Kennedy Space Center, NASA, announced U.S. Bureau of Sport Fisheries and Wildlife would administer for NASA the entire buffer zone Merritt Island lowlands, nearly 40,000 acres. (KSC Release 34–64)
- Dr. Norbert Wiener, 69, died in Stockholm. Considered the "father of automation" and one of the world's ranking mathematicians, Dr. Wiener was prof. of math at MIT from 1932 to 1960. (AP, *NYT*, 3/19/64, 1)
- In the U.S. for speaking tour, Sir Bernard Lovell, Director of Britain's Jodrell Bank Experimental Station, told the *Boston Globe*: ". . . I am terribly impressed by the dynamic effort this country is making to be first in space. It appears to me the signs are extremely favorable so far as the West is concerned in the stake it has in America's space program. . . .

"This is the greatest challenge man has ever faced. If you're going to be the leading power in the world, you simply can't afford to come in second. . . ."

Sir Bernard said press reports quoting him as saying the U.S.S.R. had given up going to the moon were "completely erroneous.

"I have every reason to believe the Russians are trying to get to the moon every bit as fast as the Americans are." (Leland, *Boston Globe*, 3/19/64)

March 19: Attempt to orbit NASA Explorer Beacon satellite (S–66) failed when third stage of Delta launch vehicle burned for only 22 sec. instead of normal 40. This was only the second Delta failure and followed 22 consecutive successes. The 120-lb. satellite was designed to transmit data on the ionosphere directly to worldwide network of ground stations. In addition to making major ionosphere studies, the satellite was to have served as test bed for two geodesic experiments—reflection of a laser beam directed from Wallops Island, Va., and transmission on two frequencies permitting ground stations to study Doppler method of satellite tracking and influence of ionosphere on Doppler tracking. (UPI, *Wash. Post*, 3/20/64; *Av. Wk.*, 2/23/64, 25; NASA Release 64–60)

March 19: NASA Nike-Apache sounding rocket with Rice Univ. payload was launched from Ft. Churchill, Canada, to 94-mi. altitude. Two photometers, two scintillators, three Geiger counters, and two magnetometers made coordinated measurements of light intensity and particle fluxes. (NASA Rpt. SRL)

- Two U.S. commercial airline pilots reported seeing unidentified object explode in a "spectacular flash" as it entered earth's atmosphere over the Atlantic Ocean, about 200 mi. west of Land's End, England. Press sources reported speculation that explosion was part of jettisoned rocket used the previous day in Soviet launch of Cosmos satellite. (UPI, *NYT*, 3/20/64, 2)

- Report by Committee on Science and Public Policy of the National Academy of Sciences, *Federal Support of Basic Research in Institutions of Higher Learning*, was issued by NAS. Prepared in response to requests from American Society of Biological Chemists and other scientific societies, the report attributed success of Federal scientific support program "in no small measure to enlightened policies of several federal agencies . . . specifically to the current emphasis on support by research projects grants and by fixed-price research contracts. . . .

  "The record shows a continuous regard for the government's responsibility for the money entrusted to it by the people. And the overwhelming majority of the scientific community has throughout the record respected that responsibility. . . ." (NAS–NRC Release)

- AFCRL scientist Sheldon B. Herskovitz had created a theoretical model and operation of an all-plasma radar system. The hypothetical radar system was based on some devices already developed, some representing techniques presently available but not applied, and some that simply were theoretically promising. (OAR Release 3–64–3)

- Dr. Robert L. Barre, former NASA Scientist for Social, Economic, and Political Studies, said in address before American Orthopsychiatric Association in Chicago: "Space activities may be viewed either as a cause or as a consequence of social and cultural change. From either point of view, freedom to engage in scientific activity entails significant social responsibility. . . .

  "It is ironic that a consequence of the extension of definite verifiable knowledge, the primary product of science, should be to extend confusion. To increase knowledge, not only adds knowledge where it did not previously exist, but corrects previously held mistaken knowledge and belief. In so doing, it alters the conditions of existence; hence man's relationship, not only to his physical environment, but to other men.

  "Scientists produce the conditions from which these alterations develop. The inescapable responsibility of scientists is to continuously inform society of the knowledge they have, or are trying to create, and to explain its scientific significance.

  "The institutional locus and the social role of the scientist has traditionally fulfilled this responsibility. The scientist in the past was, not only a creator of knowledge, but an integrator, a teacher—a transmitter of knowledge. New knowledge could be incorporated in its appropriate context and transmitted more rapidly than it could be produced. Today we have altered these arrangements. . . .

"While slowing down the system for the transmittal of new knowledge and reducing the proportion of the educated populace who have access to the existing system, we are simultaneously speeding up the production of new knowledge. While ignoring the processes of evaluation, integration, interpretation, and assimilation of new knowledge, we have fixedly accelerated the processes of specialized acquisition of new knowledge.

"We have upset the equilibrium between invention and innovation. I submit that the social responsibility of the scientist is obvious: to assist society restore the balance." (Text)

March 20: Dr. Hugh L. Dryden, NASA Deputy Administrator, was presented the Robert H. Goddard Memorial Trophy for 1964, at National Space Club's annual Goddard Memorial Dinner in Washington. Other Space Club awards: Dr. Harold A. Rosen and Donald D. Williams of Hughes Aircraft Co. received the astronautics engineer award for contributions to conception and development of Syncom satellite. Tulane Univ. received a $1,500 grant to further advanced education in the field of science and astronautics; McDonnell Aircraft Corp. and General Dynamics/Astronautics received Nelson P. Jackson Aerospace Award for performance on Mercury spacecraft and Atlas booster of Astronaut L. Gordon Cooper's 22-orbit flight; editors and staff of *Fortune* magazine received the Space Club's press award for series of articles "which provided provocative insights into the motivations and accomplishments of the nation's space-exploration programs"; Robert P. Walker, mathematics and physics student at Univ. of Maryland, received $1,500 Goddard Scholarship; and Robert Cargill Hall, historian of Lockheed Missiles and Space Co., received $200 Goddard Historical Essay award for second straight year. (NSC Release)

- Dr. Edward C. Welsh, NASC Executive Secretary, delivered keynote address at the Robert H. Goddard Memorial Dinner. He compared U.S. and Soviet space accomplishments:

"1. As regards that key feature of booster power, the Soviets are ahead on an operational basis, although we are moving up rapidly, having attained the current weight-lifting record on a test shot by the Saturn I.

"2. We have placed almost four times as many payloads into earth orbit, while they continue each year to increase their absolute lead in total weight of net payloads orbited.

"3. Our numerical progress in orbiting payloads is impressive. Last year alone, the U.S. put more payloads into earth orbit than the USSR has since Sputnik I in 1957.

"4. The increasing reliability of our launchings is also impressive. During the past three years, we placed successfully into earth orbit 82% of all the payloads which we attempted to put there. The Soviet record, while not identical, is comparable.

"5. The Soviet heavy Vostok class of vehicle has accumulated some 1600 orbits, as compared with 37 for our smaller Mercury. . . .

"6. The Soviets have devoted more of their attention to far-out space than we have. As a percentage of total space launching attempts, they have devoted five times as much effort to escape-missions to the moon and planets as we have. The opposite is true as to near-earth orbits, in which the U.S. emphasis has been much greater than theirs.

"7. Both countries have made impressive strides in accumulating scientific data about space. We may even be ahead in this regard, but it is not safe to assume so.

"8. The U.S. is clearly ahead in world-wide sharing of space research in cooperative programs, and in direct application of space technology to practical use in communications, weather reporting, geodesy, and navigation.

"In summary, we have made enormous gains in the past few years in the race with our able and aggressive competitor.

". . . I want to emphasize that the Soviets have a strong orderly program, with every indication that it is continuing vigorously, apparently unaffected by adverse economic factors in their country. Wishful thinking to the contrary, we dare not slow down unless we are willing to pay the terrific price of second place. . . ." (Text)

March 20: General B. A. Schriever (USAF) addressed an Air Force Association squadron in New York City on the subject of "The Air Force in the Years Ahead":

"A major consideration in long-range planning today is the rapid pace of technical progress. Since the early days of World War II we have been caught up in a massive technological explosion. One of the most important military products of this explosion is the intercontinental ballistic missile, which today forms an increasingly important part of our nation's strategic forces. . . .

"In addition to its direct contribution to our national security the Air Force ICBM development program has created much of the technology, hardware and facilities that have been utilized in the nation's space program. Most of the nation's space shots to date have been launched by military boosters from Air Force facilities. The Titan III Standard Launch Vehicle, which is currently under development by the Air Force Systems Command, is based on the technology and experience gained in ICBM development.

"The Air Force Manned Orbiting Laboratory (MOL) program is the logical next step in military space development. The MOL is required to help us answer basic questions about man's role in space, such as the length of a time a man can safely remain in orbit, his ability to make full use of his faculties in space, and his ability to perform a useful military function there. . . ." (Text, AFSC Release 43–R–32)

- Dr. John Peter Minton of the National Cancer Institute confirmed that infrared laser rays had killed certain malignant tumors in mice. Dr. Minton had conducted several months of experiments in conjunction with engineers of the U.S. Army Missile Command engaged in laser research. (DOD Release 237–64)
- U.S. and Japanese meteorologists ended week-long conference in Miami where they formulated informal recommendations for increased international cooperation in problems of hurricanes and typhoons. Satellites and rockets were expected to play key role in the program, administered by Japanese Research Council and the National Science Foundation. (*M&R*, 3/23/64, 8)

March 21: In experiment to test reactions of astronauts under environmental conditions they would encounter outside their spacecraft, three out of four subjects suffered the bends. Capt. Eugene A. Degner, expert in body metabolism conducting tests planned on 50 volunteers at Brooks

AFB, said: "All three men with the grade one [least severe] bends had them dramatically dissipate as the altitude chamber pressure dropped below that of 30,000 feet.

"I was surprised at the results, and a little worried because the three men complained of the bends and mild discomforts early in the experiment period.

"I was afraid they would keep getting worse, but the trouble was intermittent and not severe enough to abort the [10-hour] test." (UPI, *Houston Chron.*, 3/21/64)

March 22: New laser radar system using three lasers operating in tandem was under development in Sperry Rand Corp. Electro-Optics laboratories, according to Sperry announcement. Using the three lasers in tandem to build up to peak power, the device would be the first to incorporate the full potential of laser radar, Sperry said. (*N.Y. Her. Trib.*, 3/22/64)

March 23: USAF launched Titan II on successful test flight from Cape Kennedy. The nose cone, carrying malfunction detection instrumentation to be used in Project Gemini manned space flights, impacted 5,000 mi. down the Atlantic Missile Range. (AP, *NYT*, 3/24/64; *M&R*, 3/30/64, 16)

- Two NASA Nike-Apache sounding rockets were launched from Ft. Churchill, Canada, and returned good data. The first attained 95-mi. altitude, its instrumented payload measuring light intensity and particle fluxes. Rice Univ. payload included two photometers, one scintillator, three Geiger counters, one Geiger tube, and two magnetometers. The second rocket reached 117-mi. altitude with instrumented payload to obtain data on charged particles responsible for the visual aurora and on those responsible for auroral absorption. (NASA Rpts. SRL)

- NASA Electronics Research Center Site Evaluation Committee was formalized by Deputy Administrator Dr. Hugh L. Dryden. First task of the six-man committee would be to draft proposed plan for criteria and procedures to be used in the site evaluation; after doing this, the Committee would begin evaluating proposed sites in the Greater Boston area. Col. Robert P. Young, NASA Executive Officer, was appointed Chairman of the Committee. (NASA Release 64-67; NASA Circular 311)

- European Space Research Organization (ESRO) announced in Paris its plans for satellite development program. First two ESRO satellites would be launched into orbit from the U.S. with Scout boosters; launch vehicles for later satellites would be supplied by member nations. ESRO set budget of $300 million for eight years. (*NYT*, 3/25/64, 31; *A&A*, 5/64, 9)

- Michael L. Garbacz was appointed Program Manager, Meteorological Flight Projects, in NASA Hq. Office of Space Science and Applications. Garbacz had served as Acting Program Manager since Sept. 1963. (NASA Announcement 64-59)

- By midnight deadline, 197 communications carriers had filed applications with FCC to invest in Communications Satellite Corp. (AP, *NYT*, 3/25/64, 53)

- Economic analysis of short-haul transport by Systems Analysis and Research Corp. (SARC) was presented to FAA. SARC concluded modern 20-passenger, twin-engine aircraft would provide U.S. airlines the most favorable operating results for short-haul requirements. Study was part of FAA's program to promote development of economic, short-haul transport to replace the DC-3. (FAA Release 64-25)

March 24: Mrs. Lyndon B. Johnson visited NASA Marshall Space Flight Center, where she toured the facilities, witnessed a Saturn I and an F-1 static firing, presented awards to MSFC employees, and was honored at luncheon and tea with her Alabama relatives.

At awards ceremony, Mrs. Johnson presented 30-year service pin to Michael J. Rabatin, program coordination officer, and a 20-year pin to Mrs. Mary Derryberry, engineering designer. She presented five invention awards: $300 to Robert J. Carmody for invention of highly flexible cellular structure; $400 to Klaus Juergensen for invention of bilateral energy transfer system; $400 to Lester Katz and Jack J. Nichols for force measuring instrument; and $300 to Benjamin M. Saunders for self-latching handle. Group achievement award of $500 went to Walter F. Gillespie and Alonza J. Davis for contributions to recoverable motion picture camera system developed for the Saturn I SA-5, SA-6, and SA-7 launch vehicles. Mrs. Johnson also presented a $250 Sustained Superior Performance award to Mrs. Hazel T. Atchley, contracts assistant.

In letter to MSFC Director Dr. Wernher von Braun, the First Lady described her visit as "stimulating, exhilarating, informative. . . . The significance of the Saturn has come home to me." (*Marshall Star,* 3/25/64, 1, 4; Robertson, *NYT,* 3/25/64; *Marshall Star,* 4/1/64, 1)

- NASA launched Nike-Apache sounding rocket from Ft. Churchill, Canada, to 111-mi. altitude, but telemetry transmitter failed. Five charged-particle detectors, an aspect magnetometer, and an accelerometer were included in the instrumentation, which was to have obtained data on charged particles responsible for visual aurora and those responsible for auroral absorption. (NASA Rpt. SRL)
- USAF launched Thor-Agena booster combination from Pt. Arguello with unidentified payload. (UPI, *Wash. Post,* 3/25/64)
- USAF conducted second flight test of Asset re-entry vehicle at AMR. The second stage of the Thor-Delta launch vehicle failed to ignite, so the 1,100-lb. research vehicle fell back into the Atlantic only 500 mi. away from the launch site. USAF said the Asset payload had reached 33 mi. altitude and was traveling at 10,000 mph at the time of stage seperation, but it was supposed to have reached 12,000 mph at separation time. (AP, *Houston Chron.,* 3/24/64; UPI, *Wash. Post,* 3/25/64)
- Chief engineer of Air Transport Association's Supersonic Advisory Committee, John T. Dyment, estimated it would cost half-again as much as the $1 billion estimated by the Government to develop the U.S. supersonic transport and test it adequately. He criticized plans to take steps toward manufacturing before prototype testing was completed: "We must have supersonic transports built and flying for a year or so to gain the experience and the reliability before we even start a production line. This is essential. . . ." Detailed findings of the committee's two-day meeting in New York would be reported to IATA Technical Committee May 5. (Clark, *NYT,* 3/25/64, 65)
- Export of U.S. aerospace products in 1963 totaled $1.3 billion, according to Aerospace Industries Association (AIA). (AP, Wash. *Eve. Star,* 3/24/64)

March 25: First TV transmission from Japan to the U.S. was made, using NASA'S RELAY II communications satellite. Quality of the live pictures as well as the sound was excellent. In the telecast Prime Minister Hayato Ikeda offered his country's apologies directly to the U.S. for the "unfortunate coincidence" of the stabbing of U.S. Ambassador Edwin Reischauer by a deranged Japanese youth the previous day. The eight-minute telecast was seen simultaneously on the three national networks. (NASA Release 64-63; UPI, *NYT*, 3/26/64, 3; AP, *Wash. Post*, 3/26/64)

- NASA launched Nike-Apache from Ft. Churchill, Canada, to 120-mi. altitude, its instrumented payload studying particle intensities and energy distribution of electrons with energies greater than 50 kev. Instrumentation performance was good, but there was some question as to validity of data from the plastic scintillator and further study of the data was planned. (NASA Rpt. SRL)

- House of Representatives voted (283-73) to approve NASA FY 1965 budget authorization of $5,193,810,500. It defeated the only two amendments offered: one proposed by Rep. Donald Rumsfeld (R.-Ill.) to delete the $14 million for the new Electronics Research Center in the Boston area; and one proposed by Rep. Henry S. Reuss (D.-Wis.) to eliminate the $24.7 million for aeronautic R&D on supersonic transport aircraft. (*CR*, 3/25/64, 6071-6123)

- Rep. Melvin Laird (R.-Wis.), speaking in favor of the amendment to eliminate SST funds from the NASA FY 1965 authorization, discussed his reasons for supporting the amendment:

    "Within the next 3 or 4 weeks I predict there will be announced a new single manager program by the administration as far as supersonic transport craft is concerned. It will not be a responsibility of NASA, Defense or FAA to carry out this program. A new authorization to transfer appropriations will be submitted to the Congress to carry out this program under single management.

    "As long as we are going to have a new program submitted for a Polaris-type single manager concept it is not necessary to include this authorization in the NASA authorization bill. I for one heartily endorse the plans of the administration to combine the management of this new aircraft development program in the hands of a single manager responsible to the President. This is in accordance with the Black report. . . ."
    (*CR*, 3/25/64, 6121)

- NASA Lewis Research Center granted $900,000 authorization to AEC Hanford Atomic Products Operation for R&D in connection with advanced nuclear rocket engines—fabrication, test, and evaluation of tungsten uranium oxide fuel elements for advanced nuclear rocket engine now under study at Lewis. The agreement brought value of work being performed at Hanford for NASA to more than $1.5 million. (LRC Release 64-21)

- NASA awarded two contracts to Aerojet-General Corp. for development of M-1 1.5-million-lb.-thrust rocket engine. One contract was for $190,075,635 and provided for R&D work; the other contract was for $48,550,074 and provided for necessary test facilities. Development of the liquid hydrogen/liquid oxygen engine was under technical direction of NASA Lewis Research Center. (NASA Release 64-66; LRC Release 64-31)

March 25: With the move of the Advanced Spacecraft Technology Div., NASA Manned Spacecraft Center had located 2,000 persons at the new Clear Lake site. One third of the 60 facilities at the site had been certified by the U.S. Corps of Engineers as operational or ready for occupancy. (MSC Release 64-59)

- Senators Albert Gore (D.-Tenn.) and Ralph Yarborough (D.-Tex.) said DOD negotiations to lease channels from ComSatCorp were grounds for repeal of the Communications Satellite Act of 1962. Labeling the law a "giveaway" of a national resource, Senator Gore said it would be "remarkable and bizarre" for the Government to lease the system. "We will be paying for using what we have given away." (*Wash. Post*, 3/26/64)
- Minuteman ICBM exploded shortly after launch from silo at Vandenberg AFB, and USAF was investigating cause of the explosion. This came five days after USAF conducted 50th launch in its Minuteman development program, a successful AMR flight. (*M&R*, 3/30/64, 16)

March 25-26: Government and airline technical officials met in Washington to review findings in their independent evaluations of designs proposed for the U.S. supersonic transport aircraft. On the basis of further, detailed consideration of airline and Government evaluation findings, a decision would be made on the future course of the SST program: (1) selection of an airframe and an engine proposal for development, (2) selection of two airframe and two engine proposals for one-year detailed design competition, or (3) determination that the program should be terminated or redirected if it is decided that none of the proposals meets established criteria. (FAA Release 64-26)

March 26: Mockup of the Apollo Lunar Excursion Module (LEM) was shown to the press at Grumman Aircraft Engineering Corp. after three days of design inspection by officials of NASA Manned Spacecraft Center. Objective of review was to establish design freeze on most areas of LEM design. Under NASA contract, Grumman would build 9 LEMS for ground testing and 11 for flight. (Clark, *NYT*, 3/27/64, 7; *M&R*, 4/6/64, 26)

- Soviet satellite COSMOS XXIII burned up in the atmosphere between the Great Lakes and the northeastern Canadian coast, according to NORAD. Observers in Pendleton, Ore., and Duluth, Minn., reported seeing fiery streak in the sky, and Smithsonian Astrophysical Observatory officials at Cambridge, Mass., said they assumed the streak was the reentering COSMOS XXIII. (AP, Balt. *Sun.* 3/28/64; AP, *NYT*, 3/29/64, 31)
- At close of 18-day session of U.N. Outer Space Committee's legal subcommittee in Geneva, U.S. delegate Leonard C. Meeker praised the "business like" manner in which the subcommittee had studied the draft treaties presented by the U.S. and the U.S.S.R. The subcommittee would meet again later this year. (*NYT*, 3/27/64, 9)
- Space Science Board of National Academy of Sciences concluded that Project West Ford dipoles placed in orbit last year had not interfered with either optical or radioastronomy observations. Board Chairman Dr. H. H. Hess said in report that the board's conclusion "should not be taken either as an endorsement of the experiment or as tacit agreement to the launching of another similar belt without further discussion." Consideration of the West Ford dipoles would continue under Space Science Board's newly established Committee on Potential Contam-

ination and Interference from Space Experiments. (NAS–NRC Release)

March 26: Joint Congressional Committee on Atomic Energy voted $14 million in AEC's FY 1965 budget for flight test of Snap–10A reactor. This action followed Budget Bureau letter to the Committee acknowledging that "reasonable and well informed men can differ" on such issues as the Snap flight test and stating that the Administration would conduct flight test if Congress authorized the funds, as long as the Administration's $2,651,542,000 request for AEC was not exceeded. Late last year the Administration had canceled orbital flight-test because of lack of specific mission for the Snap; leaders of the Joint Congressional Committee, Sen. John O. Pastore (D.–R.I.) and Rep. Chet Holifield (D.–Calif.), had protested the decision, urging President Johnson to approve the flight test. (*NYT*, 3/26/64, 24)

- Dr. Albert J. Kelley, Director of Electronics and Control in NASA Office of Advanced Research and Technology, said in address before Aero Club of New England, in Boston: " . . . the NASA effort is not confined to a single objective such as landing men on the moon. Rather, it is a broad-based research and development effort which is designed to meet the needs of any agency of the government having work to do in space. And the vigor with which we pursue this activity will determine how effectively those agencies can meet their responsibilities, and cope with their problems in space, in the years ahead. . . .

    "With respect to such a large, complex, and unknown environment as space, and the still not precisely defined characteristics of the earth's atmosphere, this nation would be be oblivious to the lessons of history if it, in every case, required that exploratory research and development be matched to a completely defined mission. . . ." (Text)

- Rep. Roman C. Pucinski (D.–Ill.) spoke on House floor urging passage of H.R. 8104, his bill to prohibit flight of civil supersonic aircraft generating sonic boom overpressures above 1.5 lbs. per sq. ft. Rep. Pucinski cited FAA's current sonic boom experiment in which flights producing overpressure of 1 to 1.5 lbs. per sq. ft. had prompted some 2,000 complaints. "The tests in Oklahoma City can indeed be a big help in this whole subject, but only if the FAA has the courage to increase the booms to the overpressures which we know the supersonic jets will have to produce to be efficiently and economically operational. . . . I am firmly convinced that the 2,000 complaints . . . would skyrocket if the FAA produced sonic booms approaching 3 pounds or more overpressure per square foot. . . .

    "Mr. Speaker, I submit that this whole [SST] project must be held in a state of abeyance until after the FAA has had the courage to generate sonic boom overpressures in excess of 2 pounds per square foot. . . ."

    He then cited FAA and NASA statements to House Interstate and Foreign Commerce Committee: "The FAA admits that sonic booms from 2 to 3 pounds overpressure per square foot will do damage to glass and plaster in isolated cases and that sonic booms creating overpressure of 3 to 5 pounds per square foot will cause widespread window and plaster damage. But . . . [NASA] states that tests carried out by NASA in cooperation with the U.S. Air Force show that the lowest overpressure at which house windows were broken was 23 pounds per square foot, and most of the windows tested survived pressures exceeding 56 pounds per square foot."

He said NASA was "way off the pad in its analysis" and added: ". . . here we have so vital an issue as the health and welfare of millions of Americans and the safety of property throughout the country involved and yet these two agencies cannot agree on standards for damage due to sonic booms. . . ." (*CR*, 3/26/64, 6193–97)

March 26: U.S. manned space flight plans were outlined in *Krasnaya Zvezda* (*Red Star*) article with Col. G. Terent'yev byline. Saying that the U.S. was "not leaving a stone unturned in trying to outstrip the Soviets in the conquest of space with manned space vehicles," the writer gave particular attention to Gemini and Apollo plans, then went on to claim that U.S. policy "in the conquest of space is far from a peaceful one; it has a clearly evident military (more accurately, aggressive) nature." (*Krasnaya Zvezda*, 3/26/64, 3, ATSS–T Trans.)

- Wilbur L. Pritchard, Group Director of ComSat Systems, Aerospace Corp., before Military Operations Subcommittee of the House Committee on Government Operations, was first witness to speak out against military usage of commercial comsat system. Speaking both for himself and for Aerospace Corp., Pritchard said military requirements would be so different from commercial requirements that either the national security would be sacrificed or ComSatCorp would have to charge such high rates that there would be no savings to the Government. He proposed alternative system based on simple spin-stabilized satellite, capable of growth and expansion to accommodate increased needs of military satellite communications. (*Av. Wk.*, 3/30/64, 15; *SBD*, 3/27/64, 479)

March 27: ARIEL II (UK–C) scientific earth satellite, the second in series of three U.S.–U.K. satellites, was orbited by NASA Scout booster launched from Wallops Island, Va. Initial orbital data: 843-mi. apogee, 180-mi. perigee, 101-min. period, and 52° inclination to the equator. Under overall administrative responsibility of the British Office of the Minister for Science, the three onboard experiments were designed and built by the U.K. to measure vertical distribution of ozone in the upper atmosphere, measure galactic radio noise, and detect micrometeoroids encountered by the satellite. The experiments were chosen by the British National Committee on Space Research, chaired by Sir Harrie Massey, in consultation with NASA. NASA Goddard Space Flight Center built and tested the spacecraft and contracted with Westinghouse Electric Corp. for subsystems. (Wallops Release 64–30; NASA Release 64–56)

- COSMOS XXVII satellite was launched into orbit by U.S.S.R. Announced orbital data were: 237-km. (147-mi.) apogee; 192-km. (119-mi.) perigee; 88.7-min. period; and 64.8° inclination to the equator. Soviet news agency Tass said the satellite carried scientific instrumentation intended for the "further study of outer space in accordance with the program announced by Tass" March 16, 1962, as well as radio transmitter and telemetering systems. All onboard equipment was functioning normally. (Tass, *Pravda*, 3/28/64, 1, ATSS–T Trans.)

- X–15 No. 1 flown by USAF pilot Maj. Robert Rushworth to maximum speed of 3,920 mph (mach 5.76) and altitude of 101,000 ft., in flight near Edwards AFB, Calif. For the flight a special camera was installed in the plane's fuselage to test optical degradation caused by temperatures of up to 1,200° F. This was the 103rd flight of the rocket research aircraft. (NASA X–15 Proj. Off.; UPI, *Chic. Trib.*, 3/28/64)

March 27: Dr. Hari K. Sen, research scientist in Space Physics at USAF OAR's Air Force Cambridge Research Laboratories, had shown that evaporation of solar corona was largely responsible for origin of the solar wind, OAR announced. Dr. Sen evolved simple physical theory explaining how electrons and protons of solar winds escape solar corona, then compared his results with observed interplanetary electron densities and with experimental data obtained from spacecraft on solar winds near the earth. His theory agreed, in order of magnitude, with experimental data and gave estimate of interplanetary electron density beyond observed range out from the sun. He set critical altitude for escape at about four solar radii, about 1,300,000 mi. above the visible solar disc. (OAR Release 3-64-5)

- World Book Encyclopedia announced 14 new NASA astronauts had been added to the 16 astronauts with four-year contracts for newspaper distribution rights to their life stories. World Book—a unit of Field Enterprises Educational Corp.—also said the astronauts had signed separate contracts with *Life* magazine. (AP, *NYT*, 3/28/64, 20)
- Severe earthquake caused major damage in southern Alaska, and President Johnson declared a major disaster area in Alaska. There was no apparent damage at NASA tracking station near Fairbanks, although tremors were felt. Cape Kennedy rose and fell about $2\frac{1}{4}$ in. because of seismic wave from Alaskan earthquake, U.S. Coast and Geodetic Survey said. Earth movement at Cape Kennedy was greatest in U.S. (*NYT*, 3/29/64, 1; AP, *Kansas City Star*, 5/10/64)
- NASA Lewis Research Center announced award of $797,200 contract for construction of concrete-lined shaft for the 500-ft. gravity facility, to Patrick Harrison, Inc. The underground shaft would be first phase of construction of largest zero-gravity facility in the U.S. Completion of the facility was expected this year. (LRC Release 64-30)

March 28: In conjunction with dedication of the Center of Science and Industry in Columbus, O., NASA Lewis Research Center sponsored a comprehensive exhibit of U.S. space program. The new center was designed as a museum of science, industry, health, and history. (LRC Release 64-25)

March 29: President Johnson told a Texas news conference that he would establish a seven-man advisory board for the U.S. supersonic transport (SST) development. Members of the board would be: Defense Secretary McNamara, Treasury Secretary Dillon, Commerce Secretary Hodges, CIA Director John McCone, NASA Administrator Webb, FAA Administrator Halaby, and New York banker Eugene Black. Asked whether he had given the Supersonic Transport Advisory Board a specific assignment, the President replied:

"The executive order will do that. It is being drawn. I wanted you to know that we are doing that. It has not been finalized yet. It may be changed in the details. But we want to have the broad spectrum of the Government interested in it, to get the best judgments of all of our people in an advisory capacity."

Asked if we were setting a deadline on the Board's report back to him, the President answered:

"No. This will be advisory. I am not asking for a report. I am asking them to advise in connection with the contracts and all the matters covered by the Black report and by the report that I made in connection with the testimony before the Congress. The Congress has already appropriated $60 million. We have already had an evaluation of the various proposals. But we just want to get this senior group of officials to sit in and counsel with us." (Transcript, *NYT*, 3/30/64; *WSJ*, 3/30/64)

March 29: Discovery by two astronomers of what they believed to be the most distant object identified to date was announced. Thought to be several billion light-years away, the object—named 3C–147—was considered to be from 10 to 20 per cent farther away than 3C–295, previously considered the most distant object. Both objects were "quasi-stellar radio sources," which produce 100 times the brightness and energy of an entire galaxy of 100 billion stars. Dr. Thomas A. Matthews located the new object by observing its radio emissions with Cal Tech's twin antennas. Then Dr. Maertin Schmidt photographed the object from Mt. Palomar Observatory. (Sullivan, *NYT*, 3/30/64, 27)

- Educator Robert M. Hutchins, writing in the Omaha *World-Herald*, said of U.S. project to land a man on the moon: "I think it will not be argued that the primary object of this venture is to advance knowledge. It seems rather to be designed as a huge public relations stunt. The 'knowledge industry' will restore the prestige we lost when the Russians outdistanced us in space. . . .

  "If we were dedicated to advancing knowledge, we would have had some serious debate about the best way to do it. Is it to improve our colleges and universities or to put on scientific spectaculars?" (Omaha *Sun. World-Herald*, 3/29/64)

March 30: Czechoslovakian news agency Ceteka reported Czechoslovakia would issue series of eight postage stamps in April bearing faces of 10 Soviet cosmonauts and American astronauts. (Reuters, *Houston Post*, 3/30/64)

- NASA awarded $158,466,800 contract to North American Aviation's Rocketdyne Div. for production of 76 F–1 rocket engines for the Saturn V launch vehicle. The action defined and detailed the procurement action initiated with letter contract in June 1962. Delivery of static-test engines to NASA Marshall Space Flight Center had already begun, and the first flight F–1 would arrive at MSFC in November. (NASA Release 64–48; *Marshall Star*, 4/1/64, 1)

- NASA's newly created Science and Technology Advisory Committee for Manned Space Flight convened for first time at John F. Kennedy Space Center, NASA. Formed to advise Dr. George E. Mueller, NASA Associate Administrator for Manned Space Flight, the committee met for briefings by NASA officials and tours of NASA facilities at Cape Kennedy and Merritt Island. (NASA Release 64–72)

- MSC operations at Cape Kennedy, renamed MSC–Florida Operations, was reorganized as "part of a broad NASA organizational realignment aimed at strengthening Gemini and Apollo management structures at Washington, Houston, and Florida." The organization was headed by G. Merritt Preston, responsible for MSC operations at the Cape since 1961. (KSC Release 43–64; MSC *Roundup*, 4/15/64, 8)

- U.S.S.R.'s unsuccessful attempts to send payloads to Venus Feb. 26 and Mar. 4 were reported in the U.S. press. Failures were attributed by U.S. military officials to difficulties with upper-stage rocket supposed to send the spacecraft out of parking orbit and onto its interplanetary course. (*Av. Wk.*, 3/30/64, 15; Finney, *NYT*, 4/1/64, 1, 10)

March 30: USAF announced the nine major modifications to Titan II changing it from ICBM to a man-rated launch vehicle for NASA's two-man Gemini spacecraft: addition of malfunction detection system; backup flight control system; redundant electrical system with changes for additional launch vehicle equipment; substitution of radio guidance for inertial guidance; elimination of retrorockets and verniers; new structure on second stage to hold new flight guidance and control equipment; new second-stage skirt assembly to join spacecraft; simplification of trajectory tracking; and redundancy in hydraulic systems. (AP, Balt. *Sun,* 3/31/64)

- MSC announced first meteorite ever found in Fisher County, Texas, was turned over to MSC for analysis by Mr. and Mrs. Bernard W. Neeper who found the four-pound meteorite two years ago on their farm. Meteorite investigations were made at MSC in cooperation with the U.S. National Museum. (MSC Release 64-61)

- George A. Fuller and Warrior Constructors, Inc., jointly received a $1,967,868 contract for construction of electronics, instrumentation, and materials laboratory at NASA Mississippi Test Facility. The Army Corps of Engineers awarded the fixed-price contract. (DOD Release 257-64)

- Former astronaut Lt. Col. John H. Glenn, Jr. (USMC) announced his decision to withdraw from the Democratic primary campaign for Senate candidacy in Ohio. Colonel Glenn was still convalescing from severe injury to his vestibular system in the inner ear sustained when he accidentally slipped and struck his head on a bathtub Feb. 26. In Houston, NASA Manned Spacecraft Center Director Dr. Robert R. Gilruth said: "The news of John Glenn's continued illness is upsetting. Our first concern is for his recovery. . . . We will be happy to discuss with John and the commandant of the Marine Corps his future if John so desires. . . . We wish him a speedy recovery." (AP, *NYT,* 3/31/64, 1, 18, 19; Burkett, *Houston Chron.,* 3/30/64)

- Dr. V. A. Bailey, prof. emeritus of physics at Univ. of Sydney, Australia, stated in the British journal *Nature* his theory that the sun has a large negative electric charge. He proposed that two satellites be launched around the sun, one clockwise and the other counterclockwise, to learn whether or not the sun had electrical charge and whether such a charge was negative or positive. Most physicists believed sun had no electrical charge. (Sci. Serv., *NYT,* 3/31/64, 21)

- Possibility that lasers might someday be used in preventing dental cavities and in repairing existing ones was reported by researchers Dr. Ralph H. Stern and Dr. Reidar F. Sognnaes of UCLA School of Dentistry at International Association for Dental Research meeting in Los Angeles. The investigators emphasized that their preliminary study dealt with effects of lasers on dental tissue and standard dental repair material. (Sci. Serv., *NYT,* 3/31/64, 24)

- Nicolai P. Erpilev, scientific secretary of the Astronomical Council of the Soviet Academy of Sciences, announced in Havana that the first Soviet satellite tracking station in Cuba would be installed in Havana in a special observatory and another one would probably be established later at Santa Clara. (*NYT,* 3/31/64, 26)

March 31: Unnamed "high State Department official" told *New York Herald Tribune* that talks between the U.S. and U.S.S.R. regarding Soviet cooperation in worldwide comsat network had been scheduled for June in Geneva. The official said the talks came in response to Soviet government's recent expression of interest in such negotiations to the U.S. Embassy in Moscow. Other sources indicated Soviet representatives had visited Communications Satellite Corp. in Washington and FCC inquiring about procedural arrangements. (Loory, *N.Y. Herald Trib.*, 4/1/64; *Av. Wk.*, 4/6/64, 15)

- USAF fired advanced Minuteman ICBM down the Atlantic Missile Range and 30 seconds later followed it with a Nike-Javelin sounding rocket to study the Minuteman's exhaust. This was first in 16-test series. (UPI, *NYT*, 4/1/64, 60; *M&R*, 4/6/64, 10)

During March: NASA Manned Spacecraft Center received design recommendations for 24-man Large Orbital Research Laboratory (LORL) that could be operational by 1968—the six-month study conducted by Douglas Aircraft's Missile & Space Div. and IBM's Federal Systems Div. The cylindrical space station would operate under zero gravity conditions in a 260-mi.-high orbit. (Maloney, *Houston Post*, 3/21/64)

- NASA Marshall Space Flight Center accepted the first of three new F-1 rocket engine test stands at NASA's engine test site at Edwards AFB, Calif. (*M&R*, 3/30/64, 17)

- Article in *Fortune* magazine said:

    "The U.S. is spending a lot of money—but it is also getting a lot of results in technological breakthroughs, any one of which would have been regarded as wondrous only a few years ago. The milestones passed since last summer indicate how fast the space program is now moving. . . .

    "The Soviet Union isn't going to collapse if it loses the race. If the U.S. loses, it isn't going to collapse either. For both countries, there are other goals in space that may yield rich rewards in both prestige and scientific advance. Meanwhile, NASA has much to be proud of. . . .

    "The superiority of liquid hydrogen as a rocket fuel has been proved and the U.S. has definitely overcome the Soviet lead in big boosters." (*Fortune*, 3/64)

- Results and new research of X-15 project were detailed in *Astronautics & Aeronautics* article by Thomas A. Toll, Chief of Research Div., NASA Flight Research Center, and Jack Fischel, head of FRC Manned Flight Control Branch.

    ". . . In the area of aerodynamics, problems resulting from the heat generated by high-speed flight were of most concern to designers at the time of the X-15 feasibility studies. . . .

    "At the time thermodynamic design of the X-15 was first contemplated, it was thought likely that fairly extensive regions of laminar flow might be achieved in flight, thereby alleviating the problem of heat protection as it would exist for an all-turbulent condition. This reasoning was based largely on rocket-model and wind-tunnel results available at the time. These results indicated a stabilizing effect on the boundary layer at hypersonic speeds for low ratios of the surface temperature to the stagnation temperature. As the design of the airplane progressed, it became evident that the probability of obtaining laminar flow would be greatly reduced by the presence of unavoidable surface irregularities.

As a basis for the aerodynamic heating design, a relatively low value of the transition Reynolds number ($R_r = 100,000$) was therefore stipulated. Flight results have generally confirmed this design approach, inasmuch as it has been found that at the lower altitudes, where the heating rates are highest, the boundary layer is largely turbulent, due primarily to the surface irregularities. Regions of laminar flow generally are limited to the immediate vicinity of the leading edges and are quite small. The conditions found to exist on the X-15 are consistent with results of refined wind-tunnel techniques now being used. . . .

"Maximum structural temperatures have been experienced during X-15 operation at high speeds and comparatively low altitudes, with accompanying normal-acceleration values much lower than those experienced during recovery from high altitudes. . . .

"By way of assessing the [structural] failures experienced, there appear to have been no cases in which the structure deteriorated to such a degree as to seriously endanger the integrity of the airplane. It is entirely probable, however, that had a less-conservative design been followed the problems noted might have resulted in greater deterioration with more serious consequences. . . ." (*A&A*, 3/64, 20–28)

During March: Martin-Denver revealed design concept of its Astrorocket, a proposed reusable aerospacecraft. Launched vertically, the vehicle would have two stages—both manned—equipped with auxiliary fan jets as well as rocket engines. After stage separation, first stage would re-enter atmosphere and land on airstrip like conventional airplane. Second stage, carrying payload, would also land at air base after completing mission. Astrorocket concept employed parallel staging rather than conventional bottom-to-top staging. Martin said advantages of the Astrorocket were its reusability and its capability to operate with existing air bases for space operations. (*A–N–AF J&R*, 3/7/64, 38)

- FCC ruled AT&T must confine its use of channels in undersea cables to voice and related service, and leave the services of non-voice and combined voice/non-voice to smaller message companies. Although the decision applied to undersea cables, there was speculation that it could affect future commercial communications satellites. As one FCC aide said, "The rationale of the decision would seem to extend to any type of facility." (Denniston, Wash. *Eve. Star*, 3/20/64)

- Brig. Gen. Joseph S. Bleymaier (USAF) was appointed to newly established position of Deputy Commander for Manned Systems, AFSC Space Systems Div. In this capacity he would be responsible for all aspects of the Manned Orbiting Laboratory program. (*SBD*, 3/13/64, 399; *M&R*, 3/23/64, 9)

April 1964

April 1: Fourth anniversary of TIROS I, first meteorological satellite, orbited with a Thor-Able booster from AMR. The spectacular accomplishments by the experimental TIROS weather satellites were to develop into a global cloud-cover reporting system.

- NASA launched Aerobee 150 sounding rocket from White Sands, N. Mex., with two cameras and two spectrographs to obtain spectra of nebulosites of certain star fields. Rocket attained 118-mi. altitude and instrumentation performed properly, but failure of the attitude control system prevented the rocket from pointing at the targets and there were no experimental results. (NASA Rpt. SRL)

- Five men emerged from 30 days of isolation in sealed chamber where they conducted first U.S. experiment integrating all systems for self-contained life support in space. The test was made in Seattle by the Boeing Co. for NASA. NASA Director of Biotechnology and Human Research Dr. Eugene B. Konecci called the successful experiment a "major breakthrough" in research on life support systems for extended space flights.
    "The basic principles and techniques of life-support systems have been proved workable and sound. Now we refine and advance them. We are already looking and preparing far beyond." (AP, Balt. *Sun*, 4/2/64; AP, *Houston Chron.*, 4/2/64)

- NASA Lewis Research Center was investigating several techniques for producing flexible, thin-film solar cells. One of the techniques being investigated was sputtering, which required absolutely clean atmosphere and a low pressure of inert gas. Sputtering apparatus in which controls were fully automated—rigidly controlling pressure, temperature, and voltage in relation to each other—had been built at Lewis. With this equipment, researchers were attempting to increase efficiency of thin-film cells to that of silicon cells. (LRC Release 64-28)

- Vice Adm. Rufus E. Rose (USN, Ret.) was sworn in as Director of Policy Planning Div., NASA Office of Technology Utilization and Policy Planning, by NASA Deputy Administrator Dr. Hugh L. Dryden. The most recent assignment during Admiral Rose's 44-year career was as Commandant of the Industrial College of the Armed Forces. (NASA Release 64-73)

- Picket lines were set up at the entrances to Cape Kennedy and the Merritt Island Launch Area (MILA) by United Plant Guards Workers of America. The union charged that Wackenhut Corp., subcontractor to TWA at the Saturn/Apollo site, was employing nonunion guards. 1,886 of 2,597 construction workers at the Saturn V complex refused to cross the picket lines, and about 700 of 2,500 workers at the USAF Titan III site were absent. William Simkin, vice chairman of the President's Missile Sites Labor Commission, requested that the building trades unions return to work. (UPI, *Wash. Post*, 4/2/64; AP, Balt. *Sun*, 4/2/64; UPI, *NYT*, 4/3/64)

April 1: AEC conducted test of Snap-10A reactor to see what would happen if it were to fall into the sea after an aborted space launch attempt. The Snap-10A reactor was placed in concrete water tank and the mechanism to absorb neutrons—to control the chain reaction—was removed, causing a "runaway." A wall of water 16 ft. wide erupted 40 ft. into the air, due either to sudden heat increase in the "runaway" or the contact of water with the reactor's liquid metal coolant (sodium potassium). AEC spokesman said core of the reactor was damaged as expected, and test was completed with no hazard to personnel or the public. Another test would be made to investigate consequences of a ground landing. (AP, Balt. *Sun*, 4/2/64)

- NASA Administrator James E. Webb addressed the American Assembly Student Conference at the Air Force Academy, Colorado Springs:

    ". . . in assessing the future one can always learn from the past, and if the thoughtful student of history learns anything from a review of the march of human progress, it is that the ultimate potential of contemporary undertaking is almost never perceived.

    "So, I believe, it is with research and exploration in space. . . .

    "Today, 500 years after Prince Henry and his use of the caravel to initiate the Oceanic Age, 400 years after Copernicus presented his theories about the solar system, 300 years after Newton worked out the formula for space flight, but less than 40 years after Goddard devised the vehicle to make it possible, man has left the earth, and widened his horizons to include the entire solar system.

    "Science today has at its disposal a modern caravel, the chemical rocket, which has already traveled to the moon and Venus. And, as with Henry and his explorations, we have improved navigation systems to guide them out and bring them back to make steadily more accurate measurements of the environment of space—that environment through which Gargarin and Glenn and others traveled, and through which our own planet, Mother Earth, is also traveling around the sun at the relative speed of 67,000 miles per hour.

    "And to complete the parallel, just as Prince Henry inaugurated an Age of Discovery which led man to all the seas and continents of the earth, and into the atmosphere up to twice the height of his tallest mountains, so have we today embarked on a new Age of Discovery in which there is no foreseeable horizon. . . ." (Text; *CR*, 4/6/64, 6686)

- Atlas ICBM launched by USAF from Cape Kennedy sent new deceptive warhead 5,000 mi. downrange and carried two piggyback rockets which were fired from a 7-ft. pod attached to the Atlas. The small rockets carried radiometers to measure radiation in the Atlas engine exhaust. Another attached pod contained scientific experiments to study the ionosphere, magnetosphere, and radio frequency noises. (AP, *Wash. Post*, 4/2/64; *M&R*, 4/6/64, 10)

- According to Hebrew Univ. sources, Dr. Menahem Steinberg of Hebrew Univ. in Jerusalem had developed new oxidizer for use in solid-fuel rocket engines, said to be superior to those in general use because the gases produced by combustion would be of a low molecular weight and so would increase propulsion power. The sources said two years would be required to develop and confirm the new discovery. (Reuters, Balt. *Sun*, 4/2/64)

April 1: At zero hour GMT, astronomers in 10 nations, including the U.S., turned back their atomic clocks a tenth of a second to adjust them to time measured optically by the passage of the stars. (Plumb, *NYT*, 4/1/64, 33)

April 2: U.S.S.R. launched ZOND I space probe "for the purpose of developing a space system for distant interplanetary flights." Final stage of "improved booster rocket" placed earth satellite into parking orbit; then a rocket took off from the satellite and propelled the ZOND I to escape velocity, sending it into flight trajectory "close to the computed one." There was no indication as to what the computed course was, and no details of the probe's weight, size, or destination were given. (Tass, *Krasnaya Zvezda*, 4/3/64, 1, ATSS–T Trans.; CSFC *SSR*, 4/15/64)

- In letter to chairmen of Senate and House space committees—Sen. Clinton P. Anderson (D.–N. Mex.) and Rep. George P. Miller (D.–Calif.)—NASA Administrator James E. Webb detailed findings of committees investigating failure of RANGER VI lunar probe. Although the committees had not yet identified any specific cause for failure, Mr. Webb's letter outlined five major faults in the spacecraft:

    1. The two onboard TV systems "were more complex than required and were not completely redundant. They included a number of common components in which a single failure would lead to disablement of both television systems."
    2. "Possibilities of failure . . . increased as a result of practices employed in the design and construction of the spacecraft. . . ."
    3. Preflight ground testing "may have obscured potentially dangerous situations which could have enhanced accidental triggering of critical control circuits."
    4. The antenna for transmitting the lunar photographs had never been tested with the TV transmitters.
    5. "Because of reluctance to risk possible damage to the space vehicle, pre-launch systems verification was not complete. . . ." (Loory, HTNS, *Wash. Post*, 4/3/64, 8)

- Boilerplate Apollo spacecraft was mated to Saturn I SA–6 at Cape Kennedy in preparation for flight later this spring. (*Marshall Star*, 4/8/64, 1)
- Labor dispute continued at Cape Kennedy, with 1,973 of 4,068 construction workers at the Cape and MILA refusing to cross picket lines of the United Plant Guards Workers Union. National Labor Relations Board began investigating the situation, which was further complicated by failure of Iron Workers Union and Patrick AFB Contractors Association to work out a new contract, causing about 400 ironworkers to walk off the job. (UPI, *NYT*, 4/3/64)
- With the publication of the name change from Cape Canaveral to Cape Kennedy in the official decision list of the U.S. Board on Geographic Names, the legal procedure involved in renaming the Cape was completed. (AP, *NYT*, 4/3/64)
- Vice Adm. William F. Raborn, Jr. (USN, Ret.), told industry delegates to conference at Georgia Tech that primary purpose of nuclear and space exploration was "the maintenance of national defense to guarantee a continuation of our way of life." Technology assures "not so much a better texture for our bread but the fact that we will have that bread to eat," he maintained. Conference was sponsored by Southern Interstate Nuclear Board and Georgia Tech's School of Nuclear Engineering in cooperation with AEC and NASA. (Rutherford, *Atlanta Const.*, 4/3/64)

April 3: U.S.S.R. announced several radio communication sessions were held with the ZOND I probe, which appeared to be functioning normally. When the probe attained predetermined position a special power guidance system was switched on to impart additional velocity to the probe. Programing data relative to direction of ZOND I's orientation of its axes and on the continued operation of the powered guidance system were transmitted to the probe by radio from earth when it reached 560,000 km. (347,967 mi.). The probe was said to be in a flight trajectory corresponding "with a high degree of accuracy" to the predetermined program. (Tass, *Krasnaya Zvezda*, 4/5/64, 1)
- NASA awarded $5,944,000 supplemental contract to the Boeing Co. for additional R&D and planning in the manufacture of Saturn V first-stage rocket at Michoud Operations, La. (New Orleans *Times-Picayune*, 4/4/64; *Marshall Star*, 4/8/64, 10)
- DOD announced Army Corps of Engineers had awarded $4,397,000 fixed price contract for construction of propellant facilities at NASA Mississippi Test Facility to Broadway Maintenance Corp. and Glantz Contracting Corp.; and $3,573,025 fixed price contract for construction of high-pressure water heating facility at MTF to Leslie Miller, Inc., Power Engineering Co., and Donover Construction Co. (DOD Release 278–64)
- Dr. Hugh L. Dryden, NASA Deputy Administrator, received honorary degree in mechanical engineering from the Polytechnic Institute of the University of Milan. (NASA Release 64–74)
- Lt. Col. John H. Glenn, Jr., left the hospital and returned home to Seabrook, Tex., where he would convalesce from inflammation of his inner ear. (AP, *Wash. Post*, 4/4/64)
- Secretary of the Air Force Eugene M. Zuckert said in address at sixth Air Force Academy Assembly:

    "It is not possible to predict the course of events, but there are certain points I want to make tonight that might serve as benchmarks.

    "First, the nation has no choice but to pursue the almost limitless possibilities of space with energy and imagination.

    "Second, we must do everything possible to prevent exploitation of space for aggressive purposes, and be prepared to defend freedom in space.

    "Third, the emerging defense responsibilities in space exemplify and extend the requirements for professionalism in the military, focusing sharply on the mutual responsibility of the military and civilian professionals in science, industry, and government.

    "Fourth, despite the appeal of space exploration, we must continue to increase our investment of brains and money in the effort to build a a decent life on earth if this nation is to continue as leader of an expanding free world on the earth." (Text)
- Sen. William Proxmire (D.–Wis.) said GAO study showed that USAF could save more than $1 million a year in administrative costs at AMR through consolidation of supply management functions. (AP, *Houston Chron.*, 4/3/64)
- Atlas F ICBM exploded on launch pad at Vandenberg AFB, Calif. (*M&R*, 4/13/64, 11)

April 4: U.S.S.R. announced routine launching of COSMOS XXVIII. Initial orbital data: apogee, 395 km. (245 mi.); perigee, 209 km. (130 mi.); period, 90.38 min.; and inclination, 65° to the equator. Scientific equipment onboard the satellite was "intended for the further exploration of outer space in accordance with the program announced by Tass March 16, 1962." In addition, the satellite contained radio transmitter and telemetering system. All instrumentation was functioning normally. (Tass, *Krasnaya Zvezda*, 4/5/64, 1, ATSS–T Trans.)

* At 18 hours Moscow time ZOND I reached 837,000 km. (520,097 mi.) from the earth, Tass announced. At that time the probe was in the following position: right ascension, 5 hrs. and 56 min.; inclination, minus 4° 22 min. Experimentation with ZOND I continued. (Tass, *Krasnaya Zvezda*, 4/5/64, 1; ATSS–T Trans.)
* Temporary injunction forbidding United Plant Guards Workers of America to picket at Cape Kennedy and Merritt Island was issued by U.S. District Judge George C. Young at request of the National Labor Relations Board. (UPI, *Wash. Post*, 4/5/64.)
* Reporting its findings after series of hearings, Senate Subcommitte on Employment and Manpower forecast "a leveling off or actual decline" in the present $55 billion annual expenditure for weapons and space programs. Among the report's predictions: thousands of scientists and engineers may have to seek jobs outside the aerospace industry; thousands of skilled workers may have to learn now trades; at least 10 states should plan now for converting to non-government markets. Sen. Joseph S. Clark (D.–Pa.), subcommittee chairman, said studies showed that "nearly one-tenth of the Nation's work forces is employed in defense-related enterprise." (Haakinson, AP, *Wash. Post*, 4/5/64)
* Industry sources revealed only one more Ranger lunar probe could be launched before NASA's contract with the Jet Propulsion Laboratory expired in December. If Ranger 7 failed, the sources said NASA might remove JPL from Cal Tech's management, might remove Director William H. Pickering from his post, and might hire a "hard-headed businessman" to manage the facility. (AP, *Boston Globe*, 4/6/64)

April 5: English-Russian astronautical dictionary was issued in Moscow, Tass reported. (AP, *NYT*, 4/6/64)

April 6: NASA launched Aerobee 150 sounding rocket from White Sands, N. Mex., with two cameras and two spectrographs to obtain spectra of nebulosities of certain star fields. Abnormal coning motion of the rocket and subsequent pitch-roll coupling caused the rocket to reach only 46-mi. altitude; and the attitude control system had no chance to point at the programed targets, so no experimental data were obtained. (NASA Rpt. SRL)

* First large batch of unusual stones arrived from Iowa at NASA Goddard Space Flight Center for analysis in Project Moon Harvest. Residents of six Iowa counties were gathering the specimens, which GSFC scientists were analyzing to determine if they were of lunar origin. Dr. John A. O'Keefe, project scientist, theorized that pieces of the moon's surface were chipped off the moon by meteorite impact and attracted by gravity to the earth. (*Goddard News*, 4/20/64, 4; GSFC Release G–13–64)
* About 450 iron workers stayed off the job at Cape Kennedy because union and management negotiators had failed to work out a new contract. Most of the other 3,500 construction workers returned to their jobs after

picket lines kept workers away the previous week. (UPI, *Chic. Trib.*, 4/7/64)

April 6: By this date USAF had made 21 manned drop-tests of the ballute at El Centro, Calif., from C-130 flying at altitudes up to 35,000 ft. Basically a drag balloon, the ballute was under development by Goodyear Aerospace Corp. with NASA funds. To be installed in the ejection seat system of Gemini spacecraft, the self-inflating ballute would stabilize the astronaut during free fall down to 10,000 ft., where final recovery parachute would deploy. (Yaffee, *Av. Wk.*, 4/6/64, 40–45; MSC *Roundup*, 4/15/64, 8)

* Researchers at Lankeneau Hospital in Philadelphia had concluded that orthostatic hypotension such as occurred in Astronauts Schirra and Cooper after orbital space flights could be caused as much by dehydration as by weightlessness, *Missiles and Rockets* reported. (*M&R*, 4/6/64, 25)

* Glide parachute achieved glide ratio of up to 2:1 in wind-tunnel tests at NASA Ames Research Center, it was reported. Northrop Ventura, the parachute developer, was proposing it to NASA for later phases of Project Gemini. (ARC; *M&R*, 4/6/64, 25)

* Of the formerly operational Thor missiles that were being converted to space missions, NSAF spokesman said: "This program has proved that it is possible to take a complex vehicle that was designed as a weapon and convert it to peaceful uses at a substantial saving and without sacrificing reliability. We can expect that there will be more than this." AFSSD estimated converted booster cost up to $150,000 less than new Thor booster. (*M&R*, 4/6/64, 31)

* USAF was considering developing new 40,000-lb. thrust third stage for Titan III launch vehicle, *Aviation Week* reported. Burning high-energy fuels, the new stage would replace the 16,000–17,000-lb.-thrust transtage. Meanwhile USAF asked six firms for proposals for a back-up and/or replacement rocket engine system for the transtage, said to be experiencing development difficulties. But USAF was not asking for upgrading of the 16,000-lb.-thrust system at this time. (*Av. Wk.*, 4/6/64, 13; *SBD*, 4/10/64, 59)

April 6–8: U.S. officials met in London with representatives of six European countries for further talks on investing in the global communications satellite network. Spokesman said there was "substantial progress" but no concrete agreements were reached, and another meeting would take place in about a month. (Kohlmeier, *WSJ*, 4/6/64; *NYT*, 4/9/64)

April 7: President Johnson ordered prompt removal of NASA tracking station from Zanzibar shortly after President Abeid Karume requested that the U.S. remove it before the end of this month. President Karume made the request to Frank Carlucci, U.S. Charge d'Affaires in Zanzibar, and reportedly told Carlucci that the action was a result of statement by U.S. Ambassador-to-Kenya William Atwood at New York press conference that Red China and East Germany were trying to turn Zanzibar "into a kind of non-African state to be used as a staging base for political maneuvers" on the African continent. NASA announced in Washington that the loss of the station would have no adverse effect on the U.S. manned space program. The station had been on a stand-by basis since end of Project Mercury, would have been of "minimum use" in Project Gemini and not used at all in Project Apollo. (Simons, *Wash. Post*, 4/8/64; Smith, *NYT*, 4/8/64, 11; Sehlstedt, *Balt. Sun*, 4/8/64)

April 7: NASA anounced closing the command and control station for Relay communications satellites at Nutley, N.J., and moving the operation to the Space Tracking and Data Acquisition Network (Stadan) at Blossom Point, Md. The move was made to consolidate personnel and equipment and save operation costs. (NASA Release 64-77)

- Soviet news agency Tass announced that a human being had stayed for 25 days in pressurized cabin, breathing a combination of oxygen and helium. Test results were said to be good, although a test subject with a bass voice became a tenor and scientists were still testing for the right proportion of helium to use. Reason for using the mixture was to enable future space travelers to "better endure pressure fluctuations and high temperatures in a space ship cabin," Tass said. (AP, *NYT*, 4/8/64)
- House Committee on Government Operations' Military Operations Subcommittee heard testimony from NASA witnesses regarding military communications satellites. In response to questioning, Leonard Jaffe, Director of Communications and Navigation Programs in NASA Office of Space Science and Applications (OSSA) and Robert F. Garbarini, Director of Applications in OSSA, said NASA had not participated in current DOD-ComSatCorp negotiations for possible DOD use of commercial comsat channels. ComSat Act of 1962 designated NASA as technical advisor to ComSatCorp. "We are aware of the negotiations and we expect to be informed so that we can respond under the act," Jaffe said. (*Av. Wk.*, 4/13/64, 32)
- Striking United Iron Workers forced contractors to lay off 500 other workers at Cape Kennedy. Army Corps of Engineers spokesman said the walkout was "definitely crippling some of the most crucial of our projects and three good size jobs are completely shut down." (UPI, *NYT*, 4/8/64)
- Lt. Col. Yuri A. Gagarin said in Moscow press conference that he and other Soviet cosmonauts were training for new space flights. "When we go and who will go—that we do not know. We will wait and see." (AP, *NYT*, 4/8/64, 10; AP, *Wash. Post*, 4/8/64, C6)

April 8: Titan II launched unmanned Gemini spacecraft into orbit in first Project Gemini flight (GT-1), a test of Titan II launch vehicle system, Gemini spacecraft structural integrity, and spacecraft-launch vehicle compatibility. After an uninterrupted countdown at Cape Kennedy, the Titan II lifted off and placed the spacecraft in orbit of 204-mi. apogee, 99.6-mi. perigee, and 89.27-min. period. Walter C. Williams said all systems functioned "well within manned tolerances." Only imperfection in the flight was Titan II's 14-mph excess speed (17,534 instead of 17,520 mph), sending spacecraft 21 mi. higher than planned. However, this was within tolerance and, on manned flight, crewmen could easily correct the extra height. The 7,000-lb. spacecraft was not separated from the spent rocket casing, and NASA officials estimated the orbiting assembly would re-enter the atmosphere and disintegrate in about 3½ days. (NASA Release 64-70; MSC *Roundup*, 4/15/64, 1, 3)

- X-15 No. 1 flown by Capt. Joe H. Engle (USAF) to maximum speed of 3,477 mph (mach 4.8) and altitude of more than 33 mi. in test to prepare the pilot for future flights at higher altitudes. The flight also conducted optical degradation studies. (FRC Release; UPI, *Wash. Post*, 4/9/64)

April 8: Test of parasail for space craft landings was conducted by MSC officials in Galveston Bay. The test was termed only a partial success. A model Gemini spacecraft was dropped from C–119 flying at 10,200 ft., and the drogue parachute opened perfectly. But parasail failed to deploy properly after its nylon lines snapped, and engineers had to trigger the reserve ringsail parachute to lower the vehicle safely to the water's surface. (*Houston Post,* 4/9/64; MSC *Roundup,* 4/15/64, 8)
- Soviet scientist Sergei Vernon said at Moscow press conference that the ZOND I space probe was launched to study radiation hazards in space. The probe would "provide information about more remote areas of space we know little about." (Balt. *Sun,* 4/8/64)
- NASA Technology Utilization Division and the Food and Drug Administration reported that preliminary experiments by FDA's Dr. Jacqueline Verrett indicated that a piezo-electric transducer invented at NASA Ames Research Center could, for the first time, monitor the heartbeat of a chick embryo quickly, continuously, quantitatively, and without egg damage. Instrument was invented to measure impact of micrometeoroid dust in space by Ames' Vernon L. Rogallo, who suggested its potential in chick embryo research. Through its Technology Utilization program, NASA informed FDA of the innovation. FDA had been conducting research during past four years using chick embryo technique, correlated with animal tests, to study effects of drugs, pesticides, and food additives. (NASA Release 64–76)
- Ironworkers who had walked off the job at Cape Kennedy and Merritt Island Launch Area (MILA) when their contract expired March 31 began returning to work at the request of the Missile Sites Labor Commission. Contract negotiations continued in Washington. Construction projects that had been delayed by the walkout included three MILA buildings of the Kennedy Space Center, NASA, the Saturn Vertical Assembly Building, as well as facilities for USAF Titan III, Army Corps of Engineers said. (UPI, *Wash. Post,* 4/9/64; *M&R,* 4/13/64, 10–11)
- Gov. Nelson Rockefeller said in speech at Johns Hopkins Univ.:
    "I believe strongly that it is time for the American people to take a long, hard look at the political, military and scientific merit of the 'one-shot,' crash program to place a man on the moon by 1970.
    "I think we ought to put a man on the moon. But we must make sure that our program is sound, that our priorities are proper—that we are, in short, taking maximum advantage of the virtually unlimited opportunities which the scientific and technological revolution of our times presents us—that we are not warping our entire scientific effort for this one purpose.
    "I fear that the present moon-shot program of the Democratic administration in Washington does not meet the criteria for a soundly conceived program of scientific research and development in the national interest...." (Text)
- Soviet military newspaper *Krasnaya Zvezda* (*Red Star*) announced that Lt. Col. Yuri Gagarin had been assigned head of its new "Cosmonauts' Desk" (AP, *Wash. Post,* 4/9/64)

April 9: EXPLORER IX re-entered the earth's atmosphere and disintegrated after more than three years in space. The lightweight, 12-ft.-diameter balloon was NASA's most effective satellite for measuring atmospheric density and temperature. Placed in orbit Feb. 16, 1961, EXPLORER IX traveled more than 340 million mi. during its 14,000 orbits of the earth. It provided a better understanding of upper atmosphere characteristics, determining more precisely the relationships between air density and solar radiations. Measurements of the satellite's atmospheric drag effects were correlated with those of a similar inflated sphere, EXPLORER XIX, orbited Dec. 19, 1963. EXPLORER IX was the first satellite orbited by a solid-fuel launch vehicle (Scout) and the first satellite launched at Wallops. (NASA Release 64–81; Wallops Release 64–35)

April 9: NASA announced it would negotiate with RCA Data Systems Div. for purchase of 19 additional ground computer systems for checkout, static tests, and launch of Saturn IB and Saturn V launch vehicles. This would complete the required 26 computers for the Saturn/Apollo program; seven systems were ordered last year. Total cost was expected to exceed $47 million. (NASA Release 64–79)

- Titan II ICBM was launched on successful USAF test flight down the Atlantic Missile Range. This was the 33rd and final R&D firing of the Titan II. (UPI, *Wash. Post*, 4/10/64; Witkin, *NYT*, 4/12/64, 31)
- French Foreign Ministry announced France would establish observation station in the Azores under an agreement with Portugal. The installation would give France a clear Atlantic firing range of 1550-mi. for its missiles and rockets launched near Biscarrosse on the coast of southwest France. (Reuters, *Wash. Post*, 4/10/64)
- Dr. Jeanette Piccard, widow of balloonist and atmospheric scientist Dr. Jean Felix Piccard, was appointed a consultant to NASA Manned Spacecraft Center in formulation of effective program to provide information to U.S. scientific community. (MSC Release 64–66)
- Tass reported monument to mark man's first space flight would be erected in Moscow. The monument would consist of bust of Maj. Yuri A. Gagarin and a meteorite of white metal, both the work of Lev Kerbel, attached to metal pylon about 80 ft. high. (Reuters, *NYT*, 4/10/64)

April 10: NASA launched 250-lb. instrumented Ram (Radio Attenuation Measurement) spacecraft on a ballistic trajectory from Wallops Island, Va., on a three-stage rocket system. Experiment was one of series to investigate the problem of communications through the ionized plasmas created when a spacecraft re-enters earth's atmosphere at high velocities. Spacecraft was to attain top speed of 18,400 fps. Preliminary data indicated the vehicle and payload performed satisfactorily; the flight lasted about 10 min. and impact occurred approximately 650 mi. east of Bermuda. Flight measurements, telemetered to ground receiving stations would be reduced and analyzed at Langley Research Center, which designed and built the spacecraft. Ram series was under LaRC direction for NASA's Office of Advanced Research and Technology. (NASA Release 64–65; Wallops Releases 64–29 and 64–34)

- Dr. John Strong, Director of Johns Hopkins Astrophysics and Physical Meteorology Laboratory, announced definite evidence of water vapor around and on the planet Venus. His determination was made as the result of a Feb. 21 daytime balloon flight to study Venus by means of a unique robot telescope system. The study was made under research grant for USAF Cambridge Research Laboratories. "Since it is known that carbon dioxide exists on Venus, proof of water vapor forces us to re-examine every previous calculation made concerning the possibility of some sort of life existing on the planet," Dr. Strong said. Amount of water vapor in the air above the Venusian cloud level was established with a relative precision of 5%. Launched from Holloman AFB, N. Mex., the balloon carried the telescope to altitude of 87,000 ft. (OAR Release 4-64-2)

April 10: Iron meteorite landed in two pieces in Muzzaffarpur, India. One piece was later sent to U.S. for analysis by E. P. Henderson, curator of meteorites for the Smithsonian Institution, and by the Brookhaven National Laboratories. (*Wash. Post*, 5/22/64)

- United Plant Guards Workers of America members who had picketed Cape Kennedy and Merritt Island Launch Area agreed to six-month suspension of picketing while negotiators continued trying to settle the dispute. Informed of the agreement, District Judge George C. Young dismissed the temporary anti-picketing injunction against the guards which had been obtained by the National Labor Relations Board on behalf of NASA. (AP, Wash. *Sun. Star*, 4/12/64)

- Airport towers at 277 airports reported record 31 million takeoffs and landings during 1963, FAA statistics showed. This was almost double the 16.8 million operations 10 years ago and was a 10% gain over the 28.2 million operations in 1962. (FAA Release 64-33)

April 11: President Johnson announced at White House press conference:
"The world record for aircraft speed, currently held by the Soviets, has been repeatedly broken in secrecy by the United States aircraft A-11. The President has instructed the Department of Defense to demonstrate this capability with the procedure which, according to international rules, will permit the result of the test to be entered as a new world record.

"The Soviet record is 1,665 miles an hour. The A-11 has already flown in excess of 2,000 miles an hour." (Transcript, *NYT*, 4/12/64, 64)

- NASA Nike-Apache sounding rocket launched from Ft. Churchill, Canada, with Univ. of Alaska instrumented payload attained 107-mi. altitude and obtained data on the visible aurora and on vacuum ultraviolet emissions. This information was supplemented by data from two electron particle detectors. Three photometers and two vacuum ultraviolet detectors also were included in the payload. (NASA Rpt. SRL)

- Former President Dwight D. Eisenhower said in *Saturday Evening Post* article:
"As a Republican I believe in what has been called the 'conquest of space,' but which I prefer to think of as a thorough exploration of a new scientific frontier to improve our living here on earth. This is an exploration we should pursue vigorously with a step-by-step program in line with our means and our needs. That was the aim of the space program as initially undertaken during my Administration: On the advice of eminent scientists we conceived a long-range effort, not a stunt, with costs to be stabilized around two billion dollars a year. That hardly could be called picayune by any reasonable standards.

"But now, under the Democrats, this program has been blown up out of all proportion. With hysterical fanfare our space research has been presented as a crash effort, as a 'race to the moon' between the United States and Russia which we must win at all costs. And the costs are tremendous: They now are running well over five billion dollars a year. The Government now has more than 73,000 engineers and scientists working on the nonmilitary space program, either on the federal payroll or employed under contract. This swollen program, costing more than the development of the atomic bomb, not only is contributing to an unbalanced budget; it also has diverted a disproportionate share of our brainpower and research facilities from other equally significant problems, including education and automation.

"We are breezily assured that the cost and dislocation brought about by this moon race are worthwhile for the new 'prestige' they will bring us. There is no way of telling how true that may be, but we can be sure of one thing: The voyage to the moon will set a new record for a trip taken on borrowed money." (*Sat. Eve. Post*, 4/11/64, 17–19)

April 12: Cosmonautics Day in the Soviet Union, commemorating the day when Yuri Gagarin made the first space flight in 1961.

U.S.S.R. announced steerable spacecraft POLET II had been placed in orbit. Purposes of the launching were said to be to "further improve space vehicles capable of maneuvering extensively in all directions, and to work out questions concerning the problem of space rendezvous." After being placed in initial orbit, POLET II was maneuvered in various directions by means of special engine system. "As a result of one of these maneuvers in the area of the equator, the space vehicle substantially changed the angle of inclination of its orbital plane." After the programed space maneuvers, POLET II was in the following orbit: apogee, 500 km. (311 mi.); perigee, 310 km. (193 mi.); period, 92.4 min.; and inclination to the equator, 58.06 degrees. (Tass; Shabad, *NYT*, 4/13/64, 4)

- Gemini spacecraft launched into orbit in unmanned test April 8 (GT–1) re-entered the earth's atmosphere and disintegrated midway between South America and Africa. Attached to the 11-ft.-long spacecraft was 20-ft.-long cylindrical rocket stage. They had been orbiting the earth at altitudes from between 100 and 190 mi. (AP, *Wash. Post*, 4/14/64; AP, *Houston Post*, 4/14/64)

April 13: At Manned Spacecraft Center press conference, MSC Director Dr. Robert R. Gilruth announced Astronauts Gus Grissom and John Young would be prime pilots for the first manned Gemini space flight (GT–3), with Astronauts Walter Schirra and Tom Stafford as backup pilots. Also at the press conference it was disclosed that Astronaut Alan B. Shepard, Jr., was suffering from labyrinthitis. MSC's Dr. Charles Berry emphasized that Shepard's inner ear inflammation was not caused by space flight but by infection that spread into his inner left ear. (MSC Press Conf., MSC Historian)

- NASA announced it would negotiate contract with Space Technology Laboratories, Inc., for two additional Orbiting Geophysical Observatories (Ogo). Under existing contract, STL was designing, developing, fabricating, and testing three Ogo's, the first of which was scheduled for launch this year. Contract for additional two Ogo's was expected to total about $17 million. (NASA Release 64–83)

- RCA announced it had received $22 million subcontract from Grumman Aircraft Engineering Corp. for communications system to be used in Project Apollo lunar landing. The system would transmit continuous voice and telemetry data to earth, provide communications between the astronauts on the moon and Lunar Excursion Module (LEM) and Apollo command module. (Phil. *Eve. Bull.*, 4/14/64)

April 13: President's Advisory Committee on Supersonic Transport, under chairmanship of Defense Secretary Robert S. McNamara, held its first meeting. The Committee reviewed status of U.S. supersonic transport program and began discussing major issues involved in the program. (DOD Releases 295-64, 303-64)

- Lockheed-California Co. announced it had developed concept for space shuttle vehicle that could be available in 1966. The "space tug" was designed primarily as spacecraft-to-spacecraft cargo transportation system and was result of study sponsored by AFSC. (*SBD,* 4/14/64, 75; *Av. Wk.,* 4/20/64, 59)
- Invention of pocket-size x-ray unit, smallest in the world, was announced by IIT Research Institute. Device was invented by Dr. Leonard Reiffel, vice president of the Institute, and developed by James J. Ezop, Institute physicist. Besides uses by doctors for home calls, Dr. Reiffel said the instrument would be useful in testing mechanical parts in aerospace hardware. (*Chic. Trib.,* 4/14/64)

April 14: Atlas D launch vehicle sent NASA's 200-lb. Project Fire spacecraft more than 500 mi. into space from Cape Kennedy in test to provide data on re-entry heating of spacecraft returning from the moon. As the spacecraft fell back to earth, a solid-fuel Antares II motor ignited for 30 sec. to hurl it through the atmosphere at the speed of 37,800 fps (about 26,000 mph). With exterior heating at estimated 20,000 degrees, instruments in spacecraft interior relayed data back to earth while cameras and other instruments recorded the "flaming meteor" descent. About 32 min. after launch the Project Fire craft impacted in the Atlantic more than 5,200 mi. from the launch site. Dr. Raymond L. Bisplinghoff, NASA Associate Administrator for Advanced Research and Technology, commented that the experiment, conducted by NASA Langley Research Center, "recorded the highest speed ever reached by a manmade object in free flight." In conjunction with the Project Fire test, NASA Nike-Apache sounding rocket was launched to 97-mi. altitude from Ascension Island. Its instrumented payload, provided by Univ. of Michigan, measured pressure, temperature, and density in the equatorial atmosphere using pitot-static probe technique. (NASA Release 64-69; AP, *Wash. Post,* 4/15/64; *Cinc. Enqr.* 4/16/64, NASA Rpt. SRL)

- NASA Aerobee 150 sounding rocket was launched from White Sands, N. Mex., with JPL payload to measure ultraviolet light and fluctuations in earth's magnetic field. Rocket started large yaw maneuvers soon after 30 sec. and reached only 18.2-mi. altitude; no new scientific data were obtained. (NASA Rpt. SRL)
- A Delta rocket's third-stage motor (X-248) inexplicably ignited in checkout building at Cape Kennedy, burning 11 engineers and technicians, three critically. The rocket stage had just been mated with Orbiting Solar Observatory (Oso) in preparation for prelaunch spin testing, when the motor suddenly ignited, rose from the alignment stand, hit the ceiling knocking off the Oso, and landed in a far corner. NASA and USAF immediately began investigating the cause of the ignition. USAF said this was the "first serious incident of its kind in 14 years of operations and over 1,400 missile launchings" at the Cape. (AP, Wash. *Eve. Star* 4/14/64; AP, *Wash. Post,* 4/15/64; *Goddard News,* 4/20/64, 3; *Marshall Star,* 4/29/64, 1, 3)

April 14: NASA had developed lightweight and inexpensive shock absorber that had no recoil. Built for space use, the shock absorber would break on impact rather than bounce back. (Phil. *Eve. Bull.*, 4/14/64)

- British Postmaster General Reginald Bevins said in House of Commons that, under the right conditions, "the United Kingdom will be willing to provide up to £15 million of the capital cost of a world system" for satellite communications. He said U.K. would share in design of comsat system, share in its ownership, and play full part in providing satellite launch vehicles and other equipment. He reported that in recent meeting of representatives of European, U.S., and Canadian governments on problems of establishing a global comsat system, "good progress" was made. (UPI, *NYT*, 4/16/64, 9)
- Dr. Smith J. DeFrance, Director of NASA Ames Research Center, was one of 10 recipients of the 1964 Career Service Award of the National Civil Service League. Director of Ames since 1940, Dr. DeFrance was recommended for the award on the basis of his 43-year career as an outstanding administrator and leader of aeronautical and space research. (NASA Release 64-75)
- Brig. Gen. Thomas J. Hayes, III, Assistant to the Chief of Army Engineers for NASA Support for the past two years, received certificate of appreciation from NASA Administrator James E. Webb. General Hayes supervised such Engineer Corps activities as design and construction of NASA facilities, lunar mapping, and advance engineering related to future programs. He was cited by NASA particularly for his "skillful direction" in the "planning, design and construction of the massive and complex structures required for the manned space flight system." (NASA Release 64-84)
- House Committee on Science and Astronautics was briefed on comparative U.S. and U.S.S.R. space programs by National Aeronautics and Space Council Executive Secretary Dr. Edward C. Welsh and NASC staff members. Committee Chairman Rep. George P. Miller (D.-Calif.) said the executive session gave the committee a better appreciation of Soviet scientific space exploration and reaffirmed his belief that the U.S. "must keep up its effort to stay in the space race." (*Av. Wk.*, 4/20/64, 30)
- Maj. Gen. Benjamin D. Foulois (USA, Ret.) was presented a special Medal of Recognition for his more than 50 years of dedication and service to the development of aviation, in ceremony at the Pentagon. He had been a member of the NACA while Chief of the Army Air Corps (1932-1936). (DOD Release 309-64)
- FAA announced Braniff Airways had reserved two delivery positions for U.S. supersonic transport aircraft and Trans-American Aeronautical Corp. of Washington had increased its delivery reservations for SST from one to two. (FAA Release 64-35)
- Sen. Barry Goldwater (R.-Ariz.), candidate for Republican Presidential nomination, released a campaign paper regarding missile reliability:

    "I have raised, and will continue to raise until all the facts are in, fundamental questions about the reliability of our intercontinental ballistic missiles. It is not a question of theoretical accuracy. The fact is that not one of our advanced ICBM's has ever been subjected to a full test under simulated battle conditions."

    Later in the day, DOD issued listing of U.S. missiles, bombers, and missile-bearing nuclear submarines and compared them with Soviet strength in each category. DOD said it was acting to prevent another "missile-gap myth" from circulating. (*NYT*, 4/15/64, 1, 6)

April 15: NASA launched modified four-stage solid-propellant Javelin (Argo D–4) sounding rocket from Wallops Island, Va., carrying 120-lb. instrumented payload to 452 mi. Approximately 15 min. of data were telemetered to ground stations before fourth stage and payload impacted in the Atlantic Ocean 642 mi. from launch site. Primary purpose of flight was to test X–258 solid-propellant rocket motor, and secondary purpose was to conduct scientific investigations in the ionosphere. Test was part of GSFC's development of the X–258 for NASA use in satellite launch vehicles and possibly in sounding rockets. (Wallops Release 64–36)

- NASA Nike-Apache sounding rocket launched from Ft. Churchill, Canada, with Univ. of Alaska instrumented payload attained 96-mi. altitude. Experiment objective was to obtain data on visible aurora in three wavelengths and data on vacuum ultraviolet emissions. This information was to be supplemented by data from two electron particle detectors. Payload also included three photometers and two vacuum ultraviolet detectors. Good telemetry signal was received for 383 sec., but door covering instrumentation apparently came off at about 20,000-ft. altitude and most of the sensors were damaged by the severe environment. (NASA Rpt. SRL)

- NASA Nike-Apache sounding rocket launched from Ascension Island carried Univ. of Michigan instrumented payload to 98-mi. altitude to measure atmospheric pressure, temperature, density, and winds using the pitot-static probe. All objectives appeared to have been met. (NASA Rpt. SRL)

- NASA announced selection of 23 scientific and technological experiments to be flown aboard Gemini spacecraft, selected by NASA Manned Space Flight Experiments Board from almost 100 proposals made by scientists and engineers in NASA, DOD, and the scientific community at large. 13 of the experiments—eight scientific and five technological—were sponsored by NASA; ten experiments, all technological, were sponsored by DOD. (NASA Release 64–78)

- At NASA Hq. news conference, NASA Administrator James E. Webb said new contract with Jet Propulsion Laboratory would specify changes to improve JPL's organization and management. "NASA is not going to try to impose rigid methods of organization and management on the Jet Propulsion laboratory. We believe they know a great deal about how to succeed at this business. But neither are we going to abdicate our own responsibility for this expenditure of public funds and for the success of the missions.

"We are in a process by which we are both working to solve the problem. If this process produces the kind of organizational changes and the kind of ability to fix organizational responsibility that . . . [is desired], then we will go forward. If they don't both we and JPL will be looking for some other arrangement. . . .

"I think all of us see the total picture better. Each of us tended to see one part of the picture rather than the total picture." Under the new contract, JPL would be organized more like the NASA Centers. ". . . the United States government cannot find any arrangement under which the high quality of work which JPL has done and is capable of doing in the future can be available to the country better than the arrangement that we have negotiated. I hope and believe it will be signed. But I still feel it is my responsibility to make very sure that the contract represents a full meeting of the minds."

Mr. Webb stressed that he was not singling out anyone for blame for Ranger lunar spacecraft failures. "Our objective is success on the next flight and not to find a scapegoat for this kind of thing." Of the next Ranger lunar probe, he said, "We are still striving to get it off about June, but there are lots of problems that have to be settled. Some of us feel quite strongly that we must have a very strong assurance of success before we would be ready to launch. But we also run into the problem if we don't launch then we may have difficulty fitting into the schedule later."

Asked about effect of the delay in Congressional appropriation of $141 million supplemental requested for FY 1964 manned space flight, he said: ". . . this delay is at least reducing what the President gave us—a fighting chance. I mean, with each passing chance we have a little bit less of a fighting chance to do the job within this decade. . . . I would say that costs are going to increase for the same amount of work with each passing month, and that each month roughly delays one month on the other end, or maybe more, and that this does certainly reduce the fighting chance." (Transcript)

April 15: NASA Flight Research Center formally accepted from Bell Aerosystems Co. the first of two Lunar Landing Research Vehicles (LLRV), to be used in extensive flight research program conducted by FRC in support of Project Apollo. The program would investigate possible operational and piloting problems that might occur during final phase of manned lunar landing and initial phase of lunar takeoff. (FRC Release 5-64)

- Soviet news agency Tass said 25 communication sessions had been held with ZOND I space probe since April 2, and a great deal of "interesting experimental data" on the operation of onboard systems and individual units had been obtained. By 6:00 p.m., Moscow time, ZOND I had reached 4,250,000 km. (2,640,823 mi.) from earth. (Tass, *Komsomolskaya Pravda,* 4/16/64, 1, ATSS–T Trans.)

- Dr. George E. Mueller, NASA Associate Administrator for Manned Space Flight, said in *Space Business Daily* interview that use of Centaur as a third stage for Saturn IB launch vehicle was now a "matter of study with a serious approach to the various alternatives" it could serve. He mentioned one such use might be as a booster for logistics support vehicle called Stay Time Excursion Module (STEM) for establishment of lunar bases. Dr. Mueller also disclosed that payload for Saturn SA-10 flight had been changed from an Apollo boilerplate to a micrometeoroid detection satellite (MDS), supplementing those for Saturns SA-8 and SA-9. Added MDS was to further confirm that skin thicknesses on boosters and spacecraft can be reduced, thus trimming weight requirements for Project Apollo. (*SBD,* 4/15/64, 83)

- Based on spectrographic studies of about 100 stars comparable to the sun, Carnegie Institution of Washington astronomers concluded the sun, like these stars, was radiating less heat than it used to. Dr. Olin Wilson of Carnegie staff found that the sun must have completed its period of greatest radiation activity long before life on earth began. The earth and other planets must have experienced much stronger radiation.

Report said this radiation was "of more than passing interest for hypotheses about the atmosphere and other earth features which ultimately affected the origin of life." (Henry, Wash. *Eve. Star*, 4/15/64)

April 15: Thomas E. Dolan, NASC staff member, proposed at AIAA Symposium on Space Stations and Their Logistics Support in Pasadena that "saddlebags" be attached to LEM to increase its carrying capacity for use in equipping a lunar base. 3,000-5,000 lbs. of supplies would be attached to manned Lunar Excursion Module. The system would make it possible to set up a lunar base with a 7- to 10-day staytime. He said saddlebags mode would increase total Apollo spacecraft weight by 15,000 lbs., but Saturn V/Apollo system theoretically had growth option of 100% and 15-20% growth was entirely possible. "The LEM saddlebag concept would avoid multiple launches, hardware development programs for unmanned landing systems, dependence on successful unmanned landing system performance prior to manned landings, and dependence on surface rendezvous. As such, the post-Apollo exploration phase could be carried on with higher mission success and with less total investment than that estimated for the other systems." (*M&R*, 4/20/64, 15; *SBD*, 4/17/64, 98)

- FCC approved application of ComSatCorp to build communications satellite for its planned experimental-operational synchronous-orbit network. The authorization was for construction only; further authorization would be required for placing the satellite in orbit and using it in experimental or operational service. (AP, *NYT*, 4/16/64)

- NASA announced appointment of Ray Hooker, Chief of Research Models and Facilities Div. at Langley Research Center, as NASA's Senior Scientific Representative in Australia. Hooker would replace Edwin P. Hartman, first NASA Senior Scientific Representative in Australia, now completing his second two-year term. Hooker would leave in June for Melbourne, where his primary function would be liaison with the Ministry of Supply of the Commonwealth of Australia, the cooperating agency which builds, staffs, and operates NASA facilities there. (NASA Release 64-85)

- Dr. Raymond L. Bisplinghoff, NASA Associate Administrator for Advanced Research and Technology, said at University of Cincinnati: "We have reached a state where we can do a lot more things [in space exploration] than we may be allowed to do. . . .

    "The work of my office is to develop the concepts. I can only say what can be done. It may never be approved. . . .

    "The scientist and engineer cannot answer what is next. He can only answer by outlining the options which science will permit as the next step.

    "What is in fact done next must be determined by an assessment of the scientific, social, economical and political implications of these options. . . ." (Text)

- Brig. Gen. Joseph S. Bleymaier (USAF), USAF Space Systems Div. Deputy Commander for Manned Systems, said at AIAA space station symposium in Pasadena that USAF planned to assign to a single contractor the task of integrating the Gemini spacecraft and the manned orbiting laboratory (MOL). He said astronaut transfer from the Gemini to the MOL was a major problem and described several alternative methods of transfer being studied. (*Av. Wk.*, 4/20/64, 39)

April 15: Large manned space station design was described before Orange County, Calif., Section of AAIA by Edward H. Olling, head of MSC Space Station Program Office. The 24-man station would be 150 ft. in diameter with three radial modules, rotate at maximum speed of 4 rpm. Lifetime would be from one to five years. Saturn V could launch it into orbit and Saturn IB would launch logistics vehicles. (*M&R*, 4/20/64, 22)

- The planet Jupiter, more than 10 times the size of the earth, has density about that of the sun and major elements probably consisting of hydrogen and helium, according to Office of Naval Research. The analysis was based on observations made with a balloon-carried telescope launched into the stratosphere last November from Palestine, Tex. (NANA, *Houston Chron.*, 4/15/64)
- Prof. Colin O. Hines of Univ. of Chicago told national committee of the International Scientific Radio Union meeting in Washington that an unexpected high "jet stream" was to be blamed for disruption of long-distance radio communications. The layer of strong winds occurred at 50-mi. altitude, where shearing disturbed the ionospheric layers which in turn disrupted radio communications, he said. (*Wash. Eve. Star*, 4/15/64)
- NASA Assistant Administrator for Technology Utilization and Policy Planning Dr. George L. Simpson, Jr., said in *Space Business Daily* interview that "we are beginning to feel that progress is really being made in this [technology utilization] program." In many instances, he said, use of technology gained through various projects had "advanced the state-of-the-art" when the various data were "pulled together"—for example, the advancements in fuel cell programs, to be made available to industry shortly. (*SBD*, 4/15/64, 86)

April 16: NASA launched Nike-Apache sounding rocket with ionosphere experiment payload from Wallops Island, Va., to peak altitude of 105 mi. The payload carried instrumentation to measure electron and ion density and solar radiation in the D and E regions of the ionosphere; it impacted 114 mi. downrange in the Atlantic Ocean. Provided by Univ. of Illinois and the Geophysics Corp. of America, the experiment was an investigation of the International Year of the Quiet Sun (1964–65) (IQSY). (Wallops Release 64–37)

- Japan's first television transmission to Europe by communications satellite was received in Pleumeur-Bodou, France, and fed into France's television network for European viewing. Reception was excellent. The relay was made by TELSTAR II communications satellite. (AP, Balt. *Sun*, 4/17/64)
- NASA–AEC Space Nuclear Propulsion Office announced successful completion of cold-flow experiments on a Nerva (nuclear engine for rocket vehicle application) experimental reactor, designated the NRX–A1, by the Aerojet-General and Westinghouse Electric Corporations. The experiments were designed to check the structural stability and flow dynamics of the NRX–A reactor under nonpower conditions. All test objectives were met. (SNPO)
- Asked for his opinion about developing a new manned bomber, President Johnson said at White House news conference that USAF Chief of Staff General LeMay had requested funds be reprogramed to study plans for a new bomber: "I told him I would give consideration to his proposal. I understand that proposal has been formulated and is now going through channels, and will shortly come to the President. When it gets to me, I

will study it as best I can and make the decision that I think is in the national interest." (Transcript, *Wash. Post*, 4/17/64, A18)

April 16: New three-year labor stabilization agreement, including no-strike clause, was signed in Washington by representatives of AFL-CIO and Patrick AFB Contractors' Association. Agreement covered MILA and Cape Kennedy as well as Patrick, was retroactive to April 1 when previous two-year agreement had expired. (AP, *NYT*, 4/17/64)

- D. Brainerd Holmes, vice president of Raytheon Corp. and former director of NASA manned space flight program, was interviewed in the *Boston Globe*. Asked if he still thought the 1970 manned lunar landing goal was realistic, he replied, "The timetable now is tougher than it was, but still possible." (White, *Boston Globe*, 4/16/64, *CR*, 4/21/64, A2002-03)

April 17: NASA launched three Nike-Cajun sounding rockets from Ft. Churchill, Canada, and Wallops Island, Va., two with grenade payloads to investigate winds, temperature, density, and pressure in the upper atmosphere. The first, a Nike-Cajun was launched from Wallops; it sent payload to about 57-mi. altitude, where aluminum-coated mylar plastic sphere was ejected and inflated to 26-in. diameter. The sphere continued up to about 78.4-mi. altitude before floating down to earth. Radar tracking of ascent, descent, and drift provided data for calculating densities at various altitudes. Less than two hours later the Ft. Churchill rocket was launched, sending 82-lb. payload which ejected grenades detonating at intervals in region of 23 to 58 mi. 20 min. later the final rocket—the second carrying grenades—was launched from Wallops Island, with grenades exploding in about the same altitude range. NASA Goddard Space Flight Center conducted the flights to observe characteristics of the upper atmosphere and to measure variations occurring through the period of increased heating. The grenade experiments were an extension of series conducted from Jan. 23 to Mar. 6 involving 13 Nike-Cajun flights which "yielded excellent data on the characteristics in the upper atmosphere during winter conditions," according to project scientist Wendell S. Smith. Follow-up experiments were planned for late summer when solar heating would be near maximum, for comparative data. (Wallops Release 64-38)

- Sidney J. Dagle, 29, Ball Brothers technician, died of injuries caused by ignition of X-248 rocket motor three days earlier at Cape Kennedy. This was the first death among the 11 injured personnel. Next day, John W. Fassett, 30, succumbed to burns sustained in the accident. An engineer in GSFC Field Projects Branch, Fassett was the only NASA employee among the 11; the others were employees of Douglas (booster contractor) and Ball Brothers (Oso satellite contractor). One other man remained in critical condition. (AP, *NYT*, 4/18/64, 12; UPI, *NYT*, 4/19/64, 80; *Goddard News*, 4/20/64, 3; GSFC PIO, 5/13/64)

- House Appropriations Committee reported $46,785,867,000 military appropriations bill for FY 1965. Report said: "The committee explored in detail the question of missile dependability. . . . [The committee] is inclined to conclude with the Secretary [of Defense] that the missile force we have programmed can be depended upon to carry out its military mission under all of the conditions we can foresee." The committee also proposed appropriation of $47 million for initiating development of a new manned bomber. (Raymond, *NYT*, 4/18/64, 1)

April 17: USIA communications exhibition to be shown in U.S.S.R. this summer was previewed by Government, industry, and press representatives in New York. Exhibition covered U.S. communications in space satellites, TV and radio, telephone, and business and industry. The space portion featured models of nine satellites and live TV images of visitors bounced off working model of Telstar satellite. (Shanley, *NYT*, 4/19/64, 47)

- NASA Lewis Research Center announced seven researchers had won awards for four inventions, the awards totaling $1,400 under provisions of the Government Employees Incentive Awards Act. Winners were: S. Stephen Papell, $500 for low viscosity magnetic fluid obtained by collodial suspension of magnetic particles; Paul D. Reader and Harold R. Kaufman, $500 for electrostatic ion engine with permanent magnetic circuit; Robert J. Branstetter and Allen J. Metzler, $300 for black-body furnace; John R. Jack and Paul F. Brinich, $100 for improved heat exchangers for electrothermal rockets. (LRC Release 64–38)
- Geraldine Mock, Columbus, O., housewife, became the first woman to fly around the world when she landed at Port Columbus Airport, O. Mrs. Mock completed her 23,103-mi. solo flight in 29 days, 11 hours, 59 min. The flight was monitored by National Aeronautic Association. (NAA Release)

April 18: Second meeting of the President's Advisory Committee on Supersonic Transport was held in the Pentagon. The committee heard views of airline representatives regarding economic aspects of SST, then continued its discussion of major issues involved in the SST program. (DOD Release 320–64)

- His Majesty Hussein I, King of Jordan, visited NASA Kennedy Space Center and toured facilities at the Cape. (NASA Off. Int. Aff.)

April 19: NASA Technology Utilization Report (SP–5005) on the retrometer, a new system of voice communications transmitted on a beam of light, was published. Invented at NASA Langley Research Center by Numa E. Thomas, the retrometer differed from previous light beam-communications systems in that the originating station required no power other than the human voice. The T–U Report described the system, gave instructions for building experimental retrometers, and suggested many possible applications of the system. (NASA T–U Rpt. SP–5005; NASA Release 64–82)

- Soviet V–8 helicopter, a 25-passenger single-rotor vehicle, covered 1,530 mi. in 12 hrs. 3 min. 34 sec., to claim new world nonstop records for distance and speed, *Pravda* reported. (AP, Wash. *Eve. Star*, 4/20/64; *NYT*, 4/22/64, 77)

April 20: NASA announced it had selected IBM as lead contractor for development and fabrication of instrument units for Saturn IB and Saturn V launch vehicles. IBM was selected last October to design and manufacture the data adapters and digital guidance computers and be responsible for integration and checkout of the units. Now, as lead contractor, IBM would assume additional responsibility for structural and environmental control systems and integration of all systems. NASA would supply telemetry system and ST–124M stabilized platforms. (NASA Release 64–89; KSC Release 52–64)

April 20: NASA Deputy Director Dr. Hugh L. Dryden said in speech before Economic Club of Detroit:

"It has been our experience at NASA, in dealing with contractors, that they often fail to appreciate that their plants do not possess the type of reliability standards and quality control standards which we demand. Once we have convinced him that our standards do outstrip his normal assembly line standards, the contractor himself often encounters difficulty in convincing his supervisors and work force that what's good enough for the public just isn't good enough for the national space program.

"NASA has set high standards of reliability and quality control for the boosters that send manned and unmanned spacecraft into space, the spacecraft themselves, and for their related ground support equipment. . . ." (Text)

- NASA and DOD signed agreement concerning DOD contract administration services for NASA within the Philadelphia contract administrative services region, the pilot test intended "to avoid duplication of effort and to achieve the most effective and economical utilization" of DOD and NASA resources in contract administration services for NASA. (Agreement, NMI 2-3-31)
- Western Union announced it had completed a microwave system for the next lunar spacecraft's communications with earth. The system was in test operation, linking Goldstone Tracking Station with Jet Propulsion Laboratory, 164 mi. away. For the next lunar probe, as well as subsequent lunar and planetary probes, data obtained by the spacecraft would be sent through Goldstone to JPL for computer analysis; commands to the spacecraft would be sent from JPL through Goldstone to the probe. (*Wash. Post,* 4/21/64)
- Brig. Gen. T. J. Hayes, Assistant to Chief of Army Engineers, told Space Congress at Cape Kennedy that U.S. should begin immediately to plan building a lunar base. A lunar base would be the next logical step after Project Apollo, he said, and such an installation would "require solving unprecedented engineering problems imposed by delivery-system limitations and the harsh environment about which there is little information.

"A significant program is needed to develop the equipment, techniques, and reliability necessary for this difficult task." (AP, Balt. *Sun,* 4/21/64)
- President Johnson announced U.S. would reduce production of enriched uranium over a four-year period. Addressing AP in New York, he said: "When added to previous reductions, this will mean an overall decrease in the production of plutonium by 20 per cent, and of enriched uranium by 40 per cent." Soviet Premier Khrushchev simultaneously announced in Moscow that U.S.S.R. would discontinue construction of two atomic reactors for production of plutonium; reduce production of uranium-235 in the next several years; and allocate more fissionable materials for peaceful uses of atomic energy. (Texts; *NYT,* 4/21/64, 14)
- At White House ceremony, with President Johnson participating, Dr. W. Randolph Lovelace II was sworn in as Director of Space Medicine, NASA. (Wash. *Eve. Star,* 4/21/64)
- JPL had recommended use of a steerable entry vehicle or steerable parachute system for 1969 Mars probe, *Missiles and Rockets* reported. JPL wanted to land the complete vehicle and achieve telemetry from the Martian

surface, notably on whether or not signs of life existed. Chief area of interest was Syrtis Major, prominent green area on Martian equator that might be one of the more likely spots to support life. (*M&R*, 4/20/64, 11)

April 20: Interviewed in *Missiles and Rockets*, Presidential Science Adviser Donald F. Hornig replied to question about U.S. space goals after Project Apollo:

"There are clearly a series of successively more difficult scientific objectives which we will pursue with unmanned spacecraft. It's very difficult, though, to document a convincing case for a national commitment on major space goals that are far in the future. In any case, I can't believe that the momentum gained by our space program is going to be allowed to be lost when we achieve the first major objective." (*M&R*, 4/20/64, 35)

- Problem of nylon parachutes' deteriorating in space was suggested by tests conducted by Northrop Ventura. Strength of nylon, after two weeks in vacuum chamber, reduced up to 20%; after seven minutes exposure to normal atmosphere the nylon returned to within 2.3% original strength. (*M&R*, 4/20/64, 21)

April 21: NASA launched Aerobee 150 sounding rocket from White Sands, N. Mex., with instrumental payload to obtain evidence of hypervelocity micrometeoroid impacts, samples of low-velocity cosmic dust, and measurements of electron densities. Flight failed when rocket's propulsion system malfunctioned and the rocket attained only 6.25-mi. altitude. (NASA Rpt. SRL)

- NASA announced Walter L. Lingle, Jr., Deputy Associate Administrator, was leaving NASA May 15 to return to private industry. Lingle would be succeeded by Earl D. Hilburn, Deputy Associate Administrator for Industry Affairs. (NASA Release 64–90)

- Rep. Joseph E. Karth (D.–Minn.) addressed the National Space Club in Washington:

"The great promise for the future of the space and defense industry is to turn some of the tremendous imagination, some of the talents and energies of your management, your scientists and engineers, to the major social and economic problems of our country.

"Your industries, I submit, are eminently qualified to cooperate in this entirely different kind of takeoff because of your unique experience in solving inter-disciplinary problems through the technique of systems analysis. . . ."

Emphasizing that "space is not a WPA program," he said he would not support accelerated space expenditures merely to stimulate the economy depressed by reduction in defense spending because of disarmament. (NSC Release)

- Rep. Roman C. Pucinski (D.–Ill.), Chairman of House Subcommittee on National Research Data Processing and Information Retrieval Center, wrote in Washington *Daily News* article:

"The greatest challenge in this changing world is the solution of the problem of handling scientific and technical information.

"I am deeply concerned with the staggering acceleration of scientific and technical information which is the life-blood in the heart of our research and development effort. . . .

"I am convinced somewhere in the mountainous scientific literature there is a cure for cancer, a solution for harnessing limitless energy from nuclear fusion and a way to solve the food requirement of a growing world population." (Wash. *Daily News*, 4/21/64)

April 21: Rep. John Brademas (D.-Ind.) inserted in *Congressional Record* an editorial on Rep. J. Edward Roush's proposal that NASA establish a Midwestern office, broadcast by station WOWO: ". . . . We agree with Congressman Roush that if it is in the best interests of NASA to operate liaison offices to aid potential contractors in securing NASA business, then every area of the United States should qualify for this aid. . . ." (*CR*, 4/21/64, A1998–99)

April 22: NASA launched Nike-Apache sounding rocket from Ft. Churchill, Canada, with Univ. of Alaska payload to obtain data on the visible aurora and on the vacuum ultraviolet emissions. This information was to be supplemented by data from two electron particle detectors. Payload also included three photometers and two vacuum ultraviolet detectors. The rocket attained peak altitude of 103 mi., and the flight appeared satisfactory. (NASA Rpt. SRL)

- Michael I. Yarymovych, Director of NASA Manned Earth Orbital Mission Studies, said before Canaveral Council of Technical Studies, Cape Kennedy, that NASA was coordinating fully with DOD in exploring four different types of orbital systems: Extended Apollo, Apollo Orbital Research Laboratory (AORL), Medium Orbital Research Laboratory (MORL), and Large Orbital Research Laboratory (LORL). The first three concepts were designed to be orbited by Saturn IB; LORL, by Saturn V. Yarymovych said that it was "becoming increasingly clear that the Extended Apollo is an essential element of an expanding Earth orbital program. . . .

    "In the initial stages, it would be used as a laboratory and later it could be converted to a logistics system." AORL, MORL, and LORL were in competition with each other, he said, and a decision would have to be made among them. USAF's Gemini B/Manned Orbiting Laboratory was being studied as possible integral element of Orbital Research Laboratory program. (*M&R*, 4/27/64, 16)

- NASA announced Flight Research Center's selection of Norair Div., Northrop Corp., for design and fabrication of two "lifting body" research gliders. The fixed-price contract was expected to amount to about $1 million. The research gliders would be used by FRC in studying problems of piloting advanced spacecraft during landing. (FRC Release 7–64; NASA Release 64–93)

- World's Fair opened in New York, its U.S. Space Park featuring such exhibits as full-scale mockup of Saturn V's first stage boattail, Gemini-Titan, Mercury-Atlas, Delta, and the X–15. (*Goddard News*, 5/4/64, 3)

- USAF information office was quoted by Stuart H. Loory in *New York Herald Tribune* as saying that since 1959, 12 SAC operational training missions of Atlas ICBM had been failures. USAF had called in General Dynamics Corp. engineers to help solve the problems. (Loory, *N.Y. Her. Trib.*, 4/22/64)

- AFSC/BSD commander Maj. Gen. W. A. Davis, asked at Cape Kennedy press conference to comment on the recently reported Atlas ICBM failures at Vandenberg AFB, said:

"There are several possible reasons for the failures. The Atlas is a complex piece of machinery that has a habit of using up some of its reliability with age. SAC has been taking Atlases off of operational sites, where they have been for a year or more, placing them on the stand at Vandenberg and firing them.

"I feel that this type of test is not operational realism as SAC calls it. You take the missile out of its environment, load it on a plane, fly it 1,000 or more miles, load it on another stand and fire it. This requires disconnecting and reconnecting wires and fuel lines and installing destruct packages and some instrumentation for the training launches." (AP, *Wash. Post*, 4/23/64)

April 23: NASA and NBC conducted one-hour test of the SYNCOM II communication satellite hovering in synchronous orbit near the West Coast, testing the satellite's capability to relay video coverage from Tokyo's Olympic Games next fall. In the test, signals were transmitted from an antenna at Fort Dix, N.J.; signals returned from the satellite were received at Andover, Me., and relayed to NBC studios by land lines. Quality of the pictures was considered below standard commercial quality but adequate for brief broadcasts of select events. NBC, which had purchased exclusive rights to TV coverage of the 1964 Summer Olympics in the U.S. and Central and South America, would now decide whether to attempt live coverage with satellites or to rely on tapes. (Witkin, *NYT*, 4/24/64, 55)

- USAF launched Atlas-Agena D booster combination from Point Arguello, Calif., with unidentified satellite payload. (AP, *Houston Post*, 4/24/64; *M&R*, 5/4/64, 8)
- AFSC announced that 36 projects initiated under the canceled X-20 program would be continued in support of USAF manned orbiting laboratory (MOL) and other programs. Examples of projects to be continued: X-20 full pressure suit, search and rescue receiver/transmitter, studies of pilot control of booster trajectories, nose cap concept, X-20 heat protection system, coated molybdenum panels, high temperature bearings, sensors/transducers, test instrumentation subsystem ground station, flight simulation instruments and controls. (AFSC Release 43-10-44)
- Dr. Robert M. White, Chief of U.S. Weather Bureau, said in address before joint American Meteorological Society-American Geophysical Union banquet in Washington:

  "Man has always had to contend with hurricanes, tornadoes, seismic sea waves, floods, earthquakes, and the like. His scientific, technological, and social advances have now brought him face to face with new environmental hazards. I am thinking of communications blackouts, the effects of atmospheric turbulence on high-speed aircraft, radioactive fallout, and radiation in space. . . .

  "We are now developing revolutionary new methods of acquiring global information about our environment. The space satellite, although still in its embryonic state, offers a potential for acquiring environmental information over the entire globe in a manner hitherto unimaginable. However, satellites are extremely costly, and their cost will require that a system be devised for multiple uses. A satellite observational system for environmental purposes will have to serve navigational and geodetic functions, make observations of the weather, probe the oceans, and sound the ionosphere.

"Any observational platform should be put to multiple uses if it is expensive to build and operate. I am thinking not only of space satellites, but also of ships and aircraft. We can no longer ignore the necessity of using expensive platforms to collect data for many geophysical purposes. The concept of an oceanographic survey without a simultaneous atmospheric survey must be abandoned by force of economics alone. . . ." (Text)

April 24: NASA investigative committee recommended series of tests to determine whether static electricity caused accidental ignition of Delta rocket motor in the Spin Test Facility at Cape Kennedy last April 14. In its preliminary report, the fact-finding committee said there were four possible causes for the ignition: heat, shock, outside radio signals, and electrical impulse. The first three possibilities seemed unlikely, but they would be studied further. Committee considered it "possible, however, that static electricity could have built up in such a way as to cause ignition." Until the committee completed its study and made its final report, NASA would not fly the X-248 motor. (NASA Release 64-98; KSC Release 55-64)

- USAF OAR announced determination by Cambridge Research Laboratories scientists R. A. Skrivanek, R. K. Soberman, and Tom Ryan that noctilucent clouds were composed of meteoric particles or of residue of burned out meteors. The determination was made after extensive examinations of noctilucent cloud particles gathered by specially designed rocket nose cones sent aloft on Nike-Cajun rockets in joint AFCRL–NASA–Univ. of Stockholm program. The researchers expected that another series of noctilucent cloud experiments to be conducted in Sweden this summer would confirm these results. (OAR *Release* 4-64-3)

- Douglas DC-7 aircraft crashed at airport near Phoenix, Ariz., in controlled test conducted by Flight Safety Foundation under FAA contract. Damage was worse then expected, and researchers were delighted with the "perfect crash." Main purpose of test was to obtain data on fracture affecting passenger survival, evacuation, and fire following impact. A NACA/NASA developed system to protect passengers was tested in the crash: plastic bag which can be inflated rapidly to hold the person in his seat and thus provide cushion against vibration, buffeting, high acceleration and deceleration forces, hard landings, and crash shocks. (NASA Release 64-88; FAA Release 64-30; *NYT*, 4/25/64, 1)

- NASA announced Georgia Institute of Technology would receive $1 million grant for construction of research facilities for projected Space Science Technology Center. Grant would build two-story building containing about 51,000 sq. ft. to accommodate laboratory research in atmospheric sciences, materials and structures, transport phenomena, and systems engineering. To complete the Center, Georgia Tech would build a matching structure with funds from other sources. (NASA Release 64-98)

- NASA Flight Research Center's selection of Cornell Aeronautical Laboratories, Inc., for contract negotiations on general purpose jet airborne simulator system was announced. The cost-plus-fixed-fee contract was expected to amount to about $1 million. The new system would be installed in Lockheed Jetstar aircraft (C-140A), purchased last year for use in extensive flight research directed toward problems of a variety of aircraft types, including supersonic transport. (NASA Release 64-97)

April 24: NASA selected six firms for competitive negotiations leading to contract for launch support services at Kennedy Space Center, NASA: Aerojet General Corp., American Machine and Foundry Co., Bendix Corp., Chrysler Corp., General Dynamics Corp., and Westinghouse Electric Corp. The contract would be cost-reimbursement type with incentive-award fee provisions, and would be awarded for one year with renewal options for two more years. Total cost of three-year contract was expected to be more than $20 million. Most of the work would be performed at new facilities now under construction on NASA Merritt Island Launch Area. (NASA Release 64–95)

- Dr. Gerard P. Kuiper, head of Lunar and Planetary Laboratory of the Univ. of Arizona, said at meeting of American Geophysical Union in Washington that Mars did not have enough water to support any but microscopic life forms. He described recent observations from McDonald Observatory in Fort Davis, Tex., that showed Mars to be enveloped in sheath of atmospheric ice crystals. Dr. Kuiper said in Martian spring the frost covering its pole during the wintertime evaporates directly into the air rather than melting. He estimated Martian air to be two parts nitrogen to one part carbon dioxide, with small amount of argon and even less water vapor and oxygen. Dr. Hyron Spinrad of Univ. of Calif. (Berkeley) described recent observations of Mars from Mt. Wilson Observatory. He said Mars was so cold that water melted only in the afternoon and when Mars was nearest the sun; amount of water in the Martian air was a thousandth of that in sea-level air in the arid southwestern U.S. The speakers rejected the idea that Martian air contained nitrogen-oxygen compounds, proposed last December by Georgetown Univ. astronomer Dr. Carl C. Kiess and chemist Dr. Sebastian Karrer. (Sullivan, *NYT*, 4/25/64, 8)

April 25: President Johnson said in Washington press conference:

"I have sent a memorandum to Mr. Halaby of the Federal Aviation Agency that states:

" 'I realize that you had hoped to select the contractors to proceed with the development of supersonic transport by May 1 of this year, as a result of the preliminary design competition conducted over the past several months among a number of airframe and engine manufacturers.

" 'As you have reported to me, however, the 210-member Government evaluation group, after analyzing the proposals in depth, found that none of the proposed airframe designs met the minimum range-payload requirements of the F.A.A. request for proposals of carrying a 30,000-pound payload for a distance of 4,000 statute miles.

" 'Moreover, none of these designs met what you properly emphasized as a basic requirement, namely that the aircraft be capable of economic operation.

" 'As you have also emphasized, it has been the objective that the development stage of the supersonic transport be financed 75 per cent by manufacturers. The F.A.A. request for proposals pointed out that "the Government's decision to proceed with the supersonic transport program is based on manufacturers' participation in an amount equivalent to 25 per cent of the total cost of the development through certification of the transport."

" 'I remain convinced that it will be possible to develop an American supersonic transport which will be economic to operate, will find a substantial market among the airlines of the world and will help to maintain American leadership in the air.

" 'In view of the current situation, however, I recognize that it is no longer appropriate for me to hope for a recommendation by May 1. Difficult and complex issues are now presented for resolution so that I can determine how best to proceed.

" 'I have asked the members of the President's Advisory Committee on Supersonic Transport, of which you are a member, to study this program thoroughly, after which I will expect recommendations from the committee and from you.'

"So the call for bids on May 1 is off and we cannot expect a recommendation by then." (Transcript, *NYT*, 4/26/64, 64)

April 25: U.S.S.R launched COSMOS XXIX into orbit from unspecified site. Soviet press agency Tass said the satellite was in the following initial orbit: 309-km. apogee (192 mi.) ; 204-km. perigee (127 mi.) ; 89.52 min. period; and 65°4' angle of inclination to the equator. Tass said scientific equipment, radio systems, and onboard instrumentation were functioning normally, and ground station was processing incoming information. (Tass, *Komsomolskaya Pravda*, 4/26/64, 1, ATSS-T Trans.)

- First industry-produced Saturn booster (S-I) arrived at NASA Marshall Space Flight Center after nine-day barge journey from Michoud Operations. To be used on SA-8, the booster was undergoing static firings and other tests at MSFC before being sent to Cape Kennedy where it would be launched. (*Marshall Star*, 4/29/64, 1)
- Harold Goldstein, Assistant Commissioner of U.S. Bureau of Labor Statistics, discussed changes in the labor force at symposium in New York. "We now have about 1.2 million engineers and scientists. Their numbers may rise to somewhere in the neighborhood of 2 million by 1970.

"There productivity is also rising rapidly. New inventions can be made faster, because as science and technology have progressed the techniques available for analyzing a problem, the knowledge of scientific principles that can be brought to bear, the different types of hardware that are readily available to be adapted in a new invention are increasing at a geometric rate." (Lissner, *NYT*, 4/26/64, 85)

April 26: NASA Associate Administrator Dr. Robert C. Seamans said Soviet spacecraft ZOND I apparently was traveling on a Venus trajectory. Dr. Seamans was speaking on Mutual Broadcasting Systems, Inc., radio program "Reporters Roundup." Asked why he thought ZOND I was a Venus probe, he said that he understood U.S.S.R. had identified it as such and that "this payload is traveling in that direction." Also, he added, the firing occurred "at the right time for a Venus launch." (AP, Wash. Eve. Star, 4/27/64; AP, *Wash. Post*, 4/27/64)

- NASA Deputy Administrator Dr. Hugh L. Dryden said in address before National University Extension Association in Washington:

"The exploration of space is a continuation of the geographical exploration by man of unknown areas of the earth from the days of the Phoenician mariners three thousand years ago. The New World, the polar regions, the depths of the ocean, the limits of the atmosphere—have each in turn been the temporary goal. Space is the new frontier. . . .

"We have as a nation accepted the challenge of the new frontier, and this year are spending a little more than five billion dollars on the exploration of space for peaceful purposes. This represents an expenditure of approximately 50 cents per week by each of the 200 million inhabitants of our country. . . ." (Text)

April 26: NASA announced 1964–65 Alfred P. Sloan Fellowships in executive development were awarded to three NASA employees: Walter C. Scott, Chief of Chemical and Solar Power Generation Branch in Office of Advanced Research and Technology; Edward J. Lievens, Jr., member of Technical Staff, Plans and Analysis, in Office of Manned Space Flight; and Neal A. Holmberg, Aerospace Technology Program Manager at NASA Western Operations Office. (NASA Release 64–94)

April 27: NASA announced Langley Research Center would conduct tests with Boeing 707 air transport prototype to find ways to lower landing and takeoff distances required for large jet transports. NASA pilots and engineers would study performance and handling characteristics of the aircraft in three-month program. (NASA Release 64–100)

- House Committee on Science and Astronautics' Subcommittee on NASA Oversight began hearings on Project Ranger. Subcommittee Chairman Rep. Joseph E. Karth (D.–Minn.) remarked that "the action that precipitated these hearings are letters which Mr. Webb addressed to the chairman of the two space committees" (March 31 letters to Senator Anderson and Representative Miller outlining problems with Ranger spacecraft). (Wash. *Eve. Star*, 4/27/64)

- Dr. Homer E. Newell, NASA Associate Administrator for Space Sciences and Applications, said before House Subcommittee on NASA Oversight about Project Ranger:

". . . From the outset, the concept was advanced. It was conceived to seize the initiative in space exploration from the Soviet as well as to obtain important information about the moon. Had it succeeded, it certainly would have been a major first for this country, and it may still obtain that objective. The spacecraft itself is quite complex. . . . It is comparable to the most sophisticated launch vehicles in terms of electromechanical system complexity. Launch vehicles developed in this country have typically taken 20 or more flights to achieve 50% reliability. Some have not done much better than this after many more firings. Indeed, we have read recently that last year there were 13 consecutive failures of the Atlas. In addition, the Discoverer project took 13 attempts to recover its first payload from orbit. Polaris experienced 5 initial failures before its first success. Skybolt experienced 5 flights without achieving complete mission success. In each of these projects, partial success in early flights helped improve later flights. Although I believe Skybolt was dropped for other reasons, Atlas, Discoverer, and Polaris all went on to become vital elements in the defense posture of this country.

"Ranger is going through this same process. We have solved some of our launch vehicle problems to the point where the reliability is improving appreciably. The basic spacecraft bus has been retained and has now demonstrated its capability to deliver a payload to a precise spot on the lunar surface. We now must make the payload work. The payload also is pushing the state of the art in high power television telemetry. This was necessary to obtain high resolution photographs. . . .

"Our reviews of the Ranger system continue to convince us of its capability of performing the mission. While we know we cannot have

100% reliability with launch vehicle, spacecraft, or payload, we expect to achieve at least one complete success with the remaining three flights. . . .

"Ranger is a tough job. Although success has eluded us, it is within our grasp. We must have the fortitude to proceed. Space will only be conquered by those who do have such fortitude. It might be well to point out that our competitors in this area of exploration are not lacking in this respect. It is reported that the Soviet Union has failed in from 15 to 20 consecutive lunar and planetary mission attempts since Lunik III photographed the back side of the moon in 1959.

"Even should subsequent Rangers fail through unforeseen causes, we must move ahead. We must look at the totality of the space program. . . . The over-all picture is very good and is most convincing of this country's capability to successfully complete these undertakings. The trends are all in our favor. . . ." (Testimony)

April 27: Oran W. Nicks, Director of NASA Lunar and Planetary Programs, presented history and status of the Ranger project to the Subcommittee on NASA Oversight, detailing in particular the recommendations of NASA Board of Inquiry (Kelley Board) prior to RANGER VI and the Hilburn Board following RANGER VI. Hilburn report outlined: (1) design changes effected after analyses of RANGER VI flight data; (2) changes agreed to be desirable "if properly evaluated and qualified"; and (3) board recommendations "not entirely resolved by OSSA and JPL." Notable change in the first category: NASA would closely monitor Ranger test program at JPL, would have to concur in spacecraft's flightworthiness before it could be shipped. (Testimony)

- USAF launched Thor-Agena D booster combination from Vandenberg AFB, Calif., with unidentified satellite payload. (UPI, *Wash. Post*, 4/28/64)
- Four-day Conference on Space Nutrition and Related Waste Problems, sponsored by NASA and organized with assistance of National Academy of Sciences' Space Science Board, opened in Tampa, Fla., with some 200 engineers and scientists attending.

In introductory remarks at Conference, NASA Associate Administrator for Advanced Research and Technology Dr. Raymond L. Bisplinghoff said that in both the life sciences and the physical sciences, "when we strip away the unessentials, we are really dealing with the conversion and control of energy. There is the muscle with its tiny fibrils on the one hand and the chemical fuel energy converters on the other. There are the electrical activities of the brain and nerve impulses on the one hand and the computer and feedback loop on the other.

"The state of man's technology in every age has depended on his ability to convert and control energy. The Industrial Revolution brought on an economic structure that could not be operated without knowledge of the equivalence of different forms of energy. . . .

"The opportunities that are now presented for advancements in energy conversion and control are breathtaking. New methods of energy conversion from solar, chemical, and nuclear energy sources which will be useful to all of mankind are being pioneered in the space program. . . . Perhaps the most dramatic energy conversion devices in the NASA program are the propulsion devices that convert chemical or nuclear energy to kinetic energy. The Nerva rocket engine converts nuclear energy to kinetic vehicle energy by passing liquid hydrogen through a graphite reactor at 6000° R and ejects it through a nozzle. In these devices we will convert as much energy as is converted by the Grand Coulee Dam in a device about the size of an American motor car. . . ." (Text)

Dr. Charles S. Sheldon of NASC staff said in address at the Conference:

". . . As of April 24, the United States had orbited 203 payloads, and the Soviet Union 65 since 1957. But weight comparisons are quite different. They have put up about three times as much net payload as we have. What is more discouraging, every year for the last five years, the lead of the Soviet Union in that year has grown larger than it was for the year before. . . .

"My best assessment is that the Soviet [space] program has close to the same variety of goals as our own, including undertaking very complex missions. Let us look at just the main outlines.

"*Scientific Exploration in Earth Orbit:* . . . The Soviet Union is clearly interested in more than spectaculars. In the Earth-orbital regime. it has put up 28 Kosmos class scientific and engineering-test satellites with a minimum of fanfare, and these have included a range of experiments. . . . Additionally the pair of Elektron satellites are providing synoptic measurements of the radiation belts out to 40,000 miles.

"*Space Applications:* When it comes to practical applications of unmanned satellites, the United States has the lead in demonstrated accomplishments. . . . But similar flight operations are not beyond Soviet capabilities. . . .

"*Lunar and Planetary Probes:* . . . In absolute number of launchings the Soviet Union has made a larger commitment to such flights than we have. As a percentage of total flight effort, their commitment for that purpose is running about five-fold our own. And they use larger vehicles by far. Take a part of the comparative planetary effort for example. Of our two launch attempts to date, the 447-pound Mariner II carried about 40 pounds of instruments near Venus and returned some data. We are rightfully proud of this accomplishment.

"The Soviet pattern has been to orbit 14,300-pound platforms from which a probe weighing up to 2,000 exclusive of the rocket casing can be launched on the interplanetary trajectory. Between 1960 and 1962, ten such launchings were made. Typically, the comparison is a difficult one to make. We met with a great success in Mariner II. Not one of the Soviet craft attained its full objective. . . .

"While neither country should be overly proud of its lunar probe record, the pictures taken by Lunik III back in 1959 are still a highwater mark.

"*Manned Space Flight:* . . . Through the 1963 season each country had made six manned launchings, and neither side had killed an astronaut or cosmonaut. Stories to the contrary are fabrications. . . .

"*Human Safety:* . . . The Russians sent up without recovery Laika, four years before Gagarin. The Vostok ship itself completed 100 orbits of fully controlled flight, including successful recovery of four dogs, before the first man was sent. . . .

"*Future Manned Missions:* . . . To the best of our knowledge, the Russians have not yet demonstrated a launch system capable of convenient manned flight to the Moon for a landing and return. But then, if we chose to practice Soviet-style security around our launch sites and factories, there would be no hard evidence available to foreigners that the United States is building a lunar capability either. . . .

"Because we are not permitted to inspect their facilities, we have to do our best by inference to judge the truth of their repeated statements that they intend to be first in ambitious manned flights. . . . With all this circumstantial evidence I feel it is entirely possible and even probable that they are planning officially to make long-duration and distant manned flights. . . .

"I feel we have the knowledge, the manpower, the resources, and the facilities, to mount the finest space program in the world. I feel it is worth undertaking on its own merits. I am also aware that we are in a race with a competent and tough-minded competitor. We must recognize the nature of this race and organize to meet it, not by year-by-year reactions, but with sound, consistent planning. . . ." (Text)

April 27: At American Physical Society Meeting, the planet Jupiter was reported to be rotating slower. Prof. Alex G. Smith said in 1961 radio sources showed an abrupt change in rotational rate; observations made by radio observatories in Florida and Chile showed that the rotation had slowed by 1.3 sec. Dr. Smith said this was "almost as surprising as if the city of Washington had begun to drift across the surface of the earth at a rate of 10 degrees longitude a year." Studies were continuing to determine whether Jupiter itself had changed its rotation or whether the change was limited to radio sources. (Finney, *NYT*, 4/28/64, 19)

- The "complete success" of Project Vela high-altitude nuclear-detection satellites led DOD's Advanced Research Projects Agency (ARPA) to expand and revise the project. Future launchings had been rescheduled farther apart so that subsequent satellites could incorporate any major improvements between launchings, according to ARPA Director Dr. R. L. Sproull. (*M&R*, 4/27/64, 11)

- Neutron star, detected by x-ray instruments aboard sounding rocket launched last year from White Sands Missile Range, was discussed in paper delivered to National Academy of Sciences by NRL astrophysicist Dr. Herbert Friedman. The invisible star, located in constellation Scorpio, emits about 10 billion times more x-ray intensity than visible light, he said, and density of the star was probably about one billion tons to the cubic inch. Star's surface was "so hot that practically all its luminosity is in the form of X-ray." Another experiment to detect a suspected neutron star in the Crab Nebula was being planned for July 7. (Hines, *Wash. Eve. Star*, 4/27/64; Haseltine, *Wash. Post*, 4/28/64; Sullivan, *NYT*, 4/28/64, 19)

- AFSC announced new graphite refractory coatings for use in aerospace vehicles had been developed by Research and Technology Division's Air Force Materials Laboratory. Materials research engineer Khodabakhsh S. Mazdiyasni applied for patent on the process of coating graphite with zirconium oxide. (AFSC Release 43-13-64)

- Nine-member Republican study group on space and aeronautics, chaired by Rep. Charles S. Gubser (Calif.), said President Johnson had shown a "shocking lack of confidence in United States aircraft manufacturers" with his decision to re-examine the U.S. supersonic transport program. Rep. Gubser charged the decision would "doom the United States to second place in the high-performance transport field." (AP, *NYT*, 5/3/64, 68)

April 27: Remarks of Dr. James R. Killian, Jr., president of MIT, to an MIT alumni group, appeared in *Aviation Week:*

"The growth of our innovation industry—shorthand for 'research, development, test and evaluation'—is a measure of the progress of the U.S. toward becoming a research-based society. From the very earliest days of the Republic, it has been recognized that knowledge is power and that the nation grew stronger by taming nature than it could by framing ideologies. So today we have the research society, affluent in the production of knowledge as well as dollars, with planned discovery, innovation, and experiment constantly renewing its vitality and harnessing the future for the benefit of the present. . . ." (*Av. Wk.*, 4/27/64, 21)

- Prospect that U.S. supersonic transport development program management would be transferred from FAA to DOD reported in *Aviation Week.* "It now appears that Defense Dept. will be assigned over-all management responsibility for development, with FAA acting as a contracting agency. . . ." (*Av. Wk.*, 4/27/64, 38)
- Plans for formation of National Academy of Engineering were drawn up by committee of 25 leading engineers appointed by NAS president Dr. Frederick Seitz. The committee, which would seek Congressional charter for the proposed academy, elected Dr. Augustus B. Kinzel, Union Carbide Corp.'s Vice President for Research, as chairman. (NAS-NRC Release)
- Ray Romatowski, NASA Deputy Assistant Administrator for Legislative Affairs, was among the 12 Federal employees selected by Woodrow Wilson School of Public and International Affairs to study at Princeton Univ. during 1964–65 academic year. (*Wash. Post*, 4/27/64)
- Astronomer Dr. Willem J. Luyten of the Univ. of Minnesota received the James Craig Watson Medal of the National Academy of Sciences at NAS Annual Meeting in Washington. The medal, the Academy's oldest, recognized Dr. Luyten's contributions to understanding of white-dwarf stars. (NAS-NRC Release)
- DOD renamed the A–11 supersonic aircraft the YF12A. A–11 had been the code-designation by manufacturer Lockheed Aircraft Corp., while YF12A would be a military designation. DOD called it an "interceptor aircraft." (UPI, *Wash. Post*, 4/28/64)

April 28: Aerojet General Corp. successfully test fired Apollo service module engine in long duration test at Sacramento, Calif. The multiple-restart, 22,000-lb.-thrust engine was activated four times for total of 421 sec. (UPI, *Houston Chron.*, 4/29/64)

- Use of lasers to measure craters on the moon was reported by J. S. Courtney-Pratt, of Bell Telephone Laboratories, to Society of Photographic Scientists and Engineers in New York. Based on computations made with laser beams sent to crater lips and then to crater bottoms, lunar craters were found to be as deep as 20,000 ft. (Devlin, *NYT*, 4/29/64, 15)
- NASA Administrator James E. Webb, addressing Texas Technological College, said that the role of NASA was "essentially the same as that of its predecessor, the National Advisory Committee for Aeronautics. Both were created in an atmosphere of chagrin and deep concern because prior to World War I and following World War II the nation had forfeited early leadership in a new field of technological opportunity—aero-

nautics in 1915 and space in 1957. In recommending the creation of NACA in 1915, a committee of the Smithsonian Institution presented a Memorial to the Congress which began: 'This country led in the early development of heavier-than-air machines. Today it is far behind. . . .'

"This was at the outset of a war in which, despite the fact that the Wright Brothers had first demonstrated powered flight at Kity Hawk in 1903, the United States had no aircraft design or manufacturing capability of its own. No aircraft of American design or manufacture flew in Europe during World War I, and it remained for NACA, in the years which ensued, to provide the basic research and development upon which achievement of ultimate U.S. air supremacy, civil and military, was based. . . .

"The advent of Sputnik I in 1957 revealed that once again, this time in space, the United States had neglected a clear opportunity for technological leadership and to overcome the deficiency, the President and the Congress turned once again to the formula which had proven so effective in aeronautics for more than 40 years. It created the National Aeronautics and Space Administration as a research and development agency charged with restoring the nation's leadership in space by developing the rockets, the spacecraft, the facilities, the scientific knowledge, and the operational skill required. And like NACA, NASA has fulfilled its mission to the extent that we have now reached a point of parity in space with the Soviet Union. . . ." (Text)

April 28: At annual meeting of National Academy of Sciences, in Washington, Dr. Lloyd V. Berkner, President of Graduate Research Center of the Southwest, was elected to a second four-year term as Treasurer.

Also at the meeting, NAS elected 35 new members on the basis of distinguished and continuing achievements in original research, among them: George Howard Herbig, astronomer and assistant director of Lick Observatory; and Clark Blanchard Millikan, professor of aeronautics and director of the Guggenheim Aeronautical Laboratory.

Election of six distinguished foreign scientists also was announced, among them: Marcel Gilles Josef Minnaert, Professor Emeritus at the Univ. of Utrecht and co-inventor of the stellar curve of growth, one of the most significant developments in modern astronomy; and Maurice Roy, Directeur General de l'Office National d'Etudes et de Recherches Aerospatiales and currently President of the International Committee on Space Research (COSPAR). (NAS–NRS Releases)

* Quasars (quasi-stellar objects) were subject of discussion at American Physical Society meeting in Washington. Too large to be stars, too small to be galaxies, quasars emanate tremendously strong radio waves even though they are billions of light-years away.

Dr. Peter Bergmann of Yeshiva Univ. said pulsing quasar 3C–273 must have been shining for at least a million years, because it must have taken at least that long for the jet of luminous gas protruding from one side to reach its presently observed length. Dr. Bergmann presented his theory that quasars shine because of gravitational collapse—transformation of mass to energy by gravitation rather than stellar fusion.

Dr. Louis Gold and Dr. John W. Moffatt of Martin Co.'s Research Institute for Advanced Studies rejected theories that quasars operate on stellar fusion principle, theorized that "ordered oscillation" was operating inside the vast gaseous envelopes rimming quasars. Laser effect, or coherence of particles, occurred in the gaseous envelope, which was about 1 to 10 light years thick. Dense core of quasar must act as fusion furnace, providing particles which acquire coherence on their way to the outside. (Sullivan, *NYT*, 4/29/64, 15; Haseltine, *Wash. Post*, 4/29/64; Hines, Wash. *Eve. Star*, 4/29/64)

April 29: X–15 No. 1 was flown by Maj. Robert Rushworth (USAF) to altitude of 102,000 ft. and speed of 3,903 mph (mach 5.72) in test to determine effect of shock waves on photography. Both the test and the 105th flight of the X–15 were successful. (UPI, *Wash. Post*, 4/30/64; NASA X–15 Proj. Off.)

- First succesful, full-scale controlled test of parasail was conducted by NASA Manned Spacecraft Center. "It could not have been more successful," Lee Norman, parasail project engineer said. Model of two-man Gemini spacecraft was dropped from C–119 aircraft from 11,000-ft. altitude. Small drogue parachute immediately deployed, and then the parasail quickly deployed and stabilized the spacecraft's fall. After 7.8 min. the spacecraft hit the water of Trinity Bay. Previous attempt to drop-test the parasail was unsuccessful because of deployment problems. (Maloney, *Houston Post*, 4/30/64)

- Unnamed NASA officials were quoted as saying U.S. had detected Soviet failure within the last week in an attempt to launch a probe to the moon, by John Finney in *New York Times*. Failure was said to have occurred within minutes after rocket booster rose from launch pad near the Aral Sea. (Finney, *NYT*, 4/30/64, 9)

- NASA announced $1 million facilities grant to Texas A&M Univ., which would build an activation analysis laboratory for use in its space-related research program for remote analysis of materials through nuclear activation. (NASA Release 64–102)

- Philip Donely, Chief of Flight Mechanics and Technology Div., NASA Langley Research Center, received 1964 Laura Taber Barbour Air Safety Award of the Flight Safety Foundation. Citation said that Donely's "diverse and unique contributions to air safety during his service of over 30 years with the NACA and NASA have made him an internationally recognized authority in the areas of aircraft loads, operating problems, and airworthiness requirements. . . . The counsel he has given to industry, operators, and government agencies on matters such as airworthiness aspects of new aircraft, assessment of design criteria, and airworthiness problems related to aircraft accidents has contributed significantly to the development and operation of safe aircraft." (NASA Release 64–92)

- NASA Deputy Administrator Dr. Hugh L. Dryden said in keynote address at Fourth National Conference on the Peaceful Uses of Space:

  "We have chosen to go to the moon because manned exploration of the moon involves every facet of overall space capability this nation must develop if we are to become a leading spacefaring nation. Also, the moon's pockmarked surface, untouched by water or wind erosion since it has no atmosphere, bears the traces of everything which ever occurred there. . . .

  "We are carrying forward an active national space program, not limited to the moon, encompassing science, advanced engineering, practical applications, including manned space flight.

"We are building toward pre-eminence in every phase of space activity—all the way from microscopic electronic components to skyscraper-tall rockets.

"We are building a network of large-scale engineering facilities, spaceyards, proving grounds, and space ports to assemble, test, and launch the space vehicles we need now and in the future.

"We are creating new national resources of lasting value in these facilities; in the industrial and managerial capabilities we are developing; and in the growing number of scientists and engineers who are learning about space and space technology.

"We are filling the pipelines of hardware and knowledge, and as measured by the financial resources required, will be halfway toward our first manned lunar mission by mid-1965.

"We are accumulating, in space, the basic scientific knowledge about the earth, the solar system, the universe, and about man himself.

"We are bringing benefits not only to the United States but to all the world through the use of space and space technology, employing such new tools as weather, communications, and navigational satellites, and applying space-based techniques, equipment, and materials to improve industrial products, processes and services.

"We are providing a much-needed stimulus to the energies and creativity of people everywhere, particularly to the minds and aspirations of young people.

"We are bringing about increased economic activity at a time when the effects of automation on our society are beginning to be felt.

"And we are making certain, through our sustained efforts, that the realm of space now opening up to us shall be a domain of freedom.

"It is for these reasons that we have mounted the greatest peacetime undertaking in the history of mankind. . . ." (Text)

April 29: Dr. George E. Mueller, NASA Associate Administrator for Manned Space Flight, discussed Project Apollo before the Conference on the Peaceful Uses of Space:

"The overall time phasing . . . is quite conservative. The Apollo spacecraft is being developed on a schedule four years longer than was needed for the Mercury spacecraft, and two years longer than was taken to produce the B-58 bomber. The Saturn IB and Saturn V launch vehicles are being developed on a schedule two years longer than that of the Atlas missile, and a year longer than was required for the Titan. The total duration scheduled for the Apollo program is longer than that of any previous United States research and development effort.

"The Apollo job, of course, is a big one and we will need all of the time allotted. . . . We found . . . that if the remaining six years of work [in Project Apollo] were stretched out over 12 years the total cost of the presently approved manned space flight program would increase by about 30 percent, or about six billion dollars. The economic considerations, therefore, support the maintenance of the present schedule. It is six billion dollars cheaper to continue on the course we are now following than to set out on a new course at this late date. . . .

Among the benefits from Project Apollo is "the rapid advancement of United States capability in space—the ability to undertake whatever space activities the national interest may require.

"There are seven major elements in this capability—people, industrial base, ground facilities, launch vehicles, spacecraft, operational know-how, and the ability to manage research and development. Together, they add up to space power, which provides this country with freedom of action in this new medium. . . ." (Text)

April 29: Delta Air Lines had reserved three delivery positions for U.S. supersonic transport, FAA announced. (FAA Release 64-37)

- Britain's VC-10 long-range jet aircraft made its first commercial passenger flight—a BOAC flight from London to Lagos, Nigeria. The aircraft featured rapid lift, allowing it to use shorter runways than its U.S. counterparts, the Boeing 707 and Douglas DC-8. (Farnsworth, *NYT*, 4/30/64, 57)

April 29–May 1: Air Transport and Space Meeting and Production Forum held in New York, sponsored by Society of Automotive Engineers and American Society of Mechanical Engineers. At the meeting, NASA Ames Research Center scientists reported results of a study of a manned Mars mission: Nova-class launch vehicles and nuclear-propelled spacecraft, combined with extensive aerodynamic braking techniques at Mars and Earth, would allow manned Mars mission to be made with a single, direct launch from Earth. Use of aerodynamic braking at Mars and Earth would reduce total velocity requirements 50 per cent. (*M&R*, 5/4/64, 17)

- Forum on Developments and Techniques for Air Traffic System of the Future, sponsored by Radio Technical Commission for Aeronautics, held in Washington. At the forum, details of Westinghouse Electric's concept of navigation satellite system were disclosed by E. S. Keats, Westinghouse project manager for navigation and surveying systems. The single, synchronous-orbit satellite could provide navigation, communication, and traffic service to aircraft over North Atlantic or Pacific. The satellite would interrogate each aircraft periodically, measuring time to determine navigational position. (*Av. Wk.*, 5/4/64, 69)

April 30: Full-scale paraglider was successfully deployed at Edwards AFB in test conducted by North American Aviation's Space and Information Systems Div. for NASA to demonstrate the method of maneuverable recovery of manned spacecraft. After an air drop from USAF C-130 transport at 33,000 ft., the 31-ft.-long inflatable wing deployed and assumed the programed trimmed glide. (NAA S&ID *Skywriter*, 5/1/64, 1; *Av. Wk.*, 5/4/64, 27)

- NASA Deputy Administrator Dr. Hugh L. Dryden and Astronaut L. Gordon Cooper revealed at luncheon meeting with newsmen in Washington that scientists would be invited to volunteer for training for lunar missions. Plans had been finalized at meeting between NASA officials and members of the National Academy of Sciences' Space Science Board. Up to 50 scientists would be selected to train with the Gemini/Apollo astronauts at NASA Manned Spacecraft Center during the summer months. Then, perhaps two years before a lunar landing, NASA would select some persons from the reserve to begin full-time training as crew members for later lunar flights. (AP, 4/30/64)

- Harold Gold, head of Spacecraft Branch of NASA Lewis Research Center, discussed Sert I project at the Air Transport and Space meeting in New York. Scheduled for launch on a ballistic flight this summer, Sert (Space Electric Rocket Test) would be NASA's first electric engines in space. The two electric engines aboard Sert I would differ: (1) built by Lewis was electron-bombardment ion engine using mercury propellant; (2) built by Hughes Aircraft's Hughes Research Laboratory was contact-ionization thrustor using cesium propellant. Hughes engine would operate for first half of flight, then Lewis engine would take over. "If Sert I data indicate that beam neutralization is completely effective the development of ion thrustors for space missions can continue in vacuum chambers. However, if beam neutralization is of limited effectiveness a new program of vacuum chamber and flight tests may be undertaken," Gold said. (NASA Release 64–101; LRC Release 64–39)

April 30: Development of highly precise laser plasma diagnostic probe by General Electric under contract to Cambridge Research Laboratories was announced by USAF OAR. The probe showed promise as an in-flight instrument to measure electron density distribution in the flow field of a re-entering hypersonic vehicle and simultaneously measure electron temperature of the plasma. (OAR Release 4–64–5)

- FAA announced Spain's Iberia Air Lines had reserved three delivery positions for the U.S. supersonic transport aircraft. This brought total reserved positions to 84 by 17 airlines. (FAA Release 64–40)
- USAF annual report of unidentified flying objects said there were 382 sightings in 1963 compared with 469 in 1962 and 1004 in record year of 1957. During past 16 years, report said, USAF had investigated more than 8,000 sightings, all but 7.7 percent of which had been explained by natural phenomena. (UPI, *Houston Chron.*, 5/1/64)

During April: Aluminum pressure suit designed for astronaut use on the moon was undergoing evaluation at MSC. Officially called a "constant volume, rigid articulated, anthropomorphic protectic suit," it featured increased mobility and constancy of pressure. Rigid suit concept was first introduced by Litton Industries in 1956 for use in pressure chambers. Advanced version for space use, with adaptations and improvements, was proposed to NASA by Litton last year. (MSC Roundup, 4/15/64, 8)

- DOD tracking station was established and began operating in the Seychelles Islands in the western Indian Ocean, some 1,000 mi. northeast of Zanzibar, where NASA tracking station had been closed. (*M&R*, 4/13/64, 9)
- NASA Goddard Space Flight Center's Space Operations Control Center established that Soviet LUNIK IV spacecraft, launched April 21, 1963, was in a barycentric orbit—i.e., orbiting around the center of gravity (4,600 km. from earth's center) of the earth-moon system. This unusual orbit was apparently a coincidence, because final Tass announcement about the probe had said it would become a satellite of the sun. (GSFC, and MSFC *SIN*, 4/64, 15–16)
- First production J–2 rocket engine was accepted by NASA Marshall Space Flight Center from NAA's Rocketdyne Div., then turned over to Douglas Aircraft Co. for use in ground captive tests associated with Saturn S–IVB stage work. (*Marshall Star*, 5/6/64, 1)
- Shipments of nearly six million pounds of large components for NASA Langley Research Center's $12.5 million Space Radiation Effects Laboratory began arriving. Beginning in mid-1965, the new facility would be operated by Virginia Associated Research Center (VARC), comprised of representatives of the College of William and Mary, U. Va., and VPI. (LaRC Release)

During April: NASA awarded contract modification exceeding $12 million to North American Aviation's Space and Information Systems Div. for further work on Saturn V's S-II stage. (*Marshall Star*, 4/22/64, 4)

- Postmaster General John A. Gronouski revealed plans for issuance of Robert H. Goddard air mail commemorative stamp this year. The stamp would honor 50th anniversary of Dr. Goddard's first patent registrations in 1914 for solid/liquid fuel methods and multi-stage rocket designs. (NSC Ltr. 4/1/64; *Goddard News*, 4/20/64, 3)
- Test firing of Bell Aerosystems Co. 40,000-lb.-thrust fluorine-hydrogen rocket engine showed combustion efficiency exceeded 95 per cent of theoretically obtainable maximum. (*Av. Wk.*, 4/27/64, 69)
- Dr. Raymond L. Bisplinghoff, NASA Associate Administrator for Advanced Research and Technology and AIAA Technical Director, said in *Astronautics & Aeronautics* editorial on research exploitation:

  "Those of us who seek specific scientific or technological goals in our free-enterprise system too often lose sight of the fact that the resources made available for this purpose cannot continue to increase indefinitely. In the years to come, it will be increasingly incumbent upon the individual to improve the effectiveness of his R&D activities, and so conserve the nation's resources. This is not just a matter to be dealt with at the national level, but one which will require understanding and discipline at all levels of our profession." (*A&A*, 4/64, 18–19)
- Rep. Emilio Q. Daddario (D.-Conn.) said in *Air Force and Space Digest* article on technological goals:

  "A creative idea may be born in the mind of one man. Of primary importance is the basic and fundamental research necessary to spell out this idea. Here the financial investment is probably the smallest, limited to the man, his instruments, and the time. The amounts of money are relatively modest, except for capital facilities. Indeed, sometimes this money is used and reused, for basic research is often done in the university climate where its effects nourish graduate students as well.

  "The planning, the adaptation, and the oversight of applied research and development is a more resource-consuming operation. By skilled and thoughtful planning project managers can avoid difficulties which tend to drag on important programs. Here the important resource is people, for the decisions are difficult. The history of many important programs in the United States in recent decades affords case studies of matters where waste might have been avoided if top-level people could have anticipated and eliminated specific problems.

  "One of the most difficult problems facing the United States is to determine the standards by which we can exercise the necessary judgment to avoid moving into the most expensive areas of technology—development, testing, and production—at an appropriate time to ensure that we do not become prisoners of high costs and unnecessary use of resources. We need to exercise more rigorous standards, more decisive judgment, and action." (*AF Mag.*, 4/64, 53–56)
- Space Technology Laboratories, Thompson-Ramo-Wooldridge, selected by JPL to provide throttleable vernier engines for Surveyor lunar spacecraft being developed by Hughes Aircraft Co. under JPL supervision. (*Av. Wk.*, 4/27/64, 30)

During April: European Space Research Organization (ESRO) was planning to orbit its first two satellites by late 1967 or early 1968, from U.S. Pacific Missile Range. Using purchased Scout booster, ESRO would launch the 200-lb. satellites into polar orbits. (*Av. Wk.*, 4/27/64, 31)

- Joint Congressional Committee on Atomic Energy voted unrequested funds for flight-testing Snap-10A reactor in the AEC authorization bill for FY 1965. In voting to include the requested amount for Rover nuclear propulsion, the Committee called for review of the program after data from certain critical tests were obtained. "The committee believes that the Rover project is of great importance to the space program . . . [and] the committee therefore intends to follow very closely the developments of the coming year with the hope that this program will lead to successful reactor design and operation." (*M&R*, 4/20/64, 9; *M&R*, 4/27/64, 18)

- National Science Foundation issued study of scientist and engineer manpower. According to the study, "fewer than 765,000 newly trained scientists and engineers will become available to fill more than 1 million openings for them between 1960 and 1970." (*Newsweek*, 4/13/64)

- President Charles de Gaulle approved France's plan to establish satellite launching center on east coast of French Guiana as an alternative to the Hammaguir complex in Algerian desert. Hammaguir site would come under Algerian control in July 1967 unless new treaty could be concluded.

  Later on, Gen. Robert Aubiniere, director-general of France's CNES, said in London that by 1966 France's space program would stabilize at $100 million per year. By 1966 France annually would launch two Diamant-boosted satellites, about 100 sounding rockets, and 300 balloons. (*Av. Wk.*, 4/27/64, 30)

- Col. Francis X. Kane (USAF) said in *Fortune* article that too much emphasis was placed on "computerized planning," thus handicapping our ability to prepare for the unpredictable. ". . . discovery nearly always comes by chance," he said. He proposed "meta-planning," an approach to "bring the whole of experience and of the human personality to bear on planning. . . ." (*Fortune*, 4/64, 147ff)

- United Aircraft Corp. Chairman H. M. Horner told New York Society of Security Analysts that development problems of supersonic transport engine were being underestimated. U.A.'s Pratt & Whitney estimated costs of $1 billion in its proposal to FAA, the figure based on estimated development costs of $500 million and another $500 million to bring the engine up to airline standards of reliability. (*Av. Wk.*, 4/27/64, 47)

- DOD accelerated construction of two Thor launch sites and ground support equipment at Johnston Island, part of DOD's commitment to maintain adequate safeguards against U.S.S.R.'s sudden resumption of nuclear testing. (*M&R*, 4/20/64, 16)

- Spacecraft Technology and Advanced Re-entry Tests (Start) established in Advanced Planning Directorate of the Deputy for Technology, Air Force Space Systems Div. Start included two activities: Asset re-entry vehicle program and unnamed glide re-entry vehicle, still under study. (*Av. Wk.*, 4/27/64, 30)

May 1964

May 1: NASA Associate Administrator Dr. Robert C. Seamans said in speech before Fourth National Conference on the Peaceful Uses of Space:

"We are searching for a type of contract that will motivate the contractor to become more deeply involved in performing work of high quality with maximum speed and minimum cost.

"In this way we hope to reduce the number of persons presently required to carry out what are essentially policing actions. These would be largely unnecessary if we could place more of the responsibility for basic decisions of performance, time, and cost in the hands of industry management. Only through such an approach can we hope to reverse the constant escalation of costs that stems from adding persons on both sides of the equation. That is, as the government adds people, the contractor has to add people to respond to our people, and the result is not satisfactory to either side. . . .

"Let me emphasize . . . that there are three factors we want the contractor to be involved in—cost, time, and performance. We want contracts to be written in such a way that industry management will carefully weigh and consider what any change in his operation will do to all three items. Obviously, there cannot be an alteration in any one of them without a concomitant effect on the other two. We feel that only in this way can the present performance be improved at the contractor level. . . ." (Text)

- President's Advisory Committee on Supersonic Transport held its third meeting in the Pentagon, where it heard executives of the three aircraft and three engine manufacturers that had submitted preliminary designs. (Clark, *NYT*, 5/2/64)
- NASA announced it would award grants to Maryland Univ. and Rice Univ. for construction of aeronautical and space science research facilities. Maryland's $1.5-million grant would be used for a building to provide additional space for research in such areas as chemistry, physics, fluid dynamics, astronomy, computer sciences, and mathematics. Rice's $1.6-million grant would be used for a space science building to accommodate further research in space-related scientific areas. (NASA Release 64/104)

May 2: Five-segment 120-in. rocket motor for Titan III exploded after 55 sec. during static firing at Edwards AFB, Calif. Explosion caused extensive damage to new $10-million solid-motor test facility, dedicated just three hours earlied by SSD commanding officer Maj. Gen. Ben I. Funk, who had called the new stand "the most sophisticated and highly instrumented facility in the world." Stand was operated for USAF by United Technology Center (UTC), and further tests were shifted to UTC's test facility at Coyote, Calif. Cause of explosion was thought to be failure of casing weld when a "boot" in the bottom of the motor was removed. (*M&R*, 5/11/64, 16)

May 2: More than 2,000 high school juniors and seniors attended Student Space Conference at Northeastern Univ., Boston, where they heard remarks by Astronaut L. Gordon Cooper, Bruce T. Lundin of NASA, John D. Nicolaides of NASA, Nancy G. Roman of NASA, and Isaac Asimov, writer and biochemist. (*Boston Sun. Globe,* 5/3/64)

- 150th anniversary of U.S. weather observation network, now composed of more than 13,000 stations providing records of U.S. climate. (Commerce Release WB 64–8)

May 2–3: Lewis Research Center conducted drop tests down a main elevator shaft at Cleveland's Terminal Tower, dropping model experimental packages from heights of up to 400 ft. to gather data for design of decelerator for LRC's new zero-gravity facility. (LRC Release 64–40)

May 4: Appearing at Project Ranger hearings before Subcommittee on NASA Oversight, House Committee on Science and Astronautics, NASA Administrator James E. Webb cited recent management changes by Cal Tech and JPL in the continuing evolutionary process of management relations with NASA: "These changes and continuing improvements are the result of management cooperation and responsiveness over a wide range of complex problems; the new contract we hope to sign soon is a further example of this effort. In this contract, NASA will endeavor to add to the existing system an incentive fee approach, a separate facilities management contract, a joint semi-annual review with CIT of progress and problems, and a revised overhead determination method. We will discard the concept of mutual agreement at the level of every task and replace it with a contractual statement of our current practice, i.e., laboratory acceptance of NASA program direction, after, of course, a full presentation of its views.

"This represents the position we have achieved in our negotiations looking toward the new contract for the period beginning January 1, 1965. However, NASA has made it clear that it wishes to see the agreed upon relationship put into action prior to final execution of the contractual document. This is a prudent way to avoid overlooking some aspect or approach which might yield better results than those we have reduced to contract language in our negotiations. . . ."

Mr. Webb discussed the report on RANGER VI he had sent to Chairman Miller and Chairman Anderson: "This was an important flight; the entire country desired its success. When it did not meet its mission objectives, I felt it was our responsibility to examine the reasons therefore and to report these to the Chairmen of the House and Senate Committees. The report made was the assessment of Dr. Dryden, Dr. Seamans and myself of the technical problems we have encountered and of the possible causes of failure. The Ranger 6 failed because something did not work, not because someone did not work. There had been unstinting effort dedicated to its success; this effort will be combined with lessons learned from the Ranger 6 failure to provide the best assurance of Ranger 7 success. We have invested the energy of some of the best minds of the country in Project Ranger and we expect to capitalize on that investment in both the near and the far future." (Testimony)

- At conclusion of Ranger management hearings by Subcommittee on NASA Oversight, House Committee on Science and Astronautics, Subcommittee Chairman Rep. Joseph E. Karth (D.–Minn.) remarked that testimony "indicated to me that there was a serious loss of NASA liaison or NASA management. This, I think, is primarily the question that is before the subcommittee, having arrived at the conclusion . . . that the professional, technical people at Cal Tech are extremely able people. . . ." (*M&R*, 5/11/64, 15)

May 4: General B. A. Schriever, AFSC Commander, announced organization for National Range Division (NRD), commanded by Maj. Gen. Leighton I. Davis. Air Force Missile Test Center, Patrick AFB, Fla., was redesignated Air Force Eastern Test Range (ETR), and Air Force Space Test Center (Provisional) at Vandenberg AFB was redesignated the Air Force Western Test Range (WTR). Both ranges would be assigned to NRD. General Davis also was assigned responsibility for Satellite Control Facility (SCF), Sunnyvale, Calif., which has operated DOD R&D satellite tracking and control facilities. With Hq. at Andrews AFB, General Davis would function as both NRD Commander and AFSC Deputy Commander for Global Range. (AFSC Release 45–R–50)

- Inserting in *Congressional Record* a *Saturday Evening Post* article entitled "The Great A–11 Deception," Sen. Gordon L. Allott (R.–Colo.) remarked that the A–11 had "caused me a great deal of concern. . . . I still believe it is an outrageous imposition on our system when $500 million or more is spent on an airplane which may be an interceptor or may not be, but which very few—and only a few—of the people in the Government knew about, including even some of the highest ranking military people." (*CR*, 5/4/64, 9579–81)

- President Johnson announced formation of 32-member FAA Women's Advisory Committee on Aviation, following ceremony at the White House in which he presented FAA Decoration for Exceptional Service to Mrs. Jerrie Mock for her recent 22,858-mi. global solo flight. Mrs. Philip A. Hart was named committee chairman and Mrs. Mock, vice chairman. (FAA Release 64–41)

- H. J. High Construction Co. received $1,012,800 NASA contract for construction of third-floor addition to Operations and Checkout Building, John F. Kennedy Space Center, NASA (Merritt Island). Army Corps of Engineers awarded the fixed-price contract. (DOD Release 357–64)

- Unique cost-plus-incentive-fee (CPIF) contract used by USAF for Titan III launch vehicle imposed "strict penalties on the prime contractor [Martin Co.] for deviating from the target cost, schedule, or flight-performance objectives," *Missiles and Rockets* reported. DOD officials said Titan III was "on schedule and within the planned funding." (*M&R*, 5/4/64, 13)

May 4–7: Tenth annual meeting of American Astronautical Society held in New York. Honored by AAS were: Maj. L. Gordon Cooper (USAF), the AAS Flight Achievement Award; Dr. Hubertus O. Strughold, the Melbourne W. Boynton Award; Col. Charles Yeager (USAF), the Victor A. Prather Award. (*Av. Wk.*, 5/18/64, 21)

- At AAS meeting in New York, Dr. Thomas Gold, Cornell Univ. Director of the Center for Radio Physics and Space Research, said the porous, crushable material of the lunar surface extended "to a depth of tens of centimeters or meters," according to radio thermal measurements.

"It seems very probable that a few per cent of the lunar surface is hard and rough, probably in the form of many young craters, while the rest is covered with material considerably less dense than solid rock. . . .

[We must be prepared] to find a crushable surface material many meters deep and smoothly deposited, covering most of the moon, and a few per cent of rough hard rock regions covered only with a thin layer of the fluff. . . ." (*Av. Wk.*, 5/18/64, 83)

May 4–7: At same meeting, Dr. Duane O. Muhleman of JPL criticized "fairy castle" theory of lunar surface composition. Dr. Muhleman said the concept was consistent with lunar measurements, but that "a clever man in the laboratory could build other models of surfaces that would be equally consistent." (*M&R*, 5/18/64, 23)

May 5: Senate Committee on Aeronautical and Space Sciences, meeting in executive session, favorably reported with amendment S. 2446, authorizing $5.246 billion for NASA in FY 1965. This amount restored $52 million of the $110 million cut by the House in its authorization bill in March. (NASA LAR III/89; *Av. Wk.*, 5/11/64, 24)

- Soviet "automatic interplanetary station" ZOND I was 10,137,000 km. (6,298,838 mi.) from the Earth, according to Tass, and more than 60 radio communication sessions had been held with the spacecraft since its launch more than a month ago. The spacecraft had sent scientific information, data on systems operation, and "information relating to the assigned program of investigations." (Tass, *Komsomolskaya Pravda*, 5/6/64, 1, ATSS–T Trans.)

- House Committee on Science and Astronautics, Subcommittee on Space Sciences and Applications, held hearing on geographical distribution and indirect costs of Federal research and development. NASA Deputy Associate Administrator for Industry Affairs, Earl D. Hilburn, said it was NASA's policy to "locate and fully utilize the best technical competence available wherever it might be." It was not NASA policy to award contracts on a geographical basis, Hilburn said, but NASA tried to "provide opportunities for and encourage all qualified industrial firms, universities and non-profit organizations to participate in our program." He added: "NASA makes every effort to award negotiated procurements to contractors who will perform a substantial proportion of their work within labor surplus areas." (NASA LAR III/89; *SBD*, 5/6/64, 30–31)

- Subcommittee on Science, Research and Development, House Committee on Science and Astronautics, opened hearings on problems of the Government-Science relationship. Rep. Emilio Q. Daddario (D.–Conn.), Subcommittee Chairman, said the Subcommittee would inquire into (1) geographical distribution of Federal R&D contracts and grants, and (2) allowance of indirect costs by those performing basic research for the Government. (NASA LAR III/85)

- Lt. Gen. William J. Ely (USA), Deputy Director of Defense Research and Engineering (Administration and Management), said in testimony before House Science and Astronautics Committee's Research and Development Subcommittee:

"For the past six years, over 80% of the net value of RDT&E contracts of the Department of Defense has been to companies concerned with aerospace and electronics. Actually, 11 states received 88.3% of the awards in FY 1962; 18 states received 95.3% of the total. Five major Defense RDT&E complexes, centered around Los Angeles, the San Francisco Bay area, the New York City-northern New Jersey area, Boston, and the Washington, D.C., area accounted for about 58% of the total net value of RDT&E awards.

". . . we recognize that Defense R&D is concentrated geographically and that this concentration will continue for the foreseeable future. The Department of Defense, of all the agencies of government, is the one that must feel itself most constrained to the purpose of getting the best results in RDT&E at the least cost. We cannot speak for the remainder of the government; such agencies as the National Science Foundation and the National Institutes of Health were created in part to support research, and therefore have legitimate additional criteria for their operations. We tend to follow competence where we find it and therefore consider that greater uniformity in the geographical distribution of our R&D contracts and grants can only be achieved to the extent that more uniform competence is developed." (Testimony)

May 5: Membership on the NASA Policy Planning Board, established November 1, was formally designated, with Langley Research Center Director Dr. Floyd L. Thompson as Chairman and NASA General Counsel Walter D. Sohier as Vice Chairman. (NASA Circ. 317)

- On third anniversary of his historic space flight, first American in space Astronaut Alan B. Shepard, Jr. (Cdr., USN) received the Langley Medal, highest award of the Smithsonian Institution, in ceremonies at the Smithsonian, Washington, D.C. (*CR*, 5/5/64, 9767–68)
- L. D. Gable, of Longmont, Colo., died at Cape Kennedy of injuries sustained in the April 15 accidental ignition of Delta X–248. The Ball Bros. technician was third victim of the accident, which was still being investigated. (GSFC PIO)
- NASA Director of Biotechnology and Human Research Dr. Eugene B. Konecci said in keynote address, Fifth Annual Conference on Human Factors in Electronics, in San Diego:

  "There is a great need today for simplification through a more fundamental understanding of phenomena we take for granted. Mathematical theories to aid in simplification and a better understanding of the cosmos around us are needed. Specifically in Electronics we need to re-examine our traditional complex circuitry and the logic behind it, so we can make even more rapid strides in solid state physics and electronics without imposing old circuit theory and wiring diagrams on new active elements. In 50 years an electron tube as we know it will be an archaic museum piece.

  "The greatest need in Human Factors research is a better understanding of man, his capabilities and limitations. The study of human brain mechanisms promises the greatest rewards to science and technology and in turn to the betterment of mankind. We need more cross fertilization between neuro- and psycho-physiologists and electronic engineers, through a new educational tool 'Systems Engineering' that includes the human element. . . ." (Text)
- AIAA 1964 James H. Wyld Memorial Award was presented to Brig. Gen. Joseph S. Bleymaier (USAF) head of USAF Titan III development program, for his work on Titan III launch vehicle. AIAA 1964 Propulsion Award was presented to David Altman, United Technology Center vice president, for his contributions to science of propulsion. (*A&A*, 6/64, 87; AIAA "Honors & Awards")
- Dr. Alexander H. Flax, Assistant Secretary of the Air Force for Research and Development, addressed AIAA in Cleveland: ". . . the number of new systems which can be fully developed is limited. The future, on

the other hand, is highly uncertain. The inevitable result . . . is that the number of major weapon systems being developed for the inventory is relatively small. The number of concepts and potential weapon systems being explored in other categories of R&D are very large but still not all inclusive. The approach which we are taking to assure the availability of proven technology and prototypes of critical system elements is through exploratory and advanced development programs; this does not usually include full vehicle or system prototypes although in some cases it may. By working on the critical technical, cost or operational problems of a potential new system intensively while ignoring the other elements of the system, we are placing ourselves in a better position to decide to buy that system and to get it in reasonable time, in the event that either our own national policy, the international situation (especially the actions of our adversaries), or obsolescence arising from new technological developments in counterweapons make necessary the rapid acquisition of a new system, and we are doing this at considerably lower cost than the cost of developing all the candidate systems even if these developments were at a very low rate. For example research and exploratory development on high energy fuels for both solids and liquids support potential needs both for larger and improved ICBMs and for new space boosters. The 120" and 156" solid rocket under development by the Air Force and the 260" solids being supported by NASA provide at the prototype level the necessary components of much larger ballistic missile and space launch vehicle systems over a wide range of payload capacities. The application of the 156" solid technology to a new missile in the Minuteman tradition would provide a payload capability many times that of Minuteman. Studies are under way to examine the payoffs and tradeoffs relating to the application of high energy fuels to a new upper stage suitable for application to both DOD and NASA launch vehicles. The aim of these programs is to reduce the degree of uncertainty and the element of risk which would be associated with the decision to proceed with such systems. Yet there is no assurance that these developments will, in fact, be used in their present form. They may simply be stepping stones to a later generation of propulsion systems in advanced development. The coupling of requirements for economy, reliability and commonality has led to the dictum for technical advances that many are called, but few are chosen. . . ." (Text, DOD Release 361–64)

May 6: MSC conducted roughwater egress test of Gemini spacecraft using boilerplate version of the spacecraft in the Gulf of Mexico. In waves running from six to eight feet high, Astronaut James A. Lovell, Jr., and Gordon Harvey of MSC Flight Crew Support Div., practiced egress maneuvers with the boilerplate spacecraft, which was loosely tethered to NASA Motor Vessel Retriever. (MSC *Roundup*, 5/13/64, 1)

- AFSC announced experimental arc-jet engine for space use had completed a 500-hour endurance test at Air Force Aero Propulsion Laboratory, Wright-Patterson AFB, Ohio. The 10-lb. regeneratively cooled, hydrogen-propelled engine had specific impulse greater than 1,000 sec. and efficiency of 55 per cent. (AFSC Release 43–94–48)

- At American Astronautical Society meeting in New York, MSC's Dr. Charles A. Berry said Apollo astronauts would be kept in isolation longer than earlier space crews, to prevent infection and disease. Families of crewmen would be given intensive medical care to ensure the good health of the astronauts. If one member of the three-man crew becomes ill, the entire crew would be replaced, he said. (*Houston Chron.*, 5/6/64)

May 6: ComSatCorp filed registration statement with Securities and Exchange Commission seeking registration of 10 million shares of common stock, one half to be offered to communications common carriers authorized by FCC and the other half to the general public, with 11 managing underwriters of the public offering. (ComSatCorp Release 64-5-4; Prospectus)

- NASA announced selection of RCA Missile and Surface Radar Div. for negotiations to procure C-band radars for support of Project Apollo. These radars would be part of instrumentation on the tracking ships being designed and built as joint DOD–NASA effort. (NASA Release 64-107)

- Westinghouse engineer Harry L. Thurman, Deputy Director of laser program in Westinghouse's Defense and Space Center in Baltimore, told AIAA meeting in Los Angeles that laser may someday be used to cook meals in microseconds. Citing many present and future applications of laser, Thurman said industry spent $1,000,000 on laser research in 1960 and estimated Government laser contracts alone in 1964 would amount to $27,000,000. (Roberts, Balt. *Sun*, 5/7/64)

May 7: EXPLORER XVIII, the Interplanetary Monitoring Platform (IMP) satellite, survived sub-zero temperatures during eight-hour flight in the earth's shadow, a duration record for satellites. IMP remained in the shadow for so long because of its highly eccentric orbit, carrying it out 122,000 mi. at apogee. As it entered the shadow, IMP shut off automatically, and temperatures on IMP's surface fell to more than 400° below zero. After it emerged once again, IMP's transmitter signal was picked up by NASA tracking station at Santiago, Chile. Preliminary evaluation of telemetry tapes at NASA Goddard Space Flight Center indicated data from the IMP experiments were of good quality. (NASA Release 64-111)

- First anniversary of TELSTAR II communications satellite. AT&T reported its second communications satellite was "functioning normally in every respect" and had completed 2,340 orbits of the earth. Nearly 60 public demonstrations, most of them overseas TV transmissions, were conducted with TELSTAR II. Radiation data collected by TELSTAR II enabled scientists to determine that the very high intensity of electrons in inner region of the Van Allen belt had decayed slowly during the past year. Built with stronger shielding against radiation than its predecessor and equipped with specially developed transistors to better withstand the space environment, TELSTAR II was orbiting in higher apogee than TELSTAR I, making possible communications experiments of greater length. (AT&T Release)

- At NASA background briefing for news media, Doctors W. Randolph Lovelace II, Orr E. Reynolds, and Frank B. Voris made presentations on space medicine, biosciences, and effects of space environment on man. Dr. Lovelace described the close working relationship between the offices of Space Medicine, Biotechnology, and Space Science and Medicine. He pointed out Center responsibilities as well as the "detailed task-level technical coordination" completed at the NASA–DOD level "which ensures most effective use of all existing facilities" throughout the Nation. In question period, Dr. Reynolds discussed timing of life-detector experiments for Mars: ". . . we had hoped that it would be possible to have a landing capsule on Mars for the 1966 mission. Just as the plans for that were being developed, questions were raised by recent astronomical data about the density of the atmosphere of Mars that made it very difficult, if not impossible, to design the proper landing system.

"At the present time we are hoping to have a landing vehicle for Mars for the 1969 opportunity. This is not an improved program. But we are working on the development of plans in that direction. It has already been decided that the biological sciences payload, that is, the life detector instruments, or whatever else the biological community thinks are most appropriate for that flight, will have high priority on any landing mission.

"So I think the best guess now would be that certainly our first opportunity would be for 1969, and we certainly hope we will have such instruments aboard."

Discussing space radiation, Dr. Reynolds said: ". . . a number of committees and meetings that have been held by the radiobiologists in this country have recommended that an experiment be done below the radiation belt carrying a known source of radiation. The reason for this is that there have been reports from organisms, living organisms flown in the natural radiation belt, of a higher than expected effectiveness of the radiation. There is no good theoretical basis for knowing why this should occur. . . ." Experiment for checking this would be flown in first flight of Biosatellite, carrying known source and dosage of radiation below the area of the Van Allen belts. Asked about the previous data, Dr. Reynolds said:

"There are experiments both by United States experimenters in piggyback flights, and the Russian experiments that show a higher than expected effectiveness of the ambient radiation in genetic effect and in systemic effect. These experiments we strongly suspect had this result due to lack of adequate dosimetry, measurement of the actual dosage that was encountered. For this reason the radiobiologists who have been advising us have recommended this experiment with a known source so as to eliminate any question of whether the dosimetry was wrong. . . ."

Dr. Voris discussed "our studies into the biological effects of extremely low magnetic fields and into alternating and high magnetic field forces. Because of a recently expounded theory that individual neurons possess individual and isolated electromagnetic fields of their own, it is conceivable that the functions of the human nervous system may suffer impairment as a result of exposure to electromagnetic fields that deviate markedly from ambient terrestrial levels. . . ."

Asked about slow rotation for manned space stations to provide partial gravity, Dr. Voris said that it presented "a problem in disorientation and in the effects of mal de mer. Slow rotation is fine if you can keep your head immobilized and become acclimated to it and stay in one spot. If you move around, move with the turn or away from the turn, or move from the periphery of a circular station into the center, you are changing the field force there, you are developing what is called a Coriolis effect, and you then disturb the balance mechanism or vestibular mechanism of the body and you develop recurrent effects such as seasickness." (Transcript)

May 7: Anglo-French manufacturers of the supersonic airliner Concorde revealed a long-range version of the Concorde was in development stage. British Aircraft Corp. and France's Sud-Aviation said the aircraft would have longer body and greater wingspan, greater fuel capacity and greater reserves, increasing passenger load from 100 to 118. (UPI, Phil. *Eve Bull.*, 5/7/64; *Av. Wk.*, 5/11/64, 36)

- NASA Director of Manned Lunar Mission Studies Thomas C. Evans discussed at AAS meeting the systems that might be used for post-Apollo exploration of the moon. The concepts, making maximum use of the investment in Saturn and Apollo, included: Apollo itself, with separate missions to each point of interest; Apollo supplemented with Stay Time Extension Module (STEM), with increased payload capacity developed in Apollo system to transport "saddle bags" of supplies or surface equipment attached to outside of LEM; Apollo supplemented with Apollo Logistic Support System (ALSS), with two flights—one to transport two astronauts, the other to deliver support equipment and supplies probably including lunar surface vehicle; and, Apollo supplemented with Lunar Exploration System for Apollo (LESA), with 2–3 flights and payload to include shelter-laboratory and roving vehicle. (Text)

- In paper on manned lunar scientific operations, prepared for AAS meeting in New York, NASA's Dr. Paul Lowman (Office of Space Science and Applications) and Donald A. Beattie (Office of Manned Space Flight) said that ". . . it is apparent that the moon is a virtual Rosetta stone that, if properly read, may permit us to learn how the solar system, the earth, and the continents on which we live were formed. . . .

  "Since the scientific investigations of highest priority are in the fields of geology and geophysics, it appears that there is definite value in extending manned lunar exploration by the use of modified Apollo equipment. Experience has shown that geophysical and geological mapping investigations of large areas are continuing tasks; like the painting of the Golden Gate Bridge, they are never really finished, because new concepts, questions, and methods evolve during the program. The occasionally-expressed view that the major questions about the moon could be settled by a few landings and the return of a few samples ignore the experience of several centuries of terrestrial geology and geophysics. . . ." (Text)

- Maj. Gen. O. J. Ritland (USAF), Deputy Commander for Space, AFSC, discussed in AAS luncheon address the role of USAF in U.S. space program, tracing history of ICBM development:

  "The task for the Air Force in those early days was a tremendous one. It had full responsibility for the development, acquisition, and deployment of an ICBM force capable of deterring any missile threat the Soviets might develop. This meant that we had to develop reliable high thrust boosters, materials that could withstand extreme heat caused by re-entry friction, guidance components, and a host of subsystems for use in our ICBM's. We had to develop command, control and tracking systems, and a wide range of support and test facilities for launch and retrieval. We had to recruit and train a tremendous force of highly skilled specialists and technicians. In addition we had to set up whole new management structures and apply expedited management methods under which these tasks could be accomplished with the greatest possible speed.

"Time was the critical factor. The Soviets had a very substantial advance in technological lead times over us. For this reason it was essential that we have the fullest cooperation and support from all segments of our society. The response to these needs was characteristic of the American people. . . .

"We had tremendous support from many government agencies. The National Advisory Committee for Aeronautics, which became the National Aeronautics and Space Agency in 1958, gave unfailing support. For example, NACA's Moffett Laboratories did a major portion of the high speed wind tunnel testing for our nose cone development program. NACA's Cleveland Propulsion Laboratories provided vital support to the Air Force in the rocket engine field. These are just two examples of the highly successful partnership which has evolved between the Air Force as the primary military agency in space and the NASA as the primary civilian agency in space. The partnership continues to be one of the most vital and dynamic aspects of the national space effort. . . ." (Text, AFSC Release 45–R–51)

May 7: Martin Marietta recently completed NASA-funded, DOD-concurred-upon study of Centaur upper stage mated with various Titans. Most promising configuration—particularly for heavy Surveyor spacecraft and for advanced Mars missions—was using Centaur as trans-stage for Titan III–A (without the strap-on solids). (*SBD*, 5/7/64, 34)

- FAA adopted safety regulation requiring that cockpit doors on all commercial aircraft be locked throughout flights. (FAA Release 64–45)

May 8: NASA and Smithsonian Astrophysical Observatory had begun operating "Prairie Network" in midwestern U.S., a network of 16 automatic camera stations to photograph falling meteors and aid in tracking and recovering meteorites, NASA announced. Photographic measurements would assist scientists in determining how much mass is lost by bodies during atmospheric entry and in calculating origin of meteors. (NASA Release 64–106)

- NASA signed $80-million incentive contract with Boeing Co. to build five Lunar Orbiter spacecraft. Beginning in 1966, Atlas-Agena boosters would launch Lunar Orbiters from Cape Kennedy on missions to take close-range photographs of moon's surface for scientific study and to help select landing sites for Project Apollo. (NASA Release 64–109)

- Four Soviet scientists reported to Committee on Space Research (COSPAR) in Florence that "a distortion of sex ratio" occurred among offspring of fruit flies that bred on board manned spacecraft during four-day space flights. The abstract prepared for COSPAR preliminary program did not explain the change in sex ratios, did say that experiments' purpose was to study effects of space flight—primarily weightlessness and space radiation—on the fruit fly and its cells. (AP, *Balt. Sun*, 5/9/64; HTNS, *Wash. Post*, 5/9/64)

- Soviet Academician Anatoli Blagonravov, chief Soviet delegate to COSPAR meeting in Florence, indicated to newsmen in press conference that the next major Soviet space flight might be a year away: "We are trying to open a way to the nearest planet. It will take a long time." Asked if this meant that no launch was planned for the present, he replied: "We are preparing projects. The preparation might take a year." (Levin, AP, 5/9/64)

May 8: NASA Manned Spacecraft Center announced scientists were investigating use of chewing gum for astronauts on long-duration space flights, "to stimulate dental tissues and freshen the astronaut's mouth. . . ." (MSC Release 64–69)

- Patent award to Bell Aerosystems Co. for electronic stethoscope, a by-product of space research, was announced. The 20-oz. transistorized stethoscope enabled physicians for the first time to eliminate all unwanted heart sounds and focus on only those desired. Bell was offering manufacturing rights to medical-electronics companies. (Jones, *NYT*, 5/9/64, 31)

May 9: U.S. delegation to international Committee on Space Research (COSPAR) presented annual report of U.S. activities in space. Presented by Richard W. Porter of National Academy of Sciences, the chief of U.S. delegation, the report said 900 small rockets and 100 large rockets for weather and scientific studies were fired by U.S. in 1963, and 56 satellites were successfully launched by the U.S. Porter said Project West Ford copper dipoles placed in space in May 1963 for communications experiments had not interfered with astronomy as had been feared; however, "no further experiments of this kind are now contemplated by the United States." (Levin, AP, 5/9/64)

- Soviet annual report to COSPAR listed 21 satellite and 114 rocket launchings for 1963.

    Omitting any mention of when U.S.S.R. might launch major spacecraft, report said past manned space flight had "set some new problems before scientists requiring immediate solution for successful preparation and realization of more prolonged space missions. . . ." A major question to be answered was whether it was "possible for an astronaut, after two or three weeks stay under conditions of weightlessness, to endure satisfactorily stresses acting on an organism during re-entry." Also studies must be made of methods to safeguard astronauts' cardiovascular systems. Longest section of report was on medical and biological problems of manned space flight; it offered evidence that U.S.S.R. was paying increasing attention to weightlessness and other problems of manned space flight.

    Soviet report was first such version ever submitted in English and was thought to be strikingly free of propaganda. Possibly significantly, report referred to U.S. space research in correlation with Soviet research; for example, findings of MARS I were said to correlate with findings of U.S. EXPLORER X and MARINER II. (AP, 5/9/64; Simons, *Wash. Post*, 5/13/64)

May 11: XB-70 supersonic aircraft was unveiled at North American Aviation plant, Palmdale, Calif. The 2,000-mph intercontinental-range plane would be used to investigate feasibility of long-range high-speed flight. Speaking of achievements in the program thus far, Brig. Gen. Fred J. Ascani (AFSC Aeronautical Systems Div.) said: ". . . Not all of these and other accomplishments can be termed radical breakthroughs (although the XB-70 program has produced about a thousand patent disclosures), but taken together they have helped to advance a technology which constitutes a base for a new generation of civil and military aircraft which will operate at sustained supersonic speeds." (DOD Release 374–64; Witkin, *NYT*, 5/12/64, 1; AF Pol. Info. Ltr., 6/1/64)

May 11: NASA announced appointment of Dr. Alfred J. Eggers, Jr., Assistant Director of Ames Research Center, as Deputy Associate Administrator for Advanced Research and Technology. In the newly created position, Dr. Eggers would work closely with staff members in planning NASA's advanced and supporting research programs. Specialist in problems of hypersonic and space flight research, Dr. Eggers made many contributions to theories of hypersonic aerodynamics and atmospheric entry problems of motion and heating. Among his past awards were Arthur S. Flemming Award in 1956 and Sylvanus Albert Reed Award (1962) of the IAS. (NASA Release 64–110; NASA Announcement 64–95)

- Dr. George E. Mueller, NASA Associate Administrator for Manned Space Flight, told group touring Saturn-Apollo industrial facilities at NASA Michoud Operations that astronauts on flight to the moon would not be harmed by radiation from Van Allen belts or solar flares. Project Apollo astronauts would be "safer than test pilots in industry have ever been," he declared. (Maloney, *Houston Post,* 5/12/64)
- MSFC Director Dr. Wernher von Braun discussed Saturn IB/Centaur launch vehicle at Michoud tour and said that "there will be money to develop the booster in the FY '66 space agency budget." Dr. von Braun said he believed Saturn IB/Centaur would compete on cost-effectiveness basis with Titan III; with the contemplated missions, cost of $20 million per launch was possible. Dr. von Braun also disclosed NASA studies indicating manned circumlunar flight with Gemini spacecraft was feasible. Such a mission was not being seriously considered at this time, however. (*M&R,* 5/18/64, 17)
- AEC announced a satellite carrying small Snap-9A atomic generator had been launched on Thor-Able-Star booster from Vandenberg AFB last April 21, but it did not go into orbit. Preliminary data indicated the satellite burned up upon re-entry into earth's atmosphere. (AP, *NYT,* 5/12/64; *Av. Wk.,* 5/18/64, 38)
- Walter L. Lingle, Jr., NASA Deputy Associate Administrator, said in address to the Missouri Bankers Association in St. Louis:

    "I cannot overemphasize the importance of incentive contracting as a force for improving NASA's own planning and decision making processes, as well as the contractors'.

    "We expect to write more than $500 million of incentive contracts in the present fiscal year.

    "As you know, President Johnson has taken a very personal interest in the cost effectiveness and efficiency of companies which are contractors to the Government. We regard incentive contracting as a primary means for achieving the added efficiency which President Johnson seeks.

    "More than incidentally, we think that they can become a primary means for improving contractor profit levels. It is NASA's policy to let a contractor substantially improve his profit ratios if he can achieve the incentive targets which we mutually establish. Some recent targets which we have established will permit the contractor to more than double the profit ratio . . ., if all incentive targets are achieved." (Text, *CR,* 5/12/64, 10308–11)
- Sen. J. W. Fulbright, Chairman of Senate Foreign Relations Committee, said at Copenhagen celebration of 50th anniversary of Denmark-America Foundation: "The most urgent need of this troubled world is new

thoughts and attitudes toward the old problem of human relations, as distinct from excessive preoccupation with traveling to the moon." (AP, *Kansas City Times*, 5/12/64)

May 11: Dome of ancient rock at Vredefort, South Africa, was said to be one-mile-dia. meteor which impacted earth at 36,000 mph about 250 million years ago. Willian I. Manton, research geologist of Johannesburg's Univ. of the Witwatersrand, said: "It is almost certain that a meteorite struck the earth at Vredefort and completely changed the geological structure for 60 miles around the point of impact." (AP, *Phil. Eve. Bull.*, 5/11/64)

May 11–14: 35th annual meeting of Aerospace Medical Association held at Bal Harbour, Fla. At the meeting, USAF School of Aerospace Medicine scientists led by Dr. Billy Welch reported on SAM's two 30-day experiments with 100% oxygen atmosphere and concluded that 100% oxygen was acceptable for 30-day period. USAF doctors also reported on tests with oxygen-nitrogen mixture at 7 psi for Manned Orbiting Laboratory Problem of bends in test subjects, occurring during change from MOl atmosphere to 3.5 psi atmosphere in Gemini extravehicular spacesuit caused testers to extend denitrogenization period to 4½ hours. (*M&R*, 5/18/64, 16)

- At Aerospace Medical Association conference, Dr. Dietrich E. Beischer and James C. Knepton, Jr., of Naval School of Aviation Medicine said high magnetic fields, such as those under study for spacecraft shielding, may have noticeable effect on behavior of heart. Squirrel monkeys subjected to high magnetic fields (20,000 to 70,000 gauss) showed up to ¼ decrease in heart rate and marked change in heartbeat's rhythm. Dr. Beischer also reported sea urchin eggs exposed to 140,000 gauss were retarded in their development, but fruit fly showed no effects after two-hour exposure to high gauss. High magnetic fields were being investigated for use in shielding astronauts against cosmic radiation and other space hazards, and Dr. Cornelius Tobias and N. M. Amer reported that magnetic fields around 3.5 kilogauss appeared to have protective effect against high doses of ionizing radiation. (*M&R*, 5/18/64, 16)

- At Aerospace Medical Association meeting, Maj. Gen. Marvin C. Demler, Commander of AFSC Research and Technology Div., said USAF was studying use of "space lifeboats" for astronauts who might have to abandon their spacecraft. The self-contained unit would be designed to sustain astronauts until they were rescued. (UPI, *NYT*, 5/12/64, 43)

May 12: X–15 No. 3 flown by NASA pilot Jack McKay to speed of 3,000 mph (mach 4.55), the aircraft's skin reaching recorded temperatures of 800°. With sharp leading edge on top portion of vertical tail altering airflow over the tail surfaces, the X–15 was fitted with special sensors to measure effects of this altered airflow on skin friction and heat transfer rates. Also, special microphone mounted on aircraft's side measured noise level of boundary layer flowing over surfaces. (NASA X–15 Proj. Off.; FRC Release)

- Soviet physicians' report on behavior of human heart in weightlessness of space said increase in gravity when space traveler re-enters earth's atmosphere and returns to earth would place heart muscle under heavy stress. Through Moscow broadcast from Tass, Drs. Roman Bayevsky and Oleg Gazenko reported cosmonauts' pulse rate slowed as did electrical conductance of the heart. The doctors believed that the "lower level of the functioning of the heart and 'readjustment' of the cardiovascular system to a new level should cause a certain weakening of the heart muscle. As far as was known, not one of the Soviet cosmonauts suffered any untoward permanent effects in the state of weightlessness. However, for future prolonged flights, it is extremely important to establish the degree of 'weakening' of the heart. . . ." (AP, 5/12/64)

May 12: FAA announced Panagra and Aeronaves De Mexico had each reserved two delivery positions for U.S. supersonic transport airliner, bringing total reserved positions for SST to 88 by 19 airlines. (FAA Release 64–46)

- District Judge issued temporary restraining order to halt FAA's sonic boom flight-tests over Oklahoma City. The order was issued in suit filed by Attorney George Miskovsky alleging that sonic boom shock waves were "destroying, damaging and weakening the structures of private homes and business properties." Hearing was set for May 19, after which the District Judge refused to halt the sonic boom tests (see May 21). Sonic booms were said to have cracked 147 windows in downtown office buildings. (Miles, *Wash. Post*, 5/12/64; AP, *Wash. Post*, 5/13/64)

May 13: Successful test of Apollo escape system was conducted at White Sands Missile Range by NASA Manned Spacecraft Center. Boilerplate model of Apollo spacecraft (command and service modules) was mounted on Little Joe II. At altitude of 17,000 ft. the Little Joe II was exploded. Instantly the rockets in the escape tower ignited and propelled the Apollo command module away from the exploding booster—to maximum altitude of 24,000 ft.; then the escape tower was jettisoned automatically. Drogue parachute opened from nose to stabilize the spacecraft as it descended. Then at about 7,500 ft. three large parachutes were deployed from the nose. First chute broke loose from the spacecraft, but the spacecraft landed safely with two chutes at speed of about 30 fps. The 7½-min. test was termed successful, with all objectives achieved. (NASA Release 64–108; Freelander, *Houston Post*, 5/14/64; Witkin, *NYT*, 5/14/64, 18)

- Los Alamos Scientific Laboratory successfully operated a Kiwi B–4 reactor, designated the Kiwi B–4D, at near full design power for approximately one minute, at Jackass Flats, Nev. The test used liquid hydrogen as a propellant and coolant with a liquid-hydrogen-cooled nozzle as would be required in a flight system. All principal test objectives were met. Operating time at the power plateau had to be reduced from the planned eight minutes to one minute because of nozzle leaks, but reduced test duration did not compromise the meeting of test objectives. Reactor performance was excellent, and the test was considered a major milestone in the NASA–AEC program to develop technology for using nuclear propulsion in space missions. (Finger, 5/29/64; SNPO)

- Disclosed at press conference in Florence, Italy, in connection with COSPAR meeting, that U.S.–U.S.S.R. cooperative space negotiations would be held in Geneva within two weeks. Heading U.S. team would be NASA Deputy Administrator Dr. Hugh L. Dryden, meeting with Soviet team led by Anatoli Blagonravov. (Simons, *Wash. Post*, 5/14/64)

- Aerospace Medical Association honored Dr. W. Randolph Lovelace II and Col. William K. Douglas at its meeting in Miami. Dr. Lovelace, President of Lovelace Foundation for Medical Education and Research and recently appointed NASA Director of Space Medicine, received the 1964 Theodore C. Lyster award, and Dr. Douglas, Assistant Deputy for Bioastronautics at AFMTC and formerly personal physician to the original seven Mercury astronauts, received the Louis H. Bauer Founders award. (Wash. *Eve. Star*, 5/14/64)

May 13: NASA and DOD announced Astronaut M. Scott Carpenter (Lt. Cdr., USN) would participate in USN Project Sea Lab I off coast of Bermuda in July. The submerged lab would collect and evaluate scientific data on physiological and psychological reactions to the undersea environment. (NASA Release 64–112; DOD Release 383–64)

- Launch of Ranger 7 lunar probe was postponed from late June date to third quarter of the year, to ensure most careful possible check and review of the many tests being made on the spacecraft. (NASA Release 64–114)

May 14: ZOND I interplanetary spacecraft made a course correction maneuver 13 million km. from earth, according to Soviet news agency Tass. Radio signals from ground command station brought the spacecraft "into an assigned position," and onboard engine "boosted the station's speed by about 50 meters per second." Tass said the maneuvers of the "test session" were performed successfully, "thereby allowing the solution of problems connected with the precise correction of trajectory of the automatic cosmic apparatuses." (Tass, *Pravda*, 5/19/64, ATSS–T Trans.)

- Completely assembled four-stage NASA launch vehicle was transported for first time in an airplane, the 72-ft.-long Scout flown from NASA Wallops Station to Langley AFB, Hampton, Va., in C–133 transport. Flight was initial test of procedures being considered to improve efficiency and reliability of launch vehicle operations. (NASA Release 64–117; Wallops Release 64–42)

- NASA announced Collins Radio Co. and Philco Corp. had been selected for final competition for prime contract to provide Apollo Network S–Band Systems. Estimated $5-million contract called for basic purchase of two complete tracking and data acquisition systems with 30-ft.-diameter antennas and three electronic subsystems. The band would handle such functions as tracking, telemetry, and TV and voice communication. (NASA Release 64–116)

- 25-hour high-altitude balloon flight was completed when 100-lb. instrument package was dropped from the balloon near Oakwood, Tex., where it was "recovered in good shape." Released from the Palestine, Tex., National Center for Atmospheric Research station, the balloon carried instrumentation to study cosmic rays to altitude of 120,000 ft. (*Houston Post*, 5/15/64)

- JPL dedicated its new Space Flight Operations Facility at Pasadena, Calif., the facility to act as command center for future unmanned space flights to the moon and planets. Participating in dedication ceremonies were Sen. Clinton P. Anderson, Rep. George P. Miller, Dr. Robert C. Seamans, Jr., Dr. Homer E. Newell, Dr. Lee A. DuBridge, and Dr. William H. Pickering. At dedication ceremony, radar signal was bounced off planet Venus, the 83 million mi. round trip taking seven minutes and 25 sec. (*SBD*, 5/14/64, 75–76; AP, *Houston Post*, 5/14/64)

- At the dedication of the JPL Space Flight Operations Facility, Dr. Homer E. Newell, NASA Associate Administrator for Space Science and Applications, remarked:

"Often in our discussions of the Space Program we refer to the 'manned program' and the 'unmanned program.' We even talk about manned space science and unmanned space science. Strictly speaking, this is loose talk. There is, in fact, no such thing as unmanned science. Man and his thinking are the prime ingredients of science.

"In the Space Program, it becomes simply a matter of where the man is in relation to his instruments and measuring devices. For those space missions that we call unmanned, man is back on earth while his eyes and ears and other senses are extended electronically and mechanically far out into space by the spacecraft and its instrumentation. From his remote position, he must monitor, issue commands, receive and record data, make routine or emergency decisions as required in a continuing interchange with his inanimate partner out in space.

"It is to make this partnership effective and productive of data, measurements, observations, and information that a centralized facility like the SFOF is necessary. . . ." (Text)

May 14: Langley Research Center announced three key management appointments: Mr. Eugene C. Draley was appointed Assistant Director for Flight Projects; Dr. John E. Duberg was named an Assistant Director in charge of the research activities of the Dynamic Loads Div. and the Structures Research Div., and Mr. Francis B. Smith was appointed Assistant Director responsible for the Analysis and Computation Div., the Instrument Research Div., and the Space Mechanics Div. (LARC Release)

- NASA announced it would negotiate contract with General Electric Co.'s Valley Forge Space Technology Center to design and build gravity-gradient stabilization system for Advanced Technological Satellite. Contract was expected to exceed $5 million. The GE devices would be flown on three satellites; in addition, GE would supply two prototype and two engineering gravity-gradient system models plus ground operations and data analysis. (NASA Release 64–115)

- Tenth anniversary of directive to accelerate Atlas ICBM development program. Ten years ago General Thomas D. White, then USAF Vice Chief of Staff, directed that Atlas ICBM program proceed with highest USAF priority. (AFSC Releases)

- Explosion and fire destroyed Atlas F ICBM site near Frederick, Okla. The missile was not armed, and there were no injuries in the explosion. This was fourth such event in less than a year and prompted renewed inquiry into the situation by Senate Preparedness Subcommittee. (UPI, *Wash. Post*, 5/15/64; *M&R*, 5/25/64, 11)

- Federal Judge Stephen Chandler refused to reinstate State District Court's temporary restraining order against FAA's sonic boom tests at Oklahoma City, and FAA resumed its schedule of eight flights per day. (AP, *Wash. Post*, 5/14/64; AP, Balt. *Sun*, 5/15/64)

May 15: Soviet cosmonauts Andrian Nikolayev and Pavel Popovich (VOSTOK III and IV) had orthostatic hypotension following their space flights, Soviet physiologist Prof. Vasily V. Parin revealed. The condition was noted in the cosmonauts "even two days after the flight." In address at COSPAR—prepared and delivered in English—and in the discussion which followed, the Soviet Director of Institute of Normal and Pathological Physiology, Soviet Academy of Medical Sciences, said that the cosmonauts "showed no disturbances in their state of health" after their prolonged space flights, and the "transient functional reactions" were said to have "disappeared in seven or 10 days. . . ." He added that "post-flight disturbances were largely related to the central nervous system, the cardiovascular systems and metabolism."

Prof. Nello Pace, head of Physiology Dept. at Univ. of California (Berkeley), closely questioned Prof. Parin about his statements. Prof. Pace considered that hypotension was "among the most significant facts" disclosed at the conference. He noted that the condition had been observed in U.S. Astronauts Schirra and Cooper, and that NASA had said that implications of blood pressure responses "will have to be given very serious consideration as larger missions are undertaken." (Text; Hillaby, *NYT*, 5/16/64, 8; Burkett, *Houston Chron.*, 5/17/64)

May 15: USAF grounded its fleet of more than 500 F-105 aircraft because of recent crashes. The F-105 had been involved in 18 major accidents this year, DOD spokesman said, and four pilots had been killed in 10 major accidents since March 1. (Raymond, *NYT*, 5/16/64, 8)

- Hand-held navigational device called a "stadimeter," conceived by USAF scientists at Air Force Avionics Laboratory, was displayed by AFSC. To be tested in Project Gemini manned space flights, the device used only manual and visual alignment and required no electronics. Astronauts would use stadimeter to determine optically the range of the spacecraft from the earth. (AFSC Release 44-14-56)

- General order from AFSC established Eastern and Western Test Ranges of the National Range Division. WTR would assume range support for ICBM and satellite programs on a phased basis, gaining full responsibility some time between Mar. 1 and Jun. 30, 1965. Until then, overall responsibilities for these functions continued with PMR; after that time, PMR would continue to operate as a national sea test range under the Navy. It would comprise facilities in Pt. Mugu-San Nicholas Island area and certain downrange stations in the area of Hawaii, and would be responsible for Pacific Ocean-based launches. (AFSC Release 45-R-61; NASA Announcement 61-161)

- First meeting of NASA Historical Advisory Committee, chaired by Prof. Wood Gray of George Washington Univ.

May 16: New and more detailed information about Van Allen belts gleaned by HITCHHIKER I satellite was reported to COSPAR by Ludwig Katz, AFCRL project scientist. HITCHHIKER satellite was launched pickaback on USAF satellite and separated last July 1. In the AFCRL measurements the count rate for protons between one and five million electron volts (mev) indicated that for this energy region there were two distinct Van Allen belts. The count was significantly lower in the middle region between two belts. A second feature, found in measurements between 15 and 100 kev, was that more electrons were found in energy region of the radiation belt, although there was decrease in total flux of electrons from the inner to the outer areas of earth's magnetic lines of force. (USAF OAR Release 5-64-2)

- In speech to COSPAR on the NASA Biosatellite Program, Dr. Dale Jenkins, Bioscience Programs Div., NASA Office of Space Science and Applications, remarked:

"The United States Biosatellite Program is a second generation series of carefully planned and selected experiments, including some highly sophisticated experiments which have required several years of baseline study and equipment development. This program has been designed to study the biological effects of: (a) zero gravity or weightlessness, (b) the effects of weightlessness combined with a known source of radiation, and (c) removal of living organisms from the Earth's rotation. These orbiting recoverable Biosatellites provide an opportunity to critically test major biological hypotheses in the areas of genetics, evolution, and physiology. The Biosatellite studies will delineate hazards to astronauts, and determine and define effects on degradation of human performance. Prolonged manned flights, for example, may involve changes observed in ten days of bed rest such as decalcification of bones (particularly the vertebrae), loss of muscle tone and physical capability, and certain cardiovascular changes. The effect of continued sensory deprivation on behavior and performance is unknown." (Text)

May 16: Results of Soviet biological experiments reported to COSPAR meeting. Discussing fruit fly experiments in space [see May 8], report said fruit flies mated in space apparently normally, but when the eggs hatched in the laboratory after the flight there was excess number of females. Furthermore, report said, average weight of male fruit flies was lower than of flies bred in the laboratory. Soviets disclosed that Cosmonaut V. Bykovsky had conducted fixed biological experiments during his five-day VOSTOK V space flight, killing specimens to stop their development at various intervals. In another biological experiment, Soviets reported that normal cell division in spores of Tradescantia plant was upset in space environment, with rearrangements of plant chromosomes as well as development of freak cells. (Simons, *Wash. Post*, 5/17/64)

- Eight Tiros meteorological satellites had taken more than 350,000 photographs from altitudes above 400 mi. The spectacularly successful weather-forecasting satellites did not live up to the one-time charge of "spying," for of all the pictures, only two showed any indication of existence of intelligent life on earth—and both of these pictures required corroborative knowledge to identify them. (UPI, Phil. *Sun. Bull.*, 5/17/64)

- Temperature on planet Jupiter was reported to be higher than previously thought. Geology-astronomy team at Cal Tech that studied the planet with heat-sensitive detector fitted to 200-in. telescope on Mt. Palomar found sunlit atmosphere of Jupiter averaged 230 degrees below zero F. But in shadow cast by one of its moons—Ganymede—temperature jumped to minus 117 degrees. Dr. Bruce C. Murray, associate professor of planetary sciences, and Dr. Robert L. Wildey, research fellow in astronomy and geology, gave two possible explanations for the unexpected rise in temperature: "Perhaps in sunlight the atmosphere is photochemically stable, and the shadow removes the stability. Then the atmosphere may start to break down into more transparent material, permitting the heat detector to observe much deeper into the atmosphere where temperatures are higher. Another possibility is that a breakdown of the shaded atmosphere may result in an upwelling from deeper, warmer zones of the atmosphere." Study also showed that another of Jupiter's moons—Callisto—radiated twice as much heat as a normal moon would be expected to do. (Dighton, AP, *Wash. Post*, 5/17/64)

May 17: Herbert Friedman of Naval Research Laboratory reported to COSPAR a new picture of space cosmic rays, based on results of experiments with rocket-borne x-ray telescope. New results showed that less than one per cent of all cosmic rays could be produced in intergalactic space. Therefore, Friedman said, all cosmic rays would appear to come from exploding stars. He said these findings showed that there were too many x-rays to support theory that x-rays are produced by cosmic rays' colliding with starlight, and too few x-rays to support theory that they come from the near-collision of electrons and protons (formed by neutron's disintegration in so-called "hot universe" theory). Further experiments were planned to gather more evidence. (Simons, *Wash. Post*, 5/18/64)

May 17: Dr. N. A. Kozyrev of Pulkovo Observatory, U.S.S.R., told conference on geological problems in lunar research in New York that volcanic activity was evident in two of the moon's craters—Alphonsus and Aristarchus. Overall, he said, "the Moon had an internal energy which can be not less than the internal energy of the earth." And from analysis of certain light wavelengths, he believed there was "significant volcanic activity on Venus and only slight activity on Mars." (AP, *Houston Post*, 5/18/64)

- At COSPAR, Harvard Univ. scientist Dr. Carl Sagan said acetaldehyde gas, believed to exist in atmosphere of Mars, turns to sugar on contact with ultraviolet sunlight. Martian life, which probably is microorganic, possibly gives off the acetaldehyde gas and then feeds on the sugar that falls from the atmosphere, Dr. Sagan said. (AP, *Houston Post*, 5/18/64)

- Dr. Thomas Gold, Chairman of Cornell Univ.'s Astronomy Dept., told COSPAR that laboratory experiments to simulate conditions on lunar surface pointed "toward a surface of soft, crushable material, between 6 and 10 times less dense than solid rock. It would have a texture something like cake. A man walking on the moon might sink pretty far into the soft surface. How far we don't know." (AP, *Houston Post*, 5/18/64)

May 18: House Appropriations Committee voted to approve $5.2 billion in funds for NASA in FY 1965, $245 million less than the requested $5.445 billion ($5.304 billion for FY 1965 and $141 million FY 1964 supplemental combined by the Committee). The appropriation was contained in FY 1965 Independent Offices Appropriations Bill. (Finney, *NYT*, 5/19/64, 1)

- U.S.S.R. announced routine launching of COSMOS XXX satellite. Initial orbital data, according to Tass: 383.1-km. apogee (238 mi.), 206.6-km. perigee (128 mi.), 90.24-min. period, and 64.94° inclination to the equator. (*Pravda*, 5/19/64, ATSS–T Trans.)

- Discovery of relatively cool region inside three-million-degree solar corona was announced at COSPAR meeting by Dr. Armin Deutsch of Mt. Wilson and Palomar Observatories and Dr. Guglielmo Righini, Director of Astrophysical Observatory at Arcetri, Italy. The corona clouds were believed to contain ionized calcium at relatively cool 20,000 degrees. They were discovered in analyzing spectrogram taken while the two astronomers were in DC–8 "flying observatory" during solar eclipse last July. (Reuters, *Houston Post*, 5/19/64)

- Sen. Hubert H. Humphrey introduced in the Senate a bill to establish an Office of Solar Energy in the Dept. of Interior "to foster and coordinate the research, development, and utilization of solar energy to serve many terrestrial power needs." The bill (S. 2853) was referred to Senate Committee on Aeronautical and Space Sciences. (*CR*, 5/18/64, 10800–02)

May 18: A. O. Tischler, Director of Propulsion in NASA Office of Advanced Research and Technology, said $4 million added by Congress for advanced propulsion work in FY 1965 would allow NASA to investigate a number of new possibilities, including all the amines and select hydrocarbons. Also there would be heavy concentration on materials technology and cooling techniques. (*M&R*, 5/18/64, 9)

- Sen. A. S. (Mike) Monroney (D.-Okla.) wrote in *Washington World:*
"Space research and development may be the glamour girl right now, but the lady-who can do the work for us is aviation. . . . The market exists in the field of aeronautics, while it is still too soon to start printing tickets for round trip flights to the moon.
"Government itself is in danger of being too yielding to the glamour girl. My principal complaint is about NASA's division of the $5.5 billion it will spend in Fiscal year 1965. Only $37 million is budgeted for aeronautical research. The percentage for aeronautics is only seven-tenths of one per cent. Space exports will develop as we excel in this field, but by products—not systems—will constitute the commodities. Aviation has immediate economic potential, and we must pursue it vigorously." (*Wash. World*, 5/18/64)

- Noting there were 262 USAF officers assigned to NASA and that NASA had requested another 55, and pointing out areas of NASA–USAF mutual research and test support, *Missiles and Rockets* editorialized that U.S. military space effort was inadequate: ". . . We would like to see somewhat less worry about coordination with NASA and considerably more concern with staying ahead of the Soviet Union in space technology." (*M&R*, 5/18/64, 46)

- Aviatrix Jacqueline Cochran claimed new speed record for women—1,429 mph (mach 2.2) at 37,100-ft. altitude, flying in F-104G at Edwards AFB, Calif. (AP, *Wash. Post*, 5/19/64)

- GSFC announced engineers Barbara Lunde and William D. Hibbard had invented a valve with no moving parts for controlling flow of liquid propellants to satellite reaction jets. Valve freezes and thaws the propellant in the feed line. (*Goddard News*, 5/18/64, 2)

- Seventeen of the buildings programed for the John F. Kennedy Space Center, NASA, in the Merritt Island spaceport area had been completed or were about ready for occupancy, KSC announced. (KSC Release 69–64)

- USAF advanced development program to investigate laser's intelligence-gathering capabilities in space was reported. Called "Lariat" (Laser Radar Intelligence Acquisition Technology), program was outgrowth of three-year effort by AFSC Aeronautical Systems Div. to develop satellite surveillance system at Cloudcroft, N. Mex. (*Av. Wk.*, 5/18/64, 104)

- Three persons were killed and eight injured when small missile exploded during preparations for test-firing from an airplane at Tyndall AFB, Fla. (AP, *Wash. Post*, 5/19/64)

May 18–22: Review of recent accomplishments in advanced aeronautical and space research was held during Inspection Week at NASA Langley Research Center, with Ames Research Center and Lewis Research Center also participating. More than 2,000 guests from aerospace industry, Congress and Government agencies, and educators and civic officials attended the inspection. At the tour, LaRC scientists described newly developed material for spacecraft heat shields—silicone elastomer—

containing hollow glass and plastic spheres in plastic honeycomb matrix to provide structural strength. Lunar Landing Research Facility and lunar mission simulators also were among the featured attractions. In conjunction with the inspection was the LaRC Open House on May 23. (Program; NASA Release 64–99; ARC, LARC, LRC Releases; AP, 5/18/64)

NASA Administrator James E. Webb remarked on May 18:

"Years ago, the National Advisory Committee for Aeronautics, one of NASA's ancestors, annually conducted field inspection trips so that the nation's leaders in aviation could have a first-hand look at the progress in research made by the scientists and engineers of NACA.

"Today, we are gathered for a first-hand look at advanced research and technology not only in aeronautics, but also in the new realm of space.

"It is most fitting that the revival of the annual field inspection trip takes place first at the Langley Research Center, for Langley was the first research center of the original NACA, and is today, as it was in 1917, working at the frontiers of scientific and technical knowledge. . . .

"It is important to remember that NASA is not an operating agency, in the sense that the Weather Bureau and the Air Force are. Rather, NASA's role is much closer to that of the NACA, a research and development organization. . . ." (Text)

May 19: X–15 No. 1 flown to 196,000-ft. altitude and 3,477-mph speed (mach 4.85) by USAF pilot Capt. Joe Engle in flight to serve as altitude buildup for Captain Engle. The flight also conducted optical degradation studies. (NASA X–15 Proj. Off.; FRC Release)

- Two successful tests of parasail landing gear were conducted by NASA and U.S. Army at Ft. Hood, Tex. Army helicopter dropped one-third scale model of Gemini spacecraft from altitudes of about 10,000 ft., the attached parasail maneuvered by radio signals from the ground. (UPI, *Houston Chron.*, 5/20/64)

- USAF launched an unidentified satellite aboard an Atlas-Agena D booster from Point Arguello, Calif. (AP, 5/19/64)

- MSC began 10-day series of short weightlessness tests with Gemini spacecraft installed inside a KC–135 jet aircraft. Astronaut John Young took part in the tests along with four USAF and McDonnell Aircraft Corp. test pilots. (*Houston Chron.*, 5/19/64)

- U.S. National Academy of Sciences and Soviet Academy of Sciences announced signing of exchange agreement providing for each academy to send 55 scientists to the other country for inspection visits, lectures, and cooperative research. This was third such agreement since the General Cultural Exchange Program was begun in 1959. (Finney, *NYT*, 5/20/64, 4)

- General B. A. Schriever, AFSC, said in address to National Space Club:

"There are three facts about the ballistic missile and space effort which I would like to emphasize: first, it has given us today a substantial strategic superiority; second, it has built a base of experience, facilities, and hardware for the national space program; and third, it provides convincing evidence that trained and dedicated people, who are properly motivated, can work together to achieve something that skeptics consider to be 'impossible.' . . .

"These five factors . . .—awareness of the competition, rapid advancement of technology, acquisition of modern research and test facilities, employment of skilled and dedicated people, and use of streamlined methods of management—all of these were vital to the achievements of our ballistic missile and space efforts during the past ten years. The same factors are vital to any future national effort.

"One very clear lesson emerges from our ICBM experience, and it is this—we cannot afford to rest on our past accomplishments. Technology is not standing still. It keeps moving rapidly ahead, and we must keep pace with it. We must make use of our past experience as we tackle the challenges of the future." (Text, AFSC Release 45-R-62)

May 19: DOD announced Deputy Secretary of Defense had directed military departments and Defense agencies to establish central pool of funds and manpower spaces to be used for long-term education and training of key civilian DOD employees. (DOD Release 395-64)

May 20: White House announced new development plan for U.S. supersonic transport, authorizing technical design competition between two airframe manufacturers (Boeing and Lockheed) and two engine contractors (General Electric and Pratt & Whitney). Thus North American Aviation and Curtiss-Wright were eliminated from airframe and engine competition respectively. New "program of action" recommended by FAA and President's Advisory Committee on Supersonic Transport and endorsed by President Johnson called for: Commerce Dept. studies and forecasts of overall economy of the 1970's; National Academy of Sciences advice on sonic boom studies; and Boeing and Lockheed examination of purchase cost, operating cost, and sonic booms as related to various designs, sizes, and ranges of the plane. This phase would be completed in November. (Clark, *NYT*, 5/21/64, 1; *Av. Wk.*, 5/25/64, 22)

- FAA Administrator N. E. Halaby announced National Academy of Sciences had assumed advisory capacity in connection with sonic-boom research program, at invitation of FAA. First step in NAS program would be review of sonic boom study now being conducted at Oklahoma City. (FAA Release 64-50)
- During annual meeting of international Committee on Space Research (COSPAR), which ended today, U.S. and Soviet scientists said manned space flights had raised the question of whether a man could tolerate more than five days in space, with present equipment. Scientists said astronauts showed effects of space flights, and last year's five-day manned space flight by Soviet cosmonaut was about the longest that could be attempted safely now. (Levin, AP, 5/20/64)
- Independent Offices Appropriations bill (H.R. 11296) was introduced in the House and discussed, with no resolution reached. The bill provided $5.2 billion for NASA in FY 1965. (*CR*, 5/20/64, 11063-66)
- NASA Marshall Space Flight Center selected Bendix Corp. and Boeing Aerospace Div. for negotiations of separate identical nine-month $1.5 million studies of lunar exploration payloads for Apollo Logistics Support System (ALSS). Payload would be carried to moon by Saturn V/Apollo using modified LEM ("LEM truck") to deliver from lunar orbit to lunar surface. (*Marshall Star*, 5/20/64, 1)
- NASA Administrator James E. Webb said in speech to Armed Forces Communications and Electronics Association, Washington:

"At the National Aeronautics and Space Administration, it has been estimated that about 40 percent of our booster costs, 70 percent of our major spacecraft dollars, and 90 percent of our tracking and data acquisition funds go into electronics. Unfortunately, as Dr. Albert Kelley . . . recently pointed out in this same context, a vast majority of our flight failures, not to mention flight delays, arise from electronic failures. Success in this area is obviously a major factor in overall mission accomplishment. . . .

"In space activity the United States is now at the mid-point of a 10-year program. It might be said, in a sense not entirely figurative, that we have reached the 'mid-course maneuver point.' Just as the Mariner II Venus fly-by mission would have missed by almost a quarter of a million miles instead of coming within 22,000-mile target area, we can fail to achieve our full potential if we do not seize every opportunity to correct our course as we go along. . . ." (Text)

May 20: In address to the Wayne State University College of Engineering Honors Convocation in Detroit, Mich., Dr. George E. Mueller, NASA Associate Administrator for Manned Space Flight, said:

"In the manned space flight program, we have tried to profit from the experience of the past. We are creating this very great capability in advance of the formulation of many of the specific operational requirements, which inevitably will emerge when it becomes clear what can be done. There are many elements to this capability. It includes trained people, a base of industry throughout the United States, ground facilities, launch vehicles, spacecraft, operational experience, and the ability to manage large research and development efforts. All of these things will enable the United States to undertake whatever exploration of space the national interest may require.

"In the three years since 1961, the manned space flight effort has built up at a rapid rate until there are now 250,000 persons at work throughout the country—more than 90 percent in private industry—and their numbers will increase to 300,000 when we reach the peak activity of the presently approved programs next year. About 15 percent of these are scientists and engineers. The program has become a major element in the economic life of the country. Prime contractors assign work to subcontractors, who in turn engage second and third-tier subcontractors.

"As the program progresses, we are now beginning to realize some of the benefits. Whereas six years ago we were extremely limited in the power available to place objects in space, we are now approaching a period when we will have the rockets available to do things beyond those being used in present programs.

"The time is approaching when we must give more thought to the exploitation of the capabilities provided through these efforts. To avoid the pitfall into which we have fallen in the past, we must plan for the future even while we press vigorously to accomplish present programs." (Text)

- President Johnson presented President's Safety Award for 1963 to NASA at White House ceremony. NASA Associate Administrator Dr. Robert C. Seamans, Jr., accepted the award on NASA's behalf. (NASA Safety Off.; *Marshall Star*, 5/20/64, 2)
- LRC announced two technical staff members, Richard A. Terselic and Dr. Louis Rosenblum, had applied for a patent on a portable electron-beam welder. (LRC Release 64-44)

May 20: Resolution approved at final session of COSPAR recommended that no spacecraft be landed on Mars if there were more than one chance in 10,000 that it was carrying earthly micro-organisms that could contaminate Mars. Resolution also said no nation should attempt an "unsterilized fly-by" of Mars if there were more than three chances in 100,000 of its accidentally hitting the planet. (AP, *Tulsa Daily World,* 5/21/64)

- NASA Administrator James E. Webb was presented with Armed Forces Communications and Electronics Association's Distinguished Service Honor Gold Medal Award and Certificate at the Association's annual banquet. (*A-N-AF J&R,* 5/23/64, 2)

May 21: X-15 No. 3 flown by NASA pilot Milton O. Thompson to 63,000-ft. altitude and 2,045-mph speed (mach 3.10). The X-15 rocket engine burned 41 sec. before Thompson cut back power to planned 40 per cent. Then engine unexpectedly cut off. Thompson could not restart it, and he had to make emergency landing on Cuddeback Dry Lake. Cause of premature shutdown was malfunction in engine control system. This was fourth forced landing in 107 X-15 flights. (NASA X-15 Proj. Off.; AP, Balt. *Sun,* 5/22/64)

- First two-way telephone call between U.S. and Japan via a communications satellite was conducted by personnel of NASA Goddard Space Flight Center, Greenbelt, Md., and Radio Research Laboratory Station, Kashima, Japan, using RELAY II satellite in orbit about 4,000 mi. above the Pacific Ocean. (NASA Release 64-122)

- House passed H.R. 11296 by voice vote, making FY 1965 appropriations for independent offices but excluding the $5.2 billion for NASA. Because Senate had not yet acted upon NASA authorization legislation for FY 1965, Congressional rules permitted any member to block appropriations on a point of order. Rep. John W. Wydler (R.-N.Y.) objected to the construction and administrative operation appropriation—because of $60 million included for electronics research center in Boston—and his objection was sustained. Then Rep. Olin E. Teague (D.-Tex.), Chairman of Manned Space Flight Subcommittee of House Committee on Science and Astronautics, objected to remainder of the NASA appropriations, thus blocking any attempts to cut them further. (NASA LAR III/102; *CR,* 5/21/64, 11241 ff)

- Following House debate of Independent Offices Appropriations Bill, Rep. Olin E. Teague (D.-Tex.) told press interviewers that cuts from FY 1965 space appropriations came "very close to being the straw that broke the camel's back" on goal to land man on the moon by 1970. $245 million less than requested, the $5.2 billion appropriation would leave "no place to cut this budget." Without the $141 million requested supplemental, which was included in the $245 million cut, Rep. Teague said he feared manned lunar landing goal was lost. (AP, 5/21/64)

- Rep. Louis C. Wyman (R.-N.H.) submitted his separate views on NASA FY 1965 appropriations, which he had filed with Independent Offices Subcommittee. In his report, he said: "Two things are seriously wrong in our space spending: First, the 1970 target date for a man on the moon; and second, advance contracting for Apollo before we know that Gemini is going to work. . . .

  "We should slow down in our spending on space, reassess the situation and bring the programs back to a more realistic, down-to-earth basis. . . . We should cut back the present space appropriation by at least a billion dollars, or to an appropriation for fiscal 1965 of $4.2 billion. This is not going to wreck anything except the best-laid plans of a few space contractors. But it will help a great deal in providing additional funds for some of the pressingly important things we must do now on earth." (*CR*, 5/21/64, A2704–05)

May 21: Dr. Eugene Fubini, Deputy Director of Defense Research and Engineering, told House Military Operations Subcommittee of Government Operations Committee that Titan III space booster was potentially the most desirable and economical launch vehicle for medium-altitude communications satellite system. ". . . if we could use the Titan there would be a tremendous money saving" over the Atlas-Agena. He said a shared commercial military comsat system would have "undeniable advantages" over separate military system. (NASA LAR III/102; *Av. Wk.*, 5/25/64, 32)

- NASA said major contractor on Project Apollo, North American Aviation, Inc., Space and Information Systems Div., would soon reduce its Apollo personnel by about 1,000. The action was a cost-reduction measure, and was not indicative of either current Congressional debate or change in Apollo scheduling. "We are merely trying to keep costs down as much as we can," NASA spokesmen said. (AP, *WSJ*, 5/22/64; Finney, *NYT*, 5/22/64, 9)

- Dr. Donald F. Hornig, Director of President's Office of Science and Technology, testified before House Committee on Science and Astronautics' Subcommittee on Science, Research and Development in its hearings on geographical distribution and indirect costs of Federal R&D. Dr. Hornig said Congress was cutting administrative overhead funds in Federally financed research at universities, and universities were forced to divert funds from other needs to help pay for Federally sponsored research. "Since the university has many scholarly functions to perform, while Federal funds are devoted generally to the sciences or to areas within science, it does not seem to be in the national interest to pursue a policy which diverts private funds from the education of students or the performance of scholarly research in non-scientific areas." Dr. Hornig stated Federal support of research in past decade had produced "a dispersal of university strength," adding that Southeast and Mountain regions now were producing proportionately more doctorates than 10 years ago. (NASA LAR III/102; *NYT*, 5/22/64, 8)

- Addressing American Ordnance Association in Washington, Dr. Edward C. Welsh, Executive Secretary of National Aeronautics and Space Council, said:

    "It is my thesis that mutual disarmament—and prudence dictates that there should be no other kind—would greatly increase the flow of funds and effort into space exploration and into the building of greater space capability. This does not necessarily mean a full transfer of all available resources to space applications. But, just for illustration, suppose that there were a 10% decrease in weapon systems spending. With good planning at least half of it could be diverted smoothly and efficiently into an increased space effort. With greater difficulty, the remainder might be directed toward mending other inadequacies in our society. . . .

"I am an avid supporter of the private competitive enterprise system. I also am an avid supporter of a responsible government which creates a climate in which the private system can work efficiently. My proposal to accelerate space activity to offset decreases in weapon system activity does not conflict with my support of private enterprise. In fact, it supports it, even though the funds would pass through government hands in the process. The basic objective is to strengthen our country. . . .

"In conclusion, let me emphasize:

"First. Disarmament, in the form of decreased production of weapon systems, does not need to cause an economic depression.

"Second. Resources made available as a result of decreased production of weapon systems could most effectively be applied to accelerated development of space systems.

"Third. While it is important to keep private industry healthy, government financing of research and development is not in conflict with such objectives. In fact, it may well stimulate increased production and increased employment by private companies.

"And finally, I do not suggest that there should be a sharp increase in spending for space. But I do believe space spending should increase at a rate at least consonant with the increase in our gross national product, not only because increase in the GNP makes resources available for more space activities, but also because space exploration itself stimulated the gross national product." (Text)

May 21: U.S. District Judge refused to halt supersonic flight tests over Oklahoma City in FAA sonic boom study, after 19 Oklahoma City residents filed suit attempting to halt the flights. (AP, *NYT*, 5/22/64, 70)

- NASA announced that Joseph T. Dickerson, Jr., was appointed Special Assistant to Dr. George E. Mueller, Associate Administrator for Manned Space Flight. Mr. Dickerson would assist Dr. Mueller in the areas of management, labor relations, and Government-industry relationships. (NASA Release 64-119)

May 22: NASA Deputy Administrator Dr. Hugh L. Dryden stressed to Scientific and Technical Subcommittee of the U.N. Committee on the Peaceful Uses of Outer Space, at opening of its third session in Geneva, that U.S. scientists shared the concern of those from other countries over possible contamination of Mars and other planets. (*NYT*, 5/23/64)

- Dr. W. Randolph Lovelace II, NASA Director of Space Medicine, said in Dallas: "Physiologically, women would make excellent astronauts."

  However, he added, "They lack the experience in flying high-speed planes and handling the complicated equipment used in the program.

  "All of our astronauts are seasoned test pilots or men with much experience in combat flying. It would be very difficult for any woman to gain such experience." (*Dallas Morning News*, 5/23/64)

- FAA Administrator Najeeb E. Halaby said FAA had prepared new proposal for Government-industry sharing of supersonic transport development costs. Details were not given, but Halaby said the plan "spreads the risk of the high expenditures and protects the taxpayers as well as the stockholders" of manufacturers and airlines. Government share would be larger than original 75% formula. (Toth, *Wash. Post*, 5/23/64)

- Sen. Kenneth B. Keating (R.–N.Y.) discussed in the Senate NASA's program of research in aircraft noise abatement. Senator Keating introduced in the *Congressional Record* letters from NASA on its past and present noise research and said: "NASA spending on aircraft noise abatement research has increased by 628 percent since 1962, rising from $148,000 in that year to $247,000 in 1963, to $931,000 in 1964. The Space Committee . . . has tentatively approved a sum of $1.4 million for this program next year, a decision which I strongly support. . . ." (*CR*, 5/22/64, 11344–46)

May 22: Capt. Robert E. Jasperson (USN, Ret.) recently received patent for invention of "satellite alert," a hand-held computer for informing ships' officers when to expect radio reports from navigation satellites and meteorologists when to expect readings from weather satellites. (Jones, *NYT*, 5/23/64)

May 23: AEC spokesman said about 2.2 lbs. of plutonium 238 had been lost over the west coast of Africa last month after USN satellite carrying it had failed to orbit (see May 11). Onboard as fuel for Snap–9A generator, the plutonium apparently vaporized on re-entry into earth's atmosphere and now was dispersed in very fine particles at the edge of space, where it would gradually lose its radioactivity. It posed "no health hazard to the world's population," AEC spokesman said. (UPI, *Wash. Post*, 5/24/64)

- Dr. Ted Foss, geologist at NASA Manned Spacecraft Center, described model of the moon to be constructed at MSC for conducting time and motion studies with astronauts in their space suits. To be covered with slag rocks and ash, the area would be about 340 ft. in diameter and feature several craters about 50 ft. in diameter and about 15 ft. deep, as well as ridges as high as 12 ft. U.S. Army Corps of Engineers would perform basic work on the $50,000 construction job. (*Houston Chron.*, 5/24/64; AP, 5/23/64)
- NASA's Goddard Space Flight Center held Open House with more than 4,200 visitors attending. (*Goddard News*, 6/1/64)
- Open House at Langley Research Center, in conjunction with field inspection during the past week, featured specially prepared exhibits and other items of interest for citizens of Virginia and nearby areas. (NASA Release 64–99; NASA Announcement 64–89)
- The U.S. Fighter Aces Association awarded Astronauts Walter Schirra, Alan Shepard, Virgil Grissom, and Donald Slayton honorary memberships. (*Houston Chron.*, 5/14/64)
- New program combining study of science with study of U.S.S.R. at Columbia University's Russian Institute was announced by Institute director Prof. Alexander Dallin. (*NYT*, 5/24/64, 21)
- George C. Marshall Research Library dedicated by Presidents Eisenhower and Johnson at the Virginia Military Institute, Lexington, Va. (*Wash. Post*, 5/24/64)

May 24: Interviewed in Washington *Sunday Star*, NASC Executive Secretary and Acting Chairman Dr. Edward C. Welsh was asked how far U.S. now was on its way to the moon.

"I suppose we went a good way toward the moon objective when the decision was made. Consequently, I would say that we are much more than a third of the way to the moon at the present time."

Asked about Administration policy to classify Soviet space failures, Dr. Welsh replied:

". . . The policy of classifying Soviet space failures was developed largely to protect sources of information, many of which in those days were individuals actually in the Soviet Union.

"Most of the information we get today comes from sources outside the Soviet Union which need no protection. As a consequence, there is less justification at the present time for classifying Russian space failures than there was in the past. I am not saying that there is no justification. . . .

"I think we will continue to examine this policy to see whether or not it can be relaxed, to see whether there is any justification at all for keeping from the American people information in regard to Soviet failures." (Wash. *Sun. Star*, 5/24/64)

May 24: Recently published testimony by NASA Administrator James E. Webb before a House subcommittee indicated for the first time that there was more than just presumption to indicate U.S.S.R. might be developing larger space booster to compare with Saturn I. In response to question, Mr. Webb said that "there is some evidence, although it is not sufficient to draw a firm conclusion, that they [the Soviets] are proceeding with larger developments. . . . There is some evidence that they are not stopping with their present [space] vehicle and are expecting to move into larger categories. It will take a longer period of time to evaluate that evidence clearly and be sure." (Finney, *NYT*, 5/25/64, 1, 4)

May 25: Third anniversary of President Kennedy's address to Congress on urgent national needs in which he said: ". . . I believe this nation should commit itself to achieving the goal, before this decade is out, of landing a man on the moon and returning him safely to earth."

- Douglas Aircraft Co. Missile and Space Systems Div. summarized study on Manned Orbital Research Laboratory (MORL) made for NASA Langley Research Center. Douglas said it would be possible for U.S. to have a six-man operational space station within the next five years. MORL would be launched with Saturn IB, would remain in orbit just over one year. (*Houston Post*, 5/25/64)

- Major improvements in performance of supersonic transport shapes had come from aerodynamic refinements at NASA Langley Research Center, NASA announced. Applied to the wing design of supersonic transport concept known as "Scat 15," the refinements achieved much better flight characteristics and more than 15 per cent increase in range. The improved concept—Scat 15F—was now regarded more favorably in comparison with the two established concepts, Scat 16 and 17. (NASA Release 64-120)

- Compilation by NASA Procurement Office showed that two thirds of the $1 billion in subcontract work assigned by NASA's 12 largest contractors during 1962 and 1963 went to 2,697 different subcontractors in 46 states. (NASA Release 64-118)

- Delegates from U.S., Canada, Australia, and Japan met with the European Conference on Satellite Communications—a body representing 15 European nations—to further negotiate formation of a world-wide communications satellite system. Francis Trew, secretary of the European Conference, said at the London meeting he was hopeful an agreement "would be ready for ratification in the 'near future.'" (Farnsworth, *NYT*, 5/26/64, 11)

- NASA announced it would negotiate with Douglas Aircraft Co. for S–IVB stage mockup to be used with Saturn V launch vehicle simulator at NASA Marshall Space Flight Center. Expected to cost about $5 million, the mockup would be used in study of prelaunch, launch, and orbital checkout of third stage. (NASA Release 64-123)

May 25: United Technology Center (UTC) had developed hybrid rocket combining best features of solid and liquid propellants and 20% more powerful than any solid-propulsion rocket operational today. Rocket used solid fuel—butadiene rubber—and liquid oxidizer—liquid oxygen. UTC Assistant Manager of Advanced Technology Douglas D. Ordahl said the rocket also was safer than solid-fueled rocket motors and was simpler and more reliable than liquid-fueled rockets. (*WSJ*, 5/26/64)

May 26: NASC Executive Secretary Dr. Edward C. Welsh said at dedication of Aerojet General's Dade County (Fla.) plant:

"Leadership in propulsion is . . . synonymous with leadership in space—for both military and non-military purposes. We cannot afford to relax in this area. There is no room for complacency, only room for urgency. . . .

"In supporting the thesis that there is real danger from complacency, I list the following points:

"1. The USSR is having some economic difficulties, particularly in agriculture. Yet, their space program is expanding.

"2. The USSR is far ahead of the United States in manned flight experience. Even their female cosmonaut has flown more orbits than all of our astronauts combined.

"3. The USSR has flown their heavier Vostok spacecraft in manned and unmanned flights some 45 times as many earth orbits as we have flown our much smaller Mercury.

"4. Practically all of the competences developed for space could be used for aggression as well as for peaceful purposes.

"Please do not misunderstand what I am saying. We have made great strides, particularly in the past three years. We are actually ahead of the Soviets in some applications of space, such as communications and meteorology. We are, however, not yet ahead overall. But, our rate of progress is very impressive.

"Probably what I am warning about more than anything else is the danger that we might slow down our rate of progress—that we might heed the voices of defeatism and impracticality. This is the one thing we cannot afford. We cannot afford to lose the momentum which stems from a growing program. To slow down is to be content with second place. This we should not do. In fact, we dare not do it." (Text)

- Soviet scientists had established existence of cosmic-dust cloud averaging 437 mi. deep surrounding the earth, Tass said. Cloud was believed to be composed of residues from burned up meteorites after they entered earth's atmosphere. (Reuters, *Wash. Post*, 5/27/64)

- General Bernard A. Schriever, AFSC, said in speech to the Aviation/Space Writers Association in Miami, regarding the USAF/USN F-111:

"The outstanding characteristic of the F-111 is its versatility. The F-111A—the Air Force version—can penetrate supersonically on the deck for interdiction or low altitude tactical reconnaissance. For purposes of close support, it must be able to maneuver well at low altitudes and be able to carry a variety of conventional weapons. It must also be able to loiter for extended periods while awaiting target assignments from the local ground commander. Combined with this low altitude capability, the F-111 requires a very high altitude capability for reconnaissance flexibility and air superiority missions. It must be capable of high altitude deployment between theaters without tanker support.

"The F-111A will significantly increase the Air Force's capability in support of the Army. Its great ferry range will increase Tactical Air Command's capability to react even more rapidly than now to meet contingencies anywhere in the world. But even more importantly, the F-111A, with its variable geometry wing design will operate effectively over a wide range of speeds and altitudes in any conservative combat environment.

"Basically the F-111 will be the first truly multi-purpose all-weather tactical fighter in TAC's inventory with a capacity to operate from short austere fields. This means greater dispersal and flexibility and, of course, quicker responsiveness to support Army requirements for immediate and effective close air support, air superiority and interdiction. In short, the F-111 represents a quantum step forward in the development of a tactical air weapon system providing optimum combat effectiveness at the lowest cost and with the greatest possible flexibility.

"The F-111B—the Navy version—has a primary role of air superiority. It will be armed with the Phoenix missile system and will be able to loiter at the outer defensive perimeter of the fleet for much longer periods of time than present day Navy fighters. It will also be able to accelerate out of a subsonic loiter condition and maneuver for supersonic attack.

"It is easy to see that both the Air Force and Navy versions of the F-111 call for a number of conflicting inflight performance characteristics. The requirements for high subsonic cruise efficiency and for short takeoff capability call for maximum lift from the wing area. In other words, you need wings swept fully forward. On the other hand, supersonic low altitude flight calls for swept back wings, since the body design itself provides sufficient lift to support flight; and additional wing area only induces instability. At altitude more wing area is required, but the higher speeds obtainable dictate either very thin straight wings or else a higher degree of sweep.

"How do you answer these conflicting demands? The answer is obvious—develop a variable sweep wing aircraft. Contrary to some impressions, the variable sweep design is not new. A variable sweep design was used in the Bell X-5, which was developed under Air Force sponsorship and flown at Edwards in 1951. A similar principle was utilized in the Grumman F-10F prototype which flew in 1952. But there is a major difference between those designs and the one to be used in the F-111. The early designs required a bulky internal mechanism to move the wings. That mechanism is not required for the newer designs, and consequently the use of a variable sweep wing has become practical." (Text, AFSC Release 45-R-64)

May 26: Kingman Brewster, Jr., president of Yale Univ., testified before the House Subcommittee on Science, Research and Development of the Committee on Science and Astronautics:

". . . if we would do our best to see to it that the Government's support of scientific activity distorts as little as possible the competitive allocation of talent in the scientific market place, we must not penalize the universities in competition with industrial and consulting firms; and we must not penalize the less well endowed or supported universities in competition with their richer rivals. Anything less than full reimbursement of indirect costs [of research] does exactly this."

He continued: ". . . it seems to me wholly unwise to permit considerations of political geography to control the choices of faculties and students alike, to keep them from working wherever they think they would be most productive. Because the excellent are so few, and because modern science cannot very often be pursued in isolation from a community of one's peers, a productive science policy is bound to lead to a high degree of concentration in relatively few centers of advanced education and research." (CR, 6/17/64, A3307)

May 27: NASA launched Nike-Apache sounding rocket at White Sands, N. Mex., the rocket serving as a test round for a series of rockets to be flown in Sweden in July, and containing micrometeoroid-particle collector as follow-up to Aerobee 150 experiment three years before. Despite extremely low peak altitude (about 30 mi.—the second stage broke up after about 4.5 sec. of burning), instrumentation functioned until impact except for aspect sensor and Dovap. No useful scientific data were obtained. (NASA Rpt. SRL)

* NASA announced the large air-filled bag safety system performed successfully in DC-7 crash experiment conducted April 24. Results indicated that if a man had sat in for the dummy, he would have survived the crash. Martin Co. designed the airbag system under NASA contract, for possible use by both airline passengers and astronauts. (NASA Release 64-121)

* Brig. Gen. Joseph S. Bleymaier (USAF) told Aviation/Space Writers' Association that USAF would seek bids for the Manned Orbiting Laboratory (MOL) by Dec. 1. DOD had released $6 million for the project May 1, opening the door to preliminary study. Under present schedule, he said, hardware development could begin in about a year, with first unmanned flight late in 1967 or early 1968. (Benedict, AP, 5/27/64)

* Maj. Gen. Leighton I. Davis (USAF), Commander of National Range Div., presented to Aviation/Space Writers Association meeting at Patrick AFB, Fla., a "status report" on newly organized NRD: "The activation of the NRD and the renaming of its organizations has already been well publicized. Basically we have brought the old Atlantic Missile Range and the ICBM and satellite test range portion of the Navy's Pacific Missile Range under single management. Additionally, my job as the DOD representative in support of NASA's Projects Gemini and Apollo has moved with me to the new organization. . . ." (Text, AFSC Release 45-R-61)

* ComSatCorp disclosed its initial stock issue (5 million shares) had been oversubscribed by 27% by 163 firms in the communications industry. Of the 5 million shares, 2,895,750 were bought by AT&T. The other 5 million shares were scheduled for sale to the general public June 2. (AP, *Wash. Post*, 5/28/64; MacKenzie, *Wash. Post*, 5/28/64)

May 28: Saturn I SA-6 flight was conducted successfully from Cape Kennedy, the huge two-stage launch vehicle boosting into orbit an Apollo boilerplate spacecraft attached to the S-IV second stage. The one malfunction in the flight was premature shutdown of one of the eight first-stage engines, cutting off 24 sec. early. Preliminary analysis of telemetry data indicated that the remaining seven H-1 engines burned additional two sec. to compensate for lost thrust of the eighth engine, and the Saturn I's guidance system corrected course deviations caused by the shut-off engine, so that orbital parameters (140-mi. apogee, 123.9-mi. perigee,

88.5-min. period, and 31.8° inclination to the equator) came very close to those predicted. This was first Saturn I flight test of active guidance system to inject upper stage and attached Apollo spacecraft into orbit. Eight movie cameras mounted on the Saturn I to photograph propulsion and fuel operations were ejected and recovered. Main purpose of the flight was further qualification of the Saturn I launch vehicle and continued development of technology necessary to build the more powerful Saturn IB and V launch vehicles. (*Marshall Star*, 6/3/64, 1; NASA Releases 64–113 and 64–130; *M&R*, 6/8/64, 10)

May 28: Seventh test in paraglider series was conducted by NASA Manned Spacecraft Center at Edwards AFB, Calif. Simulated Gemini spacecraft was dropped from 33,000-ft. altitude and paraglider deployed according to plan and lowered the spacecraft to 15,000 ft., where parachute opened for remaining portion of descent. (AP, *Phil. Eve. Bull.*, 5/29/64)

• Plans for first operational weather satellite system, to be called Tiros Operational Satellite (TOS) system, were announced. NASA had selected RCA's Astro Electronics Div. for negotiations leading to manufacture of five Tiros operational weather satellites, at cost expected to exceed $9 million. Dept. of Commerce/Weather Bureau would finance TOS system, provide overall management, and be responsible for its operation. NASA was responsible for spacecraft procurement, launch, and initial checkout after satellite is in orbit. After system begins in late 1965, two meteorological satellites would be in orbit at all times, photographing cloudcover of the entire earth every day. (NASA Release 64–125; Commerce Dept. Release WB 64–9)

• Successful static firing of 156-in. solid propellant rocket motor was conducted at Lockheed Propulsion Co., Potrero, Calif. The 110-sec. firing was considered successful in all respects. Brig. Gen. Joseph Bleymaier, Deputy Commander for Manned Systems, AFSSD, said the test signified " a completely new era with regard to the size of solid motors for ballistic missiles, and in particular, for space boosters." (*M&R*, 6/8/64, 25)

• USAF destroyed Athena missile about 70 sec. after launch from Green River, Utah, when "it appeared to be progressing out of the flight corridor." (*M&R*, 6/8/64, 10)

• NASA signed $237 million contract with General Dynamics/Astronautics for development of Centaur launch vehicle stages. The contract, a confirmation of letter contract issued January 1963, called for delivery of 14 flight stages, exclusive of inertial guidance system being developed by Honeywell Co., and liquid hydrogen-fueled RL–10 engine being built by Pratt & Whitney Div. of United Aircraft Corp. (NASA Release 64–127)

• NASA announced it would negotiate with Douglas Aircraft Co. for purchase of eight additional S–IVB stages for Saturn IB launch vehicle. The eight stages would be flown on vehicles number 5 through 12. Estimated at more than $110 million, the contract would call for building the eight stages, supplying automated ground support equipment for Sacramento test site, providing launch support services at Cape Kennedy, and furnishing spare parts for entire Saturn S–IVB program. (NASA Release 64–126)

May 28: NASA awarded $100,000 contract to North Carolina Science and Technology Research Center for establishment of experimental program for transfer of technology developed in U.S. space effort to civilian industry. Other participants in the program would be Univ. of North Carolina, Duke Univ., North Carolina State College, and Research Triangle Institute. (NASA Release 64-124)

- Republican Critical Issues Council of the Republican Citizens Committee, chaired by Dr. Milton Eisenhower, issued statement calling upon U.S. to abandon the 1970 goal of a manned lunar landing. It also called for increased international cooperation in space exploration: ". . . The exploration of our universe is a goal too vast, too hazardous, too costly, and too important to all mankind to be financed and conducted by one country alone, and least of all in an atmosphere of unfriendly competition." Council also deplored claims for military value of space exploration: "We are aware of no compelling argument thus far for the development and deploying of weapons in space. . . ." (Text, *CR*, 5/28/64, 11775-77)

- NASA Associate Administrator for Manned Space Flight, Dr. George E. Mueller, addressed banquet of Aviation/Space Writers' Association in Miami. Discussing possible hazards to Apollo astronauts, he said that based on data from EXPLORER XVI satellite it was "not anticipated at this time that meteoroids will constitute a major problem in the planning or scheduling of the first manned lunar exploration." Radiation did not present a hazard either: ". . . In the case of solar flares, we compared the allowable safe dose with what would have been produced within the Apollo command module by the worst flares on record, that of July 1959. If an Apollo spacecraft had been in flight on a lunar mission at this time, the astronauts would have received only 15 percent *of the safe dose,* calculated on the basis of a report by a group established by the Space Science Board of the National Academy of Sciences. . . .

  "Very recently there have been suggestions that space flight—even relatively short duration flight—is harmful to man. There is no evidence from Project Mercury to support these fears. Based on our experience in Mercury, which included a flight of 34 hours by Gordon Cooper, there is no reason to believe that flights of two weeks or less will be harmful. . . ."

  Dr. Mueller then outlined nine medical experiments to be conducted during the two-man orbital space flights of Project Gemini. (Text; NASA Release 64-128)

- Three researchers from North American Aviation, Inc., stepped out of a rotating capsule after three days of rotation and one with the capsule at rest. The capsule, about twice the size of an Apollo capsule, was on the end of a 23-ft.-radius arm which traveled at a rate of five rotations per minute, producing a g force of 1.08. The test was preliminary to more extensive ones intended to study living and working conditions in a manned orbital laboratory. While in rotation, the researchers shot darts, maneuvered on a teeter-totter, worked puzzles and mathematical problems. Conclusion was that by end of second day, performance was back to normal. (NAA S&ID *Skywriter,* 6/12/64, 2)

- B. K. O. Lundberg, Director General of the Aeronautical Research Institute of Sweden, predicted at Aviation/Space Writers Association in Miami that more than 15,000 passengers would be killed each year by the year 1990 unless aviation safety was improved. Deploring the aviation industry's emphasis on supersonic transport development ("we are now apparently on the verge of plunging ourselves into a supersonic adventure"), he stressed that the "No. 1 challenge to aviation and the aeronautical scientists" was aviation safety, and urged the world to adopt "new, radical and long-term aviation safety policy." (Morrow, *NYT*, 5/31/64, 78)

May 29: At Aviation/Space Writers Association meeting, Harold B. Finger, Manager of AEC-NASA Space Nuclear Propulsion Office and Director of NASA Nuclear Systems and Space Power, said that in the Kiwi B-4D reactor test last May 13 the "planned power and temperature of the reactor were exceeded. . . . [Also,] there were no flashes in the jet as had been experienced in the Kiwi B-4A reactor test in November 1962, none of the vibrations that had caused damage to that reactor were detected in this test, and extensive data . . . [were] recorded on the reactor performance.

". . . I think we can safely say that our structural problems have been overcome and nuclear rockets can be developed for the performance of future missions beyond Apollo." (Text)

* NASA appointed Breene M. Kerr as Deputy Assistant Administrator for Technology Utilization, effective June 1. As deputy to NASA Assistant Administrator for Technology Utilization and Policy Planning Dr. George L. Simpson, Kerr would have operating responsibility for NASA's Technology Utilization program. (NASA Release 64-129)

* Soviet Premier Khrushchev was quoted by former U.S. Senator William H. Benton (now U.S. representative to UNESCO in Paris) as saying: "If you wish, I can show you photos of military bases taken from outer space. I will show them to President Johnson if he wishes." Mr. Benton said Premier Khrushchev suggested that U.S. could maintain adequate inspection of Cuba with satellites and therefore should discontinue its aerial reconnaissance flights over the island. (Middleton, *NYT*, 5/30/64, 1)

May 30: NASA Administrator James E. Webb addressed Missouri Cotton Producers Association in Sikeston, Mo.:

"Whether the Russians are attempting to build a rocket and spacecraft to match Saturn V and Apollo, we simply do not know. When they start flight tests of such a rocket, we should be able to identify it. Flight tests of our Saturn V will not begin until the 1967-68 period, and it is quite possible that the Russians may also be testing a super rocket by then.

"This much we do know: It would be very foolish on our part to sit idle or to delay the Saturn V until we could confirm that the Soviets were also building super rockets. By that time, it would probably be too late for us to match them again in this century.

"In our national space program we are moving rapidly and vigorously to do the things we need to do as a nation to prevent technological surprise, to prevent another major Soviet initiative based on space superiority.

"We believe that peace and security and the possibility of useful cooperation with the Soviet Union all depend on our making doubly clear that they can hope to have no monopoly of space achievement to exploit for the purposes of communism.

"Some people seem to think that our present space effort can be justified only if we are in a neck-and-neck race to the moon. These people miss the point. Our aim is not to race the Soviets, but to thoroughly discourage them from making an all-out effort to use space as a political, propaganda, or military weapon for greater power on earth.

"I don't believe the American people are investing $20 billion in the Saturn-Apollo program of manned space flight in this decade just to keep even with the Soviets. I think the American people want to lead, and lead without doubts and excuses. . . ." (Text)
- At Missouri Valley College commencement exercises, NASA Administrator James E. Webb said: "For some time we in NASA have been concerned with doing our part to help build up the nation's supply of scientific and engineering manpower, and to ensure that our national competence grows rather than diminishes. Accordingly, in 1962, the agency initiated its present program of training graduate students in space-related fields, and nearly 900 students are already in training.

"This fall the number of schools participating will grow from 88 to 131, located in 47 states—with four of these colleges here in Missouri. . . ." (Text)

May 31: Saturn I SA-6 satellite, consisting of boilerplate Apollo spacecraft and S-IV second stage of the Saturn I vehicle, re-entered earth's atmosphere and disintegrated over the Western Pacific Ocean about 8:30 p.m. EDT, during its 50th orbit of the earth. (NASA Release 64-130)
- NASA Administrator James E. Webb said in remarks at Univ. of Alabama: "In the $5\frac{1}{2}$ years since the launching of Explorer I in January of 1958, scientists in the nation's universities and in government laboratories, supported by the engineering competence of American industry, have made an impressive record. In this $5\frac{1}{2}$ years 42 different university sponsored experiments have been orbited on 22 NASA satellites. Twelve of these experiments on five satellites are still operating. There are pending at the present time a total of 123 experiments representing 36 different experimental concepts contributed by nineteen universities. These experiments have been approved and will be flown on 19 NASA satellites including the orbiting geophysical observatories, the orbiting astronomical observatories, the orbiting solar observatories, and the Explorers. Of great significance is the fact that some of these experiments will be flown on future international satellites, particularly the United States–United Kingdom series." (Text)
- Reported that John Stack of Republic Aviation believed U.S.S.R. probably would beat the West with supersonic transport. Stack revealed Soviets had developed 1,200-mph delta-wing bomber, "the Bounder," which could quickly be converted into transport "because the Soviets have lower performance standards than the U.S. for their aircraft and are not worried about the effects of sonic booms on their cities." (*Parade*, 5/31/64)

During May: Sylvania Electric Products, Inc., recommended to NASA that a data collection satellite system could be provided by a single satellite in polar orbit between 500 and 700 n. mi. altitudes. Satellite would collect meteorological and oceanographic data from about 1,200 sensors per day, store the information on magnetic tape, then relay it upon command to ground station. Sylvania's study was performed under $97,000 contract awarded last year by NASA. (*M&R*, 6/1/64, 11; *SBD*, 5/13/64, 68)

During May: Sert I (space electric rocket test) payload began final series of tests in large vacuum tank at NASA Lewis Research Center, in preparation for space flight later this summer. (*Lewis News*, 6/5/64, 2)
- David B. Clark Co. began building eight flight suits for Project Gemini astronauts. MSC suit engineer Henry Friloux said the pressurized suits would protect astronauts up to 150-mi. altitudes. (*M&R*, 5/25/64, 31)
- Flight-rating test for F–1 rocket engine was delayed from mid-year to end of the year, and tests leading up to the qualification would begin in September. (*M&R*, 5/25/64, 9)
- On 20-day shakedown cruise in the Chesapeake Bay, tracking ship *Rose Knot* conducted tests with instrumented Lockheed Super Constellation aircraft equipped to simulate systems of Gemini and Agena. (*M&R*, 5/25/64, 9)
- USAF antisatellite system, utilizing Thor missile, completed successful operational testing and became operational. (Transcript, Defense Secy. McNamara News Conference, 9/18/64)
- Bell Aerosystems study for NASA concluded that design studies of lunar surface vehicles could not be made in detail until data were obtained from Ranger and Surveyor lunar probe programs. General results of the lunar surface analysis indicated the best possibility for surface locomotion would be wheeled vehicle. (*SBD*, 5/21/64, 117)
- Results of meteoroid-protection research were presented at AIAA Structures and Materials Conference in Palm Springs. North American Aviation said its research suggested that current protective requirements against impact hazard might be reduced. GM Defense Research Laboratories reported on hypervelocity impact studies that showed meteoroid bumper concept may do more harm than good at lower impact velocities. NASA Ames Research Center research indicated use of honeycomb material for meteoroid protection would increase damage from impacts, because honeycomb walls channel sprayed particles into highly damaging clusters of pulverized material that blast through rear sheet of target material. Investigation of fiber metals by IIT Research Institute showed that fiber metal bumpers were more efficient than solid bumpers in preventing damage to target material. (*M&R*, 5/25/64, 29)
- USAF began awarding small study contracts to industry to explore various aspects of the Manned Orbiting Laboratory. Directed toward the orbiting of a 4-ton laboratory in 1968 to help DOD determine the military role of manned space flights, MOL would be launched by a Titan III–C rocket and would orbit for approximately one month. Contracts were let last month to the Martin Co. to study the role of Titan III–C in relation to launching the laboratory. Hamilton Standard Div. of United Aircraft Corp. and the AiResearch Div. of the Garrett Corp. got contracts for study of environmental control systems. (Sehlstedt, *Balt. Sun*, 6/5/64)
- USAF Aerospace Research Pilot School at Edwards AFB, Calif., acquired 12 more students, bringing enrollment to 35. The school's one-year course was designed to produce pilots, project managers, or consultants for future manned space programs. (*A–N–AF J&R*, 5/30/64, 17)
- In *Astronautics & Aeronautics* article on Germany's space program, Dietrich E. Koelle discussed recoverable sounding rockets being studied. Preliminary research and development as well as testing had been begun on the new guidance system for these rockets, which would make

guided descent with assistance of Rogallo wing. Largest German space project was production of third stage for ELDO launch vehicle, which would have its first live flight-test in 1966. (*A&A*, 5/64, 66 ff)

During May: At annual meeting of Soviet Academy of Sciences, Academician L. A. Artsimovich, physicist, observed that U.S.S.R. did not have a leading role in astronomy because of inadequate facilities. He said that astronomy was more important than high-energy physics, but "our expenditures on high-energy physics many times exceed those on astronomy." Head of the Academy M. V. Keldysh replied:

"There are some areas of science in which it is necessary for us under all conditions to maintain a high level. But we don't have to try to maintain such a level in absolutely all sectors. In scientific development, it is necessary to base oneself on international scientific cooperation.

"I think that it is from this aspect we must examine the problems of astronomy's development and in this area broadly utilize international cooperation. It is not obligatory to try to have in our country the world's largest telescopes and radio telescopes, and telescopes in balloons, and on sputniks, etc." (Schwartz, *NYT*, 6/1/64, 2)

- Richard L. Schleicher of North American Aviation, Inc., received Royal Aeronautical Society of Great Britain's George Taylor Gold Medal for 1963 for paper on X–15 structural design which he presented to the Society. (*Av. Wk.*, 6/1/64, 13)
- USAF's "Summary of Lessons Learned from Air Force Management Surveys" found that cause of management deficiencies was absence of three essential elements: program-oriented management structure; comprehensive written procedures spelling out guidelines in all foreseeable situations; and effective discipline. (*Armed Forces Mgmt.*, 5/64, 37)
- Canadian Air Marshal C. R. Slemon retired as Deputy NORAD CinC at Colorado Springs to become Director of Aerospace Educational Center at USAF Academy. (*A–N–AF J&R*, 5/16/64, 4)

June 1964

June 1: In an interview with *U.S. News & World Report*, Dr. Wernher von Braun, Director of NASA Marshall Space Flight Center, defended the U.S. program to land a man on the moon by 1970 and stated that it was his belief that at the present time the chances were "better than 50%" that the project would be successful by the target date. In commenting on reports that the Russians had forsaken a race to the moon, von Braun ventured that it was possible that the Russians had "picked another focusing point besides the moon for their manned spaceflight program" such as the development of a large manned space station orbiting earth. In addition, he commented: "I don't think the Russians have ever said they aren't going to the moon. But, even if we are going alone, this would not hurt one bit. The purpose is not so much to race the Russians—or, for that matter, even to land on the moon—but to design a hard-hitting program to develop an American space-flying capability." (*U.S. News*, 6/1/64)

- Dr. George E. Mueller, Associate Administrator for Manned Space Flight, NASA, asserted that "we learned we had more work to do" in commenting on the fact that one engine in the Saturn I booster (SA-6) had shut down prematurely. Because the Saturn I booster was designed to operate successfully with one engine out, no damage was done by the early shutoff and the test was successful. (*Houston Post*, 6/2/64)

- NASA awarded a $1.2 million contract for the design, development, and construction of a new simulator system for installation in a subsonic jet transport to Cornell Aeronautical Laboratories, Buffalo, N.Y. The new system, called the General Purpose Airborne Simulator (GPAS), was to provide airborne simulation of advanced aircraft with particular emphasis on the proposed supersonic transport. The system should allow inflight evaluation of specific configurations for a wide variety of flight conditions of speed, altitude, and attitude. (FRC Release 13-64)

- *Missiles and Rockets* reported that NASA planned two Mariner-type flights to Mars in 1969 and a Voyager mission to Mars in 1971. *M&R* reported that D. P. Hearth, of the office of Advanced Lunar and Planetary Programs in the NASA Office of Space Sciences, said that he expected both new projects to be funded in the FY '66 budget. According to Hearth, neither project had won program approval by top NASA officials, but the chances for both appeared good. Both 1969 and 1971 would offer attractive launch windows because propulsion requirements to complete the missions would be the smallest of any period in the next 15 years. (*M&R*, 6/1/64, 15)

- NASA contributed $177,000 to the Univ. of Houston to support 10 predoctoral students studying space-related sciences and technology. The fellowships were for a three-year period and brought the total to 20 in the NASA-sponsored study program. With this grant, the space agency raised its total contribution to $354,000 in fellowship grants to the university. (*Houston Post*, 6/1/64)

June 1: *New York Times* reported that AEC and USAF were jointly supporting work on a relatively simple nuclear rocket engine (Project Poodle) that was likely to be flyable in the next five years. The rocket would be clustered to propel an upper rocket stage, and it was reported that the newly designed device might save more than $300 million on space missions which had already been planned and for which conventional upper stages had been contemplated. The novelty of the new engine stemmed from the fact that its power would be derived not from a controlled nuclear reaction but from the decay of radioactive polonium 210. Propulsion would be provided in gentle acceleration of about $1/4$ lb. thrust sustained for periods of weeks. (Witkin, *NYT*, 6/1/64)

- In a letter to *Aviation Week*, S. B. Kramer suggested that, on the basis of his computations, the U.S.S.R.'s ZOND I probe was on a trajectory that would bring it on the sunlit side of Venus around July 24, 1964. His calculations were based on bulletins on ZOND I issued by Tass, and he asserted that they led him to two conclusions: first, that ZOND I was on a trajectory that would lead it to approach Venus closely; second, that the trajectory was such that the probe could return to the vicinity of earth during the latter half of March 1965. (*Av. Wk.*, 6/1/64, 102)

- The policy committee of the Cornell Guggenheim Aviation Safety Center of New York urged the Government to begin immediately to develop procedures for the selection and training of crews for the supersonic transports now in the design stage. The group urged the initiation of intensive studies of the "operating environment" of supersonic aircraft now in use by the military, and the development of joint management-union programs for selection of crews based upon physical and technical competence rather than upon the seniority principle. (AP, *NYT*, 6/2/64, 63)

- Marilyn Link, Executive Secretary of the Link Foundation of New York, was named winner of the national Frank G. Brewer Trophy for 1963. The trophy, awarded by the National Aeronautic Association, honors "distinguished and outstanding meritorious service to the Air Youth of America in Aero Space Education." (NAA Release)

- Miss Jacqueline Cochran flew the Edwards AFB 100-km. circular course at 1,302 mph to recapture the international woman's record held by Jacqueline Auriol. Miss Auriol had averaged 1,266 mph on June 4, 1963, to capture the 100-km. record from Miss Cochran. (AP, *Wash. Post.* 6/5/64)

June 2: At his news conference, President Johnson stated, in reply to a question about reconnaissance of Cuba by means of planes and satellites: "I will say this: that we, at the time of the Cuba missile crisis, attempted to work out an agreement for inspection and that was refused. This nation, in order to protect its people, must have a knowledge of what is taking place and we propose to keep informed. As to what offers the best method, we will have to determine that." (*NYT*, 6/3/64, 25)

- The Senate Committee on Aeronautical and Space Sciences approved authorization of $5,246,293,250 in the next fiscal year for NASA. This figure would be $52,482,750 more than was voted by the House, but still $57,706,750 less than the budget requests. The Senate Committee's recommendations included restoration of budget cuts by the House from the unmanned Surveyor program, the Orbiting Geophysical Observatory program, and funds for tracking network operations. (*Wash. Post*, 6/3/64; *Av. Wk.*, 6/15/64, 30)

June 2: In Senate debate, Senator Frank Lausche (D.-Ohio) criticized NASA's refusal to divulge to congressional space committees its documents concerning the selection of Boston as the site of the Electronic Research Center. He called NASA's consideration of requests of other states for the installation "a mere front" hiding the fact that the decision had already been made. Senators Keating and Javits (R.-N.Y.), and Case (R.-N.J.) joined Lausche in criticizing NASA's refusal to divulge the contents of its site selection studies. (*CR*, 6/2/64, 11951-52; UPI, *Boston Globe*, 6/2/64)

- NASA named two companies to enter competitive negotiations for providing instrumentation support services at the Merritt Island Launch Area and the NASA facilities located within the Cape area. The two companies were Federal Electric Corp. of Paramus, N.J., and Chrysler Corp., Florida Missile Operations, Cape Kennedy, Fla. The total period of performance would be for three years, at a cost of $4.5 million. (KSC Release 80-64)
- NASA engineers at the Goddard Space Flight Center were puzzled by the question of how a colony of fleas managed to infiltrate a special chamber which was supposedly protected against invasion even by bacteria. The chamber had been designed and installed to test, under conditions simulating those of space, optical equipment to be placed on board the Orbiting Astronomical Observatory. One possible explanation for the presence of the fleas suggested that they entered concealed in the fur-lined parkas worn by technicians adjusting equipment inside the low-temperature chamber. (*Wash. Daily News*, 6/2/64)
- NASA Flight Research Center awarded a $1.2 million contract to the Norair Div. of the Northrop Corp. of Hawthorne, Calif., for the design and construction of two full-scale lifting body research vehicles. The two research vehicles (the M-2 and HL-10) were to be used initially by NASA for full-scale wind tunnel testing and for free flight tests as manned subsonic gliders. (FRC Release 14-64)
- Representative H. R. Gross (R.-Iowa) told the House that Breene M. Kerr, son of the late Sen. Robert S. Kerr (D.-Okla.) and recently named Deputy Assistant Administrator for Technology Utilization of NASA, would have to exercise great care to avoid conflicts of interest in his new job. He warned that Kerr's new assignment "will require some serious thinking on his part if he is to avoid actions that may tend to confuse the Kerr financial interests with the interests of the government." (Mollenhoff, *Des Moines Register*, 6/3/64)
- In a House debate over the patent policy for the proposed Water Resources Research Center, Rep. Chet Holifield (D.-Calif.) criticized current DOD and NASA patent policies. He urged that the policy be revised to permit any industry to "use any kind of a device or any kind of a patent procedure which is paid for by federal funds." (*CR*, 6/2/64, 12045)
- The FAA awarded contracts totaling $6 million each to the Boeing Company and Lockheed Aircraft Corp. for the design of a supersonic transport. In addition, contracts totaling $5 million each were awarded to the General Electric Co. and the Pratt & Whitney Division of United Aircraft Corp. for designs for the engines of the transport. Boeing and Lockheed would do extensive design work, wind tunnel testing, and structural studies aimed at refining designs to conform to the technical and economic requirements established by the FAA. The two engine contractors would develop detailed preliminary layout designs. (AP, *NYT*, 6/3/64, 16; FAA Release 64-55)

June 2: Alexander Kuzin, Soviet member of U.N. Scientific Committee on the Effects of Atomic Radiation, said that contaminations by short-lived radioactive substances were disappearing. By the end of the current year, it would be practically impossible to trace any such contamination of the earth, Kuzin said in a speech reported by Tass. (AP, *NYT*, 6/4/64, 4)

- A $100-million issue of Communications Satellite Corp. stock was put on the market and almost immediately sold out. Five million shares were involved in the stock issue. The maximum allotment was 50 shares per customer from each security house, but the average was only about 15 shares because of the great public demand. The par value of the stock was $20 per share, and it closed at around $23 after having reached a high of $27 in the first few minutes of trading. (*NYT*, 6/3/64, 59)

June 2–3: New, improved solar cell was described at Conference of Photovoltaic Specialists by Joseph Mandelkorn, head of group at NASA Lewis Research Center that developed the solar cell. The new cell was made of pure silicon with small amount of aluminum added. Before 1960, solar cells were made of silicon with small amount of phosphorus added. In 1960, Mandelkorn discovered that substituting boron for phosphorus made cells with greater radiation resistance; these cells were used on TELSTAR I, whose orbit passed through Van Allen radiation belts. The newest cells, even more radiation resistant, were expected to have useful lifetime of 25–50 times that of silicon-phosphorus type. (LRC Release 64–51)

June 3: NASA launched Nike-Apache sounding rocket from Wallops Island, Va., to 86.6-mi. altitude, but nose cone failed to release. Using CW propagation transmitter, Gerdien condenser probe, and ion trap, experiments were to have measured ambient electron and ion density, monitored integrated ultraviolet flux from the sun, and tested two 12.267-mc shroud antennas. The propagation experiment obtained good data from 83 sec. to peak, but no other experimental results were obtained. (NASA Rpt. SRL)

- A new division was formed within the NASA Office of Advanced Research and Technology. The new Programs and Resources Division of OART would be headed by Merrill H. Mead who came to Hq. from the Ames Research Center. Dr. R. L. Bisplinghoff, Associate Administrator for Advanced Research and Technology, said the new division was created to "improve the program planning capability and provide common objectives toward which all of the elements of the office can be directed." (NASA Release 64–132)

- NASA revealed that it spent a record $1,939,000,000 in the first half of the current fiscal year, July 1, 1963, to December 31, 1963, for goods and services. The total sum was $15 million more than the previous high, for the six months ending June 30, 1963. 44 states participated in direct awards of NASA prime contracts of $25,000 or more. Of these 44, California received by far the largest amount in money terms, more than $700 million. (Sehlstedt, Balt. *Sun*, 6/4/64)

- A $700,000 contract was awarded to the Defense and Space Operations Division of the Westinghouse Corp., Baltimore, Md., Operations Center by the Goddard Space Flight Center. The contract called for the complete integration, checkout, and testing of the ground communications system for NASA's Rosman, N.C., Stadan (Space Tracking and Data

Acquisition Network) station's second 85-ft. antenna system which would support the new Advanced Technological Satellite (ATS). (GSFC Release G-16-64)

June 3: Sh. P. Darchiya, an astronomer from the Pulkovo Observatory, succeeded in photographing details of the lunar surface on the unexposed side of the lunar disk, i.e., under conditions of earthlight. Astronomers had previously recorded the luminescence of individual details of the lunar surface on the unexposed side of the lunar disk, but according to A. A. Mikhailov, Director of the Observatory, this phenomenon had never been confirmed by instrument observation. (*Izvestia*, 6/3/64, 4, ATSS-T Trans.)

- It was reported that U.A.R. planned to soon launch its own earth satellite—a 40–70 lb. satellite boosted by a three-stage rocket. A source close to German rocket scientists in U.A.R. said the launching would probably be purely a prestige gesture, having little or no scientific value. (UPI, *Wash. Post*, 6/4/64)
- HF-24 fighter-bombers designed by a team of 18 German aeronautical engineers and manufactured in Bangalore at the Hindustan Aircraft Co. were reportedly being delivered to the Indian government. The chief of the team designing the HF-24 was reported to be Dr. Kurt Tank, principal designer for Germany's World War II Luftwaffe. (NANA, Wash. *Eve. Star*, 6/3/64)
- Miss Jacqueline Cochran claimed her third jet speed record in less than a month, flying a Lockheed F-104 fighter aircraft. Flying at 1,135 mph over a 500-km. closed course at Edwards AFB, Calif., Miss Cochran bettered her own international record of 680.75 mph which she established over the same course on September 17, 1961. (NAA Release; AP, *Wash. Post*, 6/5/64)
- USAF launched unidentified satellite payload aboard a Scout vehicle from Pt. Arguello, Calif. Purpose of the shot was not announced. (*M&R*, 6/15/64, 10)

June 4: The House Republican Conference's task force on space and aeronautics charged that NASA was disregarding the safety of astronauts in its haste to place men on the moon. The report specifically criticized the agency for not developing means of rescuing astronauts stranded in orbit and called for a slowdown in the program until the feasibility of "rescue ships" could be explored. The criticisms were directed at the Project Gemini flights. The report suggested that NASA's reluctance to develop a Gemini space rescue system resulted from a fear that it would slow the Apollo project. (Finney, *NYT*, 6/5/64)

- Dr. Hugh L. Dryden, Deputy Administrator of NASA, said he hoped for cooperation on space flights between the U.S., the U.S.S.R., and other countries at some future time. He asserted that such cooperation was not possible at the current time because of the secrecy that surrounded Soviet space efforts. (*N.Y. Her. Trib.*, 6/5/64)
- A $1,313,000 contract was awarded by the Army Corps of Engineers to Carpenter Brothers of Dallas, Texas, for the construction of a test maintenance building at the NASA Mississippi Test Facility. (DOD Release 438-64)
- USAF and USN astronomers planned special telescopic observation of the moon on June 4 and 5 in an attempt to identify the mysterious red spots first observed by Sir William Herschel in 1738. The spots,

located in the crater Aristarchus, were observed in October 1963 by astronomer J. C. Greenacre from the Lowell Observatory in Flagstaff, Ariz. They indicate the possibility of volcanos below its barren surface. (Ubel, *N.Y. Her. Trib.*, 6/4/64; AP, Balt. *Sun*, 6/5/64)

June 4: Conference on New Technology at Lewis Research Center was opened by LRC Director Dr. Abe Silverstein: "The vehicles and spacecraft launched from Florida are the visible results of our nation's space effort. Less evident is the body of technology that makes these and future flights possible. Almost half of NASA's 30,000 employees are among those creating this technology." Attending the two-day conference were some 350 representatives of industry; area development groups; universities; Federal, state, and local government; labor; banking; and NASA Hq. and Centers. (*Lewis News*, 6/6/64, 1)

- It was announced that the British government was spending $2.8 million to enlarge its space communications facilities at Goonhilly Downs. Future communications satellites would require more powerful receivers because of higher orbits. (AP, Phil. *Eve. Bull.*, 6/4/64)

- Soviet scientists disclosed that they were planning to grow fresh vegetables in future Soviet spacecraft. *Izvestia* reported that scientists had already grown cabbage, beets, tomatoes, and carrots in synthetic soil in simulated spacecraft conditions. (AP, *Chic. Trib.*, 6/5/64)

- France and Spain signed an agreement to set up a joint satellite tracking station in the Spanish-owned Canary Islands. (UPI, *Wash. Daily News*, 6/5/64)

- New FAA *Statistical Handbook* showed substantial increases in aviation activity. Since 1950, domestic and international air passenger traffic had more than tripled. The same was true for air cargo traffic. Domestic flights carried 56 million passengers in 1962. International and territorial flights carried 6.6 million passengers. Revenue ton miles flown by all U.S. airlines, scheduled and nonscheduled, exceeded 6.6 billion in the same year. Employment within the aeronautical industry increased from 668,500 in 1961 to 707,300 in 1962, an increase of six per cent. (FAA Release 64–57)

June 5: The U.S. and the U.S.S.R. reached further agreement during negotiations conducted in Geneva on cooperative efforts in space research. A joint weather satellite program would go into operation by the end of this year. The weather program would consist of the U.S. and the U.S.S.R. launching meteorological satellites on a coordinated basis with the data being exchanged via a special telegraphic link between Moscow and Washington. In addition, preliminary agreement was reached for extension of scientific cooperation in space matters to cover biology and medicine. Although details of this agreement were not available, it was reported to cover a wide exchange of information on the biological and medical problems raised by prolonged manned space flights. The negotiations were conducted by Dr. Hugh Dryden, Deputy Administrator of NASA, and A. A. Blagonravov, a member of the Soviet Academy of Sciences; both agreements were subject to the formal approval of their governments. (*Wash. Post*, 6/6/64; *NYT*, 6/6/64; UPI, *Wash. Daily News*, 6/6/64)

- NASA and the Swedish Space Committee agreed to extend for another year their cooperative sounding rocket studies of the upper atmosphere and noctilucent clouds. Under the new agreement, the U.S. would provide the eight payloads and part of the ground equipment on a loan basis. Sweden would provide four of the eight sounding rockets and most of the ground instrumentation. Launching, payload recovery, and other support operations would be carried out by Swedish personnel. The joint studies began in 1961. (NASA Release 64-131)

June 5: USAF Secretary Eugene M. Zukert said that the Titan III rocket would make its maiden flight during the coming summer. Eventually, Titan III would launch the USAF's Manned Orbital Laboratory, probably sometime in 1967 or 1968. In addition, Zuckert announced that three firms, Douglas Aircraft Co., General Electric Co., and Martin Co., had been authorized to go ahead with work on orbital space-station studies. (UPI, Louisville *Courier-Journal*, 6/6/64)

- Prof. A. A. Blagonravov, chief delegate of the U.S.S.R. to the U.N. Committee on Peaceful Uses of Outer Space, told a news conference in Geneva that the Soviet Union had no set plan for landing a man on the moon "within the present decade." (AP, *Houston Post*, 6/6/64)
- USAF launched an unidentified satellite aboard a Thor-Agena D booster from Vandenberg AFB, Calif. (*M&R*, 6/15/64, 10)
- Britain successfully launched a Blue Streak rocket from the Woomera Range in Australia. The rocket reached an altitude of 110 mi. and a top speed of 6,400 mph. Because the engines cut off seven seconds earlier than had been planned, the rocket fell 382 mi. short of impact area. The rocket traveled 620 mi. northwest of Woomera and disintegrated over Western Australia. (*NYT*, 6/6/64)
- Dr. I. M. Levitt, Director of the Fels Planetarium in Philadelphia, displayed in Tel Aviv a clock designed for lunar explorers. The clock, designed by Dr. Levitt, accounted for the fact that a lunar day and night would be more than 28 times as long as the earth's 24-hour cycle. His clock showed Greenwich time on earth as well as local time for any point on the moon. (*NYT*, 6/7/64, 78)
- Astronomers at Lowell Observatory, Flagstaff, Ariz., failed in an attempt to sight and photograph red spots in the vicinity of the Aristarchus region of the moon. Conditions were about the same as on Oct. 29 and Nov. 27 when Lowell observers had sighted the red spots. (AP, Balt. *Sun*, 6/6/64)
- A spokesman for Stanley Gibbons Ltd. of London, one of the world's leading stamp dealers, cast doubt on recent reports that the U.S.S.R. had issued a commemorative stamp honoring three cosmonauts lost in orbit. He stated, "We get all new issue stamps from the Soviet Union. We know nothing about this one." (AP, *Chic. Trib.*, 6/6/64)

June 6: U.S.S.R. launched earth satellite COSMOS XXXI. The orbit had the following initial parameters: period of rotation, 91.6 min.; apogee, 508 km. (309 mi.); perigee, 228 km. (141.67 mi.); inclination to equator, 49°. (*Izvestia*, 6/7/64, 1, ATSS-T Trans.)

- Amateur astronomers in Riverdale, the Bronx, saw a red spot on the moon between the craters Aristarchus and Herodotus at 4:20 a.m. It vanished at 5:10 a.m. before the moon had come into view of the western observatories which had observed spots near Aristarchus last Oct. and Nov. (*NYT*, 6/18/64)
- It was reported that U.S. and Soviet officials would meet in Geneva on June 15 and 16 to discuss a possible Russian role in a global space communications system. The U.S. was already in the process of working

out an agreement with the Europeans, and it was believed that the system would be greatly strengthened by Russian participation. The Geneva meetings, which would involve the U.S., European nations, and the U.S.S.R., would include a general discussion of recent developments in space communications. It was also reported that before the Geneva meeting a negotiation session in London between the U.S. and European nations would work on a preliminary agreement on an interim organization for a space communications system. (Farnsworth, *NYT*, 6/7/64, 6)

June 6: Brig. Gen. Joseph Bleymaier, Deputy Chief of Staff for Manned Space Flight, AFSC, said that USAF would sharply increase its participation in NASA's manned space flight program; USAF astronauts might be assigned for training as crew members on some Gemini and Apollo flights, and 100 officers would help staff Integrated Mission Control Center (IMCC) at MSC. [George Low, MSC Deputy Diricetor, denied this claim three days later (see June 9).]

General Bleymaier disclosed that: (1) flight crews of Manned Orbiting Laboratory (MOL) would come from nine Dyna-Soar pilots now at Aerospace Research Pilot School, Edwards AFB; (2) USAF would begin discussing specific flights aboard Gemini and Apollo when Sec. of Defense Robert S. McNamara gave approval for production of MOL; (3) 17 Titan III-C launches were planned, two allocated to MOL; (4) payloads of 6,000 lbs. would contain scientific research projects, with NASA experiments to be accommodated. (*Houston Post*, 6/10/64; *M&R*, 6/15/64, 12)

- Speaking at commencement exercises at New Mexico State University, University Park, N. Mex., Dr. George Mueller, NASA Associate Administrator for Manned Space Flight, noted that the NASA program for manned space flight has recorded 16 successes in 16 attempts. He explained this record: "A rather simple formula has contributed to the success to date. We fly a spacecraft, manned or unmanned, only after a comprehensive set of ground and flight tests has been completed and every deviation from expected performance has been explained. We are continuing to follow that policy." (Text)
- Dr. Joseph Shea, manager of MSC Apollo Spacecraft Office, said that U.S. would land men on the moon before Dec. 31, 1969. He stated that in his judgment "we have a much better than even chance" of landing the first two astronauts on the lunar surface ahead of the 1970 deadline set by the late President Kennedy. (UPI, *Wash. Daily News*, 6/6/64)
- Lt. Gen. James H. Doolittle (USAFR) was awarded the Pennsylvania Military College Engineering Centennial Medal. In his acceptance speech, General Doolittle spoke on USAF Project Forecast report, which considered "national policy, technological opportunities, and capability" in its 10-year predictions. Based on the technological opportunities evaluated (including materials, propulsion, aerodynamic development, special weapons, and guidance), there would be eight important possible technical advances: Advanced Manned Precision Strike System (AMPSS), very accurate ICBM's, very precise air-to-ground missiles, a Vtol fighter, a Vtol light transport, a large cargo aircraft (CX-X), Mobile Air Defense, and a hypersonic mach 6 reconnaissance aircraft. General Doolittle concluded: ". . . our national security and our national welfare, in future, will depend very largely on military technology and, of course, as always, on the spirit and will of our citizens.

Our Nation must become and remain superior in science and technology. We must have more and better scientists and engineers. This is one of the very important challenges to Pennsylvania Military College and to all of the other fine educational institutions of our land. . . ." (Text, CR, 8/31/64, 20404–06)

June 6–7: NASA Manned Spacecraft Center near Houston held its first public open house, with more than 52,000 visitors attending. Visitors walked among the 27 buildings and viewed models of the Gemini and Apollo spacecraft set up inside the auditorium, as well as viewing a film on NASA. (MSC Roundup, 6/24/64, 2)

June 7: Speaking at the commencement of the Univ. of Vermont in Burlington, James E. Webb, Administrator of NASA, said that "historians and psychologists see a universal interest and participation in space efforts as providing mankind with a moral and creative substitute for war." He remarked that the U.S. was ahead of the U.S.S.R. in weight-lifting power in space. He stated that the U.S. was building a national capacity for operating in the new environment of space that would serve the national purpose for many years to come. (Text; NYT, 6/8/64, 22)

- *Parade* magazine's Intelligence Report said the U.S. had 50 Samos surveillance satellites in orbit photographing Soviet military installations. Photographs were reported to be parachuted to earth between Calif. and Hawaii. Russians were said to have similar satellites spying on U.S. (*Parade, Wash. Post*, 6/7/64)
- It was reported that the Army was seeking ways to increase the forward speed of helicopters. To speed its rotor-borne craft, and thus lessen their chances of being shot down by enemy air or ground forces, the Army was reported to be interested in adding short conventional wings and small jet engines. The resultant craft, called a compound helicopter, was considered capable of flying above 200 mph. Research contracts were awarded to the Lockheed-Georgia Co. and the Kaman Aircraft Corp. for development of helicopter modifications. (*NYT*, 6/8/64, 58)
- The USAF was reported pushing the development of a new material which offered promise of revolutionizing the manufacture of planes, missiles, and many other products. The material was a combination of boron fibers and epoxy resins to bind the fibers together, reported to be far stronger and stiffer per unit weight than any existing structural material. According to Gen. Bernard Schriever, Commander, AFSC, if the development program were adequately funded, the technology could be perfected in five to seven years. (Witkin, *NYT*, 6/7/64, 63)
- Dr. George E. Mueller, NASA Associate Administrator for Manned Space Flight, answered GOP critics, complaining of a "crash program" to reach the moon, with a "Not so. The eight years devoted to the Apollo Project is a longer period than the duration of any previous United States research and development program." He continued: "Apollo's time scale for development is four years longer than the time scale for the Mercury spacecraft, two years longer than the B-58 bomber and one year longer than the X–15." (*Houston Post*, 6/7/64)
- Dr. S. Burman Long, Worcester, Mass., minister, delivered address on religion in the space age at the 73rd convention of the General Federation of Women's Clubs in Atlantic City, N.J.:

". . . My religion in the space age confirms my faith in a spiritual heaven as real as my humble earthly home. The sooner that mankind accepts an orderly and unlimited universe, the earlier we will have faith in truth. Our cosmology has changed because our knowledge has changed. It has changed before. Once a new continent was discovered. We no longer limit the universe to hell below, heaven above, and some of both between. The God of creation is in control now as He always has been. . . ." (Text, *CR* 7/24/64, A3801–02)

Early June: Air Force Space Systems Div. announced 10-ft. delta wing glider able to travel almost around the world as goal of improved Asset program. Glider would be ready for launch in two years. This improved Asset program was said to be more "realistic" prospect for manned reentry vehicle than canceled Dyna-Soar project. (*M&R*, 6/15/64, 10)

June 8: NASA Deputy Administrator Hugh L. Dryden announced to newsmen the bilateral agreement negotiated between U.S. and U.S.S.R. in June 1962 (the "Dryden-Blagonravov Agreement"), by which the two countries would cooperate in space programs for communications, meteorology, and geophysics, was discussed by him and Academician Blagonravov at Geneva. Cooperation in Soviet-American communications for global meteorology was to be implemented by the two countries' sharing costs equally, each supporting the program in alternate months; a bilaterally financed communications link between Moscow and Washington was to be solely for the purpose of exchanging cloud pictures. The 1962 agreement grew out of correspondence between President Kennedy and Chairman Khrushchev in Feb. and March 1962. Dryden and Blagonravov also discussed American-Soviet cooperation in space biology. Agreement was reached, subject to review before final approval, by which American and Soviet space biologists would prepare bilingual reports on similar subjects.

Soviets had proposed cooperation in biology and medicine, both of which had great bearing on manned space flight when exploration extended beyond the moon. (*Wash. Post*, 6/9/64; NASA Transcript, 6/8/64[2] "Background Information," "News Conference.")

- NASA announced industry design competition for Automated Biological Laboratory (ABL) for detecting possible life on Mars. An ABL would be landed by Voyager spacecraft to be flown in 1970's. Proposals resulting from preliminary studies would lead to contract for a one-year effort to review scientific objectives, define an experimental payload suitable for landing and operation on Mars, and establish spacecraft and launch vehicle requirements. ABL would incorporate several life-detection devices still in developmental stages and would be equipped to analyze and describe any organisms discovered by such devices. (NASA Release 64–134)

- It was reported that technicians had experienced some difficulty returning to their spacecraft couches during test designed to see how well astronauts would be able to perform certain functions in weightless space flight. The tests were performed inside a mockup of a Gemini capsule mounted in a KC–135 flown to produce short periods of weightlessness. The difficulty was caused by pressure suits swollen by pressurization and couches designed to fit the astronaut only when his suit was not pressurized. Dr. M. Debrovner of MSC said that the difficulty would soon be overcome. (*Houston Post*, 6/9/64)

June 8: Officials of NASA's Saturn program expressed interest in a miniature version of the Rover nuclear rocket engine. Called the "Poodle," the liquid-propellant engine was developed under an AEC contract with Space Technology Laboratories of Thompson-Ramo-Wooldridge. (*M&R*, 6/8/64, 22)

- Federal investigators and an attorney from the NLRB were dispatched to Cape Kennedy to seek a solution to labor troubles which halted work on America's programs for landing on the moon and for orbiting a manned space laboratory. An estimated 80% of workers on the Vertical Assembly Building for Project Apollo and the launch complex for the USAF's Titan III refused to cross picket lines thrown up by the Brotherhood of Maintenance of Way Employees in protest to use of a Government-owned railroad spur which had been operated since Jan. 23, 1963, by nonunion personnel. Cost of the projects affected by the work halt was estimated at $215 million. (*Chic. Trib.*, 6/9/64; *Balt. Sun*, 6/9/64)
- ComSatCorp said it would award three additional study contracts for design of communications satellites systems. AT&T and RCA would design a system of 18 satellites to be placed in random orbits. TRW Space Technology Laboratories would engineer a system of 12 satellites in controlled orbits. Hughes Aircraft would plan an improved synchronous-orbit communications satellite. Based on these preliminary studies, a choice was to be made for use in the initial comsat system to begin operation in 1966–67. (ComSatCorp Release; *Wash. Post*, 6/9/64; *Balt. Sun*, 6/9/64; *NYT*, 6/9/64, 25)
- A daughter was born to Soviet Cosmonauts Valentina Nikolayeva-Tereshkova and Andrian Nikolayev. She was first child to be born of parents who had flown in space, and no ill effects were noted. (*Houston Chron.*, 6/10/64; *Balt. Sun*, 6/11/64)
- Nuclear submarine *Daniel Webster* completed a perfect series of tests by firing two Polaris A–3 missiles from beneath the surface of the Atlantic. (*Wash. Post*, 6/9/64)

June 9: George M. Low, NASA MSC deputy director, said all pilots in Projects Gemini and Apollo would be astronauts trained by NASA for its own manned space projects. Brig. Gen. Bleymaier (USAF), in interviews at Houston, had previously said he thought pilots selected by USAF, whose MOL project he heads, would participate in Apollo and Gemini in order to acquire experience for the MOL project. Without NASA training for its astronauts, USAF would have to orbit its men without previous space flight experience or launch its own Gemini training flights. (Maloney, *Houston Post*, 6/10/64; *Houston Chron.*, 6/10/64)

- Managing underwriters of ComSatCorp's 10-million-share public offering at $20 a share announced the offering was oversubscribed and the books closed. Thus ComSatCorp had enough money to launch the fleet of satellites that was expected to be handling global communications by 1967. (*WSJ*, 6/9/64; *NYT*, 6/12/64, 47)
- A new British jet, the Hawker-Siddeley DH–125, was expected to be the first of five new models of aircraft to get government approval for use as a private passenger carrier. The new small jets were being built for the business executive market. The British plane was to cost $750,000 and fly about 500 mph. (*NYT*, 6/9/64)

June 9: Soviet Ministry of Defense said Pentagon was spying on Russian military and space installations in Moscow, via radio and radar monitoring stations on borders of U.S.S.R. and other Socialist nations and satellites. Allegation called for increased Soviet vigilance against such "espionage." American embassy in Moscow called statement "nonsense." (*NYT*, 6/11/64, 10)

- AFSC reported completion of a portable hydrogen-fire detector, for use with the volume of hydrogen anticipated with future space boosters. The unit detected ultraviolet radiation emanating from invisible hydrogen flame. (AFSC Release 43–89–67)

June 10: NASA launched an Aerobee rocket from White Sands Missile Range, N. Mex., to a height of 95 mi. The payload, an inflatable paraglider equipped with mylar-sandwich-type meteoroid penetration sensors, was flown in a meteoroid shower to investigate meteoroid hazards in space. Experiment performed as planned, vehicle worked successfully, and payload was recovered. (*M&R*, 6/15/64, 10; NASA Release 64–141, 64–135)

- Nike-Apache launched from Wallops Island, Va., to 90.6-mi. altitude, in flight to detect quiet-day ionospheric electric currents using for first time a very small rubidium vapor magnetometer designed for use in small rockets. Experiment was project of NASA Goddard Space Flight Center. (NASA Rpt. SRL)

- Tass reported Soviet Union had launched COSMOS XXXII, latest satellite in series of launches initiated in March 1962. Initial orbit: 333-km. apogee (206.9 mi.), 209-km. perigee (130.5 mi.), 89.78-min. period, and 51.28° inclination to the equator. COSMOS XXXII re-entered earth's atmosphere June 18. (*Pravda*, 6/11/64, in MSFC *SIN*, 7/64; GSFC *SSR*, 6/30/64)

- Two-stage French Rubis rocket reached altitude of 1,800 km. (1,118.49 mi.) in launch from Hammaguir base. First stage of Rubis was Agate rocket and second stage was third stage of France's satellite launcher Diamant. (*M&R*, 6/29/64, 12)

- A Japanese astronomer of the Kurashika Astronomical Observatory observed a new comet. It had been earlier sighted by an observer at Tokyo Astronomical Observatory on June 7, as well as by an Australian. On June 9, Soviet astronomer Anatoli Bakharev discovered what was described as a 9th-magnitude comet moving rapidly with respect to the stars. (AP, *NYT*, 6/15/64; *Pravda*, 6/13/64, 6, ATSS–T Trans.)

- American Institute for Biological Sciences expressed concern that NASA was not giving sufficiently high priority to the investigation of extraterrestrial life. The Institute, a federation of biologists, said NASA should begin planning a project for detection of possible life on Mars when the planet comes into favorable position in 1969. Search for extraterrestrial life also was advocated by NAS Space Science Board nearly two years ago. NASA originally had hoped to perform the Mars biology mission in 1964, but technical, scientific, and political obstacles delayed the attempt. (Finney, *NYT*, 6/11/64, 2; *Wash. Post*, 6/11/64)

- NASA announced replacement of Dr. Richard B. Morrison by Vincent L. Johnson as Director of Launch Vehicle and Propulsion Programs Div. of Office of Space Science and Applications, effective June 14. Dr. Morrison was to return to Univ. of Michigan where he had been on leave of absence as Prof. of Astronautical Engineering since 1962. (NASA Release 64–136)

June 10: Rep. Roman C. Pucinski (D.-Ill.) called attention to CBS broadcast by correspondent Stuart Novins in which he discussed Soviet efforts to provide an efficient system for the retrieval of data from scientific research. In the Soviet Union's Institute of Scientific and Technical Information, thousands of specialists read the scientific literature and abstract the contents of scientific publications from around the world. Last year they abstracted the contents of 770,000 reports, and it was decided to publish an annual summary of world-wide scientific results, an achievement probably of the widest scope yet attained in retrieval of scientific information. (*CR*, 6/10/64, A3174)

- Rep. Roman C. Pucinski (D.-Ill.) called attention to an article in the *Chicago Sun-Times* on the Midwest's dissatisfaction with the distribution of Federal R&D projects. President's science adviser Dr. Donald F. Hornig said two Congressional committees were investigating whether research contract funds had been concentrated in too few areas of the country to the detriment of other regions. Dr. Hornig also said that this investigation showed that science can no longer ignore the political process. (*CR*, 6/10/64, A3138)

- House and Senate received from NASA Administrator reports on a grant to Cornell Univ. of $1,350,000 for construction of facilities for research. Reports were referred to the two Congressional space committees. (NASA LAR, III/116)

- In answer to Air Transport Assoc. president Stuart G. Tipton's disapproval of FAA's policy of publicizing its test plane crashes, FAA Administrator N. E. Halaby said FAA would continue informing the public on its safety research. Tipton had said the publicity "scared the daylights out of people." (*Wash. Eve. Star*, 6/11/64)

- American scientists began a two-year experiment at the South Polar ice cap in which they would study cosmic rays of solar and galactic origin. Located at the South Pole so that the same background of stars would be present at all times, experiment was to investigate any regularities in solar outbursts of cosmic rays, as well as reason for cosmic rays from sun and galaxy. Cosmic rays could be lethal to astronauts. (*Wash. Eve. Star*, 6/10/64)

- U.S. Army Missile Command awarded contract to Perkin-Elmer Co. for Optical Re-entry Instrumentation System, Project Glow, which would gather "signature" data from vehicles re-entering earth's atmosphere, in various wavelengths. Program to use system was being sponsored by ARPA, as part of Project Defender, and by Army Materiel Command's Nike-Zeus Project Office. (*Marshall Star*, 6/10/64, 7)

- First of improved H-1 engines, whose thrust had been upped from 188,000 lbs. to 200,000 lbs., were delivered recently by North American Aviation's Rocketdyne Division to NASA Michoud Operations. Cluster of eight H-1 engines was to improve thrust of Saturn booster, augmenting thrust of 1.5 million lbs. for Saturn I to 1.6 million lbs. for Saturn IB. (*Marshall Star*, 6/10/64, 1)

- Gen. B. A. Schriever (USAF), Commander AFSC, at a 10th anniversary commemoration of USAF ballistic missile and space programs, said Communist drives toward world domination presented challenge to U.S. to maintain technological superiority as war deterrent. At same time we should develop military flexibility to deal with conflict levels below all-out war. In future such flexibility would be provided by manned aircraft of novel design, to be used for support as well as actual fighting.

Next step in space would be to investigate military strategy of manned space flight. This was a major purpose of USAF's recently acquired mission of developing Manned Orbiting Laboratory (MOL).

Another possibility to investigate would be recovering used boosters to save new construction costs. "This would be a very important building block for future operations in space." (AFSC Release 46-R-72)

June 10: Soviet aviatrix Marina Popovich, wife of Cosmonaut Pavel Popovich, flew a jet trainer at speeds of 600 km/hr (373 mph), 5 km. (3.1 mi.) above the ground. This was said to be an excellent result for such a light aircraft as the L-29. (*Izvestia,* 6/12/64, 4, ATSS-T Trans.)

June 11: NASA announced plans to let contracts for improving thrust of Delta launch vehicles. Boosted energy was to have been achieved by enlarging 2nd-stage fuel tanks. Some Deltas to be modified would use three solid-propellant motors strapped to the Thor main stage, being called Thrust Augmented Deltas (TAD). With larger tanks, TADs could place 1,200 lbs. into earth orbit or 225 lbs. into escape velocity, as compared to 100 lbs. for standard Delta. (NASA Release 64-133)

- Goodyear Aerospace Corp. developed solar collector for USAF. The 4-story-high mirror, to be used in tests of electricity sources for manned space station, was capable of setting fire to buildings or blinding people with a single flash of sunlight. (*Wash. Eve. Star,* 6/11/64)
- 1.5 million-lb.-thrust S-I-8 Saturn booster was successfully fired in a full-duration static test at MSFC. Built by Chrysler Corp. at Michoud Operations plant near New Orleans, the booster was the first to be produced by private industry. (AP, 6/12/64)
- Cook Electric Co., Chicago, unveiled a $2 million observatory to be included in NASA's first Oao. Cook built the equipment for NASA under contract to Univ. of Wisconsin, where astronomers were planning observations in the far ultraviolet region of the spectrum. (*Houston Post,* 6/12/64)
- In an address to Plans for Progress Seminar, NASA Administrator James E. Webb said American success in space technology represented several accomplishments and potentialities: we can make benefits available to other nations; space achievements are vital in maintaining technological balance of world power; and it gives opportunity for utilizing all our human resources in carrying out the broad-gauge programs necessary for making significant inputs into space knowledge, regardless of race, creed, or color. (Text)
- A Federal court ban on picketing of railroad workers at Cape Kennedy that was issued June 10 went into effect, and construction of the Saturn and Titan III launch facilities resumed. 3,824 members of building trades union effected the work stoppage. The District Judge who handed down the temporary injunction also scheduled a hearing with NLRB. (AP, 6/11/64; UPI, *Houston Chron.,* 6/11/64)
- NASA MSC officials announced first Gemini spacecraft for a manned spaceflight (GT-3) had passed design and engineering inspections; project manager Charles W. Mathews declared it would perform its mission in satisfactory fashion. (*Houston Chron.,* 6/11/64; MSC *Roundup,* 6/24/64, 8)

June 11: House Appropriations Committee cut off developmental funds for the AEC's Pluto low-flying, nuclear-powered missile, proposing the missile be mothballed until DOD decided whether it needed the weapon. Having gone through nearly 10 years of development costing $196 million, Pluto was nearing the flight stage. The committee provided $1 million to mothball it. AEC had requested $8 million additional funds to get the missile to operating stage. (Finney, *NYT*, 6/12/64, 1, 23)
- NASA announced appointment of two directors in Advanced Manned Missions Program, Office of Manned Space Flight. M. J. Raffensperger assumed duties last week as Director of Manned Earth Orbital Missions Studies. F. P. Dixon would become Director of Manned Planetary Mission Studies on June 15. (NASA Release 64–137)

June 12: It was reported that the Atlas ICBM program had cost $5.4 billion for 129 Atlases. In the closing year of his administration President Eisenhower had objected to the $35 million cost of each Atlas on the launching pad. (*Wash. Eve. Star*, 6/12/64)
- Soviet Defense Ministry newspaper reported a trainee cosmonaut had narrow escape recently when he jumped with defective parachute. (AP, 6/12/64)
- NASA intended to award contracts for studying feasibility of improving the weight-lifting capacity of the Saturn V by more than one third in case the needs of the moon program dictated such an increase. Preliminary studies were to cost about $2 million. Proposals were being sought by MSFC. With its 7.5-million-lb.-thrust booster, Saturn V would be able to launch either 120 tons into earth orbit or 45 tons to the moon. (AP, 6/12/64; NASA Release 64–140)

June 13: USAF launched Thor-Agena D booster combination with unidentified satellite payload into a polar orbit. (AP, *Chic. Trib.*, 6/14/64; UPI, *Wash. Post*, 6/14/64)
- Space editor James J. Haggerty, Jr., compared Saturn IB and Titan III–C launch vehicles: "... In any head-to-head clash between the two programs, Saturn IB must be considered the favorite. Titan III–C's quick reaction capability may be handy some day, but not in the next few years, since there are no military manned spacecraft to do the reacting. Saturn IB starts out with a 25% payload advantage and it has considerably more 'stretchability.' For instance, adding the same 120-inch solid motors in the Titan III–C design to the Saturn IB's lower stage would permit boosting 65,000 pounds into a 100-mile earth orbit. Other improvements being considered could give IB a 120,000-pound earth orbit capability.

    "Thus, Saturn IB poses a threat to continued development of the only heavy payload military launch vehicle. (*A–N–AF J&R*, 6/13/64, 15)
- It was reported that former astronaut John Glenn had recovered sufficiently from his inner ear ailment to work in his yard and walk around the neighborhood. (*Phil. Eve. Bull.*, 6/13/64)
- Resolution of the policy committee of Cornell-Guggenheim Aviation Safety Center, headed by former FAA Administrator Elwood R. Quesada, was inserted in *Congressional Record*:

    "(1) That the Government agencies and the broad aviation industry recognize and avoid the danger and economic waste that could result, if the exacting technological task of developing a sound supersonic transport were permitted to become a race to be first or a psychology of haste.

"(2) That the pressures of prestige and competition not be permitted to overcome the more worthy considerations of public safety and economic logic.

"(3) That all design and economic factors be deliberately tested and evaluated to assure maximum effectiveness of the solutions selected to insure that the end product be first in point of excellence rather than first in point of time." (CR, 6/13/64, 13226)

June 13: Johnson Administration was expected to postpone until next year a decision as to whether the U.S. should establish antimissile defenses around the country in a $15 billion program, as advocated by Gen. Maxwell D. Taylor (USA). Such a defense system would work in conjunction with civil defense facilities. Question seemed to be whether proposed system's effectiveness would warrant the cost. (*NYT*, 6/14/64, 72)

June 14: Space Park at New York World's Fair, sponsored by NASA and DOD, had not turned out to be a prominent drawer of crowds, although those who visited found it one of the most instructive and rewarding exhibits at the Fair. (*NYT*, 6/14/64, 79)

- Sen. Gaylord Nelson (D.-Wis.) said General Accounting Office had, in spot audits of military financial records, discovered almost $500 million of waste. Included in the alleged waste was Navy's ordering of 24 aircraft to be built before satisfactory model had been developed and contract monies paid by military to builders of planes and missiles who, doing no work themselves, made profits on work done by subcontractors. (*Houston Post*, 6/15/64)

- Dr. Abe Silverstein, Director of NASA Lewis Research Center, was awarded honorary Doctor of Applied Science degree by Fenn College, Cleveland, O. (*Lewis News*, 6/19/64, 2)

June 15: FRC test-flew modified X-15 No. 2 with Maj. Robert Rushworth (USAF) as pilot, the aircraft remaining attached to B-52 mother craft for airborne checkout of the aircraft systems. Flight was first for this X-15 since Nov. 1962 landing accident. Modifications, completed by North American Aviation, included four astronomical cameras mounted on fuselage, designed to take stellar photographs above ozone layer of earth's atmosphere, and external propellant tanks which would enable engine to increase top speed by at least 1,000 mph, to a maximum speed over 5,000 mph. (FRC Release 15-64; NASA Release 64-138)

- NASA's leading 12 contractors issued $1,027,000,000 in subcontracts during 1962 and 1963 to 2,697 firms in 46 states, *Aviation Week* reported. (*Av. Wk.*, 6/15/64, 71)

- FAA fire tests began in cooperation with USAF at FAA's National Aviation Facilities Experimental Center to determine whether rotor downwash from helicopters can disperse flames and cool the fuselage of a crashed and burning plane. Five surplus USAF Stratocruisers were burned in the tests. (FAA Release 64-62)

- NASA granted nine companies 16 waivers of U.S. commercial rights to inventions that resulted from work performed on NASA contracts, bringing to 90 the number of patent waivers granted by NASA. (NASA Release 64-141)

- NASA Summer Faculty Fellowships, a unique 10-wk. summer program, began at Lewis Research Center. The 12 Fellows, all young professors and instructors in engineering and the physical sciences, would work on space research projects at LRC and would attend a special graduate level course in chemical rocket technology at Case Institute of Technology. (LRC Release 64-53)

June 15: William G. Gisel, President of Bell Aerosystems Co., said before N.Y. Society of Newspaper Editors, Niagara Falls: "I cannot contemplate the space exploration project . . . without a tremendous surge of pride in the human race. We are going to the stars, gentlemen. The human race will travel to the planets and, in time, to other solar systems and in doing so will fan the spark of greatness which in spite of ourselves has carried us from the cave to the threshhold of space. Man can go to the moon in this decade. The important question is, will it be an American?" (NASA R&SC Div.)

June 15–16: U.S. began exploratory talks in Geneva with U.S.S.R. on the use of artificial satellites in a global communications system. This was first talk between American and Soviet representatives to discuss possible Russian participation. Informed sources in Geneva said the Soviets were friendly but showed no current interest in a global satellite communications scheme. In joint communique at end of two days of exploratory talks, Soviet and American officials said only that "both sides recognize the usefulness of the exchange of opinions." No future meeting was scheduled. (*NYT*, 6/16/64; Simons, *Wash. Post*, 6/17/64; *M&R*, 6/22/64, 10; *NYT*, 6/18/64)

June 16: NASA ordered a third Oao (Orbiting Astronomical Observatory) and took an option on two more, from Grumman Aircraft Engineering Corp., prime contractor for the current Oao program. If only one satellite were purchased the contract would be more than $20 million, and if all three were purchased the contract would total more than $50 million. (NASA Release 64-142)

- House Subcommittee on NASA Oversight of House Committee on Science and Astronautics issued a report on its hearings on the repeated failures in the Ranger spacecraft program. The committee concluded that NASA provided too little direction on Ranger, while JPL was too resistant to following NASA orders when they were issued.

  To strengthen management over JPL, the committee recommended: (1) JPL should be restricted to doing the work of a center rather than that of an industrial contractor; (2) a general manager with industrial experience should be appointed as deputy to JPL director to be in charge of complex projects such as Ranger; (3) NASA should streamline its relationship to JPL to achieve JPL responsiveness to NASA direction. (Text; Finney, *NYT*, 6/17/64; *Wash. Post*, 6/17/64)

- Scientist R. M. Lemmon told lunar exploration symposium at MSC that he would recommend officially that astronauts returning from trips to the moon be quarantined for at least three weeks "for the protection of everybody." He said that astronauts might pick up disease-causing organisms during their stay on the moon "which may be totally different from anything we have here on earth." (*Wash. Post*, 6/17/64)

- Dr. Edward C. Welsh, NASC Executive Secretary, dedicated J-4 propulsion engine test cell at Arnold Engineering Development Center. He said: "In recent years, there has been much talk about a 'space race'. There is such a race and we should not forget it for a minute. However, we must not let the pace of our national space program be determined just by the actions of others. Actually, the real opponent in the space race is time." We must continue "to push technology vigorously," and "recognize the importance to the national space program of adequate technical facilities."

In answer to reporters' questions following dedication, Dr. Welsh said the U.S.S.R. still led the U.S. in the space race and was speeding up its program. (Text; Clark, *NYT*, 6/17/64)

June 16: Western Materials Co. was awarded a $1,178,517 contract by the Army Corps of Engineers for construction of roads at NASA's Mississippi Test Facility. (DOD Release 464-64)

- FAA announced plans to celebrate the formal beginning of postwar international cooperation in civil aviation twenty years ago. It was in November 1944 that the U.S. convened the International Civil Aviation Conference in Chicago, resulting in the Convention of International Civil Aviation signed by the 52 attending nations. (FAA Release 64-60)

June 17: Space and Information Div. of North American Aviation, Inc., announced that it would lay off workers at a rate that would total 2,500 firings by July 1, 1964, as a result of "continuing studies of changes in the moon program." But a NASA spokesman said the reduction was caused by termination of the tooling-up phase and start of the production phase in Project Apollo. (Wash. *Eve. Star*, 6/18/64)

- FAA announced award of $1,659,460 contract to Raytheon Co. to design, develop, and manufacture data-filter equipment to give the air traffic controller a means of selecting out desired information for display on his radarscope. Delivery of the equipment would be completed in March 1965. (FAA Release 64-59)

- NASA announced it had selected Collins Radio Co. for procurement of a major portion of the S-band tracking, data acquisition, and communications system to be used in Project Apollo. Nine systems with 30-ft.-diameter parabolic antennas were to be provided, at a cost of approximately $20 million. (NASA Release 64-146)

- NASA purchased hingeless rotor helicopter, the XH-51N Research Vehicle, from Lockheed Aircraft Corp. for use in flight research at LaRC. The cost would be $556,158. It would be used as a flying laboratory to assess the loads and flying qualities problems that are likely to be encountered by operational aircraft of the hingeless rotor type. (NASA Release 64-147; LaRC Release)

- XC-142A V/Stol aircraft, built by Ling-Temco-Vought for the Army, USN, and USAF, was unveiled in Dallas. The aircraft's four turboprop engines act like helicopter rotors for takeoff and landing, when the aircraft's wings are tilted. Once aloft, the aircraft flies with wings and propellers in conventional position and attains top speed of over 400 mph. (*U.S. News*, 6/29/64)

- NASA announced Purdue Univ. had been awarded grant of $840,000 to finance expansion of Purdue's Jet Propulsion Center. The construction would enable Purdue to conduct studies of ultra high pressure combustion in rocket thrust chambers. (NASA Release 64-145)

- Scientists from the U.S. and Norway sent three gigantic balloons across the North Atlantic above the Arctic Circle to Alaska during the summer to measure cosmic radiation and other phenomena at an altitude of 115,000 ft. (AP, *NYT*, 6/19/64, 16)

June 18: USAF launched Thor-Agena D booster from WTR with unidentified satellite payload. It was later disclosed that two satellites were orbited with one launch vehicle. (HHN-48)

June 18: Lewis Research Center completed wind tunnel tests to study the aerodynamic forces which would occur on the launch vehicle when Mariner spacecraft was launched toward Mars. (LRC Release 64–55)
- Edward Z. Gray, Director, Advanced Manned Missions Program, Office of Manned Space Flight, addressed the National Space Club:

    "From [the objectives of the Space Act of 1958], the objectives of the Manned Space Flight Program are . . . derived. They are:

    "1. The conduct of manned scientific space flight missions . . . as an integral part of a total program to expand our knowledge of the universe.

    "2. The development of launch vehicles, spacecraft, lunar and planetary surface equipment . . . and a broad foundation of industrial support, all of which are required to support manned space flight.

    "3. The utilization of the space environment to provide direct benefits to the earth. . . .

    "4. The application of manned space flight activities to promote the national security and to stimulate the national economy, scientific achievement, technological growth, and educational capability." (Text)
- USAF fired Atlas–D 5,000 miles downrange from Vandenberg AFB as part of an advanced re-entry systems program, and indicated shot was successful. (UPI, Wash. *Eve. Star*, 6/18/64)
- NASA granted Dr. George Wharton, Ohio State Univ. scientist, nearly $70,000 for a three-year research program to determine the least amount of water necessary to sustain earth life. (*Wash. Daily News*, 6/18/64)
- Dr. Edward C. Welsh, NASC Executive Secretary, delivered an address on occasion of the Martin Co. Honors Night Banquet. He suggested that private industry was not doing its share in the job of educating "the public to the value, the significance of space progress, space leadership, and space benefits. It should not be possible for anyone to take seriously the public diatribes against the national space program. Anyone who knows anything about 'crash programs' or the space program would never talk [about 'moondoggle']. Crash programs have unlimited funds, unlimited overtime, and parallel approaches to the end objective. None of these aspects applies to the space program or even to the lunar project."

    Dr. Welsh said that in educating the general public, the contribution of the space program to education, national security, technological and managerial innovations, international status, and the national economy should be emphasized. He concluded: "The national space program is a creative force in our society. As we marshal our progressive strengths, let us use them to stamp out the evils of ignorance and misunderstanding. . . ." (Text)
- AFSC revealed a device known as RMU (Remote Maneuvering Unit), which is designed to be launched from a manned orbiting mother craft and carries a television camera to examine other nearby objects in space. The RMU would televise its image to the crew of the mother ship. Built by Ling-Temco-Vought, the device weighed less than 125 lbs. and measured less than 2½-ft. long. It is propelled by a nitrogen reaction gas ejected from one or more of 16 nozzles on the RMU. (AFSC Release 42–43–76)
- NASA issued list of its Technology Utilization publications, designed to acquaint the public with useful innovations from space research. (NASA Release 64–149)

June 19: TIROS VII completed one year in orbit with all components working as well as the day it was launched. During the past year, the satellite transmitted about 70,000 pictures to earth, 90 per cent of them usable. and spotted 16 hurricanes and typhoons. (NASA Release 64-144)
- President Johnson was shown lunar landing research vehicle and X-15 during visit to Edwards AFB. It was first public showing of the lunar craft. The President said: "No one can come here and doubt that America is, and will remain, first in the world in the strength, the diversity, and the genius of its defense." (*N.Y. Her. Trib.*, 6/20/64)
- USAF launched Thor-Agena D booster from WTR with undisclosed satellite payload. (HHN-48)
- Navy dedicated 61-inch astrometric reflector telescope at Naval Observatory, Flagstaff, Ariz. Its quartz primary mirror was largest quartz reflecting surface manufactured to date. It would be used primarily to study white dwarf, red dwarf, and sub-dwarf stars. (*Wash. Post*, 6/18/64)
- Labor Dept.'s new edition of government handbook on job classifications was to include 6,000 job titles not in existence when the last edition was published several years before, a reflection of the technological upheaval which had occurred in science and industry. (*Houston Post*, 6/19/64)

June 20: NASA announced that AC Spark Plug Div. of General Motors Corp. would become the prime contractor for production of the complete guidance and navigation systems for Project Apollo. Formerly, in the design and development phase of the systems, AC shared responsibility with Kollsman Instrument and Raytheon Missile and Space Div. The latter two firms now become subcontractors to AC. In this realignment of contractor relationships, Dr. George E. Mueller, NASA Associate Administrator for Manned Space Flight, said: ". . . it has been determined that the most effective management of manufacturing complete systems can be achieved through a prime contractor." (NASA Release 64-148)
- First full-scale development firing of uncooled M-1 engine was cut short after 1.6 seconds instead of the planned three seconds because of malfunction in the liquid oxygen system. The engine reached 800,000-lb. thrust in test programed for 1,000,000 lbs. (*Av. Wk.*, 6/29/64, 25; *M&R*, 6/29/64, 11)
- Third international working conference on satellite communications ended in London after three days of consultation between delegates of U.S., U.K., Canada, Australia, Japan, and European nations. Next conference would be held in Washington in July. (*NYT*, 6/22/64, 45)
- USN commissioned the guided missile ship *Norton Sound* at Baltimore, Md. For 12 years previously *Norton Sound* had served as a seagoing test platform for various missile systems, particularly for numerous sounding rocket and "Rockoon" launches. (DOD Release 468-64)

June 21: Ranger 7 arrived at Cape Kennedy after cross-country trip by van from JPL. Spacecraft was to undergo series of tests with Atlas-Agena launch vehicle before the planned launch in late July or August. (NASA Release 64-152)
- France planned to launch a series of constant-level balloon weather stations in the southern hemisphere and a satellite to "interrogate" them on their findings. Named Project Eole, it was scheduled for 1967. Up to 256 balloons were to be sent aloft, each carrying four pounds of weather and communications equipment, to remain at fixed altitudes between

16,500 and 33,000 ft. for at least ten days. The French satellite would interrogate each balloon twice during a single pass overhead. (Simons, *Wash. Post*, 6/21/64)

June 21: Among 49 new projects voted by U.N. Special Fund was $582,000 grant to help India establish a satellite tracking station and perform research in space communications. (L.A. Times, *Wash. Post*, 6/22/64)

June 22: U.S. Senate passed H.R. 10456, authorizing $5,246,293,250 budget for NASA in FY 1965. Included in this amount was $52,482,750 restored by Senate Space Committee from the $110,189,500 reduction made by the House. The bill would authorize $4,354,150,000 for research and development; $262,880,500 for construction of facilities; and $629,-262,750 for administrative operations. Bill was passed by voice vote, was reintroduced two days later for record vote (see June 24). (*CR*, 6/22/64, 14069-78)

- NASA announced initiating Atlas flox program, involving addition of about 30% fluorine to the rocket's oxidizer to obtain "flox," which would significantly increase booster performance. In first phase of program possible hazards and operation problems of flox would be studied, including component compatibility tests and rocket engine static-firing, at estimated cost of $8 million. Second phase, not yet funded, would conclude with R&D launch of flox Atlas with dummy upper stage in 1966. NASA Lewis Research Center would negotiate contracts with GD/A for work on the Atlas and with Rocketdyne Div. of NAA for work on the engines. (NASA Release 64-150)

- First manned Gemini space flight (GT-3) was scheduled for late December, MSC public affairs official Ben Gillespie told reporters at weekly press briefing. (Maloney, *Houston Post*, 6/23/64)

- NASA Wallops Station announced selection of General Electric Co. to negotiate 21-month $1.5-million contract for real-time impact prediction and data processing system. The system would enable Wallops Station to more accurately predict impact points for launch vehicles at any time during flight as well as to process telemetry, radar, and general purpose data in larger quantities and at greater speeds. (NASA Release 64-151; Wallops Release 64-50)

- Radiation hazards to Project Apollo astronauts bound for the moon were described at American Medical Association conference by Dr. Hermann J. Schaefer, scientist at U.S. Naval School of Aviation Medicine. Dr. Schaefer predicted that in 1969 proton streams from the sun could range from deadly "downpour" to very light "shower." He calculated chance of lethal-dosage proton stream was three in 10,000; chance of moderate but incapacitating "rain" would be one in 1,000; and chance of light "shower" would be one in 100. Dr. Schaefer emphasized his forecast was based on predictable solar behavior and that the weather of the sun, like that of the earth, could be unpredictable. (Justice, *Houston Post*, 6/23/64)

- Communications Satellite Corp. announced it had filed application with FCC for authority to modify AT&T ground station at Andover, Me., so it could be used with ComSatCorp synchronous satellite scheduled for launch in March 1965. The application requested FCC approval of technical characteristics of the modified station. ComSatCorp also filed proposed agreement with AT&T for modification and operation of the station. (ComSatCorp Release)

June 22: NASA Flight Research Center Director Paul Bikle said in *Missiles and Rockets* that rocket engines to power M-2 and HL-10 lifting bodies would accelerate the manned mockups to mach 2 for investigation of their handling characteristics. "The maximum altitude they reach will not be very high, maybe 70,000 or 80,000 ft." FRC planned to select the engine by mid-July. (*M&R*, 6/22/64, 20-21)

- Dr. Uri Shafrir, head of cosmic dust and hydrodynamics labs of UCLA, conducted drop tests under NASA-NSF grants program. Tests proved Dr. Shafrir's conviction that spherical objects descend in irregular corkscrew pattern of spirals rather than straight down. In the three drops at Pt. Arguello, aluminum spheres 16-in. in diameter and 10 lbs. in weight were dropped from USN helicopter at 10,000-ft. altitude. As the spheres descended, plume of red smoke from sphere marked 400-ft. width in the spirals. Two skydivers falling within 40 ft. of the spheres photographed them during descent, and ground observers tracked them with radar, survey instruments, and cameras. (*L.A. Times*, 6/23/64; AP, Balt. *Sun*, 6/23/64)

- *Aviation Week* reported that NASA and JPL were seeking a compromise to House subcommittee recommendation that NASA appoint a general manager to act as deputy to the director of JPL. James E. Webb, NASA Administrator, opposed the suggestion that NASA make the appointment. It was reported that JPL was seeking someone mutually acceptable to the committee, NASA, and JPL. (*Av. Wk.*, 6/22/64, 25)

- Hearings on S. 2602, to establish a national air and space museum in the Smithsonian Institution, were held by Subcommittee on the Smithsonian Institution of the Senate Committee on Rules and Administration. (NASA LAR III/125)

- In letter to *New York Times*, FAA Administrator Najeeb E. Halaby endorsed suggestion in *Times* editorial that two or more aircraft manufacturers join together to develop U.S. supersonic transport. Halaby said he would welcome such a partnership and would recommend to Justice Dept. that it "interpose no objection on antitrust grounds" because of "the clear national interest factors in the program." (Clark, *NYT*, 6/23/64, 66)

- Based on reports circulated at recent COSPAR meeting, *Missiles and Rockets* said that U.S.S.R. probably would not attempt another manned space flight for about a year but that flight probably would be an orbital rendezvous mission. (*M&R*, 6/22/64, 33)

June 23: COSMOS XXXIII launched into orbit with the following initial parameters: apogee, 293 km. (182.06 mi.); perigee, 209 km. (129.87 mi.); period, 89.38 min.; and inclination to the equator, 65°. Tass said the satellite carried scientific apparatus designed for investigation of space in accordance with program announced March 16, 1962, and onboard instrumentation was operating normally. (Tass, *Komsomolskaya Pravda*, 6/24/64, 1, ATSS-T Trans.)

- Army Map Service announced completion of its topographic study of the moon. Prepared under agreement with NASA, the lunar map was the most complete ever made, the first to show variations in heights over all the visible surface of the moon. Map covered area of about 8,000,000 sq. mi., and more than 5,000 surface features were identified by name. The topographers established that the moon's highest peaks reach about 35,000 ft. above its lowest points—making the lunar mountains about 6,000 ft. higher than Mt. Everest. (NYHT, *Wash. Post,* 6/24/64; Sehlstedt, Balt. *Sun,* 6/24/64)

June 23: At request of Senator Wayne Morse (D.-Ore.) and by unanimous consent to motion of Senator Mike Mansfield (D.-Mont.), Senate requested House to return H.R. 10456, NASA authorization bill for FY 1965. Senator Morse was requesting record vote on the bill; action of the previous day was taken without quorum call and with only voice vote. (NASA LAR III/126; *CR,* 6/23/64, 14234-35 and 6/24/64, 14392)

- NASA announced formal award of $24 million contract to General Electric for development of six flight model Biosatellites plus additional spacecraft for ground testing. GE was selected for the contract last August and had been working under preliminary $5 million contract, now included in formal contract. (*WSJ,* 6/24/64)
- NASA Group Achievement Award presented to Centaur "E" Stand Project personnel at NASA Lewis Research Center's Plum Brook Station for their "significant contributions resulting in the successful completion of research testing of the Atlas 116 D booster under simulated loading and environmental conditions and [for] providing data vital to the first flight of the Centaur vehicle." (LRC Release 64-56)
- Speech by Peter Hackes of NBC News inserted in *Congressional Record* by Rep. Emilio Q. Daddario (D.-Conn.) advocating a progressive space program. Hackes said:

  "Some of those who have studied space age economics tell us that by the mid-1970's our space spending will return to the American economy a dime for every nickel we're spending now. Aside from the very obvious race with Russia to be first to land on the moon; in addition to advancing science to the point where we may locate life on another planet and might discover the secrets of the beginnings of mankind; and besides the use of space to defend ourselves militarily—besides all these reasons, the U.S. space program will bring us two other exceedingly important dollars and cents benefits: a sizable advance in industry and employment. And an as-yet uncalculated fallout in new civilian products. . . ." (Text, *CR,* 7/2/64, A3690-92)
- Nuclear submarine *John Adams,* submerged off Florida coast, successfully fired Polaris A-2 missile about 1,500 mi. to its target in the Atlantic Ocean. This was last American missile-firing submarine to be equipped with the A-2 model. (UPI, *NYT,* 6/24/64, 38)

June 24: Senate passed (78-3) H.R. 10456 authorizing funds for NASA for FY 1965. Senate adopted the committee amendments en bloc. Amendment introduced by Sen. Fulbright to reduce R&D funds for Project Apollo by $267,750,000 was rejected by vote of 42-38. Bill would be sent to House-Senate conference committee. (NASA LAR III/127; *CR,* 6/24/64, 14367-84)

- NASA Manned Spacecraft Center made its final move from all leased facilities, with more than 700 employees moving to the Clear Lake location between now and June 30. (MSC *Roundup,* 6/24/64, 1)
- 15 NASA astronauts began three days and two nights of jungle survival training in the Panama Canal Zone, under supervision of the USAF Tropic Survival School. (MSC *Roundup,* 6/24/64, 8)
- Theory that moon's surface resembled cotton candy and was only about 19% solid was expounded by David D. Cudaback of Univ. of California's radio astronomy laboratory at Berkeley. Based on his four years of

research with microwaves from the moon, Dr. Cudaback said moon had highly porous surface extending down to a depth of "at least" nine or 10 ft., and "very possibly" deeper. He said only one other American scientist had obtained the same figures on high porosity, but that "quite a few Russians" had produced findings similar to his own. (Justice, *Houston Post*, 6/25/64)

June 24: It was reported DOD may coordinate a geodetic survey program combining Secor satellite, a planned Transit satellite, ECHO II, and a planned Anna satellite. A ground station would correlate the flashing light of Anna, the optical system of Secor, and the doppler radar system of the Transit, using ECHO II as a reference point. (*SBD*, 6/24/64)

June 25: Maj. Robert Rushworth (USAF) flew X–15 No. 2 to 83,000-ft. altitude and 2,966-mph speed (mach 4.49). Engine of the rocket research aircraft burned 76 sec. This was the first flight of the rebuilt No. 2 aircraft, and performance was successful. (NASA X–15 Proj. Off.)

- NASA launched companion Nike-Apache sounding rockets from Wallops Island, Va., with instrumentation to detect quiet day ionospheric electric currents and to detect and study configuration of ionospheric currents under disturbance conditions. Each payload contained a dual-cell rubidium-vapor magnetometer to measure magnitude of the magnetic field. Both rockets performed and transmitted excellently, the first (launched at 10:57 a.m.) reaching 92.5-mi. altitude and the second (launched at 8:52 p.m.) reaching 96.7-mi. altitude. (NASA Rpts. SRL)
- First tracking station to be built for Project Gemini was dedicated at Carnarvon, Australia, a site close to the antipodal point to Cape Kennedy, the launch site. Carnarvon was first land station over which Gemini spacecraft would pass as it went into orbit. NASA Administrator James E. Webb participated by telephone in the ceremonies, and his remarks initiated the new station into "SCAMA," the NASA 13-country network of operational voice communications connecting the manned space flight tracking network. (NASA Release 64–155)
- Scout rocket launched by USAF from Pt. Arguello, Calif., exploded in flight. (*M&R*, 7/6/64, 8)
- NASA announced award of three-year, $3.6-million contract to Federal Electric Corp. for instrumentation support services to the Merritt Island Launch Area of the John F. Kennedy Space Center, NASA. (NASA Release 64–156)
- First four divisions of NASA Lewis Research Center had completed moving into new Development Engineering Building, LRC announced. Occupancy of the building by LRC units temporarily housed in a Cleveland shopping center would be completed by end of summer. (LRC Release 64–58)
- Dow Chemical Corp. selected by Kennedy Space Center, NASA, for contract to provide engineering support services at Merritt Island Launch Area facilities. Cost-plus-fixed-fee contract was estimated at $1.4 million for the first year, with two one-year options for extensions. (*SBD*, 6/26/64, 315)
- DOD announced Aeronca Manufacturing Corp. was receiving $1,809,724 contract from Office of Naval Research for design, fabrication, and installation of an advanced antenna tracking system for NASA Wallops Station. (DOD Release 485–64)

June 26: NASA announced it had requested proposals from scientists for experiments on Advanced Technological Satellites to be launched from Cape Kennedy in 1966–68. Primary mission of ATS would be to evaluate advanced communication techniques, meteorological components, and gravity gradient stabilization systems, but additional space for scientific experiments was available. (NASA Release 64–153)

- NASA Lewis Research Center program of metal fracture toughness testing was helping make supersonic transport possible, according to William F. Brown, Jr., Chief of LRC Strength of Materials Branch. Co-author of paper on fracture toughness test methods with John E. Strawley, Head of Fracture Section at LRC, Brown said listening to amplified sounds of metal cracking under stress was one of the techniques of testing fracture toughness. (LRC Release 64–59)
- Curtiss-Wright X–19 Vtol transport was flown successfully for the first time in series of short-duration tests. (*Av. Wk.*, 7/6/64, 278)

June 27: Analysis of findings by Stratoscope II balloon-astronomy investigation last November was announced by National Science Foundation and Princeton Univ. astronomers. Dr. Martin Schwarzschild, director of Project Stratoscope II, reviewed first findings of the telescope and instrument package carried aloft by balloon. Studies of the moon showed lunar surface material reflected more solar light in the infrared band than in the visible spectrum. Other findings included confirmation that water vapor exists near red giant stars—for example, there is "about as much water vapor above one square mile of Mira as there is above one square mile of Earth." (*NYT*, 6/28/64, 56; Burkett, *Houston Chron.*, 6/28/64)

June 28: NASA Administrator James E. Webb said in open message to AIAA, convening in Washington, that Saturn V space booster, used in Project Apollo, would enable U.S. to "catch up with and pass the Russians" in space. "But the development and flight testing of this rocket have been delayed by a reduction of $600 million in the NASA budget for this fiscal year and efforts are now being made to further delay it through reductions of up to $500 million for the next fiscal year. This effort to slow up and hold back the progress both President Kennedy and President Johnson have recommended is a matter of important and immediate concern as the AIAA meets.

"Members of AIAA will do well to study and understand the very large foundation of space operational capability which is now being put in place by this country and consider the long-range future of space in terms of the proper use of a cap........ ..... will be very large. . . ." (Text, *Wash. Post*, 6/28/64)

- New maps of the moon made by USAF Aeronautical Chart and Information Center, St. Louis, were termed "the best lunar maps ever made" by E. A. Whitaker, astronomer at Lunar and Planetary Laboratory in Tucson, Ariz., and Dr. Eugene M. Shoemaker, lunar geologist of U.S. Geological Survey Observatory in Flagstaff, Ariz. Maps were on scale of one inch equals eight miles. (Burkett, *Houston Chron.*, 6/28/64)

June 29: New research project would study feasibility and design concept of advanced hypersonic ramjet engine, NASA announced. The work would extend ramjet propulsion technology which NASA and NACA had been developing for more than 15 years. To be conducted under direction of Langley Research Center, the project was being organized to determine potential of the ramjet as a propulsion system for hypersonic flight. Project would use hydrogen as fuel. (NASA Release 64–158)

June 29: George Friedl, Jr., was appointed NASA Deputy Associate Administrator for Industry Affairs, succeeding Earl D. Hilburn, who was recently named Deputy Associate Administrator. Since 1962 a management consultant in Santa Ana, Calif., Friedl served from 1960 to 1962 as Chairman of the Board and President of Matthews Corp. and from 1956 to 1960 as Corporate Vice President of Litton Industries, Inc., and General Manager of its Electronic Equipment Div. (NASA Announcement 64–140; NASA Release 64–157)

- GSFC reported that NASA was considering development of a satellite-borne radiotelescope to investigate low frequency radio emissions from galactic regions, distant planets, and stars. Long antennas would be required to receive the very-low-frequency radio emissions. The program would study the development and effects of sporadic radio noise bursts from the sun and the sporadic decameter radio emission from Jupiter, determine the low frequency spectra of galactic and extra-galactic radio sources, and map the spatial distribution of galactic and extra-galactic radio sources at low frequencies. (*Goddard News*, 6/29/64, 3)
- Appointment of Maj. Gen. Alvin R. Luedecke (USAF Ret.) as Deputy Director of Cal Tech's Jet Propulsion Laboratory was announced by JPL Director Dr. William H. Pickering. General manager of AEC since his retirement in 1958, General Luedecke would replace Brian Sparks effective August 1. (NASA Release 64–159)
- NASA awarded $16,975,000 contract to Blount Brothers Corp. for construction of three major elements of space environment facility known as Space Propulsion Facility, to be built at Plum Brook Station of NASA Lewis Research Center. The facility would simulate temperatures and pressure conditions at altitudes up to 100 mi. from earth, permitting evaluation and development testing of spacecraft, chemical propulsion systems, and nuclear electric power generation and propulsion systems. (LRC Release 64–61)
- Studies of two-stage recoverable lifting body space vehicle, by NASA Marshall Space Flight Center were reported in *Aviation Week*. Envisioned primarily as a passenger/supply space ferry, the system could be launched by track-mounted rocket sled or take off horizontally similar to conventional aircraft. Both stages would be manned, the first with only a flight crew and the pickaback second stage with 10–14 men and supplies. In this concept, both delta-shape stages would be recoverable. (*Av. Wk.*, 6/29/64, 38)
- NASA Deputy Administrator Dr. Hugh L. Dryden, interviewed in *Washington World*, replied to interviewer's question:

    "Yes, Gemini is really necessary to the Apollo program. As a matter of fact, Gemini would be necessary as the next major step in manned space flight even if there were no Apollo mission. Gemini provides the United States large increments of manned space flight capability. Gemini is vital to the Apollo program since our astronauts must learn— and perfect—the techniques of rendezvous and docking in space before attempting the trip to the moon and back. . . ." (NASA)
- NASA completed negotiations with Honeywell Aeronautical Div. for contract to furnish three spacecraft for Project Scanner horizon definition

experiment. The $1,224,616 contract included design, fabrication, integration, qualification, environmental testing, and certain launch support services. Project Scanner payloads would be flown on suborbital trajectories from Wallops Island, Va. (LaRC Release)

June 29: Future Delta launch vehicles with larger propellant tanks, the Thrust-Augmented Delta (TAD), would be used to launch payloads in the Biosatellite, Pioneer, and meteorological satellite families, GSFC reported. (*Goddard News*, 6/29/64, 1)

- Senate's narrow margin of defeating amendment to cut NASA's FY 1965 Apollo authorization by 10% was subject of editorial by Robert Hotz in *Aviation Week:*

    ". . . There should be little satisfaction for supporters of the U.S. space program in the narrow margin by which the Fulbright amendment was defeated. It indicates clearly what a long, hard pull still lies ahead to educate the people of this country fully on the long-term needs to maintain this nation's pre-eminence in the vital area of space technology. There will never be any lack of criticism of any legislative program that has progress as its goal. But it should be the genuine concern of the supporters of the U.S. space program to state their case with such force and clarity that this conservative carping from Sen. Fulbright and his ilk will produce only the hollow echo in history that it merits." (*Av. Wk.*, 6/29/64, 11)

- *Aviation Week* reported that President Johnson had inspected USAF-Lockheed YF-12A (A-11) supersonic aircraft at Edwards AFB during his recent tour of West Coast defense installations. (*Av. Wk.*, 6/29/64, 15)

- John H. Boynton of MSC discussed pilot's role in Mercury space flight systems failures, in paper to AIAA first annual meeting. He summarized: "For the two unmanned missions, MA-4 and MA-5, control system difficulties made impossible the successful completion of a three-pass orbital flight, which was the design mission for the Mercury spacecraft. But in both cases, the presence of a pilot and the capability to control the spacecraft attitudes manually would have made the three-pass mission possible. This fact was demonstrated in the MA-6 mission in which a similar malfunction occurred but in which the astronaut completed the flight by using his manual control modes. . . .

    "For the MA-6 mission, . . . the flight would undoubtedly have been terminated before the planned time because of an excessive rate of fuel consumption in the automatic control mode. In the successful MA-7 flight, manual initiation of the retrorocket-ignition sequence prevented the horizon scanner malfunction from significantly delaying this critical event. Although an error in yaw angle caused the spacecraft to land somewhat beyond the planned landing area, this error would have undoubtedly been much greater had the pilot not been present. The six-pass MA-8 mission was successful, although a problem of an abnormal increase in suit temperature could have terminated the flight early had no means been available for reducing this temperature. The pilot, however, was able to correct this situation by manual adjustment of his coolant flow valve. Short circuits in the automatic control circuitry during the MA-9 flight made it essential for the pilot to control the spacecraft manually during retrofire and reentry." (Text)

June 29: C. A. Syvertson and David H. Dennis of NASA Ames Research Center, in paper for AIAA first annual meeting, reviewed problems of atmospheric entry, atmospheric exit, and atmospheric cruise, suggesting that "future vehicles in each of these areas will differ significantly from their current-day counterparts. For example, entry vehicles for use in the interplanetary missions which may follow the lunar mission will enter the Earth's atmosphere at speeds up to twice that of Apollo. At these speeds, radiative heating predominates and dictates the use of more slender configurations than the very blunt Apollo. The interplanetary entry vehicles will also require very precise guidance and control both when approaching the Earth and when flying within the atmosphere. . . .

"For cruise aircraft that follow the supersonic transports, the most characteristic feature will be the probable use of hydrogen fuel. This high-energy fuel provides excellent payload and range capabilities for transports flying up to about twice the speed of the supersonic transport. The low density of the fuel results in aircraft with very large fuselages which may be twice the size of current-day subsonic jet transports." (Text)

- At AIAA meeting and press conference, Dr. Bernard M. Wagner, Chairman of New York Medical College's pathology dept., reported Soviets had decided not to send any more women cosmonauts into space. Soviet doctors were not satisfied with performance of Valentina Tereshkova, first woman cosmonaut. She had been too excited during the space flight, as seen by wide variations in her pulse rate, and she did not recover from effects of space flight as quickly as the male cosmonauts. Dr. Wagner visited in Moscow earlier in June at invitation of Prof. Vassily V. Parin, Director of Institute of Normal and Pathological Physiology. Dr. Wagner reported Prof. Parin, prominent in Soviet space program, believed artificial gravity would be necessary for astronauts on long space flights, and Soviet space planners were considering possibility of equipping Vostok spacecraft with centrifuge providing periodic spin of $1/3$ earth's gravity for the crew. Dr. Wagner added that Soviets were preparing for another manned space flight before the end of the year. (Loory, *N.Y. Her. Trib.*, 6/30/64)

- First meeting of the AIAA Technical Committee on History of Aeronautics and Astronautics held in Washington, NASA Historian Eugene M. Emme as Chairman.

- Employment in aerospace industry during the past year reflected overall downward trend, according to *Aviation Week* survey. Exceptions to the general trend were McDonnell Aircraft Co. and Douglas Aircraft Co., the former mainly due to work on Gemini spacecraft and F-4 Phantom II aircraft and the latter due to increased work for Saturn launch vehicle program and commercial programs. (*Av. Wk.*, 6/29/64, 20)

- Reported that legislation was being drafted that would apply system of U.S. agricultural extension service to industry to help solve problems of economic dislocations due to decreased defense spending. President Johnson said such an industry extension service would "do for urban America what the agricultural extension service has done for rural America." (*Av. Wk.*, 6/29/64, 21)

- Pilots of South African Airways had been warned to fly at least 9,500-ft. above sea-level altitude when over Muden on the route between Johannesburg and Durban, because large citrus estate at Muden was shooting rain-making rockets into storm clouds. The estate had been plagued by hailstorms and was experimenting with rockets to reduce the fruit-damaging hailstones. With eight launch pads on the estate, it was possible to launch 50 Italian rockets at once to a height of 6,000 ft. (AP, *NYT*, 6/29/64, 30)

June 30: Atlas-Centaur 3 (AC-3) was launched from ETR in partially successful development test which NASA officials termed "highly successful from an engineering point of view." Five of the six primary objectives were fully achieved: nose fairing and insulation panels withstood flight-loads and jettisoned properly; structural integrity of Atlas and Centaur stages during all phases of flight were verified; Atlas-Centaur separation operated satisfactorily; operation of the guidance system was demonstrated; and capability of Atlas-Centaur to be launched at scheduled time was demonstrated, the vehicle lifting off only four minutes after pre-planned launch time. The sixth, partially-achieved objective was ignition and burn of Centaur stage's two RL-10A3 liquid hydrogen engines: They ignited properly but cut off 127 sec. before programed 380 sec. burn time. Four seconds after engine ignition, hydraulic pump to engine actuators had failed, so that hydraulic system did not actuate the engine swiveling mechanism to maintain control of the stage during powered flight. The Centaur began to roll. Increasing roll motion forced propellant to side of propellant tanks, uncovering feed-line outlets; engine cut off after 253 sec. because of lack of fuel. Stage maintained its trajectory, however, and at programed cutoff time the stage attitude control system regained roll control and remaining inflight events occurred as planned. Because of the shorter engine burn, AC-3 reached speed of only 11,425 mph instead of the 17,400 mph which would have put it, incidentally, into orbit. Spent stage re-entered and impacted in Atlantic Ocean 2,706 mi. from Cape Kennedy. (NASA Release 64-143; *CR*, 7/1/64, 15086-87; AP, Balt. *Sun*, 7/1/64; Wash. *Eve. Star*, 7/2/64, A5)

- X-15 No. 1 piloted by John B. McKay (NASA) reached only 98,000-ft. altitude and 3,375-mph speed (mach 5.10) in flight near Edwards AFB, Calif. Within seconds after launch from mother ship, all guidance equipment on the rocket aircraft failed, forcing Pilot McKay to cut back speed and altitude. Flight had been planned to reach 180,000-ft. altitude and 3,550-mph speed in test for future X-15 flights to photograph the stars from above atmospheric interference. The rocket engine burned 83 sec. (NASA X-15 Proj. Off.; *Wash. Post*, 7/1/64)

- House and Senate conference committee agreed on $5,227,506,000 as compromise between the two versions of H.R. 10456, the NASA Authorization Act for FY 1965. This was $33.7 million more than the House had voted and $18.8 million less than the Senate had approved. (NASA LAR III/129-131; Confr. Rpt., House Rpt. 1529)

- *New York Times* editorialized on the four-vote margin of the Senate defeat of Sen. Fulbright's amendment to reduce FY 1965 Project Apollo funds by 10%. ". . . The narrowness of the margin may be illustrated by the fact that if there had been abstentions by the Senators from even three of the states having direct economic interest in the most rapid prosecution of the moon race—say, California, Texas and Florida—Senator Fulbright's amendment would have been adopted.

". . . [Senator Fulbright] lost his battle on the authorization bill, but he will have another chance when the bill actually appropriating the funds is considered. The new mood in Congress suggests that this battle may produce a decision to scrap the artificial 1970 deadline. . . ." (*NYT*, 6/30/64)

June 30: NASA Associate Administrator Dr. Robert C. Seamans, Jr., said at AIAA Luncheon in Washington: ". . . the concept of the lunar landing was not new when NASA was founded in 1958. Research was continuing on the various technical aspects of such a mission. In 1959, in-house work began on mission definition. In the fall of 1960, three six-month feasibility design study contracts were awarded. In January 1961 NASA received these interim findings. There were two basic vehicle approaches to be considered—direct ascent and rendezvous. In May of 1961, the study contract results were available, and supported our interim findings, and the recommendation was made to the Congress that the nation proceed with the mission. . . .

"There will be missions after Project Apollo. We are already examining their feasibility, cost, and potential return. We are carefully establishing a variety of future space missions, both manned and unmanned, that exploit the near earth, lunar, and planetary environments. No decisions have been made and we are not ready to recommend them yet. As the information becomes available and as the results are analyzed, we will be able to present the country with a solid spectrum of achievable, well-conceived mission alternatives and options. The selection will be a national, not an agency, decision." (Text)

- Self-sealing structural walls for spacecraft application were demonstrated at AIAA by Arnold J. Tuckerman, Hughes Aircraft Co. materials scientist. Tuckerman and his colleagues fired .22 pellets at pressurized chamber with walls in which polyurethanes and polyamide prepolymers were impregnated within the laminations of the walls. As the pellets pierced the chamber, they triggered chemical reaction causing the sealant to foam instantly, hardening into an airtight plug that would last up to two weeks without major leakage. (Haseltine, *Wash. Post*, 7/1/64)

- A. O. Tischler, NASA Director of Propulsion and Power Generation, said in paper for AIAA:

"At this time the United States stands on the threshold of transport in the space environment. Until now, with the exception of a relatively few payloads such as Syncom, payloads launched into space were put on their trajectory by the launch vehicle. Only minor path corrections were applied by the on-board propulsion systems. Even the Mercury capsule returns were made by a relatively minor slowing of the orbiting capsule, with aerodynamic braking providing the rest of the kinetic energy change. However in Gemini which may fly with men this year, and which will later effect rendezvous of separate payloads in space, and in Apollo, which will perform an intricate series of space maneuvers, we recognize systems that carry significant propulsion capability into space for use entirely in that environment. The time is rapidly coming when space is no longer something we throw darts into but is an environment in which working propulsion systems will maneuver and transport spacecraft payloads and eventually convey these payloads back to the earth's atmosphere for descent to the surface to be used again. . . .

"[Space] . . . booster engines have passed, or in the case of the Saturn V engines, are approaching their preliminary flight rating tests, while the more conservative spacecraft propulsion systems have not yet arrived at that milepost. While these spacecraft engines may be expected to come along at a faster pace than the big booster engine developments, the fact is that they are presently in a tail chase situation. I think we must observe that at this moment the status of spacecraft propulsion systems under development lags the development status of launch vehicle engines and is additionally behind in performing sophistication. That is the area that warrants propulsion attention." (Text)

June 30: Paper surveying space nuclear propulsion by Harold B. Finger, James Lazar, and James J. Lynch of NASA and AEC was presented to AIAA.

". . . a major forward step has been taken in the milestone established by the Los Alamos Scientific Laboratory in its design, development, operation, and analysis of the KIWI-B4D nuclear rocket reactor experiment. This test provides good reason for confidence in the successful execution of the tests to be conducted this year and next and provides a good basis for confidence in the availability of nuclear rockets when they will be required for the performance of advanced space missions. The availability of these nuclear rocket propulsion systems will give this country a propulsion capability far advanced over any other rocket propulsion system available. . . .

"Progress has been made in electric propulsion, particularly in the thrustor area, but important research data and technology are also beginning to be provided in the difficult area of nuclear reactor electric generating systems required for prime electric propulsion in space.

"Beyond these systems, other advanced nuclear propulsion concepts are not yet well enough understood to justify undertaking significant development efforts. Careful and systematic accumulation of research information to evaluate the feasibility of these concepts and to arrive at a real understanding of them appear appropriate." (Text)

- Appearing before Special Subcommittee on Air and Water Pollution of the Senate Committee on Public Works, NASA Assistant Associate Administrator for Advanced Research and Technology John L. Sloop said:

"The NASA activities having a bearing on air and water pollution are almost entirely in the area of nuclear and chemical propulsion. Even in this area, however, only a fraction of the work offers a pollution threat.

"With regard to water pollution, it is standard practice at NASA Centers and in the rocket industry to capture water used in test operations and, if necessary, treat it before re-use or discharge.

"In considering air pollution, the NASA uses two approaches:

"(1) Isolated locations and atmospheric dispersion, and

"(2) Containment and neutralization. . . .

"In . . . our nuclear propulsion activities, neither routine reactor development testing and flight operations nor conceivable accidents are expected to release sufficient radioactivity into the atmosphere to constitute a hazard to public health and safety.

"In chemical propulsion, there are some propellants that are toxic prior to combustion and some rocket exhausts that contain toxic products. Fortunately, the NASA space program is primarily concerned with non-toxic propellants. In our largest program, the Apollo effort, for example, the giant Saturn I and V boosters both burn oxygen and kerosene in their first stages and oxygen and hydrogen in their upper stages. The Apollo spacecraft uses toxic nitrogen tetroxide and a mixture of hydrazine and unsymmetrical dimethylhydrazine, but the quantity is small compared to that used by the Saturn boosters . . . [and is not used] until the Apollo is well into space under normal operation. . . .

"In most rocket testing and in vehicle launches, the rocket exhausts directly into the atmosphere. The very high exhaust temperature and velocity serve to rapidly disperse the exhaust products. In some rocket testing we must simulate the low pressures encountered at altitude. In these facilities, the rocket exhausts into a duct where there is an opportunity to scrub the products with water before discharging into the atmosphere. A unique facility, an example of the containment and treatment method for use in populated areas, was built at the Lewis Research Center in Cleveland, for rockets using fluorine. This facility, put into operation in 1956, was designed to handle hydrogen-fluorine rockets up to 20,000 pounds thrust. . . ." (Testimony)

June 30: Vernon G. MacKenzie, Chief of Public Health Service's Div. of Air Pollution, testified before Senate Committee on Public Works' Special Subcommittee on Air and Water Pollution that investigation should be made of "suitable means" to insure that rocket-fuel development tests by private organizations be conducted under "suitably controlled conditions." He said PHS had "full confidence" that Federal agencies would apply appropriate precautions of health protection of rocket fuels "insofar as these are under the direct control of Federal agencies." He named beryllium among "the most toxic and hazardous of the non-radioactive metallic substances used in industry today," and cited also fluorine as a potentially dangerous component. (*M&R*, 7/6/64, 11)

- Museum of Modern Art in New York opened exhibition titled "20th Century Engineering," which featured 195 spectacular and dramatic structures combining beauty with utility. NASA Langley Research Center wind-tunnel was represented among the dams, antennas, observatories, skyscrapers, and other structures. (Huxtable, *NYT*, 6/30/64, 29)
- Arthur Kantrowitz delivered 1964 von Kármán Lecture at AIAA annual meeting.

During June: Analysis of infrared radiation measurements of TIROS II meteorological satellite was detailed by R. S. Hawkins of AFCRL. Hawkins suggested that when discrepancies between data from satellite and from other sources are understood and accounted for, infrared data from meteorological satellites could provide an invaluable tool for plotting and predicting course of cold fronts as well as for obtaining information on structure of frontal zones and regions of convective activity. (OAR *Research Review*, 6/64, 3)

- Total number of contractor and civil service personnel at NASA Michoud Operations passed the 10,000 mark. Of the total, 281 were employees of NASA. (*Marshall Star*, 6/24/64, 9)
- Water-cooled space suit, employing plastic tubing sewn into long underwear, was in research and development at NASA Manned Spacecraft Center. Begun by Royal Aircraft Establishment in 1962, suit was undertaken by MSC in 1963 when Dr. John Billingham left RAE to join MSC. Dr. Billingham said, "In operation and in theory the suit looks acceptable. But we must test it much more for reliability." (*M&R*, 6/22/64, 28)

During June: Detailed report on global satellite weather prediction system, "Swami" (Stanford Worldwide Acquisition of Meteorological Information), was completed by 31-man team of Stanford Univ. students. Three 1,200-mi. circular polar orbit satellites orbiting 120° apart would receive meteorological data from global network of weather balloons, transmit the stored information after each two-hour orbit to a single ground command station. (*M&R*, 6/22/64, 25)

- NASA-DOD-AEC committee studying space power requirements of U.S. space program through 1975 presented its report to Aeronautics and Astronautics Coordinating Board (AACB). (*M&R*, 6/22/64)
- NASA Reliability and Quality Assurance Office published "Quality Assurance Provisions for Government Agencies, June 1964 Edition" (NPC 200–1A), a complete revision, reflecting experience gained by NASA and DOD agencies at supplier operations on behalf of NASA. (NASA Release 64–154; NASA Announcement 64–129)
- First of two main experiment units for Orbiting Astronomical Observatory (Oao) was delivered to NASA by Univ. of Wisconsin. Designed and built by Cook Electric Co. under contract to the university, the experiment consisted of clustered 16-in. nebular photometer telescope, four 8-in. stellar photometer telescopes, and two scanning spectrometers. (*Av. Wk.*, 6/29/64, 55)
- At Institute of Navigation meeting in New York, panel discussion on proposed civilian navigational satellite system reportedly indicated only mild interest by potential users of the system. Unnamed NASA officials were said to be optimistic that NASA-FAA-Commerce-DOD-Interior representatives would agree "in the near future" on setting up an ad hoc committee to study the requirements of such a system. (*M&R*, 6/22/64, 17)
- A 60-ft. pad was rapidly fabricated at Moffet Field, Calif., for full-scale evaluation using the Bell X–14 vertical take-off aircraft. In this test neither the 1,500° F temperature nor the wheel-loading adversely affected the rapidly prepared pad, AFSC Research and Technology Div. reported. The instant-landing pad, a ground-hardener of polyester resin, was being developed by Air Force Aero Propulsion Laboratory for rapid preparation of landing sites for future jet vertical take-off and landing (Vtol) aircraft. Operational Vtol aircraft would drop or spray the semi-liquid preparation in forward combat areas, land on it 15 min. later. (AFSC RTD *Biennial Rpt.*, 7/62–7/64, 23)
- Douglas Missile & Space Systems Div. concluded in its manned Mars exploration study for NASA Marshall S$_{\text{r}}$. Flight Center that future interplanetary space travelers should be no taller than 5'7" and no heavier than 145 lbs. Six of these "20 percentile" men—smaller than approximately 80 per cent of American men—on a manned Mars flight would require 1⅓ lbs. less food per day than larger men, thus saving nearly 500 lbs. of food-cargo weight on a one-year trip. Study said 70,000 lbs. could be saved in gross weight of Mars spacecraft if the six astronauts were 20 percentile men. (DM&SSD *Apogee*, 6/64)
- Dr. George E. Mueller, NASA Associate Administrator for Manned Space Flight, described in *Astronautics & Aeronautics* the future space capabilities being developed for the U.S. through Project Apollo. The Saturn IB, Saturn V, and Apollo manned spacecraft would provide capabilities including extended lunar explorations, orbiting laboratory re-

search, manned escape missions, and long-duration operations—including satellite inspection and repair. (*A&A*, 6/64, 26–29)

During June: Dr. Raymond L. Bisplinghoff, NASA Associate Administrator for Advanced Research and Technology, said in *Scientific American* article on the supersonic transport that NASA Langley Research Center and Ames Research Center had "devised several aerodynamic designs for what they call a SCAT, which is an acronym for 'supersonic commercial air transport.' Four of these designs, known as SCAT's 4, 15, 16, and 17, are the products of eight years of research. Each design has characteristics that make it somewhat different from the others. . . .

"NASA engineers conducted extensive wind-tunnel tests of the four SCAT models at the Langley and Ames centers to evaluate and compare their aerodynamic characteristics. . . . It is evident that we must discover how to combine the supersonic L/D [lift-drag ratio] characteristics of SCAT 4 with the subsonic performance of SCAT 16. We are optimistic that this can be done. . . .

Discussing special structural problems of SST designs, Dr. Bisplinghoff said the metal fatigue would be accelerated under exposure to high temperatures of supersonic flight, but at the same time it was essential that the airframe provide safety and serviceability for more than 10 yrs. or 30,000–50,000 hrs. flying time.

"Prolonged screening of materials by NASA's Langley and Lewis laboratories has identified a titanium alloy that appears to meet the unique and stringent requirements for strength, stiffness and high temperatures of an SST. The alloy contains 90 percent titanium, 8 percent aluminum, 1 percent molybdenum and 1 percent vanadium. . . ." (*Sci. Amer.*, 6/64)

- Dr. Wernher von Braun, Director of Marshall Space Flight Center, was awarded an honorary degree by Iowa Wesleyan College. (*Marshall Star*, 6/17/64, 1)
- Article written by Sen. Barry Goldwater, Republican nominee for President, in which he laid down his views on the form the U.S. space program should take, appeared in *Science and Mechanics:* "The concept that speeding men to the moon is the same as jockeying trotters over a muddy track must be abandoned, regardless of Kremlin propaganda. The idea that we can cooperate with the Russians to do the job quicker and cheaper is too ludicrous for comment. We are spending entirely too much money on the manned moon program, when a carefully plotted program using unmanned lunar landing equipment could steadily build up a solid body of scientific knowledge about the lunar environment that would increase the safety factor for astronauts much later. Automatic landing equipment could be designed at minimum cost. The program should expand in easy steps, each one leading to the next. The billions of dollars thus saved could be profitably invested in other aspects of space research."

Speaking of the military role in space, he said: "All manned space research should be directed by the military, with national security and control of the access to space as primary goals. The threat from space now and in the foreseeable future is from spatial regions within a thousand miles of Earth and not from lunar distances. Routine daily surveillance of these regions must be established, and the additional capabilities of positive identification, interception, and destruction of hostile enemy space vehicles will be necessary. This would be the greatest deterrent to an enemy's military space aspirations. It would also give us the means of developing experienced space pilots, scientific and technological methods, and new scientific knowledge which could be used later in sending men to the moon and the planets. The hardware, on a routinely reliable basis, would be available off-the-shelf.

"It might be added that almost every successful NASA launch to date, manned and unmanned, has been made with military support—with missile rocket-boosters, launch facilities, technical know-how, communications (including air transport of boosters) or safety procedures—or with all five. . . .

"Finally, I believe that an immediate initial step of any effective military space program must be a manned station in near orbit about the Earth, with necessary support equipment. . . ."

Senator Goldwater also stated that in his opinion there was the need for two crash programs in space research. The first would develop the Nike-Zeus antimissile missile, and the second would develop the potentiality of lasers for use as military weapons. (*Science and Mechanics*, 6/64; *Wash. Post*, 7/21/64)

During June: France's Office National d'Etudes et de Recherches Aerospatiales (ONERA) revealed it had successfully flight-tested hybrid rocket engine in a rocket launched April 25 from Ceres sounding rocket facility on the Ile du Levant in the Mediterranean Sea. Engine employed combination of liquid nitric acid oxidizer and unidentified solid fuel. (*Av. Wk.*, 6/29/64, 24)

- Trustees and President of Clark Univ. announced Clark's proposed new library would be named the "Robert Hutchings Goddard Library" in honor of the distinguished rocket pioneer, who was an alumnus and professor of physics at Clark. (Clark Announcement)
- Pan American Airways' Guided Missile Range Div. was awarded one-year, $112 million contract from USAF for continued operation of Air Force Eastern Test Range. (*Av. Wk.*, 6/29/64, 25)

During May and June: Project Taper (Turbulent Air Pilot Environment Research) was conducted by FAA and NASA to study turbulent air hazards for jet aircraft and ways for pilots to deal with them. Project involved flights into light-to-moderate turbulence at high altitudes by FAA Convair 880 flown by FAA and NASA pilots. Data on motion effects experienced in the cockpit, gathered in the 800 flights, were used in NASA laboratory tests on motion simulator. In second series of flight tests, using FAA Boeing 720 airliner, instruments recorded pilot control actions and aircraft response in turbulent air, also tested autopilot and yaw damper for their ability to maintain control. (Hudson, *NYT*, 5/1/64, 58)

July 1964

July 1: Referring to Centaur rocket (AC-3) which splashed into Atlantic on June 30 after failing to orbit, NASA officials pointed out that "from an engineering point of view the flight was highly successful. . . ." Five of six test objectives were achieved, and the sixth—liquid-hydrogen engine start and run—was partially achieved. Grant Hansen, vice-president of General Dynamics/Astronautics, which built Centaur, said examination of instrument readings from the flight indicated failure of a nylon coupling which joined two shafts in hydraulic system associated with one of Centaur's two engines. Coupling seemed to have failed about four seconds after Centaur stage ignited. (Wash. *Eve. Star*, 7/2/64, A5)

- Soviets launched COSMOS XXXIV artificial earth satellite equipped for scientific study of outer space. It was part of program announced by Tass on March 16, 1962. Period was 90 min., apogee was 360 km. (223 mi.), perigee was 205 km. (127 mi.), and equatorial inclination was 64.96°. (Tass, *Komsomolskaya Pravda*, 7/2/64, 1, ATSS-T Trans.)

- President Johnson congratulated American Institute of Aeronautics and Astronautics (AIAA) on its first annual meeting and technical display, held June 29–July 2, Washington. He said: "I am particularly pleased that you have selected this occasion to honor the highly respected and capable Dr. Hugh L. Dryden, of the National Aeronautics and Space Administration." (AIAA *Booster*, 7/2/64, 1)

- NASA's George E. Mueller reported to AIAA that "beginning in 1958 with the first United States satellite launching, the Explorer-Jupiter C, and continuing through the Thor, Atlas, Thor-Delta, and Thor-Agena to the Saturn V in 1968 through 1970, the cost of payload in orbit will be reduced by about three orders of magnitude.

    "This remarkable decrease in costs results from the growing reliability and operational effectiveness of our launch vehicles and from the trend towards larger and larger weight-lifting capacity. The relative increase has been greater ·· ·' · ge vehicles because of the economies associated with their operation."

    Regarding the extension of capability of existing launch vehicles, Mueller reported: "Basically, our launch vehicle technology is sound, and it is advancing in an orderly, evolutionary fashion. We have mastered the design, manufacturing, transporting, checking out, and launching of small, medium, and large launch vehicles.

    "Existing space boosters have done a creditable job. Engine reliability has been exceptionally good; the clustering of relatively large numbers of engines is now a proven technology; uprating of engines in thrust and efficiency has been possible; and further uprating can be accomplished to meet additional payload demands.

"We are developing a basic launch vehicle industrial and technological capability that can be extended and expanded for both future space exploration and exploitation.

"For the future, we are building on today's firm foundation. New high-energy propellants and nuclear rockets have great promise. Large solid-propellant motors are being employed. Much progress is being made in correlating high-energy propellants and engine materials and engineering technology.

"Increased emphasis will be placed on management and the development of reliability. Management responsibilities will continue to grow more complex, while cost effectiveness will continue to be stressed. A smaller inventory of vehicles will be developed which have greater reliability and broader utility.

"All in all, we can look forward to a new and challenging era of progress in launch vehicle technology." (Text)

July 1: Paul G. Johnson of AEC–NASA SNPO said at AIAA meetings that early explorers of Mars may pass by planet Venus on way back to earth, to slow down spacecraft to safe re-entry speed. Side trip would lengthen voyage by a few months. Johnson also said that nuclear rockets might make Mars trip possible in 1980's. Mission could involve six to eight crewmen and take from 400 to 500 days. (NASA Release 64–165)

- Dr. Eugene G. Fubini, Deputy Director of Defense Research and Engineering, called on defense industry at AIAA meetings to stop what he called unjustified and potentially dangerous "gloom and doom talk" about future defense spending. He said Pentagon R&D spending, currently at a $5.5 billion annual level, should remain relatively stable in coming years. Such pessimistic fears on part of industry, based on recent very small decrease in defense budget, Fubini said, could prompt top engineers and scientists to move into other industries and thereby weaken U.S. military posture by reducing industry's capability of generating new ideas for DOD and developing new weapons. (AIAA *Booster*, 7/2/64; *NYT*, 7/2/64, 32M; *Av. Wk.*, 7/13/64, 19)

- President Johnson transmitted NASA's *Ninth Semiannual Report to the Congress* on progress and projects for period from January 1, 1963, to June 30, 1963. President's covering message stated that "the breadth of performance promises subsequent periods of even greater accomplishment in meeting the challenge of space." (NASA LAR III/132)

- Richard H. Nelson, Boeing's project manager for production of Saturn V rockets, said production was about 10 weeks behind schedule because of a parts shortage. However, first rocket scheduled for flight was to be delivered on schedule in late summer or fall of 1966. Present delay affected Saturns designed to test design and engines in ground firings at MSFC. Last 10 of Saturn V's were expected to be delivered to Cape Kennedy by late 1968. Nelson was addressing Aviation Writers Association in Houston. (*Houston Chron.*, 7/2/64)

- Dr. Hugh L. Dryden, NASA Deputy Administrator since 1958, received the AIAA premiere award, the Louis W. Hill Award for 1964, for the Outstanding Contribution to Space Transportation. Dr. Dryden was honored for his fluid mechanics research leading to space era, technical leadership of NACA and NASA teams which made civilian space program a reality, and personal contribution to research in the X–15 program. He was called "a symbol of our aerospace success."

In his address at the AIAA Honors Night Banquet, Dr. Dryden said that those involved with our space program must consider broad national goals in addition to scientific details. Other goals mentioned by Dryden were "to obtain maximum benefit to the welfare of the Nation and to all mankind" and "to provide not only material benefits to free man in a peaceful world, but also incentives for mental and spiritual growth and accomplishment." While recalling the day about 40 years ago, "when an airplane was essentially the product, from concept to final machine, of one man, the designer with the aid of two or three assistants," Dr. Dryden said today's space technologist ". . . must search for knowledge of the facts, conduct our work with devotion to intellectual honesty and objectivity, and realize in our activities the highest moral aspirations and ideals of which we are capable."

Dr. Dryden expressed confidence that the Nation would continue to support comprehensive space exploration program because of sum total of goals space planning strove for. These included national prestige, national defense, human knowledge of space, development of efficient space vehicles, and utilization and international exchange of technological information. He added: "I believe that the rate and scale of the program are not excessive in view of the great resources of this country." (AIAA, *The Louis W. Hill Award* pamphlet, 7/1/64; AIAA *Booster*, 7/2/64, 1; Speech Text; NASA Release 64–161)

July 1: At AIAA banquet, G. Edward Pendray Award was presented to Andrew G. Haley for his "pioneering contribution to the analysis of governmental and international legal questions arising from the rapid development of space travel and space exploration, as exemplified by his book, *Space Law and Government*, published in 1963." AIAA Lawrence B. Sperry Award was presented to Daniel M. Tellep, Manager of Launch and Entry Thermodynamics Research and Engineering at the Lockheed Missiles and Space Co., for Tellep's "contributions to re-entry technology through participation in the United States' first re-entry flight test program and for advancements in the field of heat transfer as related to aerospace programs." Also presented were the 1964 Graduate and Undergraduate Student Awards to Richard C. Lessman and Lawrence S. Iwan. (AIAA Annual Mtg. Program; AIAA *Booster*, 7/2/64)

• At AIAA former American Airlines' vice president William Littlewood said problems of sonic boom and economics might postpone introduction of supersonic airliner from 1970 target date to some time between 1975 and 1980. Problems of practicability and noise would have to be solved before supersonic transport became reality, Mr. Littlewood said: (Finney, *NYT*, 7/2/64, 28)

• NASA awarded $1,125,040 contract to Ling-Temco-Vought for construction of Dynamic Crew Procedures Simulator which would allow astronauts to simulate orbital, lunar, and interplanetary flights in model spacecraft. Simulator's completion was scheduled for May 1965, when it would be delivered to MSC. (MSC PAO; *Denver Post*, 7/5/64)

• NASA announced award of contract to North American Aviation's Rocketdyne Div. to provide four RN–6 liquid-hydrogen-cooled nozzles to be used in tests. Three nozzles were to be used for Nerva program, the fourth for Lewis Research Center's Advanced Nuclear Rocket Systems Analysis Program. (NASA Release 64–162)

July 1: NASA awarded contract to Documentation, Inc., of Bethesda, Md., for $4.9 million to operate NASA Scientific and Technical Information Facility for another year. Maintaining world's largest collection of aerospace literature, facility was to operate selective dissemination of information program during coming year, using automatic equipment to notify individual scientists and engineers of new developments in their special fields. (NASA Release 64–163)

- Dr. Oscar E. Anderson, Jr., was appointed Director of International Organizations Div. of NASA's Office of International Programs. He was previously Project Officer in Cooperative Projects Div. of same office. Before coming to NASA in 1962, Anderson was a historian at AEC. (NASA Announcement 64–135)
- John H. Glenn, Jr., discussed with NASA Administrator James E. Webb the possibility of the former astronaut's returning to NASA in advisory capacity, according to NASA spokesman, but nothing had been decided by the two. (UPI, *Wash. Post*, 7/2/64; AP, Wash. *Eve. Star*, 7/2/64, A8)
- Dr. Allen E. Puckett, vice president of Hughes Aircraft, said world's first commercial communications satellite, Early Bird, being built for ComSatCorp, would be completed in less than one year and ready for launch a month later. Early Bird was a first step toward global telecommunications and was a commercial version of Hughes' Syncom satellite. (Miles, *L. A. Times*, 7/1/64)
- NASA–USAF Memorandum of Understanding defined responsibilities of NASA and various USAF organizations which loan aircraft to NASA or provide logistical support for such aircraft. (NMI 2–3–34A)

July 1–2: NASA's Science and Technology Advisory Committee for Manned Space Flight was holding three days of meetings with MSC officials on problems related to Project Apollo. Members of Committee, established by NASA Administrator James E. Webb in December 1963, and which functioned in advisory capacity to NASA Associate Administrator for Manned Space Flight, Dr. George E. Mueller, delved into problems posed by launch vehicle systems, spacecraft systems, launch operations, science program, and space medicine.

MIT physicist Dr. Charles Townes said at MSC meeting of the Committee that he believed astronauts, not scientists, should be first men to explore moon. Scientists, he said, would concentrate on their own specialties in exploring the moon's surface, but astronauts trained in scientific principles would bring back important information on the moon as a whole. (MSC *Roundup*, 7/8/64, 1; *Houston Chron.*, 7/2/64)

- Manpower needs of civil aviation were discussed in Washington by FAA's Aviation Human Resources Study Board. (FAA Release 64–61)

July 2: House adopted by voice vote the conference report on NASA FY 1965 authorization bill (H.R. 10456) and sent bill to Senate. Senate adopted conference report, clearing bill for President's signature. Bill provided NASA authorization of $5,227,506,000 of which $4,341,100,000 was for research and development, $622,880,500 was for construction of facilities and $623,525,500 was for administrative operations. (NASA LAR III/133)

- Sen. J. W. Fulbright (D.–Ark.) called for 10% cut in NASA's Apollo FY 1965 budget, in statement submitted to Senate Appropriations Committee. He stated 1970 Apollo goal had already been missed, so that cut in funds would have no serious effect, and that NASA's past estimates of costs and time were not reliable, so that little confidence could be placed in NASA's meeting its 1970 deadline. (Wash. *Eve. Star*, 7/2/64, A8)

July 2: USAF launched Thor-Agena D space booster from WTR with undisclosed satellite payload. (HHN-48)

- At AIAA, William G. Huber of MSFC outlined NASA's plans for space projects beyond Apollo project. According to one set of plans for the 1970-1990 period, space program costs would be $6,970,000,000 per year for 20 years. This figure was highest yet publicly discussed by a NASA official. Huber was outlining possibilities for the future in his discussion of large post-Saturn launch vehicles. (Wash. *Eve. Star*, 7/3/64)
- Paul D. Reader, Head of Lewis Research Center's Electrostatic Thrustor Systems Section, presented test results of experimental ion engine which was six times larger than any other electric engine. Large engine ejected beam of charged particles 20 in. in diameter with power of 30 kw. Development of large, high-powered ion engines, which may provide thrust for extended deep-space exploration, was step in direction of solving weight problem posed by electrical circuitry for smaller ion engines. (LRC Release 64-60)
- Columbia Univ. Electronics Research Laboratories announced development of plasma-jet generator that produced continuous jet of hot plasma—8,000 to 20,000° F—and that required no confining jacket or chamber. (*A&A*, 9/64, 97)
- In satellite communications demonstration at AIAA Washington meetings, Lt. Cdr. J. F. Debold, Commanding Officer, U.S. Naval Research and Development Satellite Communications Group, spoke via SYNCOM II satellite with his second in command on board USNS *Kingsport*. *Kingsport*, which can acquire, track, control, and communicate via a satellite while cruising at sea, was built by Navy to complement two Army-developed communications terminals on East and West Coasts. (*AIAA Booster*, 7/2/64, 2)
- Engineers proposed development of intercontinental ballistic troop transport capable of carrying 1,200 troops to any trouble spot on earth in 45 minutes. Such an Ithacus rocket transport would be ideal for helping stamp out "brush fire" conflicts, according to engineers Philip Bono and George C. Goldbaum of Douglas Aircraft who spoke at AIAA annual meeting. If funds were committed to project, they considered the ICBM transport system might be operational by 1975. (*Houston Chron.*, 7/3/64)
- Final Polaris A-3 in test program was launched from Cape Kennedy. Advanced Polaris' 2,500-mi. flight was 41st in program begun in summer 1962. (*M&R*, 7/13/64, 14)

July 3: In preparation for launching of seventh Saturn I rocket (SA-7), NASA had completed major phase of evaluation of flight records of SA-6. Review of SA-6 data, in form of magnetic tapes, strip charts, oscillograms, photographs, etc., was made in collaboration with major vehicle contractors. Analysis reaffirmed the success of previous flight on May 28 despite loss of one engine 23 sec. before engine programed cutoff. (NASA Release 64-160)

July 5: NASA extended contract with Mason-Rust Co. for provision of support services at Michoud Operations near New Orleans. New $6,839,069 modification awarded by MSFC, to continue in force till end of 1964, increased total Mason-Rust contract to $17,751,806. (*Marshall Star,* 7/8/64, 1)

July 5: Library of Congress had distributed to 30 large city and state libraries copies of Braille translations of books on space exploration. These informations aids for the blind had been published by Library of Congress under a NASA grant. (Las Cruces *Sun-News,* 7/5/64)

- Effective this date, USAF Western Test Range's facility at Point Arguello became part of Vandenberg AFB, and its facility at Kwajalein Island was turned over to the Army. Organization plans for National Range Division with its Eastern Test Range and Western Test Range had been announced May 4. (*L.A. Times,* 7/2/64)

- During 1,000th session in U.S.S.R. on radar probing of planets, Tass said, Soviet scientists used newly designed equipment and high-speed computer to successfully gather data on topography of planet Venus. (*Pravda,* 7/5/64, ATSS-T Trans.)

- Second U.S. manned rocket flight was made 20 years ago today, when Northrop MX-324 flying wing was towed aloft and propelled by 200-lb.-thrust Aerojet rocket engine and flown for four minutes by test pilot Harry Crosby over Harper's Dry Lake, Calif. (Clark, *NYT,* 7/5/64, 44; *A&A, 1915-60,* 47)

- Reported that new 125-ft. radiotelescope opened at England's Jodrell Bank Radio Astronomy Station early in June could give insight into creation of universe. Radiotelescope was to detect and analyze high-frequency radio pulses emitted by stars 6 billion light-years distant. (Reuters, *Chic. Trib.,* 7/6/64)

During week of July 5-11: Unmanned paraglider wing designed to make air-to-ground cargo deliveries to forward battle areas successfully completed series of three flight tests at U.S. Army Yuma Proving Ground, Ariz. The paraglider, also designed to be a spacecraft recovery system, slipped three 1,000-lb. cargo packages to earth without mishap from 3,100 ft. Demonstration was performed by Space and Information Div. of North American Aviation under program called Precision Aerial Delivery System (PADS). (NAA S&ID *Skywriter,* 7/17/64, 1)

July 6: USAF launched Atlas-Agena D space booster from Vandenberg AFB, Calif., with unidentified payload. (*M&R,* 7/13/64, 14; *SBD,* 7/8/64, 33)

- European Space Research Organization (ESRO) made first launching, a British Skylark sounding rocket, to altitude of 125 mi. from Salto di Quirra range in Sardinia. (*NYT,* 7/11/64, 5)

- NASA selected International Business Machines Corp. Federal Systems Div. to supply technical services for support of Goddard Space Flight Center's computer facilities. The computation services and allied systems permitted GSFC to acquire high-speed raw data from spacecraft and to convert it almost instantaneously into intelligible information concerning the spacecraft. Contractor was to staff and operate computer facilities related to launch monitoring system at Cape Kennedy, the early flight tracking station on Bermuda, and GSFC control center. (NASA Release 64-166)

- MSC announced approval of improved Gemini spacesuit for use by Apollo astronauts in the earlier flights in earth orbit. Adoption of Gemini spacesuit would allow more time for development of Apollo suit, which would allow astronauts greatest mobility for walking on surface of moon. (*Houston Post,* 7/7/64)

July 6: An airlines pilot who used a weather map based on information transmitted to earth by a Tiros meteorological satellite on a flight from New York City to Dakar, Africa, said ". . . the map was so accurate as to be almost unbelievable. . . . in my estimation we have found the answer, let's send up more Tiros." Weather maps based on Tiros information were being given to pilots flying international routes, as a matter of course. (Las Cruces *Sun-News*, 7/6/64)

* In *Aviation Week* editorial, Robert Hotz said that until the past two years most nations in the world considered space technology to be monopoly of the competing U.S. and U.S.S.R. Now they were beginning to participate in space programs themselves. Two primary reasons given were technical success of weather and communications satellites and "enlightened and effective" program in international cooperation operated by NASA. Hotz saw international space cooperation as the most powerful force binding nations of the world if early promise of international space technology is realized.

"Perhaps the most heartening aspect of the international expansion of space activity is the manner in which the hard technical and financial facts are persuading nations to work together on cooperative projects ranging from global satellite systems to booster development and scientific exploration.

"Part of this has been achieved through the NASA international cooperation program for which Dr. Hugh L. Dryden, Deputy NASA Administrator; Arnold Frutkin, head of the international programs office of NASA; and Dr. Harry Goett, director of Goddard Space Flight Center, deserve special credit.

"The other portion has been achieved through Communications Satellite Corp., currently organizing global communications system with heavy international participation, and through the U.S. Weather Bureau's leadership in developing a global weather satellite system. . . ." (*Av. Wk.*, 7/6/64; *CR*, 7/31/64, A4040–41)

* McDonnell Aircraft Corp. celebrated its 25th anniversary. Today with facilities worth $90 million, a backlog of $1 billion in contracts, and 35,000 employees, the company looked back to the day in 1939 when James S. McDonnell's office had a typewriter, one employee, and no contracts. (AP, New Orleans *Times-Picayune*, 7/6/64)

* Scholars on UCLA faculty, having judged that space age overemphasized science in education at the expense of the humanities, sought support for proposed independent National Humanities Foundation. Foundation was to be structurally similar to NSF. Proposal, drawn up by 20-man Commission on the Humanities, was already in the hands of President Johnson. (Hartt, *L.A. Times*, 7/6/64)

July 7: Secretary of Defense Robert S. McNamara reported to President Johnson that DOD's cost-reduction efforts had saved Government $2.5 billion, instead of previously estimated $1 billion, during FY 1964. In next three years DOD's annual savings would total at least $4.6 billion. McNamara singled out Titan III contract with Martin Co. as one of best-managed contracts DOD had, as result of incentives applied to program. Spending on Titan III program was 1% below cost estimate. McNamara assured reporters that savings were not being achieved at expense of defense posture. (*NYT*, 7/8/64, 1, 3; *M&R*, 7/13/64, 11)

July 7: U.S. Army presented two pioneer communications satellites, Score and Courier, to Smithsonian Institution during opening of Museum of History and Technology's Hall of Electricity. Both satellites were developed at present Army Electronics Command, Ft. Monmouth, N.J. (Balt. *Sun,* 7/8/64)

- In study financed by NASA, a Mexico City matador had his heart beat, respiratory rate, and body temperature monitored during the height of a bull fight. Readings from sensors fixed to his skin were telemetered to instruments inside bull ring. Doctors at MSC used information in study to determine how much physical stress human body could withstand and what effect stress had on performance. It had been determined previously that many athletes undergo more physical stress than do astronauts in flight. (AP, *NYT,* 7/8/64, 12; MSC *Roundup,* 7/22/64, 7)
- Yevgeny F. Loginov, head of Soviet Union's airline, Aeroflot, said Soviet designers were working on supersonic airliner. He indicated U.S.S.R. hoped to have one in operation before the U.S. (AP, *NYT,* 7/8/64, 58)

July 8: X-15 No. 3, piloted by Capt. Joe Engle (USAF), flew 3,511 mph (mach 4.94) and gained altitude of 170,000 ft. in first test of infrared horizon scanner designed to give sharper outline of earth to astronauts re-entering earth's atmosphere. Flight also enabled Engle to make pilot checkout of adaptive control system which had ability to sense atmospheric conditions surrounding the craft. Flight was postponed twice before successful effort. (NASA X-15 Proj. Off.; AP, *Wash. Post,* 7/9/64)

- Scientists from the Air Force Cambridge Research Laboratories, the Swedish Space Committee, and NASA announced they would investigate the nature of noctilucent clouds during a period starting the last week in July and extending through Aug. 24. This would be the second in series of rocket-borne experiments to be conducted from Kronogård Range in Northern Sweden. Principal purpose of experiments was to confirm results of experimental flights completed in August 1962, when it was tentatively determined that noctilucent clouds were composed of meteoric particles or the residue of burned out meteors. Program calls for the firing of four Nike-Apache rockets for collection purposes and four Nike-Cajun rockets to measure temperatures and winds associated with the presence or absence of the clouds. (OAR Release 7-64-2)
- NASA announced grant of $1,125,000 to Univ. of Illinois to finance construction of research laboratory facilities as wing to university's Coordinated Science Laboratory. To be called Aerospace Research Center, the facilities were to be for investigations of ionosphere properties, propulsion, molecular densities, rendezvous problems, and plasma physics. (NASA Release 64-169)
- After two previous failures, USAF successfully launched four-stage, solid-fuel Athena missile. Rocket was fired from Green River, Utah, on estimated 15-min., 475-mi. flight to White Sands, N. Mex., carrying one of series of subscale experimental re-entry vehicles being tested as part of program to develop advanced re-entry vehicles for space travel and ICBM's. No other details of flight were given. (AP, *Hartford Courant,* 7/9/64)
- Senate Independent Offices Subcommittee of Committee on Appropriations concluded its hearings on H.R. 11296, FY 1965 appropriations for independent offices (including NASA). (NASA LAR III/134-135)

July 8: European Space Research Organization launched second British Skylark sounding rocket with scientific payload to altitude of 125 mi. from Salto di Quirra range in Sardinia. This launch and identical one the previous day were initial attempts of the 10-nation organization. (*NYT,* 7/11/64, 5)

- NASA Administrator James E. Webb told Senate Appropriations Committee's Independent Offices Subcommittee that 15 Saturn V flights were planned for Project Apollo program, the lunar landing to be accomplished in the last three months of 1969 with the fifteenth flight. Webb also said that if the proposed $5.3 billion NASA budget for fiscal year 1965 were cut so that the Apollo program had to be slowed down, the cost of the program would increase by a billion dollars with each added year of delay. (Text)
- Rocketdyne Div. of North American Aviation received four NASA contract modifications worth $22,378,626 on existing research and development production contracts for the F–1 engines, which provide thrust in first stage of Saturn V launch vehicle. NASA also awarded contract modifications to Douglas Aircraft Co. totaling $31,471,836 for additional work on the S–IV and S–IVB rocket stages. (*Marshall Star,* 7/15/64, 1, 9; *Nashville Tennessean,* 7/9/64)
- Donald K. Slayton, Assistant Director for Flight Crew Operations, MSC, announced specific assignments for astronauts. Included in assignments was that of Alan Shepard, Jr., who was named chief of Astronaut Office, taking place of Slayton, who had been acting in the position in addition to other duties. (MSC *Roundup,* 7/8/64, 1)
- Sir Bernard Lovell, Director of Jodrell Bank Experimental Station, reported in *New Scientist* that technical difficulties, and not a deliberate slackening of effort, were hampering Soviet activities in getting men to the moon. He noted that the Russians apparently had not made a probe work in space successfully for more than a few weeks and concluded that the Soviets were failing where "once they shone—in basic rocketry techniques." (Simons, *Wash. Post,* 7/9/64)
- FAA experiments conducted over Oklahoma City during prior five months showed that the magnification of a sonic boom under certain atmospheric conditions occurred more frequently and with more force than had been predicted. The unexpected finding led FAA Administrator Najeeb E. Halaby to declare at a news conference that the sonic boom problem presented by a supersonic airliner appeared to be "more serious" than had been supposed. Halaby rejected charges that nation's supersonic airliner program was too hasty. He said that those most concerned with the plane's development felt that they were "going about the right rate." Further sonic boom tests would be made, he said, probably over desert areas. (Finney, *NYT,* 7/9/64, 12; *Wash. Post,* 7/9/64)
- ComSatCorp announced that six persons had been nominated by its board of directors for election to the new board by public stockholders. They were: Leo D. Welch, Dr. Joseph V. Charyk, David M. Kennedy, George L. Killion, Leonard H. Marks, and Bruce G. Sundlun. The corporation announced that the first annual meeting of the shareholders was tentatively planned for Sept. 17 in Washington. (ComSatCorp Release, 7/8/64; *NYT,* 7/9/64, 43C, 45C)

July 8: Dr. John P. Meehan, physiologist at Univ. of Southern California, announced he would begin experiment in which 15 young men would lie flat on their backs for 30 days, then be put in a centrifuge and be subjected to forces eight times normal gravity while attempting to operate simulated space capsule controls. Purpose of the experiment was to duplicate effect of long periods of weightlessness in space on body's heart-blood system and to devise ways of overcoming it. (AP, Balt. *Sun,* 7/9/64)

- Teamster President James R. Hoffa ordered end to picketing that had tied up vital construction work at Cape Kennedy. (*Wash. Daily News,* 7/8/64)

July 9: NASA Nike-Apache vehicle launched at 12:13 a.m. EDT from Wallops Station carried 65-lb. instrumented payload to 104-mi. altitude in experiment to measure light intensity and charged particle flux as functions of altitude. Payload and spent second stage, which were not separated during flight, impacted 98 mi. downrange in Atlantic Ocean. Experiment was fifth in series to measure and study the nature and causes of aurorae and airglow in the night sky during the dark of the moon, project of Rice Univ. under NASA grant. (Wallops Release 64–53)

- Dr. Hugh L. Dryden, NASA Deputy Administrator, discussed utilization of space technology for commercial application before American Management Association in New York:

"In the present century, the lag from basic discoveries to applications has become progressively shorter. The Space Age is still further reducing the lag.

"Two major programs are under way to meet the Congressional mandate for the widest possible dissemination of the results of space activities. The first aims to make available the primary technical data to aerospace scientists and engineers. The second repackages the information and moves space-generated technology out of the agency's laboratories and the plants of its contractors into the non-space industrial complex.

"NASA collects, organizes, and disseminates scientific and technical information on its in-house and contractor work by means of the latest in computer, microfilm, and other methods. Decentralization with local access is the keynote.

"To be of most use this information must be identified and repackaged to meet the special needs of nonaerospace industries. This is the purpose of the Technology Utilization Program, the first national program of its type. Staffs at each NASA Center are systematically organizing the reporting of new or improved ideas, materials, methods, and technology developed in the course of research and development activities. The more promising reports go for analysis to a group of industrial research institutes familiar with current industrial needs and requirements. If the analysis indicates promise, reports are published and indexed so that they are readily available to any segment of the economy that may find them useful.

"NASA is making significant contributions to the growth of science and technology in most major fields. More important, NASA is dealing with the organization and utilization of expanding knowledge in a highly advanced and, we believe, very effective manner." (Text)

July 9: Dr. Robert R. Gilruth, Director of MSC, was one of featured speakers at opening of Houston Museum of Natural Science and Burke Baker Planetarium in Houston. Dr. Gilruth said MSC was negotiating with museum officials for part time use of planetarium for Apollo project. (MSC *Roundup,* 7/22/64, 1)

- George Miskovsky, Oklahoma City attorney, said he would seek reinstatement of Okla. court injunction against sonic booms later reversed by a U.S. District Court. He termed FAA plans to continue sonic boom testing over Oklahoma City while admitting it was a more serious problem than originally believed as "more bureaucratic double talk." (AP, Balt. *Sun,* 7/10/64)
- Lt. Col. Floyd J. Sweet (USAF, Ret.), member of NASA Administrator's staff, became the twenty-ninth member of the Helms Hall Soaring Hall of Fame. He was cited for sustained leadership in U.S. soaring activities as a pilot, instructor, club organizer, and officer of the Soaring Society of America. Colonel Sweet's first glider competition was in 1932. (Helms Release)

July 10: Asked at news conference if he thought Congressional budget-cutting would preclude lunar trip in this decade, President Johnson said: "I think the funds requested are necessary. I hope the Congress will act on this appropriation soon. The committee has not acted on the appropriation yet. I don't know how to gauge the exact way it will exercise its judgment. I have every reason to believe that it will be sympathetic to our national objectives, and I would hope that they will be generous in this response to our request for funds." (Transcript, *NYT,* 7/11/64, 6)

- Dr. Winston Edward Kock, vice president-research for Bendix Corp., was named Director of NASA's Electronics Research Center to be located in the Boston area, effective Sept. 1. As director, he would have charge of fundamental work in space and aeronautical electronics. (NASA Release 64–172; UPI, *Wash. Post,* 7/11/64)
- USAF launched unidentified satellite toward polar orbit from Vandenberg AFB with Thor-Agena D launch vehicle. (UPI, *Wash. Post,* 7/11/64; *NYT,* 7/12/64, 2)
- Astronaut Neil Armstrong addressed Aerospace Workshop at Drake Univ.: "We don't feel at this time it is particularly important whether an astronaut is masculine or feminine. Astronauts are picked on the basis of experience and training. At present there are no women with the proper qualifications available. Some day we will have qualified women and I'm sure they'll be included in the program." (*Des Moines Register,* 7/11/64)
- Gen. Bernard A. Schriever, AFSC, said at a Los Angeles Rotary Club luncheon: "We have identified and produced experimental quantities of revolutionary new high strength, lightweight, high temperature materials" among which is one formed of boron fibers and a new plastic binder. "The advantages of such a material for construction of aircraft and space vehicles are obvious," pointing out that it would allow a high increase in the operating temperatures of most power plants. This would permit the use of fuels generating higher thrust levels for greater payloads and speed. (Text; AP, *NYT,* 7/12/64, 5)

July 10: Univ. of Southern California's human centrifuge, only facility of its kind on a college or university campus in America, was to be modernized and its office areas enlarged to accommodate expanding biomedical and aerospace research programs under two NASA grants totaling $280,000. (*L.A. Times*, 7/10/64)

- James E. Webb, NASA Administrator, talked and dined with Astronaut John H. Glenn, Jr., at the home of Dr. Robert R. Gilruth, Director of MSC, in Houston. Afterwards, Col. Glenn told an interviewer: "We talked about a number of different areas in which I could work. I want a part-time job, a consulting job." (*NYT*, 7/12/64)
- Reception honoring Dr. Abe Silverstein, Director of NASA Lewis Research Center, was held on his 35th anniversary of service with NACA–NASA. (*Lewis News*, 7/17/64, 3)

July 11: U.S.S.R. announced successful orbiting of two satellites, ELECTRON III and ELECTRON IV, with a single launch vehicle. Soviet news agency Tass said the satellites were studying the earth's magnetic field and radiation belts, radiation arriving from deep space, and the physical conditions in the upper atmospheric layers. ELECTRON III was orbiting at 7,040-km. (4,375-mi.) apogee, 405-km. (252-mi.) perigee, 2-hr., 48 min. period, and 60.86° inclination to the equator. ELECTRON IV was orbiting at 66,235-km. (41,158-mi.) apogee, 459-km. (285-mi.) perigee, 21-hr., 54-min. period, and 60.86° inclination to the equator. (Tass, *Izvestia*, 7/12/64, 1, ATSS–T Trans.; AP, Wash. *Eve. Star*, 7/11/64)

- Gemini-Titan II (GT–2) rocket which was to launch unmanned Gemini spacecraft arrived at Cape Kennedy. It arrived by C–133B cargo plane from Martin Co. plant in Baltimore where it had undergone testing. (AP, *Orlando* (Fla.) *Sentinel*, 7/12/64)
- USAF Titan III transtage was successfully static-test-fired for more than six minutes to verify compatability of various subsystems, in operation to simulate injection of MOL into 100-mi. orbit. Officials said firing went as expected, including length of burn time, although no mention was made as to whether it reached 312-second specific impulse demanded by Air Force. (*SBD*, 7/14/64, 66; *M&R*, 7/20/64, 8)
- Japan successfully launched three-stage Lambda III, its largest and most advanced research rocket, on a flight to collect data on the upper layer of the atmosphere. The rocket attained an altitude of 600 mi. in the 17-min., 20-sec. flight from Kagoshima Space Center in Kagoshima Prefecture, Japan. (AP, *Wash. Post*, 7/12/64; NASA Off. Int. Aff.)
- The Federal government was critized by National Academy of Sciences committee for plunging into such major undertakings as a moon expedition or a supersonic airplane without assessing the effects on the Nation's limited scientific and engineering manpower resources. The study, made public by President Johnson, said that since the Government supported about three fifths of all those engaged in research and development, it had a "massive influence" on the deployment and utilization of scientists and engineers. This, in turn, imposed an "entirely new order of responsibility" upon the Government to prevent "malutilization" of the manpower and to achieve a "balanced allocation of scientific and engineering talent." Two of the committee's principal recommendations:

(1) Steps be taken by the Executive Office of the President to develop an integrated program for collecting and analyzing information about scientific and engineering manpower.

(2) "Before the government reaches a decision to undertake a great technological program, it should make a careful assessment of the impact of the decision on the deployment and utilization of scientists and engineers."

Nowhere in its report did the committee suggest that the lunar expedition had unduly diverted manpower from other fields; in fact, the report showed that slightly less than 30 per cent of the Nation's scientists, engineers, and secondary school teachers of science and mathematics were engaged in defense, space, and atomic energy research, and that space accounted for less than 5 per cent.

The study, an outgrowth of a request by President Kennedy in 1961 for a review of the Nation's scientific and engineering manpower resources, was prepared by a 17-man committee headed by Dr. James R. Killian, Jr., science adviser to former President Eisenhower and now chairman of the corporation of MIT. (Finney, *NYT*, 7/12/64, 1, 33; Carey, AP, *Wash. Post*, 7/12/64)

July 11: *Journal of the Armed Forces* noted that Philip Bono, with Missile and Space Systems Div. of Douglas Aircraft Co., had redeveloped Ithacus (formerly Icarus), his study concept of a spacecraft which could move 132 tons of cargo or 1,200 troops anywhere in the world in 30 or 40 min. Ithacus was formerly conceived for launch from fixed land site, but under the new concept, there would be seaborne launch from nuclear aircraft carrier of *Enterprise* class. This would reduce vulnerability of the system. (Haggerty, *J/Armed Forces*, 7/11/64, 2)

- Council of Defense and Space Industry Associations (CODSIA) formed under sponsorship of Aerospace Industries Association, National Security Industrial Association, and Electronic Industries Association. Council would operate solely as coordinating and communicating vehicle between Government and industry, not as master association. (*J/Armed Forces*, 7/11/64, 15)

July 12: NASA announced that Europeans in nine countries would be able to watch the Republican National Convention via NASA's RELAY II communications satellite. Television networks in England, France, Belgium, Germany, Denmark, Sweden, Norway, Switzerland, and Italy were to use RELAY II five times during the four-day convention, July 20–23. Similar coverage would be provided during the Democratic National Convention, beginning August 24. (NASA Release 64–170)

- NASA announced that four flight tests of liquid-hydrogen stages in Centaur and Saturn rockets had established liquid hydrogen as a new high-energy fuel for space vehicles. Liquid hydrogen, when mixed with liquid oxygen, produced about 35 per cent more thrust than the kerosene-type fuels normally used. (NASA Release 64–164)

- It was probable that the Soviet space probe ZOND I, was nearing Venus and would try to send back the first close-up television pictures of that planet, reported Sir Bernard Lovell, Director of Jodrell Bank Radio Experimental Station. (UPI, *Wash. Post*, 7/13/64)

- AEC disclosed decision to discontinue Project Pluto, attempt to build low-flying, atomic-powered missile, despite ten years of effort and $200 million investment. Livermore Radiation Laboratory, initiator and developer of project, had been notified of decision about ten days earlier. Cause of discontinuation was reportedly combination of indecision, indifference, and lack of support by DOD. (Finney, *NYT*, 7/13/64, 11)

July 12: Abraham Hyatt, formerly NASA Director of Plans and Program Evaluation, was appointed Jerome Clarke Hunsaker Professor of Aeronautical Engineering at MIT. (*Boston Sun. Globe*, 7/12/64)

* Lockheed Missiles and Space Co. proposed 100-ton space station that would support up to 24 men on long-duration earth-orbiting missions. Known as Olympus, the station would be 63 ft. long and 118 ft. in diameter with central hub and three radiating spokes. Each of spokes would carry cylindrical module 18 ft. in diameter and 40 ft. long. (*Tulsa Daily World*, 7/12/64)

July 13: President Johnson signed NASA Authorization Act (H.R. 10456) authorizing $5,227,506,000 for NASA FY 1965. Largest part of authorization was for Project Apollo, $2,677,500,000. The bill was $76,494,000 less than the Johnson Administration had requested. (AP, *NYT*, 7/15/64, 15)

* The largest contract awarded to date for work on the NASA Mississippi Test Operations facility was awarded to Koppers Co., Inc., Malan Construction Department, New York City, for the construction of the first test position on the S–IC dual test stand. The contract was for $17,280,157, and was awarded by the U.S. Army Corps of Engineers, Mobile District. Work on the stand would include construction of three concrete tiers each about 158 ft. high, a tower approximately 105 ft. high, a steel flame deflector, and other equipment. (*Marshall Star*, 7/15/64, 1)

* DOD announced it was abandoning plans to join with ComSatCorp in a space communications system meeting both military and commercial needs. Despite the economies of participation, the factor of international participation in the satellite system and the control foreign nations would have over a communications network that DOD wanted for emergency command control over its forces finally tipped the balance against DOD cooperation. (Finney, *NYT*, 7/14/64, 15)

* MSFC planners believed a post-Saturn V booster would be needed if the U.S. were to undertake a manned planetary exploration program, *Missiles and Rockets* reported. They also favored the so-called Block II vehicle, which would have two stages using new propulsion techniques and high-pressure engines. First stage would be recoverable. (*M&R*, 7/13/64, 9)

* Mercury capsule (MA–9) in which Astronaut L. Gordon Cooper flew 22 earth orbits was viewed by more than 20,000 people when on display in Nebraska State Capitol at Lincoln. (*Omaha Eve. World-Her.*, 7/13/64)

* Spatial orientation and stabilization of Early Bird, ComSatCorp's first experimental satellite, was to be achieved using a hydrogen-peroxide reaction jet control system. Developed by Walter Kidde and Co., and originally intended for SYNCOM II, newer version would be capable of at least a 10-yr. lifetime. (*M&R*, 7/13/64, 27)

* NASA announced that it would participate in about 200 one- to three-week aerospace workshops and teacher institutes at colleges and universities in 48 states during the summer. NASA spacemobiles, films, exhibits, and lectures would be used to acquaint teachers with the latest in space activities. (NASA Release 64–168)

July 13: Paul Haney, MSC Public Affairs Officer, said he would make a study of facilities used for dissemination of news at Republican National Convention in effort to develop efficient techniques for disseminating news during manned space flights controlled by MSC. (Maloney, *Houston Post*, 7/14/64)

- Secretary of Defense Robert S. McNamara announced that the week of July 20 would be Defense Cost Reduction Week as an occasion to honor civilian and military personnel who had helped trim $2.5 billion from defense costs and to stimulate efforts to reach the projected goal of $4.6 billion savings over the next three years. (Phillips, *NYT*, 7/14/64)
- Largest monolithic solid grain ever poured was well into its 20-day curing cycle at Aerojet-Dade Div. of Aerojet-General Corp. The 120-in., 37-ft.-long "Subscale 1" would be test-fired in the same cast-cure pit in mid-August. Both batch and continuous mix processes were used with the continuous mix working for 55 straight hours in the 2½-day pouring job. The 100-ton grain was to produce about 600,000 lbs. thrust in checking out the nozzle, materials, and processes developed for the upcoming 260-in. large solid rocket test at Aerojet. (*M&R*, 7/13/64, 11)
- Yelena Andrianovena Nikolayeva, infant daughter of Russian cosmonauts Valentina Tereshkova and Andrian Nikolayev, was to be thoroughly examined for next few years to determine whether weightlessness or radiation of her parents had any effect. (Snider, CDNS, *Houston Chron.*, 7/13/64)
- Observations by University of Bern (Switz.) scientists for several years in the Simplon Tunnel showed that effects of ionizing radiations also existed in a radiation-free environment. In the tunnel, which had proven to be free of any traces of cosmic radiation, artemia eggs showed only a 60–70% rate of offspring, Hordexum seeds became sterile after six months, and algae did not grow after a few weeks. (*M&R*, 7/13/64, 27)

July 14: Rocketdyne Div. of North American Aviation and Douglas Aircraft Co. received three contract increments totaling more than $8 million for work on S–IVB stage of Saturn IB and Saturn V launch vehicles from NASA MSFC. Additional funding of $3.6 million went to Rocketdyne to equip their plant for production of J–2 engines and for providing static test site and propellant storage area. Douglas received two incremental contracts. One, for $2.3 million, was for purchase of long-term items on recently ordered additional eight S–IVB stages. Other, for $2.2 million, was for purchase of additional set of automatic checkout equipment. (*SBD*, 7/14/64, 63; *Marshall Star*, 7/15/64, 9)

- NASA announced it had selected General Atomics Div. of General Dynamics Corp. to perform preliminary work, estimated to be worth about $1.7 million, on development of advanced type of nuclear reactor for use in powering spacecraft. Company would conduct studies on use of tungsten to contain uranium fuel in compact high-powered reactor that might power a rocket. Work would be performed for Space Nuclear Propulsion Office operated jointly by NASA and AEC. (NASA Release 64–173; *WSJ*, 7/15/64)
- Pratt & Whitney received $685,000 contract from NASA MSFC for study and integration of an advanced liquid-oxygen–hydrogen rocket engine. Objective would be to integrate existing feasible concepts into one engine best able to meet future requirements. Work would be coordinated with other advanced mission studies such as Saturn V improvement, post

Saturn, and the reusable orbital shuttle transport. Configuration would be an engine with a single chamber turbopump unit with variable mixture ratio control. It would be throttleable and capable of multiple restarts and with a thrust range of 300,000 to 350,000 lbs. One-year study would use data produced under several separate NASA engine technology contracts at Pratt & Whitney with inputs from Aerojet-General Corp., Rocketdyne Div. of North American Aviation, and General Electric Co. (*M&R*, 7/20/64, 8; *SBD*, 7/14/64, 63)

July 14: Missile and Space Systems Div. of Douglas Aircraft Co. has submitted study to NASA which concluded that six-man space research station, capable of orbiting for one year, could be orbiting earth within five years. Crew, serving on staggered schedule, would travel to and from station on modified Gemini or Apollo spacecraft. Station itself would provide small degree of artificial gravity by rotating slowly and would include centrifuge to simulate re-entry forces. (AP, *N.Y. Her. Trib.*, 7/14/64)

- Dr. Charles A. Berry of NASA's MSC said that some communicable diseases could develop on space missions of up to 14 days if astronauts were exposed during preflight preparation period. Therefore NASA plans included turning the spacecraft around and returning to earth in severe cases and equipping vehicles with emergency medical supplies. Also, doctors might be trained as astronauts for 30- to 60-day missions. (*Phil. Eve. Bull.*, 7/14/64)

July 14–15: NASA's Wallops Station, Wallops Island, Va., launched seven scientific experiments in a period of less than 10 hours. All of the experiments involved sounding rocket experiments carried aloft by Nike-Apache rockets. Four of the payloads carried sodium vapor cloud experiments for GSFC to provide data on wind directions and speeds and at various altitudes and the changes which occurred during the period from sunset to dawn (peak altitudes: 118 mi., 119 mi., 119 mi., and 119 mi., respectively). The other three were ionosphere experiments for the Univ. of Illinois and the Geophysics Corp. of America (peak altitudes: 97 mi., 99 mi., and 106 mi., respectively). The main purpose of these tests was to observe and measure structural changes in the ionosphere from nighttime to daytime, as radiation from the sun penetrated downward. All seven experiments were successful. The sounding rocket experiments were part of the third International Year of the Quiet Sun (IQSY) "quarter day" in the coordinated solar study program. (Wallops Releases 64–54, 64–55, 64–56; GSFC Historian; NASA Rpts. SRL)

July 15: The U.S.S.R. launched COSMOS XXXV into orbit. The satellite's initial orbital parameters were: period, 89.2 min.; apogee, 268 km. (161 mi.); perigee, 217 km. (130 mi.); inclination, 51.3°. The satellite, carrying scientific equipment for space studies, was reported to be functioning normally. (Tass, *Izvestia*, 7/17/64, 1, ATSS–T Trans.; AP, *NYT*, 7/16/64)

- NASA announced that it had selected 22 scientific experiments for the fifth Orbiting Geophysical Observatory (Ogo-E), scheduled to be launched in 1966. 54 scientists from four countries (U.S., U.K., France, and the Netherlands) representing universities, government, and private industry would take part in designing and constructing the experiments. The satellite was scheduled to be launched into an eccentric orbit, with an apogee of 92,000 mi. and a perigee of 170 mi. The Ogo-E experiments would study energetic particle radiation, earth and interplanetary magnetic fields, effect of solar gases on the earth's magnetic fields and radio communication, and the effect of solar radiation on the earth. (NASA Release 64-171)

July 15: DOD announced that since early July the Army had operated a special radar at White Sands Missile Range, N. Mex. The radar, the Multifunction Array Radar (MAR) was part of the Nike-X anti-ICBM missile system. Called a phased array radar, MAR worked on the theory that radar beams could be "bent" electronically by playing one beam against another. This called for a delicacy and timing far exceeding man's capabilities. It took an extremely fast computer to play the beam. The MAR was to replace three radars used in the Nike-Zeus Antimissile Missile System: acquisition radar for detecting missile warheads at long ranges; the target tracking radar for pinpointing the location of the warheads; and the discrimination radar for sorting live warheads from decoys. (DOD Release 519-64)

- In a speech before the Briefing Conference on Government Contracting for Research and Development sponsored by the Federal Bar Association and the Bureau of National Affairs, Inc., in Boston, Ernest W. Brackett, NASA Assistant Deputy Associate Administrator for Procurement Policy, spoke of NASA's policy towards incentive contracts: "Incentive contracts are not the answer to everything, but much good can and will come from them. They will effect more realistic cost proposals. When a company's incentive fee or profit is based on its cost proposal, it will not be on the low side. This means less cost overruns which are subject to so much criticism and which disrupt Government budgets. . . . There will also be more careful and detailed work definitions because a contractor wants to know what he is expected to do and be able to price it out if his incentive fee depends on performance and cost. It takes longer to negotiate incentive contracts and negotiations are harder. . . . Government personnel will be more careful, particularly in the technical areas, when incentive provisions are included. Another advantage should be fewer controls by Government over contract administration details, such as approval of overtime or sub-contractor selection. NASA would like to see contractors make higher profits but wants to have proof they have earned them. The health of the aerospace and electronics industries depends on their realizing a reasonable return for what they do. Incentive methods of contracting appear to be the best way this can be done." (Text)

- The platform adopted by the Republican National Convention contained a number of statements pertaining to the U.S. space program. The platform stated: "This administration has delayed research and development in advanced weapons systems and thus confronted the American people with a fearsome possibility that Soviet advances, in the decade of the 1970's, may surpass America's present lead. Its misuse of cost effectiveness has stifled the creativity of the Nation's military, scientific, and industrial communities.

    "It has failed to originate a single new major strategic weapons system after inheriting from a Republican administration the most powerful military force of all time. It has concealed a lack of qualitative advance for the 1970's by speaking of a quantitative strength which by then will be obsolete. It has not demonstrated the foresight necessary to launch as many as five satellites. The system, called Multiple Orbit-

"It has endangered security by downgrading efforts to prepare defenses against enemy ballistic missiles. It has retarded our own military development for near and outer space, while the enemy's development moves on. . . .

"We pledge . . . replanning of the present space program to provide for a more orderly, yet aggressively pursued step-by-step development, remaining alert to the danger of overdivision of skilled personnel in critical shortage from other vital areas such as health, industry, education, and science." (Text, *CR*, 7/28/64, A3934, A3939)

Mid-July: The U.S. Weather Bureau prepared a letter to the World Meteorological Organization advising that Automatic Picture Transmission (APT) pictures from TIROS VIII were no longer usable. Useful APT pictures totaled 4,067. (GSFC Historian)

July 16: NASA Wallops Station launched a sounding rocket with instrumented payload to determine the effects of the ionosphere on delicate low-frequency radio radiation receiving sensors. A Nike-Apache rocket boosted the payload, designed to test instrumentation for possible use in the space astronomy program, to an altitude of 85 mi. Preliminary data indicated the experiment was successful. (Wallops Release 64–57)

- U.S. Army's XV–5A "lift-fan" research plane, a craft with horizontal fans buried in its wings and nose, made its first vertical takeoff and landing, at Edwards AFB. The horizontal fans provided downward thrust for the vertical takeoff. The fans were powered during takeoff by the exhaust from the craft's jet engines, which was then diverted straight behind the plane for horizontal flight. This technique had the advantage that no more engine power was required for vertical flight than for normal subsonic cruising, and hence the plane had more economical fuel consumption. (DOD Release 521–64; Witkin, *NYT*, 7/18/64)

- White House announced that the U.S. and the U.S.S.R. had reached an informal agreement on limited scientific cooperation in developing methods for desalting sea water, including possible use of atomic energy. The first step in the cooperative program was to be an exchange of inspection visits by technical experts of the two countries to laboratories and experimental desalinization plants in the two countries. In addition to the reciprocal visits, the informal agreement called for an exchange of scientific reports, including results obtained by the two nations from work on pilot and demonstration desalting plants. (Finney, *NYT*, 7/16/64, 1, 6)

- It was reported that NASA was actively considering methods to be used for exploration by astronauts on the moon. Two devices were under consideration: a mobile laboratory (Molab) in which men could work and also go out in spacesuits to gather rock and soil samples, and a "lunar hopper," a helicopter device for short flights to sites that could not be reached on foot or by Molab. Parallel studies of the Molab concept were being undertaken by Boeing Co., Seattle, and Bendix Corp. of Ann Arbor, under contracts of $800,000. Bell Aerosystems Co., Buffalo, had been awarded a $199,333 contract to study the lunar hopper concept. (AP, *Balt. Sun*, 7/17/64)

- It was reported that Douglas Aircraft Co., Santa Monica, had submitted a proposal to NASA for a system which would enable a single booster to launch as many as five satellites. The system, called Multiple Orbit-

Multiple Satellite (MOMS), consisted of a number of satellites fastened together by explosive bolts. The system was designed for use with a Douglas Delta launch vehicle. Several successful launches of two and three satellites together had been accomplished before, but never as many as five as proposed in the new system. (*Houston Post*, 7/16/64)

July 16: In a speech before a briefing conference on contracting for research and development sponsored by the Federal Bar Association and the Bureau of National Affairs, General Bernard Schriever, AFSC, stated that in his opinion "when understanding is fostered and encouraged among the various sectors of national life, such as industry, the professions, the scientific community, and government, then rapid progress can be made toward national objectives." He cited the USAF ballistic missile program as an example of this principle at work and stated that the long-range USAF planning study (Project Forecast) set up in 1963 attempted also to put this principle to work, by employing broad participation by all the services, several Government agencies, and personnel from universities, corporations, and non-profit organizations. (AFSC Release 47–R–91)

- Astronaut Scott Carpenter was involved in a motor bike accident in Bermuda and suffered a fracture of his lower left arm. Carpenter was in Bermuda to participate in a Navy test in which he would have submerged in Sea Lab I diving chamber with four Navy divers to a depth of 192 feet, but his injury precluded his participation in the test. He was reported to be resting comfortably. (*N.Y. Her. Trib.*, 7/17/64; AP, Balt. *Sun*, 7/17/64)

- Brig. Gen. David Sarnoff, Chairman of the Board of RCA, addressed National Automation Conference of the American Bankers Association at the New York World's Fair, and predicted: "Over the next 20 years, I am convinced, computers will touch off an explosion in the social sciences comparable to that which we witnessed during the past half century in the physical sciences. . . .

 "A global linkup of computers will be accomplished through communications satellites, high-capacity transistorized cables, microwave conduits, as well as standard telephone and telegraph links. Data will move through broadband channels at speeds of up to 20 million bits per second, thousands of times faster than in today's systems. Ultimately, laser beams may transmit vast quantities of information instantaneously in the form of light through hoselike cables. . . .

 "Such advances will inevitably bring about basic transformations in fields far beyond those in which computers function today. Matching the technical prospects to the human environment, I see society making profound adjustments in many directions. The five main areas affected are likely to be: Work, leisure, education, health, and politics. . . ." (Text; *CR*, 8/3/64, A4051–53)

July 17: USAF launched Atlas-Agena D booster from Cape Kennedy with triple-satellite payload. Press sources reported the booster orbited two DOD Nuclear Radiation Detection Satellites and a 4.5-lb. tetrahedral satellite. The three satellites were reported to be injected initially into elliptical orbits ranging from 120 to 65,000 mi. altitudes; ground command signal maneuvered one NRDS into 65,000-mi. circular orbit, and similar maneuver was planned for the other NRDS two days later. The "Pygmy," designed to measure electrons in the Van Allen belts, would continue in the elliptical orbit. Unofficially nicknamed "Sentries," the twin radiation detectors were said to be similar, with minor exceptions, to the first pair, reported to have been successfully launched last October. (AP, Wash. *Sun. Star*, 7/19/64; *Av. Wk.*, 7/20/64, 24; *M&R*, 7/27/64, 22)

July 17: The U.S. Weather Bureau reported that a hurricane having winds of 75 mph had been located by TIROS VII satellite 1,000 mi. southwest of San Diego. (*San Diego Union*, 7/18/64; U.S. Weather Bureau)

- British scientists at the Jodrell Bank radio astronomy station were reported searching the skies for the Russian space probe ZOND I which was believed to be approaching Venus. The Russians had never revealed the precise mission of the probe and refused to supply the Jodrell Bank scientists with the radio frequencies on which it was operating, but it was believed that the probe would closely approach Venus during the 72-hr. period beginning July 17. (Reuters, *Minneapolis Morning Trib.*, 7/18/64)
- NASA's Marshall Space Flight Center awarded 29 contracts involving studies of advanced manned missions. The 29 studies, which were designed to look beyond the time when American astronauts would have landed on the moon, covered the subjects of launch vehicle development, earth orbital operations, lunar exploration, and planetary exploration. (NASA Release 64-175)
- In speech accepting Republican nomination for President, Sen. Barry M. Goldwater said: "I can see, and I suggest all thoughtful men must contemplate, the flowering of an Atlantic civilization, the whole world of Europe reunified and free, trading openly across its borders, communicating openly across the world.

  "This is a goal far, far more meaningful than a moon shot. . . ." (Text, *NYT*, 7/17/64, A10)
- Two attorneys in Oklahoma City sought a Federal court injunction to stop unlimited testing of supersonic aircraft over the city. The suit alleged that the tests being conducted by the FAA were causing undue mental and physical harm to Oklahoma City residents because of sonic booms. (AP, *NYT*, 7/19/64)
- The British Aviation Ministry was reported to have recently agreed to pay the equivalent of $840 to a Welsh farmer, David Lloyd-Davies, who contended that his mare died of fright when a sonic boom hit his farm. (UPI, *NYT*, 7/17/64, 12)
- A report from GAO to Congress charged that USN had spent $13.3 million to put rocket launchers on fighter planes and then had taken them off again because they created too many problems. The rockets had caused the fighters to pitch upward, and in many cases a plane would overtake and collide with a rocket it had launched. The GAO report said that the Navy could have saved about $4 million if it had gotten information available before the launchers were ordered. (UPI, *Houston Chron.*, 7/18/64)
- An article in the Russian publication *Krasnaya Zvezda* (*Red Star*) reported a speech delivered by Russian Premier Khrushchev before a group of professors and students in Mishkol'tse in April 1964. In the speech Khrushchev noted that in the U.S. "the scientists to a considerable degree are not Americans . . . the U.S. has pirated its scientific staffs from the representatives of various nations." The article went on to

call attention to British criticism that the U.S. was drawing scientists out of Britain and detailed the activities of the American army at the end of World War II in attempting to capture German rocket scientists. The article concluded, "We can justifiably be proud that all our achievements in the field of rocketry and cosmonautics, to be sure in all fields of science and technology, are due to the brains and the talent of the Soviet scientists alone, without any help from the outside." (*Krasnaya Zvezda*, 7/17/64, 4, ATSS–T Trans.)

July 18: It was reported that NASA was considering diverting money from some of its other programs in an effort to save the National goal of landing men on the moon in the 1960's. Dr. Robert C. Seamans, Jr., Associate Administrator of NASA, said in an interview that the agency "faced a real uphill battle in the next five years" to achieve a lunar landing before 1970. He stated that the probability of success was "not as high as we'd like it to be, since this is a national commitment." Dr. Edward C. Welsh, Chairman of the National Aeronautics and Space Council, said in interview that any further schedule slippages would inevitably throw the lunar landing into the 1970's. Space officials were still holding to the 1969 goal, and, he added, "We haven't given this up yet." At the current time NASA was $217 million short of the bare minimum it had requested for FY 1965. (Hines, *Wash. Sun. Star*, 7/19/64)

- A Naval Research Laboratory team under Dr. Herbert Friedman announced that it had disproved the theory that the Crab Nebula in the constellation Orion had a superdense neutron star at its center. The experiment was carried out July 7 by means of x-ray detectors flown to an altitude of 144 mi. over New Mexico by an Aerobee sounding rocket. (Hines, *Wash. Eve. Star*, 7/18/64)

- It was reported that USAF was planning to orbit the third in the Oscar series of satellites. These satellites were built by radio hams in California and had no military significance. It was designed to receive and re-transmit a single word in Morse—"Hi." The 25-lb. satellite was to be launched as a pickaback payload on some USAF launch from Vandenberg AFB. (*J/Armed Forces*, 7/18/64, 8)

- A report in the *Journal of the Armed Forces* noted that during FY 1964 199 new objects were placed in orbit. NORAD Space Detection and Tracking Center (SPADATS), which tracked the objects, reported that as of the end of FY 1964 there were 438 items in orbit. The items included not only satellites and probes, but also "junk" items such as upper stages of rocket boosters. As of the date of this report there were 101 satellites or probes in space. The U.S. had been responsible for 89 of them, the U.S.S.R. for 9, the U.K. for 2, and Canada for 1. Of the 199 objects launched in FY 1964, 122 were pieces of debris, and the other 77 were payloads. 58 were U.S., 18 were U.S.S.R., and 1 was from U.K. (*J/Armed Forces*, 7/18/64, 8)

- Clarence L. Johnson, Lockheed Aircraft Corp. designer of the F–12A (A–11) aircraft, said that the airplane could easily outfly Russian aircraft. He stated that the F–12A could capture all three absolute world flight records currently held by the Russians. (AP, *Kansas City Times*, 7/18/64)

- It was reported that DOD had decided to invest $200 million in a military communications satellite system. The first launching was said to be scheduled for late 1965 or 1966. Plans called for seven or eight satellites to be placed in orbit by a single Titan III booster. The satellites, perhaps as many as 24 simultaneously, were to be placed in orbit at around 6,000 mi. altitude (see July 13). (UPI, *NYT*, 7/19/64)

July 18: Hanson Baldwin, military analyst for the *New York Times,* reported that the U.S.S.R. had developed and was presumably producing an improved ICBM. The missile was said to have a range of more than 6,000 mi. and could carry a multimegaton warhead to its target. In addition, its accuracy was believed to be greater than the first Soviet long-range missiles. It was reported that it was designed for use against the below-ground silos of U.S. Minuteman missiles. (Baldwin, *NYT*, 7/18/64)

- An experiment designed by youthful rocket scientists ended in tragedy when "Gus," a two-ounce mouse, was killed in a mishap. Several youths from Schenectady had designed the rocket, which was to carry Gus to an altitude of 6,000 ft. The rocket went off at an angle upon launch and smashed into the ground killing its passenger. The launch, which had been protested by local humane societies, was viewed by a crowd of 400. (AP, *L.A. Herald Examiner,* 7/19/64)

July 19: The second of two DOD satellites, reported by press sources to be one of two Nuclear Radiation Detection Satellites, maneuvered into station 65,000 mi. above the earth. On signal from the ground a rocket motor aboard the satellite fired, moving it from an eccentric orbit with an apogee of 65,000 mi. into a circular orbit at that altitude. The motor aboard the companion satellite, boosted by the same launch vehicle, was fired successfully on July 17. (AP, *NYT*, 7/20/64; *Wash. Post,* 7/20/64)

- NASA announced it had begun a pilot program designed to help scientists and engineers cope with the greatly increased amount of literature on new discoveries and developments. The program, called Selective Dissemination of Information (SDI) was being developed for NASA under a contract to the Advanced Systems Development Division of IBM, Yorktown, N.Y. The SDI system was designed to notify those enrolled in the program of fresh information within their respective fields. (NASA Release 64–174)

- Leo D. Welch, Chairman of the Communications Satellite Corp., reported that Soviet officials had refused to answer a question as to whether they were developing their own system of communications by satellite. Welch said that the Soviets would say only that they were conducting experiments in the space technology. The questions were asked when the Soviets attended a meeting in Geneva in June at which they received a briefing on the formation of the 18-nation communications satellite network. (*Wash. Post,* 7/20/64)

- Preliminary findings from a six-year study by the National Merit Scholarship Corp. found that many top students who as college freshmen wanted to work as engineers frequently changed their minds by the time they were seniors. The report also stated: "The interest of able students in physical sciences and engineering has been decreasing during the period covered by this study and interest in the social sciences and humanities has been correspondingly increasing." For both sexes biology, history, and mathematics had become more popular. At the start of the study 13% of the males and 27% of the females said they expected to major in the social sciences, history, English, and foreign languages. In 1963, students desiring majors in these fields rose to 19% for males and 34% for females. (Janssen, *Phil. Inquirer*, 7/19/64)

July 19: In editorial *New York Times* applauded the agreement between U.S. and U.S.S.R. aimed at improving methods for desalination of sea water. The *Times* lauded the agreement both because of the seriousness of the increasing shortage of fresh water on earth, and also as a step in the "trend toward wider common efforts by Soviet and American scientists in areas where progress will bring mutual benefits." The editorial continued, ". . . as Americans and Russians get accustomed to cooperation for common benefit in scientific fields, they may generate habits and attitudes that could ultimately spread to the political field, again with mutual gains." (*NYT*, 7/19/64, E8)

July 20: Space flight of NASA's Sert I (Space Electric Rocket Test) spacecraft marked first successful operation in space of an electric rocket engine. NASA launched the Sert I on a suborbital trajectory from Wallops Island, using four-stage Scout booster rocket which lofted the Sert I to 2,500-mi. altitude before impacting in Atlantic Ocean about 2,000 mi. downrange from Wallops. One of the two ion propulsion engines aboard the 375-lb. Sert I was operated twice, for total run of 30 min.; this engine was built by NASA Lewis Research Center. The other engine, built by Hughes Research Laboratories, did not produce thrust. Primary purpose of the experiment was to verify that electrostatic (ion) engines can produce thrust in space, and all technical objectives were achieved. NASA later announced the test was so successful that a second test would not be necessary (Oct. 16). (NASA Releases 64–167, 64–180; LRC Release 64–63; Wallops Release 64–58)

- NASA Administrator James E. Webb reported to the President that NASA had accomplished cost reductions totaling $128,783,000 in FY 1964. This greatly exceeded the agency's target of $81,780,000. (NASA Announcement 64–162)

- AEC scientists were reported to be planning even more sophisticated nuclear radiation detection satellites to follow the two identified by press sources as having been maneuvered into patrol positions during the preceding weekend. The advanced Sentries were reported to carry optical devices to monitor not only space nuclear blasts but also those in the earth's atmosphere. (AP, Wash. *Eve. Star*, 7/20/64)

- Sen. Stuart Symington (D.–Mo.) discussed national space program on Senate floor: "If our nation is to remain strong and a leader of men, then it is necessary that we maintain a position of leadership in the greatest adventure of our time.

"Not only would a slowdown in this program have an adverse effect on U.S. leadership, but it would cost the taxpayer more money. NASA estimates that for every year the program is stretched out a billion dollars would have to be added to the program. That is not surprising. Any major R&D program requires a tight timetable if it is to be an efficient program and provide the necessary incentive. Without a tight timetable the work drags on through repeated changes and indecision. As in most human endeavors the establishment of a tight schedule in fact provides assurance that the program will in fact be accomplished.

"For the present, however, there is the fact that the huge space industry complex—and the space psychology itself—has found a place in the Nation's consciousness. The United States is more than six years away from its first puny satellite, and in that time the acceleration of progress has been on the edge of the fantastic. The space age has barely been born but we cannot see how anyone can misread the promise of adventure and discovery in the years to come." (CR, 7/20/64, 15720–21)

July 20: It was reported that NASA planned to cancel its lease for a Lockheed JetStar at the KSC in Florida and purchase the plane outright. This came out in a report by NASA to a Senate appropriations subcommittee. The report indicated that NASA planned to operate a total of 80 aircraft in FY 1965, an increase of 4 planes. NASA officials told the committee that 37 of the 80 would be used for research, 20 for proficiency and flight training, and 23 for administrative aircraft. (*Av. Wk.*, 7/20/64, 18)

- Premature shutdown of the U.K. Blue Streak space launcher engines during its first flight for ELDO at Woomera, Australia, in June had been traced to gain settings of the autopilot system, it was reported. Calculations showed that a simple change in autopilot settings could prevent future recurrences of the problem. (*Av. Wk.*, 7/20/64, 24)
- *Aviation Week* recounted an exchange between Defense Secretary Robert McNamara and Sen. J. William Fulbright (D.-Ark.) in which McNamara stated that the Apollo program had psychological value but no direct military worth. In response to a question about the effect of the Apollo program on the mission of DOD, McNamara said: "... if a psychological victory that might result from the Soviets arriving at the moon first leads to a shift in attitudes in the world, tending to reduce the belief in the strength of the West, and in increasing the Soviet's view of the weakness of the West, this is bound to have a military effect. This is exactly what occurred in connection with our space programs, and it led to a very costly response by this country." (*Av. Wk.*, 7/20/64)
- *Aviation Week* stated that a technological war plan task force under the cognizance of AFSC had been established at Space Systems Div., to investigate a variety of subject areas as a continuation of the USAF's Project Forecast. Participants were expected to include key personnel from industry and NASA as well as AFSC. Areas to be studied included boron fiber technology for application to structural materials, operational requirements of biological satellites, manned hypersonic vehicles to investigate thermodynamic and aerodynamic regions, and laser applications and requirements. (*Av. Wk.*, 7/20/64, 23)
- USAF was reported continuing to seek industry support for avionics subsystems studies for DOD's manned orbiting laboratory (MOL) program. The Space Systems Div. of AFSC asked avionics companies to submit bids for a study of an image velocity sensor subsystem, which would enable orbiting astronauts to acquire and track space targets visually and to take fixes on terrestrial bodies for navigation. SSD also asked Honeywell to submit a sole-source proposal for a study of the laboratory's attitude control system. (*Av. Wk.*, 7/20/64, 24)
- It was reported that contractors were accusing NASA of naming a Mars exploration idea "Project Beagle" just to appeal to President Johnson. NASA officials denied this and stated that the project was named after the ship in which Charles Darwin sailed around the world. (*Av. Wk.*, 7/20/64, 15)

July 21: NASA invited scientists to propose research experiments and studies for manned and unmanned space flights during the last half of the 1960's. The announcement of deadlines and guidelines for the submission of proposals was distributed by NASA's Office of Space Science and Applications to domestic and foreign scientists in universities, industries, and Government installations. The agency gave the world scientific community detailed descriptions and timetables covering a wide range of space research opportunities, including placing scientific experiments in available space on Gemini and Apollo manned flights, weather satellites, sounding rockets, the X-15 research plane, orbiting observatories, and balloons. (NASA Release 64-177)

- Col. Robert P. Young (USA), Executive Officer of NASA since 1961, was awarded NASA's Medal for Outstanding Leadership. In making the award, NASA Administrator James E. Webb cited Young for "his outstanding leadership and dedicated service displayed in all aspects of performance while serving as executive officer of NASA, and for his significant and noteworthy contributions in co-ordinating internal management of the agency." (NASA Release 64-181; *Wash. Post*, 7/22/64)
- The appointment of Dr. Jerome Spar, Professor of Meteorology and Oceanography, NYU, as Director of Meteorological Research for the U.S. Weather Bureau was announced. (Commerce Dept. Release WB 64-12)
- Commenting on Republican Presidential candidate Goldwater's recently published views on the space program, Warren Burkett, science writer for the *Houston Chronicle*, pointed out that a good deal of research which can be of use to the military services was being carried on as part of NASA's Project Apollo. He suggested that an Apollo Orbital Research Laboratory, based on components being developed in the Apollo program, could be used for military aspects of the space program such as space patrol and interception. The article suggested that the Apollo program was generally developing a group of "off-the-shelf" space systems which could be adapted for military use should this appear desirable. (*Houston Chron.*, 7/26/64)

July 22: NASA and ESRO announced signing Memorandum of Understanding for project involving two scientific satellites. Under the agreement, NASA would launch into orbit two satellites, ESRO 1 and ESRO 2, for the nine-nation West European organization. Both satellites would be launched on NASA Scout rocket boosters into near-polar eccentric orbits and were tentatively scheduled during 1967 from Western Test Range. (NASA Release 64-178)

- USAF successfully sent a delta-winged Asset glider through a 12,000-mph re-entry test in an experiment designed to gather data on the conditions that future manned spacecraft might encounter when returning to earth from space missions—including skin heating as high as 4,000° F. The 1,150-lb. spacecraft was lifted to an altitude of 44 mi. aboard a Thor-Delta booster launched from Cape Kennedy, then was fired downward for a 1,650-mi. glide through the atmosphere. The glider, the third in this series of experiments, was recovered in the Atlantic Ocean by frogmen. (UPI, *NYT*, 7/23/64, 46; *Wash. Post*, 7/23/64; *Wash. Eve. Star*, 7/22/64)
- The White House and Communications Satellite Corp. announced television coverage of the Olympic Games from Tokyo to the U.S. and Canada would be attempted via communications satellite. Project

would be carried out with cooperation of NASA and DOD and with Government of Japan and Japan Broadcasting Corp. NASA's Syncom 3 experimental communications satellite, planned for launch into synchronous orbit during August, would be made available for transmission of TV coverage of the Games, which would take place Oct. 10–24. (ComSatCorp Release; Finney, *NYT*, 7/23/64, 42)

July 22: Details of method to be used in first U.S. attempt to rendezvous a manned and unmanned spacecraft in space were disclosed by NASA Manned Spacecraft Center. Launch of Atlas-Agena into orbit would be followed 95 min. later by launch of Titan II with two-man Gemini spacecraft. Five hours after Titan II launch, rendezvous and docking Gemini spacecraft and Agena stage would be accomplished. (AP, Wash. *Eve. Star*, 7/22/64)

- Results of experimental drop tests down elevator shaft of Cleveland's Terminal Tower showed that styrofoam could be used successfully as a deceleration medium, according to NASA Lewis Research Center design engineer Terrence E. Russell. Purpose of the deceleration tests, conducted by LRC last May, was to verify use of styrofoam as filler in cart used to "catch" free-falling test packages at LRC's new 500-ft. zero-gravity facility. (LRC Release 64–64)

- NASA announced Univ. of Arizona would receive $1,200,000 NASA grant to construct Space Sciences Center on the campus at Tucson. The four-story structure would "further the University of Arizona's multi-disciplinary program that is proving invaluable to the national space effort," NASA Administrator James E. Webb said. (NASA Release 64–183)

- NASA selected Philco Corp. for negotiation of contract for feasibility and preliminary design studies for automated biological laboratory (ABL) to detect and describe indigenous life on the planet Mars. The one-year study would define life-detection payloads for Voyager-class missions, which were under study by NASA Office of Space Science and Applications. (NASA Release 64–182)

- Japan launched its first meteorological sounding rocket from the National Space Development Center's testing range on Nii Jima Island. The two-stage rocket reached 68.2-mi. altitude. (AP, *NYT*, 7/23/64, 46; AP, *Wash. Post*, 7/23/64)

- House passed H.R. 11611, a bill to establish a National Commission on Technology, Automation, and Economic Progress. Bill was then referred to Senate Committee on Labor and Public Welfare. (NASA LAR III/140)

- William B. Bergen, president of Martin Baltimore, said U.S. had closed the so-called "space gap" of Russian competition "without fanfare, and almost without knowing it ourselves." Addressing National Space Club in Washington, Bergen said that, with all the "Hollywoodian propaganda" about enormous Soviet satellites, "there seems little doubt that the objective part of the scientific world is now convinced that our modest-sized couriers into space have brought back far more useful information than has been collected by the Russians' more massive messengers." (AP, *Houston Post*, 7/23/64)

- FAA awarded $975,000 contract to Bunker-Ramo Corp. for study of all-weather landing instrumentation of modern jet passenger aircraft. (FAA Release 64–72)

July 23: Aerobee 150 sounding rocket launched by NASA from Ft. Churchill, Canada, reached 144.5-mi. altitude in experiment to study very low energy cosmic ray heavy nuclei. Rocket performance was good and instrumentation performance was satisfactory. Two of three trays containing experimental emulsions did not completely retract and were damaged by heat and upon impact by water. Preliminary estimate was that 100% good data were obtained by the emulsion that functioned properly and 60% by the two which did not retract. (NASA Rpt. SRL)

- NASA Langley Research Center asked industry for proposals on the kind of launch vehicle it would take to put a small spacecraft into orbit around the sun. The purpose of the program would be to put an instrumented satellite into position for a clear view and study of the sun's plasma corona. (Newport News *Daily Press*, 7/24/64)
- Senate passed S. 2602 to establish National Air and Space Museum under the Smithsonian Institution. The bill was then sent to the House. (NASA LAR III/141; *CR*, 7/23/64, 16138–39)

July 24: At international conference in Washington, two interrelated agreements for establishment of global commercial communications satellite system were initialed by the negotiators from 13 participating countries. Five other nations planned to join the agreement on or before Aug. 19, date of formal enactment. Under these agreements, ownership in the $200 million enterprise would be shared among 18 nations and Vatican City, with the United States' Communications Satellite Corp., owning 61% of the shares and serving as manager of the system on behalf of all participants. Agreement would be open for six months beginning Aug. 19 to any nation belonging to International Telecommunications Union. (ComSatCorp Releases; Finney, *NYT*, 7/25/64; N.Y. Her. Trib., 7/25/64; *WSJ*, 7/27/64)

- 1,000-hr. operating test at full power of Snap-8 experimental reactor was completed in a ground test at Atomics International's Nuclear Field Laboratory, contractor to AEC for the reactor. During the test, the reactor was operated at 600 kw. of heat output and 1,300° F coolant outlet temperature, thus exceeding the current Snap-8 system requirement of 450 kw. of heat. Snap development was joint program of NASA and AEC. (*Marshall Star*, 9/23/64, 8)
- At his Washington press conference, President Johnson announced successful development of an advanced, long-range reconnaissance aircraft to be employed by Strategic Air Command:

    "This system employs the new SR-71 aircraft and provides a long-range, advanced strategic reconnaissance plane for military use capable of worldwide reconnaissance for military operations.

    "The Joint Chiefs of Staff, when reviewing the RS-70, emphasized the importance of the strategic reconnaissance mission. The SR-71 aircraft reconnaissance system is the most advanced in the world.

    "The aircraft will fly at more than three times the speed of sound. It will operate at altitudes in excess of 80,000 feet. It will use the most advanced observation equipment of all kinds in the world.

    "The aircraft will provide the strategic forces of the United States with an outstanding long-range reconnaissance capability. The system will be used during periods of military hostilities and in other situations in which the United States military forces may be confronting foreign military forces.

"The SR-71 uses the same J-58 engine as the experimental interceptor previously announced, but it is substantially heavier and it has a longer range.

"The considerably heavier gross weight permits it to accommodate the multiple reconnaissance sensors needed by the Strategic Air Command to accomplish the strategic reconnaissance mission in a military environment.

"This billion-dollar program was initiated in February, 1963. The first operational aircraft will begin flight-testing in early 1965, and deployment of production units to the Strategic Air Command will begin shortly thereafter.

"Appropriate members of Congress have been kept fully informed on the nature of, and the progress in, this aircraft program.

"Further information on this major advanced aircraft system will be released from time to time at the appropriate military secret classification levels." (Transcript, *NYT*, 7/25/64, 6)

July 24: McDonnell Aircraft Corp. was awarded a $1,189,500 fixed-price contract for study of Gemini B spacecraft for the Manned Orbiting Laboratory (MOL) program. The contract was awarded by the USAF Space Systems Division. (DOD Release 545-64)

- NASA announced small businesses received three out of every four subcontracts awarded by NASA contractors in the first three quarters of FY 1964. Of 1,337 subcontractors, 983 (or 74%) were small companies. The small businesses received $119 million, or 23% of the $526 million awarded in the time period. (NASA Release 64-179)
- Governors and senators from 12 midwestern states meeting in Washington held under consideration a suggestion from Gov. James A. Rhodes (R.-O.) and Sen. Gaylord Nelson (D.-Wis.) that a regional bloc be formed in Congress to obtain Federal contracts for the Midwest: "When we refuse as a unit to support east and west coast projects until a sensible policy of government spending is established, then we'll get a hearing," Senator Nelson said. "We aren't going to support the others and make a wasteland of the Midwest." (Dodd, CTPS, *Chic. Trib.*, 7/25/64)
- Cape Kennedy celebrated its 14th anniversary as a missile and space vehicle launching base. The first missile from the Cape was a hybrid Bumper (V-2) booster with an Army Wac Corporal second stage, launched on a 200-mi. trajectory on July 24, 1950. Since then more than 1,400 launches had been made. Since its establishment in 1958, NASA had been responsible for 84 flights. (*Marshall Star*, 7/29/64, 9)
- It was reported that USAF was planning to permanently preserve the launch complex from which the first American astronauts were launched into space and to develop the area around it into an Air Force Space Museum. (Strothman, *Miami Herald*, 7/24/64)
- An exhibition of U.S. communications equipment—including a model of the Telstar satellite—opened in Leningrad. (AP Wire, 7/24/64)
- NASA Goddard Space Flight Center announced researchers had developed first known high-temperature-resistant ceramic potting compound in which space electronics could be imbedded for protection against extreme heat. The ceramic potting compound can be cured at temperatures as low as 150° F, permitting it to be cured in place around the electronics without subjecting the delicate equipment to damaging heat. The ceramic potting compound combined protection against heat up to 2,500° F with protection against vibration. (GSFC Release G-18-64)

July 24: Results of space poll of 2,000 randomly selected AAAS members, 1,134 of whom responded, were published in *Science.* To the statement, "The vital national interests of the United States require that a high priority be given to landing a man on the moon by 1970," 31% of the respondents said "Yes," 62% "No." To the statement, "A reasonable objective would be a lunar landing by 1970, 1975, 1980, 1990, [or] Never," 20% chose 1970; 23%, 1975; 19%, 1980; 9%, 1990; and 7%, Never. Asked to select "the most important justification for manned exploration of the moon" from prepared list of alternatives, respondents said: military, 7%; scientific, 58%; resources on moon, 1%; national prestige, 13%; technological fallout, 6%; spirit of adventure, 7%. Asked to rank following fields "in order of their potentiality for producing new knowledge," respondents said: earth-based astronomy, 2%; manned lunar research, 10%; biomedical research, 51%; oceanography, 6%; physics research, 15%; chemical research, 9%; other, 7%. Asked to choose "what fraction of the total federal $15 billion" R&D funds "should be devoted to the space program," 30% chose 0–10%, 31% chose 10–20%, 14% chose 20–30%, and 5% chose 30–40%.

Commenting on the poll, Robert C. Toth of *Los Angeles Times* noted that *Science* editor Philip Abelson was "an outspoken opponent of the moon program" and that the survey did not directly ask the key question: Should the existing program be slowed down now? *Journal of the Armed Forces* space editor James J. Haggerty, Jr., pointed out that "it all depends on who you ask." Of the 1,134 respondents, he said, "it turned out that 86% were not connected with the space program. As a matter of fact, there is a strong suspicion that they are all medical researchers, since 51% of them indicated that biomedical research is the scientific field with the potential for producing the greatest amount of new knowledge, where only 10 percent thought manned lunar research would contribute anything of value. . . ." (*J/Armed Forces,* 8/8/64, 12; *Science,* 7/24/64, 368; Toth, *Wash. Post,* 7/27/64)

July 25: NASA launched Aerobee 150 sounding rocket from Ft. Churchill, Canada, with instrumented payload to study very low energy cosmic ray heavy nuclei. The rocket reached 133.7-mi. altitude, ejected recovery package. After land recovery of the payload, examination revealed the emulsion trays all had fully extended and fully retracted properly, so that very good data would be obtained. (NASA Rpt. SRL)

- According to press reports, the twin satellites launched July 17 by USAF continued to send back clear signals from space. Press sources had indicated the satellites were Nuclear Radiation Detection Satellites, said one satellite was in almost perfectly circular orbit ranging from 63,314 mi. to 64,915 mi. altitude, while the other was in slightly higher orbit ranging from 58,686 mi. to 69,387 mi. (AP, Balt. *Sun,* 7/26/64; *Wash. Sun. Star,* 7/26/64)
- Soviet space probe ZOND I apparently was unsuccessful in its mission to land on or fly by the planet Venus, according to unidentified official in Washington. Western scientists had predicted ZOND I would reach Venus between July 18 and 25, but no Soviet announcement had been made. (AP, *Wash. Post,* 7/26/64)
- Drew Pearson wrote in his syndicated column that there were about 500 Germans in Egypt working on military rockets and other arms programs. Among more important: Prof. Wolfgang Pitz, former Austrian, and

expert on rockets; Prof. Paul Goercke, expert on radar; Dr. Hans Kleinsaechter, expert on automatic pilots. (Pearson, *Wash. Post*, 7/25/64, B11)

July 26: Details of a report financed by NASA Manned Spacecraft Center delving into the question of why so few Negroes were qualified for jobs with MSC were made public. The report on manpower available from minority groups indicated that Negroes seeking scientific and technical jobs in the space age would have to change their attitudes toward education, shifting from traditional subjects to those needed in the space age, such as physics. It also indicated a need for changes in curricula and in educational emphasis by Negro schools, as well as improved placement provisions in many schools. The report, financed by a $38,000 contract from MSC, was prepared by Drs. J. A. Pierce and B. A. Turner of Texas Southern Univ. (*Houston Chron.*, 7/26/64)

- Prime Minister Tunku Abdul Rahman of Malaysia toured John F. Kennedy Space Center. Rahman stated after his tour: "I can't say what impressed me most. Everything impressed me. It was so scientific." (AP, N.O. *Times-Picayune*, 7/27/64)

July 27: NASA Manned Spacecraft Center named Capt. James A. McDivitt (USAF) and Capt. Edward H. White II (USAF) as pilot and copilot for the second manned space flight of Project Gemini (GT-4), a four-day earth-orbital mission in early 1965. Astronauts named as backup crew for the flight were Maj. Frank Borman (USAF) and Lt. Cdr. James A. Lovell, Jr. (USN). All four were "second generation" astronauts selected for the NASA program in Sept. 1962. (MSC *Roundup*, 8/5/64, 1)

- NASA announced modifications to X-15 No. 1 rocket-powered aircraft that would enable it to carry on scientific experiments above much of the earth's atmosphere were being made at NASA Flight Research Center. The aircraft would have new inertial guidance system, originally designed by Minneapolis Honeywell Regulator Co. for the USAF X-20, and streamlined compartments for experiments at each wing tip. One compartment would contain device to collect micrometeoroids and extraterrestrial dust and special spectrometer to measure radiation and polarization of daytime sky-light; other pod would contain densitron to measure air densities at high altitudes. (NASA Release 64-184; FRC Release 19-64)

- Dr. Robert H. Goddard was named posthumously as winner of the 1964 Daniel Guggenheim Medal. (*Wash. Eve. Star*, 7/27/64)

July 28: NASA's RANGER VII spacecraft was launched from Cape Kennedy on its flight to the moon. After an almost perfect countdown, Atlas-Agena launch vehicle boosted the 806-lb. spacecraft into a parking orbit around the earth, and then the Agena engines refired to place RANGER VII on its trajectory to the moon. The spacecraft carried six television cameras designed to take and relay to earth closeup pictures of the lunar surface. Flight to moon would take about 68 hours, with the pictures to be taken during the last 15 min. before craft impacted on the moon. The launching was so accurate that, even without the standard midcourse correction maneuver, RANGER VII would hit the moon. However, scientists wanted pictures of the "Sea of Clouds" region and planned to execute midcourse correction to ensure impact and pictures in this area. (*NYT*, 7/29/64; NASA Release 64-176)

July 28: NASA Aerobee 300 sounding rocket with Univ. of Michigan payload to measure electron and neutral particle temperatures in the atmosphere was launched from Ft. Churchill, Canada, to 200-mi. altitude. Instrumentation included electrostatic probe, omegatron gage, Bayard-Alpert ion gage, as well as a solar aspect sensor to provide proper attitude. Instrumentation performance was satisfactory, although the omegatron made no measurements. (NASA Rpt. SRL)

- Unspecified "informed sources" in Paris said that the first foreign base on U.S. territory—a satellite tracking station in Alaska—was to be established by the European Space Research Organization. It was reported that the establishment of the station had been agreed to in principle by the U.S. State Department, and that a formal agreement was expected to be signed in the near future. ESRO had plans for launching about 400 sounding rockets, 6 small satellites, 8 medium satellites, 7 space probes, and an orbiting astronomical observatory during the following 8 years and would use the station to help track, control, and communicate with these spacecraft. ESRO planned a network of similar stations extending from Alaska across North Pole to the Shetland Islands, Belgium, and Italy. (Simons, *Wash. Post*, 7/29/64)
- The USAF successfully fired a Minuteman ICBM from Vandenberg AFB, Calif. (*M&R*, 8/3/64, 8)

July 29: RANGER VII spacecraft successfully completed its midcourse correction maneuvers to put the spacecraft on its final collision course to the moon. RANGER VII carried out series of orders involving switching from the directional to the omnidirectional antenna, folding the directional antenna out of the way of the midcourse motor, pitch and roll maneuvers to properly orient the spacecraft, firing the midcourse motor, and reorientation by earth and sun acquisition after the midcourse motor had been fired. Telemetry indicated that everything aboard the spacecraft was normal after the midcourse correction activities had been completed. (NASA Release 64–187)

- USAF Capt. Joe Engle flew the NASA-AirForce X–15 No. 3 experimental plane from Edwards AFB, Calif., in a test of the heat-resistant qualities of special materials attached to the craft's fuselage. The X–15 reached maximum speed of 3,511 mph (mach 5.32) and peak altitude of 78,000 ft. (NASA X–15 Proj. Off.; AP, Wash. *Eve. Star*, 7/30/64)
- NASA conducted two successful test-flights of paraglider vehicles at Edwards AFB, Calif. In the first test, a simulated Gemini spacecraft manned by North American Aviation, Inc., test pilot E. P. Hetzel was towed aloft beneath a pre-inflated paraglider landing system. The tow test vehicle (TTV) was towed to 6,000-ft. altitude by CH–46C helicopter. While airborne, Hetzel manually controlled the TTV through preplanned maneuvers to its landing. In the second test, a similar but unmanned system was dropped from a C–130 aircraft at 32,000-ft. altitude. The paraglider was automatically deployed and remotely controlled in flight from 22,000-ft. to 12,500-ft. altitude, where it was separated from the Gemini test vehicle. The tests were part of NASA's continuing evaluation of the paraglider system as a means of landing and recovering manned spacecraft (MSC *Roundup*, 8/5/64, 1)
- NASA announced its Ames Research Center would negotiate with J. W. Fecker Co., division of American Optical Co., for satellite attitude control simulator. Contract would amount to about $1 million. The

simulator would be used in vacuum chamber to evaluate performance of systems and components being developed for attitude control of spacecraft. (NASA Release 64-188)

July 29: At MSC press conference, Kenneth S. Kleinknecht, Deputy Project Manager of Project Gemini, stated that an experiment assigned to the second manned flight (GT-4) in Project Gemini would involve the first exposure of an astronaut to the hazards of outer space without full space vehicle protection. He said that the experiment would involve "stepping into space," but modified this later to say that it might involve nothing more than the crewman's opening up the spacecraft hatch and standing up in his seat. In addition to this experiment, the second flight (scheduled for early 1965) also would include other scientific assignments, including medical tests, radiation measurements, and measurement of the earth's magnetic field. (AP, *NYT,* 7/30/64)

- Christopher C. Kraft, Jr., Assistant Director for Flight Operations of NASA Manned Spacecraft Center, was appointed flight operations director for the second Gemini-Titan flight (GT-2), an unmanned suborbital flight scheduled for autumn. Kraft had been responsible for flight direction in Project Mercury and in the first Gemini-Titan flight. For the GT-2 mission he would replace Dr. Walter C. Williams, who resigned from NASA last March to become Vice President and General Manager, Manned Systems Div., Aerospace Corp. (NASA Release 64-185)

- Communications Satellite Corp. announced its Board of Directors had elected John A. Johnson as Vice President-International. Mr. Johnson had been Director of International Arrangements of ComSatCorp since Dec. 9, 1963. From October 1958 until he joined ComSatCorp in 1963 he had served as General Counsel of NASA. (ComSatCorp Release)

- His Excellency Philibert Tsiranana, President of the Malagasy Republic, toured NASA Goddard Space Flight Center and received briefings on the Goddard mission by GSFC Director Dr. Harry Goett. In his official party were his wife, the Malagasy Foreign Minister, their Ambassador to the U.S., and officials of the U.S. State Dept. and NASA Hq. (*Goddard News,* 8/10/64, 1)

- Dr. Norman W. Rosenberg, of the Air Force Cambridge Research Laboratories, was presented the DOD Distinguished Civilian Service Award, highest decoration for civilian employees. Since joining AFCRL in 1959, Rosenberg designed and executed a number of important experimental studies in upper atmosphere physics, most of which were part of Project Firefly. (OAR Release 7-64-3)

- West Germany's vertical take-off aircraft successfully passed the speed of sound in a test flight. The two-jet VJ101 achieved its vertical rise by tilting its jets downward and then went into forward flight by swinging them up to a horizontal position. The plane was being developed by the West German firms of Boelkow, Heinkel, and Messerschmitt. (AP, *Wash. Post,* 7/31/64)

- USAF successfully fired an Atlas ICBM from Vandenberg AFB, California. (*M&R,* 8/3/64, 8)

July 30: The U.S.S.R. launched another earth satellite, COSMOS XXXVI. The initial orbital parameters were as follows: period, 91.9 min.; apogee, 503 km. (300 mi.); perigee, 259 km. (155 mi.); inclination, 49°. (*Krasnaya Zvezda,* 7/31/64, 1, ATSS-T Trans.)

July 30: At his news conference, President Johnson noted the first anniversary of the nuclear test ban treaty: ". . . a year ago this week the nuclear test ban treaty was signed and agreed upon. Today, a year later, more than 100 nations have joined the three original signing countries. We have also seen a United Nations resolution banning weapons of mass destruction in outer space and steps to cut back production of fissionable materials. . . . A year without atmospheric testing has left our air cleaner. This is a benefit to every American family, and to every family everywhere, since all radiation, however small, involves some possibility of biological risk to us or to our descendants. . . ."

President Johnson said that the problems of economic conversion posed by shifts and cutbacks in military activities had been "greatly relieved by our general prosperity." In a report delivered at his news conference, the President called the problem of economic conversion "hard" and "complex" and said that it "will not be solved in one month or one year." But he stressed that the main element in a solution was to keep the general economy prosperous. (Transcript, *NYT*, 7/31/64, 8; Dale, *NYT*, 7/31/64)

- The Senate Committee on Appropriations marked up and ordered favorably reported with amendments H.R. 11296, FY 1965 appropriations for independent offices. The bill included a NASA appropriation of $5.3 billion. (NASA LAR III/143-46)
- The USN's Sea Lab I, with four Navy divers aboard, began ascent to the surface after 10 days on the ocean floor at a depth of 192 ft. The experiment was terminated 11 days ahead of schedule because of an approaching storm. Astronaut Scott Carpenter was to have participated in the experiment, but was forced to withdraw because of the broken arm he suffered recently. (DOD Release 558-64)
- USAF's Vice Chief of Staff General William F. McKee was presented his third Distinguished Service Medal by President Johnson. Retiring after 35 years in the armed forces, General McKee was joining NASA Sept. 1, as Assistant Administrator for Management Development. (Wash. *Eve. Star*, 7/28/64; AP, *NYT*, 7/31/64, 5)
- Professor Wallace Osgood Fenn of the Univ. of Rochester School of Medicine was named recipient of the Daniel and Florence Guggenheim Award for 1964 of the International Academy of Astronautics. This was the first time the award had been given to a physiologist. Prof. Fenn was Secretary General of the International Union of the Physiological Sciences. Previous winners of the award included Sir Bernard Lovell, Dr. James A. Van Allen, and Prof. Marcel Nicolet. (IAA Release 21)
- French Premier Georges Pompidou declared that France would take every possible precaution to prevent harmful fallout in forthcoming atomic tests in the Pacific. He said that a special team would be sent to the Polynesian test site area on Mururoa Island to ensure that waters and air were not contaminated by the tests. (AP, *Wash. Post*, 7/31/64)
- The nuclear submarine *Daniel Boone* test-fired a Polaris A-3 missile with a dummy warhead more than 2,000 mi. down the Eastern Test Range while lying submerged 30 mi. off the Florida coast. This marked the eighth success in nine undersea shots with the advanced-model Polaris. (UPI, *Wash. Post*, 7/31/64)

July 31: RANGER VII successfully completed its historic mission of taking and relaying to earth closeup pictures of the lunar surface. The spacecraft sent back 4,316 high-quality photographs of the moon before it crashed on the surface of the moon in an area northwest of the Sea of Clouds. All six cameras aboard the spacecraft worked properly, sending back photographs of the moon that improved the resolution of lunar detail as seen from the earth by a factor of 1,000. The closest pictures were snapped 2.3 sec. before impact, at 10.7° south latitude and 20.7° longitude, 68 hrs. and 36 min. after takeoff.

At news conference on preliminary results from RANGER VII, held at JPL, principal experimenter Dr. Gerard P. Kuiper of Univ. of Ariz. Lunar and Planetary Laboratory described sampling of the lunar photographs as they were shown to the public. In reply to a question, he said: "... the thing that struck me most was that the Moon, the region of the Moon, the dimensions which we have been able to observe ... this range, tremendous range of new knowledge has not resulted in any totally unforeseen problems....

"The Moon has shown—the records show a continuity of features from the large craters observable with telescopes down to smaller sizes.

"Think that the rounded features of the large numbers of secondary craters is new. Also their large number is new....

"It is also clear that primary craters exist, little pits which have very much the same shape.... They are smaller, but the shape, the form, is the same as the bigger craters....

"There are some ridges shown. These ridges in the maria are rather of the same nature, of the same shape, sort or form, as the larger ridges.

"So I would say, in summing up, that the new region of knowledge is in a true sense an extension of the earlier knowledge. I think it shows that the craters continue to occur down to sizes of about three feet in diameter, about one foot depth.

"This clearly implies that we are not dealing here with enormous layers of dust, of very loose dust, that is, very loose material, where one would not expect any small features to be visible at all. It is not like sand, in other words....

"This is a great day for science, and this is a great day for the United States.... the amount of information that has been gained about the lunar surface is truly remarkable." (NASA Transcripts)

- President Johnson personally congratulated scientists and technicians responsible for the successful flight of RANGER VII. He congratulated Dr. William Pickering, Director of JPL, and Dr. Homer Newell, NASA Associate Administrator for Space Science and Applications, and then issued a statement on the experiment. He called the flight "a basic step forward in our orderly program to assemble the scientific knowledge necessary for man's trip to the moon.... The pictures obtained of the lunar surface should prove extremely useful. ... They will be a guide in constructing the lunar excursion module and in planning the trip. Now we shall be able to better map out our descent route. We'll be able to build our lunar landing equipment with greater certainty and knowledge of the conditions which our astronauts will encounter on the moon. I recognize that this great success has come only after a number of failures and partial failures in our efforts to send probes to the moon. This success should spur us on to added effort in the future. . . . On behalf of a grateful nation, let me again congratulate you on this magnificent achievement. All of you today have helped further the peaceful exploration of space." (*NYT*, 8/1/64)

July 31: The U.S.S.R. announced a new program of Pacific rocket firings, to begin August 4 and to last to the end of 1964. The announcement, describing the impact areas south and west of Hawaii, said "test flights of new types of rocket carriers for space objects" would be conducted. The Soviet government requested that ships and aircraft stay out of the impact areas during announced periods. (Balt. *Sun*, 8/1/64; Tass, *Pravda*, 8/1/64, ATSS–T Trans.)

- W. S. Bellows Construction Corp. and Peter Kiewit and Sons, Inc., received a $2,061,358 fixed-price contract for construction of the Contractor's Support Facility at NASA Manned Spacecraft Center. The contract was awarded by the Army Corps of Engineers. (DOD Release 563–64)

- A six-month study was concluded of public acceptability of sonic booms that had been conducted by the FAA in the Oklahoma City area since February 3, 1964. The study was designed to gather data on public reaction to the noise that would follow the supersonic airliners of the 1970's. During the test approximately 8 booms a day (total in test, 1,253) were created. During the 6 months the agency received 12,588 telephone calls, most of them complaints. In addition to these direct manifestations of opinion, the Univ. of Chicago conducted a detailed public opinion survey in conjunction with the experiment. Gordon M. Bain, FAA Deputy Administrator for Supersonic Transport Development, said in announcing the completion of the study: "Sonic boom presents the most imposing single challenge to design and operation of a supersonic transport plane. The people of Oklahoma City have been helping explore this challenge as, in effect, major participants in the nation's supersonic transport development program." (FAA Release 64–74; Clark, *NYT*, 8/1/64)

- Alvin H. Parker soared his Sisu 1A sailplane from Odessa, Tex., to Kimball, Neb. (644 mi.) to claim new world soaring distance record over the previous 544.331-mi. distance record. (SSA Release)

During July: Two 10-story buildings had been completed in KSC Merritt Island Launch Area and were turned over to NASA. They were the $7.8 million operations and checkout building for the Gemini and Apollo spacecraft, and the $1.3 million spacecraft pyrotechnic installation building. (AP, *Orlando Sentinel*, 7/22/64)

- Sen. Clinton P. Anderson, Chairman of Senate Committee on Aeronautical and Space Sciences, said in *Pageant* article:

"Our [U.S.] drive since Sputnik I has led skeptics to call the manned lunar program only a race to beat Russia. This is a fallacy. The moon is not an advantageous spot from which to shoot nuclear-tipped missiles. We already have rockets that can hit the Soviet Union's interior in 20 minutes. Polaris missiles can hit Russia in one or two minutes, maybe less. It would take two and a half to three days for a moon-based rocket to reach Russia.

"But the Soviet Union's design to rule the world makes it essential that we reach the moon. This goal isn't, as skeptics argue, just a couple of weeks of good publicity. Sure, we have a signed United Nations agreement forbidding any nation from claiming the moon or space as a colony. But Russian domination of the moon would equate, in world opinion, with the dominance of all space. We can't afford that false impression. The nation that retains space superiority holds an immeasurable psychological advantage. Space feats are an index of a country's strength. If Russians were the only earthmen on the moon, you'd hear people around the world saying, 'If they can do that, no telling what they can do to us.'" (*Pageant*, 7/64)

During July: It was reported that nine payloads had been chosen for flight aboard the research and development launches of the USAF Titan III. Five of the experiments were said to be experimental communications satellites built by the MIT Lincoln Laboratory, and the sixth a backup communications satellite from Lincoln Lab. The remaining three were scientific satellites using the modified Arents spacecraft shell. To avoid confusion with the defunct Arents program, the latter spacecraft were to be called Orbiting Vehicle 2 (OV-2). (*M&R*, 8/3/64, 8)

- The Boeing Co. was awarded the annual *Missiles and Rockets* New Product of the Year Award for its development of a microminiature welder. The award was presented to Boeing officials by Dr. Edward C. Welsh, Executive Secretary and Acting Chairman of the National Aeronautics and Space Council. Awards of Honor went to Kollmorgen Corp. (for compact pointing interferometer), and to Control Data Corp. (for its 6600 computer). (*M&R*, 8/3/64, 8)

- Maj. Gen. Ben I. Funk, AFSSD, said perfect record of reliability of the Atlas (LV-3) during 14 consecutive flights in FY 1964 had resulted in savings of $81 million. Three additional flights, made since that period, had given the Atlas 100% reliability record for 17 straight firings; five of these were for NASA. General Funk made the statement in letter to J. R. Dempsey, President of General Dynamics/Astronautics, prime contractor for Atlas. (*Av. Wk.*, 8/3/64, 19; *Missile/Space Daily*, 7/27/64)

- First stage of Titan III-A had been installed at Launch Complex 20, Cape Kennedy. The second stage, transtage, and an aerodynamic nose shield were expected to arrive at Cape soon. Assembled vehicle, to form core of Titan III-C, was scheduled for launch in late summer. Launch of III-A would be first from Cape Kennedy of vehicle specifically designed for a military space system. (*M&R*, 7/20/64, 8)

- U.S.S.R. submitted brief, nontechnical report of its space activities to the United Nations and donated magazines for new U.N. "space library." This was first such Soviet report submitted to the U.N. and was interpreted as evidence that U.S.S.R. was willing to cooperate in space research and share its findings with others. (Teltsch, *NYT*, 7/18/64)

- One of U.S.S.R.'s top test pilots, Col. Sergei G. Brovtsev, was killed in accident while testing helicopter. (*Av. Wk.*, 8/3/64, 15)

- Kenneth W. Gatland, Vice-President of British Interplanetary Society, described space progress in Great Britain in *Astronautics & Aeronautics*. He pointed out that Britain not only played a leading role in establishing ELDO but also was contributing to ELDO "the 300,000-lb.-thrust Blue Streak . . . as the booster for a three-stage satellite launcher [Europa 1]." Other activities—not counting the ELDO contribution—had made Britain, in cooperation with Australia, "easily the biggest participant in space research outside the United States and the Soviet Union.

"Upper-air research with Skylark rockets began at Woomera in February 1957 and has continued to the present time. Black Knight rockets have reached heights of 500 mi., and with it a number of interesting investigations have been made on both re-entry and stabilization and control of rocket vehicles at high altitudes. The growth of the Weapons Research Establishment at Woomera as a launching, tracking, and communications center under joint investment by the British and Australian Governments, and latterly by ELDO and NASA, has made this facility the finest and best-equipped overland rocket range in the Western World. . . .

"With the advent of Sputnik 1 in October 1957, Jodrell Bank [radio-astronomy facilities] quickly established itself on both sides of the Iron Curtain as an important center for space tracking and communications. . . .

"In addition, the General Post Office in 1961–62 built an experimental ground station at Goonhilly Downs, Cornwall, for relaying telecommunications signals between the United States and the United Kingdom via Telstar and Relay. There are plans to develop this station into a ground terminal for a global telecommunications network. . . .

"In July 1963, it was announced in Washington that Black Knight would be used by the United States, Great Britain, and Australia in a joint program of re-entry research [Project Dazzle] to be conducted over the Woomera rocket range. . . .

"[British] universities and industry . . . [supplied] scientific instruments for the satellites S–51 and S–52 built and launched by the United States. A third scientific satellite, S–53, wholly engineered in the United Kingdom, will be launched by Scout in 1966. . . .

"In the small rocket field, excellent results are currently being obtained with the Skua meteorological rocket . . . . a cheap and simple tool for the meteorologist to conduct routine research measurements in the upper atmosphere. . . .

"Finally, . . . [studies] are currently being made in the United Kingdom into the concept of the Aerospace Plane, or space Transporter. . . ." (*A&A*, 7/64, 18–26)

August 1964

August 1: The day after RANGER VII's historic photographs of the moon climaxed a "textbook flight," President Johnson was briefed on the mission by NASA Associate Administrator for Space Science and Applications Dr. Homer E. Newell and Jet Propulsion Laboratory Director Dr. William H. Pickering. Discussing the lunar surface photographs, Dr. Newell said the principal investigators (Dr. Gerard Kuiper and Dr. Eugene M. Shoemaker) had tentatively concluded that "because there exists such sharp little craters here on the surface, this indicates that there is not a deep layer of dust in this area. There would be expected to be a layer of dust on the lunar surface . . . there has been much speculation about how deep it would be. . . . Some people had speculated as much as a thousand feet. Many as much as 30 or 40 feet. Others had said no, it can't be. And those who said no, it can't be, Dr. Shoemaker now thinks were right. At least as far as this area is concerned."

Asked by the President whether the RANGER VII success confirmed "that you folks had known what you were doing," Dr. Pickering replied: "Yes, I think it does.

"First of all, it confirms we know what we are doing technically when we design something to do this job. Secondly, that as far as the Apollo Program is concerned, it confirms that the basic assumptions that they were making about the sort of surface we are going to have to land on is probably correct."

The President asked, "What similar achievements, scientific achievements, can be compared to this?" and Dr. Newell replied: "Dr. Kuiper was asked this question yesterday. He stated he felt this was comparable to the photographing of the sun in the ultraviolet light by the Naval Research Laboratory a number of years ago.

"Someone else said that this amounts to a big jump in lunar science, equivalent to the jump that occurred when Galileo turned the telescope on the heavens."

President Johnson said: "If we were to conclude—if I were to conclude, if the Budget were to conclude, or if the American people should conclude—that we want to effect a savings here of a few billion dollars, would it be your opinion, Doctor [Pickering], that we would be pennywise and pound-foolish?"

Dr. Pickering replied: "It would indeed, sir, because I believe that this is truly an investment not only which is needed to demonstrate our leadership in the relatively near future, but it is a long-term investment for the future." (NASA Transcript)

- Concluding his White House briefing by scientists on RANGER VII, President Johnson stated:

    "I want to say that all Americans are very proud of you today. We are proud of this historic extension of man's knowledge. We are proud of our scientists, our engineers and all the great team under the leadership of one of the greatest of all Americans, Jim Webb, who are responsible for this success. We can be duly proud of our free and open society, our system of government.

"We started behind in space. We were making many apologies just a few years ago. We had our failures, but we kept our faith in the ways of freedom, and we did not follow the easy or the inexpensive course. We know this morning that the United States has achieved fully the leadership we have sought for free men. . . .

"I think we can say this morning that this is a victory for peaceful civilian international cooperation in this hour of frustration, when so many people are getting upset at some minor disappointments.

"I think we can all take great pride in this development. More than 60 countries all around the world work for us and work for peaceful progress and work for peaceful uses of outer space. . . .

"These men don't wear a DSM this morning, and we are not presenting them any Congressional Medal of Honor. But they do have, and all of their associates from Mr. Webb down to the fellow who sweeps out the dust in the remote test laboratory, the gratitude and the admiration of all Americans of all faiths, of all parties, of all regions.

"You are welcome to the White House. The people who live here are mighty proud of you." (NASA Transcript)

August 1: In article on RANGER VII flight, Soviet newspaper *Pravda* concluded: "Soviet scientists are happy to note that the launching of 'Ranger 7'— a new national achievement of the USA in its program to investigate the moon—represents a contribution to the study of outer space." *Izvestia* published photograph of the lunar surface taken by RANGER VII as illustration of interview with Soviet astronomer Boris Levin, who said the RANGER VII flight was "definitely an important contribution to the study of the composition of the lunar surface." (*Pravda*, 8/1/64, 1, ATSS-T Trans.; USIA Research Results, *CR*, 8/11/64, 18370–73)

- In lead editorial on the RANGER VII flight, *New York Times* lauded the success of the experiment but expressed the opinion that the flight once more emphasized the possibilities for international cooperation in such efforts: ". . . Proud as this country has reason to be of Ranger 7's accomplishment, we can only conjecture how much more rapidly and cheaply its photographs could have been obtained had Soviet and American scientists been able to pool their efforts and resources years ago, rather than being required to pursue their efforts in competitive—and costly—isolation. The case for a unified international effort to reach the Moon—as urged by Presidents Kennedy and Johnson—is strengthened by every consideration of the needless waste and duplication that have marked humanity's space exploration to date." (*NYT*, 8/1/64, 20)

- *Washington Daily News* editorialized that the success of RANGER VII's mission represented "a significant milestone on a historic road to manned exploration of the moon." The lunar flight was termed "a demonstration of man's ceaseless, boundless quest for knowledge—and one which, in the end, is likely to pay off as well as similar past voyages into the unknown." (*Wash. Daily News*, 8/1/64; *CR*, 8/4/64, A4114)

- Baltimore *Sun* editorialized on RANGER VII: ". . . we hear much talk of what this means in international prestige, and to what extent we have given the Russians (as well as the moon) one in the eye. Yet surely the thing itself is the thing itself, and is wonderful in the full sense of that word. To see actual photographs close-up of the surface of the moon, whose mystery throughout the history of mankind has but been enhanced by its proximity and by the fact that *something* could always be seen there, is a matter of awe and wonder; and that is the main thing about this moonshot." (Balt. *Sun*, 8/1/64)

August 1: Soviet satellites ELECTRON III and ELECTRON IV were successfully conducting scientific investigation of upper atmosphere and outer space, according to Soviet newspaper source. Since their orbiting last July 11, 50 communication sessions had been held with ELECTRON III and 19 with ELECTRON IV, and valuable scientific and engineering data had been obtained. (*Krasnaya Zvezda*, 8/2/64, ATSS-T Trans.)

- Clifford B. Harmon Trust announced that Astronaut L. Gordon Cooper, Jr. (Maj., USAF), would receive the Harmon International Aviator's Trophy for 1964 for his 34-hour orbital space flight (MA-9) in a Mercury spacecraft May 15-16, 1963. Others to receive Harmon Trophy would be Aviatrix Betty Miller for her solo transpacific flight and Lt. Col. Fitzhugh Fulton (USAF) for 11,000-lb. payload of B-58 to 85,318-ft. altitude. The Harmon awards for 1963 and 1964 would be presented to recipients in White House ceremony September 14. (*NYT*, 8/2/64, 78)

- Expressing congratulations on the success of RANGER VII lunar mission, Chancellor Ludwig Erhard of West Germany said in telegram to President Johnson: "We are happy with the American people over this tremendous technical and scientific achievement." Other response to the successful lunar mission included statement by President of Mexico Adolfo Lopez Mateos, who said that RANGER VII was "without doubt, up to the present moment, the most brilliant page registered in the scientific history of modern times." British astronomer Sir Bernard Lovell praised the success of RANGER VII and said: "The results will be of . . . critical importance to the American plan for a soft landing of instruments on the lunar surface next year and for Project Apollo, involving a manned landing at the end of the decade." (AP, *Wash. Sun. Star*, 8/2/64)

- French Dragon rocket was successfully fired to an altitude of 275 mi. from a base in Iceland, National Center of Space Research announced in Paris. Fired in cooperation with Icelandic scientists, the sounding rocket carried instruments to measure energy and direction of particles from Van Allen radiation belts. (AP, *Wash. Sun. Star*, 8/2/64; *M&R*, 8/10/64, 11)

August 2: Giving his Sunday blessing to the public, Pope Paul VI referred to the successful U.S. RANGER VII lunar flight: "We also would unite ourselves in the admiration of all for a happening that concerns the life of all humanity and that shows how far the man of science and technology has progressed. . . ." (AP, *Houston Post*, 8/3/64)

- In article from Houston, UPI reported on first manned space flight in Project Gemini (GT-3), quoting "informed sources" as saying: "We cannot possibly make it this year. It will be early 1965." NASA's officially announced planning date was in last quarter of 1964. (UPI, *L.A. Times*, 8/2/64)

- Soviet newspaper *Krasnaya Zvezda* published erroneous map of the target area in the Pacific for its rocket tests which it had previously announced would begin August 4. The map placed the impact areas within 100 mi. of populated islands, but the targets as previously defined were several hundred miles from nearest inhabited areas and in another part of the Pacific. (Shabad, *NYT*, 8/3/64)

August 3: Rep. Carl Albert (D.–Okla.), speaking on the House floor, praised the team of scientists and technicians that made the successful RANGER VII lunar flight possible, and said that the "validity of the space program has been reconfirmed. . . ."

Also in the House, Rep. Louis C. Wyman (R.–N.H.) said the transmission of lunar photographs was "a great tribute to the scientific progress of our experts in America's space program," and RANGER VII was "a highly technical success . . . in the right direction; namely, the use of machines" rather than man in space exploration. He added that Project Apollo need not be called off, but he felt Congress should "hold back on the appropriations for Apollo space vehicle procurement until we have established that man can rendezvous in space by completing successfully the present Gemini program. . . ." He also advocated elimination of a target date for U.S. manned landing on the moon. (*CR*, 8/3/64, 17203)

- NASA announced Col. Lawrence W. Vogel (U.S. Army) replaced Col. R. P. Young as NASA Executive Officer. Colonel Vogel was detailed to NASA from U.S. Army Corps of Engineers. (NASA Announcement 64–172)
- Results of decompression experiments with mice in atmospheres of helium-oxygen, nitrogen-oxygen, and argon-oxygen were reported in *Missiles and Rockets*. Performed by group at Purdue Research Facility, experiments showed mice survived decompression markedly longer in helium-oxygen atmosphere than in air. In one experiment, eight mice decompressed in helium-oxygen atmosphere survived an average of 6.1-min. longer than six mice decompressed in air. Extrapolation of these results might indicate desirability of helium-oxygen atmospheres in spacecraft for prolonged manned flights; the extra minutes of survival would provide safety margin, in case of accidental decompression, for crew to take necessary emergency measures. (*M&R*, 8/3/64, 32)
- Discussing President Johnson's July 24 announcement of the SR-71 mach 3 aircraft under development for SAC, Robert Hotz editorialized in *Aviation Week*: ". . . The 'SR-71' was, of course, the A-11 [YF-12A] with still another designation pasted on its titanium skin. But many congressmen were fooled, and unthinking daily newspapermen and wire service reporters failed to catch the deception and spread the news across the nation of the 'new' billion-dollar aircraft program.

  "The inception of this program also was conveniently shifted from 1959 in the Eisenhower Administration to 1963 in the Kennedy Administration. Anybody familiar with aircraft development cycles knows that if this program were really started in 1963, it would be impossible to deliver operational aircraft to Strategic Air Command in 1965, as President Johnson stated. . . ." (*Av. Wk.*, 8/3/64, 11)
- Preliminary survey of public reaction to sonic booms in Oklahoma City, disclosed by FAA, indicated that 10–20% of those interviewed expressed "varying degrees of negative reactions" and about 4% made formal complaints. In the six-month test program which ended July 30, 1,253 sonic booms were created over the city. (Clark, *NYT*, 8/4/64)

August 4: Appointment of Dr. Albert J. Kelley, Director of Electronics and Control Div. of NASA OART, as Deputy Director of the new NASA Electronics Research Center, was announced by NASA Associate Administrator for Advanced Research and Technology Dr. Raymond L. Bisplinghoff. Appointment was effective Sept. 1. Dr. Kelley formed NASA research program in electronics and control when he was assigned to his position in November 1961; he joined NASA in March 1960 to establish development program for Agena launch vehicle. (NASA Release 64-199)

- Sen. Hubert H. Humphrey (D.–Minn.) discussed responses he had received to bill he had introduced for establishment of Office of Solar Energy, among them a letter from George Lof, Acting Director, Solar Energy Laboratory of Univ. of Wisconsin. After visiting U.S.S.R., Mr. Lof related that the solar development program there was "well directed, well staffed, and exceptionally well financed." There were about 10 institutions in various parts of the U.S.S.R. engaged in this work, with 80–100 full-time scientific personnel. (*CR*, 8/4/64, 17471)
- Rep. Lionel Van Derelin (D.–Calif.) said in a telegram to J. R. Dempsey, president of General Dynamics-Astronautics: "Spectacular success of the Ranger moon shot should not overshadow the equally spectacular record of consecutive successful launchings by Atlas boosters. Ranger shot was the 17th straight. Such amazing reliability is the backbone of our space program. . . ." (*CR*, 8/4/64, 17368)
- The "strange object" that fell from sky near remote Venezuelan village was identified by two U.S. Army officers as a U.S. satellite, according to U.S. Embassy spokesman in Caracas. The 175-lb. satellite had been charred by re-entry heat, damaged by impact, and partly dismantled by local residents. (UPI, *Wash. Post,* 8/7/64)

August 5: RANGER VII briefing to members of Congress, with Harris ("Bud") Schurmeier, Ranger Project Manager, JPL, and Dr. Gerard P. Kuiper, Director of Lunar and Planetary Div., Univ. of Arizona, as principal speakers. Dr. Kuiper said: "I do not think it is an exaggeration to say that a new era has begun in the exploration of the Moon. . . .

"Here, in one step last week a gain of a factor of 2,000 was made—2,000 in the effective magnification of the image of the lunar surface [earlier reported as 1,000]. . . ."

Discussing preliminary results of the lunar photographs, Dr. Kuiper said: ". . . down to the size of approximately two or three feet, the primary craters caused by direct impacts on the moon covered . . . an area amounting to about one percent of the lunar surface. So this is a very great help in landing operations, either by unmanned soft landings, such as Surveyor, or manned landings as planned for Apollo.

". . . the innocent-looking ray systems, such as one has around Copernicus, Tycho, Kepler, a number of craters really are danger spots. When you look at them closely, . . . one sees them resolved into multitudes, clusters of secondary craters. . . . I wouldn't be surprised if slopes of the order of 15 degrees would be found. This would be somewhat difficult. This is a rather uncomfortable inclination on which to land.

". . . the surface between these craters is smoother than I had thought. . . . I have . . . many times hypothesized that if you would go to higher and higher magnification, higher and higher resolution, that there would be a large number of cracks on the moon, fissures similar to those one finds on lava fields such as on Mauna Loa, Hawaii, or recent lava fields which have not yet been eroded.

"This apparently is not the case. This was wrong . . . these cracks are not present. It looks as if the surface of the moon has been buffeted, has been simply pounded on, and has been sort of smoothed out, and the cracks filled in the process. . . .

"Now the question is, what about the surface? What is the nature of the surface. . . . It is not just a loose pile of sand, dust. It has considerable stiffness. And . . . while we know from other data that the top structure is open, it is sort of like snow or a biscuit, a cracker. It has an open structure; nevertheless, it looks as if the moon may have considerable bearing strength. That, however, is something which must be measured separately, with separate equipment.

"However, we can say that there are not enormous depths, several feet, ten, or a hundred feet of dust, which would simply not have permitted formation of the sharp little craters as we observe them." (NASA Transcript)

August 5: Success of RANGER VII spotlighted NASA's preparation for further lunar and planetary exploration, including two more Rangers (early next year), Surveyor soft-landing spacecraft, Lunar Orbiter photographing spacecraft, and Project Apollo manned lunar expedition, as well as two Mariner spacecraft to gather data from the vicinity of Mars (last quarter of 1964). (NASA Releases 64–196, 64–205)

- Senate passed H.R. 11296 (FY 1965 appropriations for independent offices), after rejecting (69–20) amendment sponsored by Sen. William Proxmire to reduce NASA funds by $68,594,000 for Research and Development; by $17,880,500 for Construction of Facilities; and by $13,525,500 for Administrative Operations. As passed by the Senate, NASA appropriations for FY 1965 amounted to $5.3 billion, divided as follows: R&D, $4,413,594,000; CofF, $262,880,500; and AO, $623,525,500. Bill was sent to conference. (*CR*, 8/5/64, 17554–77)

- Boilerplate Gemini spacecraft was dropped from C–119 aircraft at 1,500-ft. altitude, sixth in series of such tests to study landing on solid ground. The tests were being conducted by Landing and Recovery Div., NASA Manned Spacecraft Center. (AP, *Houston Post*, 6/8/64; MSC *Roundup*, 8/19/64, 3)

- At NASA Hq. awards presentation ceremony, Joseph Mandelkorn of Lewis Research Center and William R. Cherry of Goddard Space Flight Center jointly received NASA Contributions Award for development of improved (n-on-p) radiation-resistant solar cells. Mandelkorn received $5,000 cash award and Cherry, $1,000. William W. Smith, formerly of Jet Propulsion Laboratory, and Bruce W. Schmitz of JPL received $1,000 award for their contribution of Trajectory-Correction Propulsion System, used recently in RANGER VII lunar photography mission. Wilmer H. Reed III of LaRC received $1,500 award for his contribution of a Test Unit Free-Flight Suspension System. Howell D. Garner and Henry J. E. Reid, Jr., of Langley Research Center, received $1,700 award for their contribution of Dynamic Precession Damper for Spin Stabilized Vehicles. Awards were presented by NASA Deputy Administrator Dr. Hugh L. Dryden. (Program)

August 5: USAF announced launching of Thor-Agena D space booster from Vandenberg AFB with unidentified satellite payload. (UPI, *Wash. Post*, 8/6/64; *M&R*, 8/17/64, 9)
- Senator John Stennis (D.–Miss.) said on the floor of the Senate:
  "In exploring space, our Nation as well as any other nation is straining at the state-of-the-art and technology. We are moving into a medium where relatively little is known. As we move into such a medium we are confident that we can stay abreast in the field of scientific discovery and its application to space with any other nation. In this medium it is not a case of being able to defend ourselves by building ten planes to every five planes of a potential aggressor nation or five ships to two ships or three tanks to one tank in order to establish our superiority. By staying abreast of advancements in science and technology our Nation will be able to prevent any surprise in this new medium. Our national space program is geared to this accomplishment. As for the present, our Nation subscribes to the philosophy that space should be explored for peaceful purposes, and we will do so. However, should an aggressor attempt to utilize this medium for warlike activities in the future, I am confident that our Nation will be prepared to defend itself." (*CR*, 8/5/64)
- National Academy of Sciences announced establishment of 10-man committee to study effects of sonic boom phenomena as related to development of supersonic transport. Chaired by Dr. J. R. Dunning, dean of School of Engineering and Applied Science, Columbia Univ., the committee would assist Government in planning an expanded sonic boom test program and evaluating the data obtained in the program. FAA would continue to manage the test program operation, with support from NASA and USAF. NAS organized the committee upon request from President Johnson. (NAS-NRC Release)
- Dave W. Lang, Chief of MSC Contracts and Procurements Div., said in address to Gulf Area Development Committee of the Houston Chamber of Commerce that MSC "ranked first in the agency in terms of dollar and percentage awards" to contractors for the first half of FY 1964. MSC awards totaled $629.1 million, or 32% of total NASA spending in that period. (Maloney, *Houston Post*, 8/6/64)

August 6: NASA and Dept. of Health, Education and Welfare announced cooperative one-year research project to study anti-cancer, carcinogenic, and anti-radiation potential of certain chemical compounds closely related to plant growth regulators. Project would extend studies made at North American Aviation's Space and Information Systems Div. in NASA research program to protect man in space. To be conducted at S&ID under $198,185 contract with HEW's Public Health Service, project would be directed by PHS National Cancer Institute and financed through a transfer of funds by NASA as part of its Technology Utilization program. (NASA Release 64–190)
- Nike-Cajun sounding rocket with upper atmosphere grenade experiment was launched from Wallops Island to 73-mi. altitude, and series of 12 grenades was ejected and detonated at intervals from 23- to 58-mi. altitude. Experiment was coordinated with similar grenade shot from Sweden's Kronogård Range (81-mi. peak altitude) and with another experiment launched there on a Nike-Apache rocket to measure noctilucent clouds (77-mi. peak altitude). These launchings marked

beginning of third series of grenade experiments conducted by NASA GSFC from Wallops Island, Fort Churchill Research Range, and Ascension Island, with Sweden joining the group for this series. The Fort Churchill and Ascension Island launchings were canceled because of weather conditions. (Wallops Release 64–59; NASA Rpt. SRL)

August 6: Black Knight rocket was fired at Woomera Rocket Range in U.K.-U.S.-Australia Project Dazzle. Minister of Supply Allen Fairhall said the rocket reached 370-mi. altitude, where second stage rocket ignited to drive instrumented re-entry package back through the atmosphere. (Reuters, *Houston Post*, 8/7/64)

• Research in extrasensory perception as a means of communication with astronauts on space flights was advocated by Harold Sherman, President and Executive Director of ESP Research Associates Foundation, Little Rock, Ark. Speaking in Los Angeles, Sherman said: "It is entirely feasible for the Air Force to experiment in the training of astronauts to transmit and receive thoughts and, in this connection, a battery of 'sensitives,' who can communicate with them on a two-way basis while they are traveling through space." (L. A. *Herald-Examiner*, 8/6/64)

August 7: NASA launched Nike-Cajun sounding rocket from Ft. Churchill with grenade payload to obtain atmospheric temperature, wind, density, and pressure data. The rocket attained 86.5-mi. altitude and all but the first of the 12 grenades detonated. (NASA Rpt. SRL)

• First manned free-flight test of Gemini paraglider landing system was unsuccessful at Edwards AFB, Calif. The paraglider, with pilot E. P. Hetzel of North American Aviation in mockup spacecraft slung beneath it, was towed to 13,000 ft. by helicopter, then was cut loose. Pilot Hetzel lost control and at 3,000 ft. was forced to bail out. The paraglider crashed. (Wash. *Eve. Star*, 8/7/64)

• NASA and Netherlands Organization for the Advancement of Pure Research announced agreement for joint scientific sounding rocket project. Under the agreement, four Nike-Apache sounding rockets carrying sodium vapor payloads would be launched from Coronie, Surinam, near the equator next autumn. Netherlands would provide scientific payloads and launch site, as well as other support facilities in Surinam. NASA would provide the rockets and launching device and train Dutch launching personnel. (NASA Release 64–197)

• USAF Titan III solid-fuel rocket motor was fired in static test at United Technology Center's Coyote, Calif., test facilities. A joint in the nozzle skirt failed during last 10 sec. of the 112-sec. test, causing a fire that did only minor damage to the test stand, but all test objectives were met. AFSC Space Systems Div. Deputy Chief of Staff for Manned Spaceflight Brig. Gen. Joseph Bleymaier termed the test an absolute success. (AP, Wash. *Eve. Star*, 8/7/64; *M&R*, 8/16/64, 8)

• Research in use of laser radar to map air turbulence in the skies was described by Drs. Ronald T. H. Collis and Myron G. H. Ligida of Aerophysics Laboratory at Stanford Research Institute. The laser equipment would be installed in aircraft. Dr. Collis and Dr. Ligida reported their research in *Nature*. (*NYT*, 8/7/64, 8)

• Discussing U.S. space costs in terms of the gross national product, Sen. Stuart Symington (D.–Mo.) said the GNP was nearly $2 billion a day, "and I guess we can afford a couple of days of that to get to the moon." (AP, Louisville, Ky., *Courier-Journal*, 8/7/64)

August 7: AFSC announced fabrication of ultrahigh strength, lightweight steel cases for solid-propellant rockets. Solid-rocket case constructed of 18% nickel maraging steel was as strong as similar-size conventional steel case but was 25% lighter. The steel was developed under contract by Curtiss-Wright Corp. (AFSC Release 44-129-94)

- Michael V. Lighthill, director of British laboratory involved in developing Anglo-French supersonic airliner Concorde and reputable expert on sonic booms, spoke optimistically in interview about outcome of recent U.S. sonic boom experiments. Lighthill said that although there was not yet enough information from the Oklahoma City studies, preliminary reports were "very encouraging." Damage from sonic booms was apparently less than many experts had feared. (Simons, *Wash. Post*, 8/8/64)
- USN successfully launched Polaris A-3 from submarine submerged 30 mi. east of Cape Kennedy. (*M&R*, 8/17/64, 9)
- Common stock of Communications Satellite Corp. had risen as high as $35 since it was put on sale to the public at $20 last June 2, surprising many Wall Street experts. (Vartan, *NYT*, 8/7/64, 33)

August 8: President Lyndon B. Johnson said in press conference at LBJ ranch, Stonewall, Tex.: ". . . it seems to me that these last seven days deserve very special consideration and contemplation of every thinking citizen. All week long the Americans have been doing what Americans do best—working together.

"The results have been highly gratifying. The week has been deeply reassuring. Wherever we have faced them, we have been meeting our challenges—at the Gulf of Tonkin, the halls of Congress, in distant space of our universe and all sections of our Nation.

"Only a week ago, we saw a steady, stable, straight-forward national course yield an important national success in the mission of Ranger 7. . . ."

The President announced:

"The Air Force will proceed immediately with the program to orbit 24 satellites for an interim, independent defense satellite communications system. This system will provide reliable, worldwide circuits, highly resistant to jamming and physical attack, for carrying essential military communications in times of crisis." (Transcript, AP, *Wash. Post*, 8/9/64, A12)

- DOD statement on military communications satellite system, announced earlier in the day by President Johnson, described the satellites as an "interim system" designed to provide "extremely reliable duplex high-quality voice circuits." The 24-satellite network would be placed into near-equatorial orbits, eight at a time, by three Titan III-C launch vehicles, beginning with the first launching of eight satellites in early 1966. This interim system would be replaced after about three years by an "advanced" military comsat system. (*NYT*, 8/9/64; UPI, *Wash. Post*, 8/9/64)
- Addressing Aerospace Industries Association Council in Seattle, Dr. Edward C. Welsh, Executive Secretary of the National Aeronautics and Space Council, summarized future aerospace trends. He foresaw no immediate substantial decrease in dollar amount of defense spending,

but said industry should be able to shift emphasis from production of weapon systems to "other useful types and forms of economic activity"—which he said should be toward an expanding space program and an increase in R&D of advanced weapons technology. Of space program values he said:

"The space program has been a catalyst, a stimulus to *education* at all levels, with particular attention to science and engineering. How much is it worth to have raised the educational sights of our young people and at the same time to have increased significantly the wealth of knowledge with which to condition them? I cannot put a price on it, but I believe its value will exceed the total cost of the space program.

"The contribution of our space program to our *national security* is also considerable. How much more secure are we, due to improved communications, more accurate navigation, more complete weather information, and better world-wide mapping? How much is it worth to be better informed about potential sources of danger? How can we assess the advantage of developing competence to detect and offset possible aggression from space? I cannot put a price on these contributions to national security, but I am confident that their value also exceeds the total cost of the space program.

"The space program stimulates the development of new products, new productive processes, and new managerial techniques. I cannot place a precise value upon such *innovations*, but I would estimate that such investment will repay itself many times over.

"A substantial difference in influence in *world affairs* evolves from whether a country is in a first position or a second position in power. To a significant degree, a nation's relative position depends on how it stands in advanced technology. . . .

"Combining, as the space program does, the best talents in management, in engineering, and in science, with the most modern facilities available, the net result is the production of *progress*. A program which stimulates education, expands research and development, augments total productivity, increases employment, and improves our international relations, is a program of the greatest economic significance. . . ." (Text)

August 8: Brazilian sources reported that Brazil was negotiating with British firm for purchase of unspecified number of Seacat ship-based guided missiles. (Miller, Balt. *Sun*, 8/9/64)

August 9: Smithsonian Institution, under grant from NASA, had purchased collection of meteorites from estate of Arthur R. Allen who, before he died in 1962, collected 45 meteorites and several hundred other rock specimens. (*Wash. Post*, 8/9/64, B9)

August 10: Paraglider landing system definitely would not be used for the scheduled Project Gemini space flights, MSC Public Affairs Officer Paul Haney announced. Development of paraglider system was too far behind for the 12 scheduled flights. After many rumors and much speculation, this statement was first official announcement of decision not to use the paraglider. It followed by three days the first manned free-flight test of the system, which was unsuccessful. Development and testing of paraglider by North American Aviation would continue, Haney said. (Maloney, *Houston Post*, 8/11/64)

August 10: Ernest W. Brackett, NASA Assistant Deputy Associate Administrator (Procurement Policy), addressed National Association of Professional Contract Administrators in Orlando, Fla. Discussing some NASA contract matters of current interest, he outlined ways in which NASA worked similarly to and differently than DOD contract administration, then said:

"Perhaps the outstanding difference between NASA and DOD in the field of contract administration is in the area of quality control and inspection. NASA's contracts are almost wholly for research and the development of satellites, space vehicles, engines, etc., the like of which have never been made before. There are no production type contracts and these probably will be very few. The system of inspection and quality control as envisaged by NASA is quite different from the sampling type of inspection used on a production line.

"We buy only a few of each item and a failure may mean the compromise of a program. Also, development items are very expensive and the loss of a shot may mean the loss of millions of dollars. There have been failures where the cause has been traced to the failure of a small component costing a hundred dollars or so which might have been caught with proper inspection. The Saturn is an example of a project where very high degrees of quality control have been applied and very detailed inspection has taken place. The six Saturn shots have all been successful and it is difficult to criticize success.

"NASA has had over 100 space shots. There have been successes and failures. In 1960 only 47% of the shots were successful, but in 1963 successes had risen to 85%. You can expect continued emphasis on quality control from NASA. Someone described that degree of reliability that NASA is looking for as being the equivalent of 1,000 TV sets made to run continuously for a year without a single one failing. . . ." (Text)

- The newest 14 NASA astronauts began week-long desert survival course in Nevada, under direction of USAF 3637th Survival and Special Training Squadron, Stead AFB. (MSC *Roundup*, 8/19/64, 2)
- Interviewed in *Missiles and Rockets*, Rep. Emilio Q. Daddario (D.–Conn.) discussed the work of his research and development subcommittee of the House Committee on Science and Astronautics. One current activity: "The National Academy of Sciences is undertaking several studies for us through a 14-man committee headed by Dr. George Kistiakowsky, science advisor to former President Eisenhower. It will look into such questions as what is the level of basic research that should be supported Federally to maintain the nation's security, economy and industrial strength. . . ." (*M&R*, 8/10/64, 26)
- Orbital tanker study by Lockheed Missiles & Space Co. for NASA Marshall Space Flight Center was discussed in *Aviation Week & Space Technology*. Tanker would be unmanned and would be used to transfer propellants to other unmanned or to manned spacecraft in earth orbit. Concepts in which orbital refueling capability was underscored were future manned planetary flights and potential reusable lunar transport ferries. (*Av. Wk.*, 8/10/64, 55–57)
- U.S. Samos (Satellite and Missile Observation Systems) satellites, undiscussed by DOD since secrecy policy was adopted in late 1961, had been orbited "periodically" ever since then, Albert Ravenholt said in New Orleans *Times-Picayune* article. Even though Samos was de-

veloped as successor to U–2 high-altitude photographic airplane, Ravenholt said, U.S.S.R. had "made no threatening noises about Samos," probably because some Russian Cosmos satellites are equipped with cameras; U.S.S.R. has no missile capable of shooting down a Samos; and, if it could down one, the action would open whole new vista of legal problems about control of outer space. (Ravenholt, N.O. *Times-Picayune*, 8/10/64)

August 10: According to unnamed DOD official, U.S.S.R. had yet to demonstrate "even the beginnings of U.S. solid rocket technology." The official said in *Missiles and Rockets* that this fact, combined with the fact that Soviet operational ICBM's were liquid-fueled vehicles stored above ground rather than in hardened silos, had greatly influenced defense planners' viewpoint that U.S. could mount a second strike and destroy a number of Soviet ICBM's before they were fired. (Trainor, *M&R*, 8/10/64)

- Soviet newspaper *Izvestia* singled out Marina Popovich, wife of Lt. Col. Pavel R. Popovich, as a possibility to become woman space traveler, saying the cosmonaut's wife had "finished the higher aviation school and has now left for a distant business trip." *Izvestia* quoted Pavel as saying: "The machine with which she is now familiarizing herself is considerably more powerful than the one in which she established a new record for speed two months ago." (UPI; *NYT*, 8/11/64, 3; *Wash. Post*, 8/12/64)

August 11: Echo, Relay, and Syncom communications satellites were presented to the Smithsonian Institution in ceremony at Museum of History and Technology. Among the NASA officials participating in the presentation were Deputy Administrator Dr. Hugh L. Dryden, Langley Research Center Director Dr. Floyd Thompson, and Goddard Space Flight Center Director Dr. Harry Goett.

Carl T. Rowan, Director of U.S. Information Agency, said in address at the ceremony:

"As we view the unfolding drama of the American space program it can be clearly seen that one of its most significant aspects is that it has been carried out in the full, open view of an interested and anxious world. When it was written into the National Aeronautics and Space Act of 1958 that the activities of the new space agency were to be open to the widest practicable and appropriate spread of information, the Congress cast into law one of the deepest instincts of the American people. It was an act of faith to specify that this enterprise should be carried on openly before the gaze of those who wished us well and those who hoped otherwise; it was truly an expression of faith in freedom and the ability of Americans to meet the challenges and opportunities of an era when each day unfolds new perils and new promises.

"If openness to public scrutiny was a basic concept in the entire space program, the decision to carry out a satellite program dedicated to freedom of communications on a global scale represented that concept converted into a highly meaningful fact. . . .

"It is obvious that the over-riding quality of communications satellites is their internationalism. A practical system based on communication satellites is inherently universal, for it would be of little use if it did not mean communicating between the peoples of widely separated lands. Its basic premise involves the inclusion of as many countries as possible in the interchange that the system allows.

"This presentation of communication satellites to our national museum of science and technology is, therefore, most truly an international occasion. For it is their international utility in the cause of world understanding, peace and progress which gives them their deepest meaning." (Text)

August 11: NASA Aerobee 150 sounding rocket launched from White Sands, N. Mex., reached 107-mi. altitude with four spectrographs to obtain spectra of planets, but no data were obtained because attitude control system's despin valve failed and the rocket spun out of control. (NASA Rpt. SRL)

- USAF Titan II rocket was launched successfully from an underground silo on flight down Western Test Range. Same day, faulty relay box in Thor-Agena launch vehicle postponed launch of NASA's Nimbus meteorological satellite into polar orbit. (*Wash. Post,* 8/12/64; *M&R,* 8/17/64, 9)

- Communications Satellite Corp. said DOD's last-minute demands on ComSatCorp to change the terms of the international comsat agreement were the reason why DOD-ComSatCorp negotiations fell through. ComSatCorp President Joseph Charyk said DOD had insisted on changes which would have barred foreign members from any voice in letting contracts to build the military portion of the network. DOD's demands were presented in July, shortly before the conference that created the new consortium and after all technical details of the proposed military-commercial sharing plan had been agreed upon in June. (Toth, *Wash. Post,* 8/12/64)

- Four Soviet parachutists at world parachute jumping championships, Leutkirch, West Germany, set new world record for men's team jumping from 1,000 meters, landing an average of 66.5 centimeters from center of target. Previous world record of 89.5 centimeters was set two days earlier by U.S. team. (Reuters, *NYT,* 8/12/64, 35)

August 11–12: Four NASA sounding rockets were launched in coordinated studies of the atmosphere from Wallops Island, Va.: Ft. Churchill, Canada; and Kronogård, Sweden.

Nike-Cajun launched from Ft. Churchill reached 77.5-mi. altitude with 12 grenades ejecting and detonating during rocket ascent. Ground-based microphone array recorded time and direction of detonation sounds' arrivals to obtain atmospheric temperature, wind, density, and pressure data.

Nike-Cajun launched from Wallops Island ejected 12 grenades during its ascent to 72-mi. altitude, with similar ground-based microphone array to obtain data.

Nike-Apache launched from Kronogård reached 75.5-mi. altitude with micrometeoroid-particle-sampling payload. Although sampling cans did not seal completely, preliminary evaluation indicated that data would be recovered. Experiment was cooperative effort of USAF Cambridge Research Laboratories and Meteorological Institute of Stockholm Univ.

Nike-Cajun launched from Kronogård, with grenade payload functioned properly, but no usable data were obtained because of loss of Dovap near Cajun ignition. For same reason, peak altitude was unknown. (NASA Rpts. SRL)

August 12: X–15 No. 3 flown by NASA pilot Milton O. Thompson in flight to test and measure heating effect of atmospheric friction on certain aircraft structural components. Aircraft reached altitude of 81,000 ft. and traveled at 3,511 mph (mach 5.32) in the 7-min. flight near NASA Flight Research Center, Edwards, Calif. (NASA X–15 Proj. Off.; *CR*, 8/20/64, 20046)

August 12: Senate-House Conference Committee on H.R. 11296 filed its Conference Report on the Independent Offices Appropriation Bill for FY 1965. Committee recommended $5,250,000,000 NASA appropriation, which was $50,000,000 less than Senate-passed bill had provided and $50,000,000 more than House Committee had recommended. One amendment, regarding use of funds for a manned lunar landing by U.S. and another country, was reported in disagreement. (House Rpt. 1781)

- Fourth birthday of ECHO I passive communications satellite, still orbiting the earth in its visible path. G. T. Schjeldahl Co., builders of the 100-ft. sphere for NASA, said ECHO I's supply of gas had long ago leaked out through holes made by micrometeoroids, but even so, ECHO I had "demonstrated the practicability of employing large inflatable and erectable reflectors to rebound intelligent signals beamed at it." (Schjeldahl Release; *Wash. Post*, 8/11/64)

- NASA Goddard Space Flight Center announced selection of GE Command Systems Div. to negotiate contract for additional spacecraft tracking system for Stadan (Space Tracking and Data Acquisition Network) station at Fairbanks, Alaska. The new system, known as Goddard Range and Range Rate System because it was originally developed by GSFC engineers, was designed to provide improved tracking data for space probes, satellites, and launch vehicles traveling in elliptical or synchronous orbits. Two previously completed systems installed at Rosman, N.C., and Carnarvon, Australia, had proved highly successful in tracking EXPLORER XVIII (IMP). (GSFC Release G–21–64)

- Assistant Secretary of Defense for Public Affairs Arthur Sylvester denied statement made yesterday by Communications Satellite Corp. President Joseph Charyk that DOD had demanded, at the last minute, unmeetable conditions for military-civilian comsat cooperation. Sylvester said: "The Department of Defense stated at the outset of negotiations that it would require assurance that the international agreement be so drafted as to protect the technical and security requirements of the national communications system." (AP, *NYT*, 8/13/64, 17)

- IBM Corp. said newly developed mathematical methods would permit great increase in use of computers to design and test aircraft, missiles, and spacecraft. Designers now could "simulate flight 15 to 20 times faster than ever before possible on a general-purpose digital computer." IBM had used the new technique to help design and evaluate guidance computer for Gemini spacecraft. (*NYT*, 8/13/64)

- General B. A. Schriever, AFSC, said in keynote address at Ninth Symposium on Ballistic Missile and Space Technology, U.S. Naval Training Center, San Diego:

    ". . . technology is a key to our nation's future. This is true not only in the military area but also in the economic area. Technology is vital not only to our security and survival, but also to our growth and general well-being. Technology must be utilized to support national objectives.

    "Those of us who are responsible for the national security must ensure that technical progress does in fact serve national military policy effectively. To meet this responsibility there are three things we must do.

    "First, we must advance the state of the art as the necessary first step toward new capabilities. . . .

"Second, we need to identify the technological opportunities during the next ten to fifteen years and relate them to national security objectives. . . .

"[Third, we need] a means of presenting alternatives clearly so that decisions can be made promptly and with a high degree of confidence. . . ." In this regard, he discussed new AFSC Hq. activity to facilitate decision-making process: PROM (Program Management, Resources Management, and Objectives Management). (Text)

August 13: House passed conference report on H.R. 11296 (Independent Offices Appropriation), providing NASA FY 1965 appropriation of $5,250,000,000. House also voted to recede from its disagreement with Senate-sponsored amendment which provided that no part of the NASA appropriation be used "for expenses of participating in a manned lunar landing to be carried out jointly by the United States and any other country without the consent of the Congress." (*CR*, 8/13/64, 18672–81)

- Rep. Emilio Q. Daddario (D.–Conn.) spoke in the House of recent report by Subcommittee on Science, Research, and Development which was a preliminary investigation of how existing Congressional resources of scientific and technical information can be made more effective. ". . . The critical need for Congress to study and evaluate its uses of scientific information cannot be minimized. It is imperative that we be prepared to assemble the necessary information for Congress to make complex policy decisions in order to meet national goals. As I have indicated we are more concerned with the policy of the decision than its technical aspects. In making these decisions the translation and conversion of scientific and technical information into policy terms is essential. . . ." (*CR*, 8/13/64, 18753)

- ComSatCorp announced in its first report to shareholders that there were more than 130,000 public stockholders in the Corporation. President Joseph Charyk said: "It is good to be able to report that the foundations of the Corporation have been solidly placed—in technical preparations, in the recruitment of key personnel, in the development of a program to create a commercial communications satellite system, and in other ways.

"It gives even greater satisfaction to report that the program to create a satellite system is under way and on schedule and that the documents setting upon the arrangements for foreign participation in the system have recently been initialed by the United States Government, the Corporation, and a number of foreign governments and telecommunication entities." (ComSatCorp Release)

August 14: X–15 No. 2 made its second flight since major modifications were made on the aircraft, with Major Robert A. Rushworth (USAF) at the controls. Flight was part of series of flight requalification tests being conducted on the modified aircraft, which attained maximum speed of 3,580 mph (mach 5.29) and altitude of 99,000 ft. The nose wheel popped out as the X–15 was flying 3,068 mph at 88,000-ft. altitude, but the wheel performed normally when the aircraft landed after nine-minute flight at Rogers Dry Lake. (NASA X–15 Proj. Off.; FRC Release; AP, *Milwaukee Journal*, 8/15/64)

- Two satellites were orbited with single Atlas-Agena D booster, launched by USAF from WTR. (HHN–48)

August 14: Senate agreed to report of Conference Committee on H.R. 11296 and cleared for the White House the Independent Offices Appropriation, including $5,250,000,000 for NASA in FY 1965. (NASA LAR III/159)

- NASA announced Syncom C communications satellite would be assisted into proper orbit later this month by its sister satellite Syncom II, which would provide a reliable relay point in space for communications between ground stations and NASA Goddard Space Flight Center. NASA Syncom Project Officer H. N. Stafford said: "As far as we know, this is the first time one satellite has been used to help in the launch of another. Tiros satellites have been used to help predict weather before manned space flights. The Relay satellite carried pictures of the Cooper flight. But this appears to be the first case of direct use of a satellite in an operational role during a launch." (NASA Release 64–204)

- U.S.S.R. launched COSMOS XXXVII into earth orbit: 300-km. (186.42-mi.) apogee, 205-km. (127.39-mi.) perigee, 89.45-min. period, and 65° angle of inclination to the equator. Tass said the satellite contained scientific equipment intended for carrying on space research according to program announced March 16, 1962. (Tass, *Krasnaya Zvezda*, 8/15/64, 1, ATSS–T Trans.)

- NASA announced selection of the name "Pegasus" for its micrometeoroid detection satellite, to be launched into orbit with Saturn I rocket booster. Named for the winged horse of ancient mythology, the satellite has a wing-like panel 96-ft. long and 14-ft. wide which would record and transmit the penetrations of meteoroids. (NASA Release 64–203)

- ComSatCorp petitioned FCC for rule to give it exclusive eligibility for ownership and operation of the initial ground stations for the global commercial comsat network. Under Communications Satellite Act of 1962, FCC must decide whether ComSatCorp or other commercial communications carrier, or both, shall build and operate the ground stations. (ComSatCorp Release)

- DOD announced award of $2,750,000 initial increment to Martin-Marietta Corp.'s Space Systems Div. on a letter contract for work on an advanced re-entry test program for AFSC Space Systems Div. Press sources identified the program as Start (Spacecraft Technology And Re-entry Test), an expansion of USAF Asset re-entry program, and said this contract was just one of 10–14 areas which had been identified in the program. (DOD Release 596–64; *M&R*, 8/24/64, 10)

- Moscow Radio quoted Cosmonauts Andrian Nikol'ayev and Pavel Popovich as saying their "standbys" were ready for space flight and that their scientists were "planning new experiments in space exploration." (*Wash. Post*, 8/14/64)

- Sen. Hubert Humphrey (D.–Minn.) discussed in the Senate H.R. 11611, act to establish National Commission on Technology, Automation, and Economic Progress. He said the legislation would "enable the essential fact-gathering" about technological change to be made. He continued: "I have long been interested in legislation of this kind. I have introduced S. 2427 to establish a Hoover-type Commission to work broadly in the area of investigating the impact of automation, technology, and employment. I have also joined in the sponsorship of S. 2274 which focuses primarily on problems in the defense industry. . . ." (*CR*, 8/14/64, 19018)

August 15: In memo to heads of Executive Departments and Agencies, President Johnson said that in the past, U.S. had not done enough to help international organizations "secure the services of highly qualified men and women from private life and from government agencies. . . ." The President presented ways in which all Executive Departments and Agencies should take "affirmative and continuing steps" to assist international organizations in gaining access "to talented citizens of this country who are qualified for positions in the international agencies." (White House Memo, 8/15/64)

- In White House press conference, President Johnson announced Secretary of Defense McNamara had approved "a program for the development of an aircraft designed specifically for air support of counter-insurgency, and limited war operations, and the Department of the Navy to contract for the building of seven prototype aircraft at a cost of about $18 million.

  "This counter-insurgency (COIN) aircraft will be an airborne equivalent of the 'jeep.' It will be able to perform peacetime emergency functions such as disaster relief, medical missions or riot control, as well as military missions to include light armed reconnaissance, helicopter escort and attack, and support of ground troops." (Transcript, *Wash. Post,* 8/16/64)

- Patent for manned space station design was awarded to its inventor Emanuel Schnitzer, technical assistant at NASA Manned Spacecraft Center. The space station was designed to erect itself in space after being lofted into orbit. Crewmen could be sent up in an auxiliary spacecraft or launched to it later in a ballistic "ferry." Until two weeks ago, Schnitzer was at NASA Hq. in Advanced Manned Missions Program Office. (Justice, *Houston Post,* 8/15/64; NYTS, Cleveland *Plain Dealer,* 8/15/64)

- Astronauts' contracts with Field Enterprises Educational Corp. and Time Inc. were discussed in *Editor and Publisher* editorial, which stated that the "principle of letting the astronauts sell their stories is all wrong." Before granting personal interview, editorial said, "astronaut must get permission in writing from Field and Time which will grant approval 'in an instance where the acceptance of such outside offers will not materially affect the value of any of the rights granted.' . . ." It concluded by saying that reporters could not interview these "national figures" unless two "competitors say it's okay." (*E&P,* 8/15/64)

August 15–16: Coordinated series of four NASA sounding rocket launches was conducted at Wallops Island, Va.; Ascension Island; and Kronogård, Sweden. A Nike-Cajun was launched from each range with exploding grenade payloads to provide temperature, wind, density, and pressure data. Peak altitudes: rocket launched from Wallops, 75.9 mi.; from Ascension, 77 mi.; and from Kronogård, 84 mi. The fourth experiment was a Nike-Apache launched to 74.3-mi. altitude with payload to collect micrometeoroid particles during a weak noctilucent cloud display above Sweden. Meteorological Institute of Stockholm Univ. participated in in the Kronogård experiments. (NASA Rpts. SRL)

August 16: President Johnson announced he was sending a set of lunar photographs taken by RANGER VII to leaders of 110 nations. "The success of the Ranger 7 exploration has been greeted with enthusiasm and interest around the world. Men of all nations recognize this is one of the greatest extensions of human knowledge about the lunar surface to occur in many centuries.

"The American people can be justly proud. We started behind in the space age. But we placed our trust in our open society and free system. Our achievements in peaceful exploration of space give us all cause to be proud, grateful and confident....

"More than 60 countries now work together voluntarily in this pursuit. We continue to hope that the extent of such international cooperation will be enlarged and that all nations will join through the United Nations to place the peaceful realms of space off limits to the designs of aggressors on earth.

"I am sure the American people will continue to support what is necessary to assure leadership. In return, they will rightfully expect that our program hold to the orderly and responsible course which has brought such outstanding success in such a short period." (Spivak, UPI, *Wash. Post*, 8/17/64)

August 16: Scientists at AF Cambridge Research Laboratories discovered definite relationship between phases of moon and cloudiness around the earth, OAR announced. Statistics covering 21,184 days from 1905 to 1962 indicated sunshine was slightly below seasonal averages for a few days following each new moon and slightly above average between last quarter and the new moon. (*S.F. Chronicle*, 8/17/64; *SBD*, 8/18/64, 241)

- Analyzing the U.S. aerospace industry Richard Rutter said in *New York Times:* "There have been some reductions in employment here and there and some production cut-backs. On the whole, however, the aerospace-defense industry remains not only very healthy but still growing....

  "Change, of course, has been the hallmark of the aerospace industry almost since its very beginning. This characteristic continues. At present, for instance, there is some shift in production and procurement from aircraft and missiles to space vehicles and related space projects. This is not yet a sharp trend, but it may be indicative of the industry's future course...." (Rutter, *NYT*, 8/16/64, F1)

- USAF reported experiments to develop chemical to be spread on aircraft and spacecraft windows for protection of pilots' eyes against a nuclear explosion's damaging flash showed promising results. (AP, *NYT*, 8/16/64, 92)

- Stuart G. Tipton, head of Air Transport Association of America (ATA), was quoted as advocating development of U.S. supersonic transport "because in and of itself, this new transportation tool is good for the United States.... The advance here is speed and speed in transportation is good in itself. Time, after all, is our one irreplaceable resource....

  "Every time that an airline or airliners have offered to the public a faster trip between two points, the results have been clear and positive.

  "When an airline ordered an aircraft which would cut the flying time from New York to Chicago from 10 hours to eight or five to two, that airline got the business...." (Clark, *NYT*, 8/16/64, 94)

August 17: Two NASA sounding rockets were launched from Kronogård, Sweden, in early morning coordinated launchings. Nike-Apache with micrometeoroid-particle-collector payload reached 77-mi. altitude in conditions of moderate noctilucent cloud display. The payload impacted in a lake and was recovered safely. USAF Cambridge Research Laboratories and Stockholm Univ.'s Meteorological Institute cooperated in the experiment. Twenty minutes later, Nike-Cajun was launched with grenade payload to 80-mi. altitude, providing good data on atmospheric temperature, wind, air density, and pressure. Stockholm Univ.'s Meteorological Institute cooperated with NASA Goddard Space Flight Center in this experiment. (NASA Rpt. SRL)

August 17: NASA Nike-Cajun sounding rockets launched within 10 min. of each other to provide data on evening atmospheric temperatures, winds, densities, and pressures from Ft. Churchill, Canada, and Wallops Island, Va., using explosive-grenade technique. Ft. Churchill rocket reached 77.5-mi. altitude, ejecting and detonating all 12 grenades. Wallops Island rocket reached 74.9-mi. altitude, ejecting and detonating all but the ninth grenade. (NASA Rpts. SRL)

- NASA launched Nike-Cajun sounding rocket from Ascension Island to peak altitude of 75.8-mi., in flight to measure wind and temperature and derive density and pressure from 12 grenade explosions during rocket ascent. All grenades ejected and detonated as programed, and ground microphones recorded the sounds. (NASA Rpt. SRL)

- Lightning damaged certain electrical systems in Titan II launch vehicle and ground checkout equipment at Cape Kennedy, where the booster rocket was being prepared for launching of unmanned Gemini spacecraft (GT-2). NASA spokesman said: "The over-all effect of this condition may result in a delay of the first manned flight [GT-3] into next year." (*Houston Post*, 8/19/64; UPI, *NYT*, 8/19/64, 15)

- U.S.S.R. was developing an "aerospace craft" capable of taking off and landing on a conventional air strip as well as flight in outer space, Chief Air Marshal Konstantin Vershinin announced in Moscow. "Our generation will definitely witness the flight of such an aerial-spacecraft," Vershinin said. (UPI, *NYT*, 8/18/64, 17)

- Numerous reports of an "unidentified flying object" observed in the skies over northeastern Ohio early in the morning were recorded, but R. A. Thompson, Assistant Director of the Sohio Tracking Station, identified the UFO as ECHO I. (Cleveland *Plain Dealer*, 8/18/64)

- Plans for military communications satellite system were described in *Aviation Week & Space Technology*, which reported that latest DOD plan for interim military comsat network called for two or three Titan III-C boosters to place 16–24 satellites into near-equatorial orbit. First launch was scheduled for February 1966. (*Av. Wk.*, 8/17/64, 19)

- USAF was reported to be interested in possible use of air-cushion vehicles (similar to "hovercraft") for transporting heavy missiles and space boosters over land, and scientists from Republic Aviation Corp. were planning presentation to USAF on this adaptation of the transporters. (Getler, *M&R*, 8/17/64, 18)

- Dr. Winston R. Markey, Associate Professor of MIT Dept. of Aeronautics and Astronautics and Director, MIT Experimental Astronomy Laboratory, was named to succeed Dr. Robert W. Buchheim as Chief Scientist of USAF. (DOD Release 600–64)

August 18: NASA launched re-entry heating experiment aboard Scout booster at 2:06 a.m. EDT from Wallops Island, Va., in test of low-density charring ablator materials' ability to withstand intense heat during spacecraft re-entry into earth's atmosphere. In addition to the four stages of the Scout launch vehicle, 17-in. spherical rocket was attached to the payload as velocity package. First two Scout stages lofted remainder of the vehicle to altitude of about 130 mi.; then the vehicle arced over in descending path and remaining stages fired the payload back into the atmosphere, attaining maximum speed of about 19,500 mph. Payload impacted about 1,265 mi. downrange after 11-min. flight. No attempt was made to recover it. Experiment was devised by NASA Langley Research Center scientists. (Wallops Release 64–62; NASA Release 64–202)

August 18: Results of five "Sammy" experiments sent aloft with NASA Nike-Apache sounding rockets were announced in Houston by Rice Univ. team of experimenters, which assembled the Sammy payloads and interpreted the data. After studying data, scientists concluded that airglow and aurora were of different origin. Data from experiments probing aurora, launched last March from Fort Churchill, showed that aurora originated when electron particles from space bombarded oxygen and nitrogen atoms in the upper atmosphere. However, data from two experiments launched from Wallops Island, Va., to probe airglow showed "there weren't enough particles around to account for airglow" by electron bombardment. (AP, Wash. *Eve. Star*, 8/19/64, C20; *Houston Post*, 8/19/64)

- U.S.S.R. launched into orbit three earth satellites—COSMOS XXXVIII, COSMOS XXXIX, and COSMOS XL—with a single booster rocket "of a new type." Tass said the three satellites were orbiting in close orbits with initial parameters as follows: 876-km. (544-mi.) apogee, 210-km. (130-mi.) perigee, 95.2-min. period, and 56.16° angle of inclination to the equator. (Tass, *Krasnaya Zvezda*, 8/19/64, ATSS–T Trans.)

- Addressing National Space Club in Washington, ARPA Director Dr. R. L. Sproull said the nuclear detection satellites orbited last month were operating "without a hitch." He described the satellites as "experimental", said they were being used "to determine the most appropriate sensors and the most attractive modes of processing information from them." He said the success of first launch (October 1963) enabled ARPA to revise the program with fewer launches. The satellites were still operating "excellently" and providing information on radiation. (*SBD*, 8/19/64, 246; NSC Release).

- Thomas M. Canning, chief of Hypersonic Free-Flight Branch of NASA Ames Research Center, described aerodynamic shape being studied for spacecraft of the future, at Conference on the Role of Simulation in Space Technology, VPI, Blacksburg, Va. Canning said laboratory tests were being made of high-angle (wing-shaped craft) as possible optimum design for manned spacecraft entering atmosphere at speeds ranging from 50,000–55,000 fps returning from interplanetary journeys. (AP, *Wash. Post*, 8/19/64)

- Professor Verner E. Suomi of Univ. of Wisconsin was appointed to the newly created post of Chief Scientist of the U.S. Weather Bureau. (Commerce Dept. Release WB 64–15)

August 19: NASA's SYNCOM III synchronous-orbit communications satellite was launched into preliminary orbit high above the earth by a TAD (Thrust-Augmented Delta) launch vehicle, its premiere satellite launching. Preliminary orbit: 23,675-mi. apogee, 695-mi. perigee, 11 hrs. 35 min. period, and 16° inclination to the equator. Four hours after launch from Cape Kennedy, the satellite received and sent back a test recording of music of "The Star Spangled Banner" as well as test voice and teletype message. Further maneuvers during the next 10 days would move SYNCOM III into a synchronous position over the Pacific, where it would be able to communicate with surface stations at Clark AFB in the Philippines, at Guam, and at Camp Roberts, Calif. Work was underway to make it possible to transmit the 1964 Olympic Games from Tokyo via SYNCOM III in October. (NASA Release 64-193; NASA Transcript of Press Conference; AP, *NYT*, 8/20/64, 51)

August 19: NASA GSFC launched 201-ft.-diameter balloon carrying 160-lb. experiment package from Moberly, Mo., at sunrise. The balloon flew at altitude of about 24 mi. over 300-mi. course, ejecting its payload on signal about 12 hours after launch. The experiment package contained instruments to record and store data on cosmic rays in the earth's magnetic field, data which would be used to help map radiation fields around the earth and in space. Experiment payload was designed and built by NASA Goddard Space Flight Center; the balloon was launched under ONR's "Skyhook" balloon flight program. (GSFC Release G-22-64; AP, *NYT*, 8/19/64; *K.C. Times*, 8/20/64)

- Rep. Peter W. Rodino (D.-N.J.) said on House floor: ". . . Hardly any problem seems beyond solution by our scientists and engineers. We marvel at our achievement of landing a vehicle on the moon, transmitting pictures of its progress to the very instant of impact. Yet, we are doing practically nothing to solve a major air safety problem, or, worse yet, to discover its cause. I refer to the increasingly serious jet age problem of clear air turbulence—CAT—a danger to every passenger flying commercial or military transport aircraft. . . .

  "Our scientific and industrial communities must be given adequate funds promptly to initiate work on the problem; and that can best be done through a coordinated research effort jointly undertaken by Government and private industry. . . .

  "Our failure to implement without delay a national effort to develop an effective CAT detection and avoidance system may—in fact, probably will—cost the lives of many future air passengers. With every gain in air traffic, with every increase in aircraft speed, with every additional mile flown by every additional plane, the problem becomes more serious. . . ." He introduced into the *Record* an article on CAT by James A. Fusca in *Space Aeronautics*, August 1964. (*CR*, 8/19/64, 19715-20)

- Western Devices, Inc., was awarded $1,615,111 NASA contract for work at Launch Complex 39, John F. Kennedy Space Center Merritt Island Launch Area. Contract was awarded by Army Corps of Engineers. (DOD Release 605-64)

- Thomas D. Morris, Assistant Secretary of Defense (I&L), advised Army, Navy, Air Force, and DSA that it was DOD policy "to provide quality and reliability assurance services to NASA in a manner fully responsive to NASA's technical requirements." Memo clarified use of NASA Quality Status Stamps by DOD personnel and NPC 200-1A as the document establishing NASA's quality assurance requirements. (NASA Proj. Off.)

- Communications Satellite Corp. stock reached $48, highest price since it went public June 2 at $20. (Hammer, *NYT*, 8/20/64, 37)

- Cal Tech announced new superconducting alloy of gold and germanium was made by a new ultra-fast cooling technique that opened the way toward development of other such alloys. (SciServ., *NYT*, 8/19/64, 13)

- Lockheed-California Co. engineers disclosed concept for a reusable rocket plane to transport 10 passengers between earth and orbiting space station. Prepared in a study for NASA Marshall Space Flight Center, the design concept was a winged, two-stage carrier with two-man crew in each stage and capable of hauling three tons of cargo in addition to the passengers. Lockheed engineers said if the vehicle system development program were begun in 1967, the transport could be operational in 1975. (*Huntsville Times*, 8/19/64)

August 19–22: 50 amateurs competed in national model rocket championship meet, sponsored by National Association of Rocketry, at NASA Wallops Station, Va. Astronaut William A. Anders made trophy presentations on the 22nd. (Wallops Release 64–61; AP, *Wash. Post*, 8/19/64)

August 20: SYNCOM III communications satellite, high above Sumatra in the Indian Ocean, was boosted into near-circular orbit above the equator by its apogee motor. The motor burned for 19.6 sec. at the command of ground station at Salisbury, Australia, which based its command on computer calculations at NASA Goddard Space Flight Center. The satellite was now in orbit of 23,550-mi. apogee; 21,235-mi. perigee; $\frac{1}{4}°$ inclination; and 1,449-min. (24 hr. and 9 min.) period. It was drifting westward at rate of about $3\frac{1}{4}°$ per day. (NASA Press Conf. Transcript; NASA Release 64–193)

- NASA Administrator James E. Webb informed Mayor Edward A. Crane of Cambridge, Mass., that 29-acre site offered by Cambridge had been selected for NASA's new Electronics Research Center. Decision was result of four-month study of 160 proposed sites in the greater Boston area by NASA committee. (NASA Release 64–208)

- NASA announced $4.6-million contract had been awarded by MSC to Space Technology Laboratories for mission analyses and trajectory simulation work in Projects Apollo and Gemini. Contract would be a continuation of Gemini spacecraft analysis project begun in September 1963 and an extension of similar activities in Apollo. (NASA Release 64–206)

- Speaking on the floor of the House, Rep. George P. Miller (D.–Calif.) pointed out:

  "In our preoccupation with our national space program, we often overlook the fact that aeronautical research and development is a very important and productive part of the whole NASA program. . . .

  "I should point out that the technology that has evolved from the X–15 program has made possible the creation of the A–11 and the RS–71, hypersonic aircraft in the 2,000-mile-an-hour range, which offer great potential to our Air Force as versatile, high-flying reconnaissance vehicles. NASA, with the X–15, is contributing significantly to the supersonic transport program through new materials, construction techniques, guidance and control systems, and aerodynamic designs that are available now because the NACA had the foresight and determination to undertake the X–15 program in 1954, in association with the Air Force and the Navy.

  "In 1959, the first flight was made. Since then, the X–15 pilots have rocketed out into near space, where 99 percent of the earth's atmosphere is below them, have reached speeds 6 times the speed of sound that have made the plane glow red hot.

  "Now, an advanced X–15 is ready to reach for new heights and even greater speeds. Our investment in the X–15 program and our confidence in the men and pilots who have made it so successful, have paid off handsomely for the individual taxpayer and for the country. Our world leadership and prestige in aeronautics have been enhanced manifold by the X-15. It has played an indispensable role in making sure that America's place in the search for new knowledge of flight will never be less than it is now." (*CR*, 8/20/64, 20046)

August 20: Dr. Raymond L. Bisplinghoff, NASA Associate Administrator for Advanced Research and Technology, spoke at Conference on the Role of Simulation in Space Technology, at VPI:

"The history of applied science and engineering is really a history of simulation—a history of searching for methods of observing apriori the behavior of engineering systems and of varying the parameters of such systems in order to seek optimum results. . . .

"Our ability to develop the engineering systems of atmospheric and space flight depends in turn upon our ability to simulate in the laboratory and thereby study new concepts as well as systems and their components prior to flight. In general, two kinds of simulation facilities are required. Those which are used to develop new concepts and physical understanding . . . [and] the larger simulation facilities needed for the ad-hoc testing and development of subsystems and systems in preparation for flight. In addition to facilities, however, there is needed a certain level of knowledge of the arts and sciences of simulation. . . ." (Text)

- Delegates from 10 nations and the Vatican formally signed in Washington an agreement for a global communications satellite system. Previously initialed by the negotiators, the agreement designated Communications Satellite Corp. of the U.S. as the partnership's "manager in the design, development, construction, establishment, operation of the space segment." (Ward, Balt. *Sun*, 8/20/64)
- FAA's publication "General Aviation Aircraft Owners Survey—1962" showed more than 20% increase in number of general aviation airplanes since 1957, approximate increase of 32% in hours flown, and 38% in miles flown. (FAA Release T 64-43)

August 21: A basic principle in bearing technology, discovered by NASA Lewis Research Center engineers, was hailed in 53rd Annual Report of the Fafnir Bearing Co. as "carrying bearing technology a step beyond conventional upset forgings." The underlying principle was first reported by the LRC engineers in 1957; now the principle was developed into a manufacturing process. Walter T. Olson, assistant director of Lewis, said this was one instance where results from exploratory research had been adapted to production techniques, an important aim of NASA. (LRC Release 64-76)

- NASA announced revision of its patent policy and regulations for waivers of commercial rights to inventions resulting from space program work. Effective September 28, the revised patent waiver and related procurement regulations implemented Presidential patent policy statement of last October which established guidelines for a more unified approach among Federal agencies regarding allocation of patent rights to inventions produced in performance of Government contracts. The new regulations also reflected more than three years of intensive study, including public hearings, on ways to improve NASA patent policies. NASA officials anticipated that they would encourage and accelerate contractors' efforts to identify and report technological advances, thereby increasing the flow and dissemination of technical information derived from the space program. (NASA Release 64-209)

August 21: USAF announced it had launched Thor-Agena D satellite booster from Vandenberg AFB. (AP, 8/21/64)

August 22: SYNCOM III was in near equatorial orbit over the Indian Ocean, with mean altitude of 22,395 mi. and $\frac{1}{4}°$ inclination to the equator, while drifting westward at the rate of about 3° per day. On ground command, the satellite ejected two bursts of steam, correcting its position to 21,890 mi. mean altitude (22,540-mi. apogee and 21,240-mi perigee) and inclination of less than 1° to the equator. It then began drifting eastward; the drift would be halted by onboard control jets, fired by ground command, once SYNCOM III reached the International Date Line. (Hines, Wash. *Sun. Star*, 8/23/64; NASA Release 64–193 & Press Conf. Transcript)

- U.S.S.R. launched COSMOS XLI, COSMOS XLII, and COSMOS XLIII earth satellites into orbit, the last two with a single rocket booster. COSMOS XLI was in initial orbit of 39,855-km. (24,775-mi.) apogee, 394-km. (245-mi.) perigee, 11 hr. 55 min. period, and 64° inclination to equator. COSMOS XLII and XLIII were in initial orbit of: 1,099-km. (683-mi.) apogee, 232-km. (144-mi.) perigee, 97.8-min. period, and 49° inclination to equator. (*Izvestia*, 8/25/64, 6, ATSS–T Trans.)

August 23: Democratic Party Platform completed, containing space plank which stated:

"In four vigorous years we have moved to the forefront of space exploration. The United States must never again settle for second place in the race for tomorrow's frontiers.

"1. We will continue the rapid development of space technology for peaceful uses.

"2. We will encourage private industry to increase its efforts in space research.

"3. We will continue to ensure that any race in space is won for freedom and for peace." (Text, AP, Balt. *Sun*, 8/24/64)

- In videotaped interview broadcast on NBC–TV, Mrs. Lyndon B. Johnson was asked to name those accomplishments of her husband of which she was most proud. The First Lady named, in the following order, the President's work on behalf of a TVA-type project for Texas in the 1930's, the National Aeronautics and Space Act of 1958, and the Civil Rights Act of 1957. (NASA Memo)

August 24: MSC Public Affairs Officer Paul Haney said it was doubtful that there would be a manned space flight this year. The two-man Project Gemini flight (GT–3) had previously been planned for late 1964. However, lightning damage at Cape Kennedy where Titan II launch vehicle was being prepared for GT–2 unmanned flight would cause minimum delay of three weeks in GT–2, Haney said. Length of time GT–3 would be delayed would depend upon how long GT–2 would be delayed. (UPI, *NYT*, 8/25/64; Maloney, *Houston Post*, 8/25/64)

- SYNCOM III might be in position to relay television pictures from Japan as soon as the first week in September, NASA said. The satellite was drifting eastward on the equator at rate of about 483 miles per day. (AP, Balt. *Sun*, 8/25/64)

- NASA announced it would purchase 102 additional J–2 rocket engines for Saturn IB and Saturn V launch vehicles at cost of about $165 million. Contract would be negotiated by NASA Marshall Space Flight Center with Rocketdyne Div. of North American Aviation, Inc. (NASA Release 64–211)

August 24: NASA Goddard Space Flight Center announced Sylvania Electronics Systems-East had been selected to negotiate $1.8 million contract for transportable ground station ("Advanced Technological Satellite Transportable Ground Station"). Sylvania would design, fabricate, install, test, and check out the system, which featured two extremely low noise amplifiers: one a maser and the other a parametric amplifier. These amplifiers operate at temperatures in the range of −440° F and prevent spacecraft signals being overshadowed by heat-induced noise generated inside the receiver. (GSFC Release G-23-64)

- NASA Associate Administrator Dr. Robert C. Seamans, Jr., said in an address before Veterans of Foreign Wars, Cleveland:

  "Today, because of vigorous United States space effort, the early Soviet leadership has been largely overcome. We now must demonstrate our determination to drive ahead toward unchallenged and enduring pre-eminence in space. . . .

  "To be secure in the world today a great nation must maintain a posture of strength. . . .

  "Space research and national security are also intertwined because of the potential uses to which the fruits of our research can be put. Because the consequences of nuclear war are potentially so disastrous for the human race, economic competition has assumed increasing importance as a measure of the strength of competing ideologies. Moreover, scientific and technical research and development have become the basic elements in this competition. . . .

  ". . . for nations which are today still floundering amidst conditions less tolerable than those of Colonial America, great hope for more rapid progress lies in the new ideas, techniques and methods which are arising from our research and development in space.

  "If we can employ the scientific knowledge and technical ideas developed in our space program to help the impoverished nations help themselves, and thus reduce their requirement for direct financial aid, how much better for them and for ourselves.

  "Our greatest hope for national security, of course, lies in achieving in the world a degree of international amity which can lead to enduring peace. Here, again, the exploration of space offers great promise, for in the acceptance of this greatest of all challenges, man can join in shooting rockets into space, instead of shooting them at each other.

  "Our national leaders have recognized that if we are to arrive at an understanding with the Soviet Union which will permit the exploration of space to become a mutual project, for the benefit of all mankind, we must maintain a driving national effort in space to insure a technological balance that will make this possible. . . ." (Text)

- Dr. Orr E. Reynolds, Director of NASA Biosciences Div., told *Missiles and Rockets* that it probably would be 1971 or 1973 before U.S. landed Automated Biological Laboratory (ABL) on Mars: "I think it would be awfully optimistic to think that we could fly a fully integrated laboratory system to the Martian surface in 1969. I would like to hope such a mission would be feasible in 1971, and the 1973 launch window would also provide a good opportunity." Other unnamed NASA officials were said to have considered launch of ABL with Saturn V booster in 1971 as improbable, citing the high cost and "problems with sterilization methods which have to be solved." ABL missions would be part of the Voyager planetary program. (*M&R*, 8/24/64, 15)

August 24: FAA announced purchase of runway traction device to enable engineers to predict stopping distances for aircraft on wet or slippery runways. Designed and developed by J. I. Thompson Co., the equipment consisted of hydraulically operated mechanism mounted on aircraft tire and towed behind specially equipped truck; special controls permitted technicians to duplicate actual speed, skid, and braking effects of aircraft landing on slick runways. (FAA Release T 64-44)

August 24-28: International Council of the Aeronautical Sciences held its fourth Congress in Paris. At the Congress, Harold B. Finger, Manager of AEC–NASA Space Nuclear Propulsion Office and Director of NASA Nuclear Systems and Space Power, made presentation on nuclear space propulsion systems:

". . . Space propulsion using nuclear energy offers a capability for accomplishment of high energy increment, high payload missions in space beyond the capability of the chemical combustion propulsion systems when considering practical operating limitations. Work now underway in the United States indicates that nuclear rockets can be anticipated for earliest use in the space program. Reactor tests being conducted during this year should provide a firm technical basis for system development. Electric propulsion using the nuclear reactor energy source offers promising storable propellant performance if light-weight, long-life power supplies can be developed. Technology investigations are now underway to evaluate the feasibility of achieving the required performance. This work will simultaneously provide the information that is required to provide large amounts of electric power for non-propulsive purposes in space. Beyond these systems, a host of new and advanced concepts have been proposed. These are not well enough defined or evaluated to assure that their high performance potential can actually be achieved. Some research work is underway on these systems." (Text)

August 25: NASA launched EXPLORER XX scientific satellite into orbit from Western Test Range, Calif., using Scout booster rocket. Initial orbit: 625-mi. apogee, 540-mi. perigee, 104-min. period, and 80° inclination to the equator. The ionosphere Explorer satellite, nicknamed "Topsi," immediately began mapping the ionosphere and sending data back to earth. The findings of this satellite would be compared with those obtained by ALOUETTE, the Canadian-developed "topside sounder." (NASA Release 64-207; *Wash. Post,* 8/26/64; AP, *Chic. Trib.,* 8/26/64; *Av. Wk.,* 8/31/64, 25)

- Team of NASA space scientists and engineers, including Astronaut Walter Cunningham outfitted in a pressurized spacesuit, began week-long mobility tests at McKenzie Pass lava area in central Oregon's Cascade Mountains, where they investigated man's ability to perform on surfaces thought to resemble that of the moon. (AP, *Balt. Sun,* 8/26/64)

- About 1,600 astronomers representing every country except Communist China opened week-long assembly of the International Astronomical Union at Hamburg Univ., Hamburg, Germany. Soviet astronomer Viktor A. Ambartsumian, president of IAU, said in address that the accelerated rate of discoveries in 20th century astronomy would continue to challenge astronomers for a long time. He cited two developments since the last IAU assembly three years ago which deserved special attention: study of x-ray spectrum with use of rockets to carry instruments beyond the earth's atmosphere; and discovery that at least some galaxies have cores that seem to have tremendous mass and a nature unlike stars. (Sullivan, *NYT*, 8/26/64)

August 25: USAF XB–70A aircraft made fourth of series of taxi runs at Palmdale, Calif., reaching for first time the planned ground speed of 140 mph. Flight tests of the 185-ft. experimental bomber would be made after one or two more taxi checkouts down the runway. (AP, Balt. *Sun*, 8/31/64; *Av. Wk.*, 8/31/64, 25)

August 25–27: Highlights of Democratic National Convention in Atlantic City, which selected Lyndon Johnson and Hubert Humphrey as Presidential and Vice Presidential candidates, were broadcast to 15 countries in western Europe's Eurovision network via TELSTAR II communications satellite. (AP, *Wash. Post*, 8/19/64)

August 26: X–15 No. 3 was flown by John B. McKay (NASA) to altitude of 87,000 ft. and speed of 3,716 mph (mach 5.58), in flight near Edwards, Calif. During the flight, heat transfer and noise level measurements were made on the aircraft's skin. (NASA X–15 Proj. Off.; AP, *NYT*, 8/27/64)

- NASA Nike-Apache sounding rocket, launched from Wallops Island, Va., reached 95.3-mi. altitude in satisfactory flight to measure ambient electron density, measure ion densities, and monitor solar ultraviolet radiations. Instrumented payload, provided by NASA Goddard Space Flight Center, contained CW propagation experiment, probe with Gerdien condenser, planar ion trap, and planar secondary emission detector. Another objective of the flight was to test efficiency of the 12.267 mc/s shroud antenna at D-region altitudes; one of the antennas performed unsatisfactorily. (NASA Rpt. SRL)

- Radio Corporation of America announced receiving $27-million contract from NASA to build 19 ground computer systems for checkout and launch of Saturn IB and Saturn V space boosters. (AP, *Wash. Post*, 8/26/64)

August 27: Dr. Robert C. Seamans, Jr., NASA Associate Administrator, visiting NASA Manned Spacecraft Center, said extended manned exploration of the moon "does make considerable sense" as follow-on program to Project Apollo. He said such a program was one of several which NASA would present as possible post-Apollo goals to the President "in approximately a month's time." Other possibilities would include such programs as landing unmanned spacecraft on other planets. (Maloney, *Houston Post*, 8/28/64)

- Venus was found to be rotating backward with a period of between 248 and 258 earth days, Dr. Gordon H. Pettingill of Cornell Univ. reported at IAU assembly in Hamburg. The new value was a refinement of one made in late 1962 by JPL studies, which indicated a rotation of between 239 and 293 (266 mean period) and was obtained by radarastronomers using new 1,000-ft.-diameter antenna at Arecibo, Puerto Rico. (Hines, *Wash Eve. Star*, 8/27/64; Sullivan, *NYT*, 8/28/64)

- Hurricane Cleo battered Cape Kennedy rocket launching site with gale winds up to 65 mph. Technicians dismantled second stage of Titan II for GT-2 Gemini flight and stored it in hangar, lashed first stage of the Titan II to the pad. No damage was done to the Titan II or other lashed-down rockets. (*Wash. Post*, 8/28/64; AP, Balt. *Sun*, 8/29/64)

August 27: NASC Executive Secretary Dr. Edward C. Welsh said in address at dedication of North American Aviation, Inc., Science Center, James Howard Kindelberger Memorial Laboratories, Thousand Oaks, Calif.:

"The whole complex activity of exploring space is important not only because the many private and governmental organizations contributing to space flight are continually uncovering new facts, new products, and new processes which bring immediate benefits to mankind. These benefits are only the beginning. Looking beyond them, we see vast unknown frontiers opening up, with unguessable discoveries yet to be made. Step by step we are conquering the hostile environment of space. We are accumulating new knowledge about the heavenly bodies, and disproving old beliefs about them.

"I would emphasize several points about the space program:

"First, it is here to stay;

"Second, it promises a profitable return on investments in basic research;

"Third, it will be an expanding activity, with no finite limits to its potential;

"Fourth, it will enhance our national security, stimulate our accumulation of knowledge, improve our standard of living, and further the chances of world peace. . . ." (Text)

August 28: NASA's NIMBUS I meteorological satellite was launched into polar orbit from Western Test Range with Thor-Agena B launch vehicle, and on its sixth orbit began transmitting weather photographs which were called "the best pictures ever" from space. Orbit of the 830-lb. satellite had been planned as circular 575-mi. altitude orbit; Agena B engines burned for only .84 sec. instead of the 3.8 sec. expected for injection burn, so the satellite was orbiting with 578-mi. apogee, 252-mi. perigee, 98-min. period, and 81° inclination to the equator. NIMBUS I, unlike the Tiros meteorological satellites, was earth-oriented and stabilized in all three axes. It contained Advanced Vidicon Camera System and Automatic Picture Transmission System, both of which operate during daylight portion of each orbit; High Resolution Infrared Radiometer, first to be flown on NASA meteorological satellite, provided first high resolution night-time cloud-cover pictures and cloud-top temperatures taken from a satellite. About 62 APT stations, including those in 12 foreign countries and 4 owned privately, participated in NIMBUS I picture reception. The APT system was designed for Nimbus and was first tested—successfully—on TIROS VIII launched last December. Rep. George P. Miller (D.–Calif.) said at NIMBUS I news conference: ". . . This much is certain: We are way ahead of the rest of the world. I am certain that if nothing came out of the space effort but the communications satellite, the weather satellite, and the navigational satellites, which will soon be flying, in a limited number of years, it would more than pay back all of the investment that has been made in space." (NASA Release 64–195; GSFC Press Briefing Transcript; Wash. *Eve. Star,* 8/29/64; NASA R&SC Div.)

- Firing of hydrogen peroxide gas jets aboard SYNCOM III communications satellite changed its drift rate from 7° east per day to 3.3° per day. (NASA Release 64–217)

- NASA presented RANGER VII press briefing, essentially the same presentation as that planned for the IAU three days later, in which refined flight data were announced. RANGER VII impacted on the moon within 19 sec. and 6 mi. of the planned time and target. It landed in one of the rays from the crater Tycho; Dr. Eugene M. Shoemaker and Dr. Gerard P. Kuiper expressed opinion that the crater rays were lava flows. Dr. Kuiper stated the impact area was not typical of the lunar surface. Of the RANGER VII lunar photographs, Dr. Kuiper revealed that the photographs resolved objects down to 10–15 inches, and the resolution was considered 1,000 times better than the very best earth-based observation and 5,000 times better than routine earth-based observation. 200 more of the lunar photographs were released.

On the question of the nature of the lunar surface, no firm conclusion was drawn. Dr. Kuiper, who estimated erosion on the moon to be about 5 ft., said it was "purely a guess" that there was "fairly solid stuff there." He added that it was not possible to obtain such data as surface bearing strength from the RANGER VII photographs. Dr. Harold Urey, whose paper was read in his absence, also expressed his belief that the photographs did not answer the question of surface strength; however, he said that a "spongy," "fairy castle structure" could be prevalent many feet down. Dr. Shoemaker, who expressed view that moon loses more material than it gains, suggested erosion could be on the order of 50 ft. He said U.S. Geological Survey was converting the photographic data into topographic profiles, and the results were showing that moon's surface was "smoother than I had dared hope." He said average slope in the area being studied did not exceed 10° and pointed out that this was reassuring for Project Apollo, whose Lem spacecraft had been designed to land safely on slope of 15°. (NASA Press Conf.; Finney, *NYT*, 8/29/64; Toth, *Wash. Post*, 8/29/64; Watson, Balt. *Sun*, 8/29/64)

August 28: U.S.S.R. announced launching of COSMOS XLIV earth satellite into the following orbit: 860-km. (534-mi.) apogee, 618-km. (384-mi.) perigee, 99.5-min. period, and 65° inclination to the equator. The satellite was said to be carrying on the Cosmos program announced March 16, 1962. (Tass, *Pravda*, 8/29/64, 1, ATSS-T Trans.)

- NASA Aerobee 150 sounding rocket launched from White Sands, N. Mex., reached 110-mi. altitude in experiment to collect data on celestial x-ray sources. Despite slight underperformance of the rocket, preliminary experimental results were considered good. American Science and Engineering, Inc., provided instrumented payload consisting of four Geiger counters, one photoelectric detector, two scintillation counters, and two star sensors. (NASA Rpt. SRL)

- First Apollo spacecraft preflight acceptance test station was declared operational. Built by GE Apollo Support Dept. with associate contractors Control Data Corp. and Radiation, Inc., the station would be used at North American Aviation's Downey, Calif., facility in Apollo preflight testing. Three others would be located at Grumman Aircraft Engineering, Bethpage, N.Y.; NASA Manned Spacecraft Center, Houston, Tex.; and MSC Florida Operations, Merritt Island, Fla. (NASA Release 64–214)

- Los Alamos Scientific Laboratory successfully completed eight-minute power experiment with the Kiwi–B4E reactor, the eighth and final reactor built for power testing under the Kiwi phase of the NASA–AEC nuclear rocket program. All planned test objectives were met and reactor performance was smooth. Duration of the test was limited by the liquid hydrogen capacity of the Jackass Flats, Nev., facility. (SNPO; *Wash. Post,* 8/29/64)

August 28: Dr. George E. Mueller, NASA Associate Administrator for Manned Space Flight, said at Apollo electronics panel, IEEE Western Electric Show and Convention, Los Angeles:

"The United States is now completing the first decade of the exploration of space. During the last five years major strides have been made in exploring, understanding, and exploiting space. Today we have allocated almost half the resources needed to carry out the greatest adventure in the history of mankind—the manned lunar landing which is the goal of the Apollo Program. This goal could not have been conceived, much less achieved, without the amazing growth of knowledge, techniques, and technology in the field of electronics which has been ever accelerating since the last war. Communications, guidance, power, control, display, all these have passed and surpassed the maximum expectations of a decade ago.

"In the Apollo Program we are developing a set of capabilities which will provide for manned exploration, understanding man's proficiency in space and the vicinity of the earth. This is a capability which can be extended to meet the needs of this Nation for science or technology, for transportation or observation, for defense or commerce.

"Although the end is manned exploration of space, the means, including the production of space vehicles, is being created here on earth. It is based in large measure on the expansion and growth of the electronics industry. It depends on that electronics industry to meet and exceed new goals for reliability and quality as well as performance in their products. . . ." (Text)

August 28–September 7: In conjunction with Cleveland's Parade of Progress exhibit commemorating opening of new Cleveland Convention Center, NASA exhibited Centaur space vehicle, full-scale spacecraft such as Ranger and Mariner, Sert spacecraft, RANGER VII moon pictures, Gemini two-man spacecraft, and other exhibits. Space lectures and motion pictures were presented. NASA Lewis Research Center coordinated the display. (LRC Releases 64–72, –73, –75, –79, and –80)

August 29: Soviet Premier Nikita Khrushchev expressed his pleasure with success of U.S. lunar-photography probe RANGER VII, in letter to President Johnson replying to letter and photographs sent to him by Mr. Johnson. (Tass; Reuters, *NYT,* 8/30/64)

• USAF announced launching four-stage Blue Scout, Jr., rocket combination from Vandenberg AFB, Calif., with undisclosed payload. (UPI, *NYT,* 8/30/64)

August 30: FY 1965 Independent Offices Appropriation Bill containing $5.250 billion NASA appropriation, was signed into law by President Johnson.

NASA Administrator James E. Webb issued statement on the effect of the appropriation, saying that NASA would "maintain the basic momentum and direction of the program as presented to Congress but will make a number of adjustments to meet the Congressional reduction of $195 million, to effect those specific changes directed by Congress, and to take advantage of experience gained from operations conducted during the nine months since the budget was submitted.

"Although these adjustments will affect launch schedules and program milestones, Dr. Dryden, Dr. Seamans, and I are convinced that it is best to make a maximum effort to gain operational leadership in manned space flight and that the manned lunar landing should continue to be targeted within this decade. This will stretch NASA and its contractors to the limit, but we have advised the President that we are not at this time willing to give up this goal which is so challenging and which is so important as a focus for the efforts we must continue to make in all our programs. . . .

"The target date is only five years away. Our assurance that it can be met is less than under President Johnson's budget, but we still hope that with hard work, dedication, continued successes, and continued support we can meet it. We are going to make a hard try.

"This decision does not involve the transfer to manned flight of funds from space science and applications programs or advanced research and technology programs. These programs will require some adjustments, but will not be drastically reduced as would be necessary if funds were reprogrammed to benefit manned space flight.

"In the program for unmanned planetary exploration, the combination of a heavy workload at the Jet Propulsion Laboratory, the short lead time available, and the importance of applying our resources to a major advance beyond the limited Mariner make it unwise to undertake a Mars mission in 1966 with the current Mariner-type spacecraft. The development of a spacecraft with much greater scientific promise for launch to Mars in 1969 is being initiated. . . ." (NASA Release 64–221)

August 30: NASA announced Franklyn W. Phillips would be Assistant Director for Administrative Operations at NASA's new Electronics Research Center, Cambridge, Mass., effective Sept. 1. A veteran NACA–NASA executive, Phillips had been serving as Director of NASA Northeastern Office, which would be absorbed within the new Center. As Assistant Director, Phillips would be responsible for administrative matters, including procurement, personnel affairs, management analysis, administrative support, and technical information. (NASA Release 64–218)

* Library of Congress announced it was creating Science Policy Research Div. within its Legislative Reference Service, to advise Congress and congressional committees on scientific and technological developments. This step was seen as first significant action by Congress to establish advisory apparatus in R&D subject area, a topic of growing concern and debate. (Finney, *NYT*, 8/31/64)

August 31: RANGER VII team of scientists presented lunar photographs to IAU in Hamburg, Germany, and discussed data obtained from them. Dr. Thomas Gold of Cornell Univ., known as an outspoken proponent of the "fairy castle" theory of lunar surface structure, said there was nothing in the RANGER VII photographs that would give him reason to revise his hypothesis. Dr. Gold also disputed hypothesis of Dr. Gerard P. Kuiper, of Univ. of Arizona, who identified what he interpreted as lava flows on the moon (crater rays, essentially); Dr. Gold stated none of the photographs showed evidence of volcanic lava on the moon. (*NYT*, 9/1/64)

* Placing a 100-in. telescope in orbit by 1975–1980 for studying the stars was "not an unreasonable goal," Dr. Leo Goldberg told IAU in Hamburg. The Harvard Univ. astronomer summed up arguments for telescope in space by saying that he could not agree with those "who prefer to look at the universe through a hazy, dirty, semi-opaque atmosphere." Cost of such a project could be justified, he said, only if

expected lifetime of such an instrument was long. However, he cited Soviet and U.S. progress toward space rendezvous techniques and expected increases in rocket booster power, and said: "The future looks terribly exciting." (Sullivan, *NYT*, 9/1/64)

August 31: Labor Dept. said industry employed 885,000 engineers and scientists in 1963, about 4% more than in 1962. 715,000 of the 885,000 were engineers and 170,000 were scientists. (AP, *NYT*, 9/1/64, 22)

During August: NASA Manned Spacecraft Center received five Northrop T–38 aircraft on loan from USAF for use as space-flight readiness trainers to maintain proficiency of astronauts. (*Av. Wk.*, 8/31/64, 25)

- Douglas Aircraft Co. Missile and Space Systems Div. received nine-month study contract from NASA Marshall Space Flight Center to study methods of uprating Saturn IB space vehicle and S–IVB Saturn stage. (*Av. Wk.*, 8/31/64, 23)

- NASA Administrator James E. Webb said in article on NASA and USAF relations in *The Airman:*

    "At a time when possible military missions for manned spacecraft have not yet been clearly defined, the NASA Gemini and Apollo programs and the Air Force MOL program are developing American manned space flight technology and operating experience at a rapid rate.

    "We believe that the progress called for in the Gemini, Apollo, and MOL will enable us to demonstrate clear superiority over the Soviet Union in manned space flight capabilities during this decade.

    "Our accomplishments in these three programs will prepare us much better for whatever further steps may be necessary to meet our national defense needs in space. . . ." (*Airman*, 8/64, 6–11)

- Dr. Ernst Stuhlinger, Director of Research Projects at NASA Marshall Space Flight Center and AIAA Board Member, said in *Astronautics & Aeronautics* editorial:

    "For some years past, it has been asked whether electric propulsion meets a 'requirement' in our space program. It may be well to remember that the airplane, the atomic submarine, the radio, and the first satellite were not developed in fulfillment of 'requirements.' An artificial atmosphere is certainly a requirement for an astronaut in space, and must be developed for that reason. But the astronaut does not go into space because it is his requirement, but because our country has decided to excel in the exploration of outer space with the most appropriate means. Electric propulsion promises to be a very appropriate method of vastly expanding our exploration capability in space beyond that of chemical and nuclear systems. For that reason, its development should be pursued with high priority and great vigor." (*A&A*, 8/64, 14–15)

- B. K. Heusinger of NASA Marshall Space Flight Center said in *Astronautics & Aeronautics* article about present and future improvements in Saturn launch vehicle propulsion:

    "First and foremost, the Saturn–I testing program has already made possible an additional payload capability of approximately 4300 lb. without major redesign efforts. A substantial portion of this gain has come from altered operational and procedural techniques developed during the normal course of static and flight testing. The impact of major redesigns will not be felt, as a result of the lead times involved, until the first of the Saturn–IB vehicles. . . ." (*A&A*, 8/64,/ 20–25)

During August: J. S. Butz, Jr., said in *Air Force and Space Digest* article on the future of aviation:

"Aeronautics in the 1970s and 1980s promises to eclipse anything of the past in both scope and the performance of individual aircraft. Although missile and space systems will continue to furnish strong competition for national support, aviation activity in the decades ahead will be much more extensive and important than it is today. . . .

"The revolutionary effect of the new technology on future aircraft can be best illustrated by discussing two key areas. One is propulsion. The other is materials. Light, more powerful engines, and lighter, stronger structures always have been key design objectives. This is still true. . . ."

He quoted Dr. Alexander H. Flax, Assistant Secretary of the Air Force for Research and Development: " 'It is possible today to begin the development of lightweight gas-turbine engines with a thrust-to-weight ratio of better than 10 to 1.' Dr. Flax points out that installed thrust-to-weight ratios apparently can rise to higher than 20 to 1 on the second generation of future engines.

"Ten years ago such a statement would have got him laughed out of most scientific and engineering meetings. A few people may still argue that the estimate is high. But the main argument now is over how best to achieve this truly stunning performance rather than over the possibility of achieving it." He detailed some of the R&D areas with great potential for future aircraft, then said:

"Technical evidence abounds to prove that an aeronautical revolution is at hand and that the first step in that revolution can have great economic importance. It clearly is time for more resolute action. Britain and France, and possibly the Soviet Union, have approximately the same opportunity as the US to capitalize on what is known now. In western Europe, at least, engine and materials technology is well advanced. . . ." (*AF Mag.*, 8/64)

- Nuclear physicist Dr. Edward Teller criticized DOD for failure to proceed with flight-testing the nuclear-powered Low Altitude Supersonic Vehicle (LASV). Speaking at Air Force Institute of Technology, Wright-Patterson AFB, Dr. Teller said: "For the sake of an economy that amounts to less than 1% of the Air Force budget this has been canceled. I believe this is the biggest mistake we have made since the years following World War 2 when we failed to develop the ICBM. . . ." (*Av. Wk.*, 8/17/64, 65)

- DOD established 22 information analysis centers to supply specific answers to technical questions asked by managerial, scientific, and engineering personnel employed by DOD and by DOD contractors. (*M&R*, 8/10/64, 10; *Missile/Space Daily*, 8/5/64 and 8/6/64)

- AFSC was conducting research with quick-setting plastic to provide "instant landing sites" for V/Stol aircraft. (*Aviation Daily*, 8/10/64)

- Terence Prittie said in *Atlantic Monthly* article that the "Egyptians, with the vital help of German and other foreign scientists and technicians, have been working on a crash program for the production of long- and medium-range rockets at their Factory 333 at Heliopolis, a few miles east of Cairo." He described the three main types of rockets being developed: Al Zafar ("the Victor") single-stage liquid-fuel rocket; Al Kahar ("the Conqueror") single-stage liquid-fuel rocket; and Al Raid ("the Explorer") two-stage rocket. (*Atlantic*, 8/64)

September 1964

September 1: NASA Electronics Research Center was formally activated in Cambridge, Mass., and Dr. Winston E. Kock took the oath of office as Director. Electronics Research Task Group from NASA Hq. and NASA North Eastern Office in Cambridge were absorbed in the new Center to form initial nucleus of 80 personnel. Design and construction of the facility would begin late this year and be completed in four or five years. (NASA Release 64–219; NASA Announcement 64–189; NASA Circulars 320, 321)

* USAF launched Titan III–A space booster from Cape Kennedy, and the first two stages burned perfectly; however, because transtage engines shut down prematurely, the Titan III–A fell slightly short of orbiting its 3,750-lb. dummy lead payload. AFSC's Brig Gen. Joseph Bleymaier said preliminary investigation showed malfunction of onboard helium pressure valve was to blame for the shortened burn time from programed 406 sec. to about 391 sec. He also said the test was 95% successful—all primary objectives and most secondary objectives had been met. (DOD Release 625–64; *M&R*, 9/7/64, 11; *Av. Wk.*, 9/7/64, 32)

* Aerobee 150A sounding rocket launched from Wallops Island, Va., reached 96.5-mi. altitude in successful flight to measure brightness of stars in the ultraviolet. Experiment instrumentation for the NASA rocket was provided by Space Astronomy Laboratory of the Univ. of Wisconsin (NASA Rpt. SRL)

* General William F. McKee (USAF, Ret.) was sworn in as Assistant Administrator for Management Development, NASA. In this position, General McKee would be responsible for review and development of recommendations concerning major management matters and would report directly to NASA Administrator James E. Webb. (NASA Release 64–216; NASA Announcement 64–190)

* MSFC Michoud Operations began mass movement of about 5,000 contractor personnel and supplies and equipment from temporary quarters in the New Orleans area to the Michoud facility. About 4,400 employees would move into the new Engineer and Office Building, and about 600 would move into existing administration and engineering buildings. Being performed after normal working hours, the move was to be completed by Nov. 1. (*Marshall Star*, 9/2/64, 1, 12)

* In honor of RANGER VII, the International Astronomical Union meeting in Hamburg, Germany, designated the region where the lunar probe impacted on the moon as "Mare Cognitum." (Sullivan, *NYT*, 9/2/64, 14)

* USAF Captains Albert R. Crews and Richard E. Lawyer emerged from two-week stay in simulated-space cabin at GE Space Technology Center, Valley Forge, Pa. Officials said test showed that man can perform more tasks in extended space flight than previously supposed. (*Wash. Post*, 9/2/64)

September 1: One-year pilot test for engineering data-retrieval system for use by defense contractors and DOD components began. Test was being conducted to determine feasibility, cost, and user acceptance of mechanized system for interchange of engineering and technical data. It utilized present data-retrieval engineering system file of Army Missile Command and additional specially assembled file to be placed at about 25 defense contractor activities engaged in R&D and production engineering. (DOD Release 636-64)

- U.S.S.R. disclosed it had succeeded in producing first nuclear-reactor system to convert atomic energy directly into electric power, small enough to fit aboard a spacecraft. Report was made to Third International Atoms-for-Peace Conference in Geneva. (L.A. Times-Wash. Post Serv., Louisville *Courier-Journal*, 9/2/64)

September 2: At press conference, NASA Goddard Space Flight Center officials said NIMBUS I meteorological satellite was performing successfully and sending back cloud cover photographs of unexpectedly good quality. First nighttime pictures of cloud cover, made with the satellite's infrared sensor, were comparable in quality and resolution to the TV pictures made by Tiros meteorological satellites. Unplanned orbit—eccentric rather than circular—reduced the satellite's earth coverage from 100% every 24 hours to about 70%. (GSFC Press Conf. Transcript; Finney, *NYT*, 9/3/64, 13)

- NASA's SYNCOM II and SYNCOM III crossed paths at the equator about 22,000 mi. above the Pacific Ocean at about 162° east longitude. SYNCOM III, launched into synchronous equatorial orbit Aug. 19, was drifting toward International Date Line at rate of 3.3° per day; SYNCOM II, launched into orbit July 26, 1963, was moving in figure-8 pattern 33° north and south of the equator. (NASA Release 64-217)

- DOD announced $5,550,012 contract for integrated navigation systems for three Project Apollo tracking ships had been awarded by Naval Bureau of Ships to Sperry Rand Corp. (DOD Release 633-64)

September 3: NASA-modified Arcas sounding rocket, launched at Wallops Island, Va., carried white rat in specially designed, instrumented payload compartment in nose cone of the rocket to 114,000-ft. altitude, where the payload separated from the booster, a parachute deployed, and the container floated down to the surface of the ocean. Package was recovered by helicopter a few minutes later about 20 mi. offshore, and the rat was returned to NASA Wallops Station for observation and analysis. During the flight, biomedical data such as heartbeat and body temperature of the rat were telemetered to ground stations. Launch was part of Bio-Space Technology Training Program. (Wallops Release 64-65)

- Reorientation maneuver was made with SYNCOM III communications satellite, high above the Pacific Ocean, to align the communications antenna atop the spacecraft for optimum coverage of the earth. This positioned the antenna along a north-south axis perpendicular to the orbital plane. Side effect of this maneuver, plus an orbital adjustment made Aug. 31, was to reduce the eastward drift rate from 3.6° per day to less than 3° per day. (NASA Release 64-230)

- X-15 No. 3 flown by Milton O. Thompson (NASA) to altitude of 77,000 ft. and speed of 3,545 mph (mach 5.37), the flight studying heat transfer rates and boundary layer noise on specially installed surfaces on the bottom of the aircraft. (NASA X-15 Proj. Off.; FRC Release)

September 3: Three NASA astronauts began flights in T-33 trainer jets simulating approaches to the lunar surface. Maj. Edwin E. Aldrin, Jr., Capt. David R. Scott, and Elliott M. See, Jr., sent their planes into dives from altitudes of about 15,000 ft. toward the large, rugged lava flow in southern Idaho; cameras in the planes recorded the approaches, and the astronauts made their own observations of the surface. Planes were leveled out at about 3,000-ft. altitude. (*Houston Post*, 9/3/64)

* NASA announced agreement had been reached by Malagasy Republic and U.S. on the relocation and expansion of NASA tracking station on Madagascar. Previously located at Majunga, the station was being relocated near Tananarive and expanded to include instrumentation for Project Gemini monitoring and communicating and for scientific satellite tracking and recording data. (NASA Release 64-222)
* Lockheed Constellation "crashed" near Phoenix, Ariz., in test conducted for FAA by Flight Safety Foundation. The airliner, with 21 passenger "dummies" and several thousand pounds of cameras and measuring instruments onboard, traveled down the runway at 115 mph to crash into barriers of wood, earth, and steel; it slid another 500 ft. and stopped with its fuselage upright but cracked in two places, one wing sheared off, and the other wing twisted. Aircraft remained in such good condition that passengers apparently would have survived the crash, unlike similar test of DC-7 last April, in which the aircraft hit the barriers at 160 mph and was destroyed. (Clark, *NYT*, 9/4/64, 53)

September 4: NASA's OGO I (Orbiting Geophysical Observatory) was launched from Cape Kennedy by Atlas-Agena B rocket booster into highly elliptical orbit (92.721-mi. apogee, 175-mi. perigee, 621½-hr. period, and 31° inclination to the equator). The huge 1,073-lb. satellite carried 20 experiments, inaugurating new "space bus" series of standardized observatories capable of conducting many related space experiments simultaneously in space. However, one of OGOI's long booms, one short boom including the omnidirectional radio antenna did not deploy. This resulted in abnormal operation of the automatic control system, and most of the control gas was used in attempting to lock the satellite into its earth-stabilized orbit. Project officials attributed the satellite's inability to lock onto the earth to the fact that its earth-seeking sensor was obscured by one of the undeployed booms. Scientists decided not to attempt turning on experiments for several days while calculating contingency operations for a spin-stabilized satellite. (NASA Release 64-213; GSFC Release G-26-64; AP, *NYT*, 9/6/64, 26; Hines, *Wash. Eve. Star*, 9/8/64)

* NASA announced signing of nine-month contract extension for Project Apollo spacecraft with North American Aviation's Space and Information Systems Div. The $496-million extension called for five additional command and service modules, three additional flight boilerplate spacecraft, a full-scale mockup, and nine adapter sections to house Lunar Excursion Modules. (NASA Release 64-227)
* NASA issued supplemental list of technology utilization publications, designed to acquaint the public with useful innovations from space research. (NASA Release 64-225; TU List)
* Cornell Univ. astronomer Dr. Thomas Gold said in *Science* article that absence of surface rock outcroppings in RANGER VII close-up lunar

photographs supported his thesis that the moon is covered with dust. Erosion of the moon's surface at rate of one micron (1/25,000 in.) per year would be enough, "over geological time," to smooth out a once-rocky surface. This erosion, caused by cosmic dust and energetic solar particles, would have deposited dust in low areas and gradually filled them. However, Dr. Gold said that even if in the craters and seas the dust were a mile deep, it was no cause for concern; dust on the moon, like that on earth, would not necessarily be loose and powdery at depths below the surface. (Wash. *Eve. Star*, 9/4/64)

September 5: President Johnson announced in his news conference that Polaris A-3 missile "will soon become part of our strategic missile force. The new A-3 will be deployed for the first time aboard the nuclear submarine USS *Daniel Webster* when she begins her first patrol later this month.

"The A-3 missile, which was put into accelerated development in 1964, has a range of 2500 nautical miles, some 1000 nautical miles greater than that of the A-2 and more than double that of the A-1. . . ."

The President also announced that AEC and DOD were "proceeding with the development of a new, high-powered, long-lived reactor, which constitutes a major step forward in nuclear technology. . . .

"Two of these reactors could power an attack aircraft carrier as compared with eight reactors required for the USS *Enterprise*. . . ." (Transcript, *Wash. Post*, 9/6/64)

- Lunar photographs made by RANGER VII and interpretations of their findings were discussed in *Saturday Review* by Science Editor John Lear:

  "The net sum of the evidence here recited is that the Congress and the people have been misled about the true significance of the Ranger VII mission. The pictures sent back home by the ingeniously constructed robot dragonfly contain no more assurance of the safety of a manned landing on the moon than existed before Ranger VII took off. On the contrary, one sure danger has been discovered that had not been suspected before. The apparent disappearance of the supposed danger of impenetrable dust is accompanied by the equal danger of a landing surface that may be as fragile as a tea biscuit. Ranger VII couldn't possibly have seen what matters most of all when human expeditions of the moon are being weighed: the bearing strength of the lunar face. Given man's present limited knowledge of soil physics, no photograph could tell whether the moon's surface will hold a spaceboat or even a lone man. . . ." (*SR*, 9/5/64, 35–43)

- USAF's START (Spacecraft Technology and Advanced Re-entry Test) program was discussed by James J. Haggerty, Jr., in *Journal of the Armed Forces*. START, according to Haggerty, "appears to be more than . . . an expansion of the ASSET program. . . . START involves a dozen or more areas of investigation in the general field of re-entry. It could lead to development of a manned re-entry vehicle for use with the Manned Orbital Laboratory, one which could make a maneuvering re-entry to a landing field, as opposed to the ballistic flight path of the Gemini spacecraft which will be the original MOL re-entry vehicle. Looking farther down the road, START technology could provide the basis for eventual development of a large ferry vehicle for shuttling men and supplies to and from an orbiting space station. . . ."

First contract for this new type of spacecraft was awarded recently to Martin Marietta Corp. for ablative-type re-entry body, designated SV-5, capable of maneuvering during re-entry into earth's atmosphere at hypersonic-transonic speeds. SV-5 was said to be only one portion of broad START program, which would investigate a number of other re-entry body shapes, possibly even lifting body types such as M-2 and HL-10 being investigated by NASA. (*J/Armed Forces*, 9/5/64, 5)

September 5: More emigrating scientists and engineers settle in the U.S. than in any other country, according to Dr. Charles V. Kidd, Associate Director for International Activities of National Institutes of Health, writing in UNESCO publication. From 1949 through 1961, 43,000 scientists and engineers (average of 3,350 per year) immigrated to U.S., reaching peak of 5,800 per year in 1957. (O'Kane, *NYT*, 9/6/64, 20)

September 6: NASA announced ECHO I balloon satellite had been used this summer by French National Geographic Institute to accurately locate satellite tracking stations being built for France and ESRO across Europe and Africa. According to reports, French National Geographic Institute photographed ECHO I simultaneously from precisely located cameras in France and Algeria, in order to join the triangulation systems of France and Algeria. Its aluminum-coated mylar skin wrinkled and misshapen, ECHO I was no longer useful in radio-reflecting communications, but it was still clearly visible in the nighttime sky after more than four years of orbiting the earth. (NASA Release 64-226)

September 7: NASA Goddard Space Flight Center scientists, via Rosman, N.C., tracking station, succeeded in turning on 14 of the experiments onboard OGO I and received scientific data transmitted from the satellite. Power level in the NASA "space bus" was said to be satisfactory. OGO I was obtaining less than half the power it needed for full-time operation of its 20 experiments, because it was spinning slowly in space instead of having its solar panels constantly aimed at the sun. (Hines, *Wash. Eve. Star*, 9/8/64; AP, *NYT*, 9/8/64, 31)

• NASA announced 39 new international fellowships in space science had been granted for one year's graduate study at universities across the U.S. Initiated in 1961, program was administered by National Academy of Sciences-National Research Council, with NASA providing university costs and travel within the U.S. So far, 34 fellows had completed U.S. studies under the program and 17 countries had participated. (NASA Release 64-223)

• General Dynamics/Astronautics received $6 million contract from USAF for additional standardized Atlas space launch vehicles. Atlases produced under this contract were said to be for sending lunar-orbiting probes on photographic missions. (*San Diego Union*, 9/8/64)

September 7-12: 15th Congress of the International Astronautical Federation (IAF) held in Warsaw. Reported that 700 scientists and engineers representing 40 countries attended; largest delegation (100 persons) from the U.S., while 60 representatives from the U.S.S.R. was largest Soviet delegation ever to attend such a conference. 180 papers were given. The Polish Government issued an IAF stamp in honor of the Congress, while *Pravda* referred to Warsaw as the "space capital of the world."

Congress opened with statement by President Edmond A. Brun (France) that deliberations "should undoubtedly contribute towards increasing international cooperation in the field of astronautics." First report to the entire Congres was on the lunar mission of RANGER VII by JPL Director William Pickering and Harris Schurmeier. RANGER VII movie was shown five times during the week to packed audiences. Informative U.S. papers on progress on comsats, weather satellites (including NIMBUS I photographs), manned lunar landing program, and bioastronautics, as well as consideration of the idea of a Lunar International Laboratory, appeared highlights of the 15th IAF Congress. (NASA Hist.)

September 7: IAF paper by Soviet bioastronautics expert N. M. Sissakian, read for him in his absence, stressed conviction that artificial gravity must be created in spacecraft for long manned flights: ". . . We see the need for a new field of research called gravitational biology that would determine the general laws relating acceleration and body function." Dr. O. Gazenko and P. K. Isakov reported Soviet cosmonauts' space flights of up to five days raised the questions of whether prolonged weightlessness would: cause loss of body calcium and softening of bones; affect the inner ear and deteriorate man's ability to concentrate, or damage the brain; weaken body systems such as heart muscle so that they would be destroyed upon sudden re-entry into earth's gravitational force. (AP, Wash. *Eve. Star,* 9/7/64; AP, *NYT,* 9/8/64, 11; L.A. Times, *Wash. Post,* 9/8/64)

September 8: Hurricane Dora with her 125-mph winds approached the U.S. Atlantic coast from Cape Kennedy to Myrtle Beach, S.C., and workers at the rocket launching center dismantled five of the six rockets being prepared for launch this fall. Only the Saturn I SA-7 rocket was secured in its gantry; removed were Titan II (for GT-2 Gemini launch), Atlas-Agena (for Mars probe launch), Thor-Delta (for Imp launch), Atlas-Centaur (launch vehicle development launch), and Atlas for USAF re-entry test. (UPI, *NYT,* 9/9/64, 1; Witkin, *NYT,* 9/9/64, 23)

- Soviet scientist Prof. A. G. Kuznetsov told IAF that U.S.S.R. had conducted tests of cosmonauts spending 25 days in sealed chamber with helium-oxygen atmosphere. He said in one test, although there were no adverse effects on the subject, slight changes were noted—quickened pulse and respiration rate, faster blood circulation, and higher voice pitch to 7/10th of an octave. Prof. Kuznetsov noted that the manned Vostok spacecraft used mixture of nitrogen and oxygen; he did not disclose whether that mixture affected cosmonauts adversely, but he said that "helium offers a possible substitute for nitrogen that might prove more suitable for spacecraft requirements." Helium might reduce toxicity and provide a more hygienic atmosphere, he said.

  Dr. Eugene B. Konecci of NASA, cochairman of the bioastronautics session, said U.S. had not pressed development of helium system for three principal reasons: (1) helium's tendency to alter voice pitch; (2) helium's leak qualities, which make it difficult to contain in enclosure; and (3) helium's high thermal conductivity, which might pose possible astronaut heating problem. "We still feel strongly about nitrogen. We presently see no advantage to helium that would cause us to change. However, we will continue to study the possibility of using it in case we encounter some unforeseen nitrogen difficulty," Dr. Konecci said. (*Wash. Post,* 9/9/64; AP, *Balt. Sun,* 9/9/64)

- Dr. Eugene B. Konecci, NASA Director of Biotechnology and Human Research, in paper delivered at IAF in Warsaw, examined man's possible roles in space flight. As a transmitter, receiver, or control mechanism, he said, man's capacity and speed are limited as compared with a machine's. Boredom, he said, makes man ineffective "in tasks that call for long periods of relative inactivity." However, he said, man "is the only available computer which can solve problems by logical induction, although unfortunately he cannot keep emotions out of this thought process. He can make sense of fragmentary information." (NASA Release 64-229)

September 8: Communications Satellite Corp., authorized two years ago by law as a private corporation to establish a global commercial comsat system, was listed on the New York Stock Exchange. Incorporated Feb. 1, 1963, ComSatCorp had offered 10 million shares at $20 each, five million to communications common carriers and five million to the general public. (ComSatCorp Release)

- Sen. Mike Monroney (D.-Okla.) said in the Senate that "one of the outstanding accomplishments of this Congress is the action taken to solve the problem of obtaining professional advice on scientific and technological questions involved in the legislative process. The appropriation of funds to establish a Science Policy Research Division of the Legislative Reference Service of the Library of Congress is the latest in a series of steps taken by the Congress to improve the effectiveness of its staff. . . .

  "I join other Members of the House and Senate in hopeful anticipation that the new Science Policy Research Division of the Legislative Reference Service, will develop successfully under the direction of Dr. Edward Wenk, Jr. Dr. Wenk's experience in both the legislative and executive branches of the Government is invaluable in meeting his new responsibilities of advising Congress on science, engineering, and public policy." The division would begin operation Sept. 28. (CR. 9/8/64, 21030; *M&R*, 9/14/64, 15)

September 9: NASA's SYNCOM III synchonous-orbit communications satellite was 178.12° east longitude and 0.03° north latitude at noon, and was traveling along the equator drifting eastward at the rate of 2.8° per day. It was expected to reach the International Date Line (180°) tomorrow. SYNCOM III was inclined 0.081° to the equator, and its orbit ranged in altitude from 21,690-mi. perigee to 22,509-mi. apogee. Control maneuvers would be signaled tomorrow to decrease the satellite's apogee and increase the perigee, thus giving it a more nearly circular orbit. (NASA Release 64-230)

- Cape Kennedy, prepared for the onslaught of Hurricane Dora by advance warnings including photographs by NIMBUS I satellite, escaped the second hurricane in a month with virtually no damage. (AP, 9/9/64)

- Permanent Hall of Science was dedicated at World's Fair, New York City. Participating in dedication ceremony, NASA Administrator James E. Webb said the Hall of Science would "contribute to the understanding of those many disciplines which comprise modern scientific undertakings. . . ." Robert Moses, World's Fair president, noted particularly the "extraordinary public interest in space exploration" in his remarks. (Webb Text; NASA LAR III/172-78; *NYT*, 9/10/64)

- At IAF, NASA's Dr. Eugene B. Konecci, cochairman with a Soviet delegate of space medicine sessions, said: "For the first time, the Russians are giving us technical laboratory data on their space research. We are quite encouraged by this exchange on a scientist-to-scientist basis." (AP, *Denver Post*, 9/9/64)

September 9: British Minister of Aviation Julian Amery announced at banquet of Society of British Aerospace Companies the government's decision to develop Black Knight rocket as an experimental satellite launcher. The new rocket, designated Black Prince, would be a three-stage configuration using hydrogen peroxide and kerosene in the first two stages and a solid-propellant spin-stabilized third stage. The first stage would be powered by a cluster of eight engines. Also announced were plans to accelerate research on an ocean-going hovercraft to fill a naval role. (AP, 9/9/64; *A&A*, 11/64, 116)

September 10: Kiwi B–4E reactor, operated at nearly full power for eight minutes Aug. 28, was restarted and run for a second time at power level nearly identical to that achieved during the first experiment. Test lasted about 2.5 min. and reactor performance was excellent. This demonstration of capability to rerun reactors came much earlier in the NASA–AEC nuclear rocket program than had been anticipated; officials considered it a significant step toward economical development of nuclear propulsion technology. Harold B. Finger, Manager of NASA–AEC Space Nuclear Propulsion Office, said: "We have proved that nuclear rockets can give the kind of performance that is needed for advanced space missions. We could use essentially this type of reactor for experimental flights in the early 1970s and have an operational system by 1975." Test marked final one planned for Kiwi. (SNPO; Wash. *Eve. Star,* 9/15/64)

- NASA Goddard Space Flight Center officials successfully made the first of two final corrections in the orbit of SYNCOM III communications satellite, raising the satellite's perigee closer to perfect circle 22,289 mi. above the earth. (Wash. *Eve. Star,* 9/10/64)

- NASA Goddard Space Flight Center scientists conducted interrogation of OGO I satellite through Rosman, N.C., tracking station, and 15 experiments were turned on to confirm their operation. Signals were weaker than previously received, because OGO I's solar panels were not locked onto the sun; the satellite's solar panels were successfully commanded to turn 42° into a more favorable sun angle, thus ensuring satisfactory power levels. As of this date, 19 of the 20 onboard scientific experiments had been turned on and had transmitted data, which were being evaluated by experimenters. (NASA Release 64–232; GSFC Release G–26–64)

- In Warsaw at the luncheon of the International Academy of Astronautics (IAA), Rep. George P. Miller (D.–Calif.), Chairman of House Committee on Science and Astronautics, expressed the intent of the U.S. to engage in the conquest of space for peaceful purposes as instituted in National Aeronautics and Space Act of 1958 and implemented by NASA. Prof. C. Stark Draper of MIT was re-elected as president of IAA for the coming year.

  In *Missiles and Rockets* interview while in Warsaw, Rep. Miller said that "the U.S. would be happy to open up negotiations with Poland on space projects." A possible means of such cooperation, he said, might be incorporation of Polish scientific experiments on U.S. satellites and sharing of data they would acquire, a program such as U.S. already conducts with U.K., France, and Italy. (NASA Hist.; *M&R,* 9/14/64, 16)

- NASA announced it had assigned management responsibility for large solid rocket motor development to its Lewis Research Center.

The 260-in. solid motor development was formerly a program of the Department of Defense. Under its FY 1965 budget, NASA assumed funding responsibility for the program; it would begin management responsibility for it next July. Two contracts for the program, with Aerojet General Corp. and Thiokol Chemical Corp. and initiated by the Air Force in June 1963, had been transferred to NASA. Smaller, 156-in. solid motor program continued under Air Force development. (NASA Release 64–231)

September 10: Joint Navigation Satellite Committee (JNSC) established by agreement between Depts. of Treasury, Defense, Interior, Commerce; FAA; and NASA. JNSC would evaluate requirements for a nonmilitary satellite system for air-sea navigation, traffic control, emergency and rescue operations, and related functions. If JNSC found navsat system to be the most advantageous means of providing these services, it would recommend appropriate program to develop the satellite system. (Agreement; NMI 2–3–39)

- Sen. Hubert Humphrey (D.–Minn.) introduced in the Senate a joint resolution providing for establishment of a President's Advisory Staff on Scientific Information Management (PASSIM), to "deal forthrightly with the problem of evaluating, developing, and operating properly structured information management and decision-aiding systems. . . . Positioned at the Presidential level the staff of PASSIM will be permitted to necessary overview of information management practices within Government, and will be better situated to develop and organize data and decision-aiding systems on an interdisciplinary and interagency, intergovernmental, interindustry, or intersector basis. . . ." (*CR*, 9/10/64, 21281)

- Findings of survey on distribution of Federal research and development funds, conducted by Subcommittee on Science, Research, and Development of the House Committee on Science and Astronautics, were made public by subcommittee chairman Rep. Emilio Q. Daddario (D.–Conn.). Said to be first definitive, Government-wide survey of this subject, survey showed that during FY 1963 California received 38.4% of the total $12 billion spent by the Government on R&D, New York received 9.2%, and Massachusetts, 4.6%. Nine states accounted for 70.9% of the total, and the 41 remaining states had less than 2.5% each. Regionally, Pacific Coast states received 41.8%; Middle Atlantic states, 15.7%; Mountain region, 8.7%; South Atlantic, 8.5%; East North Central, 6.6%; New England, 6.1%; West South Central, 5.5%; West North Central, 3.6%; and East South Central, 2.8%. (Finney, *NYT*, 9/11/64)

- NASA Administrator James E. Webb said in address at Midwest Governors' Conference, Minneapolis:

    "In many ways, it seems to me that at least a part of the answer to the problems of the Midwest can be found in further development of your great universities and finding new ways through which they can work with industry in ways that will speed up the process of utilizing the technology that builds rapidly on the scientific base of university research.

    "The success of the national space program to date has stemmed directly from cooperation between industry, government, and universities. We are convinced that the maximum advance can best be achieved by close cooperation between these segments of our society. We believe that this cooperation is most fruitful when it is broadly based or regional. . . ." (Text)

September 10: U.S.S.R.'s Ivan Cheprov, speaking at symposium on international space law held in Warsaw in conjunction with IAF congress, criticized the planned global communications satellite network being organized by Communications Satellite Corp. He said it appeared "that the very use of new communications channels by other nations would depend on the whims of Comsat. . . . The right to participate in the management of the space communications system is denied those who are not willing to subsidize and to perpetuate a monopoly in the sphere of producing communications satellites and running space communications business. . . . The other participants merely form a chorus for approving any decisions dictated by Comsat." He said the U.N. and the International Telecommunication Union could organize an international communications system "on the basis of equality." (Reuters, *Wash. Post,* 9/11/64)

- Rep. Joseph E. Karth (D.–Minn.), speaking at Midwest Governors' Conference in Minneapolis, recommended establishment of a national technology commission to award Federal research projects to promote nationwide economic development. Rep. Karth said concentration of Government R&D contracts to industries and universities on East and West coasts showed that "the broad national economic development is not being served." (Higgins, *St. Louis Post-Dispatch,* 9/11/64)

- Jet-propelled back-pack, part of a Modular Maneuvering Unit (MMU) for Gemini astronauts, was described by AFSC, whose Aero Propulsion Laboratory conceived the device. The unit would enable astronauts to fly outside their Gemini spacecraft for up to an hour and travel as far as 1,000 ft. from the spacecraft, although the first flight would be made with astronaut on a 200-ft. tether to send the astronaut forward, backward, up, and down. In addition to the back-pack, MMU consisted of a chest-pack, furnished by NASA, which the astronaut would use when leaving and entering the spacecraft, before donning and after discarding the back-pack. First flight test of the unit on Gemini mission was planned for mid-1965. (AFSC Release 47–22–113; MSC *Roundup,* 1/16/64, 5)

- Col. John M. Stapp (USAF) said at IAF congress in Warsaw that pigs and bears had been used in tests to determine best position for astronauts to be in when Apollo spacecraft returning from the moon hits the earth. Human volunteers were used in 146 test runs, but anaesthetized bears and pigs were used in tests which were considered too dangerous for human subjects. Results: safest position would be landing backwards at 45° angle to the line of impact. (UPI, *Chic. Trib.,* 9/11/64)

- Rep. William Miller (R.–N.Y.), Republican Vice Presidential candidate, said he felt recent successful RANGER VII lunar flight had "no political implications" and was "completely a scientific effort, outside the realm of politics." Upon questioning, he said the success showed U.S. is ahead of U.S.S.R. in space technology and added U.S. had never been behind U.S.S.R. in accumulation of space data. (Sci. Serv., *S.F. Chronicle,* 9/10/64)

September 11: SYNCOM III was positioned into near-perfect stationary orbit over the Pacific Ocean. Final corrective maneuvers were completed, and the communications satellite appeared to hang motionless in space above the equator and International Date Line, where it could transmit radio and television signals between Japan and California. (AP, *Wash. Post,* 9/13/64)

- 1964 would be a year of development test on individual systems which provide capability to carry out Project Apollo lunar landing, according to Dr. Joseph Shea, Manager of Apollo Spacecraft Program Office at NASA Manned Spacecraft Center. He cited the test of Apollo launch escape and earth landing systems at White Sands Missile Range last May and launch of Apollo boilerplate aboard Saturn I (SA–6) also last May as two important development flight tests—milestones marking the gradual transition of Project Apollo to flight test phase. "However, there is still much ground testing to be done." (NAA S&ID *Skywriter,* 9/11/64, 4)

- Dr. Harrison Brown, Cal Tech geochemist, said in *Science* article that as many as 2,000 earth-like planets capable of supporting life may exist in the nearest corner of the Milky Way Galaxy. Based on studies of this portion of the universe—volume of 10,000 cubic parsecs—Dr. Brown theorized: "Virtually every main sequence star should have a planetary system associated with it." He calculated there were approximately 1,000 such stars in the region and 60,000 dark bodies of Mars' size or bigger, or an average 60 "dark companions" to each of the stars. He determined a typical star should have four such planets, and of these "on the average some two planets per visible star might provide suitable environment for the emergence of life processes which are based upon chemical systems similar to those on earth." Dr. Brown's paper propounded higher probability of life in the universe than previously theorized. (Hines, Wash. *Eve. Star,* 9/11/64)

- At 15th IAF Congress in Warsaw, delegates from six nations in panel session discussed possibilities of building a Lunar International Laboratory (LIL). It was agreed such a laboratory could be established in the 1975–85 time period if U.S. and U.S.S.R had conducted thorough preliminary lunar explorations by that time. (*Wash. Post,* 9/13/64; *M&R,* 9/21/64, 16)

- FAA Administrator Najeeb E. Halaby said in London that he had urged U.K. and France to conduct sonic-boom studies such as those the FAA had sponsored in Oklahoma City this year. He said the six-month test program had shown that a serious problem existed in sonic booms. U.K. and France were developing supersonic transport "Concorde." *NYT,* 9/12/64, 38)

- "Highlights of Recent Space Research," by Robert Jastrow and A.G.W. Cameron of Goddard Institute for Space Studies, NASA, appeared in *Science.* The article presented highlights of recent space research in the physical sciences: geodesy, meteorology, the upper atmosphere, magnetosphere, magnetopause, atmosphere of Venus, exploration of the moon, solar physics, x-rays and gamma rays. (*Science,* 9/11/64, GSFC Reprint)

- USAF announced it had retired two squadrons of Atlas ICBM's and would inactivate more Atlases and Titans. Total of 105 of the liquid-fueled missiles would be retired, rendered obsolete by more modern, better protected missiles such as Minuteman and Polaris. (AP, *Wash. Post*, 9/12/64)

September 11: Awards at 1964 Air Force Association national convention included H. H. Arnold Trophy to General Curtis E. LeMay (USAF CofS), who was named "Aerospace Man of the Year" for his "enlightened and aggressive leadership in behalf of the status and welfare of the military man"; David C. Schilling Trophy to Major Sidney J. Kubesch, SAC B-58 pilot, for his record Tokyo-to-London flight in October 1963; and Theodore von Kármán Trophy to Clarence L. Johnson, Vice President (Advanced Development Projects) of Lockheed Aircraft Corp., for designing the YF-12A (A-11) aircraft.

Among the citations of honor: Dr. Edward C. Welsh, Executive Secretary and Acting Chairman, NASC, was cited for his "broad understanding and vigorous pursuit of aerospace goals." Maj. Gen. Samuel C. Phillips (USAF), deputy director of Apollo program in NASA Hq., was cited for his contributions to Minuteman ICBM program in previous assignment with AFSC/BSD. Chief M/Sgt. Frank M. Davis (USAF) and Capt. Joseph H. Engle (USAF) were named outstanding airman and outstanding officer of 1964 by AFA. Davis was cited for leadership in desert survival training including program for Gemini astronauts, among other citations. Engle was cited for "his active and aggressive" role as an X-15 test pilot. (*AF Mag.*, 10/64, 33; *Wash. Post*, 9/12/64)

- Soviet astronomer Nikolai Barabashov said Kharkov Univ. observatory ultraviolet photographs might show surface of Venus. Tass quoted Barabashov as saying a large dark spot appeared on the planet in pictures taken before dawn, and he believed this indicated a clearing in the clouds normally covering Venus. "The position of the spot on the disk warrants the assumption that we have here a glimpse of a part of the surface of Venus, the closest neighbor of the earth." He added that further studies and analyses would be made. (AP, *Houston Post*, 9/12/64)

September 12: At closing plenary session of 15th IAF Congress in Warsaw, Dr. William Pickering, Director of Cal Tech's Jet Propulsion Laboratory, was named new IAF president.

At the IAF banquet, Dr. Leonid I. Sedov (U.S.S.R.) said: "When we tell people we can land on the Moon by 1971, they say to us 'why so long, what is holding you up?'" Also at the banquet, Dr. Wallace O. Fenn, thirty-five years prominent in biological sciences and Dept. of Physiology of Univ. of Rochester Medical School, and Secretary General of International Union of Physiological Sciences since 1959, was awarded the IAA's Guggenheim International Astronautics Award for 1964. (*Wash. Post*, 9/13/64; *M&R*, 9/21/64, 16; IAA Annual Rpt. 1963-64)

- Japanese Ministry of Posts and Telecommunications conducted its first TV test with SYNCOM III communications satellite from ground station at Kashima, receiving return picture with satisfactory resolution. Purpose of test was to investigate parameters of the station. Transmission tests to Pt. Mugu, Calif., via the comsat were planned for early October. (*Av. Wk.*, 9/21/64, 30)

September 13: U.S.S.R. launched COSMOS XLV into orbit, carrying scientific equipment to continue the space research program begun in March 1962. Initial parameters reported for the satellite's orbits: apogee, 327 km.

(203.067 mi.); perigee, 206 km. (127.926 mi.); period, 89.69 min.; inclination, 64°54'. (AP, *NYT*, 9/14/64; *FonF*, XXIV, 313)

September 13: Rep. Olin E. Teague (D.-Tex.) was quoted in World Book Encylopedia Science Service article as saying NASA's Apollo three-man spacecraft could be modified for military missions to: orbit earth for two-to-three months with more than 2½ tons of weapons or reconnaissance equipment by 1967; move from equatorial orbit to polar orbit, surveying earth every 24 hrs.; survey a single strategic portion of the globe from a stationary 24-hr. orbit; carry six 500-lb. rocket packages, adaptable to containing nuclear weapons. "If we do not succeed in developing . . . [such] military potential for Apollo, it will be because of lack of funding and direction, and not because of technical problems," he said. ". . . our ability to defend ourselves in the future and thus our whole future is tied up in space." (WBESciServ, *Houston Chron.*, 9/13/64)

September 14: NASA announced the next U.S. manned space flight, Project Gemini GT-3, had been delayed into 1965 because of time lost as result of lightning damage to GT-2 launch vehicle and removal of the rocket from launch pad during hurricanes Cleo and Dora. Dr. George E. Mueller, NASA Associate Administrator for Manned Space Flight, said GT-2 unmanned suborbital mission would be made in last quarter of 1964. GT-3, with Astronauts Virgil I. Grissom and John Young as crew members, was scheduled for first quarter of 1965. (NASA Release 64–233)

* NASA Goddard Space Flight Center scientists Saul H. Genatt and Edwin Reid said they had seen "red and blue bands" on the moon last Aug. 25 through a 16-in. telescope at GSFC, Greenbelt, Md. Reid first sighted the bands, which appeared at the crater Aristarchus, then he and Genatt continued to observe them for an hour. Two bands were red and one was blue, Genatt said; the two red bands lay across the southern half of the crater and the blue band lay across the northern half parallel to the other two. The colors were "very strong" at 10:00 p.m., when the men began observations, then gradually faded in intensity until the blue band disappeared at 10:30 and the red bands, shortly after 11:00. Similar sightings had been reported by Lowell Observatory scientists in October and November 1963, and prior to that time, by Soviet astronomer Nikolai Kozyrev. (Hines, Wash. *Eve. Star*, 9/15/64; Perkinson, Balt. *Sun*, 9/14/64; *A&A 63*)

* USAF launched Thor-Agena D satellite booster with unidentified payload from Vandenberg AFB, Calif. (*M&R*, 9/21/64, 11)

* President Johnson presented the Presidential Medal of Freedom, highest U.S. civilian award, to Clarence L. Johnson, designer of U-2 and A-11 aircraft, and 29 other distinguished Americans in White House ceremony. (AP, *NYT*, 9/15/64, 26)

* NASA Administrator James E. Webb said in New Orleans that progress in development of vehicles for manned lunar flight was "very good." He predicted: "If things work well, we'll be there by late 1969. This would not be possible if it were not for Michoud [NASA Michoud Operations]." At present, he said, 11,505 persons were employed at Michoud Operations and $927 million worth of contracts had been let there since it opened. Mr. Webb was in New Orleans to meet with former members of NASA New Orleans Coordinating Committee, formed in 1961 to assist

NASA in establishing Michoud Operations. (N.O. *Times-Picayune*, 9/15/64)

September 14: Communications Satellite Corp. stock had more than doubled in price since being issued to the public in June and was continuing to rise. Michael Getler, in analysis of the situation, attributed rapid price rise to "basically an amazing scarcity of shares for daily trading—in combination with some rather short-sighted short-selling of the issue based in large measure on early, perhaps overly pessimistic, views of its likely performance. . . ." He quoted security consultant George P. Edgar: " 'The instinct of the public, in this case, has operated better than that of Wall Street. For the most part, professionals have missed ComSat and the public took the ball and ran.' . . ." (Getler, *M&R*, 9/14/64, 18)

- Rep. Melvin Price (D.-Ill.), speaking before International Convention of Military Electronics in Washington, criticized cancellation of Pluto nuclear-powered, low-altitude missile project: ". . . Repeatedly, we in Congress were told that it would constitute an outstanding weapons system and would supplement our ballistic missile system, but that the main problem in the Pluto concept was proving the feasibility of the reactor portion of the system. . . .

  "We know of nothing that has happened in the last few years to change or eliminate the original requirement for Pluto. If the need for Pluto disappeared years ago, the project should have been promptly stopped; if the need still exists, the project should be carried through to completion. . . ." (Finney, *NYT*, 9/15/64, 10)

- President Johnson presented 1963 Harmon International Aviator's Trophies to: Astronaut L. Gordon Cooper, for piloting FAITH 7 Mercury spacecraft to 22 orbits and a safe return; Aviatrix Betty Miller, for making first east-to-west transpacific solo flight by a woman, flying Piper Apache from Oakland, Calif., to Sydney, Australia; and to Lt. Col. Fitzhugh L. Fulton (USAF), for piloting B–58 Hustler bomber to 85,318-ft. altitude with payload of more than 11,000 lbs. (*Wash. Post*, 9/15/64)

- Space-General Corp. was building new five-ton rocket vehicle for USAF to use in testing recovery and re-entry parachutes at El Centro, Calif., *Missiles and Rockets* reported. Vehicle, called "Hi-Dive," would be dropped from aircraft at 45,000 ft., slowed down by stabilization parachutes for firing of rockets to speed it up to mach 5. Test parachute would deploy at 30,000 ft. Delivery was expected next spring. (*M&R*, 9/14/64, 11)

- 25th anniversary of Igor Sikorsky's flight in the first practical helicopter in the Western Hemisphere. (Sikorsky Aircraft Div. of UA)

- Accidental explosion involving liquid hydrogen demolished small building at NASA Lewis Research Center's Plum Brook Facility; there were no injuries. (Cleveland *Plain Dealer*, 9/15/64)

September 15: President Johnson, in impromptu visit to John F. Kennedy Space Center, inspected Saturn SA–7 space booster being prepared for launch and space probes being checked out for interplanetary flights, received briefing by Astronaut Walter M. Schirra, Jr., on two-man Gemini spacecraft, and made two brief speeches to space program personnel. "As long as I am permitted to lead this country I will never accept a place second to any other nation in this field," the President said. "I am proud of the long, wandering road we have traveled and I look forward to more productive days ahead. . . . We cannot be second in space and first in the world." The 90-min. visit was his first at Cape Kennedy since he became President. (Yerxa, *NYT*, 9/16/64, 11)

September 15: Rep. Carl Albert (D.-Okla.) submitted in *Congressional Record* "An Accounting of Stewardship 1961–1964," which included summary of U.S. accomplishments in space:

"Since 1961 the United States has pressed vigorously forward with a 10-year, $35-billion national space program for clear leadership in space exploration, space use, and all important aspects of space science and technology.

"Already this program has enabled the United States to challenge the early Soviet challenge in space booster power and to effectively counter the Soviet bid for recognition as the world's leading nation in science and technology. . . ."

Among the achievements listed: Saturn I rocket's successful flight; Saturn IB and V development; mastery of liquid hydrogen as rocket fuel; six manned Mercury space flights; Gemini spacecraft and Titan II launch vehicle test-flight; Apollo spacecraft development; first close-up observations of another planet (Venus) with MARINER II and pictures of the moon with RANGER VII; orbiting observatory program initiation; Tiros weather satellite operation; and ComSatCorp establishment.

"In short, the United States has matched rapid progress in manned space flight with a balanced program for scientific investigations in space, practical uses of space, and advanced research and technological pioneering to assure that the new challenges of space in the next decade can also be met, and U.S. leadership maintained." (*CR*, 9/15/64, 21498)

- FAA Administrator Najeeb E. Halaby said in Washington news conference that U.S. manufacturers had not submitted any satisfactory design proposals for the small transport aircraft designed to replace the DC-3, and FAA would not award the planned $100,000 design development contracts. A likely substitute prospect was French-built Nord 262. On subject of supersonic transport, Halaby said U.S. was proceeding "very deliberately," but he hoped that FAA would be able to select the best design by Nov. 1. (AP, *Wash. Post*, 9/16/64)

- Soviet Premier Nikita Khrushchev was reported to have told visiting Japanese legislators that U.S.S.R. had developed a new "terrible weapon" capable of destroying mankind. Two days later, however, Mr. Khrushchev clarified his remark, denying he had said that the weapon could destroy mankind. "I said that scientists had shown me the terrible weapon, which shows what mankind can do. We did not say anything about unlimited power." He also said the weapon was not a nuclear bomb. He blamed the earlier report on having been "misinterpreted" in translation from Russian to Japanese and then to English. (*NYT*, 9/16/64; Shabad, *NYT*, 9/18/64)

- NASA Astronaut Donn F. Eisele dislocated his left shoulder during weightless flight training at Wright-Patterson AFB in KC-135 jet aircraft, and was placed on limited duty for about three weeks. (*Houston Post*, 9/17/64)

September 15: Two NASA-USAF memoranda of understanding providing for exchange of technical management and scientific personnel became effective. Signed by NASA Deputy Administrator Dr. Hugh L. Dryden and USAF Secretary Eugene M. Zuckert, the agreements were based on recommendation by joint USAF/NASA Military Requirements Review Group. Previously no NASA personnel had been assigned to USAF; under the new terms, NASA would assign personnel to USAF on a "when-needed" basis. NASA personnel would be GS-13's and above, and USAF officers assigned to NASA would be majors or higher unless "the position offers exceptional opportunity to the Air Force," according to agreement. Under basic agreement dating from 1959, about 100 USAF officers were now on duty with NASA, including 13 astronauts, not affected by the agreement, and 40 recent ROTC graduates. About 40 more senior officers were scheduled for assignment to NASA in the near future. (NASA Release 64-234; DOD Release 671-64)

- Dr. Floyd L. Thompson, Director of NASA Langley Research Center, said in speech before National Space Club in Washington that the problem of maintaining radio communications with astronauts during atmospheric re-entry may have been solved. He said LaRC research in Project Ram (Radio Attenuation Measurement) showed that ejection of small amount of liquid into the air "provides an effective means of dealing with radio blackout," the liquid apparently preventing buildup of an ionized sheath that cannot be penetrated by radio waves. While it may not be the "completely workable answer, it looks like a very promising method of dealing with the problem." (NSC Release)

- General Dynamics/Astronautics Corp. was receiving two contract actions, DOD announced: (1) a $15,551,509 increment to existing contract for Atlas space boosters; and (2) a $1,100,000 increment to $94,448,904 contract for work on design and development of a standard space launch vehicle. (DOD Release 666-64)

- Personal and scientific papers of Dr. Robert H. Goddard had been given to Clark Univ. by his widow, Clark president Howard B. Jefferson announced. (AP, *Houston Post*, 9/16/64)

September 16: NASA Flight Research Center announced award of $82,000 contract to General Electric Co. for study of flight instrumentation for use on a research ramjet to be carried on the X-15A-2 aircraft. Major portion of the instrumentation would be thrust/drag measuring device and appropriate sensors to measure net thrust or drag transmitted to the X-15 aircraft from the ramjet. Design of this device also would provide capability to jettison the ramjet from the aircraft. (FRC Release 21-64)

- OGO I satellite probably would obtain 50%—and perhaps as much as 75%—of planned scientific data, NASA Goddard Space Flight Center Director Dr. Harry J. Goett said at Washington news conference. The satellite was still not locked onto the earth but was spinning at rate of 5 rpm. All 20 experiments had been turned on and had transmitted data, but only 17 were sending back useful information. Although OGO I was "crippled," Dr. Goett said, "we have quite a vigorous cripple." For example, the satellite was transmitting 10 times more data bits than ever received from the most advanced Explorer-class scientific satellite. (NASA Transcript)

September 16: Rep. Joseph M. Montoya (D.-N. Mex.) said Aug. 28 test of Kiwi B-4A nuclear reactor was "the culmination of seven previous tests—first begun in 1959. The Kiwi project has had as its basic objective the development of basic graphite nuclear rocket technology. With the completion of the Kiwi project, Los Alamos [Scientific Laboratory] will be directing its future nuclear rocket efforts to development of advanced, higher power, graphite reactor technology under the project name, Phoebus. . . .

"In a great sense the groundwork for nuclear space propulsion has been laid through the success of the Kiwi project. We are on the brink of witnessing a great event—the beginnings of the nuclear space age. We might even compare the success of the Kiwi project in the space age with the combustible engine and its profound effect on the motorcar industry.

"It is altogether fitting that this revolutionary accomplishment in the space age took place in Los Alamos—the Kitty Hawk of the nuclear age. And the birthplace of the atomic bomb. . . ." (*CR*, 9/16/64, A4719-20)

- Dr. Ernst Stuhlinger, Director of Research Projects Laboratory, NASA Marshall Space Flight Center, was awarded the Hermann Oberth Medal by the West German Association for Rocket Research Technique and Space Flight at annual meeting of West Berlin's Scientific Association for Air and Space Flight. (Reuters, *Houston Post*, 9/17/64)

September 17: NASA's EXPLORER XVIII (IMP-I) satellite, silent since last June because of low power levels, resumed data transmission; it had reached a more favorable sun angle and was able to operate on power directly from its four solar panels. Data received from the satellite, based on preliminary analysis, were described as "good." (NASA Release 64-242 Addendum)

- President Johnson said in Sacramento, Calif., speech:

"To insure that no nation will be tempted to use the reaches of space as a platform for weapons of mass destruction, we began in 1962 and 1963 to develop systems capable of destroying bomb carrying satellites.

"We now have developed and tested two systems with the ability to intercept and destroy armed satellites circling the earth in space. I can tell you today that these systems are in place, they are operationally ready, and they are on alert to protect this Nation and the free world.

"Our only purpose still is peace. But should another nation employ such weapons in space, the United States will be prepared and ready to reply. . . ."

The President also disclosed U.S. had produced "over-the-horizon" radar, which would "literally look around the curve of the earth . . . [and] give us earlier warning than ever before of any hostile launches against this country." (Text)

- In its first five years of powered flight, the X-15 aircraft amassed total of 17 hrs., 52 min., 7.1 sec. of flight, of which 4 hrs., 8 min., 16.8 sec. was powered flight. The rocket-powered research aircraft accrued 4 hrs., 13 min., 11 sec. of flight at speeds above 2,000 mph (mach 3), far more flight time at mach 3 than any other aircraft. It achieved a total of 2 hrs., 39 min., 47.4 sec. at mach 4 (about 2,700 mph), a speed impossible for any other manned aircraft; 40 min., 45.1 sec. above mach 5 (about 3,400 mph); and 11.6 sec. above mach 6 (about 4,000 mph). The

X-15 was built for top speed of 4,000 mph and altitude capability of 250,000 ft. Piloted by Joseph A. Walker (NASA), the X-15 set unofficial world speed record at 4,104 mph and unofficial world altitude record of 354,200 ft. It held current official FAI world altitude record of 314,750 ft., set July 17, 1962, with Maj. Robert M. Rushworth (USAF) as pilot. The No. 2 aircraft, modified to provide longer engine burn time, was expected to attain much higher speeds—possibly as high as mach 8 (more than 5,000 mph). In its first five years the X-15 made 116 flights. It was expected that the three aircraft would bring the total to twice that many within about five more years. (NASA Release 64-265)

September 17: NASA selected Brown and Root, Inc., with Northrop Corp. for negotiation of contract to provide operational and maintenance support services at NASA Manned Spacecraft Center. Cost for first year: about $2.5 million. Two-year extension options to the contract were contemplated, and total cost requirements for three years would be about $10 million. (NASA Release 64-235)

- Sen. John Tower (R.-Tex.) said in Houston he favored continuing exploration of space and landing a man on the moon as soon as possible, thus differing with Republican Presidential candidate Sen. Barry Goldwater. Sen. Tower said: "I believe knowledge is needed for the sake of knowledge. . . . I would continue space expenditures at their present level." (*Houston Chron.*, 9/17/64)

- Vladimir Kotelnikov, member of Soviet Academy of Sciences, proposed one-year radio search of the universe for signals from extraterrestrial civilizations, according to Tass. He proposed that network of antennas be assigned to "listen to" specific target points in space. The antennas would be connected to a central multichannel receiver. U.S. radio-astronomers at National Radio Astronomy Observatory, Green Bank, W. Va., had conducted similar experiment in 1960 (Project Ozma) without detecting messages. (Reuters, *Wash. Post*, 9/18/64)

- Communications Satellite Corp. held its first stockholders' meeting and elected its first Board of Directors—six representing public shareholders and six, communications carriers. (Finney, *NYT*, 9/18/64, 45)

September 18: Saturn I SA-7 two-stage rocket was launched from Cape Kennedy, placing in orbit a nearly-37,000-lb. payload consisting of boilerplate Apollo spacecraft command and service modules, instrument unit, and spent S-IV stage. The orbit (145-mi. apogee, 112-mi. perigee, 88.6-min. period) was similar to the interim orbit for future three-man Apollo lunar missions. All major test objectives were met: final development testing of Saturn I propulsion, structural, guidance and flight control systems; development testing of Apollo spacecraft structure and design during flight through the atmosphere; and demonstration of physical compatibility of launch vehicle and spacecraft and test-jettisoning of spacecraft launch escape system. Except for minor changes in Saturn I vehicle, test was similar to Saturn I SA-6 (May 28, 1964). For third time, elaborate system of eight motion picture cameras and one TV camera was mounted in S-I stage to record such flight events as S-IV stage separation and ignition of S-IV engines. The eight motion picture cameras were ejected following S-I powered flight, but Hurricane Gladys was located about 300-mi. from camera impact point, so the recovery ship had to leave the area the previous day and recovery was not attempted. (NASA Release 64-228; UPI, *NYT*, 9/19/64, 11; *Marshall Star*, 9/23/64, 1; MSC *Roundup*, 9/30/64, 1)

September 18: Secretary of Defense Robert S. McNamara issued statement amplifying the President's Sept. 17 announcement of U.S. systems for intercepting and destroying armed satellites. Secretary McNamara said one of the systems was operated by the Army and used Nike-Zeus missile; other system was operated by USAF and used Thor missile. In tests both systems had "intercepted satellites in space, their missiles passing so close as to be within the destruction radius of the warheads." Army system became operational Aug. 1, 1963, and USAF system May 29, 1964. (DOD Release 673-64)

- Dr. Edward C. Welsh, NASC Executive Secretary, said in address at Commercial Museum, Philadelphia: "It takes a certain type of ability to manufacture plausible-sounding myths, and unfortunately such ability has been working overtime regarding the space program. . . ." He examined "several of these confusing rumors": *national security*—"The critic who belittles what is being done in space to strengthen the nation's defense is either ignorant, irresponsible, or both. . . . Progress of a substantial nature is being made in strengthening our national security through space competence"; *crash program*—"Crash programs have unlimited funds, unlimited overtime, and parallel approaches to the end objective. None of these aspects applies to the space program or even to the lunar project . . ."; *either/or philosophy*—"Actually investments stimulate other investments; they do not deter them. . . . What is more, decreased spending in education, for example, does not automatically increase the funds available for space exploration, or vice versa . . ."; and *slowdown*—"such an approach would be of great benefit to the Soviets as it would make it easier for them to increase their lead in some aspects of space and attain the lead in others. Do not think for a moment that they are not trying hard to do just that. So far this year, the USSR has put more than twice as many spacecraft into orbit as they had put into orbit by this time last year. In a sense, they are doubling their space effort. . . . To slow down a carefully planned project, such as Apollo, would be to increase its cost and reduce its likelihood of timely success. . . . Those who propose the stretchout or slowdown in the space program are in fact arguing for inefficiency and higher costs as well as delayed accomplishments." (Text)

- Appointment of Charles W. Harper, authority on simulation and aircraft flight research at NASA Ames Research Center, as Director of Aeronautical Research, NASA Hq. Office of Advanced Research and Technology, was announced by Dr. Raymond L. Bisplinghoff, NASA Associate Administrator for Advanced Research and Technology. (NASA Release 64-239)

- Dr. Alan T. Waterman, physicist and former director of National Science Foundation, was sworn in as consultant to NASA Administrator James E. Webb. He is also a member of the NASA Historical Advisory Committee. (NASA Release 64-240)

- NASA MSC's Director, Dr. Robert R. Gilruth, announced Astronaut M. Scott Carpenter (Lt. Cdr., USN) would serve as his executive assistant. Dr. Gilruth called this a "convalescent assignment," because Astronaut Carpenter still had minor injuries from a motorbike accident several months ago. Also announced was selection of Carpenter for promotion from Lt. Cdr. to Commander. (Wash. *Eve. Star*, 9/18/64)

September 19: Static test-firing of NASA–USAF 120-in.-diameter solid-fueled rocket motor was successful at Aerojet-General facility south of Miami, Fla. The motor produced about 600,000 lbs. of thrust, met its major objective of proving the design concepts, materials, and processes to be used in building 260-in.-diameter motor. Initiated by USAF in 1963, program was being directed by joint NASA–USAF team under existing contracts until the end of this fiscal year, with project management responsibility assigned to NASA Lewis Research Center. (LRC Release 64–87; *Lewis News*, 10/9/64, 1)

September 19–20: About 5,000 persons attended Third Annual Open House at NASA Wallops Station, which featured the base and launch sites as well as special exhibits of rocket models, satellite models, and other space hardware. (NASA Announcement 64–206; Wallops Release 64–66)

September 20: Editorial in Washington *Sunday Star:*
"There can be little doubt that President Johnson has been politically motivated in revealing that the United States today has in operation weapons systems capable of intercepting and destroying armed satellites orbiting the earth at very high altitudes. . . .

"In effect, with the support of guarded detail from Defense Secretary McNamara, Mr. Johnson has lifted the lid on classified information in a way that tends to make the Republican candidate look a bit silly. . . .

"Our presidential campaigns result in too much talk about things that might better be kept under wraps in an age as deadly as the present.

"Still, . . . it is reassuring to be told that our country is equipped with devices designed to counter the most advanced Soviet weaponry, including rockets which Nikita Khrushchev claims can 'hit a fly in the sky.' Judging from the President's remarks, we appear to be at least as good as the Russians, if not better, at that Buck Rogers' game." (Wash. *Sun. Star*, 9/20/64)

September 21: USAF XB–70A aircraft took off from Palmdale, Calif., on its first flight in extensive test program. The aircraft flew about one hour, attaining 16,000-ft. altitude and subsonic speed, then landed at Edwards AFB, Calif. Despite trouble retracting the landing gear, loss of one engine's power, and blowout of two tires on left main landing gear, test was considered a "highly productive" first flight. (DOD Release 676–64; Witkin, *NYT*, 9/22/64, 1)

* At White House news conference, President Johnson announced his three appointees to the board of Communications Satellite Corp.
"I have appointed Mr. Frederick Donner, who is chairman of the board of General Motors, and has been since 1958. . . .

"I have Mr. George Meany, the president of the A.F.L.–C.I.O. I will have Mr. Clark Kerr, president of the University of California. Those are the three Presidential appointees.

"This has required a good deal of study and discussion, and I feel very pleased that these men have indicated their willingness to make a sacrifice to lend their peculiar and unusual talents to the direction of this great adventure between government and free enterprise.

"Needless to say, all of them have a good many duties at the present time, but because I pointed out the need for their experience and their broad knowledge, they accepted membership." (Transcript, *NYT*, 9/22/64)

September 21: NASA Administrator James E. Webb said in *Aviation Week & Space Technology* that U.S. space program had "entered a whole new period," in which U.S. was faced with exploring all the practical applications of knowledge suddenly gained through the space effort. "My view is to create stability, and the satellite may be one of the most important tools yet developed to create stability in the world. . . . In this effort, modern technology is going to play a very important role. The balance of technological power may be more important than any other form of power balance in the next decades ahead." He said NASA was developing a broad technological capability to conduct missions with vast international application. "This will enable our nation to say to other nations of the world, 'Join with us and we will undertake these missions together.'"

"Overt expression" of this new focus of national space program planning would be expressed in forecast report NASA was completing for President Johnson, the article continued. Report would emphasize broad range of general space capabilities available in the future and describe specific possible post-Apollo missions—such as manned lunar base, earth-orbiting space stations, Mars probes—stressing that selection of such a goal should be a national decision. (*Av. Wk.*, 9/21/64, 22)

- Editorial in *Aviation Week* by Robert Hotz:

    "The hardware for the second generation of U.S. space systems is already in the flight test stage and its feasibility already demonstrated. This is why it is necessary now to organize a new type of national space program based not on the simple concept of overtaking the Soviets but on utilizing the new riches of space technology to serve domestic and international goals of U.S. policy. . . .

    "A whole new spectrum of power has been created, ranging from the three tons of thrust per second of the huge liquid-fueled boosters through electric, nuclear and ionic power down to systems whose power can be calibrated in micromouse burps and can function for years without servicing. Harnessing these new energy sources for earth as well as space applications may be one of the most technically significant developments of the next decade.

    "NASA is picking an excellent time, aside from the election implications, to inventory the achievements of space technology and project them into a future format that will give the national space program a focus and purpose beyond the Apollo lunar landings. For the development of space technology will continue to open new vistas for man and his universe that cannot yet be perceived from this point in time with our vision still clouded by the earth's atmospheric sheath." (*Av. Wk.*, 9/21/64, 17; *CR*, 9/23/64, A4823)

- Rep. Hale Boggs (D.–La.) said on the floor of the House: "It was just a short 3 years ago when Mayor Victor H. Schiro [of New Orleans] made the first announcement about the reactivation of the Michoud plant [for production of NASA space boosters]. At that time few of us realized that this program would constitute the single most important industry that has come to south Louisiana since we were admitted into the Union in 1812. . . ." Rep. Boggs had served on the NASA New Orleans Coordinating Committee, formed in 1961 to help NASA establish Michoud operations. (*CR*, 9/21/64, A4764)

September 21: Astronaut L. Gordon Cooper (Maj., USAF) was presented the Gen. Thomas D. White Space Trophy by Air Force Secretary Eugene M. Zuckert, at the National Geographic Society, Washington. Award cited Major Cooper for "the most outstanding contribution to the Nation's progress in aerospace during 1963," orbiting the earth 22 times in Mercury spacecraft in May 1963. (*Wash. Post,* 9/22/64).

- Information about the Van Allen radiation belts, obtained from USAF satellite about a year ago, was reported by Dr. Ludwig Katz, of USAF's Space Physics Laboratory:

    "These low energy particles are not a serious hazard to men in space. Over a period of time, however, they could affect the performance of electronic equipment carried aboard a satellite.

    "The instruments carried aboard the satellite were of six types: a proton detector, plasma probe, electron detector, electrostatic analyzer, geiger counter and magnetometer. The data were gathered from more than 100 orbits." (Henry, *Wash. Eve. Star,* 9/21/64)

- Editorial in Washington *Evening Star* said postponement of manned Gemini space flight into 1965 came as "scant surprise," since "as long ago as last Christmastime a 'space forecast' story in the Star made the prediction that this would be the case, and for several months the 'last quarter of 1964' target date has been hardly more than a tongue-in-cheek fiction at NASA Headquarters. . . .

    "Of course, it is far better to delay the flights than to run the risk of tragic failure due to haste or carelessness. Wouldn't it make sense to forget about phony schedules—and phony excuses for not keeping them—and just carry on with the space program as fast and effectively as possible?" (*Wash. Eve. Star,* 9/21/64)

September 22: Boilerplate Apollo spacecraft with attached burned-out Saturn S–IV stage, launched into intentionally low orbit Sept. 18 by Saturn I SA–7, re-entered earth's atmosphere over the Indian Ocean. The orbiting package was in its 59th orbit and had been in flight for three days, 19 hrs., 37 min. (NASA Release 64–250)

- NASA announced the details of U.S. geodetic satellite program, designed to provide more precise mapping of the earth's surface and its gravitational field. Program was based on recommendation of NASA-DOD-Dept. of Commerce Geodetic Satellite Policy Board and involved three types of satellites: Beacon Explorer, equipped with reflectors to reflect laser beams from earth and with Doppler transmitter to measure effect of irregularities in earth's gravitational field; Geodetic Explorer (instrumented to conduct more experiments than the Beacon Explorer), containing flashing lights, variety of electronic beacons, optical reflectors, and when possible radar reflectors; and passive geodetic staellite, sunlight-reflecting 100-ft.-diameter Echo-type sphere. (NASA Release 64–236)

- Engine for Apollo service module fired for one second in programed 10-sec. test. Automatic cutoff device shut down the engine because of excess vibration due to unstable combustion. NASA spokesman called the test a "minimal success," said results were "not entirely unexpected. The engine had been fired three times previously in acceptance tests, but it wasn't fired in as complete a form as it was today." (AP, *L.A. Times,* 9/23/64)

September 22: NASA and Univ. of Texas signed contract for 84-in. optical telescope to observe the moon and planets. NASA would provide $2.1 million to build the telescope, and the University would provide $759,000 observatory building and $1.2-million associated shop buildings. Scheduled to be operational in 1968, the telescope would be fifth largest in the world. (*M&R*, 9/28/64, 13)

- Team of geologists and astronomers at Cal Tech made significant temperature measurements of planet Jupiter, using heat-sensitive detector fitted to Mt. Palomar 200-in. telescope and discovered that: apparent temperature of Jupiter's atmosphere increases more than 100° in areas darkened by shadows of its moons (average temperature of atmosphere is −230°F in sunlight but only −117°F in shadows, as penetrations go deeper, atmospheric temperature rises; and the moon Callisto seems to radiate about twice as much heat as might be expected. (Korman, CTPS, *Chic. Trib.*, 9/23/64)

- Electron microscope with new television attachment, 10 times more powerful than any previous microscope, was demonstrated by RCA. The system magnified objects by two million times. (AP, *NYT*, 9/23/64, 41)

September 23: NASA announced SYNCOM III synchronous-orbit communications satellite was now in such a precise orbit that its rotational speed was only five seconds slower than that of the earth. Positioned over the equator at 22,300-mi. altitude, SYNCOM III drifted less than 1/100th of a degree per day. Orbital parameters: 22,311-mi. apogee, 22,164-mi. perigee, 1,436.158-min. period, and .095° to the equator. Same day, spokesman for Communications Satellite Corp. announced SYNCOM III was successfully relaying test pictures across the Pacific and their quality was "much better than expected." The experimental transmissions between Kashima, Japan, and Pt. Mugu, Calif., had begun Sept. 21. (NASA Release 64–241; Finney, *NYT*, 9/24/64, 13)

- USAF launched Atlas-Agena D satellite booster from Vandenberg AFB with an undisclosed payload. (*M&R*, 9/28/64, 9)

- NIMBUS I experimental weather satellite ceased operating, after producing more than 27,000 remarkably sharp weather photographs. As result of several days' buildup of friction in motor that turned satellite's solar paddles, the solar paddles locked, preventing them from rotating toward sun. Lacking adequate solar power to recharge batteries, the satellite control system, three camera systems, and associated electronic equipment could no longer function. NIMBUS I surpassed all expectations both as a research spacecraft and as a storm-spotter. During its 380-orbit lifetime it provided the first satellite pictures of nighttime cloud-cover and photographed Hurricanes Dora, Ethel, and Florence, Typhoons Ruby and Sally. (NASA Release 64–243)

- Attempted flight of X–15 No. 3, with Capt. Joseph H. Engle (USAF) as pilot, was aborted 10 min. before scheduled launch from B–52 mother craft because of problems with cabin pressure. (NASA X–15 Proj. Off.)

- Dr. George E. Mueller, NASA Associate Administrator for Manned Space Flight, said in speech before Professional Group on Antennas and Propagation of the IEEE, Kennedy Airport International Hotel: "In the development of national [space] capability, NASA is carrying forward the work begun a half-century ago by its predecessor agency, the National Advisory Committee for Aeronautics. Like the NACA, the space agency is concentrating its efforts on research and development. The only important difference is that NASA also carries out operations in space and, in so doing, develops methods of operation as well as the needed technology. . . .

"The facilities, manpower, flight hardware, and experience being created in the Apollo program will serve the country long after the program itself has been completed. In fact, fully 90 percent of the work now in progress in Apollo would be done to create space power even if there were no moon and our efforts in manned space flight had an entirely different goal.

"It may be compared with the Paris objective of Lindbergh's flight. His primary purpose was not to reach Paris but to develop and demonstrate the capability of transatlantic flight. . . ." (Text)

September 23: Committee for the Preservation of Cambridge Industry, made up of representatives from each of the 94 industrial firms that would be disclocated by location of NASA Electronics Research Center on Kendall Square site, announced it would wage a legal battle to prevent NASA's obtaining the site. Committee said implementation of NASA's plan would "disrupt the jobs of 4,000 employees within this area," contended it would cost some of the industries $200,000 to $400,000 to relocate although maximum renewal funds available for business relocation were only $25,000. (Lewis, *Boston Globe,* 9/23/64)

* Five surviving men among the eight Army Air Service pilots who took part in the first flight around the world 40 years ago were honored by Explorers Club at dinner in Beverly Hills, Calif. A week later, the men were honored in ceremonies at AFSC Hq., which also honored Donald W. Douglas, Sr., designer and builder of the four Douglas World Cruiser open-cockpit biplanes. (AP, Balt. *Sun,* 9/21/64)

September 24: NASA's astronomy program was topic of background briefing for the press at NASA Hq., by astronomy program chief Dr. Nancy Roman. Dr. Roman disclosed at the briefing that NASA was studying concept of a manned orbiting telscope. The 100-in. telescope would operate in space automatically, but man would have to make final adjustments in its focus, change auxiliary instruments from time to time for varying the experiments, collect photographic film, and make repairs. NASA had invited members of the astronomical community to propose studies basic to the Mot concept. Several NASA centers and Hq. offices were supporting related engineering studies. Dr. Roman said design studies for a Radio Astronomy Observatory were "pretty well completed." Not yet an approved program, Rao would have four 750-ft. antennas extended in two V's to receive radio emissions from space. (NASA Transcript; ATPH)

* U.S.S.R. announced routine launching of COSMOS XLVI into earth orbit: apogee, 271 km. (168 mi.); perigee, 215 km. (134 mi.); period, 89.2 min.; inclination to the equator, 51.3°. (Tass, *Krasnaya Zvezda,* 9/26/64, 1, ATSS-T Trans.)

* Aerojet-General Corp. and Westinghouse Electric Corp. successfully conducted first power test of a Nerva (Nuclear Engine for Rocket Vehicle Application) experimental reactor designated the NRX-A2. The reactor was operated for six minutes, at powers ranging from 56% of full power to near full power. All test objectives were met. Nerva was a project of NASA-AEC Space Nuclear Reactor Office. (SNPO; *Lewis News,* 10/30/64, 1; Hines, Wash. *Eve. Star,* 9/25/64)

September 24: DOD announced $2,641,430 contract to Akwa-Downey Construction Co. for construction of NASA test buildings at John F. Kennedy Space Center, NASA, Merritt Island Launch Area (MILA). Contract was awarded by Army Corps of Engineers. (DOD Release 688-64)
- USAF Minuteman II "flew according to the book" in its first flight-test, USAF announced. The ICBM was launched from silo at Cape Kennedy to target site about 5,000 mi. downrange. (AP, Balt. *Sun*, 9/25/64)
- President Johnson awarded Robert J. Collier Trophy, U.S. aviation's highest honor, to Clarence L. Johnson, Lockheed Aircraft Corp. designer and developer of A-11 aircraft. (NAA Release)
- Redeye guided antiaircraft missile was publicly tested in series of four launches against drone craft, and performance was declared "technically successful." Smallest known guided missile—weighing only 28 lbs. including bazooka-like hand-held launcher—Redeye was being developed jointly by U.S. Army and USMC. (Witkin, *NYT*, 9/25/64)

September 25: DOD announced Chaney and James Construction Co., Inc., was being awarded $3,884,955 contract for test facilities at NASA Manned Spacecraft Center. Army Corps of Engineers awarded the NASA-funded contract. (DOD Release 695-64)
- James J. Kramer, research scientist at NASA Lewis Research Center, was named Project Manager of large (260-in.) solid-propellant rocket motor demonstration project at LRC. (*Lewis News*, 9/25/64, 1)
- DOD announced Aerojet-General Corp. was being obligated $1,109,000 increment to previously awarded contract for development and fabrication of Titan II propulsion system for Project Gemini. AFSC/SSD awarded the cost-plus-incentive-fee contract. (DOD Release 695-64)
- U.S. Army awarded one of its largest single contracts, worth nearly a third of a billion dollars, to Western Electric Co. for continued development and testing of Nike-X missile defense system. The cost-plus-incentive-fee contract for $309,664,200 was to cover work on the system from Oct. 1, 1964, to Sept. 30, 1965. (DOD Release 694-64)

September 26: NASA launched second modified Arcas sounding rocket with white rat as part of the payload, the rocket carrying the rodent to 21.4-mi. altitude where payload separated from booster and parachuted to ocean surface. Payload was recovered by helicopter a few minutes later about 27.8 mi. from Wallops Island, Va., launch site. Total flight time was 24.5 min. Purpose of this launch and the previous one Sept. 3 was to demonstrate feasibility of using Arcas to familiarize scientists with technological problems associated with space flight studies on small animals. (Wallops Release 64-67)
- NASA Aerobee 150A sounding rocket with University of Pittsburgh and NASA Goddard Space Flight Center instrumented payload was launched from Wallops Island, Va., to 74.5-mi. altitude. Primary objective of flight was to obtain propulsion and environmental data on the Aerobee 150A. New severance techniques also were employed. Additional experiments included low-energy gamma-ray measurements, study of yellow radiation from sodium vapor, Dovap transmitter studies for tracking rockets, nuclear emulsions for cosmic-ray studies, and attitude control system employing new techniques for pointing. Rocket performance was good, and excellent propulsion and environmental data were obtained. (NASA Rpt. SRL)

September 27: New magnetic hammer developed by MSFC team headed by Robert J. Schwinghamer had saved NASA $240,000 during the past six months, NASA announced. Magnetic hammer was used to remove distortions from huge Saturn V fuel-tank domes after they had been fabricated and assembled. (NASA Release 64-238)
- MSC Gemini Project Director Charles Mathews, asked about prospects of lengthening first manned Gemini flight, was reported as saying: "We are re-examining the possibility of a day-long mission but the probability is that it will stay three orbits [4½ hours] as planned. . . ." Question was prompted by Astronaut Virgil I. Grissom's having stated publicly a few days ago that he would like to have the flight plan extended up to as many as 18 orbits and was trying to persuade project officials to lengthen it. (AP, *Hartford Courant,* 9/27/64)
- U.K. supersonic TSR-2 aircraft was test-flown for the first time in 14-min. flight from Boscombe Down, the reconnaissance-bomber performing satisfactorily. Aviation Minister Julian Amery said the TSR-2, to be operational in about three years, would be "a formidable addition to Britain's power." (Feron, *NYT,* 9/28/64)

September 28: X-15 No. 3 flown by Capt. Joseph H. Engle (USAF) to 97,000-ft. altitude and 3,820-mph speed (mach 5.69) from Edwards AFB, Calif. During the flight, a new protective coating on underside of plane's tail was tested for aerodynamic heating, for possible future application in X-15 No. 2. (NASA X-15 Proj. Off.; AP, *Wash. Post,* 9/29/64; *M&R,* 10/5/64, 11)
- NASA instituted revised patent waivers and related procurement regulations in accordance with the late President Kennedy's directive for a more uniform patent policy among Federal agencies. New regulations gave contractors greater chance to obtain patent rights to their commercially applicable inventions made in the course of work on Government contracts. Announced Aug. 21, the new regulations were expected to accelerate contractors' efforts to identify and report technological advances, thus enhancing technology utilization from the space program. (NASA Release 64-209; *Av. Wk.,* 10/5/64, 26)
- USAF launched four-stage Athena re-entry test rocket from Green River, Utah, for a partially successful flight. Flight was aborted automatically before third-stage ignition because of an inverter malfunction. (*M&R,* 10/5/64, 11)
- DOD announced $1,245,601 contract award to B. B. McCormick and Sons, Inc., for road work at Merritt Island Launch Area, John F. Kennedy Space Center, NASA. The NASA-funded contract was awarded by Army Corps of Engineers. (DOD Release 699-64)
- Sen. William Proxmire (D.-Wis.) spoke on the Senate floor on general aviation ("all aviation activity except the airlines and the military") :
    "Three observations about general aviation are worth sharing.
    "General aviation has greatly expanded our horizons for recreation, commerce, and government for a relatively small but potent and influential number of people. It can do so for many more.
    "General aviation has sharply increased the pace of business and political competition. By reducing the unproductive travel time of costly personnel, the airplane, particularly in general aviation, has acted as a labor creator rather than as a labor saver. By making it

possible to do more things, the airplane has forced people to do more things to keep ahead of the competition.

"General aviation in the rest of the world suffers under an oppressive load of regulatory restriction to so great an extent that it barely exists. We must guard against burdening U.S. general aviation with similar controls producing the same result." (CR, 9/28/64, 22241-43)

September 28: U.S.S.R. was believed to be using over-the-horizon radar similar to type that U.S. was installing (Sept. 17). *Aviation Week & Space Technology* reported U.S.S.R. claimed that the basic technique in over-the-horizon radar—back-scattering of high-frequency radio waves from the ionosphere—was first discovered by Soviet Prof. Nikolai Kabanov and Soviets termed it "Kabanov Effect." (*Av. Wk.*, 9/28/64, 19)

- USN launched its first submarine equipped with Polaris A-3 missiles, the U.S.S. *Daniel Webster*, from Charleston, S.C. (Sehlstedt, *NYT*, 9/29/64)
- Office of Naval Research was studying submarine-based antisatellite missile system, according to *Aviation Week & Space Technology*. System would use Polaris fleet ballistic missile to supplement existing Army and Air Force weapons. (*Av. Wk.*, 9/28/64, 18)

September 29: USAF launched final Minuteman I test ICBM from Cape Kennedy. First-stage flight was satisfactory, but second stage veered out of control and had to be destroyed by range safety officer 165 sec. after launch. In launches from Cape Kennedy, the missile scored 38 successes, 8 partial successes, and 8 failures. From Vandenberg AFB, it scored 60 successes, 4 partial successes, and 2 failures. (AP, *Balt. Sun*, 9/30/64; *M&R*, 10/5/64, 11)

- President Johnson sent nomination of Lt. Col. John H. Glenn, Jr., to Senate for promotion to full colonel (USMC) in recognition of his "many accomplishments while in the service of his country." Former Astronaut Glenn said the news came "as quite a surprise" to him, since he had previously expressed his intention to retire from USMC and had requested that he not be considered for promotion. (AP, *NYT*, 9/30/64, 16; *Houston Post*, 9/30/64)
- NASA announced it would negotiate with Sanders Associates, Inc., for purchase of seven display systems to be used with computers for checkout and launch of Saturn V launch vehicles. Three of the systems would be used at NASA Marshall Space Flight Center and four at Kennedy Space Center, NASA. Cost of the cost-plus-incentive-fee type contract may exceed $5 million. (NASA Release 64-249)
- NASA Goddard Space Flight Center announced selection of Univac Div., Sperry Rand Corp., for negotiation of contract for Message Multiplexer communications processor systems. Final contract would call for Univac to supply, install, check out, program, and document two complete message handling systems to be installed in NASA communications switching centers at London and at an undetermined site in Australia. Contract was expected to amount to about $1.3 million. (GSFC Release G-64-28)
- Sen. Gaylord Nelson (D.-Wis.) spoke in the Senate about the pattern of Federal R&D expenditures which he said was "having an unfavorable effect on the Midwest, home of many of our greatest industries and our greatest universities. . . . We cannot pursue a policy, even unintentionally, which has the effect of wasting the industrial and intellectual potential of the American heartland. . . ." He pointed out that "not

all the blame for lack of Federal research and development activity can be attributed to the Government. The industries and universities of the Middle West have not in all cases been as aggressive as they should have been in seeking Federal contracts." (CR, 9/29/64, 22355-58)

September 30: X-15 No. 2 flown by Maj. Robert Rushworth (USAF) to maximum altitude of 98,000 ft. in flight that reached 3,511 mph (mach 5.19), from Edwards AFB, Calif. (NASA X-15 Proj. Off.; M&R, 10/5/64, 11)

- Dr. Gerard P. Kuiper, Director of Lunar and Planetary Observatory at Univ. of Arizona and chief experimenter on RANGER VII lunar-photographing probe, said at MIT conference that much of the moon's surface is naked lava. RANGER VII photographs showed that dark regions of the moon, particularly its seas, were of volcanic origin. Dr. Kuiper said other recent photographs of the lunar seas in various wavelengths, including infrared, showed sharp relief features in the lunar seas that were lava flows. "There cannot ever be one millimeter" of cosmic dust on the surface, he stated. (Sullivan, *NYT*, 10/1/64)

- 156-in. solid-propellant motor was fired by Lockheed Propulsion Co., producing about 1.2 million lbs. of thrust in burn time of two min., 35 sec. This was second firing of the motor and was rated "successful in all respects" by AFSC/SSD officials. It produced 30% more thrust than first firing May 28 because it contained about 30% more propellant. The test was particularly significant because it proved out a new nozzle concept; spokesman said the test showed "that we can get rid of the graphite now and use graphite tape—meaning that essentially we can go to a plastic nozzle." (*M&R*, 10/5/64, 11-12)

- Lockheed A-11 (YF-12A), fastest known jet aircraft, was displayed for first time at Edwards AFB, where the plane was flown at only 450 mph for the spectators. Also unveiled was air-to-air interceptor missile. USAF spokesman said the new aircraft-missile combination represented a "tremendous increase" in ability to intercept hostile aircraft. In announcing the aircraft last February, President Johnson had said it had been tested at speeds exceeding 2,000 mph and altitudes exceeding 70,000 ft. (Witkin, *NYT*, 10/1/64, 1)

- Senate confirmed President Johnson's nominations to Board of Directors of the Communications Satellite Corp.: Frederic G. Donner, George Meany, and Clark Kerr. (NASA LAR III/187-88)

- NASA announced New York Univ. and Polytechnic Institute of Brooklyn would receive grants for construction and renovation of space-related research facilities. NASA grant to NYU would amount to $582,000 and to Brooklyn Tech, $632,000. (NASA Release 64-255)

- Rep. J. Edward Roush (D.-Ind.), speaking in the House, refuted contention of Sen. Barry Goldwater (R.-Ariz.) in June issue of *Science and Mechanics* that " 'all manned space research should be directed by the military, with national security and control of the access to space as primary goals.'

"To turn all manned research over to the military would be contrary to the purposes and goals this Nation established even before the beginning of the space age. It is so contrary it can be considered reason for the greatest of alarm. . . .

"[It] would violate the very spirit as well as the underlying principles of the National Aeronautics and Space Act of 1958. To cause a change in such control would create distrust throughout the world about our intentions. . . .

"If we suddenly announce a change in our goal of conquering space for peaceful reasons to one of military reasons I am convinced we would lose a number of . . . [our tracking stations]. Even more important we would lose the confidence of the world in our stand that we are a peace loving nation. . . ." (*CR*, 9/30/64, 22454)

September 30: Stanford Research Institute was being awarded $1,409,346 contract for study and evaluation of Nike-X system's potential effectiveness, DOD announced. Nike-X project at Redstone Arsenal, Ala., awarded the contract. (DOD Release 708–64)

* Report of Aviation Human Resources Study Board, created by FAA last February and chaired by Dan A. Kimball, was released by FAA. Study showed that career planning and manpower training were deficient for meeting needs of civil aviation for the next 20 years. Comparatively few people under 30 were now learning to fly. Enrollment in aviation courses had declined, number of schools offering aviation training was decreasing, and only a few military trained airmen were seeking professional aviation careers. Number of professional pilots needed by airlines and general aviation was estimated at 3,650 in 1965, 4,300 in 1970, 5,700 in 1975, and 6,300 in 1980. Report urged expansion of on-the-job and apprenticeship training programs, scholarships, extensive career motivation programs, and other measures. (FAA Release 64–89)

September 30–October 1: FAA sponsored National Aviation System Symposium in Washington, which was attended by leaders of airlines and aircraft manufacturing and officials in other Government agencies. In address at the symposium FAA Administrator Najeeb E. Halaby discussed current national air space system and forecast the future of aviation. He outlined some aviation goals:

"It is in the national interest to see the products of our entire aviation industry sold to the other nations of the world. . . . [Also, we must] promote the use of our concepts, techniques and procedures of managing an airspace and an airport system.

"In my opinion, our most important domestic goal is the establishment of a truly effective air transportation system. In terms that the traveling public can understand, this means the connection of any two population centers within the United States by a door-to-door travel time of not more than 8 hours. . . .

"Finally, there is an over-riding goal which must be uppermost in the minds of everyone. None of the accomplishments I have already discussed will be possible if they are not accompanied by improvements in air safety and, here, primary reliance must be on the aviation community itself. . . ." (Text, *CR*, 10/2/64, 23043–46)

During September: After DOD installed small ground station in Saigon, military communications experiments using SYNCOM II communications satellite began between Saigon and Hawaii. Limited experiments also were made with SYNCOM III, which was in a position to relay messages from Saigon to Camp Roberts, Calif. SYNCOM II was being used on a virtually 24-hour basis. DOD had been experimenting with SYNCOM II for months—before installation of Saigon station, when SYNCOM II relayed messages between Manila and points in the U.S., and even earlier, when the satellite relayed military communications over the Atlantic Ocean. (*Bus. Wk.*, 9/12/64; Finney, *NYT*, 9/29/64, 4)

During September: Dr. Raymond L. Bisplinghoff, NASA Associate Administrator for Advanced Research and Technology, described use of nuclear energy in space propulsion and electric-power generation:

"We have observed repeatedly how a given technology has been advanced by the military requirements and how this technology is later borrowed for consumer products. However, these underlying technologies may be catalyzed with equal effectiveness by significant peacetime objectives set by the President, as exemplified by the Nation's space program.

"Therefore, the nation must support activities during peacetime which have as their purpose large-scale technological goals separated from commercial and military needs. The application of nuclear energy to space exploitation offers such a goal—one which can provide enormous subsidiary benefits in addition to opening the gateway to deep space." (*GE Forum*, 7/9/64, 15–17)

- USAF Missile Development Center reported seven-ft. monorail sled reached record ground speed of 4,200 mph in test at Holloman AFB. (*M&R*, 9/21/64, 9)
- U.S.S.R. had developed nuclear reactor for space applications, according to technical supplement of *Pravda*. Nicknamed "Romashka" ("Daisy"), it used enriched U–235 and had output of 500 watts, 88 amps, through direct control of reactor heat with beryllium reflectors. It weighed about 500 lbs., was comparable to U.S. 950-lb. Snap-10A power source in its power levels. Soviets claimed to have run it continuously 400 hrs. without failure. (*M&R*, 12/14/64, 21)
- Aerospace industry sales for second quarter of 1964 rose to $4.006 billion from $3.808 billion in preceding quarter, Securities and Exchange Commission-Federal Trade Commission reported. (*Av. Wk.*, 9/21/64, 34)
- U.S.S.R. began series of long-range missile tests from Siberian launch site to impact point in the Pacific Ocean near Christmas Island. The missiles had range of more than 7,000 mi. This series would last until Nov. 1, and a second series would extend the tests to Dec. 30, to target area near Howland Island. (Hughes, *Miami Herald*, 9/14/64)
- Interview with General John P. McConnell (USAF), Vice Chief of Staff, USAF, in *The Airman*. General McConnell gave his opinion of USAF's future role in space: "*First*, to augment the existing defense capabilities of the U.S. land, sea, and air forces by using space systems for such things as better communications, weather reporting, and navigation. *Second*, to develop a military capability for protection of U.S. interests in space.

  "I think the Air Force, by virtue of its overall mission and experience, as well as its DOD-assigned responsibilities for space developments and boosters, must seek to obtain space capabilities for our defense as quickly as they become economical and practical.

  "Our role to support the national goals of both the scientific exploration and the high-priority defense projects should continue.

  "I also visualize that in the long run, future manned vehicles will play an increasingly vital role. It is essential that the Air Force vigorously pursue space research and development as well as operations." (*Airman*, 9/64, 12–15)

During September: Writing in *U.S. Naval Institute Proceedings,* Capt. R. C. Truax (USN, Ret.) described early naval rocket development work at Engineering Experiment Station, Annapolis. Captain Truax was in charge of all rocket propulsion work for USN Bureau of Aeronautics from 1946 to 1949. (*U.S. Naval Inst. Proc.,* 9/64, 82–108)

- Frederick R. Kappel, chairman of the board of AT&T, was named recipient of the 1965 John Fritz Medal by United Engineering Trustees, Inc., for "leadership in the pioneering of space communications. . . ." (*Av. Wk.,* 9/14/64, 23)
- William J. Schulte, Assistant Administrator for General Aviation, FAA, was named 1964 winner of National Business Aircraft Association Award for meritorious service to aviation. (*Av. Wk.,* 9/14/64, 23)
- General Accounting Office told Congress that DOD was making only "negligible" progress in its program to standardize electronic parts. GAO blamed the Services' failure "to take appropriate and timely action to complete item-reduction projects initiated under the Defense standardization program" as well as lack of DOD "authoritative direction and control." (*Av. Daily,* 9/24/64)

October 1964

October 1: Sixth anniversary of NASA. In Headquarters ceremony honoring the first NASA Administrator, President T. Keith Glennan of Case Institute of Technology (Aug. 1958–Jan. 1961), and the first Deputy Administrator, Dr. Hugh L. Dryden (Aug. 1958–present), official portraits were unveiled. Administrator James E. Webb made informal remarks on the role of NASA in fulfillment of the American concept of power and purpose in human affairs. He read a letter from President Lyndon B. Johnson which said, in part:

"We must be first in space and in aeronautics to maintain first place on earth. The accomplishments of this past year show clearly that we are making rapid progress to achieve and maintain that position.

"Significant as our success has been, it is but indicative of the far greater advances that mankind can expect from our aeronautical and space efforts in the coming years. We have reached a new threshold of competence in the air and beyond it in space which opens to us the widest possibilities for the future."

Associate Administrator Dr. Robert Seamans (1960–present) reviewed the personal contributions of Drs. Glennan and Dryden in the initiation and progress of the U.S. civilian space program. Friends and relatives of those honored and official NASA guests attended the ceremony. (Program; Letter)

- NASA successfully launched Aerobee 150 sounding rocket to 89-mi. altitude from White Sands, N. Mex. Purpose was to map photon-emitting night-sky sources. Gas-filled proportional counters were used to detect and measure energy of the x-rays. The rocket performed well and experimental results were good, but 10–20% of desired data were lost because yo-yo despin mechanism failed to deploy. (NASA Rpt. SRL)

- NASA and DOD began implementing agreement on a single new system to improve communication and coordination of research and technology information between the two agencies. The agreement abolished certains forms and procedures which the two agencies had used independently for reporting on research and technology data, and instituted a new standard reporting form for common use. After the changeover is completed (July 1, 1965), all research and technology work of NASA and DOD will be covered on the new form, and any scientist or staff member from NASA or DOD will be able to obtain full information by fast machine processing on any given type of work. Signed by NASA Associate Administrator Dr. Robert C. Seamans, Jr., and Director of Defense Research and Engineering Dr. Harold Brown, the agreement was worked out in a series of conferences over a period of more than one year. It was sponsored by and would be continued under the Supporting Space Research and Technology Panel of the Aeronautics and Astronautics Coordinating Board (AACB). (NASA Release 64–258)

October 1: Dr. Gerard P. Kuiper told MIT conference on earth sciences that continuing research had verified Mars' atmosphere was too thin to allow spacecraft to land by parachute. Density of Martian atmosphere, according to Dr. Kuiper, had been found to be less than two per cent of that at sea level on earth. The new measurement was made by spectrographic studies. (Abraham, *Phil. Eve. Bull.*, 10/1/64)

- NASA Goddard Space Flight Center announced selection of Reeves Instrument Co. for negotiations of $2.5 million contract for three 30-ft. Unified S-Band Antenna Systems. The antenna systems would be installed in tracking ships for Project Apollo. (GSFC Release G-64-29)
- NASA announced award of three-year contract to Bendix Field Engineering Corp. for launch support services at John F. Kennedy Space Center, NASA. Contract was expected to cost about $2.8 million for the first year but could amount to as much as $17 million for the three-year period. (NASA Release 64-256)
- NASA announced appointment of Gilbert W. Ousley as NASA European Representative for NASA Office of International Programs. With GSFC since 1959, Ousley would be stationed within U.S. Embassy at Paris, where he would handle NASA relations with European organizations. (NASA Announcement 64-225; NASA Release 64-252)
- After examining Col. John H. Glenn, Jr. (USMC), U.S. Air Force doctors at San Antonio said the former astronaut was on the way to "virtually complete recovery" from his inner-ear injury resulting from accidental fall last February. (AP, *NYT*, 10/2/64, 62)
- Press reported that U.S. tracking network had "detected two important Soviet space failures within a month," according to "authoritative sources in Washington." (L.A. *Herald Examiner*, 10/1/64)
- FAA Administrator Najeeb E. Halaby discussed "aviation in 1975" before National Aviation System Symposium in Washington. He referred to "the coming revolution in aviation technology," a technology "just now on the threshold of developments. . . . Innovations in aerodynamics, improved propulsion systems and fuel properties and the development of new metals and other materials will produce performance characteristics and operating economies which will allow aviation solutions to entire ranges of problems where aviation has never been a factor before. . . ." (Text)
- Fifteen years ago Eastern Test Range was activated at Long Range Proving Ground. Hq. were established at Patrick AFB, then still the Banana River NAS, some 18 mi. south of Cape Kennedy (then Cape Canaveral) rocket launch site. (*A&A, 1915-60*; AP, *Tulsa Daily World*, 10/1/64)

October 2: NASA Manned Spacecraft Center conducted tests of astronauts floating in model Gemini spacecraft in the open seas, simulating astronauts awaiting recovery after splash-down from space flight. MSC project engineers considered the tests successful, despite the fact that the two astronauts became ill while the enclosed spacecraft pitched and rolled in the rough Gulf of Mexico waters. The astronauts considered the heat in the spacecraft to be cause of sickness. Tests were ended after 16 hrs. rather than lasting the planned 36 hrs. because of approaching Hurricane Hilda. Astronaut participants: Lt. Alan L. Bean (USN) and Lt. Cdr. James A. Lovell, Jr. (USN). (Maloney, *Houston Post*, 10/3/64)

October 2: USAF launched Titan II ICBM from Vandenberg AFB with pickaback instrumented capsule equipped to study heat and shock of reentry. The package containing 25 experiments was provided by General Electric Co. Re-entry Systems Dept. for DOD's Advanced Research Projects Agency (ARPA). The Titan II was launched by a SAC crew on routine training mission. (AP, Balt. *Sun,* 10/3/64; *M&R,* 10/12/64, 10)

- John Driscoll, Chairman of Massachusetts Turnpike Authority, met with James E. Webb, NASA Administrator, "to assure him we are ready, willing, and able to discuss with him and to negotiate with NASA for our turnpike location [of NASA Electronics Research Center in Cambridge] in the event the Kendall-Sq. site is not selected." After the meeting Mr. Webb repeated that the Kendall-Sq. location "better meets the requirements" of NASA than any other proposal. Meanwhile, in Boston, Dr. John J. Brennan, Chairman of the Committee for the Preservation of Cambridge Industry, said in press conference that the committee would use every available legal means to prevent location of the NASA center in Kendall Sq. (*Boston Globe,* 10/2/64)

- Sen. Everett Dirksen (R.-Ill.), minority leader in U.S. Senate, filed in the *Congressional Record* his report of the 88th Congress' activities. The portion dealing with U.S. space program stressed "what priority should be applied" to Project Apollo lunar landing program: "Republican concern over the program is engendered by the belief that far too much of our national technical talent and scientific intellect is devoted to this one project while other, more needed and worthy programs are permitted to flounder for lack of competent, imaginative direction. . . .

  "The controversy . . . [is] not whether man will ultimately reach the moon and beyond, but rather, how shall it be done, and whether other aspects of human needs should be bypassed or overlooked in the one spasmodic effort to achieve a lunar landing at once. . . ." (*CR,* 10/2/64, 23061-68)

- First meeting of Interim Communications Satellite Committee, with representatives of 15 nations, adjourned after four days. In its session, the committee approved rules of procedure, appointed Committee Chairman (John A. Johnson, ComSatCorp vice president for international affairs), and heard progress reports on development of global commercial communications system. (ComSatCorp Release; *NYT,* 10/4/64, 74)

- In presentation to NATO's AGARD, Frank E. Rom of NASA Lewis Research Center discussed fast and moderated low-power, lightweight nuclear reactors and their applications. Both one- and two-stage nuclear rockets had significantly superior performance over the best chemical rocket systems, Rom said. Further, he said, "The development of a low-power nuclear rocket would be the quickest and least costly way to achieve a practical nuclear powerplant for space use. Since early use of these smaller powerplants would be possible, operational and flight experience would aid in the development of larger nuclear rockets." (LRC Release 64-88)

- AFSC announced its Research and Technology Div. had completed tests on most effective method of heating high-pressure air to simulate aerodynamic conditions of earth's atmosphere. Wind-tunnel arc air heater produced direct power current of 5.1 megawatts [previous high was 4.2

megawatts]. During the 5.1-megawatt run, 3.85 megawatts were contained in the gas, representing higher efficiency than had been predicted. New designs based on the test would extend the probable lifetime of a projected 50-megawatt heater to be used in electrogas-dynamic facility being built at Wright-Patterson AFB. Both AFSC's Arnold Engineering Development Center and NASA Ames Research Center would use data from the test program for their work with similar heaters. (AFSC Release 42–R–119)

October 2: AEC exploded nuclear device of low–intermediate yield in underground test at Nevada site, the 16th weapons-related nuclear test announced by AEC this year. (AP, *NYT*, 10/4/64, 34)

October 3: President Lyndon B. Johnson said in his news conference that Federal non-defense agencies had reported savings of "more than $178 million on an annual basis" as result of management improvements and cost reductions initiated in July and August. "These savings are in addition to the $100 million reported last April, and to the $140 million reported in July." He listed NASA with the second-largest amount saved in July and August: $44 million. (Transcript, *Wash. Post*, 10/4/64)

- USAF successfully flew Athena re-entry test rocket from Green River, Utah, to target area at White Sands Missile Range, N. Mex. USAF said that "some test objectives were achieved." Launch was said to be second time in five launches that the four-stage missile completed the 357-mi. course. (AP, *Phil. Inq.*, 10/4/64)
- NASA Administrator James E. Webb addressed Young Presidents' Organization Eastern Area Conference in White Sulphur Springs, W. Va.: "As the national space program moves into its seventh year, the United States has reached the halfway point in the broad-based accelerated program for the present decade, a program that was planned and has been carried forward by the three Administrations.

"The National Space Program is creating for the United States new resources for whatever use the national interest may require, whether that interest is in maintaining national security, international relations, stimulating education, or implementing scientific and technological progress. . . .

"In writing the Space Act of 1958, Congress stressed that the scientific and technical knowledge acquired as the national program moved ahead should be shared promptly with all who could use it. . . .

"To be of most use this information must be identified and repackaged to meet the special needs of nonaerospace industries. This is the purpose of the NASA's Technology Utilization Program, the first national undertaking of its scope. . . ."

He detailed how the program works and enumerated outstanding examples, then concluded:

"NASA's major contribution here has been to provide a vital communications link between activities where information is developing and areas where it may be utilized.

"We are convinced that NASA's Technology Utilization Program will be of increasing benefit to the American economy. It is the mechanism through which the discoveries made in space science and technology can pay substantial dividends to American business and industry." (Text)

October 3: Col. Martin L. Raines (USA, Ret.) was named manager of NASA MSC's facility at White Sands Missile Range, Director Dr. Robert R. Gilruth announced. (AP, *Tulsa Daily World,* 10/4/64)
- Rep. F. Bradford Morse (R.–Lowell) of Massachusetts said he had written letter to NASA Administrator Jame E. Webb recommending NASA abandon plans to locate Electronics Research Center in Cambridge and, instead, select another site in a nearby community. In his letter, he said: "It is difficult to understand why the administration [NASA] persists in locating in Cambridge when there is such intense opposition. . . ." (*Boston Sun. Globe,* 10/4/64)
- James J. Haggerty, Jr., space editor of *Journal of the Armed Forces,* discussed the Agena rocket stage, calling it "a highly reliable, extremely versatile piece of hardware which serves triple duty as an orbital injection vehicle for satellites, as an upper stage booster for lunar and planetary probes, and as a spacecraft. From the standpoint of frequency of usage, Agena is far ahead of the field. In the seven years since Sputnik I went up, there have been about 290 spacecraft sent into earth orbit or deep space trajectory; Agena, in various versions, has been in on 142 of them—almost half. It is currently being employed at the rate of about 30 shoots a year and it will figure in some very important upcoming projects. . . ." Haggerty described Agena contributions, notably its remarkable accuracy in sending RANGER VII on its course to the moon. (*J/Armed Forces,* 10/3/64, 8)
- Hand-held, photoelectric sensor that vibrates when it senses changes in environmental terrain had potential as an aid to the blind, AFSC announced. Developed by Santa Rita Technology under USAF contract, the environmental sensor detects changes in light values, feeds the data to photo resistor which changes value according to amount of light, then passes it to oscillator, which creates pulse vibrations against the user's hand. AFSC said research also was underway on device transmitting speech through tactile transducer. (AP, *NYT,* 10/4/64, 46)
- The 84th Congress adjourned without acting on S. 1278, bill instructing National Bureau of Standards to study practicability of U.S. adoption of the metric system. (Library of Congress)

October 4: NASA EXPLORER XXI scientific earth satellite (IMP B) was launched into orbit from Cape Kennedy with a Thor-Delta rocket booster. Onboard instrumentation was working excellently, but the satellite's orbit (59,400-mi. apogee; 120-mi. perigee; 34-hr., 57-min. period; 33°53′ inclination to the equator) fell far short of the intended deep-space path which would have enabled the Interplanetary Monitoring Platform to measure magnetic fields, cosmic rays, and solar winds in interplanetary space. Planned apogee was 161,000 mi. Robert H. Gray, Director of Goddard Launch Operations, said minimum apogee required for mission success was about 97,750 mi. As in launch of EXPLORER XVIII (IMF A), the Delta rocket used higher-thrust third-stage motor, the X-258, to give the satellite its final boost into orbital injection. Preliminary tracking data indicated this stage had not provided the required thrust. Designed and built by NASA Goddard Space Flight Center, EXPLORER XXI contained nine experiments. Five of these were provided by GSFC researchers and four by experimenters at Univ. of Chicago, Univ. of California, NASA Ames Research Center, and MIT respectively. (NASA Release 64-242; AP, Balt. *Sun,* 10/5/64; UPI, *Chic. Trib.,* 10/5/64; *M&R,* 10/12/64, 24)

October 4: Seventh anniversary of the Space Age—orbiting of SPUTNIK I by the U.S.S.R. on Oct. 4, 1957. Soviet Cosmonaut Yuri Gagarin said in *Trud* interview that U.S.S.R. was preparing a manned space flight: "We are working and making studies and are preparing for a new launching. It is not our custom to advertise, but I'll tell you that new and more complicated research work is in store for us." (UPI, *Chic. Trib.*, 10/5/64)

- In recent interview Sir Bernard Lovell, Director of Jodrell Bank Experimental Station, had complained: "American military radar is seriously interfering with our studies of the universe." In the 406–410 megacycle band, he said, radar signals were interfering with Jodrell Bank radiotelescope's listening to radio waves from distant points in the universe. The interference had been occurring over the past 9–12 months, and "we've had to stop one program because of it." Sir Bernard and other astronomers indicated the 406–410 megacycle band was "the only relevant band" for studying quasi-stellar objects ("quasars"). U.S. Government officials acknowledged receiving similar complaints from astronomers at Cambridge Univ. However, according to William E. Plummer, acting chief of frequency management in U.S. telecommunications coordinating office, "No radar of any kind is authorized to operate on that band." (Toth, *L.A. Times*, 10/4/64)

October 5: USAF launched Thor-Agena D booster rocket from Vandenberg AFB with unidentified satellite payload. (UPI, *NYT*, 10/7/64, 19; HHN-48)

- Addressing Oklahoma Frontiers of Science Foundation in Oklahoma City during ceremonies celebrating its tenth anniversary, NASA Administrator James E. Webb pointed out that "some of our NASA projects in education and in the transfer of space technology to the general economy owe much to the philosophy and concepts threshed out in the early days of the Foundation. . . ." (Text)

- Second flight-test of XB-70 aircraft was conducted at Edwards AFB, Calif., with pilot Alvin S. White (North American Aviation, Inc.) and co-pilot Col. Joseph Cotton (USAF). The aircraft reached 600 mph instead of planned 700 mph and attained 28,000-ft. altitude instead of planned 35,000 ft. It was forced to land prematurely because of hydraulic system problems. (AP, *Balt. Sun*, 10/6/64; *Av. Wk.*, 10/12/64, 29)

- Sen. Clinton P. Anderson (D.–N. Mex.), Chairman of Senate Committee on Aeronautical and Space Sciences, said Rift program should be reactivated because "mechanical difficulties which impeded early work on the nuclear propulsion reactor for space have largely been overcome." He said success of recent nuclear reactor test for Nerva strengthened his view that NASA and AEC receive funds in the coming fiscal year "so that we can develop in a properly phased development program an operational nuclear propelled rocket." He pointed out the need for "proper consideration to nuclear energy in planning new space missions. It is obvious that this is a necessary step and one that will require restoration of target dates for an actual flight test of a nuclear rocket system." (*SBD*, 10/5/64, 179; *M&R*, 10/12/64, 17)

- USAF announced an ion rocket would be tested in 90-day orbital flight next spring, using Agena vehicle. Electrical power to operate the engine would be provided by Snap 10A reactor, the first test combining an electrical propulsion engine and a nuclear reactor in space. Spokesmen said similar rocket had been successfully tested in August in suborbital flight from Pt. Arguello, Calif. The new engine, developed by Electro-Optical Systems, was said to have "developed the unheard-of specific impulse level of 7,800 seconds—roughly 20 times greater than the best chemical rockets of today." (*Wash. Post*, 10/6/64; Hines, *Wash. Eve. Star*, 10/6/64)

October 5: NASA Director of Personnel Robert J. Lacklen said in *Aviation Week* he had ordered a "complete re-write" of NASA Management Manual section dealing with employees' acceptance of gifts and gratuities from aerospace industry representatives. "The same thing has jumped us off on a re-write that caused the Defense Dept. to issue its new orders. That is the continuing appearance of NASA names on contractor entertainment lists." (*Av. Wk.*, 10/5/64, 18)

- U.S. airmail stamp honoring Dr. Robert H. Goddard was released at Roswell, N. Mex. Stamp was first U.S. stamp with rocket motif since Fort Bliss stamp of 1946. U.S. stamps also had been issued for Project Echo (1961) and Project Mercury (1962). (NASA Historian)

- Article in *U.S. News & World Report* assessing impact of Project Apollo on U.S. industry quoted unnamed U.S. space official: " 'Project Apollo is the biggest single job in history. In terms of manpower used, it dwarfs the building of the pyramids by three times. It is five times greater in cost than the wartime crash program to build the first atomic bomb.'

"Over-all price tag on moon exploration in this decade is 20 billion dollars. Ten billion of that has been spent—or will be spent—on preliminary programs such as Projects Mercury and Gemini and for new facilities, Government payrolls, tracking stations, communications. . . .

"At a time when Pentagon spending on missiles and rockets is going down, higher spending by . . . [NASA] comes as a shot in the arm to thousands of firms. . . .

"Apollo is now entering a critical phase, one entrusted almost solely to U.S. industry. At the peak of the industrial pyramid are four major aerospace firms, supported by 66 other major companies holding NASA contracts of more than half a million dollars each. . . .

"What is bound to impress anyone is how the big companies have been pulled together on this vast project. . . . To people in industry, working together in peacetime on a 'national goal' is new and satisfying. Explains one official: 'The amazing thing is that industry is finding that space is not only good business—but it is a real adventure for us, too.' " (*U.S. News*, 10/5/64)

October 6: U.S.S.R. announced routine launching of COSMOS XLVII artificial earth satellite "for the further investigation of cosmic space." Orbital data: apogee, 413 km. (256 mi.); perigee, 177 km. (110 mi); period, 90 min.; and inclination to the equator, 64.77°. COSMOS XLVII deorbited Oct. 7. (Tass, *Pravda*, 10/7/64, 1, ATSS–T Trans.; GSFC *SSR*, 10/15/64)

- Final S–I booster in Saturn I rocket program was static-fired at NASA Marshall Space Flight Center for its second and final time. The 156-sec. test was "entirely satisfactory." This S–I stage, the second one built by Chrysler Corp. for NASA, would power the last (SA–10) Saturn I launch vehicle. First Saturn I booster was static-fired at MSFC in April 1961. Today's firing ended about three-and-a-half years of testing in which R&D and flight vehicles underwent a total of 55 firings for cumulative time of 3,867 sec. (*Marshall Star*, 10/14/64, 1)

October 6: USAF launched Thor-Able-Star rocket toward polar orbit from Vandenberg AFB, Calif., but did not disclose payload or mission. It later was revealed that the Thor-Able-Star placed three satellites in orbit. (UPI, *NYT*, 10/7/64, 19; *SBD*, 10/8/64, 202; HHN-48)

- NASA announced appointment of Jennings Pemble Field, Jr., as Director of Gemini Program Control in NASA Office of Manned Space Flight. In this position Field would be responsible for Gemini program costs and budget. Field was an executive of IMC Industries, Inc.; from 1954 to 1963 he was with Bendix Corp., and from 1940 to 1954 he was a commissioned officer in U.S. Navy. (NASA Release 64-257)

- Soviet Government had rejected the space agreement negotiated in Geneva by Dr. Hugh L. Dryden and Soviet Academician Anatoli Blagonravov last June, Howard Simons reported in the *Washington Post*. After the negotiators had returned to their respective countries, Simons reported, U.S.S.R. notified U.S. in July that it chose to reject the agreement. NASA officials had no comment on the report, but other sources (unnamed) said the tentative agreement had called for exchange of information and publication of a book about space medicine and bioscience by Soviet and U.S. experimenters. U.S.S.R. refusal was based on insistence that topics be covered by either U.S. or Soviet author, although Blagonravov had agreed to U.S. proposal that each topic be covered by a Soviet and a U.S. author. This was said to be first instance of exercising the usual option to reject the space agreement reached by negotiating teams. (Simons, *Wash. Post*, 10/6/64)

- NASA Administrator James E. Webb remarked at Univ. of Notre Dame that the establishment of its new Aero-Space Science and Engineering Department placed Notre Dame "in a vigorous and growing group of the nation's leading educational institutions. . . .

  "What is happening this fall at this university is in some ways symbolic of what is happening all over the nation. For the space program has been a catalyst to science, to engineering, to research, and to education at all levels. The power of the modern rocket has given man a new, and in many ways, an unlimited tool for probing the mysteries of the universe. . . ." (Text)

- NASA Administrator James E. Webb said in address at Annual Management Banquet of the National Management Association, in South Bend, Ind., that the "most important trend in NASA's procurement activity today . . . is our effort to use more incentive contracts.

  "Three years ago we had no incentive contracts. Today we have 52 totaling more than half a billion dollars. And we expect to raise that to one billion before the end of this year. . . .

  "So far we have experimented with a new type of incentive contract called the award fee. . . . So far our experience with this type of incentive contract has been good; and efficient contractors who appear to be saving the government money have earned extra profits for themselves. The award fee contract is especially applicable to service contracts where targets are difficult to fix. We feel that the award fee has motivated contractors to perform better, to cut costs, and to make quick corrections when they have been deficient. Contractors, in turn, have expressed their satisfaction with this new procedure. . . ." (Text)

October 7: NASA launched two Nike-Apache sounding rockets from Wallops Island, Va., with payloads to measure upper atmosphere winds and obtain measurements of earth's magnetic field. The launches were conducted within 15 min. of each other to study possible correlation of ionospheric current system and wind shears. First payload ejected trimethylaluminum vapor trail up to 107-mi. altitude, and the bluish-white cloud was visible for several hundred miles. Payload also contained instruments to measure electron temperature and density. Second experiment carried 58-lb. payload of instruments to measure earth's magnetic field. The experiments were conducted at dusk and would be followed by two similar ones at dawn the next day. (Wallops Release 64-70; NASA Release 64-260)

- NASA Argo D-4 launch vehicle launched from Wallops Island, Va., carried instrumented payload to altitude of about 625 mi. in 18-min. flight terminating in impact in the Atlantic Ocean. Designed and built for NASA by Lockheed Missile and Space Co., the experiment measured composition and density of positive ions as a function of altitude from about 150-mi. altitude to peak. (Wallops Release 64-71; NASA Release 64-260)

- Inauguration of SYNCOM III communications link from Tokyo via ground station at Kashima, Japan, and SYNCOM III hovering above the Pacific, to receiving station at Pt. Mugu, Calif., for transmission to U.S. television stations. Special program featured pre-taped appearances by President Lyndon B. Johnson, Secretary of State Dean Rusk, Foreign Minister Etsusaburo Shiina of Japan, and Communications Minister Zitsuzo Tokuyasu of Japan. President Johnson called the new comsat link an opening of "new vistas of friendship and understanding in the fields of education, cultural exchange, business and entertainment." ComSatCorp officials were said to be pleased with quality of reception. Opening ceremonies of 1964 Olympic Games Oct. 10 were scheduled for transmission via the comsat link. (ComSatCorp Release; *WSJ*, 10/8/64)

- American Institute of Aeronautics and Astronautics (AIAA) announced nomination of Richard E. Horner, Senior Vice President of Northrop Corp., as AIAA president for 1965. Mr. Horner served as the first Associate Administrator of NASA (June 1959–July 1960) and before that as Assistant Secretary of the Air Force for Research and Development. (*NYT*, 10/8/64)

- One-year program of general orientation for 60 recent university graduates beginning their professional careers at NASA Langley Research Center began with ceremony presided over by LaRC Director Dr. Floyd L. Thompson. (LaRC Release)

- NASA Administrator James E. Webb said in his remarks in Grand Rapids, Mich.: "We have a fighting chance to maintain our schedules for the manned lunar exploration mission in this decade. By so doing, we will increase our ability to operate in space for any mission the national interest may require. And we will be using space for peace, discouraging those with other ideologies from political propaganda or military threats. . . ." (Text)

- C. Leo de Orsey, attorney for the seven NASA Mercury astronauts, revealed former astronaut John H. Glenn, Jr. (Lt. Col., USMC) had turned down a $1,000,000 contract with private industry in 1963. (*St. Louis Post-Dispatch*, 10/7/64)

October 7: Celebrating its 15th anniversary, East Germany displayed two tactical surface-to-surface missiles apparently comparable to U.S. 300-mi.-range Sergeant missile. (Balt. *Sun*, 10/8/64)

October 8: NASA launched two Nike-Apache sounding rockets with instrumented payloads from Wallops Island, Va., sending them aloft 11 min. apart. The first payload ejected sodium vapor trail up to 99-mi. altitude, to measure upper atmosphere winds. The pink vapor cloud was visible for several hundred miles. Second experiment carried 58-lb. payload of instrumentation to 87-mi. altitude to obtain measurements of earth's magnetic field. Experiments were launched closely together so experimenters could study possible correlation of ionospheric current systems and wind shears; they were similar to two experiments launched at dusk the previous evening. (Wallops Release 64–71; NASA Release 64–260)

- USAF Atlas-Agena D space booster was launched from Vandenberg AFB, Calif., with undisclosed payload. (AP, *Wash. Post*, 10/9/64; HHN-48)
- Full-scale metal mockup of Lunar Excursion Module (LEM) for Project Apollo was review by NASA Manned Spacecraft Center officials at Grumman's Bethpage, N.Y., plant. Mockup's appearance was highlight of week-long analysis of the mockup design with a critique aimed at definitizing the LEM design. (MSC *Roundup*, 10/14/64, 1)
- Apollo program review was presented in special day-long session of five-day national meeting of Society of Automotive Engineers (SAE) in Los Angeles. Another highlight of the meeting was emphasis on in-flight maintenance problems associated with long manned space flights. General Dynamics/Astronautics reported its design philosophy of eliminating need for hand tools in spacecraft maintenance and repair. In address to the group, NASA Associate Administrator Dr. Robert C. Seamans, Jr., said that "virtually none" of U.S. space flight projects had been "successfully completed within their originally estimated resources and on their originally laid out schedules . . . historically, schedules have often stretched out to nearly twice their length and costs have sometimes grown by factors of three or four. . . . [However] we don't accept that management can't do something about it. . . ." He predicted fewer new projects would be initiated unless industry "can show real evidence of having overcome the point of view that says R&D cannot be controlled. . . ." (NAA S&ID *Skywriter*, 10/2/64, 1; *M&R*, 10/12/64, 18; *Av. Wk.*, 10/19/64, 32)
- NASA Administrator James E. Webb said before National Security Industrial Association meeting in Washington that many achievements had been made "in that area where science and technology are indivisible. . . . Many of the most significant *technological* advances, for example, have been made in NASA's space *science* effort.

    "Outstanding among them are our proven ability to launch payloads from earth orbit to deep space with consistent success; and the companion ability to make the necessary mid-course corrections far out in space. . . ." (Text)
- Dr. Harry J. Goett, Director of NASA Goddard Space Flight Center, said in address at centennial convocation luncheon, Worcester Polytechnic Institute:

    "The new era promised by the Space Age perhaps connotes a return to what was once called natural philosophy. The unifying element of these developments of the space program is a general spirit of inquiry into the nature of the external physical world. It represents a re-direction of interest away from the increasingly narrow specialization which has characterized the physical sciences in the last decades.

"The second distinguishing feature of the scientific research in space is the fact that the various scientists have only been able to make their observations . . . by virtue of the hardware developed by the engineer. This is in distinct contrast to the biologist who could invest in a good microscope and do research comparable with the best. . . . Generally, the engineer exploits and puts to practical use the knowledge acquired by the scientists. But in space, the scientist seems to be particularly dependent on the engineer to develop new devices and techniques. Engineering, in this context, has become a more creative and trail-blazing profession.

"The third and possibly most imposing challenge of the space age is the potential feed-back of its developments into the civilian economy. In the long run, the justification of our space budget must stand on this 'fall-out.' . . ." (Text)

October 8: Secretary of Defense Robert S. McNamara was criticized by Military Operations Subcommittee of the House Government Operations Committee in report on military communications satellite. Subcommittee said his concern with economizing had resulted in "overmanagement and underperformance" in the comsat area. "We do not believe, however, that economizing efforts should throttle programs essential to the national security. Satellite communications is one of the most vital and relatively less costly of our major defense programs." Subcommittee said Secretary McNamara's proposal last October for joint military-civilian comsat system was "ill-advised, poorly timed and badly coordinated. . . . Valuable time has been lost. Had the department moved ahead according to plans and policies laid down two years ago by the Joint Chiefs of Staff, a system could have been operating now. . . ." Subcommittee also recommended DOD begin comsat program using Atlas-Agena satellite booster instead of Titan III–C experimental launch vehicle. (Finney, *NYT*, 10/9/64, 1, 16; *M&R*, 10/12/64, 10)

- Judge Charles Fairhurst issued temporary restraining order preventing Cambridge City Council from voting on application for $663,000 in Federal funds for survey and planning the proposed area for NASA Electronics Research Center. Judge Fairhurst granted the order after six Cambridge firms requested declaratory decree that Kendall Sq. "cannot and does not legally qualify as a valid subject for an urban renewal project." (*Boston Globe*, 10/8/64)

- NASA Astronaut Coordinator Donald K. Slayton said in World Book Encyclopedia Science Service feature article that physical training of U.S. astronauts differed greatly from that of his Soviet counterparts. He pointed out: "There is a basic difference in philosophy between Soviet and United States flight crew selection and training. The Soviets select and train bodies as biological specimens for space experiments. The United States selects and trains engineer-pilots as integral operators of spacecraft. Our emphasis is on intelligence and skill as opposed to physical prowess." (WBE Sci. Serv., *Ind. Star*, 10/8/64)

October 8: DOD announced SR–71 strategic reconnaissance aircraft would be assigned to SAC at Beale AFB, Marysville, Calif. SR–71 was scheduled to become operational in 1965. (DOD Release 726–64)

October 9: NASA launched EXPLORER XXII (BE–B) beacon-explorer satellite with four-stage Scout launch vehicle from Vandenberg AFB, Calif. Orbital data: 669-mi. apogee, 549-mi. perigee, 104.7-min. period, and 80° inclination to the equator. EXPLORER XXII was last of five satellites in first phase of NASA's ionosphere exploration and first of five satellites in NASA's geodetic satellite series. Ionosphere experiment involved satellite's transmitting radio signals down through the ionosphere. Ground stations around the world acquiring the signals measured electron distribution by the Doppler shift method and the Faraday rotation method. The international scientific effort with EXPLORER XXII was the most extensive ever for a U.S. space project. Network of more than 80 ground stations was being operated by some 50 scientific groups in 32 countries. Geodetic experiment involved laser beams sent from NASA Wallops Station to the satellite, on which were mounted 360 one-inch "cube-corner" reflectors to return the light. In addition a small supporting experiment, electrostatic probe, measured electron density and temperature in immediate vicinity of the satellite. (NASA Release 64–237; NASA News Conf. Transcript)

- NASA launched second re-entry experiment for AEC, designated Reentry Flight Demonstration-2 (RFD–2), on a Scout booster from Wallops Island, Va. Nonradioactive mockup of nuclear isotopic generator was sent to 800-mi. altitude by three-stage Scout, before the payload reentered atmosphere on its ballistic trajectory and impacted about 250 mi. southeast of Bermuda. Primary objective of RFD–2 was to acquire data on generator disassembly and fuel capsule burnup rate for use in advanced generator designs. In previous flight test (RFD–1 in May 1963), nonradioactive model of Snap-10A nuclear reactor was tested in atmospheric re-entry; RFD–1 demonstrated that space reactors can be designed to melt down and disassemble upon re-entry. The nonradioactive mockup in RFD–2 was a model of an isotopic-thermoelectric generator. Earlier testing had shown that heat created during re-entry could be used to burn up isotopic generator, leaving very minute particles which then would be dispersed at high altitudes to eliminate any radiation danger. (Wallops Release 64–72)
- Four consecutive full-duration F–1 engine test firings at new Rocket Engine Test Site at Edwards, Calif., demonstrated the operational readiness of the site, which was officially accepted by NASA Marshall Space Flight Center Director Dr. Wernher von Braun from the Army Corps of Engineers. In the ceremonies, Dr. von Braun assigned site operation responsibilities to Rocketdyne Div. of North American Aviation, Inc., which would use the site for qualification firings of F–1 rocket engines. (*Marshall Star*, 10/14/64, 10; NASA Release 64–259)
- First major test of revamped NASA worldwide tracking network, part of preparation for manned earth-orbital space flights in Project Gemini. During nine-day drill, simulated flight missions were conducted which involved Goddard Space Flight Center, Mission Control Center, and eight sites in the worldwide tracking network. (MSC *Roundup*, 10/14/64, 1; *Goddard News*, 10/19/64, 1)

October 9: DOD issued rebuttal to report issued the day before by Military Operations Subcommittee of the House Government Operations Committee criticizing DOD plans for a military communications satellite system. DOD said:
"The Defense Department is being operated to provide the highest degree of combat readiness at the lowest possible cost. This policy is being followed in the development of communications satellites. As a result, the defense establishment will have a better satellite communications capability than originally foreseen at anticipated savings of $75–100 millions." (Finney, *NYT*, 10/10/64, 1, 14)

October 10: Opening ceremonies of the Olympic Games in Tokyo were telecast live in U.S. via ComSatCorp's communications link using NASA's SYNCOM III satellite. Telecast was delayed on West Coast and taped for later showing by NBC, which said it was avoiding interference with commercial programing. Japan was reported to have expressed disappointment to U.S. State Dept., and State Dept. in turn was said to be "deeply concerned" over the procedure. NBC confirmed that future coverage of Olympic Games would be taped or filmed. (ComSatCorp Release; Gould, *NYT*, 10/11/64, 1, 25)

- NASA Electronics Research Center Director Dr. Winston E. Kock announced appointment of James T. Dennison as Director of Technology Utilization, ERC. Dennison joined NASA in May 1963 as a consultant and had been serving as acting director of Technology Utilization Div., NASA Hq. Appointment was effective Nov. 1. (NASA Release 64-244)

- Dr. Gerard P. Kuiper, Director of Univ. of Arizona's Lunar and Planetary Laboratory, charged that John Lear's article in *Saturday Review* (Sept. 5) on RANGER VII's findings was "an insidious mixture of reasonably correct quotations, his own gross extrapolations, unfair allegations and personal peeves. . . .

  "One implication of Mr. Lear's article seems to be that I have given him information to the effect that scientists are under pressure to make statements that are favorable to the continuance of large government projects. This is complete fantasy. . . ." (*SR*, 10/10/64, 32)

- Douglas Missile & Space Systems Div. preliminary study of a manned Mars mission, called Project Deimos, was described by James J. Haggerty, Jr., in *Journal of the Armed Forces*. Conducted by Douglas engineers J. L. Woodworth, G. A. Ursini, and Phil Bono, the study was based on six-man crew making 830-day round trip. Key to study was chemical-fuel rocket booster of 18,000,000-lb.-thrust power capable of placing 800,000 lbs. in earth orbit. Reusable booster would be fueled in orbit (by other reusable boosters making 10 trips) before sending spacecraft into Mars trajectory. (*J/Armed Forces*, 10/10/64, 8)

October 11: Dr. George E. Mueller, NASA Associate Administrator for Manned Space Flight, appointed William B. Rieke, president of Lockheed Aircraft International, Inc., as his deputy. Rieke would succeed George M. Low, who had become Deputy Director of NASA Manned Spacecraft Center. (NASA Release 64-261)

- Sen. Hubert H. Humphrey (D.–Minn.), Vice-Presidential candidate, told rally in Washington, D.C., that the Administration would form a study group to seek "the appropriate mechanisms of development and research" by U.S. industry. (Pomfret, *NYT*, 10/12/64, 15)

October 11: Comparison of science in U.S. and U.S.S.R. was subject of article by Prof. Robert E. Marshak, Univ. of Rochester. Distinguishing between "pure science or basic research and applied science or technological development," he said U.S. was "doing exceedingly well in basic science. . . . The striking American performance in pure science in recent years results from a combination of fortunate circumstances. A primary factor is that the openness, freedom and emphasis on individual initiative that characterizes the American way of life provide a favorable climate for the practice of pure science. . . .

"On the other hand, in applied science the U.S.S.R. has demonstrated its capability of organizing large team undertakings whose favorable outcome depends as much on careful mobilization of material and human resources as on scientific brainpower. . . ." (*NYT Mag.*, 10/11/64, 59)

October 12: U.S.S.R. launched into orbit VOSKHOD I spacecraft with three-man crew: pilot-cosmonaut, Engr. Col. Vladimir Mikhailovich Komarov; scientific co-worker cosmonaut, Candidate of Technical Sciences, Konstantin Petrovich Feoktistov; and physician-cosmonaut, Boris Borisovich Yegorov. This was history's first multi-manned space flight and marked the entry into space of a scientist and a physician. Initial orbit (as given by Tass): 409-km. (254-mi.) apogee, 178-km. (110-mi.) perigee, 90.1-min. period, and 65° inclination to the equator. Tass announced purposes of the space flight were to: test new multiseat manned spacecraft; investigate work capability and interrelationship, in flight, of a group of cosmonauts consisting of specialists from various branches of science and technology; conduct scientific physico-technical research under actual spaceflight conditions; continue study of effects of spaceflight conditions on human organism; and conduct expanded medico-biological investigations under conditions of prolonged flight. Television photographs of the cosmonauts, wearing lightweight woolen suits rather than spacesuits, were received in U.S.S.R. and transmitted both taped and live by TV stations during the day. On its eighth orbit, as the spacecraft passed over U.S., the crew radioed greeting: "From aboard the spaceship VOSKHOD we convey our best wishes to the industrious American people. We wish the people of the United States peace and happiness." (Tass, *Krasnaya Zvezda*, 10/13/64, 1, ATSS–T Trans.; Tanner, *NYT*, 10/13/64, 1)

- Prof. Leonid I. Sedov, Soviet space official, said in *Izvestia* interview that success of VOSKHOD I "opens up new horizons" in space. It now made realistic any plans about building an "orbiting space platform, a flying space institute," which could serve as a "springboard for further interplanetary expeditions." And Lt. Col. Pavel Popovich, Soviet cosmonaut, said in Tass interview that flight of VOSKHOD I was of great significance "because our aim is to send space stations to distant planets." (Tanner, *NYT*, 10/13/64, 1, 18)

- Yuri M. Vorontsov, Soviet delegate to U.N. Legal Subcommittee of the Committee on the Peaceful Uses of Outer Space, said orbiting of three-man VOSKHOD I spacecraft represented "a great achievement of Soviet science and the Soviet people 'on behalf of all mankind.'" Dr. N. M. Sissakian, member of presidium of Soviet Academy of Sciences, said in Paris that U.S.S.R. hoped to present scientific results of VOSKHOD I space flight to international symposium on manned space flight in Paris next June. (AP, *Balt. Sun*, 10/13/64; Sullivan, *NYT*, 10/13/64, 18)

October 12: Yuri Gagarin, Soviet cosmonaut and first man to fly in space, said in *Izvestia* interview that Engr. Col. Vladimir Mikhailovitch Komarov, member of VOSKHOD I crew, had once been disqualified for cosmonaut training because of heart-beat irregularity, but had been reinstated when a heart specialist declared his heart to be sound. (Loory, *N.Y. Her., Trib.*, 10/13/64)

- British Labor party leader Harold Wilson said Soviet VOSKHOD I spacecraft was forerunner of missile-launching satellites that would make Polaris missile obsolete. Such space-based weapons, he said, "would mean that the all-American deterrent on which Sir Alec Douglas-Home bases his defense argument will soon be made obsolete by space missile development." British election was three days away. (Culhane, *Balt. Sun*, 10/13/64)
- Dr. Wernher von Braun, Director of NASA Marshall Space Flight Center, said in Gatlinburg, Tenn., speech to NASA–AEC–USAF symposium on radiation protection that "not too far in the future huge nuclear-propelled spacecraft will take crews of men on long voyages deep into space. . . .
    "The problems of man-made radiation connected with these space ships will prove far more challenging in the long run than those of the natural radiation in space. . . ." (*NYT*, 10/13/64, 18)
- First test patterns had been sent over special cable link between Suitland, Md., and Moscow, for exchange of meteorological data, according to U.S. Weather Bureau. The link was considered operational, could be used any time U.S. and U.S.S.R. agreed on scheduling and other details of the cooperative exchange. The cable was result of cooperative space agreement reached in 1962 by NASA Deputy Administrator Dr. Hugh L. Dryden for the U.S. and Academician A. A. Blagonravov for the U.S.S.R. (Gustaitis, *Wash. Post*, 10/13/64)
- First meeting of Joint Navigation Satellite Committee, composed of representatives from NASA, FAA, and Departments of Commerce, Defense, Interior, and Treasury. Committee would evaluate requirements for improved navigation, traffic control, and search-rescue services, then determine whether navsat system could meet these requirements and at what costs. (NASA Proj. Off.; NASA Release 64–253)
- USAF XB–70A research bomber achieved 740-mph speed and 35,000-ft. altitude in its third flight test. Pilot Alvin S. White said it hit supersonic speed three times during the 1 hr., 35 min. flight near Edwards AFB, Calif. (*Wash. Post*, 10/13/64; AP, *NYT*, 10/13/64, 31)
- Details of Phoebus-powered nuclear rocket engine, as described by Harold B. Finger, Director of NASA–AEC Space Nuclear Propulsion Office, in series of lectures sponsored by NATO Advisory Group for Aeronautical Research and Development (AGARD): The second-generation nuclear rocket would have 250,000-lb. thrust, using 5,000-megawatt Phoebus reactor. In clustered formation, the engines could launch manned spacecraft out of earth orbit on Mars trajectory. Ground-test firings of first-generation (low-power) Phoebus would be conducted in 1965–66, and tests of the heavier Phoebus reactor in 1967. (*M&R*, 10/12/64, 17)
- Safety of jet airliners had improved sevenfold since 1959, William W. Moss told 17th international air safety seminar of the Flight Foundation in New York. Capt. Moss, a Pan American World Airways pilot, said there was average of one fatal accident per 690,000 flying hours in first nine months of this year, whereas there was one per 85,000 flying hours in 1959. (*NYT*, 10/13/64, 74)

October 12–14: Entry Technology Conference sponsored by AIAA was held in Williamsburg, Va., and NASA Langley Research Center, Hampton, Va. Among the papers presented at the conference was review of Project Fire by Richard C. Dingeldein, of Langley's Flight Re-entry Programs Office. He reviewed results obtained from first Project Fire experiment, which achieved re-entry velocity of nearly 38,000 fps. (LARC Release; AIAA Bulletin)

October 13: U.S.S.R. announced three-man VOSKHOD I spacecraft had landed safely after orbiting the earth for 24 hrs., 17 min. (16 orbits). All three crewmen were reported to be in good condition after VOSKHOD I landed at the "pre-arranged spot." Indications were that the men remained inside the craft to completion of landing. Official announcement asserted that VOSKHOD I had completed its task before landing and was landing according to schedule. Soviet news report indicated the crew had requested permission to continue the flight "for another 24 hours," but Chief Designer had replied that "we shall stay within the program." (AP, 10/13/64; Tanner, *NYT*, 10/14/64, 1, 18)

- Dr. Edward C. Welsh, Executive Secretary of the National Aeronautics and Space Council, said U.S.S.R.'s successful orbiting of three-man VOSKHOD I spacecraft "shows they are actively in the race with manned spacecraft." He noted Soviet emphasis on life sciences in their space program. "They have done more in this field, I believe, than we have." Asked about U.S. chances for landing a man on the moon before U.S.S.R., he replied: "We have a 50–50 chance of getting there first." But, he added, U.S.S.R. might be first with a circumlunar flight. Citing buildup of Soviet space program, Dr. Welsh said ratio of their spacecraft launchings in 1964 to comparable portion of 1963 was about 3-to-1. He described VOSKHOD I as a "logical follow-on from a very rapidly stepped-up space program." (Sehlstedt, *Balt. Sun*, 10/13/64)

- NASA Administrator James E. Webb said in interview that Soviet three-man VOSKHOD I was a "significant space accomplishment.

    "It is a clear indication that the Russians are continuing a large space program for the achievement of national power and prestige."

    He observed that Saturn V rocket, while not yet available, represented U.S. bid for space supremacy. "It is still a good bid and our policy is to proceed rapidly to the flight stage although it is still two years away." He said Saturn V was " a vehicle that can put 240,000 pounds in orbit as against the 10,000- to 15,000-pound class the Russians have been flying." (Sehlstedt, *Balt. Sun*, 10/13/64; *Houston Post*, 10/13/64)

- Speculation about why Soviet VOSKHOD I space flight was terminated after 24 hours was proferred by experts around the world. U.S. space engineers generally indicated Soviets operated according to plan by terminating flight when they did. However, officials of other Western countries speculated VOSKHOD I was brought down prematurely, for reasons variously given as illness among the crew, faulty radio transmitter, and improper orbit. Sen. Clinton P. Anderson (D.–N. Mex.), Chairman of Senate Committee on Aeronautical and Space Sciences, said, "We knew the shot was coming and we had information it would involve a week—that's why they had the doctor along." He suggested technical problems had caused premature ending of trip. Some sources pointed to Soviets' own announcement the previous day, which referred to "conditions of prolonged flight." At least one press analysis, in retrospect, said VOSK-HOD I was perhaps ordered back to earth because of political events in the Soviet Union—Soviet Premier Nikita Khrushchev was being removed from office by Presidium of the Communist Party Central Committee, an occurrence not made public until Oct. 15. (Ubell, *N.Y. Her. Trib.*, 10/14/64; Simons, *Wash. Post*, 10/14/64; *Time*, 10/30/64, 37)

October 13: Unnamed NASA official quoted by Howard Simons in *Washington Post* as saying there was "increasing evidence" that U.S.S.R. was building powerful new rocket booster. However, it was not used in launching three-man VOSKHOD I spacecraft. U.S. Government officials were said to believe VOSKHOD I launched by "old, reliable space boosters . . . mated into a multistage rocket." (Simons, *Wash. Post*, 10/14/64)

- Analyzing impact of VOSKHOD I flight, John Finney said feeling of U.S. Government officials was that VOSKHOD I "probably was the prelude to a longer flight sometime soon and to the unveiling of a more powerful rocket for launchings." In itself, VOSKHOD I flight was considered of more psychological than technical significance. (Finney, *NYT*, 10/14/64, 18)

- Sen. Barry M. Goldwater (R.–Ariz.), Republican Presidential candidate, said in Topeka, Kans.:

  "The entry of the Soviet Union into such a commanding position in near-space [VOSKHOD I flight] also requires an urgent and full time consideration by all concerned branches of the Federal Government and, of course, the President.

  "Near space, the area in which orbital flight takes place, is the key to the military use of space. The nation that commands near space could, after developing proper weapons and maneuverable space craft, dominate the earth.

  "I have pleaded for years for a realistic emphasis on research in this area. Instead, this Administration has dedicated billions of dollars and the virtual entirety of its space efforts, to a moon shot.

  "The military possibilities of space have been downgraded by this Administration, as, indeed, has most research into the advanced weapon systems which we so urgently need to keep the peace in the face of Soviet advances scientifically, and Soviet threats politically.

  "Again, this is a full time and a full-scale problem requiring immediate attention. We need a major redirection of our space research efforts." (Text, *SBD*, 10/14/64, 236)

- Washington *Star* was informed of VOSKHOD I's safe landing 25 min. before Soviet Government made the official announcement, when Edward Baskakov of the newspaper *Soviet Russia* called from Moscow for a reaction to the flight. "Beat" of the *Star* was based upon their editorial on SPUTNIK I in 1957. (*Wash. Eve. Star*, 10/13/64)

- NASA Goddard Space Flight Center officials announced they believed they had successfully "hit" EXPLORER XXII satellite with laser beams fired one-per-second across more than 500 miles. Whether the hits were definitely achieved or not would await further analysis. (AP, *Balt. Sun*, 10/14/64; *Wash. Post*, 10/14/64)

October 14: U.S.S.R. announced routine launching of COSMOS XLVIII satellite into orbit, with the following parameters: 295-km. apogee (183-mi.), 203-km. perigee (126-mi.), 89.4-min. period, and 65.07°-inclination to the equator. (Tass, *Krasnaya Zvezda*, 10/15/64, 1, ATSS–T Trans.)

October 14: NASA Marshall Space Flight Center announced award of two contract modifications to Boeing Co., prime contractor for Saturn V first stage (S–IC): $2,918,418 modification for design, development, and manufacture of components for S–IC stage umbilical connection and related hardware; and, $11,836,000 modification for structural static-load testing program on S–IC to assure structural integrity of the 7.5-million-lb.-thrust rocket stage. These modifications increased total value of the S–IC contract to $516,206,009. (*Marshall Star*, 10/14/64, 9)

- NASA Goddard Space Flight Center announced two firms would negotiate to provide three Satellite Telemetry Automatic Reduction Systems (STARS): Beckman Instruments, Inc., and Electro Mechanical Research, Inc. Negotiations with the two firms would lead to final selection of single contractor for $2.5-million award. (GSFC Release G–30–64)
- Three Soviet cosmonauts spent the day undergoing medical checks and post-flight examinations, following their 24-hour orbital space flight. (Clymer, *Balt. Sun*, 10/15/64)
- Soviet geophysicist Academician Yevgeny Fedorov said in *Izvestia* that VOSKHOD I space flight proved Soviet space program was more advanced than that of U.S., adding: "I am sure that in the future, both Soviet and American cosmonauts will meet on the moon." *Pravda* editorial writer Victor Mayevsky said on Moscow TV that the flight of VOSKHOD I was a "Pearl Harbor in space for America." (Tanner, *NYT*, 10/15/64, 8; UPI, *Chic. Trib.*, 10/15/64)
- Speculation that U.S.S.R. timed their three-man VOSKHOD I space flight "to take the edge off the Chinese atomic demonstration which is expected very soon" was offered by Max Lerner in *New York Post*. (Lerner, *N.Y. Post*, 10/14/64)
- Discovery of a sixteenth meson particle in the atomic nucleus, named "epsilon," by scientists led by Dr. Bogdan C. Maglic at the European Center for Nuclear Research (CERN), was announced. (*NYT*, 10/15/64)

October 15: X–15 No. 1 flown by John B. McKay (NASA) to 2,932 mph (mach 4.44) and 82,000 ft. near Edwards, Calif. This was first flight of No. 1 since it was modified for high-altitude flight research (wing-tip pods containing micrometeoroid and dust collectors, spectrometer, and densitron; new inertial guidance system originally designed for use in USAF X–20). Purpose of this flight was operational checkout of new equipment and evaluation of flight characteristics with the wing-tip pods added. (NASA X–15 Proj. Off.; FRC Release)

- MSC conducted partially successful test of parasail landing system in Galveston Bay, using dummy Gemini spacecraft. (AP, 10/16/64)
- Eighty-three scientists, engineers, and operating groups received medals, citations, and cash awards totaling $7,900 for outstanding contributions to U.S. civilian space program, in NASA's 6th Annual Honor Awards Ceremony. Recipients of awards for exceptional scientific achievement: William R. Lucas, MSFC; Ernst Stuhlinger, MSFC; Frank B. McDonald, GSFC; and Daniel G. Mazur, GSFC. Recipients of award for outstanding leadership: Dr. Kurt H. Debus, Director, KSC; Dr. Harry J. Goett, Director, GSFC; and Dr. Wernher von Braun, Director, MSFC. Group achievement awards were accepted by Dr. Debus for KSC and by Dr. von Braun for MSFC. Recipients of awards for exceptional service: I. Edward Garrick, Oran W. Nicks, and Leonard Jaffe of NASA Hq.; Hans

F. Gruene of KSC; and Wesley L. Hjornevik of MSC. Public service awards went to Bernie P. Miller, of RCA, and Harris M. Schurmeier and Allen E. Wolfe, of JPL, for their work on RANGER VII lunar-photography mission. Special award was presented to Newton W. Cunningham, Hq., for his work on RANGER VII. Cash awards for inventions and contributions were presented to personnel from Ames Research Center, Lewis Research Center, Marshall Space Flight Center, and Jet Propulsion Laboratory. Personnel from Hq., FRC, LaRC, LRC, MSC, and MSFC received Presidential citations for contributions to economy and efficiency in Government. (Program; NASA Release 64–263)

October 15: 22,000-lb.-thrust rocket engine for Project Apollo service module, to return Apollo astronauts from the moon, was test-fired twice at Las Cruces, N. Mex. (AP, Balt. *Sun*, 10/16/64)

- USAF version of F–111 tactical fighter plane, designed for use by both USAF and USN, rolled out of General Dynamics' Fort Worth, Tex., plant, in ceremony attended by Secretary of Defense Robert S. McNamara, Secretary of the Air Force Eugene M. Zuckert, Secretary of the Navy Paul H. Nitze, and other officials. Secretary McNamara said: "For the first time in aviation history, we have an airplane with the range of a transport, the carrying capacity and endurance of a bomber and the agility of a fighter-pursuit plane. . . ." The technology of the variable-geometry wing configuration was developed at NASA Langley Research Center. (DOD Release 719–64; Witkin, *NYT*, 10/16/64, 1, 11; Fryklund, Wash. *Eve. Star*, 10/15/64; *Av. Wk.*, 10/26/64, 32)

- Soviet newspaper *Trud* science writer Leonid Byshko said flight to the moon was "one of the most immediate problems of cosmonautics." After lunar flight, he said, next goal would be flights to other planets in the solar system. "For this purpose, new rocket carriers based on atomic energy are needed. . . . There are no difficulties in principle for creating such nuclear rockets. Only a certain time is required for designing, building and testing." (UPI, *Wash. Post*, 10/16/64)

- U.S.S.R. announced Nikita Khrushchev had been "released" of all his official duties for reasons of "age and deteriorating health." His successors were named: Leonid Brezhnev as Secretary of the Communist Party, and Alexei Kosygin as Premier of U.S.S.R. (AP, *Wash. Post*, 10/16/64)

- Craig Breedlove bettered his own world land-speed record at Bonneville Salt Flats, Utah, driving his three-wheeled jet-propelled vehicle at an average of 526.28 mph, then survived crash when the racer veered off the course and, its braking gear failing, skidded through shallow mud before finally halting in 15-ft.-deep pond. (UPI, *NYT*, 10/16/64, 48)

- NRX–A2 Nerva (Nuclear Engine for Rocket Vehicle Application) experimental reactor, run at full power Sept. 24, was restarted in test of reactor system in the low-power, low-flow region of operation. Analysis of data from this run confirmed design predictions and indicated no unusual start-up or control problems. Test was conducted by Aerojet-General and Westinghouse Electric Corporations for NASA–AEC Space Nuclear Propulsion Office. (SNPO; NASA Press Conf. Transcript, 11/12/64)

October 15–16: NASA's RELAY II communications satellite was used to transmit television coverage of election events in Britain; it was joined by RELAY I on the 16th. (NASA Release 64–254)

October 16: NASA announced the first electric-rocket engine space flight (July 20) had been so successful that a second test would not be necessary. In Sert I (Space Electric Rocket Test), electron-bombardment engine using mercury propellant produced an ion beam developing .0055 lb. of thrust. (NASA Release 64–264; LRC Release 64–93)

- NASA Langley Research Center announced that a foam generator developed for possible use in recovery at sea of instrument packages returned from space had been patented. Attached to instrument packages, generator mixes chemicals to produce almost instant plastic foam as it parachutes to water landing, filling a bag which keeps the payload floating on the surface awaiting recovery. (LaRC Release, 10/16/64)
- U.S.-Japan telecommunications cable broke at the peak of traffic from the Olympic Games in Tokyo, including sound track for films telecast from Tokyo and sent via SYNCOM III communications satellite. Japanese officials estimated earliest repair date as Oct. 23. (*NYT*, 10/17/64, 13)
- NASA received patent for life preserver carried by Mercury astronauts on all orbital space flights. Invented by MSC engineers Mathew I. Radnofsky and Glenn A. Shewmake, life-preserver was called "Ram's Horn" because of its shape: main part fits across wearer's chest and hook-like ends fit over the shoulders. (Jones, *NYT*, 10/17/64, 35)
- Hovercraft, also known as GEM's (ground effect machines), probably will be regulated by Coast Guard, Civil Aeronautics Board, and Federal Maritime Commission, according to tentative position statements released jointly by nine Government agencies. (AP, *NYT*, 10/17/64, 46; *M&R*, 10/26/64, 45)
- Archeological relics dating as far back as 3,500 B.C. were being uncovered at future site of man's rocket launchings to the moon—Merritt Island, Fla. (NASA Space Sheet, 10/16/64)
- Communist China announced it had detonated its first nuclear device. Three minutes before the nuclear detonation an earthquake occurred in Mongolia, precluding the possibility of detecting the nuclear explosion by AFCRL's four automatic seismic stations in northeastern U.S. (OAR *Review*, 1/65, 15; Reuters, *NYT*, 10/17/64, 10)

October 17: USAF launched Thor-Agena D space booster with unidentified satellite payload, from WTR. (HHN–48)

- Under Secretary of the Air Force Brockway McMillan said in *Journal of the Armed Forces* article: "The Department of Defense and the Air Force in particular must provide every reasonable means of support to NASA and I know of no high official in DOD or the Air Force who does not share this conviction. . . ." He pointed to "a conviction shared by both DOD and NASA that there must be a maximum of cross-pollination between respective programs. To do anything less would be short sighted and wasteful and would certainly indicate that we had failed to grasp the essentiality of the national space effort as a consensus. . . ." (*J/Armed Forces*, 10/17/64, 16–17)

October 18: National Aeronautics and Space Council Executive Secretary and Acting Chairman Dr. Edward C. Welsh said on "From the People" radio program that U.S. has "at least a 50–50 chance of getting to the moon ahead of [U.S.S.R.]. . . . I'm talking about landing on the moon and taking off from the moon and returning safely to earth. I'm not talking about a [circumlunar] shot around the moon which they might be able to do even with this [present] equipment." Discussing Soviet's three-man VOSKHOD I space flight, he said the fact that the cosmonauts did not wear spacesuits was "probably one of the most significant aspects of the whole flight." (AP, *Houston Post*, 10/19/64)

October 18: NASA Administrator James E. Webb said in interview with John Finney on the subject of U.S. and U.S.S.R.'s space programs:

"The significance [of VOSKHOD I flight], it seems to me, is the clear indication that manned flight in large spacecraft is important to the Russians, and offers opportunities for them for the future that justifies the investment clearly required."

Asked if VOSKHOD I flight meant U.S.S.R. had "extended its lead over the United States in manned space flight," Mr. Webb replied:

"This is hard to answer categorically but if I had to give a quick answer, I would say 'Yes.'"

Asked when U.S. would overtake U.S.S.R. in manned space flight, Mr. Webb said:

"Well, first I think we must always keep in mind that our program including the developmental flights on the Saturn IB and the flights out to a quarter of a million miles with Saturn V are designed to meet United States needs for an operational capability out that far in space.

"If the flight capability of the Russian craft is extended to include these additional capabilities, then we will have a serious problem in catching up in the next few years. If their capability does not extend to the booster power necessary for the full range of operations out to the moon, then we will pass them in this decade, provided our present program is adequately funded and no unknown or unanticipated difficulties show up." (Interview, *NYT*, 10/18/64, E5)

- Soviet space scientists Dr. Vladimir Lebedev and Dr. Oleg Kuznetsov, writing in Soviet Air Force magazine, discussed U.S. Astronaut L. Gordon Cooper's sightings of houses and other objects from space as he orbited over Tibet in May 1963. The scientists noted that some U.S. experts had attributed Major Cooper's sightings to hallucinations caused by spaceflight conditions. However, the Soviets said, their own experiments with men in soundproof, dark chambers simulating spacecraft led them to conclude that Major Cooper's observations were "an illusion of recognition caused by insufficient information." (AP, *NYT*, 10/19/64, 3)

October 19: NASA launched four-stage Javelin (Argo D-4) sounding rocket from Wallops Island, Va., with sixth payload of Japan-U.S. experiments in cooperative program to investigate properties and characteristics of the ionosphere. Payload was sent to 521-mi. altitude before impacting in Atlantic Ocean 840 mi. from launch site. Telemetry during flight was intermittent, so preliminary results of the experiments were inconclusive. For the cooperative experiments, Radio Research Laboratory, Tokyo, supplied radio frequency (RF) resonance probe and NASA Goddard Space Flight Center supplied Langmuir probe. (Wallops Release 64-75)

- Rally in Moscow's Red Square celebrated homecoming for the three Soviet cosmonauts who orbited Oct. 12-13 in VOSKHOD I spacecraft: Vladimir Mikhailovich Komarov, Konstantin Petrovich Feoktistov, and Boris Borisovich Yegorov. This was first public appearance of new U.S.S.R. leadership, and principal address was given by CPSU First Secretary Leonid Brezhnev, who said:

"The Soviet Union has still further outstripped the United States of America in the 'space race'. . . . [We] do not regard our space research as an end in itself, as some kind of 'race.' In the great and serious cause of the exploration and development of outer space, the spirit of the frantic gamblers is alien to us. We see in this cause part and parcel of the tremendous constructive work of the Soviet people . . . in conformity with the general line of our party in all spheres of the economy, science and culture in the name of man, for the sake of man. . . ." (Tanner, *NYT*, 10/20/64, 1, 2; AP, 10/19/64)

October 19: NASA announced Dr. George E. Mueller, Associate Administrator for Manned Space Flight, assumed the role of acting director of Project Gemini. George M. Low, former Deputy Associate Administrator for Manned Space Flight at Hq. and since February 1964 Deputy Director of NASA Manned Spacecraft Center, had served as acting director of Gemini since last November. (NASA Release 64-247)

- NASA announced three-month expedition at sea using converted Navy aircraft transport to launch sounding rockets would be conducted next year as part of International Quiet Sun Year (IQSY). 40 or more scientific experiments would be carried by Nike-Apache and Nike-Cajun sounding rockets as well as Arcas and Hasp weather rockets to areas of the upper atmosphere and the ionosphere that cannot be reached by land-based rockets. Exact locations of the launches had not yet been determined, but it had been decided that a number of experiments would be made in region of the magnetic equator in eastern portion of the South Pacific Ocean. NASA Wallops Station was assigned project management of the expedition and Goddard Space Flight Center was responsible for coordinating most of the scientific experiments, being provided by universities, other Federal agencies, and NASA field centers. Project was under overall direction of NASA Office of Space Science and Applications. (NASA Release 64-246; Wallops Release 64-73)

- NASA began recruiting 10-20 scientist-astronauts for future manned space flights. Qualified applicants—with doctoral degree or equivalent experience in natural sciences, medicine, or engineering—would be screened and evaluated according to standards established by NASA and ad hoc Committee on Scientific Qualifications of Scientist-Astronauts of the NAS Space Science Board. Before final selection as astronauts, applicants would receive training in a limited space-simulation program designed to familiarize them with space environment and test their ability to withstand physical stresses of space flight. Departing from previous requirement that astronaut candidates be experienced jet pilots, NASA said selected astronauts not already qualified pilots would be given individual flight training to qualify them as pilots of high-performance aircraft and helicopters. (NASA Release 64-248)

- Maj. Edwin Aldrin, Jr., from the newest group of NASA astronauts, already was qualified as a scientist-astronaut, holding a doctorate in astronautics. Two other astronauts were nearing completion of doctoral studies: R. Walter Cunningham (physics) and Russell L. Schweikart (aeronautics/astronautics). (*Houston Post*, 10/20/64)

- Astronaut Russell L. Schweikart began week-long test of continuous living in a Gemini spacesuit. Before donning the spacesuit at Edwards AFB, he had biosensors attached to his body. Among his activities for the test: experiencing weightlessness in F-104B aircraft, going through reproduction of a Gemini space flight in Manned Spacecraft Center mission simulator, and spinning in Ames Research Center centrifuge. (Maloney, *Houston Post*, 10/20/64)

October 19: NASA Administrator James E. Webb said in interview in Chicago that he had predicted in 1961 that U.S.S.R. could launch a multimanned spacecraft with then-existing equipment and that they could send a manned spacecraft on a lunar-orbiting mission—"and this might still be attempted soon." He added: "Nothing that occurred during the recent three-man flight has caused me to deviate from the prediction I made in 1961." (*Chic. Trib.*, 10/20/64)

- IPL analysis of portion of last lunar surface photographed by RANGER VII found that median slope was about 1.6° over 30-ft. distance, not considered a problem for Appollo's Lunar Excursion Module (LEM). (*Av. Wk.*, 10/19/64, 23, 30)
- Editorial by Robert Hotz in *Aviation Week* summarized significance of Soviets' VOSKHOD I three-man space flight:

  "It establishes clearly that the Soviets have not abandoned their goal of a manned lunar landing. . . .

  "It establishes the scientific validity of the Soviet manned space program and should convince even dullards that this venture has sound purpose far beyond its propaganda value. . . .

  "It indicates clearly that the Soviets are prepared to continue with their own across-the-board space exploration program and will not be dependent on any cooperative ventures with the United States. . . .

  "It also confirms the widening Soviet lead in space medicine, which is the key to all manned space flight operations. . . .

  "It spotlights the ultra-conservatism shown by the U.S. military approach to manned space flight in starting to do now what the Pentagon planned to begin on a lesser scale in three years. . . ." (*Av. Wk.*, 10/19/64, 21)
- NASA announced appointment of Charles F. Yost, assistant director of DOD's Advanced Research Projects Agency (ARPA) since 1960, as Director of Office of Technology Utilization, NASA Hq., effective Nov. 1. (NASA Release 64-245)
- Reported that NASA was considering University Explorer program, under which universities would design and build their own scientific satellites with NASA funding. The satellites would be launched from Wallops Island with Scout boosters. (*Av. Wk.*, 10/19/64, 35)
- Univ. of Chicago president George W. Beadle said at National Electronics Conference in Chicago that communication of scientific information among industry, Government, and universities was too slow. He said two main things should be done: "(1) academic people must do better at translating their discoveries and (2) industry and government applying this basic research must do better at keeping up on new developments so they can take advantage of them." (*Chic. Trib.*, 10/20/64)
- Unnamed DOD officials said they were certain that U.S.S.R.'s Cosmos satellites were being used for reconnaissance over U.S. (Hoffman, *Wash. Eve. Star*, 10/19/64)
- British astronomer Sir Bernard Lovell, Director of Jodrell Bank observatory, told newsmen in Boston that there were "indications the Soviet economy is being strained" by their space program. He predicted that U.S.S.R. would next attempt "to achieve a rendezvous of two spaceships and to circumvent the moon." (Long, *CSM*, 10/19/64)

October 19: Royal Crown Cola Co. president W. H. Glenn announced Lt. Col. John H. Glenn, Jr., former NASA astronaut, had been elected a director of the company. (Hammer, *NYT*, 10/20/64, 49)
- Explosion-fire of fuel for USAF XB–70 aircraft at Edwards AFB, Calif., caused four injuries and one death (Ivan W. Mayo of North American Aviation, Inc.). (UPI, *Wash. Post*, 10/21/64)
- Republican six-man study group headed by former Secretary of Defense Neil H. McElroy released report charging that Kennedy-Johnson Administration had neglected research and development of new weapon systems. In reply to the charges, DOD issued statement saying it was "pursuing a vigorous and effective research and development program designed to keep this nation the strongest in the world . . ." and citing more than 200 major new projects initiated since 1961. (AP, *NYT*, 10/20/64)

October 20: U.K. successfully launched its second Blue Streak rocket booster from Woomera, Australia, in development test. Blue Streak was slated to serve as first stage of ELDO's multistage launch vehicle. (*M&R*, 10/26/64, 11)
- USAF space medicine team, in paper delivered to seminar at Brooks AFB, reported on experiment with 12 volunteers committed to "absolute bed rest" for four weeks, in attempt to simulate conditions of extended space flight. After the period, subjects showed marked increase in tendency toward fainting, but when protected with antigravity suits this tendency nearly disappeared. Heartbeat rates were faster in stand-up test after the bedrest than in similar test before it. High gravity loads in centrifuge did not adversely affect the subjects after their long rest. The volunteers lost about one-fifth of a gallon of their total blood volume, and increase in calcium loss was apparent, but was not considered hazardous. (Hines, Wash. *Eve. Star*, 10/21/64)
- Rear Adm. Francis D. Boyle, Deputy Director of Defense Communications Agency's Communications Satellite Project Office, told National Space Club in Washington of DOD plans for a military comsat system. He said Titan III–C launch vehicles would orbit up to 24 comsats during the first six months of 1966, placing six to eight satellites in orbit each launch. Orbiting at more than 22,000 mi., the satellites would be "approximately synchronous." (AP, Wash. *Eve. Star*, 10/21/64)
- Lt. Col. John D. Peters (USAF), engineer at Arnold Engineering Development Center, said in San Antonio that the four-legged landing gear designed for Apollo's Lunar Excursion Module (LEM) was "needlessly risky" and should be replaced by an inflatable "doughnut" to cushion the LEM's landing. Tube at LEM's base could be inflated just before landing, would allow a safe landing even if exact nature of lunar surface is unknown. (Hines, Wash. *Eve. Star*, 10/20/64; *SBD*, 10/22/64, 283)
- Communications Satellite Corp. requested bids from 15 companies for research data and consultant services for defining design and performance requirements of ground terminals for global comsat system. (ComSatCorp Release)

October 21: Astrobee 1500, most powerful of U.S. sounding rockets, was flight-tested from NASA's Wallops Island, Va., launch site, its second stage carrying 156-lb. instrument package to 1,212-mi. altitude before impact in the Atlantic Ocean 1,326 mi. from launch site. This was second flight test of the solid-propellant rocket; adequate performance data were not achieved in first NASA flight test (April 1963) because nose-fairing failure ended the flight only 16 sec. after launch. In today's flight, 27 min. of telemetered data were received on acceleration, vibration, temperatures, and pressures, and from preliminary examination the test appeared successful. (Wallops Release 64-76)

October 21: Mstislav V. Keldysh, president of Soviet Academy of Sciences, said in Moscow press conference that vehicle used to launch VOSKHOD I spacecraft was more powerful "than that used before to put into orbit any instruments or astronauts." Keldysh said VOSKHOD I was first of a series but revealed nothing about future flights. Konstantin P. Feoktistov, scientist-cosmonaut aboard VOSKHOD I, revealed he had made astronomical observations during the space flight, practicing "aeronavigation by means of a sextant." He predicted that on interplanetary flights "it will be possible to fix the spaceship's position autonomously on board the spaceship." Feoktistov also disclosed VOSKHOD I's attitude control system was based on ion propulsion engine. He described VOSKHOD I's return to earth as a "featherbed landing." Of the spacecraft's landing system, Cosmonaut Vladimir Komarov said after parachutes were opened, a second retrorocket was fired to achieve the soft landing. Physician-cosmonaut Boris B. Yegorov disclosed that "sharp movements of the head caused slight dizzy sensations, sensations of general discomfort" for both him and Feoktistov. (Shabad, *NYT*, 10/22/64, 15; Clymer, *Balt. Sun*, 10/22/64; AP, *Chic. Trib.*, 10/22/64)

- Dr. Hugh L. Dryden, NASA Deputy Administrator, gave the 1964 Third Bicentennial Marshall Wood Lecture, Brown Univ., Providence, R.I. Dr. Dryden predicted that by the year 2000 astronauts' trips to the moon "may be somewhat commonplace. . . . The moon will also be serving as an important scientific station for radio astronomers . . . [and] many other kinds of scientists. . . . We may be maintaining sizable installations on the moon. . . .

"Beyond the moon it is probable that manned flights will have been made to Mars and possibly to the vicinities of other planets as well. . . . In space stations orbiting above the earth we shall likely be fabricating space vehicles; it is possible that we shall be using rockets, in ballistic trajectory through space, to deliver freight shipments from one continent to another. . . .

"Let us let our imagination roam to a hundred years from now. It seems certain that we will have developed the capability to move throughout the solar system and will have sent at least exploratory expeditions to the outermost reaches. We will be considering the extension of our exploration of space to the nearest stars, but I am somewhat doubtful that we can even in this period discover new sources of energy which will make this journey possible in the lifetime of one man.

"It seems likely that within 100 years we will have established outposts in the form of stations on the moon and Mars. It is possible that we will have moved to a certain degree of colonization on Mars, but I doubt that this will take place on the moon, unless the moon turns out to contain resources of great economic value to us here on earth. The resources required for such extensive exploration of Mars will require cooperation among the nations of the earth; otherwise it is not likely to be accomplished. . . ." (Text)

October 21: FAA announced new series of about 600 sonic-boom tests late this year over White Sands Missile Range, N. Mex., creating sonic boom pressure ranging up to 30 psf above normal atmospheric pressure to measure effects on structures. NASA and other agencies would also participate in the research preparatory to development of a U.S. supersonic transport. (Clark, *NYT,* 10/22/64)

- Nomadic Lapps in northern Sweden had registered complaints about ESRO's rocket launching plans from new base being built about 100 mi. northeast of Kiruna. Beginning in 1966, ESRO would fire 50 rockets per year 120 mi. into the ionosphere every year for eight years. But the Lapps, living in the target area for the rockets, were fearful for the safety of themselves and their reindeer. (Reuters, *NYT,* 10/21/64, 26)
- NASA issued a Tech Brief describing new type of adhesive bonding technique in structural assembly, using epoxy resins as an adhesive which leaves no voids or air pockets between the two surfaces joined together. (NASA Release 64-262)
- Dr. Dwight F. Gunder, 61, special adviser on rockets and missiles to Navy Bureau of Ordnance, died in Loveland, Colo. Dr. Gunder had received citation which said he "provided the guide lines for the Polaris missile's development." (AP, *NYT,* 10/23/64, 34)

October 22: AEC detonated five-kiloton nuclear device 2,700 ft. underground in the Tatum Salt Dome, in southern Mississippi. Called Project Salmon, the test was designed to evaluate performance of seismographic equipment in detecting underground nuclear explosions. Officials estimated the blast created 120-ft.-diameter cavity in the salt dome. (AP, *N.Y. Her. Trib.,* 10/23/64)

- Secretary of Defense Robert S. McNamara said in news conference: ". . . I do not anticipate any significant increase in the Defense budget for the next fiscal year as a result of last week's events. In fact . . . , I feel quite confident that the budget will decline as a percentage of the gross national product, while supporting what is recognized as the strongest military force in the world. . . ." (DOD Release 765-64)
- John L. Sloop, Assistant Associate Administrator for Advanced Research and Technology, NASA, said in lecture to Santa Clara County Science Teachers, at NASA Ames Research Center:

    "On the momentous day of the first flight to the moon's surface, the three astronauts will be using the results of almost a decade of concentrated effort which include knowledge, skills, and equipment of 30,000 people in ten NASA Centers and over a quarter of a million people in more than 5,000 companies. They will have flight experience in space 100 times longer than the total time logged in Mercury. They will have the knowledge gained by dozens of unmanned satellites that have probed space about the moon and beyond. They will have the result of Ranger photographs of the lunar surface plus more detailed surveys made by lunar orbiting and the Surveyor lunar lander. . . ." (Text; NASA R&SC Div.)

October 23: NASA launched Javelin (Argo D-4) sounding rocket from Wallops Island, Va., with 90-lb. payload containing instrumentation to measure low-level cosmic radio noise. The experiment was timed to coincide with passage of EXPLORER XX (fixed frequency topside sounder satellite), which also was being used to investigate cosmic noise source at wavelengths which do not penetrate to earth because of shielding effect of ionosphere. The Javelin was sent up through the region of the satellite about 15 min. after EXPLORER XX had passed through the area at about 650-mi. altitude. During the flight 18 min. of scientific data were telemetered from rocket payload, which reached 660-mi. altitude before impacting in the ocean 650 mi. from launch site. Preliminary analysis indicated flight results were good. (Wallops Release 64–77)

October 23: Gemini Titan GT–2 spacecraft was mated with the Titan II launch vehicle at Launch Complex 19, Cape Kennedy, preparatory to unmanned ballistic flight later this year. (MSC *Roundup,* 10/28/64, 3)

- USAF Atlas-Agena D satellite booster combination was launched from Vandenberg AFB, Calif., with unidentified payload. It was later disclosed that three satellites were orbited. (AP, *NYT,* 10/24/64; HHN–48)

- Test of parasail spacecraft recovery system for manned space flight was conducted by NASA Manned Spacecraft Center over Trinity Bay, Tex. Unmanned boilerplate Gemini spacecraft carrying parasail system in stowed position was dropped from C–119 aircraft at 11,000-ft. altitude. Parasail deployed properly and landing rockets reduced impact to 10 fps, but because of rigging error in parachute installation the spacecraft descended straight down instead of making programed turn. MSC's John Kiker said everything functioned in proper sequence and the test was considered successful. (*Av. Wk.,* 12/14/64, 61)

- Dr. Eugene B. Konecci, Director of Biotechnology and Human Research Div. in NASA Office of Advanced Research and Technology since July 1962, resigned from NASA to join staff of the National Aeronautics and Space Council. Dr. Walton L. Jones, Jr., head of Human Research Branch, was assigned as Acting Director of the Biotechnology and Human Research Div., effective Oct. 26. Dr. Jones joined the division May 5, previously had been Director of Aviation Medicine Technical Div. of USN Bureau of Medicine and Surgery. (NASA Release 64–268)

- NASA Administrator James E. Webb said in Tyler, Tex., that the U.S., "in the short space of three years, has more than tripled its industrial, scientific, engineering, and governmental space capability. This was accomplished through the close cooperation of what might be called the 'triad' of government, industry, and the university community, in carrying out the largest peacetime research and development effort in all history. Management techniques developed in coordinating this complex endeavor are beginning to find their way into business and industrial executive processes. Experience has been gained that may prove useful in solving other difficult national and international problems...." (Text)

- NASA announced preliminary negotiations had begun to lease Convair 990 jet transport, with option to buy, from Convair Div. of General Dynamics Corp. The aircraft would be based at NASA Ames Research Center. Stripped of passenger accommodations, the aircraft would be used as a flying laboratory for research in aeronautical technology and the space sciences. (NASA Release 64–269)

- Rep. Hale Boggs (D.–La.) asserted that top management personnel would be moved from NASA Marshall Space Flight Center, Huntsville, Ala., to Michoud Operations, Louisiana, quoting NASA Administrator James E. Webb as saying that NASA was analyzing the situation. Mr. Webb told *Huntsville Times* of difficulties in hiring and keeping top executives for the Alabama installation, added: "There's nothing in this situation that can't be corrected, but it has to be worked out." (AP, Wash. *Sun. Star*, 10/25/64; Hines, Wash. *Eve. Star*, 10/26/64)

October 23: French National Assembly approved 1965 funds for U.K.-France Concorde supersonic transport. (*NYT*, 10/24/64)

- Dr. Bruno Bettelheim, professor of educational psychology at Univ. of Chicago, said in address before symposium on American Women in Science, at MIT, that women's "specific genius for nurturing, humanizing, and preserving" should be made use of in all spheres of living, in science, engineering, architecture, medicine, and all other professions. (Leland, *Boston Globe*, 10/23/64)

October 24: U.S.S.R. announced launching COSMOS XLIX artificial earth satellite into orbit: apogee, 490 km. (304 mi.); perigee, 260 km. (162 mi.); period, 91.83 min.; and inclination, 49° to the equatorial plane. Onboard equipment was said to be operating normally and ground tracking station was processing incoming data. (Tass, *Krasnaya Zvezda*, 10/25/64, 1, ATSS–T Trans.)

- President Johnson announced U.S. and U.S.S.R. had reached final agreement on exchange of meteorological data, tentatively reached in the Second NASA/Soviet Memorandum of Understanding of June 6, 1964, and approved by U.S.S.R. Oct. 23. Exchange of conventional data via the special telecommunications link would proceed soon. Formal terms of agreement would be made public next week at meeting of U.N. Committee on the Peaceful Uses of Outer Space. (UPI, *Wash. Post*, 10/25/64; NASA Off. of Int. Pro.)

- Subcommittee on Science, Research, and Development of the House Committee on Science and Astronautics recommended steps to be taken to distribute Federal funds more widely throughout the country: conference of representatives of Government and industry to explore methods of wider distribution; study by President's Office of Science and Technology to identify institutions with greatest growth potential; addition of NSF funds "for the specific purpose of insuring the existence of at least one center of excellence in research and technology in each appropriate region of the Nation"; development of long-range programs by the Administration to be carried out at under-used science centers. (Simons, *Wash. Post*, 10/25/64)

- Republican Presidential candidate Sen. Barry M. Goldwater issued report on science prepared by his Task Force on Space, Science, and the Atom, in which he had this to say of Project Apollo:

  "Man should indeed explore the moon, but I see no good reason why we must incur the large unnecessary expense of a crash program. The first man on the moon will probably not gain for us any vital military advantage." He proposed instruments be sent to the lunar surface, emphasis be concentrated on efforts to assure the "security of near-space." (*Wash. Post*, 10/25/64, A9)

- According to NORAD Commander-in-Chief Gen. John K. Gerhart (USAF), manned Soviet satellites have passed over North America 44 times. He added: ". . . strong modern aerospace defenses are indispensable; they are a vital part of deterrence—they are essential to survival and they are our insurance to victory." (*J/Armed Forces*, 10/24/64, 6)

October 24–25: Employees' Open House at NASA Lewis Research Center, with employees and their families touring the facilities and exhibit areas. Estimated 8,000 persons visited the Center. (*Lewis News,* 10/30/64, 1)

October 25: Soviet aircraft designer Artem I. Mikoyan, brother of Soviet President Anastas I. Mikoyan, remarked in Moscow: "Whatever secrets surround the earth's nearest neighbor, they will be solved by man. And one would like to believe that the first man on the moon will be our Soviet cosmonaut. We are confident of this. Confirmation of this may be seen in the flight of the spaceship 'Sunrise' [VOSKHOD]." (AP, *Wash. Post,* 10/26/64)

October 26: NASA's EXPLORER XX topside-ionosphere sounder satellite was providing new information on "the arrangement of ionospheric irregularities" and "the structure of plasma resonances," according to Central Radio Propagation of National Bureau of Standards. (*M&R,* 10/26/64, 23)

- NASA Aerobee 150 sounding rocket was launched from White Sands, N. Mex., to 119-mi. altitude in successful flight to collect data on celestial x-ray sources. American Science and Engineering, Inc., provided payload instrumentation, which included four Geiger counters, one photoelectric detector, two scintillation counters, and two star sensors. Rocket and instrumentation performance were considered good. This was the 500th launch of an Aerobee liquid-propelled sounding rocket, the booster that carried first U.S. rhesus monkeys into space on high-altitude flights in 1951 and 1952. Aerobee was outgrowth of rocket program begun in 1946 by what is now Aerojet-General Corp., which was then engaged in jet-assisted take-off (Jato) research. (NASA Rpt. SRL; Wash. *Eve. Star,* 10/28/64)

- NASA Associate Administrator Dr. Robert C. Seamans, Jr., said in *Missiles and Rockets* interview that NASA was tentatively planning to initiate program definition of Apollo-X spacecraft in FY 1965. He stressed, however, that NASA manned space station of from six-months to one-year lifetime would receive no hardware funding until the 1970's. He also said NASA's Apollo-X study did not compete with USAF Manned Orbiting Laboratory (MOL) program: "MOL is important for the military as a method of determining what opportunities there are for men in space. It is not suitable to fulfill NASA requirements to gain scientific knowledge." (*M&R,* 10/26/64, 14)

- Dr. John J. Brennan, Jr., speaking for Committee to Preserve Cambridge (Mass.) Industries, said he would petition court to decide that proposed Kendall Sq. site for NASA Electronics Research Center did not meet urban renewal standards. Dr. Brennan said he received reply from White House to his letter of protest in which Presidential aide Kenneth O'Donnell said Kendall Sq. would have to meet all legal requirements for urban renewal before NASA could locate there. (*Boston Globe,* 10/26/64)

- Astronaut Russell Schweickart removed spacesuit at NASA Ames Research Center, where he had undergone three rides on centrifuge. Schweickart, who had celebrated his 29th birthday in the spacesuit, had worn it constantly since Oct. 19. (AP, Wash. *Eve. Star,* 10/27/64)

October 26: Soviet astronomer Nikolai S. Kardashev, writing in U.S.S.R.'s *Astronomical Journal*, proposed that two sources of strong radio emissions—CTA-21 and CTA-102—were possible beacons from intelligent civilizations. First observed in 1960, the optically invisible "objects" are unique among recorded sources in that their radiations emit at peak intensities of about 900 megacycles. Kardashev said 900 megacycles would be optimum frequency for interstellar communications, since it is the frequency where background noise—cosmic noise and quantum noise—is at a minimum. (Sullivan, *NYT*, 10/26/64, 1, 36)

- Republican National Committee Chairman Dean Burch said in open letter to President Johnson that NASA Administrator James E. Webb was engaging "in the most blatant and transparent politics by suggesting he will shift the top space scientists and executives from Huntsville, Ala., to New Orleans. . . . This is political blackmail, pure and simple. . . ." (CTPS, *Chic. Trib.*, 10/27/64)
- European Space Research Organization (ESRO) announced plans to build research center at Noordwijk, the Netherlands. Center would employ 800-man international staff, would include laboratories, proving grounds, and other facilities. (*NYT*, 10/27/64, 5)
- London betting house of William Hill Ltd., announced odds of 100-to-1 against manned lunar landing by U.S.S.R. before Jan. 1, 1971 and 150-to-1 against manned lunar landing by U.S. before same date. Several weeks later the firm improved their odds by changing date to Jan. 1, 1968. (AP, Balt. *Sun*, 10/27/64; *SBD*, 11/12/64)

October 27: USAF Project Asset space glider was launched from Cape Kennedy with single-stage Thor booster to 31-mi. altitude before returning downward at 8,800-mph speed and impacting 900 mi. southeast of the launch site in the Atlantic Ocean. USAF said the spacecraft temperatures had reached about 2,000° in the 25-min. flight, which was termed "very successful." (AP, *Wash. Post*, 10/28/64; *M&R*, 11/2/64, 8)

- Red spot on the moon was observed from 12:18 a.m. to about 1:00 a.m. by three lunar observers in Port Tobacco, Md. Using 16-in. telescope, the astronomers described the spot as a "red, pinkish glow" which was "located on the base of the central peak in the Crater Alphonsus." This was said to be eighth sighting of such a spot on the moon in the last six years and the first in Alphonsus since the discovery of the phenomenon by a Russian scientist in 1958. The three observers, associated with Trident Engineer Associates of Annapolis, Md., were using equipment specially developed for detecting the red spots, in lunar observation project for NASA. (*Wash. Sun. Star*, 11/1/64; NASA Release 64-278)
- Maj. Gen. Samuel C. Phillips (USAF), Apollo program deputy director in NASA Hq. since last January, was appointed director of the Apollo program. NASA Associate Administrator for Manned Space Flight Dr. George E. Mueller had been acting Apollo director for the past year. (NASA Release 64-267)
- Balloon launched from Holloman AFB, Tex., carried telescope and instrumentation to 86,000-ft. altitude, where it focused on the planet Venus for more than three hours. Financed by USAF contract, the experiment was second such study of Venus this year by Johns Hopkins Univ. team headed by Dr. John Strong. Scientific results would be announced later this year (see Dec. 7). (Fenton, Balt. *Sun*, 12/7/64; Sullivan, *NYT*, 12/8/64)

October 27: U.S. Army's use of orbiting satellite to pinpoint exact locations of land bodies separated by large expanses of ocean was described by DOD. Using SECOR (Sequential Collation of Range) system with the satellite and overseas ground stations, Army Corps of Engineers' Army Map Service was for the first time obtaining distance measurements of up to a thousand miles to an accuracy within 30 meters. Nine unknown points were expected to be located in the Pacific during the first year of operations. (DOD Release 776–64)

- Writing about development of Vostok and Voskhod spacecraft, Soviet Lt. Gen. N. Kamanin disclosed:

  "The space flights of the six 'Vostoks' became important landmarks in the history of mankind, having pointed out and illuminated the paths into the unknown worlds of the universe. The first spaceship-satellites began to be launched in 1960. At that time, Soviet science had already progressed to a systematic preparation for manned space flights. Many difficulties blocked the way. It was necessary to have a dependable spacecraft that could be launched successfully into the stellar fastnesses and could assure the safety of a cosmonaut's flight including return to Earth. . . .

  "The flight of the 'Voskhod' spacecraft demonstrates that the time is drawing near when man can penetrate far into outer space and can realize the ancient dreams of flights to the Moon, Mars, Venus and to even more remote depths of the universe. Mankind has entered a new epoch in revealing the secrets of nature hidden in the depths of space. . . .

  "The 'Voskhod' has increased reliability, improved equipment and an expanded program. We still do not know what the next spacecraft will be named. We can already say with certainty, however, that it will be a further step along the path of progress. Just as aviation became the means for mass transport of passengers over the airways, the spaceships will become the connecting link between the planets of the solar system. . . ." (*Krasnaya Zvezda*, 10/27/64, ATSS–T Trans.)

- Communications Satellite Corp. re-emphasized to FCC its need to own and operate initial satellite terminal stations in U.S. ComSatCorp was replying to statements filed with FCC by communications common-carriers which requested FCC to deny ComSatCorp's request, originally filed Aug. 14. (ComSatCorp Release)

- Lt. Col. John H. Glenn, Jr. (USMC), received his colonelcy from President Johnson in White House ceremony. Since the Senate had not gotten to the promotion list prior to adjournment, the President, after consultation with members of Congress, had made a recess appointment for Glenn. (UPI, *NYT*, 10/28/64, 38; *Balt. Sun*, 10/28/64)

- Art Arfons set new world land-speed record of 536.71 mph in his jet-powered car at Bonneville Salt Flats, Utah. (AP, *NYT*, 10/28/64, 58)

October 28: U.S.S.R. announced orbiting of COSMOS L satellite into the following orbit: apogee, 150 mi.; perigee, 122 mi.; period, 88.7 min.; inclination to the equator, 51.3°. (Tass, *Krasnaya Zvezda*, 10/30/64, 1, ATSS–T Trans.)

- Titan II rocket booster for first manned Gemini space flight (GT–3) was turned over to the Government in ceremonies at Martin Co.'s Baltimore plant. Charles W. Mathews, Gemini project manager of NASA Manned Spacecraft Center, said parachute landing system and spacecraft's ejection seats were all that remained to be flight-qualified, and GT–3 was still planned for first quarter of 1965. Mathews disclosed MSC had been experimenting for more than a year with braking rockets to enable manned Gemini spacecraft to land on land instead of water. Parasail also would be used in such a landing. Among other officials present were GT-3 crewmen, Astronauts Virgil I. Grissom and John Young. (Hines, Wash. *Eve. Star*, 10/29/64; Witkin, *NYT*, 10/29/64)

October 28: U.S. and U.S.S.R. began exchanging weather data over special teletype, Weather Bureau spokesman said, although some operating details had not yet been worked out. Spokesman said U.S. was not yet sending to U.S.S.R. any data gathered from meteorological satellites. (AP, *NYT*, 10/29/64, 58)

- Brig. Gen. Benjamin G. Holzman (USAF, Ret.) was sworn in by Dr. Raymond L. Bisplinghoff, NASA Associate Administrator for Advanced Research and Technology, as consultant to NASA for planning and administration of its advanced research and technology program. General Holzman would work on staff organization and formation of a program for NASA Electronics Research Center. (NASA Release 64-271)

- Post-Apollo possibilities being considered by NASA were listed by NASA Associate Administrator for Space Science and Applications Dr. Homer E. Newell at luncheon meeting of Institute of Electrical & Electronic Engineers' Professional Technical Group on Nuclear Science, Philadelphia:

  "Adaptation of Gemini as an earth orbiting laboratory.
  "Adaptation of Apollo as an earth orbiting laboratory.
  "Extension of Apollo capabilities to create a six-man (or so) earth orbiting station.
  "Creation of an earth orbiting station to accommodate a score or more persons. . . .
  "Creation of an earth orbiting astronomical observatory in association with a manned orbiting station.
  "Extended manned exploration of the moon.
  "Extended unmanned exploration of the moon.
  "Establishment of a manned lunar base.
  "Establishment of a lunar astronomical facility.
  "Establishment of a lunar radio astronomy facility on the far side of the moon.
  "Unmanned exploration of the solar system.
  "Manned exploration of the planets.
  "Advanced solar studies.
  "Advanced weather satellite system.
  "Advanced communications satellite system.
  "Navigation satellite system.
  "Data collection satellites.
  "Nuclear power.
  "Nuclear propulsion.
  "Electric propulsion."

  Dr. Newell added: "The Space Science Board of the National Academy of Sciences has recently advised the Administrator of NASA that it strongly urges our moving forward with a vigorous, imaginative program of exploration of the solar system. In this regard the Board and our other scientific advisors accord much higher priority to enhancement of our unmanned planetary effort (including exobiology) than to a large-scale follow-on of the current manned lunar landing program or to manned earth orbital missions. This, I believe, reflects the general opinion of the scientific community, and is an important factor to weigh with all the other factors that must be considered in deciding on what next. . . ." (Text)

October 28: NASA Lewis Research Center announced contracts for construction of facilities at LRC's Plum Brook Station: $1,160,000 contract to Roediger Construction, Inc., and $1,386,000 contract to Mosser Construction Co. (LRC Release 64–100)

- FAITH 7 Mercury spacecraft, in which Astronaut L. Gordon Cooper made 22 orbits of the earth in final space flight of Project Mercury, went on display at NASA Hq., Washington. Nov. 2 the spacecraft would resume its tour of the 50 state capitals. (NASA Release, unnumbered)

October 29: Nobel Prize in Physics awarded to Dr. Charles H. Townes (MIT) and two Soviet physicists, Aleksandr M. Prokhorov and Nikolai G. Basov, for research in quantum electronics leading to production of oscillators and amplifiers for lasers. Dr. Townes was Chairman of NASA Manned Space Flight Advisory Committee. (*NYT*, 10/30/64, 1)

- Captive flight of X–15 No. 3 was made with Milton O. Thompson as pilot. (NASA X–15 Proj. Off.)
- NASA Administrator James E. Webb told Huntsville Industrial Expansion Committee, meeting at NASA Marshall Space Flight Center:

  "We have given . . . [MSFC] the job of managing the contractors who will assemble, test and launch giant boosters, and much of this work is handled here in Huntsville.

  "We have placed this management job with the Marshall Center because the management of industrial development contracts involving hundreds of millions of dollars per year requires the daily support by scientists and engineers intimately familiar with the many-faceted development problems. And we want to keep this arrangement. As this very large program progresses, it will be good business, and in the interests of efficiency and lowered costs, to move part of the contract management functions near our Michoud Assembly Facility at New Orleans or our industrial stage assembly plants in California.

  "We have done this in the past and we will undoubtedly do more of it in the future. The extent to which we may have to move more of these management functions out of the Huntsville area will directly depend on our ability to improve our record of attracting senior executives from industry to the Huntsville installation. . . ." (*Marshall Star*, 11/4/64, 4)

- Weber Aircraft Co. of Burbank, Calif., demonstrated the astronaut escape system that it had developed for NASA. System was designed to boost astronauts out of their Gemini spacecraft and open escape parachutes in 10 sec. if a spacecraft malfunction should develop. (Macomber, *San Diego Union*, 10/30/64)
- Minuteman II ICBM flight test from Cape Kennedy marked second success in as many launchings. USAF said the advanced missile achieved all test objectives. (AP, *Houston Post*, 10/30/64)

October 29–30: Conference on New Technology held at NASA Lewis Research Center. Part of NASA's Technology Utilization program, the conference was a means of disseminating new technology developed from aerospace research to other segments of the U.S. economy. (LRC Release 64–95)

October 30: X–15 No. 3 was flown by Milton O. Thompson to maximum altitude of 82,000 ft., the rocket-research aircraft attaining 3,051 mph (mach 4.62). (NASA X–15 Proj. Off.)

- NASA launched Aerobee 150 sounding rocket from White Sands, N. Mex., in successful flight to collect solar spectral data in the x-ray and ultraviolet regions, to test adaptation of solar spectrometer designed for Oso–C satellite, and to measure ionospheric electron density. Rocket reached 117.5-mi. altitude and data were collected in all experiments during the flight. (NASA Rpt. SRL)
- NASA pilot Joseph A. Walker flew Lunar Landing Research Vehicle (LLRV) at NASA Flight Research Center. Total free-flight time in this first test was less than one minute, the LLRV rising 10 ft. (*M&R*, 11/16/64, 8)
- NASA announced selection of scientific satellites, IMP–D and IMP–E, that will be placed in orbit around the moon in 1966–67 to study interplanetary phenomena and earth-sun relationships in cislunar space. The eight experiments selected for the Interplanetary Monitoring Platforms were designed mainly to measure magnetic fields, solar plasma, micrometeoroids, and energetic particles in the vicinity of the moon. (NASA Release 64–270)
- NASA announced appointment of Francis J. Sullivan as Acting Director of Electronics and Control Div., NASA Hq. The position was formerly held by Dr. Albert J. Kelley, who became Deputy Director of NASA's new Electronics Research Center. Sullivan was Deputy Director of the division since November 1963. (NASA Release 64–276)
- Appointment of Lt. Gen. Frank A. Bogart (USAF) as Special Assistant to Associate Administrator for Manned Space Flight announced by NASA. Appointment would become effective shortly after his retirement from USAF Oct. 31. General Bogart would be concerned with industry relationships and special management problems of program control and contracting. (NASA Announcement 64–255)
- U.S. Army's Redstone rocket, developed by NASA Marshall Space Flight Center team when it was formerly part of Army Missile Command, was officially retired from active military operation in ceremony at Army Missile Command, Redstone Arsenal, Ala. Dr. Wernher von Braun, MSFC Director, spoke on Redstone's contributions to U.S. space exploration, including serving as launch vehicle for first and second U.S. manned space flights. Redstone was replaced as military missile by Army's Pershing. (*Marshall Star*, 10/4/64, 10)
- Dr. George E. Mueller, NASA Associate Administrator for Manned Space Flight, gave address at dedication of Magic Mountain Laboratory of the Marquardt Corp., Van Nuys, Calif.:

    "There is a considerable degree of misunderstanding of the overall purposes of our national space program. Many people seem to believe that a landing on the moon is our paramount objective. This is not so.

    "Our principal goal is to make the United States first in space. The Apollo mission milestones—manned flight in earth orbit, into deep space, and to the moon and back—provide a focus for the development of the total capability that will enable the United States to compete in the race for space. . . ." (Text)

October 30: Editorial in North American Aviation's Space & Information Systems *Skywriter:*

"The company's XB-70 has now completed four test flights acquiring millions of items of information, in a highly significant beginning of its test program.

"So far, the airplane has flown at altitudes up to 45,000 feet, at speeds up to Mach 1.4; with the wing tips deflected and up, and through various yaw, flutter and stability tests.

"In 34 days, during which test flights were accomplished, the XB-70 has flown at supersonic speeds for 55 minutes. Performance at takeoff, liftoff, climb out, and angle of climb has coincided almost exactly with flight simulator estimates. . . ." (NAA S&ID *Skywriter*, 10/30/64, 1)

- DOD announced award of $1,413,890 NASA-funded contract to Smith and Sapp Construction Co. for construction of launch equipment shop at Launch Complex 39, Merritt Island Launch Area, Fla. (DOD Release 783-64)
- It was announced that in the first two-man Gemini flight, set for Feb. 1965, Astronaut Virgil I. Grissom might become the first man in space to alter the orbit of his own spacecraft. The change would be after about 90 minutes of flight in the first orbit over Guaymas, Mex. Major Grissom, the command pilot, would be sitting on the left side of the spacecraft and Lt. Cdr. John Young, pilot, would be sitting on the right side. (AP, *Sunday Oregonian*, 11/1/64)
- Colonel John H. Glenn, Jr. (USMC), was declared fully recovered from inner-ear injury by USAF doctors in San Antonio, Tex. The former NASA astronaut was expected to resign from USMC and become an active director of Royal Crown Cola Co. (UPI, *NYT*, 10/31/64, 11)
- Dr. Werner R. Kirchner, vice-president of Aerojet-General and director of Solid Rocket Operations, announced the firing of a small solid propellant motor which successfully started and stopped more than six times. In the test, Aerojet used a 4-in. motor with 6 lbs. of propellant. (*M&R*, 11/23/64, 9)
- Dr. Alvin Radkowsky, Navy nuclear scientist, received patent for design of large breeder reactor which AEC and California were proposing to use in 525-kilowatt steam-electric power plant. (Jones, *NYT*, 10/31/64, 35)
- West Germany's Ministry of Scientific Research said estimated 4,300 West German scientists had left to work in U.S. between 1949 and 1961, and number was expected to increase. (Reuters, *NYT*, 10/31/64, 8)

October 30–November 1: Inventors' Congress and Space Symposium held in Little Rock, Arkansas. NASA Administrator James E. Webb spoke on American progress and goals in space. He described "the Southern Crescent of the space program"—Cape Kennedy, Fla.; MSFC, Huntsville, Ala.; Michoud Operations, La./Mississippi Test Facility; and MSC, Houston, Tex.

"One of the central facts about this great complex of space activities . . . is the possibility that many varied research, development and production activities will gravitate in this direction. This is made likely by reasons of economics and of convenience.

"By and large, these will be activities that require heavy contributions from science and technology, and Arkansas and the other states of the South are now in a position to get ready to take full advantage of the prospect. Arkansas must be able to advance in such areas as electronics, in the development and use of new materials, in medical research related to the severe requirements of space, and the like. It must not only make provisions for suitable working and living conditions for scientists and engineers, it must also be farsighted in training its people as technicians and skilled workmen, for these are as essential in this kind of activity as the scientists and engineers. . . ." (Text)

October 30: Morris Forgash, president of United States Freight Co., gave his predictions of transportation in the year 2000, in address at National Defense Transportation Association conference in Frankfurt, Germany. Among his predictions: "By the year 2000 the hydrofoils will be the buslines of the oceans"; "people and things will be transported between all the more distant points on earth by rocket-powered craft." He concluded:

"Before you mark off anything that I have predicted for the next three and a half decades look back on the startling developments of the recent past. The breaking of the sound barrier. Space travel. Substitution of a cushion of air for the ageless wheel. The broadcasting of a message of good will to the world by a President of the United States not by cable, phone, or radio, but from a satellite circling around the globe in outer space—then the erasure of the message and substitution of another while the satellite was in orbit. The Olympic games in Tokyo brought into the homes of people on the far side of the earth through the medium of a relay station in space. Multiply these miracles of science by the rate of change which is being constantly compounded and see if my predictions do not come out on the modest side even to the point of reticence.

"I believe the most important fact for us to understand is that transportation, though it opens new worlds and changes old concepts of time and distance, will not shape the world of 2000 A.D. Transportation is the servant of the people—the genie which makes their dreams come true—the magic carpet of their future—the mobility of their desires, their aspirations, and their resources. . . ." (Text, *CR*, 1/7/65, A70–74)

October 31: Astronaut Theodore C. Freeman (Capt., USAF) was killed when his T-38 jet trainer crashed near Houston. The former experimental test-flight instructor was among third group of NASA astronauts, selected last October, and was first U.S. astronaut to lose his life. (AP, *NYT*, 11/2/64)

• Second anniversary of orbiting of ANNA IB first satellite designed for use in measuring size and shape of the earth. Johns Hopkins Univ. Applied Physics Laboratory, builder of the satellite, said ANNA IB was still flashing its beacon and relaying radio signals to earth upon command. (AP, *Houston Post*, 10/30/64)

• Report prepared for Republican Presidential candidate Sen. Barry M. Goldwater by Adm. Arthur Radford (USN, Ret.), Gen. Nathan Twining (USAF, Ret.), and Gen. Arthur Trudeau (USA, Ret.), charged the Administration with misrepresentation in its disclosure of two antisatellite defense systems last September. Republican Task Force said the systems were "situated on test sites" and "manned by test crews." It claimed all components of the systems were initiated in the Eisenhower administration. DOD retaliated to the "unfounded charges" and reiterated that the antisatellite systems were "operational and on alert status." (CTPS, *Chic. Trib.*, 11/1/64; DOD Release 786–64)

October 31: USN announced KC–130F (Hercules) transport, a four-engine turboprop aircraft, had made series of landings and take-offs from aircraft carrier *Forrestal* without aid of arresting gear or catapults. The tests were made last November, but USN spokesman said reason for delaying announcement was that "the project was being analyzed." (*NYT*, 11/1/64, 20S)

During October: NASA Assistant Administrator for International Programs Arnold W. Frutkin said in *Aerospace:*

"The prime purpose of the National Aeronautics and Space Administration's international space program is to serve broad foreign policy objectives through useful scientific and technical projects.

"But a practical consequence for the American aerospace industry is that the program has stimulated a level of space activity abroad which might not otherwise have been achieved. Purely as a by-product, then, cooperative space projects have created limited markets for American know-how and hardware overseas. . . ." (*Aerospace*, Fall 1964, 10–15)

- The French Government reported two successful firings of its four-stage Berenice rocket, a mach-12 re-entry test vehicle, from the Ile du Levant testing center in the Mediterranean. Tests were conducted by the Office National d'Etudes et de Recherches Aerospatiales (ONERA). The Berenice project was designed to supply re-entry data for use in France's strategic missile program. (*Av. Wk.*, 12/7/64, 25)

- Interviewed in *Data*, Maj. Gen. Samuel C. Phillips (USAF), director of Apollo program in NASA Hq., replied to question about "all-up" concept of testing in the Apollo program.

"The general test policy requires that all-up testing will be the basic approach to the flight test program. This requires that where practicable, all flights will be scheduled as complete space vehicles using all live stages and as much of the lunar mission flight hardware as can be made available. This policy results from the recognition that true operational suitability of the flight hardware can only be obtained by operating the systems in the space environment and that the total number of development flights and total elapsed time can be reduced by making all flights as fully representative of the operational configuration as possible." (*Data*, 10/64, 7–13)

- William C. Schneider, Gemini program deputy director, NASA Hq., discussed flight readiness of Gemini launch vehicles and spacecraft in *Data* interview:

"The launch vehicle is leading in flight readiness and this is as it should be. The Air Force Titan II has been under development for over four years, and 33 flights have been made in the program. NASA has benefited fully from the results of this extensive qualification flight test program that is now complete. The spacecraft, however, is still in the developmental phase of testing and is simply a younger program. In addition, the GLV's will be essentially identical throughout the Gemini Program. This is not true of the spacecraft. . . ." (*Data*, 10/64, 15–21)

- Vincent L. Johnson, NASA Director of Launch Vehicle and Propulsion Programs, was interviewed in *Data* on Centaur launch vehicle.

"At this point in the Centaur development program, the major portion of the ground tests have been completed, and three development flights have been accomplished. Three more devlopment flights remain before the vehicle will be considered operational in a single-burn configuration in early 1965. The two-burn capability will be certified by two additional development flights providing a proven operational two-burn capability in 1966. I would like to note, however, that lunar flights can and will be accomplished by single-burn Centaurs in 1965 and 1966, and perhaps also in future years during those lunar opportunities that allow single-burn trajectories. . . ." (*Data,* 10/64, 27–31)

During October: Lt. Cdr. M. Scott Carpenter was awarded special gold medal by International Association of Man in Space, commemorating Astronaut Carpenter's Project Mercury orbital space flight. (Dryden, Ltr. to Carpenter, 11/5/64)

- Dr. Wolfgang Priester, after a year's work at Goddard Institute for Space Studies as a senior research scientist, returned to Bonn Univ. to become Director of its Institute of Astrophysics and Space Research. (*Goddard News,* 11/2/64, 8)
- At two meetings of European Conference on Satellite Communications in Geneva, representatives proposed plans for regional comsat network independent of the U.S., while not ruling out participation with U.S. in global network. (*M&R,* 11/9/64, 12–13)

November 1964

November 1: NASA launched Nike-Apache sounding rocket from Ft. Churchill, Canada, with instrumented payload to obtain comparison of electron density profile with wind profile and to measure electron temperature. Sodium vapor ejected as scheduled and Langmuir probe functioned well. Launch was to have been one of four spaced through a single night, but high winds and subsequent cloud cover forced cancellation of the other three. (NASA Rpt. SRL)

- Experiment conducted by National Bureau of Standards' Central Radio Propagation Laboratory had found that nitrogen molecules rather than oxygen molecules destroy helium ions in the earth's atmosphere. Chief experimenter E. E. Ferguson explained that the results meant helium concentration in upper atmosphere was increasing, upsetting steady-state theory based on helium-ion—oxygen molecule reaction. However, Ferguson said, recent samplings of upper atmosphere by sounding rockets did not uphold the laboratory findings: 100-to-1,000 times less helium was present than would have been expected if helium had been continuously increasing. (Simons, *Wash. Post,* 11/1/64)

November 2–4: RELAY I and RELAY II communications satellites were used to broadcast election week previews and results to Europe. Total broadcast time amounted to slightly less than two hours. SYNCOM II synchronous communications satellite, stationed in the Pacific, relayed nearly eight hours of U.S. Presidential election returns to the Far East. Voice of America transmitted the radio signals. (*Goddard News,* 11/16/64, 3; NASA Release 64–279)

November 2: USAF announced launch of Thor-Agena D booster combination with unidentified satellite payload, from Vandenberg AFB. (UPI, *NYT,* 11/3/64, 53; HHN–48)

- FAA announced that the two airframe and two engine manufacturers participating in U.S. supersonic transport program had submitted second-round design proposals for evaluation by Government and airlines. (Airframe companies, Boeing Co. and Lockheed Aircraft Corp.; engine companies, General Electric Co. and Pratt & Whitney Div. of United Aircraft Corp.) 120-member joint Supersonic Transport Evaluation Group, comprised of specialists from FAA, NASA, USAF, CAB, and USN, would conduct Government evaluation, with aerodynamic studies of wind tunnel models submitted by airframe contractors being conducted at NASA Langley Research Center. Both Government and airline evaluations were scheduled for completion Nov. 30. Basis of the Lockheed design was still a delta wing with a small delta at front of fuselage, but the company said that a larger, thinner wing, aerodynamically "cleaner" design, and engines with 25 per cent more power gave it "major advances" in range and passenger-carrying ability, with a profit potential of 30 per cent return on investment. Three different fuselage lengths, with capacity for 170 to 250 passengers, were offered.

The Boeing Co., which initially offered two versions, said that its new design could match the economies of current jets over long ranges and would seat 200 to 250 passengers. (FAA Release 64-101; Clark, *NYT*, 11/4/64, 52)

November 2: Four-man panel of former DOD R&D directors, under auspices of Scientists and Engineers for Johnson and Humphrey, responded to Republican charges that Democratic Administration had "neglected and stifled innovation. . . ." Panel said:

"We favor maintaining a substantial base of defense research and development which is essential if we are to maintain the flow of new ideas and new devices. But support of R&D does not mean the innovation of a large number of operational systems without regard for military requirements or without ample attention to technical and operational feasibility. Senator Goldwater and his advisors have ignored the fact that many of the projects cancelled during previous administrations had proved unfeasible or too expensive in relation to the military benefits they would yield. . . ." (*SBD*, 11/2/64, 5)

- Soviet members of the U.N. Committee on the Peaceful Uses of Outer Space visited NASA Goddard Space Flight Center. They were Prof. O. G. Gazenko, Prof. Yu. D. Kalimin, and Mr. I. V. Milovidov. (*Goddard News*, 12/14/64, 2)
- *Aviation Week* editor Robert Hotz took U.S. intelligence experts and NASA officials to task for the "considerable disgruntlement [that] has been generated in the U.S. manned space flight program" since Soviets' VOSKHOD space flight.

"All of this does not mean that the U.S. manned space flight program should be slowed, diminished, or abandoned, as a variety of critics have suggested. . . . It means that the U.S. manned space flight program is not overtaking the Soviets at its present pace and must be accelerated, expanded to yield better scientific results and aimed at some longer-ranging goals beyond the lunar landing. . . .

"We always have been convinced that this nation has the technical resources to accelerate its pace and achieve the eventual leadership in man's exploration of space. What we seem to lack is the political leadership and management organization that will enable this complex partnership of taxpayer, government and industry to proceed at the pace of which it is capable toward goals that will surpass its competition. For a nation that grew to its present greatness on the strength of its competitive surge, we are coming extremely close to losing this legacy in space." (*Av. Wk.*, 11/2/64, 11)

November 3: NASA Aerobee 150 sounding rocket launched by USN from the White Sands Missile Range, N. Mex., carried a spectrograph payload prepared by Princeton Univ. to 78.8-mi. altitude before the payload returned to earth by parachute. Primary objective of studying ultraviolet radiation of three stars in the constellation Orion was achieved. Payload was recovered in good condition and the film extracted and developed. Princeton said it believed this was the first time any university had attempted to study the atmosphere and composition of stars by means of equipment contained in a rocket. (NASA Rpt. SRL; *NYT*, 11/17/64, 72)

- NASA launched a Nike-Cajun sounding rocket from Wallops Island to an altitude of 76.8, with grenade payload to obtain temperature, wind, density, and pressure measurements. Experiments were provided by Goddard Space Flight Center (GSFC) scientists. Rocket performance was excellent but instrumentation performance was unsatisfactory since grenades did not eject or explode. (NASA Rpt. SRL)

November 3: NASA announced signing Memorandum of Understanding with Norwegian Space Research Committee and the Ionospheric Research Laboratory of the Royal Technical Institute of Denmark to extend joint sounding rocket program of investigations of the ionosphere. Tentative schedule of new series called for two Nike-Apache launches this year and two more early next year from Norway's Andoeya Range and two Nike-Apache launches from U.S. facilities. Scandinavian groups would purchase the four sounding rockets to be launched in Norway, fund their own experiments, and operate Andoeya Range. NASA would supply two sounding rockets to be launched in U.S., fund its experiments, and provide use of its Wallops Island, Va., range. There had been 11 successful launchings in this program since 1962. (NASA Release 64-274)

- Appointment of Brig. Gen. David M. Jones (USAF) as NASA Deputy Associate Administrator for Manned Space Flight, effective Dec. 15, was announced by NASA Associate Administrator for Manned Space Flight Dr. George E. Mueller. Gen. Jones would be primarily concerned with major development problems in Gemini and Apollo programs, planning for advanced missions and all mission operations. His most recent assignment had been as Deputy Chief of Staff (Systems), AFSC. (NASA Release 64-277)

- Geraldo de Carvalho Silos told the U.N. Committee on the Peaceful Uses of Outer Space that Brazil would establish a rocket center near Natal and would seek U.N. sponsorship for the facility. The U.N. had been asked to sponsor a similar international facility at Thumba, India. (*NYT*, 11/4/64, 39)

- DOD announced award of $1,692,001 contract for construction of dual vertical checkout and pneumatic test station Saturn S–II facility at Seal Beach, Calif. (DOD Release 788–64)

November 4–6: The AIAA/NASA Third Manned Space Flight Meeting was held in Houston, Tex.

Franklin P. Dixon, NASA Director of Manned Planetary Mission Studies, said in a speech: "Our studies indicate that the next major phase in manned earth orbital operations should be a space station with a 6–9 man capacity. This would be specifically designed for long-duration earth orbital flights and would directly further the routine exploitation of near earth space for peaceful purposes. The studies on this concept have shown that the basic station module should have a usable total volume in the order of 8,000 cubic feet with continuous 'floor' areas of not less than 300 square feet." (Text)

Dr. Edward C. Welsh, Executive Secretary and Acting Chairman of the National Aeronautics and Space Council, spoke at the awards banquet: "There is a wonderful thing about the space business. It has a vitality and a deep sense of challenge, which most other activities do not have. If it is not a way of life, it is at least a way of looking at life. The space program is the program for the optimist, the person who is willing to look ahead with confidence, the individual who sees no limits to what can be done and no horizons to the scope of doing. One of the greatest benefits already obtained from the space program is its revitalizing effect upon our whole economy—our whole educational system—our whole attitude toward the future. Let us use all our power of persuasion against any individual or any group who would slow down the pace of this program." Referring to statements that the military possibilities of space have been "downgraded," Dr. Welsh said: "During the four-year period of 1957 through 1960, space expenditures for the DOD totaled $1 billion. On the other hand, during the last four years, that is 1961 through 1964, the DOD spent about $4.6 billion on its space responsibility. The fact is that the Department of Defense is annually spending about a billion and a half dollars in space research and development. This includes projects in navigation, communications, observation, detection, booster development, scientific instrumentation, command and control techniques, interception, and manned and unmanned space flight." He would make no prediction regarding next year's space budget, but suggested that "space expenditures over the coming years, even with occasional plateaus, would be characterized by an upward trend." Dr. Welsh confirmed the Russians were relying heavily upon unmanned Vostok spacecraft in their Cosmos program. He said the Vostoks had accumulated "more than 2500 orbits"; since the seven Soviet manned flights to date accounted for only 10% of this, the balance must have been achieved in the more than 20 Vostok flights from Tyuratam range at the 65° inclination. Unlike the Cosmos satellites, these unmanned Vostoks had been recovered after as much as two weeks in orbit. Dr. Welsh also disclosed that Soviet launching reliability was "comparable" to the 85% attained by U.S. in past three years. (Text; *A&A*, 11/64, 116; Dr. Edward C. Welsh)

Dr. Charles S. Sheldon, staff member of the National Aeronautics and Space Council, in a comparison of U.S. and Soviet space flights, said that 262 military and civilian launch vehicles had been launched by this country between 1957 and October of this year as opposed to at least 81 launched by the Russians. Of the total U.S. launch attempts, 192 were successful, and 231 payloads were put into earth orbit, eight were injected into escape trajectories, and 84 were listed as failures. (Text)

Brig. Gen. Joseph S. Bleymaier, AFSC Space Systems Div. deputy for manned systems, said that studies by Jet Propulsion Laboratory and by General Electric, under JPL contract, showed that payload capacities of both Surveyor and Mariner could be doubled if the Titan III, rather than the Atlas-Agena, were used to launch them. The GE study, he said, determined that the Voyager Mars soft-landing mission objectives could be accomplished more effectively if the lander and orbiter were launched as two separate Titan III payloads, rather than as a combined package atop a Saturn IB. (*Av. Wk.*, 11/9/64, 27)

AIAA made the following awards: 1964 Astronautics Award to Walter C. Williams, vice president and general manager of the Manned Systems Div. of the Aerospace Corp. and until April 1964 NASA Deputy Associate Administrator for Manned Space Flight Operations, for "outstanding and continuing achievements in successful aerospace systems operations"; Octave Chanute Award to Fred J. Drinkwater and Robert C. Innis, USAF pilots at NASA's Ames Research Center, Moffett Field, Calif., for their work on the V/Stol and Stol aircraft; John Jeffries Award to Dr. Eugene Konecci of the National Aeronautics and Space Council, until Oct. 1964 NASA Director of Biotechnology and Human Research, for his contributions to life sciences; Robert M. Losey Award to Col. Robert C. Miller (USAF), Kansas City Weather Bureau, for his work on turbulence. (*Av. Wk.*, 11/16/64, 19; MSC *Roundup*, 11/11/64, 1)

Joseph A. Walker, project pilot for NASA's Flight Research Center, described the first flight of the Lunar Landing Research Vehicle (LLRV). Walker said this was the first of several pilot familiarization and checkout flights. Simulated lunar missions were scheduled for early 1965. The first flight consisted of three separate take-offs and landings. Total free flight time was just under one min., with a maximum altitude of approximately 10 ft. He used only the jet engine for lift power and did not activate the lift rockets. All eight of the standard control rockets were operated briefly.

The LLRV was equipped with a jet engine that could be automatically regulated to counterbalance five-sixths of the vehicle's weight to compensate for the gravitational difference between the moon and earth. This engine was also used to provide take-off power. During forthcoming checkout flights, two hydrogen-peroxide rocket motors, capable of delivering from 100 to 500 lbs. thrust each, would be used to regulate lift. The program, in support of Project Apollo, was designed to study the piloting and operational procedures involved during the final phases of a manned lunar landing and during the initial portion of the lunar take-off. (FRC Release 28–64)

Paper on launch vehicle engines by A. O. Tischler, Director of NASA OART Chemical Propulsion Div., and Leland F. Belew, Manager of MSFC Engine Program Office, covered current status of solid- and liquid-propellant rocket engines for Saturn and Titan launch vehicles. The authors reported that the F–1 engine represented "the largest thrust engine fired to date. Started in early 1959 the F–1 engine has been under development for six years. Five F–1's will be used in the first stage of the Saturn V vehicle to produce a take-off thrust of 7,500,000 pounds. . . .

"Nominal firing duration is 150 seconds and we expect better than 15 times that life to be in the engine at qualification. . . .

"[Design refinements to correct combustion-driven oscillations] have produced an engine which in the last five months has accumulated almost as much test time as in the previous five years. . . ." (Text)

November 4: Dr. George E. Mueller, NASA's Associate Administrator for Manned Space Flight, cited during a luncheon of the AIAA Third Manned Space Flight Meeting in Houston four important decisions made in the past ten years: "The first was in 1954 when as the result of the Von Neumann Committee's recommendations we began the ballistic missile program on a top-priority basis. . . . The second decision a year later was to undertake the Vanguard program as the second priority effort, completely separate from the ballistic missile program. . . . The third decision . . . was the passage of the National Aeronautics and Space Act of 1958. . . . The fourth major decision was President Kennedy's 1961 proposal to expand and accelerate United States space activity." Dr. Mueller said that there is no such thing as unmanned exploration of space. "As sensor, manipulator, evaluator, and investigator in space, as the scope and range of our activities continue to increase, the role of man assumes increasing importance. Our challenge is to capitalize on his presence in space." (Text)

November 4: USAF launched Thor-Agena D space booster from WTR with unidentified satellite payload. (HHN-48)

- Minuteman ICBM was successfully fired from Vandenberg AFB, Calif., by a Strategic Air Command crew from Malmstrom AFB, Mont. The missile traveled 5,000 mi. down the Western Test Range. (*M&R*, 11/9/64, 10)
- Capt. Theodore C. Freeman (USAF) was buried with full military honors in Arlington National Cemetey, Va., with the entire astronaut team attending the services. (MSC *Roundup*, 11/11/64, 1)
- Controller General Joseph Campbell said that the Federal Government could have saved $60 million by buying rather than leasing electronic data-processing systems for use on Government contracts. He recommended that a central management office be set up with authority to decide on the least expensive methods of obtaining data-processing equipment. (*NYT*, 11/4/64)

November 5: MARINER III (Mars 64) was launched into interplanetary orbit from Cape Kennedy with an Atlas-Agena D launch vehicle, but spacecraft communications failed when fairing did not jettison, preventing solar panel deployment. The Agena stage pushed the MARINER III into a parking orbit around the earth and coasted to the proper position for injection into a Mars trajectory. The Agena again ignited 32 min. after launching and sent the spacecraft toward Mars.

Tracking data indicated that the second-stage Agena D booster shut down four seconds early on the second burn. MARINER III's four solar panels, designed to use power from the sun to recharge the spacecraft's batteries, were prevented from deploying by failure of nose fairing to jettison. Further, the spacecraft did not align itself properly with the sun so that the communications antennas would be properly pointed.

In addition to TV equipment for taking pictures of the Martian surface, MARINER III carried instruments to study radiation, space dust, and magnetic forces near Mars and in unmapped space after swinging past the planet. This launch marked the first use of the USAF's new Agena D upper stage and also the first U.S. space mission requiring a second igniting of the Agena D. (NASA Releases 266, 288; KSC Release 206–64; MSC *Roundup*, 11/11/64, 7; Sullivan, *NYT*, 11/6/64, 16; Chic. *Trib.*, 11/6/64; Balt. *Sun*, 11/6/64, 1; HHN-48)

- NASA launched a 110-lb. instrumented payload to measure upper atmospheric day-glow emissions from Wallops Station, Va., aboard a four-stage solid-propellant Javelin (Argo D–4) sounding rocket. Day-glow studies were extended to higher altitudes than previous observations and tests of instrumentation being considered for future orbital missions were conducted. The payload, containing a spectrophotometer, ultraviolet photometers, and a rotating filter wheel photometer, was carried to an altitude of 548 mi. as scientific data were telemetered to ground receivers for 17 minutes. Impact occurred in the Atlantic 930 mi. from the launch site. (Wallops Release 64–79; NASA Rpt. SRL)
- NASA successfully launched a Javelin sounding rocket from Ft. Churchill, Canada, to an altitude of 522 mi., with instrumented payload to measure the earth's geomagnetic field from the ground to Javelin peak altitude. Data from this experiment would be combined with that obtained from a Nike-Apache sounding rocket launched 90 sec. earlier. This was the first Javelin launched from the Churchill Research Range. (NASA Rpt. SRL)

November 5: Lockheed-Georgia Co., Dawsonville, Ga., announced completion of the nation's first facility for subjecting space hardware to the simultaneous exposure of atomic radiation and extreme cold. The company planned to produce extreme cold through the use of liquid hydrogen which would be at $-427°$ F. Equipment would consist of an insulated 1,000-gal. liquid-hydrogen tank mounted on a railroad flatcar that could be rolled by remote control into the nuclear reactor radiation target area. (AP, *NYT*, 11/8/64, 26)

- Air Force Office of Scientific Research and Massachusetts Institute of Technology announced jointly that scientists at the National Magnet Laboratory, using a giant water-cooled magnet, had produced the strongest continuous magnetic fields generated by man to date—up to 255,000 gauss. Maximum fields were maintained for periods of a minute. (AFOSR Release, 11/64, 2S)
- George H. Arthur, Deputy Director of the International Telephone and Telegraph Aerospace Laboratory at Nutley, N.J., predicted that the Russians would go to the moon before 1970 and make a manned circumlunar mission in two and one half years. He said the United States would also be able to get to the moon by 1970 if it had as much success with its developmental program in the Saturn IB and Saturn V as it had with Saturn I. (Balt. *Sun*, 11/5/64)
- Abraham S. Bass, Chief of the Technical Assistance Branch in NASA's Office of Manned Space Flight, speaking at the Liquid Missile Propellants Symposium in Hershey, Pa., pointed out that development and operation of the prime and auxiliary rocket engines required by NASA's manned and unmanned space exploration programs would push propellant production in the U.S. to unanticipated heights. NASA's liquid hydrogen requirement was expected to rise to 14 million lb. per quarter, or 28,000 tons/year, during 1967. (*Av. Wk.*, 11/23/64, 64; NASA Proj. Off.)
- Dr. Donald F. Hornig, President Johnson's Special Assistant for Science and Technology, arrived in Moscow with a six-man group that would make a two-week tour of scientific and industrial institutions in Moscow, Leningrad, Minsk, and Novosibirsk to study Soviet methods of planning and coordinating scientific research. (*NYT*, 11/6/64, 19)
- Navy Secretary Paul H. Nitze formally approved the retirement of Marine Col. John H. Glenn, Jr. The retirement was set for Jan. 1. (*N.Y. Her. Trib.*, 11/5/64; *NYT*, 11/5/64, 11)

November 5–6: NASA launched four grenade-bearing sounding rockets over an 18-hour period for meteorological studies in the upper atmosphere. Launched from Wallops Island at about six-hour intervals, the Nike-Apache two-stage solid-propellant rockets carried payloads containing 12 special explosive charges which were ejected and detonated in the four flights at altitudes ranging from 25 to 59 miles. The series of flights was designed to provide information on variations in wind directions and speeds, atmospheric densities, pressures, and temperatures over a 24-hr. period. (Wallops Release 64–80)

- NASA launched four Nike-Cajun sounding rockets within 18 hours from Wallops Island, Va., with grenade payload to obtain temperature, wind, density, and pressure measurements up to 55.9 mi. All rockets and instrumentation performed excellently. (NASA Rpt. SRL)

November 6: NASA launched 295-lb. EXPLORER XXIII (S–55c) meteoroid detection satellite from Wallops Station, Va., aboard a Scout vehicle. The satellite was a cylinder 24-in. in dia. and 92 in. long. Its orbital elements: apogee, 614 mi.; perigee 286 mi.; inclination to equator, 52°; period, 99 min. EXPLORER XXIII was expected to have a useful lifetime of one year. Its primary purpose was to provide accurate knowledge of penetration capabilities of meteoroids and the resistance of various materials to penetration, thus facilitating design of spacecraft.

The payload contained primary sensors (pressurized cells) designed to record the rate of meteoroid penetration in two different thicknesses of stainless steel; a capacitor-type penetration detector to determine effects of high-energy radiation; cadmium sulphide cells to record the size of impacting meteoroids; impact detectors capable of detecting three levels of meteoroid momentum; two separate telemeter canisters for storing experimental data and relaying it to ground stations. Also onboard the Scout rocket was an experiment to measure air loads on the structure during its ascent through the atmosphere between 25,000 and 40,000 ft.

The spacecraft was the successor to EXPLORER XVI (S–55b), launched Dec. 16, 1962, which made the first statistically significant sampling of meteoroids in near-earth space. (NASA Release 64–272; Wallops Release 64–81; *Marshall Star,* 11/11/64, 1; *NYT,* 11/7/64, 9; *M&R,* 11/16/64, 33)

- NASA successfully launched a Javelin sounding rocket from Ft. Churchill, Canada, to an altitude of 444 mi., with instrumented payload to measure the earth's geomagnetic field from the ground to Javelin peak altitude. Data from this experiment would be combined with that obtained from a Nike-Apache sounding rocket launched 90 sec. earlier. (NASA Rpt. SRL)
- X–15 No. 2 piloted by John B. McKay (NASA) in planned captive flight. (NASA X–15 Proj. Off.)
- NASA Nike-Apache sounding rocket with French sodium-vapor payload was launched from India's Thumba Range to about 192-km. altitude (about 119 mi.). Two ground camera stations photographed sodium vapor released from the rocket to obtain measurements of atmospheric winds, turbulence, and diffusivity. (NASA Rpt. SRL)
- NASA Langley Research Center announced that power supply capability tests were being conducted on two huge magnet coils at the Space Radiation Effects Laboratory. In operation, the magnet coils would be capable of creating a pulling power greater than 3,500 tons. The magnet coils were part of the 600-million electron volt synchrocyclotron which NASA would operate to study the effect on spacecraft and their systems of the particle radiation streaming from the sun or trapped in the earth's magnetic field. (LaRC Release)
- NASA announced that space scientists had adopted a new analytical method—neutron activation analysis—to help them determine the presence and quantity of oxygen in alkali metals. General Atomics Div. of General Dynamics Corp. undertook this oxygen-identification problem under a contract from Lewis Research Center. (LRC Release 64–102)
- First successful flight of a fully instrumented four-stage Athena re-entry research vehicle was achieved. Launched by USAF and USA from Green River, Utah, the payload covered approximately 417 mi. and impacted

nearly on target within the White Sands Missile Range, N. Mex. (UPI, *Denver Post*, 11/8/64; *M&R*, 11/16/64, 9, 32)

November 6: USAF fired a Minuteman missile from Vandenberg AFB. It was the 70th Minuteman to be launched from that base. (AP, Wash. *Eve. Star*, 11/5/64)

- Air Force Systems Command's Research and Technology Div. reported the High Power Facility at the Rome Air Development Center (RADC) was operational. It was designed to evaluate high-power transmitting tubes and radio frequency components, and would support research in plasma physics and engineering where brute force was needed in direct current and pulse radio frequency. (AFSC Release 44-128-139)
- Italy's largest private monitoring center said it had recorded radio signals indicating a new Soviet space launch. It speculated that the Russians might have placed a satellite in parking orbit around the earth as a base for a Mars launch. (*Phil. Inq.*, 11/7/64)
- The U.N. Committee on the Peaceful Uses of Outer Space took the following actions: (1) it invited the Scientific and Technical Subcommittee to study and submit a report on "the possibility of establishing a civil world-wide navigation satellite system on a non-discriminatory basis"; and (2) it decided to establish a working group to examine "the desirability, organization, and objectives of an international conference or meeting to be held in 1967 on the exploration and peaceful uses of outer space." (NASA Off. of Int. Aff.; U.N. Doc. A/57-79; *NYT*, 11/8/64, 12)

November 7: NASA successfully launched an Aerobee 150 to 132-mi. altitude from White Sands, N. Mex. Purpose was to measure absolute intensity of the stars in the ultraviolet with 50 Angstrom resolution. Payload was recovered about 65 mi. from the launch site. (NASA Rpt. SRL)

- Celebrating the 47th anniversary of the Bolshevik Revolution, leaders of the Communist world watched a display of the Soviet Union's newest and most powerful missiles. An antimissile missile, shown among six other rockets for the first time, appeared to be of greatest strategic significance. A four-engine missile, 60 to 65 ft. long, enclosed in a cigar-like tube about eight ft. in diameter, it was believed to be capable of reaching very high altitudes. (Clymer, Balt. *Sun*, 11/8/64; Tanner, *NYT*, 11/8/64, 1)
- West Germany's Scientific Research Ministry proposed a three-point plan to stem the flow of German scientists to the United States. The plan would seek modernization of West German research institutes, higher pay for scientists, and quicker access to the top of the academic ladder for young scientists. (Reuters, *NYT*, 11/8/28; Shabecoff, *NYT*, 11/15/64, 14)

November 8: Pravda disclosed that the three crew members of the spaceship VOSKHOD I never used seat straps during their 24-hour flight except during the launching and landing stages. These men also said that their craft was equipped with a new control system that made it possible to orient the vehicle even when it was flying on the dark side of the earth. (Shabad, *NYT*, 11/9/64, 15)

- GAO complained NASA had spent money unnecessarily by producing four basically similar films on the flight of astronaut John Glenn, three on the Alan Shepard flight, three on the Apollo lunar exploration project, two on the Ariel satellite, and two emphasizing international cooperation in the space program. The cost of the films ranged from $216,000 for one of the Glenn-flight productions to $1,500 for one of the three on Shepard's flight. (*Orlando Sentinel*, 11/4/64; *NYT*, 11/4/64, 40)

November 8: Dr. Thomas B. Weber, formerly at Aerospace Medical Center, Brooks AFB, Tex., and now manager of advanced research for space engineering at Beckman Instruments, Inc., Fullerton, Calif., said that all space-cabin simulation experiments so far have had to be called off within a few days because of accumulation of toxic elements in the closed atmospheres. "This may be the reason Russia's recent three-man space flight, which was expected to last several days, ended after only 24 hours." Dr. Weber continued, "Unless an answer is found, this is a definite threat to our own plans for a manned orbiting laboratory and it could delay the Apollo program to land men on the moon."

Working under USAF contracts, Dr. Weber had devised instruments to collect and identify more than 400 contaminants given off by the human body and by space cabin equipment. So far he had found that 20 were potentially lethal, including highly explosive hydrogen given off by water used in cooling systems; ozone, given off by electrical equipment; and ammonia, acetone, and alcohol found in breath, perspiration, and body wastes. "None of these occurs in dangerous amounts in a normal environment," Dr. Weber said, "but in a small cabin they build up to lethal concentrations in short periods of time. . . . Our longest space flight so far lasted less than two days, but even on that one we ran into trouble filtering out the carbon dioxide exhaled by the astronaut." (AP, *Denver Post*, 11/8/64; AP, Wash. *Eve. Star*, 12/16/64)

• In the Letters to the Editor column of the *New York Times*, Dr. Martin Summerfield, AIAA Vice President for Publications, took issue with sources quoted by *NYT* to the effect that Communist China would not be able to produce military nuclear-tipped rockets for another decade or even a generation. "As for rocket secrets," he said, "there are none, either in theory or design. All one needs is a reading knowledge of English, French, or German, and a good collection of recent technical journals." He said it was a serious underestimate to believe that the Chinese could not plan and execute an efficient research, development, and production program for practical missiles of medium range and a further underestimate to believe they were not close to their objective. (*NYT*, 11/8/64; *A&A*, 12/64, 10)

November 9: NASA Nike-Apache sounding rocket with French sodium-vapor payload was launched from India's Thumba Range to about 192-km. altitude (about 119 mi.). Two ground camera stations photographed sodium vapor released from the rocket to obtain measurements of atmospheric winds, turbulence, and diffusivity. (NASA Rpt. SRL)

• Sen. Clinton P. Anderson (D.–N. Mex.), Chairman of Senate Committee on Aeronautical and Space Sciences, recommended in letter to President Johnson that USAF Manned Orbiting Laboratory (MOL) and NASA Apollo-X programs be merged. Jointly operated national space station program would best utilize U.S. resources, according to his letter which *Aviation Week* cited in its Dec. 7 issue. Senator Anderson argued that $1 billion could be saved over five-year period if MOL were canceled and USAF funds applied to the Apollo-based space station. (*Av. Wk.*, 12/7/64, 16; NASA Archives)

November 9: NASA officials told *Missiles and Rockets* that Phase I of the floxed Atlas development program, which was to end this year, would be continued until mid-1965. Also delayed was the decision whether to move on to Phase II, which called for development of flight hardware and two flights in the 1966–67 period. (*M&R*, 11/9/64, 9)

- A scheduled flight of the X–15 rocket ship to test star-tracking cameras was canceled when rain soaked lakebeds used as landing fields for the plane. No new date was scheduled. (*Wash. Post*, 11/10/64)
- United Press International (UPI) reported that the Air Force would attempt pickaback launch of a secret satellite (OV–1) next month from Vandenberg AFB with hopes of proving a method which would save the taxpayers $370,000 per launch. The 300-lb. satellite, carrying its own rocket motor, would ride into space on the side of an Atlas missile making a re-entry test. Near the peak of the Atlas' flight, OV–1 would spring away, coast for a few seconds, and be launched into orbit by its own rocket. The Atlas would continue on its mission. (UPI, *NYT*, 11/10/64, 34)
- Air Force Systems Command announced that Air Force pilots would fly a high-altitude Lockheed U–2 aircraft in a research program directed by the Research and Technology Div.'s Air Force Flight Dynamics Laboratory, to measure and predict high-altitude clear-air turbulence (Hi-Cat). The U–2 would gather Hi-Cat data at altitudes above 55,000 ft. for developing a worldwide statistical model showing occurrence of clear-air turbulence. (AFSC Release 47–30–144)
- Physicists at Columbia University had produced a "photon echo" effect in a synthetic ruby crystal by shooting two successive flashes of light at it, *Physical Review Letters* reported. A ruby laser was the light source. The phenomenon, which scientists had not been sure was physically possible, was described as appearing as though a light flashed into a dark, empty closet received, after a pause, an answering wink of light. Authors of the research were Dr. Sven R. Hartmann, Norman A. Kurnit, and Dr. Issac D. Abella. The photon echo effect would seem potentially valuable in designing computers because it occurred at 186,000 mi./sec., the speed of light, and because the nature of the effect would seem readily adaptable to computer use. (Schmeck, *NYT*, 11/9/64)
- Air Force Systems Command announced plans to orbit an experimental fuel cell able to produce both electricity for space power and pure drinking water for space crews. The fuel cell would be orbited from Vandenberg AFB early in 1965. (AFSC Release 43–11–145; *NYT*, 11/10/64, 34; *Houston Post*, 11/10/64)
- California Institute of Technology radio astronomers Glenn Berge and David Morris announced that Jupiter's radiation belt is centered 30,000 or more miles from the center of that 85,000-mi.-diameter planet, making the zone of intense radiation lopsided. Berge and Morris said they couldn't explain their finding but that it did help to explain the origin of bursts of radio signals from the planet: they may result when Jupiter's rotation brought the magnetic poles into the proper angle to beam radiation in earth's direction. (AP, N.O. *Times Picayune*, 11/9/64)
- ComSatCorp announced six companies had made summary proposals to provide research data and consultant services for ground terminal stations: American Telephone & Telegraph Co.; ARCS Industries, Inc.;

Daniel, Mann, Johnson & Mendenhall; Hughes Aircraft Co.; Page Communications Engineers, Inc.; Westinghouse Electric Corp. Of these, one or more contractors would be selected. (ComSatCorp Release)

November 10: NASA launched its first pair of ship-to-shore sodium vapor cloud experiments aboard Nike-Apache sounding rockets, one at 5:25 p.m. EST from a launcher on the deck of the USNS *Croatan,* located about 75 mi. off shore south of Wallops Island, Va., the second at 5:28 p.m. EST from a pad on Wallops Island. The payloads ejected sodium vapor trails up to an altitude of approximately 123 mi. Purpose was to study variations in wind conditions occurring at different altitudes and locations and to test shipboard equipment and procedures. Ground cameras photographed dispersion of the sodium trails, permitting scientists to measure wind velocities and directions from the photographic records. (Wallops Release 64–83; NASA Rpt. SRL)

- NASA Nike-Apache sounding rocket with French sodium-vapor payload was launched from India's Thumba Range to about 192-km. altitude (about 119 mi.). Two ground camera stations photographed sodium vapor released from the rocket to obtain measurements of atmospheric winds, turbulence, and diffusivity. (NASA Rpt. SRL)

- At the conclusion today of a test in which four men whirled for five days in a centrifuge, Dr. Bernard D. Newsom, Chief of Aerospace Medicine at General Dynamics/Astronautics in San Diego, Calif., announced that they had adapted so well it appeared there may be no human problem in spinning future spacecraft to create artificial gravity. "We are extremely encouraged with the results," he said, but he cautioned that more research is necessary before decisions could be made on the basic goal of the experiment—to determine if spacemen can adapt to live and work for long periods in a rotating environment. (*L.A. Times,* 11/11/64; *Wash. Post,* 11/11/64)

- Speaking at the dedication of a monument at Cape Kennedy honoring Project Mercury and the nation's seven original astronauts, Dr. Edward C. Welsh, Acting Chairman and Executive Secretary of the National Aeronautics and Space Council said: "Mercury served to lift the self-respect of our nation when it badly needed such a boost. I know that we still have widespread public support of our national space programs in spite of the criticism of some short-sighted individuals." (*NYT,* 11/11/64, 48; AP, Balt. *Sun,* 11/9/64; Rossiter, UPI, *Miami Herald,* 11/11/64)

November 11: The second set in a series of three dual ship-to-shore sounding rocket launchings was conducted at 5:24 p.m. EST from the USNS *Croatan* and at 5:27 p.m. from Wallops Island, Va., about 200 mi. apart. Sodium vapor was ejected by the Nike-Apaches from about 25 mi. up to an altitude of approximately 125 mi. The reddish-pink clouds were visible for several hundred miles along the East Coast. (Wallops Release 64–84; NASA Rpt. SRL)

- Paul Haney, NASA Manned Spacecraft Center Public Affairs Officer, told the Radio and Television News Directors Association that television viewers would be able to watch the first astronauts land on the moon from the comfort of their living rooms. Television cameras carried by the astronauts would beam a signal back to earth which could be relayed to the major TV networks. The astronaut on the lunar surface would also take pictures as he walked along. All of these would be relayed back to earth, Haney said. (AP, Balt. *Sun*, 10/12/64; Maloney, *Houston Post*, 11/12/64)

November 11: Institute for Satellite and Space Research scientists in Bochum, W. Germany, reported monitors had picked up both radio signals and voice broadcasts on the same frequency used by the Soviet Union for manned spacecraft. No manned space launch had been announced by U.S.S.R. (Sci. Ser., *Wash. Daily News*, 11/13/64)

November 12: The third and final set in a series of dual ship-to-shore sounding rocket launchings was conducted at 5:53 a.m. EST from the USNS *Croatan* and at 5:56 a.m. from Wallops Island, about 300 mi. apart. Sodium vapor was ejected by the Nike-Apache launched from the ship up to an altitude of approximately 141 mi.; sodium vapor failed to eject from the payload launched from Wallops Island. (Wallops Release 64-85; NASA Rpt. SRL)

- Harold B. Finger, Manager of NASA–AEC Space Nuclear Propulsion Office, commented in NASA Hq. nuclear rocket background briefing that the Kiwi and NRX power reactor tests conducted during 1964 "constituted the largest advance in rocketry and in rocket performance that has been made since [Robert] Goddard started flying his pump liquid rocket systems in the '20's and '30's. . . ." (SNPO; NASA Press Conf. Transcript)

- Air Force Systems Command announced first successful demonstration of internal thrust from a "scramjet" (supersonic combustion ramjet) engine. The test was one in a series to demonstrate that scramjet engines could be developed for very-high-speed vehicles and was conducted under the direction of Air Force Aero Propulsion Laboratory. Test conditions in the 30-in.-long scramjet engine simulated a speed of approximately mach six over a brief burning period. Frontal area of the engine was approximately three-by-four inches. Scramjets differed from conventional engines in that they had moving parts only in the fuel feeding system and produced thrust by burning fuel in a supersonic air stream. (AFSC Release 49-40-141)

- Institute for Strategic Studies, an international organization for research on defense, security, and disarmament, reported U.S. had increased its strategic superiority over U.S.S.R. The report warned, however, that "this may be the last year in which the western superiority in long-range striking power is so marked." By early next year the U.S. would have 925 intercontinental ballistic missiles, compared to U.S.S.R.'s 200. American superiority in fleet ballistic missiles had increased to four to one over U.S.S.R.'s less advanced counterpart of the Polaris. Soviet Union had maintained her superiority in submarine strength. (Cerutti, *Chic. Trib.*, 11/13/64)

- In a report to Parliament, France disclosed plans to build up the French nuclear force for the next five years, including supersonic bombers, a Polaris-type submarine, and tactical atomic weapons for the army. Estimated cost of the program, which would extend over a five-year period (1965–70), was $11 billion. (*Wash. Post*, 11/13/64)

- The National Academy of Sciences' Panel on Astronomical Facilities presented a 10-year program that would approximately double the nation's ground-based astronomical facilities. Specific recommendations were contained in a 105-page report, *Ground-Based Astronomy— A Ten-Year Program*. The total cost for construction and operation of the new facilities outlined by the panel was estimated at $224 million over the 10-year period, the maximum annual expenditure of $26 million occurring during the fourth and fifth years of the program. The panel found, in analyzing "the present serious situation" with respect to ground-based astronomy: "Nearly every phase of observational astrophysics is hampered today because the rate of growth of new astronomical facilities had not kept pace with the increasing demand for fundamental data.

"If astronomy is to progress, major new facilities are required." (NAS–NRC Release, 11/12/64)

November 12: Dr. Robert L. Sproull, Director of DOD Advanced Research Projects Agency (ARPA), speaking at the dedication of the Xerox Corporation's new research and development laboratory in Rochester, N.Y., cautioned that the "spillover" from the Federal Government's own development programs would in the future be less beneficial to industry than it had in the past. Dr. Sproull explained that this would come about because of increased efficiency in Federal research programs and "good management will focus funds and attention ever more precisely on the specific mission and that mission is not the stimulation of American industry."

Dr. Sproull offered two suggestions to fill the gap created by the disappearance of Federal "spillover": (1) that a governmental extension service be set up to channel the university-scientific effort into businesses unable to provide their own basic research; (2) that there be more encouragement of industry-research institutes such as the one maintained by the textile industry at Princeton, N.J. (Frost, *NYT*, 11/13/64, 47)

November 13: NASA announced that further analysis of results of the first experiments to use the laser beam for satellite tracking indicated that scientists had accomplished precision tracking of a satellite with a laser device. Tests were conducted on 10 passes of the EXPLORER XXII satellite since the first success on Oct. 11. (NASA Release 64–285)

- NASA's Marshall Space Flight Center announced that two barnacle-encrusted capsules, each containing 100 ft. of color motion-picture film in good condition, were found, one on a beach at Eleuthera Island in the Bahamas, the other near a beach in San Salvador in Central America. The capsules had been ejected from Saturn I SA-7 booster rocket launched Sept. 18, 1964, from Cape Kennedy. (UPI, *NYT*, 11/14/64, 24)

- NASA's Lewis Research Center announced that it was testing a gas bearing, designed as a research tool, in a vacuum simulating space environment above 100 miles. Suspended from the top of a 60-ft.-long, 15-ft.-dia. tank, the bearing would provide complete freedom of motion to any payload attached to the bearing. A suspended payload would have three-axis mobility with a 10° swing in both pitch and yaw. The bearing, designed to carry an average 350-lb. payload, would enable scientists to evaluate low-thrust systems for attitude control or station-keeping in on-the-ground tests. (LRC Release 64–105)

- United Technology Center's 120-in.-dia., solid-propellant motor for Air Force's Titan III–C booster was static-fired successfully for more than 110 sec. with a new ablative nozzle-throat which probably would be substituted for the carbon throat previously used in test firings. (*Av. Wk.*, 11/23/64, 34; UTC)

November 13: Dr. George E. Mueller, NASA Associate Administrator for Manned Space Flight, at the National Editorial Writers' Conference in Cocoa Beach, Fla., said: "The Apollo development schedule is conservative. The Apollo spacecraft is being developed on a time scale four years longer than the Mercury spacecraft, two years longer than the B-58 bomber, and one year longer than the X-15." (Text)

- UNESCO decided to convene an international meeting of scientists in Paris late next year to study the use of space satellites in mass communications media. No exact date was set for the meeting. (*NYT*, 11/14/64, 29)
- Harry F. Guggenheim, president of the Daniel and Florence Guggenheim Foundation, was named by NAA as the recipient of the Wright Brothers Memorial Trophy for 1964. (AP, *NYT*, 11/14/64, 31)

November 14: NASA successfully launched an Aerobee 150 to 128-mi. altitude from White Sands, N. Mex. Primary purpose was to measure absolute intensity of the stars in the ultraviolet with 50 Angstrom resolution. (NASA Rpt. SRL)

- President Johnson appointed a national commission to study the impact of technological changes on American life. Howard Bowen, an economist and president of the Univ. of Iowa, was nominated chairman of the 14-member panel which would submit a report to Congress and the President on four areas: the role and pace of technological change, both past and future; the impact of technological change during the next ten years on the Nation's economy and social structure; unmet community and human needs to which modern technology might be effectively applied; the most effective means for channeling technology into promising directions. (*Wash. Post*, 11/15/64, A6)
- Wolfgang Cartellieri, West Germany Research Ministry state secretary, announced that preparations had begun for a space satellite program involving small research and telecommunications satellites. (UPI, *NYT*, 11/15/64, 56)

November 15: Astronomers at San Miguel Observatory near Buenos Aires, Argentina, reported seeing a red-colored, unidentified flying object, traveling at four and a half times the speed of an ordinary man-made satellite, cross the sky from west to east at 8:30 p.m. local time, from east to west five minutes later, and then from west to east at 9:05 p.m. (Reuters, *Wash. Post*, 11/16/64)

November 16: NASA launched its first Project Luster payload from White Sands, N. Mex., on an Aerobee 150 sounding rocket. Primary objectives of flight were to evaluate engineering performance of Luster micrometeoroid-sampling instrument and to collect meteoritic debris during the peak of the Leonids meteor shower. The Aerobee 150 reached 97-mi. altitude, and performance of both the rocket and the instrumentation was considered "very good." Nose cone did not fully retract, so the data collected during the flight were lost in re-entry heating and atmospheric contamination. (NASA Rpt. SRL)

- NASA launched an Aerobee 150A sounding rocket from Wallops Island, Va., to an altitude of 117 mi. The payload contained three experiments, the most important of which involved the measurement of the composition of the neutral atmosphere in the altitude region of 62-155 mi. All three experiments operated satisfactorily. (NASA Rpt. SRL)

November 16: X-15 No. 2 piloted by John B. McKay (NASA) in planned captive flight. (NASA X-15 Proj. Off.)
- NASA's Manned Spacecraft Center in Houston divulged that its post-Apollo study emphasis would be on a Y-shaped 36-man National Multipurpose Space Station that could be launched by Saturn V and be the primary space scientific and technical experimental center for at least five years in the mid-1970's. (*Av. Wk.*, 11/16/64, 51)
- An MSC investigative board announced that a snow goose, weighing about eight lbs., with a wingspan of more than 24 in., caused the fatal crash of Astronaut Theodore C. Freeman (USAF) on Oct. 31. The bird struck the left side of the airplane canopy causing both engines to flame-out on impact. Captain Freeman, however, still had control of the aircraft and attempted to make a landing at Ellington. He flew a looping pattern over the Gulf Freeway and lowered his landing gear as he approached the field. When he apparently realized that a landing was not possible, he veered to the left into an open field, probably to avoid buildings at Ellington. Seconds later, at slightly under 100-ft. altitude, he ejected, too late for his parachute to deploy fully. (MSC *Roundup*, 11/25/64)
- Dr. Albert J. Kelley, Deputy Director of NASA's newly created Electronics Research Center, said in an interview with *Missiles and Rockets* that ERC would issue between 35 and 40 contracts, averaging about $50,000 each for a total of $2 million, within the next four months. Most of the contracts were expected to run for one year. (*M&R*, 11/16/64, 10-11)
- The Sprint antimissile missile was partially displayed for the Association of the United States Army in Washington. Only the exterior shell was shown—a cone-shaped missile 27 ft. long and $4\frac{1}{2}$ ft. wide at its base. Sprint, combined with the Nike X, was designed to take off from below ground and intercept ICBM's in a matter of seconds. (Watson, Balt. *Sun*, 11/17/64)
- An announcement issued by the AEC said: "The United States today recorded seismic signals from an event in the Soviet nuclear test area in the Semipalatinsk region." This was the first apparent Soviet underground nuclear test noted by the United States since the treaty banning all but underground nuclear tests was signed in August 1963. (*NYT*, 11/17/64, 3)
- The report that the White House might establish a "space czar" cabinet post to direct DOD and NASA space programs was discounted. The White House was against it, as were Secretary of Defense McNamara and NASA Administrator Webb. (*M&R*, 11/16/64, 7)
- Col. John H. Glenn, Jr. (USMC), said during the career guidance clinic held at NASA Manned Spacecraft Center that there was a need for a national information center to make new space knowledge available to educators. (Lee, *Houston Post*, 11/17/64)

November 16-18: The Third International Symposium on Bioastronautics and the Exploration of Space, sponsored by the Aerospace Medical Div. of AFSC and organized by the Southwest Research Institute, was held in San Antonio, Tex.

Dr. James A. Van Allen, for whom the Van Allen radiation belts were named, said in a speech that radiation from solar flares would have been dangerous to space travelers only three days in the past four years. After his address, Dr. Van Allen told reporters "there would be no serious danger to travel to any point in the solar system caused by solar flares."

Dr. Fred L. Whipple, Director of the Astrophysical Observatory of the Smithsonian Institution, said that EXPLORER XVI experienced 55 meteoroid punctures during a seven-month period. However, all were in patches of the satellite's skin made very thin to assess the collision hazard. The patches, made of beryllium copper, were of three thicknesses: 1/1,000, 2/1,000, and 5/1,000 of an inch. None of the cells with the thickest skin was pierced. It was thought that the puncturing particles hit at typical meteor velocities of about 15 mi./sec. Dr. Whipple told reporters that the chances of a spacecraft being seriously damaged by meteoroids was about the same as being struck by lightning.

Dr. John W. Firor, Director of the High Altitude Observatory at the National Center for Atmospheric Research at Boulder, Colo., said astronauts on the moon would have to be exposed to solar flare radiation for "many hours" before they would suffer serious damage. Attempts to forecast solar flares accurately "have not shown any success," Dr. Firor said. (Maloney, *Houston Post*, 11/17/64; Sullivan, *NYT*, 11/17/64, 16)

Dr. Frank J. Malina, chairman of the Lunar International Laboratory (LIL) program at the International Academy of Astronautics in Paris, said: "The principal justification for the operation of a manned center on the moon on an international basis may well be a financial one." He explained that it was estimated that with present rocket technology, it would cost about $5,000 per pound of payload carried from the earth to the moon, and that one man-hour on the moon would cost as much as $80,000. Dr. Malina also mentioned that plans for LIL were begun before President Kennedy announced this country's intention of going to the moon before the end of the decade. (Maloney, *Houston Post*, 12/19/64)

Soviet scientists reported that the entire VOSKHOD I space flight was conducted by the three-man crew without spacesuits. U.S.S.R. space physiologist, Dr. O. Gazenko, said that the absence of spacesuits made it possible to perform a number of experiments using sensors directly on the skin, even during boost and re-entry. Among these were measurements of the electrocardiogram, respiration rate, electroencephalogram, eye movements, and seismocardiogram.

Dr. V. V. Antipov, a radiologist, said that VOSKHOD preliminary results indicated a substantial contribution to knowledge about the effects of space flight factors upon biological organisms. He said that significant changes revealed in tissue experiments showed that new, spherical fragments occurred in cells in all stages of division and had only been duplicated in Soviet research under extremely high g forces, such as 4,000 g. They had thus far been tied to the boost and descent phase of the flight, with the greatest number of changes occurring during the first hours of flight, and the least just before re-entry. Antipov also disclosed that in one VOSKHOD experiment, a number of cells were flown which had been flown on previous space flights. He said this indicates that the tissues could maintain viability with repeated exposures, and possibly that some selection took place. By studying the ones which "selected themselves," he said, a great advantage would be gained in calculating what happened to human cells.

Missiles and Rockets reported that during an interview Dr. Gazenko stated that due to the thermal and decompression advantages it offered, helium would eventually be used as a diluent gas for spacecraft atmosphere instead of the nitrogen gas that had been used on all Soviet space shots to date. He said the subject is a matter of scientific controversy in the U.S.S.R., but that "undoubtedly" there would be a shift to helium at some time in the future. (*M&R*, 11/23/64, 14–15)

At a press conference, Dr. O. G. Gazenko reiterated that the VOSKHOD crew did not use space or pressure suits at all. He was asked when the Russians plan to rendezvous and dock two spacecraft and if they would both be manned.

"We will, of course, try to do this as soon as possible, but I have the feeling that it would be very nice to have two humans meet in space.... It is possible one will be a man and one will be a woman." His answers were relayed to reporters by Dr. Boris Mandiovsky, a librarian from the Library of Congress.

Dr. Gazenko, giving an oral summary of his much longer paper, said medical results of seven Russian manned space flights had not shown any insurmountable barriers to space flight.

"The analogy of the physiological data obtained during manned orbital flight of up to five days duration proves that man, specially selected and trained, can satisfactorily tolerate all orbital flight stresses," Dr. Gazenko continued. "However, as can be seen in the material, important individual variations are noted. We have to devote special attention to this factor in future cosmonaut selection and training," he said.

He told scientists that there are problems still to be solved for true long space flights, and expressed the hope that between Russia and the United States there could be closer cooperation in solving these problems. (*Houston Post*, 11/18/64)

Col. Charles E. Yeager, Commandant of the USAF Aerospace Research Pilot School at the Air Force Flight Test Center, Edwards AFB, Calif., called for an end to the use of heavy work-retarding pressure suits. "We are relying entirely too much on the suit to pull the guy out of a problem," he said. (*Houston Post*, 11/19/64)

Astronaut M. Scott Carpenter called for a "new approach" in development of wearing apparel for astronauts in space. He termed the pressurized spacesuit "the greatest single handicap we have to operations in space. To do useful work is almost impossible. I'm sure there is a better way." Carpenter said that he had proposed an investigation of a "spray-on" spacesuit that would permit astronauts to move freely about the cabin of the capsule. (Webb, *Boston Globe*, 11/17/64; *Houston Post*, 11/18/64; *NYT*, 11/18/64, 24)

Maj. Herbert H. Reynolds (USAF), Chief of the Comparative Psychology Branch of the Aeromedical Research Laboratory at Holloman AFB, N. Mex., said that chimpanzees had performed so well within an almost total vacuum that there was hope space travelers could survive a vacuum caused by a puncture of their vehicle long enough for the hole to be plugged. Research was conducted at Holloman for NASA. (Sullivan, *NYT*, 11/19/64, 19)

Dr. Cyril Ponnamperuma, biochemist at NASA Ames Research Center, told newsmen that "there is a very good possibility of finding the answer to the question of the existence of life in our own planetary system by an inspection of the planets with our immediate or remote sensors." (*Houston Post*, 11/17/64)

November 17: NASA conducted two sounding rocket launches from USNS *Croatan*, near Wallops Island, Va., to determine whether or not ship's radar could track metalized-mylar spheres with sufficient accuracy to obtain significant geophysical data. Both payloads, each consisting of two metalized-mylar spheres, were provided by Univ. of Michigan. The first rocket, a Nike-Apache, reached 136-mi. altitude and two spheres ejected and inflated, but ship's radar did not acquire track. Land-based radars did obtain sphere trajectory data. The second rocket, a Nike-Cajun, reached 92.6-mi. altitude, its spheres ejecting and inflating as planned but deflating earlier than expected. Ship's radar provided questionable position data which would have to be analyzed further. (Wallops Release 64–86; NASA Rpts. SRL)

- The National Academy of Sciences released a statement from its Space Science Board recommending that unmanned exploration of the planet Mars, involving both physical and biological investigations and expressly the search for extraterrestrial life, be made the primary objective of the Nation's space effort in the ten to fifteen years following the Project Apollo manned lunar landing (1971–85). Such a program would be planned to (1) capitalize upon each stage of technological capability, (2) yield tangible, meaningful results at appropriate intervals with no potentially critical gap, (3) secure environmental data essential to manned ventures, and (4) provide time for proper development of extended manned activities in space. (NAS–NRC Release 11/17/64; Text, 10/28/64)

- NASA Langley Research Center issued requests for proposals for the conceptual and preliminary design of a hypersonic ramjet research engine to be used in the Hypersonic Ramjet Experiment Project announced by NASA last June. Since the project would ultimately include flight research with the engine mounted on the X–15 No. 2 aircraft, the guidelines included a weight limitation of 800 lbs. for the final research engine, including necessary flight-test instrumentation and dimensions compatible with mounting the engine near the lower aft fuselage of the X–15. (LaRC Release)

- U.S. Army announced that the XV–5A lift-fan jet V/Stol had made its first full transition flight, successfully converting to conventional flight after a vertical take-off. Tests were conducted at Edwards AFB, Calif. The XV–5A was said to have two big advantages over other types of vertical-flight planes: (1) it was much more economical in fuel consumption; (2) because the downward flow of air from the fans was relatively cool and gentle, the aircraft would have no trouble operating from unprepared dirt surfaces. (Witkin, *NYT*, 11/18/64, 17)

- James E. Webb, NASA Administrator, told the Council on World Affairs in Dayton, Ohio: "Space has become one of the newest dimensions of power: a dimension that ranks potentially with that of the land, the sea, and the air. And because national security is based on such dimensions, our security is heavily involved in our space achievements. Prudence demands that we conduct our program on a scale and at a pace that will permit no hostile force to employ space against us as an unchallenged avenue of aggression, politically or militarily.

  "At the same time we must never lose sight of another powerful concept, that in the eyes of the world, space has become an area of active competition with much more at stake than considerations of national pride or prestige. In the view of other nations, the relative effectiveness of the American system of free enterprise is, in a very real sense, being tested in space." (Text)

November 17: Electro-Optical Systems, Inc., reported that it had devised a high temperature, ultra high-vacuum thermionic electron emission microscope that can pictorially record critical problems associated with metal phase and grain growth change in missiles and hypersonic aircraft. The instrument was developed under contract to NASA. (L.A. *Herald Examiner*, 11/17/64)

- Maj. Gen. Marvin C. Demler, Commander, Research and Technology Div., AFSC, speaking at the Seventh Hypervelocity Impact Symposium, said: "Another example of advanced development hypervelocity in nature if you will, is the ASSET Program. ASSET stands for Aerothermodynamic/Elastic Structural Systems Environmental Tests, and involves the flight verification and demonstration of aerodynamic and structural design concepts for the lifting re-entry class of aerospace vehicles. Three ASSET vehicles have been launched to date of which two were quite successful. The ASSET Program was originally conceived and conducted by the Air Force Flight Dynamics Laboratory and has now been turned over to the Systems Command's Space Systems Div. program for even more advanced vehicles." (Text, AFSC Release)
- Army announced it had asked industry to submit proposals for a study of an advanced aerial vehicle for use as a weapons platform which might eventually replace current armed helicopters. The vehicle was conceived as an integrated aerial fire support system to include armament, avionics, and fire control equipment. (DOD Release 811–64)
- NASA announced award by Langley Research Center of $2.5 million contract to Boeing Co. for dynamic flight simulation work for the supersonic transport program. The Boeing 707 prototype would be modified by installation of major control and computer equipment to permit complete simulation of a wide variety of supersonic transport types. (NASA Release 64–286)

November 18: A Thor-Agena D booster was used by USAF to launch an unidentified satellite from Vandenberg AFB, Calif. (*M&R*, 11/23/64, 8 HHN–48)

- Flights to study the effects of sonic booms on structures and "to obtain preliminary data on sleep disturbance" began at White Sands Missile Range. The flights would be made through Dec. 17 and from Jan. 5 through Feb. 4 under the direction of the Federal Aviation Agency, as part of the program to develop a supersonic transport aircraft. (Clark, *NYT*, 11/13/64, 58)
- The Hermann Oberth Award, presented annually by the Alabama chapter of AIAA, was awarded posthumously to Hugh M. Taylor, former special assistant to the Chief of the Guidance and Control Division, NASA Marshall Space Flight Center. Taylor, who had worked at Redstone Arsenal since 1951, was killed in a plane crash July 9 near Newport, Tenn. (*Marshall Star*, 11/11/64, 1, and 11/25/64, 1)

November 19: NASA launched a Nike-Cajun sounding rocket from Wallops Island, Va., to an altitude of 78 mi., to test the prototype Nike-Cajun grenade payload produced by Zimney Corp. Instrumentation performance was satisfactory, except for failure of 12th grenade to explode. (NASA Rpt. SRL)

November 19: NASA conducted two sets of experiments with Nike-Apache sounding rockets launched from the flight deck of the USNS *Croatan:* (1) an experiment to measure pressure, temperature, and density in the region of about 20- to 75-mi. altitude for the Univ. of Michigan; (2) experiments to measure electron density in the ionosphere for the Univ. of Illinois. (Wallops Release 64–86; NASA Rpt. SRL)

- NASA Lewis Research Center undertook supervision at Lockheed Missiles and Space Company, Sunnyvale, Calif., of the design and development of a metal shroud section to replace the fiberglas section in Mariner 4. Investigations had indicated that the fiberglas fairing of MARINER III probably had structural failure when exposed to the flight environment and that this prevented shroud separation from the spacecraft. (KSC Release 206–64)

- In press interview during Third International Symposium on Bioastronautics and Exploration of Space, in San Antonio, Major Duane E. Graveline, a USAF doctor who examined each of the Project Mercury astronauts after their flights, said there was no doubt that even brief forays into space created a certain physical debilitation. "What prolonged space flights, with the constant state of weightlessness and no gravity as we know on earth might do to the human body, we don't know yet. We know that the Russian cosmonauts have suffered from the effects of prolonged weightlessness, and that U.S. astronauts Walter Schirra and L. Gordon Cooper had temporary debility for the same reason. . . . What we must do now is determine how to condition man for long-duration manned orbital missions and journeys into the universe. Frankly, we don't have the answers yet." (Macomber, Copley News Service, *San Diego Union*, 11/19/64)

- Cornell Univ. announced that scientists at the Ionospheric Observatory at Arecibo, Puerto Rico, bounced a radar signal off the planet Mars while it was about 138,000,000 mi. from earth today. Cornell said this was the farthest distance at which Mars had been detected by radar. (AP, Balt. *Sun*, 11/20/64)

- NASA announced that Francis J. Sullivan had been appointed Acting Director of the Electronics and Control Div., Office of Advanced Research and Technology. The position formerly was held by Dr. Albert J. Kelley, who became Deputy Director of the Electronics Research Center, Cambridge, Mass. (NASA Announcement 64–274)

- Dr. J. Allen Hynek of Northwestern Univ. said that he and his colleagues would submit a proposal to NASA for a 24-in. telescope to be landed on the moon by the mid-70's. Such a telescope might double the size of the optically observable universe from its present radius of about 5 billion light years to perhaps 10 billion. It would also "see" radiations from deep space that are totally excluded by earth's atmosphere. (Schmeck, *NYT*, 11/20/64)

- Defense Secretary Robert S. McNamara announced the closing of Atlas E, Atlas F, and Titan I missile installations. The inactivation would save $116.9 million annually. McNamara said he took the action because of "our now sizeable inventory of Titan II and Minuteman missiles which have far greater capabilities and can be operated and maintained at a fraction of the cost of the earlier models." (*SBD*, 11/20/64; DOD Release 822–64)

November 19: The Soviet Union marked Rocket Day, a new observance, by warning the West of the "stupendous power," great range, and pinpoint accuracy of its missiles. Marshal Nikolai I. Krylov, chief of the rocket forces repeated the Kremlin's claim that it had antimissile missiles. (UPI, *NYT*, 11/20/64, 27)

- A Commerce Dept. reprint of a Russian scientific article indicated that Soviet satellite COSMOS V had detected and reported a high altitude explosion of U.S. nuclear weapon over Johnston Island in the Pacific on July 9, 1962. (*NYT*, 11/20/64)
- A Pershing missile, fired from an Army range north of Fort Bliss, Tex., by an artillery unit from Fort Sill, Okla., as part of a training exercise, crashed 400 mi. off course somewhere in the mountains near Creede, Colo. It was not armed with explosives. (AP, Balt. *Sun*, 11/23/64)

November 20: General Creighton W. Abrams, Jr. (USA), presented the Distinguished Flying Cross to four of five Army pilots who recently broke ten world records for helicopters, including two records previously held by the Russians. The recipients were: Capt. William A. Welter, Jr.; Capt. Michael N. Antoniou; CWO Emery E. Nelson; and CWO Joseph C. Watts. The fifth pilot, Maj. John A. Johnston, was absent on an overseas assignment. The records were for speed, distance, and time-to-climb in class E–1, E–1D, and E–1E helicopters. With official recognition of the new records, the U.S. would hold 26 marks (52%), the U.S.S.R. 14 (28%), France 9 (18%), and Czechoslovakia 1 (2%). (DOD Releases 793–64 and 824–64; *NYT*, 11/7/64, 53)

- After analyzing atmospheric debris from Communist China's nuclear test, nuclear physicist Seitaro Nakamura of Tokyo Univ. Institute for Nuclear Physics said China was nearer to detonating her first hydrogen bomb than France was. Quantities of U–238 and U–237 found in atmospheric samplings Oct. 17–25 over Japan suggested detonation had been a "moderator test" of the type often made shortly before a hydrogen (fusion) test. (*Statist*, 11/20/64, 481)

November 21: Two NASA satellites—EXPLORER XXIV and EXPLORER XXV—were successfully launched into orbit from Vandenberg AFB, Calif., aboard a single four-stage Scout rocket. This marked the first time NASA had orbited dual spacecraft with a single booster. The double payload was designed to provide more detailed information on complex radiation-air density measurements.

EXPLORER XXIV (Air Density satellite) was a 12-ft.-dia., 19-lb. polka-dotted sphere identical to EXPLORERS IX and XIX. It was covered with aluminum foil to reflect both sunlight and radio waves; approximately 4,000 white spots on the surface provided temperature control. EXPLORER XXIV would provide three types of global comparative measurements: (1) high-altitude air density; (2) sources of atmospheric heat through comparison of data with Injun and other satellites; and (3) density and temperature variations of the atmosphere as a function of latitude. Orbital parameters were: apogee, 1,550 mi.; perigee, 327 mi; period, 116 min.; and inclination, 81°.

EXPLORER XXV (Injun-B), a 2-ft.-dia., 90-lb. metal satellite bearing 16 radiation sensors to measure protons and electrons in various ranges of energy, was roughly spherical in shape with 40 flat surfaces, 30 of them studded with solar cells. Orbital parameters were: apogee, 1,548 mi.; perigee, 326 mi.; period, 116 min.; and inclination, 81°.

Tracking data after one complete orbit indicated that the satellites were close to planned orbit with all instruments operating. (NASA Release 64–284; *Goddard News*, 11/30/64)

November 21: Univ. of Pennsylvania announced a team of archeologists, using a rubidium magnetometer intended for outer space, had found what may be the 2,500-year-old ruins of the fabled Greek city of Sybaris. (AP, *Omaha World Herald*, 11/22/64)

- Vice Adm. Hyman G. Rickover (USN) was selected as recipient of the 1964 Enrico Fermi Award for his contributions to development of nuclear power for submarines and power-generating plants. Rickover was the first engineer-administrator ever to receive the award. (*Phil. Inquirer*, 11/22/64).

November 22: Breene M. Kerr, who had been NASA's Deputy Assistant Administrator for Technology Utilization, became Assistant Administrator for Technology Utilization. He thus took over part of the duties of Dr. George L. Simpson, who would function as Associate Deputy Administrator and Assistant Administrator for Policy Planning. (NASA Release 64–291)

- Dr. Thomas P. Murphy became NASA Deputy Assistant Administrator for Legislative Affairs. Dr. Murphy joined NASA in November 1961 as a Staff Assistant to the Administrator and from July 1963 was Assistant to the Assistant Administrator for Technology Utilization and Policy Planning. (NASA Release 64–292)

November 23: NASA successfully launched a Nike-Apache sounding rocket to an altitude of 110.8 mi. from Wallops Island, Va. Experimental objectives were: (1) to measure daytime ion and electron densities and solar ultraviolet flux in the region from 31–124 mi.; (2) to test efficiency of loop antenna section for extending CW propagation measurements to "D" region altitudes. (NASA Rpt. SRL)

- NASA announced that USNS *Croatan* had completed its successful shakedown cruise off the Virginia-North Carolina coast during which a number of sounding rockets were launched for NASA. Electronic equipment and operational procedures were checked out in preparation for using the USNS *Croatan* as a mobile launch platform early in 1965. During this three-month NASA expedition, forty or more scientific experiments would be launched from the USNS *Croatan* as part of NASA's sounding rocket program in connection with the International Quiet Sun Year (IQSY), 1964–1965. (Wallops Release 64–86; NASA Release 64–294)

- USAF announced selection of the first six astronauts for its manned orbiting laboratory (MOL) program. An additional six would be named in early 1965. Following a six-month basic training course, the crewmen, all graduates of USAF's Aerospace Research Pilot School, would be assigned specialty areas. (*Av. Wk.*, 11/23/64, 25)

- Paul P. Haney, Public Affairs Officer at NASA Manned Spacecraft Center, asserted that the bulk of 550 communications regarding NASA scientist-astronauts had to be classified as letters of interest. Seventy-five per cent of this number, or 412, were lacking some detail and about 100 had been eliminated altogether. The number of applications was ahead of the number received for other classes of astronaut, he said. These figures clarified figures given by Julian Scheer, NASA Assistant Administrator for Public Affairs, who had said on Nov. 22 that only 20 applications had been received and that only nine of these applicants qualified. (*Houston Post*, 11/24/64)

November 23: Aviation Week reported that Soviet space-medicine specialists were considering bacterial compatibility in their selection of cosmonaut teams for the three-man Voskhod spacecraft. This was an important consideration because of the difficulty of preventing communication of bacterial flora, which might be harmless to one man and dangerous to another, in a closed-cabin environment. It was not known whether persons with similar metabolic processes would be selected for a single crew or whether those with extremely different ones might not compensate each other, nor had it been determined whether the VOSKHOD I flight crew was selected for bacterial compatibility. (*Av. Wk.*, 11/23/64, 59)

November 23–24: NASA Goddard Space Flight Center was host at the jointly sponsored Goddard-IEEE Short Term Frequency Stability Symposium to some 310 scientists and engineers from across the country, England, Canada, and Switzerland. The symposium was the first of its kind to study problems associated with generating stable frequencies in the microwave region for the control, guidance, and tracking of launch vehicles and satellites. (*Goddard News*, 11/30/64, 1)

November 24: NASA Goddard Space Flight Center announced selection of Beckman Instruments Corp. to negotiate final contract for three Satellite Telemetry Automatic Reduction Systems (Stars) at slightly more than $1,000,000. The contractor would be responsible for manufacture, installation, and maintenance of the system as well as training of Goddard personnel. This second phase of the Goddard Stars project would provide a highly automated method of preparing large quantities of data for later analysis by the scientific community. (GSFC Release G-33-64)

- At the Aerospace Medical Research Laboratory, Wright-Patterson AFB, Ohio, four college men completed a test in which they lived on astronaut food for 28 days in a spacecraft type of capsule and 14 more days in hospital isolation. They ate what was called "freeze dehydrated foods" that came in one-in. cubes of beef, chicken, etc. Liquid was distilled water, tea, and fruit juices. No ill effects were apparent. (*NYT*, 11/27/64, 31)
- Canada announced the completion of negotiations for building a 150-in. optical telescope at Mt. Kobau, B.C., in commemoration of the recent visit to Canada of Queen Elizabeth II. (Shaw, *CSM*, 11/24/64)
- *Izvestia*, the Soviet government newspaper, announced the death of Y. A. Korovin, a leading Soviet authority on international and space law. (AP, *Galveston Tribune*, 11/25/64)

November 25: James E. Webb, NASA Administrator, and Allen Fairhall, Australian Minister for Supply, announced jointly that a tracking station to support the Apollo manned lunar flight would be established near Canberra, Australia, as a part of the worldwide unified S–band Apollo tracking network. The station would be equipped with an 85-ft. diameter parabolic reflector antenna to support the lunar phases of Apollo missions. (NASA Release 64-290)

- The Lunar Landing Research Vehicle (LLRV), piloted by Joseph A. Walker of NASA's Flight Research Center, attained an altitude of 30 ft.—the highest point in five brief flights—and landed safely. Walker said he was highly pleased with the craft's performance. The flight was conducted at Edwards AFB, Calif. (AP, *NYT*, 11/26/64, 9)

November 25: Launch of Stratoscope II balloon was delayed at least 48 hours when a tube used to fill the 300,000-cu. ft. balloon with helium slipped and punctured the plastic skin. Another balloon was to be sent to the Palestine, Tex., launch site from the G. T. Schjeldahl Co. in Northfield, Minn. (*M&R*, 11/30/64, 29; *NYT*, 11/26/64, 34)

- In a speech to the National Space Club in Washington, D.C., NASA's Director of Advanced Manned Earth Orbital Missions M. J. Raffensperger said: "Our limited flight experience to date indicates that any deleterious effects of space flight can be controlled, but man's reaction to long-duration space missions is not yet quantitatively understood. Thus, the biomedical and behavioral areas must receive major attention in the earlier extended duration earth orbital flights." Other areas that Mr. Raffensperger felt showed promise in providing beneficial returns to humanity were "research and observation of meteorology from a manned spacecraft" and "earth mapping and survey." (Text)
- NASA announced it had awarded 33 supplementary or new research grants and contracts totaling $2,418,634 to 26 colleges, universities, and private research institutions. (NASA Release 64-295)

November 26: A Thanksgiving Day message from President Johnson was broadcast by RELAY I, RELAY II, and SYNCOM II communications satellites to American armed forces all over the world. (Day, Phil. *Eve. Bull.*, 11/26/64)

- The ground station at Rao, Sweden, about 225 mi. southwest of Stockholm, joined the world network of communications satellite ground stations and would receive taped television messages via NASA's RELAY I satellite. The station was the joint effort of Norway, Denmark, and Sweden. (*M&R*, 11/30/64, 20-22)
- USN announced $36,000 contract with the Convair and Electric Boat divisions of General Dynamics Corp. for analytical and design studies of a flying submarine—a small craft capable of flying to and submerging for operations in relatively closed waters. (AP, *Balt. Sun*, 11/27/64)

November 27: Dr. George E. Mueller, NASA Associate Administrator for Manned Space Flight, announced that about 100 of the 140 personnel in the Kennedy Space Center Launch Support Equipment Engineering Div. would move from Marshall Space Flight Center to KSC by Aug. 1965. (NASA Release 64-298; *NYT*, 11/28/64, 16)

- The European Space Research Organization (ESRO) Council approved its program for 1965, which would include setting up tracking and telemeasuring stations in Belgium, the Falkland Islands, and Alaska. The program would also include 41 experiments during 28 rocket firings, four experiments with the polar ionospheric satellite Esro 1 to be fired in September, and six experiments with the Esro 2 satellite for the study of cosmic rays. Payment of $17.2 million was voted and contracts worth $34.6 million were authorized. (AP, *NYT*, 11/29/64, 20)
- President Johnson named Dr. Charles Stark Draper, head of Dept of Aeronautics and Astronautics at Massachusetts Institute of Technology, as a recipient of the National Medal of Science. Dr. Draper, a leader in developing aeronautic and astronautic instruments and controls, was one of 11 scientists to be so honored. (*NYT*, 11/28/64, 16)

November 28: NASA MARINER IV Mars probe was successfully launched into interplanetary orbit from Cape Kennedy, Fla., by an Atlas-Agena D launch vehicle. The Agena D separated from the Atlas 115 mi. up and injected itself into a parking orbit around the earth. It then reignited and injected itself into a Mars trajectory. MARINER IV was detached and locked onto the sun for solar power and stabilization on one axis and opened its solar panels. Later, the probe would lock on the star Canopus for second-axis stabilization—the first time a star would be used for an attitude reference on a long space mission. In addition to equipment for televising the Martian surface, MARINER IV carried instruments to study radiation, space dust, and magnetic forces near earth, Mars and in deep space. The trajectory as presently calculated would carry MARINER IV within 151,000 mi. of Mars by mid-July. (NASA Releases 266, 288; Hines, Wash. *Eve. Star*, 11/29/64; *NYT*, 11/29/64; Miles, *Wash. Post*, 11/30/64)

November 28: USAF Athena missile was destroyed in flight over southeastern Utah when high winds threatened to force it off course. The Athena was used to test re-entry procedures for ICBM's. (UPI, *St. Louis Post-Dispatch*, 11/28/64)

- At his news conference in Johnson City, Tex., President Johnson spoke of Vice President-elect Humphrey's prospective duties. "I know he has demonstrated an intense interest in our space activities, and under the Space Act he will be the new Chairman of the Space Council. . . ." (Transcript, UPI, *Wash. Post*, 11/29/64)

November 29: Carbon 13 was observed in the comet Ikeya in about the same proportion as it occurred on the earth by Dr. Jesse L. Greenstein of the Mount Wilson and Palomar Observatories and Dr. Antoni Stawikowski of the Nicolaus Copernicus University of Poland working at Palomar Observatory. "This lends credence," said Dr. Greenstein, "to the theory that comets, and there may be millions of them, were formed in about the same region of the solar system as the earth and were blown into their present, vast, elliptical orbits. The fact that the carbon isotope ratio is similar for Ikeya and the earth argues against the theory that comets are captured, interstellar wanderers that were formed beyond the solar system. If comets were born elsewhere, one would expect them to be unlike earth." (*NYT*, 11/30/64, 11; AFOSR Release 11–64–3 "S")

- It was announced that a new ultraviolet radiation spectrograph had been developed at NASA Goddard Space Flight Center in an effort to solve the mystery of "airglow." The spectrograph would be carried aloft by a rocket and would be expected to reveal the constituents and mechanisms of the airglow. (NANA, *San Diego Union*, 11/29/64)

November 30: NASA and Pan American Airways announced that a scheduled Pan Am flight operating between San Francisco and Honolulu had successfully received error-free teletype messages for one hour from NASA's communications satellite, SYNCOM III, which was in stationary orbit over the mid-Pacific. NASA said that the relay of communications to an aircraft in flight from a station on the ground was another "first" for SYNCOM III. (NASA Release 64–297)

- MARINER IV Mars probe finally locked its sensor on the star Canopus after fixes on three wrong stars. Jet Propulsion Laboratory officials noted that picking up wrong stars was not critical in the early stages of the flight. If the MARINER IV had not dropped these wrong fixes of its own accord, the action could have been commanded from the ground. Only if Canopus were in the sensor's sights would the TV camera be aimed at the Martian surface when MARINER IV flew past the planet 7½ months from now; only with Canopus in view would scientists know the spacecraft's attitude with the precision needed for a midcourse rocket firing of maximum accuracy to refine the course of the trajectory and bring the MARINER IV within about 8,600 mi. of Mars instead of the presently anticipated distance of 151,000 mi. (Miles, *Wash. Post*, 11/30/64; Witkin, *NYT*, 11/30/64, 1; *National Observer*, 11/30/64; Witkin, *NYT*, 12/1/64, 1)

November 30: NASA launched Aerobee 150 sounding rocket from White Sands, N. Mex., with experiment to obtain optical spectra of night ultraviolet airglow. Rocket attained 113.5-mi. altitude, and the attitude control system kept the two fast spectrographs pointed accurately at the horizon, permitting four-minute exposure of the faint molecular oxygen Herzberg bands. Also included in payload was a Nikon camera to obtain star-field backgrounds for aspect determination. (NASA Rpt. SRL)

- Nike-Apache sounding rocket launched in U.S.-Pakistan project from Sonmiani reached 124-mi. altitude. Sodium vapor trail ejected during ascent was photographed by ground cameras to measure atmospheric wind directions and speeds. Launch was part of cooperative program between NASA and Pakistan's Space and Upper Atmosphere Research Committee. (NASA Rpt. SRL)

- The Soviet Union launched ZOND II probe in the direction of the planet Mars. The purpose of the launching was said to be trying of the station's systems under actual conditions of prolonged space flight and gaining of practical experience. Scientific investigations in interplanetary space would be carried out simultaneously. The probe was proceeding along a trajectory close to the computed one. (Tass, *Krasnaya Zvezda*, 12/2/64, 1, ATSS–T Trans.; Simons, *Wash. Post*, 12/2/64; Clymer, *Balt. Sun*, 12/2/64)

- X–15 No. 2 flown by pilot John B. McKay (NASA) to 86,000-ft. altitude at maximum speed of 3,000 mph (mach 4.54). The flight continued checkout of the revised landing gear system on the No. 2 aircraft and checked modifications made to prevent nose gear from coming down prematurely. Total engine burn time was 76 sec. and tests performed were successful. (NASA X–15 Proj. Off.)

- The National Center for Atmospheric Research postponed the launching of a huge balloon equipped to carry a 36-in. telescope 15 mi. into the stratosphere to photograph Jupiter and star clusters. (UPI, *NYT*, 12/1/64, 2)

- Dr. Nancy Roman, Chief of NASA Astronomy programs, told *Space Business Daily* that her office had a program under consideration for a 50–55-in. telescopic mirror manned orbiting observatory which might orbit in the mid-1970's. Although she emphasized that all plans were in the "thinking stage," Dr. Roman said the 50–55-in. configuration would provide another interim link between the unmanned Orbiting Astronomical Observatory series and the manned 100-in. configuration being considered for launching in 15 to 20 years. "The 50–55-in. interim configuration should be such," she said, "that it could be used either manned or unmanned. But it will have as its principal objective the proving of man's ability as an orbiting astronomer." (*SBD*, 11/30/64, 132)

- NASA Marshall Space Flight Center's deadline for industry proposals to design a 100-lb.-thrust rocket engine for manned space systems. Designated "C-1," the engine would be powered by monomethylhydrazine and nitrogen tetroxide, which would be hypergolic and storable. The engine could be employed for attitude control, maneuvering, and ullage control purposes. (*A&A*, 12/64, 75)

November 30: Results of five public opinion surveys conducted in 12 cities over 14 months for Thiokol Chemical Corp. and Grumman Aircraft Engineering Corp. were announced. Findings indicated that most Americans supported the national goal of landing men on the moon by 1970 but they felt no sense of urgency about achieving it. One third of those interviewed thought spending should be reduced. (*NYT*, 11/15/64, 66)

- Gen. Thomas S. Power (USAF) retired as Commander-in-Chief of the Strategic Air Command (SAC). He was succeeded by Lt. Gen. John D. Ryan, formerly SAC Vice Commander-in-Chief. Ryan would be made a four-star general. (*NYT*, 12/1/64, 46)
- Canaveral District of the Army Corps of Engineers announced award of $19,055,953 firm-fixed-price contract to the George A. Fuller Co., for construction of Pad B of NASA's Saturn V Complex 39 at Merritt Island, Fla. The contract also provided for building a two-mile section of crawlerway for transporting launch and arming towers to the pad. (DOD Release 844-64; *M&R*, 11/30/64, 22)

November 30–December 3: James T. Ramey, a commissioner of the Atomic Energy Commission, analyzed the use of cost-effectiveness techniques in a speech to the Atomic Industrial Forum in San Francisco. He said: "The cost effectiveness technique assumes end-applications or requirements, and then figures out the costs of alternative systems of achieving equal or greater effectiveness . . . it would seem that the requirements system, and the cost effectiveness techniques, tend to favor the status quo in technology, if not in economics or politics. . . .

"This leads me to the conclusion that the underlying problem is really whose judgment and recommendations should receive most weight in determining which development programs and projects should go forward. Should these decisions be made by the science advisors and coordinators, and budget specialists who have no direct programmatic responsibility and are not accountable for failure to carry out the directives of the President and the Congress? Or should the judgment of the developmental agency which has the statutory responsibility for the program and the technical depth and experience receive the most weight? . . . I have a certain affinity for giving considerable weight to the judgment of the development agency, which, in regard to atomic energy, would be the AEC and its laboratories. In regard to nuclear energy in space, it would be AEC and NASA, and to some extent the Air Force. . . .

"I am quite optimistic about achieving the great possibilities of atomic energy in space. I firmly believe the atom will provide the means for the propulsion and servicing of manned space ships for moon missions following Apollo, and to and from the planets . . ." (*Av. Wk.*, 1/11/65, 11)

During November: NASA's Flight Research Center completed a series of wind-tunnel tests of X–15 models with canard stabilizers mounted on the nose to determine if a canard surface would permit the X–15 to re-enter the atmosphere at higher angles of attack. Tests were carried out in a Jet Propulsion Laboratory tunnel. (*M&R*, 11/9/64, 9)

During November: The Titan III–C 120-in. solid strap-on engine was fired successfully at United Technology Center's Coyote, Calif., development center. The engine burned for 110 sec. (*M&R*, 11/23/64, 9)
- Astronomers Dr. Robert Kraft of the Mount Wilson and Palomar Observatories in Calif., and Dr. W. Krzeminski of the Polish Astronomical Institute working at the Lowell Observatory, Flagstaff, Ariz., reported in *Astrophysical Journal* a new concept of Novae Sagittae that may provide the first opportunity to test part of Einstein's theory predicting the existence of gravitational waves. Novae Sagittae appeared to consist of twin stars connected by a bridge of hydrogen gas flowing from the larger to the smaller star. The two stars revolve about a common center so fast— about every $81\frac{1}{2}$ min.—that they should radiate gravitational waves, if such waves exist. (*NYT*, 12/1/64, 31)
- General Curtis E. LeMay, USAF Chief of Staff, was interviewed in *General Electric Forum* and discussed aerospace power's contribution to national security. "In recent months, the area of direct technical concern to the Air Force has become much broader. For example, we must now press for technical superiority not only in aircraft but also in missiles, space systems, and command and control systems that are responsive to the compressed time scale of military operations. We cannot afford to fall behind in any of these fields, because the potential enemy can be expected to exploit all of the important technical gaps that might appear." LeMay said that "chiefly because of the cost factor, we are not exploiting anything like the total amount of technology that we think might be useful. Certainly we would like to do more, but our R&D outlay already comprises 18 percent of the Air Force's total budget and must be balanced against other needs." (*GE Forum*, 10–12/64, 14–17)
- Linde Division of Union Carbide has "grown" a new solid-state laser material for use by scientists in laser research at the Marshall Space Flight Center Astrionics Laboratory. The new material was yttrium aluminum garnet doped with neodymium. It supported solid-state continuous-wave laser action, with common tap water as the only coolant. Toughness of the garnet was comparable to that of ruby laser rods. (*Marshall Star*, 11/18/64, 2)
- In an editorial in *Astronautics and Aeronautics*, H. H. Koelle said: "Suppose . . . that a space station could be made available with working area in the station and a round-trip ticket purchased by individual countries for their scientists, and that an advanced aerospace transportation system will be developed anyway as part of the U.S. space program. Analysis of such an operation indicates that an international research station of this kind could be built in the late 1970s, and that the price of a three-month field trip to the station might well be kept in the order of a million dollars a person. Many countries . . . would be able to pay this price and would enthusiastically participate in the project. During the 1980s, . . . the charges should be reduced considerably, thus increasing international participation." (*A&A*, 11/64, 20–21)
- The House Science and Astronautics Committee Subcommittee on Science, Research and Development made recommendations to equalize the national distribution of Federal research and development funds: that additional funds be budgeted to the National Science Foundation to ensure at least one major center of excellence in research and technology in each appropriate region of the nation; that DOD and NASA emphasize

building new centers of science and technology when existing facilities in research-concentrated areas become obsolescent. The Subcommittee also called for the White House to convene a Government-industry conference to study such problems. (*M&R*, 11/16/64, 11)

During November: A team from Lockheed Propulsion Co. visited NASA Marshall Space Flight Center to brief scientists there on possible use of 156-in. solid motor as a strap-on for the first stage of both Saturn IB and Saturn V. (*M&R*, 11/23/64, 7)

- Marshall Space Flight Center awarded a $2,017,400 contract to Granite Construction Co., Houston, Tex., for construction of a vehicle component supply building and a hazardous materials storage building at its Saturn rocket production plant (NASA Michoud Operations) in New Orleans. (*Marshall Star*, 11/18/64, 4)
- North American Aviation, Inc., announced FY 1964 sales of $2,189,594,000, the largest annual total ever recorded by a company predominantly in the aerospace business. (*Av. Wk.*, 11/16/64, 29)

December 1964

December 1: First full-power static firing of the hydrogen-fueled Saturn S–IVB stage was achieved at the Douglas Missile and Space Systems Div., Sacramento, Calif., Test Center. The J-2 engines were operated at full 200,000-lb. thrust for ten seconds as planned. This marked the first use of the new NASA Test Complex Beta, consisting of two 150-ft.-high steel and concrete stands capable of holding down rockets with more than a million pounds thrust. Primary objectives of the ten-second test were to achieve engine start, advance to full thrust, stabilize main-stage operation for the duration, and shut down on schedule. (*Marshall Star,* 12/9/64, 9; *M&R,* 12/7/64, 9; DMSSD *Apogee,* 12/64)

- USAF launched a new Advanced Ballistic Re-entry System experimental nose-cone (Abres) from Vandenberg AFB 5,000 mi. down the WTR on an Atlas-D ICBM. The nosecone tests were being conducted to develop more effective re-entry systems for America's intercontinental ballistic missiles. (AP, Wash. *Eve. Star,* 12/1/64; *Houston Chron.,* 12/1/64)
- Nike-Cajun sounding rocket successfully launched from the Argentine launch range at Chamical to study the ionosphere. Payload contained continuous wave propagation, electron temperature, and ion density experiments. This was first rocket launched by a South American country under a cooperative arrangement with NASA. Project was administered jointly by the Argentine National Commission of Space Research (CNIE) and NASA. Argentina built the payloads and part of the ground station, while NASA trained Argentine technicians and engineers and provided some ground support equipment. Argentina purchased two Nike-Cajun sounding rockets in the U.S. No exchange of funds took place between the two agencies. (NASA Release 64–304; NASA Rpt. SRL)
- Nike-Apache sounding rocket reached 124-mi. altitude in U.S.-Pakistan experiment from Sonmiani. Sodium vapor trail from the rocket was photographed by ground cameras to measure direction and speeds of upper atmospheric winds. (NASA Rpt. SRL)
- Soviet Union announced that the spacecraft, ZOND II, headed for Mars in an apparent race with MARINER IV, had run into difficulties. The Soviet report said that in the first few radio contacts between ZOND II and the ground "the power supply on board the space station was about half of the expected level." Position information indicated the probe was 25,000 mi. from earth. (Witkin, *NYT,* 12/2/64, 1; Simons, *Wash. Post,* 12/2/64)
- Robert G. Deissler, NASA Lewis Research Center, received the Heat Transfer Division Memorial Award of the American Society of Mechanical Engineers in New York City. (LRC Release 64–108)
- Dr. J. E. Geake of the Manchester College of Science and Technology said that undamaged material on the moon brought to the surface by the impact of a meteorite or by possible volcanic activity could account for

patches of red light observed there. He suggested that if the impact were in a region of luminescent material, the rocks would emit light when hit by charged particles. The impact could have occurred 10,000 or even 100,000 years before the luminescence was observed. (Sci. Serv., *NYT*, 12/1/64, 19)

December 1: Former NASA astronaut Col. John H. Glenn, Jr. (USMC), completed 15-day assignment at El Toro, Calif., MCAS, where he was on TDY for flight refresher training.   (*J/Armed Forces*, 11/28/64, 5)

December 2: Henry O. Slone, Snap-8 project manager at NASA Lewis Research Center, reported a successful 76-hr. test of Snap-8's turbine alternator assembly (TAA) using mercury as the working fluid. "This test demonstrated operational capability and machine performance up to 80% of its rated power," Slone said. Snap-8 (System for Nuclear Auxiliary Power) was designed to supply electric power onboard a spacecraft and would use nuclear heat to produce 35 kw. of power. When fully developed, it would have the capability to operate unattended in space for 10,000 hours.   (LRC Release 64–111; *M&R*, 12/14/64, 10)

- NASA Manned Spacecraft Center awarded contracts totaling almost $1 million: Garrett Corp., with a $384,665 contract, would develop a portable life support system to be worn on the astronaut's back when he would leave his capsule in space, or for moon exploration; Boeing Co. was awarded a $299,746 contract for a feasibility study of placing a laboratory capsule on the Apollo spacecraft; the U.S. Naval Air Equipment Center, Philadelphia, was given $45,000 to evaluate the dependability of the water-cooled undergarment to be worn by astronauts in space flight and lunar exploration.   (AP, Balt. *Sun*, 12/4/64)
- An F–104 aircraft participating in sonic boom tests for supersonic transport program, at Alamogordo, N. Mex., inadvertently created a boom 20 times as great as those expected from the proposed airliners and 10 lbs. psf greater in pressure than the 30-lb. upper limit planned for these tests. The incident occurred during a demonstration for newsmen and recorders on measuring instruments in test structures were not turned on.   (Clark, *NYT*, 12/3/64; *Time*, 12/11/64)
- The Federal Aviation Agency (FAA), following up Defense Secretary McNamara's announced deactivation of 19 airfields, called attention to the procedures by which surplus military airfields could be transferred to state or local public agencies for development as civil airports. The procedures were established by the Surplus Property Act of 1944 and its amendments. More than 600 airports had been transferred to state and local agencies under this statute since the end of World War II.   (FAA Release 64–108)
- Four USAF Tactical Air Command F–4C (Phantom II) jet fighter aircraft landed at MacDill AFB, Fla., after completing an 18-hr. flight of almost 10,000 mi. The Phantoms refueled several times from KC–135 jet tankers. Flight, which set a new unofficial endurance record for jet fighter aircraft, was made to evaluate the capabilities of the F–4C aircraft during sustained, ultra-long-range flight and to evaluate the physiological and psychological effects on the crews under these conditions.   (DOD Release 852–64; *Wash. Post*, 12/3/64)
- USAF Electronic Systems Division announced award of $1,350,000 fixed-price contract to Space-General Corp. for design, fabrication, and testing of satellite payloads.   (DOD Release 849–64)

December 2: The New York Stock Exchange banned stop orders in shares of the ComSatCorp., hoping to prevent aggravated price swings in ComSat shares. ComSat had finished as the third most active issue for the day, having traded as high as 52¾. (Vartan, *NYT*, 12/3/64, 69)

- Dr. Raymond L. Bisplinghoff, NASA Associate Administrator for Advanced Research and Technology, spoke on "New Technology—Its Selection and Development" before the Atomic Industrial Forum conference in San Francisco. His address made the following points: (1) preeminence in space demands that a program of development of new technology be continuously supported; (2) the formulation of a responsible requirement requires the existence of an underlying body of technology; (3) the development of a new branch or extension of a branch of technology should be directed at a related class of requirements rather than a single requirement; (4) development of a new technology requires the existence of a focal point in the form of advanced hardware; (5) short and long-term goals must be related to the Nation's policies and plans; (6) our goals must be continuously assessed; (7) our work must be relevant to the goals which have been set; and (8) better processes of program management must be devised. He said in conclusion: "A balance of technologies in our favor will, however, require wisdom in the selection of new technologies in which to invest our resources and resolution in their development." (Text)

December 3: Speaking at Georgetown Univ.'s 175th anniversary convocation, President Johnson said: "For almost the first time the interdependence of nations is not a remote goal or a ringing slogan. It is a fact which we neglect at our peril.

"Communications satellites, atomic rockets and jet transports have made distant capitals into close neighbors. Our challenge is to transform this reality into an instrument for the freedom of man.

"Today the cost of failure to communicate is not silence or serenity but destruction and dissolution." (Text, *Wash. Eve. Star*, 12/3/64)

- President Johnson directed that contracts of engine and airframe manufacturers working on a projected supersonic transport plane, which expired Nov. 30, be extended. Four firms were affected: the Boeing Co., and Lockheed Aircraft Corp., which had airframe contracts; and the General Electric Co., and the Pratt & Whitney Div. of the United Aircraft Corp., which had engine contracts. (AP, *Wash. Post*, 12/4/64)
- A joint British-American group of communications engineers was preparing recommendations for a communications base in the Seychelles Islands, northeast of Madagascar. Station would fill a wide gap between Atlantic and Pacific stations in the present worldwide chain of receiving-transmitting stations for satellites. Existing network, set up by the U.S. Navy, had to rely on shipboard stations for Indian Ocean coverage. British responsibility would be limited to the real estate involved while the U.S. would build and equip the station and furnish most of the operating personnel. (Watson, *Balt. Sun*, 12/31/64)
- Eighteen top business leaders accompanied by NASA officials toured NASA Michoud Operations as part of a *Life* magazine-sponsored visit of space facilities across the country. They were briefed by Dr. Wernher von Braun, director of NASA's Marshall Space Flight Center, and Dr. G. N. Constan, Michoud manager. (N.O. *Times-Picayune*, 12/4/64)

December 3: Alamogordo, N. Mex., sonic boom test site for supersonic airliner program was repaired 24 hours after it was battered by the boom from an F-104. A spokesman for the FAA said: "The damage was not as extensive as we first feared. We are still on schedule, and should complete the tests as planned in February." (*Av. Wk.*, 12/14/64, 31)

- USN nuclear submarine *Stonewall Jackson* passed its first missile-firing test by shooting a Polaris A-3 more than 2,000 mi. to an ocean target area. The firing was the 18th success in 19 A-3 training launchings from submerged submarines. (UPI, *NYT*, 12/3/64, 83)
- It was announced that President Johnson had designated December 1964 as United States International Aviation Month to commemorate the 20th anniversary of the International Civil Aviation Organization (ICAO), a division of the United Nations. The ICAO was responsible for making the international rules and setting international standards for all civil aviation activity, including the field of general avaiation. (NAA Release)

December 4: USAF launched Atlas-Agena D from WTR with unidentified satellite payload. (HHN-48)

- Nike-Cajun sounding rocket reached 85-mi. altitude in launch from Chamical, Argentina. Instrumented payload apparently obtained good data in measurement of electron density, electron temperature, ultraviolet and solar x-radiation flux. Test was second of two Nike-Cajun launches by NASA and Argentina's National Commission of Space Research. (NASA Rpt. SRL)
- The Athena rocket was successfully fired by USAF from Green River, Utah, to a target area in the White Sands Missile Range, N. Mex. (*M&R*, 12/14/64, 10)
- NASA awarded 19 supplementary or new research grants and contracts totaling $914,889 to 17 universities, colleges, and private research institutions. (NASA Release 64-305)
- Army Corps of Engineers announced award of $8,079,003 (NASA funds) fixed-price contract to Malan Construction Co. of Koppers Co., Inc., for construction of a rocket engine test stand at the NASA Mississippi Test Operations. (DOD Release 856-64)
- AFSC Space Systems Div. announced that Lockheed Missiles and Space Co. would receive a $3,180,000 increment to a previously awarded contract for work on the Gemini program target vehicle system. (DOD Release 856-64)
- Joseph A. Darr, Jr., and John Reeves of NASA Flight Research Center received plaques from President Johnson for their dollar-saving contribution to the X-15 research rocket airplane. Darr and Reeves were cited for the design and construction of a portable engine inspection cart that was used to check the systems operation of the LR-99 rocket engine which powered the X-15. Use of the cart saved 500 man hours and about $150,000, NASA estimated. (NASA Release 64-303)
- Air Force Systems Command reported that Rome Air Development Center (RADC), sponsoring approximately $1 million annually in computer research and development, was major Air Force agency performing research in the computer field. According to Alan R. Barnum, Chief of RADC's Data Processing Section, "RADC's primary responsibility in computer research is development of high speed computer devices and special computers to meet peculiar Air Force problems. Also, methods of programming computer instructions, development of larger and faster

computer memories and research in a new area called artificial intelligence." (AFSC Release 11-64)

December 4: ComSatCorp said it had 140,000 public shareholders, with more than half holding ten shares or less, and with 95 per cent holding 50 shares or less. This total did not include an estimated 60,000 public stockholders whose shares were held for them by securities firms and banks and, therefore, not listed in their own names. ComSat stock climbed 6½ points to close at a new high of 55½ on the New York Stock Exchange. It ranked as the second most active issue with a turnover of 134,200 shares. (ComSatCorp Release; Vartan, *NYT*, 12/5/64, 39)

December 5: MARINER IV underwent a critical course change to make it possible for the craft to take useful pictures of Mars during the fly-by in July 1965. In response to commands radioed from earth, a small rocket on the base of the spacecraft, which could add anything between an eighth of a mile an hour to 188 miles an hour to the craft's speed, was fired. Initial indications were that the correction was successful and accurate.

Several days later, tracking equipment and associated computers confirmed that the rocket firing had produced three changes in MARINER IV's flight path: (1) the Mars fly-by from 151,000 mi. in front of Mars to a closest approach of 5,400 mi. behind Mars; (2) the arrival date from July 16, 1965, to July 14; and (3) the flight path past Mars from the leading edge to the trailing edge of Mars. The maneuver was followed by sun and Canopus reacquisition. A second midcourse maneuver would not be required.

The midcourse rocket used by MARINER IV was the first to be used by this country that was capable of a second firing. (NASA Release 64-310; Witkin, *NYT*, 12/6/64, 67; Miles, *Wash. Post*, 12/6/64)

December 6: NASA and General Electric Co. announced they had accomplished precision tracking of EXPLORER XXII with a laser device. Tests were conducted at NASA Goddard Space Flight Center using a ruby laser mounted on a telescope. With the telescope aimed along the predicted path, the laser was flashed several times as EXPLORER XXII, containing an array of corner reflectors, passed overhead. The reflectors, made of quartz crystal, were designed to return the beam of the laser almost exactly where it came from. The optical signals were displayed on an oscilloscope. As the laser pulse traveled from the telescope to the satellite, then back to the ground, the elapsed time measured could be converted into distance. Expected accuracy was within 10 ft. (*Chic. Trib.*, 12/7/64)

• James V. Bernardo was appointed Director of NASA Educational Programs and Services, would direct programs for developing an understanding of U.S. aerospace activities among students, teachers, and the general public, and for providing services to schools, colleges, and educational organizations. Dr. Aaron P. Seamster was named Deputy Director. (NASA Ann. 64-293)

December 7: MARINER IV lost its lock on Canopus. "Current analysis supports the theory that a dust particle reflecting the light of the sun passed through the field of view of the Canopus tracker," NASA said. Loss of lock on the star would not alter the spacecraft's course, would merely affect its attitude, which would be of importance only when the spacecraft neared Mars and attempted to photograph the planet. NASA planned to command the spacecraft to reacquire Canopus soon.

JPL, manager of the Mariner program for NASA, reported that a solar plasma probe experiment, which had functioned perfectly for eight days, had ceased to transmit useful information. (AP, *Houston Post*, 12/8/64)

December 7: NASA Nike-Apache sounding rocket launched from Wallops Island, Va., with Univ. of N.H. instrumented payload reached 90-mi. altitude, but experiment results were not obtained because of telemetry failure at 4.05 sec. Primary purpose of flight was to check out instrumentation for Nike-Apache sounding rocket experiments to be conducted on the IQSY Mobile Launch Expedition in early 1965. A second checkout launch was scheduled for mid-January. (NASA Rpt. SRL)

- Scientists at Johns Hopkins Univ. announced observations from an unmanned balloon on Oct. 27 that might indicate life on Venus. Experiments showed that the clouds covering Venus were composed of water in the form of ice particles and that the surface temperature was close to $-40°$ F, with only a few degrees difference between the light and dark sides of the planet. An earlier baloon flight in February had revealed that the air above the clouds was rich in water vapor. The scientists theorized that if there were water vapor, some of its molecules would be broken up by ultraviolet sunlight into oxygen and hydrogen; an atmosphere with oxygen and water vapor would be a propitious environment for life.

    Radiotelescopes on earth and on the MARINER II space probe, which had passed within 21,000 mi. of Venus in December 1962, indicated the surface temperature was $800°$ F. However, since interpretation of radiotelescope waves as an indication of temperature had neither been proven nor universally accepted, the Johns Hopkins scientists suggested that previous findings might only indicate a large amount of lightning-like electrical activity in the turbulent atmosphere of Venus. (Sullivan, *NYT*, 12/8/64; Fenton, *Balt. Sun*, 12/7/64; *Wash. Post*, 12/8/64)

- *Aviation Week and Space Technology* reported on a letter (dated Nov. 9, 1964) from Sen. Clinton P. Anderson (D.-N. Mex.), Chairman of Senate Committee on Aeronautical and Space Sciences, to President Johnson in which he recommended a merger of the Air Force's Manned Orbiting Laboratory (MOL) program and NASA's Apollo X program. He argued that $1 billion could be saved over a five-year period if MOL were to be cancelled and the Air Force funds applied to the Apollo-based space station. A jointly operated national space station program would, according to Anderson, best utilize U.S. space resources. Sen. Anderson felt that MOL lacked growth potential because it could not progress beyond its two-man, 30-day mission without development of a resupply system. (*Av. Wk.*, 12/7/64, 16; NASA Archives)

- Plans for adding a transmitter to the Pan American Boeing 707 for two-way communication tests using SYNCOM III as a relay point were discussed at an Air Transport Association meeting attended by representatives of FAA, NASA, Bendix, Pan Am, and Hughes Aircraft. Teletypewriter messages had been successfully transmitted in November from a ground station at Camp Roberts, Calif., to a Pan Am 707 enroute to Honolulu. (*Av. Wk.*, 12/7/64, 34)

- NASA Goddard Space Flight Center scientists would evaluate two H–10 atomic hydrogen maser clocks for possible use in high-precision satellite tracking as well as experiments testing Einstein's prediction of the effect of gravity on time. The clocks were developed by the Quantum Electronic Devices activity of Varian Associates and would use the natural oscillation resulting from the interaction of the proton and the electron in the hydrogen atom. (*M&R*, 12/7/64, 17)

December 7: USAF announced completion of the highly successful Demonstration and Shakedown Operations (DASO) series of test firings of the Titan II intercontinental ballistic missile. Five Titan IIs had been successfully launched from Vandenberg AFB by Strategic Air Command (SAC) crews and had landed on target down the WTR. The series was conducted to test launch techniques and lasted from July 30 to Nov. 4, 1964. Said Lt. Gen. Thomas P. Gerrity, Deputy Chief of Staff for Systems and Logistics, the 100 per cent successful "record . . . is unequaled in ICBM history and will provide a goal for future missile systems in the years ahead." (Sehlstedt, Balt. *Sun*, 12/8/64; AP, Balt. *Sun*, 12/9/64)

- *Missiles and Rockets* reported that a combination vacuum cleaner and fire extinguisher was being developed by Lockheed Missiles and Space Co. for use in the MOL. Dubbed VacGuard, the device would operate on the pressure difference between the cabin and outer space. It would control cabin fires, remove unwanted solids, liquids, and gases, and control sudden leaks in the space cabin wall. The device would consist of an adapter, a flexible hose, and a selector valve. (*M&R*, 12/7/64, 17)
- AF Cambridge Research Laboratories had successfully demonstrated that communications could be made through brain waves. According to Dr. Edmond M. Dewan, trained subjects could alter the pattern of their brains' alpha rhythm, the low-frequency wave related to visual perception. The alpha rhythm could be turned on and off, allowing use of a binary digit system or Morse code for communication. (*M&R*, 12/7/64, 17)
- Radar and Doppler velocity sensor system destined to assist the Surveyor spacecraft in its soft landing on the moon was in advanced test phases at Ryan Electronics, *Missiles and Rockets* reported. Flight tests of the system using helicopters at Holloman AFB and at the Hughes Aircraft Co. airport had been successfully completed and had run from 6,000 to 20 ft. above the ground. (*M&R*, 12/7/64, 17)
- In an editorial, Robert Hotz, editor of *Aviation Week and Space Technology*, said: "There is a . . . good possibility that the national space program may get its most severe shellacking to date if some plans currently finding favor within Administration executive circles are, in fact, implemented. One of these proposals would slice the scientific programs of the National Aeronautics and Space Administration to the bone and concentrate the NASA budget on the Apollo manned lunar landing program. Some portions of the scientific community have been making this charge without much justification for several years. If NASA does virtually abandon its highly successful scientific exploration program now, the howl from scientists will rise in a swift crescendo. It would be a major mistake to make such a drastic switch in what is now a pretty well-balanced NASA effort." (*Av. Wk.*, 12/7/64, 11)
- Houston "Colt .45s" National League baseball team changed its name to the Houston "Astros" to exploit the city's claim to be the U.S. space capital. (*Av. Wk.*, 12/7/64, 15)

December 7: Carl Brown Harper, expert in aeronautical engineering who had been a consultant to the Dept. of Justice since 1930, died after a heart attack. Mr. Harper had held a number of patents on the retractable gun turret and on aircraft wing designs now used on the Boeing 707, the Douglas DC-8, and other high speed aircraft. (*Wash. Post,* 12/9/64)

December 8: Apollo spacecraft (BP-23) was subjected to simulated catastrophic failure of its launch vehicle in a successful test of the abort escape system at White Sands, N. Mex. Apollo command and service boilerplate modules were boosted on a Little Joe II launch vehicle to the region of maximum dynamic pressure at 29,000-ft. altitude and launch vehicle failure was simulated. The escape vehicle, mated to the spacecraft by a connecting tower, yanked the Apollo free of Little Joe and carried it almost two miles higher. The Little Joe rocket fell to earth; the Apollo modules were returned gently to the ground by three 88-ft. parachutes. This was the third successful test of functioning of the 33-ft. launch escape tower designed to jettison the command module and three Apollo astronauts to safety in an abort before or during the launch phase of a Saturn V mission. It was the first test of the boost-protective cover subsystem, a contoured shell fitting over the command module and designed to protect the docking mechanism of Lunar Excursion Module (LEM) from excessive heating during high temperatures of first-stage flight. It also protected command module windows from exhaust of the launch escape motor. (NASA Release 64-299)

- A Saturn S-IVB stage began a 6,000-mi. land, sea, and river trip from the Douglas Missile and Space Systems Div., Calif., to NASA Marshall Space Flight Center. At MSFC, the stage would be mated with a first stage of the Saturn IB, built by MSFC and modified by the Chrysler Corp., and an Apollo spacecraft. Tests of the assembly would check the complete Saturn-Apollo vehicle for its ability to withstand vibrations and other stresses expected in flight. (*Marshall Star,* 12/9/64, 1-2)

- The U.S. Army Corps of Engineers and NASA announced the award of a contract for master planning of NASA's $60 million Electronics Research Center to the joint venture firms of Edward Durell Stone, New York City; Giffels and Rossetti, Detroit; and Charles A. Maguire Associates, Boston-Providence. The contract was awarded by the New England Division of the Corps of Engineers, which had been designated by the Electronics Research Center as NASA's design and construction agent for the buildings to be located on a 29-acre site at Kendall Sq., Cambridge, Mass. (NASA Release 64-307)

- Three contracts were awarded by NASA Lewis Research Center to Hughes Aircraft Co. The first contract, totaling $1,151,370, was a 12-mo. continuation of a previous contract for research and development of a contact ion engine using cesium as a propellant. The other two contracts were also follow-on efforts: a contract for $202,915 provided for fabrication of porous alloy and solid sintered ionizers; a contract for $243,061 covered tests and evaluation of physical, electrical, and cesium ionization properties of porous ionizers and evaluation of the electrical properties of solid sintered ionizer material. (LRC Release 64-114)

December 8: NASA Lewis Research Center awarded a $2,315,400 contract to Rocketdyne Div. of North American Aviation, Inc., for design, development, and testing of an Atlas launch vehicle sustainer engine that would use flox (liquid fluorine and liquid oxygen). The contract provided for firing a complete Atlas sustainer engine for the first time using flox as the oxidizer; RP-1, a type of kerosene, would still be the fuel. (NASA Release 64-306; LRC Release 64-113)

- ComSatCorp asked 16 companies to propose studies on how to judge the merits of launch vehicles which could be used to put medium-altitude communications satellites into orbit around the earth. ComSatCorp told FCC it planned to launch its first satellite, the Early Bird, in March as a communication link between Europe and North America. The Early Bird would be able to transmit two-way telephone conversations, television computer data, and other types of communications as an experiment in synchronous satellites, placed above the equator at an altitude of 22,300 miles. (ComSatCorp Release; *NYT*, 12/9/64)

- A team of Canadian scientists from McGill Univ., conducting Project Harp (High Altitude Research Program), was using a 250-ton, 16-in. naval gun to launch scientific rockets to heights of nearly 100 mi. The method was both reliable and economical (more than 80 shots had been fired, most of which were successful), and the launch procedure relatively simple. Objective of experiments was to find the secret of the lower ionosphere's influence on radio communications. Dr. Gerald Bull, who conceived the idea of Project Harp, said that a new technique was being developed whereby a combination of gun and rocket could place a 100-lb. payload in earth orbit. (Jeffrey, *London Daily Telegraph*, 12/8/64, 14)

- *New York Times* reported that scientists might use an infrared radiometer to predict location and time of future volcanoes. The scanner had already detected heat changes in the earth's surface in volcanic areas in Hawaii that correlated with later volcanic eruptions. (Sci. Serv., *NYT*, 12/8/64, 60)

- USAF successfully launched the fifth Asset re-entry test vehicle in its series of six from Cape Kennedy. The glider rode pickaback aboard a single-stage Thor booster. (*Wash. Post*, 12/9/64; *M&R*, 12/14/64, 10)

- The first landing by computer alone was made by a United Air Lines Caravelle jet at Dulles International Airport, Washington. With no directions from the pilot, the machine brought down the aircraft, constantly judging correct altitude, rate of descent, and speed and bringing the craft in line with its decisions. In bad weather, the system would make possible a computer landing at 100-ft. ceiling and quarter-mile visibility. Manual landing limits were 200-ft. and one half mile. (Tuck, *Wash. Post*, 12/9/64)

- NIMBUS I meteorological satellite, which had lost its electric power and stopped functioning Sept. 23, revived and was returning useful engineering data. NASA Goddard Space Flight Center said the solar paddles, which had been locked, had apparently directed themselves at the sun long enough to recharge the spacecraft's batteries but the gas for stabilization had been exhausted and the craft was tumbling at about 15 rpm. This caused photographs to be fuzzy and useless. (GSFC Release G-64-34)

December 8: Soviet Union said it continued to be in "stable" contact with ZOND II, then 1.53 million mi. from earth. No further reference was made to a partial power failure aboard the craft. (*NYT*, 12/8/64, 54)

- NASA Marshall Space Flight Center awarded a $2 million contract to Motorola's Western Military Electronics Div., to develop the Airborne Ranging Orbital Determination (AROD) system. (*M&R*, 12/14/64, 21)
- Radio Corp. of America introduced a new line of computers using tiny silicon chip circuits (monolithic integrated circuitry). A silicon chip circuit would barely cover a typewritten "o," but would contain two electronic circuits with 15 transistors and 13 resistors. The line was named Spectra 70 because RCA felt it would provide a complete spectrum of computing, peripheral, and communications equipment, and would meet the full range of system and application requirements of science and industry. (*Wash. Post*, 12/9/64)
- On the eve of his 85th birthday, M/G Benjamin D. Foulois (USA, Ret.) was honored for his achievements in aviation at a dinner at Andrews AFB. General Foulois, who was the United States' first military pilot, was inducted into the Primus Club—an elite group of aerospace pioneers— and hailed as "a living link between the age of the Wright Brothers and today's astronauts." (Raymond, *NYT*, 12/9/64, 53)

December 9: X–15 No. 3 was flown by NASA pilot Milton O. Thompson to 91,000-ft. altitude at maximum speed of 3,545 (mach 5.3) mph to check air flow around the nose, heating caused by skin friction, and boundary layer air flow at different velocities and angles of attack. (NASA X–15 Proj. Off.; *NYT*, 12/11/64, 34)

- The last in a series of 11 tests of the F–2 Apollo service module engine was successfully conducted by NASA Manned Spacecraft Center engineers at White Sands Missile Range, N. Mex. The engine was fired for 20 sec. and then shut down for a five-min. "purge" before a 100-sec. test firing. The F–2 rocket engine would be expected to bring astronauts back to earth from the moon. (Lee, *Houston Post*, 12/10/64)
- FCC postponed a decision on the question of who would own and operate the ground stations to be set up as part of the communications satellite system and called for further filings on the issue. Detailed arguments were due Jan. 5, and reply comments on Jan. 18. It was estimated it would take the FCC a minimum of six weeks to evaluate all comments. The carriers that had filed protests to exclusive ComSatCorp ground station ownership were American Telephone and Telegraph Co., International Telephone and Telegraph Corp., Radio Corp. of America, Western Union International, and Hawaiian Telephone Co. (*Av. Wk.*, 12/14/64, 27; AP, N.O. *Times-Picayune*, 12/10/64)
- A committee had been set up by Prime Minister Harold Wilson to review Britain's future in space, it was announced. The committee would examine the merits of a national satellite-launcher program based on the Black Knight rocket as well as Britain's commitment to the European Launcher Development Organization (ELDO) which would hope to start launching satellites two or three years from now. The space committee's report would be expected in February. (MacLeod, *CSM*, 12/9/64)
- A unique optical system known as a laser range finder was announced by RCA's Aerospace Systems Div. and U.S. Army Electronics Command. The unit operated by sending a very narrow beam pulse of red light toward the target at the speed of light. After reflection from the target,

the red light of the laser beam was received and sensed by a phototube in the range finder. An electronic counter measured the elapsed time between the transmitted pulse and the return reflection and converted the elapsed time instantaneously into a measure of distance.

Similar systems had suffered from loss of alignment with use between the receiving telescope, which sensed the reflected laser light, and the sighting telescope by which the receiver was aimed at the target. The new design combined these two telescopes into one unit that allowed a lighter, more precise, and more rugged system to be fabricated. (*Marshall Star*, 12/9/64, 5–6)

December 9: Soviet Premier Alexei Kosygin told the Soviet parliament that expenditures for defense would drop by four per cent or 500 million rubles ($555 million) in 1965. He also declared that his government had acted after being informed by American officials that U.S. planned to reduce its military budget in the coming year. A spokesman for the U.S. Embassy said later that the embassy had no knowledge that such information had been given to the Soviet government. In his budget message, Premier A. N. Kosygin told the Supreme Soviet that "people of science must do a lot for us to be able to provide more material benefits for the satisfaction of the population's growing requirements," complained that the opportunities for doing this "have been by no means exploited." After calling for scientific breakthroughs to the production process in chemistry, electronics, transistor technology, and biology, Kosygin said: "There are major shortcomings in scientific research planning. Our plans do not make use in good time of all that is progressive and useful in research discoveries. For this reason, scientific discoveries and technical inventions not infrequently become outdated and lose their practical worth.

"Utilization of the latest achievements of science and technology should be made one of the main criteria in the appraisal of economic plans and their fulfillment. Unfortunately, many people employed in industry continue to regard further capital investments as the main and only source of increased output and do not really strive hard for the rapid introduction of scientific and technical achievements, which could produce a considerably greater effect than additional capital investments. Scientific and technical progress opens new horizons for a radical improvement in quality and an expansion of the range of goods produced." (*Av. Wk.*, 12/28/64, 16; Tanner, *NYT*, 12/10/64; *WSJ*, 12/10/64)

December 10: Final Gemini unmanned flight test (GT–2) from ETR was postponed after a last-second engine misfire on the Titan II rocket. The trouble occurred 1.7 seconds after the Titan II booster engines ignited: loss of hydraulic pressure caused one of the two engine nozzles to swivel out of place; and automatic system sensed this and immediately extinguished both engines in the rocket's first stage one second before lift-off. A NASA spokesman said that one of the nozzle tubes received a slight crack as a result of the malfunction and that extensive checking of all systems must be completed before a decision would be made as to whether the entire first stage would be replaced.

The Titan II was to have lifted the 6,900-lb. Gemini capsule to 106 mi. in the suborbital flight and then accelerated it to faster and hotter reentry than normal. A parachute landing in the Atlantic was projected for the capsule and recovery planned. Chief goals of the flight, second in the Gemini series, would have been to test the heat shield as well as control, life support, and other systems.

NASA said the delay meant the first manned Gemini flight would not take place until the second quarter of 1965. GT-2 would be attempted again in January. (Hoffman, *N.Y. Her. Trib.* 12/10/64; Balt. *Sun.* 12/10/64; Witkin, *NYT*, 12/10/64)

December 10: USAF Titan III-A space booster was launched from Cape Kennedy and executed its first completely successful test flight. Midway in its first orbit, approximately 115 mi. above the earth, the 5,250-lb. "transtage" (a third stage with multiple start-stop-restart capability and ability to transfer a payload from one orbit to another) performed a 360° somersault to align the platform's inertial guidance system gyroscopes. At the end of its first 100-min. orbit, explosive charges automatically kicked a 3,750-lb. cylinder-shaped dummy satellite into a separate orbit. After the Titan III-A firing, Brig. Gen. Joseph S. Bleymaier, project director, said: "The success confirmed our confidence in the system. We feel that the Air Force now has a real purpose in accomplishing space missions." (AP, Balt. *Sun*, 12/11/64; Simons, *Wash. Post*, 12/11/64; AP, *NYT*, 12/11/64, 20)

- X-15 No. 1 was piloted by Capt. Joseph H. Engle (USAF) to 112,000-ft. altitude at maximum speed of 3,500 (mach 5.11) mph to test scientific instruments (a densitron and daytime sky background experiments) carried in pods recently installed on the craft's wing tips and to sample density of upper atmosphere. The flight also served as a checkout of the newly installed Honeywell inertial guidance system. (NASA X-15 Proj. Off.; *NYT*, 12/11/64, 25)

- U.S.S.R. launched COSMOS LI into orbit. In addition to scientific instrumentation, the craft contained a radio system for the precise measurement of orbital elements and a radiotelemetry system for transmitting data to earth. All equipment on board was operating normally. Initial orbital data: apogee, 544 km. (344.24 mi.); perigee, 264 km. (164 mi.); period, 92.5 min.; inclination, 48.8°. (Tass, *Pravda*, 12/11/64, 1, ATSS-T Trans.; Tass, *Komsomolskaya Pravda*, 12/11/64, 1, ATSS-T Trans.; UPI, *NYT*, 12/11/64, 43; *Wash. Post*, 12/11/64)

- An unexpected rise in the number of meteors bombarding the earth from April to October 1963 was reported in *Science* by two Canadian and two New Zealand scientists. The rise was detected by radar echoes and the number of meteors put at 300 million. Studies made by the New Zealand group after the great increase indicated that the meteor rate rise may be recurring, with decreasing strength, once every six months. (Sci. Serv., *NYT*, 12/25/64; AP, Carey, *Wash. Post*, 12/11/64)

- NASA announced that it would negotiate a one-year contract with Mason-Rust for continued support services at NASA Michoud Operations. (*Huntsville Times*, 12/11/64)

December 11: Atlas-Centaur 4 (AC-4) hurled its Centaur second stage into orbit around the earth with a mass model of the Surveyor lunar spacecraft in its nose. The first-burn performance of the Centaur rocket sent the stage into a nearly circular parking orbit of 100 mi. perigee and 106 mi. apogee. The second burn should have put the rocket into an oval path taking it out as much as 5,000 mi. from earth; however, the attempt to restart the RL-10 engines failed and the stage began to tumble and roll. After about 15 hr. in orbit, the

Centaur stage and its Surveyor dummy payload re-entered the atmosphere over the South Pacific east of Australia and disintegrated.

NASA said all prime mission objectives had been met, including the demonstration of the structural integrity of the Atlas-Centaur during powered flight, testing of the guidance system which was flying closed loop for the first time, verification of the structural and thermal integrity of the Centaur nose fairing and insulation panels, and verification of the performance of the jettison systems for the nose fairing and insulation panels.

Project officials were investigating what caused the stage to tumble, preventing propellants from reaching the pumps. (NASA Release 64–289; *Lewis News*, 12/24/64, 1; *Av. Wk.*, 12/21/64, 24; *Wash. Eve. Star*, 12/11/64)

December 11: First launch of NASA Aerobee 350 sounding rocket was conducted at Wallops Island, Va., the rocket reaching 6,000-ft. altitude in successful test. For this first test upper stage was inert, the main objectives being to demonstrate compatibility of Aerobee 350 with Aerobee launch-tower rail system and evaluate performance of Aerobee 350 during first-stage (Nike) boost. Preliminary evaluation of test indicated NASA Goddard Space Flight Center could proceed with plans to launch complete live Aerobee 350 without redesign of the rocket or the launcher rail system. (NASA Rpt. SRL)

- Developmental spacecraft landing rockets were tested using a Gemini boilerplate. Spacecraft was dropped from a C-119 aircraft at 7,000-ft. altitude. Two 12,000-lb.-thrust rockets were fired simultaneously when the spacecraft was ten ft. above the water with burn time one and one half sec. Landing speed of the craft was slowed from 27 fps to about nine to 10 fps; normal g-force of seven to eight g's was reduced to approximately two and one half g's on impact. A controllable parachute steered by radio commands from a NASA ship lowered the spacecraft into the waters of Trinity Bay, Tex. The landing rockets were not a definite part of either the Gemini or Apollo programs. (MSC *Roundup*, 12/23/64, 7)

- NASA announced that a 10-month study contract for $191,835 to assess the technological problems of designing, building, and operating a manned orbiting astronomical observatory had been awarded to the Boeing Co. Boeing would study possible ways in which such an observatory might be operated, with particular emphasis on the role of man in its scientific program. A requirement of the study was that the orbiting telescope be sufficiently adaptable to support a broad astronomical program. The study would assume that the telescope would be operated in conjunction with a proposed Manned Orbital Research Laboratory (MORL) under study by NASA Langley Research Center for NASA Office of Manned Space Flight. (LaRC Release)

- In a Gemini suit comfort test, Hoyt Maples successfully completed a 96-hr. stay in a 8-ft.-dia. chamber in the Crew Systems Laboratory at NASA Manned Spacecraft Center. The suit was not pressurized during the test, but the test chamber was a pure oxygen environment with 5 lbs. of pressure, the same environment that the Gemini spacecraft would use at orbital altitudes. During the test, Maples followed a four-hour-work, four-hour-rest cycle. He used a task board which tested his mental alertness by mathematical calculations he made on a series of lights on the panel board. Maples' food was a Gemini flight diet consisting of freeze-dehydrated, dehydrated, and bite-size food. Other tasks Maples performed during his stay in the chamber included checks of the environmental control system and exercise with a bungee cord. Although he did not take off his helmet during the test, he was able to open the faceplate and take off his gloves. With this test, the Gemini suit passed another phase of its qualification for use in the manned Gemini flights. (MSC *Roundup*, 12/23/64)

December 11: Formation of a National Academy of Engineering under the charter of the NAS was announced by Dr. Frederick Seitz, President of the National Academy of Sciences, and Dr. Augustus B. Kinzel, President-elect of the National Academy of Engineering. Among the founding members were Dr. Hugh L. Dryden, NASA Deputy Administrator, and Dr. William H. Pickering, Director of the Jet Propulsion Laboratory. (NAS-NRC Release; *Wash. Post*, 12/12/64, 20)

- The hydrogen-fueled J–2 engine that would power the upper stages of the Saturn V launch vehicle successfully demonstrated its ability to stop and restart in a ground test carried out at the Santa Susana Field Laboratory of Rocketdyne Div. The 200,000-lb.-thrust engine would operate in a cluster of five to provide one million lbs. of thrust for the S–II second stage of the Apollo moon launch vehicle, as well as operating singly in the S–IVB stage of the Saturn IB. The test engine was operated initially for 165 sec. and shut down. Following a 75-min. "coast" period, the engine was restarted. After seven seconds, it was shut down for six minutes, then restarted and operated for the full duration of 310 seconds. Among specific objectives demonstrated by the test were automatic repressurizing of the start tank during engine operation, sustained tank pressure during the orbital coast period, and reliable operation of the electrical and pneumatic systems used for restart. (*Marshall Star*, 12/16/64, 1, 12; AP, Balt. *Sun*, 12/12/64)
- A ten-second firing of the "battleship" stage of the Saturn S–II, second stage of the Saturn V moon rocket, was conducted by North American Aviation, Inc., at their Santa Susana Field Laboratory. The test demonstrated the ability of the battleship stage and accompanying facilities to operate the J–2 engine at full thrust. The S–II would be powered by a cluster of five J–2 liquid hydrogen engines, developing a total thrust of a million pounds. (NAA S&ID Skywriter, 12/31/64, 2; *Marshall Star*, 12/16/64, 1, 12)

December 11–12: MARINER IV passed unscathed through the Geminid stream, the first of three meteor streams it had to cross on its 7½-mo. journey to take television pictures of Mars. (NASA Release 64–323)

December 12: USAF and the Atomic Energy Commission (AEC) were reported to be negotiating a $13,000,000 contract with the Garrett Corp. for a 33-month program of advanced development and ground testing of the Space Power Unit Reactor (Spur) system. Spur was one of two elements in the Snap–50/Spur system—Snap–50, under development for AEC by Pratt & Whitney Aircraft, being the energy producing reactor, and Spur, the portion which converted heat to electricity. The objective would be to provide a flyable system in the 1970's. (Haggerty, *J/Armed Forces*, 12/12/64, 8)

- Thiokol Chemical Corp. Wasatch Div.'s 156-in. solid-propellant rocket motor was fired for 130 sec. and terminated within 0.5 sec. of predicted action time, although the last 30 sec. of the firing was completed without the plastic nozzle extension, which parted at the metal attachment flange. Severe erosion of the 5-ft. nozzle extension skirt, which was constructed entirely of ablative materials, weakened the structural integrity of the skirt and it blew off after about 100 sec. of motor burning. The firing continued normally after loss of the skirt with only about an 8% degradation in thrust, which had peaked at just over 1.4 million lb. This was the first demonstration of thrust-vector-control using a movable nozzle. The rocket motor case used in this feasibility demonstration was undamaged, would be used in the follow-on 156-in. program which was awaiting DOD approval. (*Av. Wk.*, 12/21/64, 26)

December 12: Secretary of Defense Robert S. McNamara held a press conference at which he responded to questions regarding the MOL project:

QUESTION: "Mr. Secretary, Senator Anderson has suggested that a billion dollars could be saved by merging the Air Force's Manned Orbiting Laboratory with NASA's Apollo. [See Dec. 7, 1964.] Can you tell us how you feel about that?"

SECRETARY MCNAMARA: "I feel very strongly that we must avoid duplicate space programs. There must be only one National Space Program. The great bulk of the expenditures for that National Space Program should be in NASA, because the bulk of the National Space Program will be oriented toward civilian scientific and other objectives. There is a place, there is a requirement for a military component. We must maintain security, a high level of security, on certain elements of the space program and that portion of the program must be handled by the Defense Department. I think Senator Anderson was simply emphasizing the absolute essentiality of fully coordinating the NASA and the Defense Department programs. With that, I agree 100 percent. . . . My principle is a very simple one . . . I believe we are a military organization, we are not interested in space except insofar as it bears directly on our military mission. If there is anything that NASA can do, that we can in effect hire them to do as our agent, I am 100 percent in favor of doing so."

QUESTION: "A related question. Could you say, specifically, whether you are backing the MOL or not?"

SECRETARY MCNAMARA: "Yes, indeed."

QUESTION: "That could go ahead under the Air Force."

SECRETARY MCNAMARA: "Yes, but I want to be sure that the MOL program, the details of it, are fully analyzed by NASA and fully taken into account when NASA establishes any portion of its Apollo Program not directly related to the Lunar Program. . . . I want to be certain that the Air Force, when it established the MOL, takes account of what NASA is required to do as part of the Apollo Program that in turn is directly related to the Lunar Program." (Text; *M&R*, 12/21/64, 20; *SBD*, 12/15/64)

- Secretary of Defense Robert S. McNamara announced that DOD would establish a Contract Audit Agency whose principal function would be to consolidate the activities and the 3,600 personnel of various units formerly engaged separately in auditing defense and space contracts. He estimated that this would result in a "manpower" saving of $1.8 million. (*NYT*, 12/13/64)

December 12: Britain's new Labor Government had reviewed plans to build the Concorde supersonic airliner and was reported to be seeking a compromise with France over the project. It was possible that the U.K. would propose to France an extension of the Concorde project over a longer period of time to stretch out cost of development. Also Britain would like France to consider the possibility of U.S. participation so that cost could be split three ways. (Farnsworth, *NYT*, 12/13/64, 4)

December 13: RELAY I began its third year in orbit and could still send clear, high-resolution television and audio signals across oceans and continents. The communications satellite had circled the globe 5,685 times and accomplished 186 demonstrations and 2,139 experiments, with a transponder-on time of 334 hours. The RELAY I project was directed by NASA Goddard Space Flight Center. (NASA Release 64–213)

- USAF launched Thor-Able-Star launch vehicle from WTR with unidentified satellite payload. It was later revealed that two satellites were placed in orbit (HHN–48)

- Translations of reports in Russian journals indicated that space flights lasting longer than 24 hours produced changes in the human body, including bladder and kidney troubles, that could not be corrected. Future Soviet flights would be aimed at detecting the beginning of these disorders and attempting to prevent them. The translations indicated that the Russians planned to send up more doctors and physiologists and that the primary reason for the 16-mo. delay between flights by the U.S.S.R. was the seriousness of the disorders affecting astronauts after flights lasting for five days. Future crews would be selected "as medical teams as opposed to individuals," said one Soviet journal. The men would also be picked on the basis of their medical compatibility— to see that one man's personal germs, bacteria, or virus did not infect other members of the team. (Burkett, *Houston Chron.*, 12/13/64)

December 14: Dr. Donald E. Hornig, scientific adviser to President Johnson, was quoted on the controversial MOL project and its possible merger with NASA's Apollo: "MOL is still under discussion, and the scope of it is still in a state of flux. The Air Force certainly has a role to play in space." Sources inside the White House said Dr. Hornig was against full funding of the MOL, for which project the Air Force wanted $175 million in FY 1966. (*Av. Wk.*, 12/14/64, 15)

- NASA designer of ECHO I, William J. O'Sullivan, said he believed the satellite still had its spherical shape after over four years in orbit. An optical investigation made by the Goodyear Aerospace Corp. for NASA Langley Research Center indicated that the satellite's aluminum surface—which made the balloon highly reflective of light and radar waves, regulated its temperature, and protected the plastic film from space hazards—still reflected like a mirror. From calculations using the known reflective characteristics of ECHO I when it was launched and his interpretation of the Goodyear measurements, O'Sullivan concluded that ECHO I "is still essentially spherical and that its very thin coating of aluminum has suffered no measurable degradation in specularity or reflectivity." (LaRC Release)

December 14: Missiles and Rockets reported that NASA was reviewing the pros and cons of Project Orion (an advanced nuclear-pulse rocket propulsion concept) to determine whether the agency could justify a mission requirement for a nuclear-pulse vehicle—such as a fast manned trip to Mars. DOD was scheduled to drop its research in January 1965, because the Air Force could not establish a mission requirement. Both DOD and AEC would join in financing a joint program if NASA established the mission requirement. Project Orion would not in any case be available in the same time period as solid-core nuclear reactors, whose development was already well underway. (Wilks, *M&R*, 12/14/64, 12–14)

- Langley Research Center aerospace engineer Dewey L. Clemmons, Jr., presented a paper at the Symposium on the Establishment of a European Geodetic Network for Artificial Satellites held in Paris, France, describing the U.S. Passive Geodetic Satellite (Pageos) planned for launch during the first half of 1966. He said: "The satellite will be an aluminum-coated sphere 100 feet (30.48 meters) in diameter, a near replica of ECHO I (1960 Iota 1) that was launched by the United States on August 12, 1960. No instrumentation will be attached to the surface of the Pageos satellite. Pageos will serve the geodesist by simply acting as a point source of light in the sky as it reflects the incident sunlight. Simultaneous photographs of this light source against the star background, taken by two or more widely separated ground-based cameras, will enable geodesists to determine the spatial coordinates of each camera position. An interconnected series of camera positions can be established that will cover the entire surface of the earth, thereby permitting geometric determination of each camera position within a single reference system. The use of this satellite for geodetic purposes will continue for at least five years during which the necessary photogrammetric observations can be made and to provide a purely geometric determination of the shape and size of the earth." (Text; AP, *NYT*, 12/15/64, 41)
- Dr. Eugene Konecci, of the NASC staff, told *Missiles and Rockets* that there had been no mention of airlocks in public or private discussions of the Vostok and Voskhod spacecraft, and no other indications of plans for extravehicular operations. Although the Soviets were aiming at rendezvous, this was expected to be strictly a vehicle maneuver. "Soviet interest in the use of helium-oxygen atmospheres for longterm space vehicles may indicate that they do not have capability for extra-vehicular experimentation," said Dr. Konecci. (*M&R*, 12/14/64, 16)
- Among the ten top scientific, medical, and technological advances in 1964 as selected by Dr. Watson Davis, Science Service Director, were:
    1. Close-up photographs of the moon taken by RANGER VII.
    2. Orbiting of the earth by three Russian cosmonauts in the VOSKHOD I spacecraft.
    3. Discovery of quasars, the brightest, most violent, and heaviest sources of light and radio waves, possibly fueled by collapse of gravity waves. (Sci. Serv., *Wash. Daily News*, 12/14/64)
- NASA announced that 142 colleges and universities would participate in NASA's graduate training program during the 1965–66 academic year. (NASA Release 64–308)

December 14: U.S. Army announced that the Lockheed-California Co.'s XH–51A compound helicopter (a helicopter having wings as well as rotor blades) had flown 242 mph—"the fastest known speed of any rotor craft in the world." The flight was made by a test pilot, Donald Segner, off the Southern California coast. (AP, *NYT*, 12/14/64, 32)

* USAF Air Defense Command 10th Aerospace Missile Sq., a fully operational unit of ADC at Vandenberg AFB which would man an antisatellite weapon, was on 5-min. alert status. (*Av. Wk.*, 12/14/64, 13)

December 14–17: Twenty-six aviation officials from 23 countries convened in Washington for activities celebrating the 20th anniversary of the formal beginning of post-war international cooperation in civil aviation after the Chicago Conference of 1944. Dignitaries were honored guests at the Wright Memorial Dinner in Washington on Dec. 17. (FAA Release 64–109)

December 15: SAN MARCO I (SM–1) was orbited from NASA Wallops Station. The NASA-trained Italian crew launched the 254-lb. Italian-designed satellite on a four-stage solid-propellant Scout to an orbit with an apogee of 490.04 mi., a perigee of 124.62 mi., an orbital period of 94.7 min., and an inclination of 37.77°. It was the first time a foreign country had designed, built, and launched a satellite in orbit in the course of NASA's international program. The satellite would measure air density and ionospheric characteristics related to long-range radio transmissions and would qualify satellite and train crew for subsequent launches from sea platform in Indian Ocean. (Wallops Release 64–91)

* At a symposium sponsored by NASA Goddard Space Flight Center, Dr. Norman F. Ness likened the earth to a kind of comet with a long magnetic tail that extended for an unknown distance. The "new look" was drawn from results of the first detailed mapping of the earth's magnetic field on the nighttime side of the magnetosphere by EXPLORER XVIII (IMP I, Interplanetary Monitoring Platform), launched Nov. 26, 1963. Until recently, scientists had believed that the earth's magnetosphere was basically spherical; however, data from EXPLORER XVIII indicate that countless magnetic lines of force stretched out like the tail of a comet to an unknown distance in space, apparently beyond the moon. Within this comet-like tail, the lines of force in the Northern Hemisphere were directed towards the sun; in the Southern Hemisphere, away from the sun. In between, there was a neutral zone.

    Dr. Ness characterized this neutral zone, which had been hypothesized but never before detected, as a thin sheet which was a permanent part of the earth's environment and virtually void of any magnetic activity. Though the neutral zone's exact role was unknown, Dr. Ness speculated that it might be responsible for formation of the auroras, creation of the Van Allen radiation belt, and presence of the gegenschein. Ness concluded that in view of the evidence from EXPLORER XVIII, "The only difference between the earth and a comet is that the earth possesses a strong magnetic field and comets do not." (Simons, *Wash. Post*, 12/16/64; Hixson, *N.Y. Her. Trib.*, 12/16/64; *Wash. Daily News*, 12/16/64; Toth, *Detroit News*, 12/16/64; AP, Wash. *Eve. Star*, 12/16/64; AP, *NYT*, 12/16/64, 30; *Av. Wk.*, 12/21/64, 25)

* Dr. Eugene B. Konecci of the NASC staff addressed the National Space Club on the Soviet bioastronautics program. He said that although

some American scientists had been critical of the amount and kinds of data released by the Soviets, Soviet scientists appeared to be desirous of further opening their research in technological areas such as life support.

On choosing the proper atmosphere for space flight, Dr. Konecci said: "Some American scientists . . . advocated a two-gas system. . . . Others chose a 5 psi 100% oxygen system. . . . Animal and human experiments to date indicate that there are no apparent physiological effects due to 100% oxygen up to and beyond 30 days.

"What is important to mention is that we had the technology as part of our aircraft system in handling 100% oxygen in low pressure-high pressure and liquid states, and we have extended this use to space cabins. The Soviets . . . have not been confronted with 100% oxygen problems. Because of the near sea-level atmosphere in their Vostok and Voskhod flights . . . they were concerned more with weightlessness and radiation effects. . . ."

Dr. Konecci said that both U.S. and the U.S.S.R. had much research work ahead. "The Soviets have still to fly their life support systems for the full designed parameter . . . in Vostok this was supposed to be ten days with later literature indicating a twelve-day capability. I expect the Voskhod has the comparable capability, that is, about twelve days, but because of the three-man crew this would in fact mean a threefold increase of their life support capability, i.e. on a man-per-day basis. I think we can also expect the Soviets to attempt manned rendezvous maneuvers. I really do not believe at this time that they will attempt manned extravehicular experiments until (1) they have a better space suit, which they appear to be working on; (2) that they have a better fix on the possible decompression sickness (bends or dysbarism) problems of decompressing from a sea-level atmosphere with nitrogen to a compatible space suit environment; and (3) until we perform the Gemini intentional decompression and extravehicular experiments." (Text)

December 15: The New York Stock Exchange banned any new margin trading in ComSatCorp stock in an effort to stop speculation. The effect of the ruling was to cut off the supply of borrowed money that had helped traders push the stock to a high of 71½ from 46, its price a little over three months ago. ComSatCorp stock had proved to be particularly volatile because many of the 5 million shares offered to the public last June at $20 were in long-term accounts and were not for sale. (Phalon, *NYT,* 12/16/64, 63)

- Col. George E. Humphries, President of the Engineering Institute of Canada, said in an interview that Canada was gaining more technically trained personnel, particularly among university engineering professors, than it had previously. "Salaries have risen here during the past 5 years, and for every engineering teacher that goes to the United States, Canada is gaining two from that country," Colonel Humphries concluded. (Canadian Press, *NYT,* 12/15/64, 79)

- Jane's *All the World's Aircraft, 1964–65,* published in London, said that the U.S. probably had led the Soviet Union for some years "in terms of quality and quantity of scientific data and military intelligence acquired in space." Jane's said the strong U.S. lead was indicated by the photographs of the moon sent back by RANGER VII last year, which it termed "one of the great astronautical achievements of 1964." (UPI, *Wash. Post,* 12/15/64)

December 16: Two NASA Aerobee 150 sounding rockets were launched from White Sands Missile Range, N. Mex. The first rocket reached 126.5-mi. altitude carrying instrumented payload which was recovered after impact. Electron measurement experiment did not work, but the other experiments apparently were successful: hypervelocity micrometeoroid collectors, cosmic dust collectors, and momentum and energy measurements as correlative data for the collectors. The second rocket reached 145.5-mi. altitude in successful nighttime test to study distribution of Lyman-alpha radiation. Scanning ultraviolet spectrometer and a photometer were used. (NASA Rpts. SRL)

- A hingeless Lockheed XH-51N rotor helicopter for advanced research in rotary wing aircraft had been delivered to NASA Langley Research Center. Studies would be made of the loads and flying qualities problems that might be encountered by operational aircraft of this type and of the extent to which their potential advantages of simplicity and improved flying qualities could be realized in practice. Rotor blades on the XH-51N were held firmly to the hub, which was attached rigidly to the engine drive shaft. On other helicopters, the blades were either hinged at the side of the rotor hub or were gimbaled atop the mast. (LaRC Release 12/16/64)

- Najeeb Halaby, Administrator of FAA, said that the four aircraft companies designing the supersonic transport (SST) had submitted proposals which showed "very substantial progress in the last 6 months." Halaby spoke of progress in the three principal problem areas associated with development of a 2,000-mph airliner: the safety and economics of its operation and the sonic booms created in its wake. The two airframe competitors were the Boeing Co. and Lockheed Aircraft Corp.; designs for the engine were submitted by General Electric and the Pratt Whitney Div. of United Aircraft Corp. (Toth, *Wash. Post*, 12/17/64, Balt. *Sun*, 12/17/64)

- USAF announced that Aerojet General Corp. would receive a $2,744,000 increment to a previously awarded contract for research and development for the Titan III transtage engine, from Ballistic Systems Div. of Air Force Systems Command. (DOD Release 888-64)

- USAF announced that Douglas Aircraft Co., Inc., Missile and Space Systems, would receive a $4,000,000 order to an existing contract for procurement of Thor space boosters from Space Systems Div. of Air Force Systems Command. (DOD Release 888-64)

- NASA used an Aerobee rocket to launch two experiments from White Sands Missile Range. The experiments were to measure cosmic dust and electron densities in the upper atmosphere. (*M&R*, 12/21/64, 8)

    NASA also successfully launched a Nike-Apache sounding rocket from White Sands, N. Mex., to peak altitude of 121 mi. Objective was to obtain data on the electron density distribution in the ionospheric D region by means of a radio propagation experiment, and to measure photoemission currents. (NASA Rpt. SRL)

December 17: MARINER IV Mars-bound space probe received two commands intended to cause the spacecraft's attitude control system to fix and remain on the star Canopus. The first required the tracker to break lock with the star Gamma-Velorum and go into a search for Canopus. Canopus was reacquired. The second command was to deter MARINER IV from losing its lock on Canopus. (NASA Release 64-323)

December 17: NASA extended until Dec. 31, 1966, its contract with the California Institute of Technology for research and development programs conducted at the Government-owned Jet Propulsion Laboratory. Work that would be performed included research and development in lunar and interplanetary programs, aeronautics, development and operation of deep space tracking facilities, providing NASA with technical advice, scientific studies, and reports of investigations, and providing supervision of related work given to contractors.

Requirements from NASA would be issued in the form of written Task Orders, and the cost of this contract would be determined by the sum of these task orders. In addition, the two parties would negotiate each fiscal year for a fixed fee for the management of the Jet Propulsion Laboratory. (NASA Release 64–320)

- Dynatronics, Inc., Orlando, Fla., was selected by NASA to negotiate an estimated $3.5-million contract award for design, manufacture, and on-site engineering support of 19 pulse code modulation (PCM) systems in NASA's worldwide Apollo Manned Space Flight Network, with options for additional systems. (NASA Release 64–319)
- NASA would negotiate with the Univac Div. of Sperry Rand Corp. for purchase of digital data processors to be used in the Apollo Manned Space Flight Network ground stations and tracking ships. The fixed-price contract would approximate $4.5 million for the ten systems with options for additional systems. (NASA Release 64–318)
- Kenneth C. McCracken, head of cosmic ray research of the Earth and Planetary Sciences Laboratory at the Graduate Research Center of the Southwest said that the Imps (Interplanetary Monitoring Platforms) scheduled to be shot into earth orbit in 1966 would be the best tool for studying cosmic rays during the years of the quiet sun. Besides lasting much longer than either the "deep" space probes or the balloons, they would have one other tremendous advantage, McCracken said. "The IMPS will have very eccentric orbits, skimming 200 miles over the surface of the earth at the closest points of their orbits, or 'perigees,' and soaring 185,000 miles out in space at their 'apogees,' or farthest points of orbit. . . . That means that in counting and analyzing cosmic rays we shall for the first time be able actually to compare the virgin, or primary, rays with the secondary particles that have had all the wear and tear of passing through the earth's atmosphere and magnetosphere." (Getze, *L.A. Times*, 12/17/64)
- Dr. Victor F. Hess, co-winner of the 1936 Nobel Prize and former Professor of Physics at Fordham Univ., died at the age of 81 in Mt. Vernon, N.Y. Dr. Hess shared the Nobel Prize with Dr. Carl C. Anderson of the California Institute of Technology for his part in the discovery of cosmic rays. (*NYT*, 12/19/64, 29)
- Boeing announced that Dr. George M. Knauf, for the past three years deputy director of space medicine in NASA's Office of Manned Space Flight, would join the company as chief of bioastronautics. He would manage groups working on life support and protection, engineering psychology, and program support. (*Seattle Post-Intelligence*, 12/18/64)
- Howard Simons of the *Washington Post* was awarded the 1964 American for the Advancement of Science-Westinghouse Science writing award. His winning entry consisted of stories on the Samos satellite, cybernetics in Russia, and the planet Jupiter. (*Wash. Post*, 12/18/64)

December 18: Tass disclosed that ZOND II was equipped with a new type of rocket engine known as a plasma accelerator and that the engine, on radio command from earth, had successfully operated the attitude-control system that oriented the spacecraft with respect to the sun. It was the first time that such a plasma engine had been tested under conditions of actual space flight. Information continued to be received on the operations of the systems and units aboard the spacecraft; radio communication with the station was stable. (Tass, *Komsomolskaya Pravda,* 12/19/64, 1, ATSS-T Trans.; Shabad, *NYT,* 12/18/64, 12)

- NASA selected 14 firms for negotiation of contracts to provide engineering, fabrication, and institutional support services to six laboratories and three offices of the Marshall Space Flight Center. Work would be in support of the Saturn/Apollo launch vehicle program with cost estimated at $68 million for one year. (NASA Release 64-322)
- More than 900 persons had applied to NASA for the 10 to 20 positions for scientist astronaut that it would fill next year. To qualify, a scientist had to be under 34 years old, a U.S. citizen, no taller than six feet, with a B.A. degree and a doctorate or the equivalent in experience. Recruitment would end Dec. 31. Recommendations would be made by the National Academy of Sciences to NASA in the spring of 1965. (AP, *Miami Her.,* 12/18/64)
- USAF successfully launched a Minuteman II "instant ICBM" from Cape Kennedy. The test warhead hit its Atlantic target area more than 5,000 mi. from the launch site. (UPI, *NYT,* 12/20/64, 41)
- Present dollar volume of contracts handled by the NASA-Michoud plant passed $1 billion during 1964. (N.O. *Times-Picayune,* 12/18/64)
- Members of Local 1685, United Brotherhood of Carpenters and Joiners of America, ended a 24-hr. strike at the Merritt Island Launch Area (MILA) and nearby Patrick AFB facilities. The walkout had tied up $246 million in construction for space projects, including construction on an assembly building for the Saturn V lunar rocket, and work at the Titan III and Saturn IB launch areas. Carpenters had protested the use of nonunion workers by the Akwa-Downey Construction Co., a subcontractor on one of the projects. It was the fourth major labor difficulty at NASA Kennedy Space Center in 1964. (UPI, *NYT,* 12/18/64, 39 and 12/19/64, 48; *Chic. Trib.,* 12/18/64)

December 18–19: Total lunar eclipse occurred. A 15-ft.-dia. dish antenna, built by Marconi Co. and designed for Queen Mary College of London Univ., was used for the first time to view the eclipse. It was operated as a radiotelescope at wavelengths of about one mm.—possibly the largest ever built to operate at such short wavelengths. The dish employed an epoxy-resins structure reinforced with glass fibers and having a zinc-sprayed surface.

During the eclipse, extensive heat-map measurements of the lunar surface were made from Egypt by team of Boeing Scientific Research Laboratories scientists. Preliminary analysis of their measurements showed hundreds of "hot spots," or surface temperature anomalies, on the lunar surface, although the experimenters had expected to find only a few dozen. The experimenters made 12 20-min. scans of the moon during the eclipse, in both the visual and infrared portions of the spectrum. Measurements were made at El Kottamia Observatory, 50 mi. east of Cairo, at invitation of United Arab Republic Astronomical Service. (*M&R,* 1/4/65, 21; *NYT,* 12/22/64; *M&R,* 1/18/65, 28)

December 19: A Gemini-type spacecraft was successfully landed at Edwards AFB in first manned test of the new paraglider recovery system. North American Aviation test pilot Donald F. McCusker guided the capsule back to earth following a four-and-a-half minute flight beneath the inflated wing, after being released from a helicopter at an altitude of 8,500 ft. McCusker was slightly injured when the paraglider landed at about 25 mph—faster than had been anticipated. (L.A. *Times,* 12/20/64; *San Diego Union,* 12/20/64; NAA S&ID *Skywriter,* 12/31/64, 4)

- Sen. Clinton P. Anderson (D.–N. Mex.), chairman of the Senate Committee on Aeronautical and Space Sciences, issued a statement related to his letter to President Johnson [See Dec. 7, 1964] in which he had proposed a merger of the Air Force Manned Orbiting Laboratory (MOL) and NASA Apollo-X programs: "Department of Defense and NASA officials have informed me of an agreement they have worked out in relation to the Air Force's proposed Manned Orbiting Laboratory (MOL) and the space administration's proposed Apollo-X.

    "The Department of Defense and NASA have gone a long way toward answering the questions I raised several weeks ago. I suggested at that time that perhaps a great savings could be achieved without sacrifice to either the NASA or Air Force responsibilities in space by harmonizing MOL and Apollo-X. I have been told that the Air Force and NASA will take advantage of each other's technology and hardware development with all efforts directed at achievement of a true space laboratory as an end goal. The review underway by both agencies will strengthen the total space effort.

    "We should take full advantage of the technology and ground-support systems developed for Project Apollo. The object of the Senate Space Committee in raising questions about MOL and Apollo-X was to accomplish the wisest use of personnel and money, not in criticism of either project."

    New York Times reporter John Finney interpreted Sen. Anderson's statement as an indication that the Administration "had gone along with his [Sen. Anderson's] suggestions. Not lost upon the Administration was the letter's implication that it would be difficult to defend the proposed $5.2 billion space budget on Capitol Hill unless the Administration could show it had eliminated duplication." (Text; Finney, *NYT,* 12/20/64, 45)

- NASA Goddard Space Fight Center announced plans for negotiation with the General Dynamics-Electronics Co. for purchase of approximately $1.25 million worth of unique ground tracking equipment to be used in the Applications Technology Satellite project. ATS Range and Range Rate electronic systems would be installed in three ground stations to provide ground trackers with extremely accurate satellite position, velocity, and range. The systems would also drive displays and digital data consoles. The contractor would design, manufacture, and test the systems under a cost reimbursable contract with incentive provisions. ATS satellites, designed to advance technology in communications, meteorology, and other applications, would require unique ground systems ready for the first launch in 1966. (GSFC Release G–36–64)

December 19: An astronomer at Mt. Wilson and Mt. Palomar Observatories, Calif., Allan R. Sandage, told the 2d Texas Symposium on Relativistic Astrophysics that an explosion in the galaxy 3C–2 may indicate the birth of a quasi-stellar force (QSF). A quasi-stellar force would be an entirely new source and kind of energy in the universe which would not follow the known rules of physical forces. (AP, *Wash. Post,* 12/17/64)

- USAF launched unidentified satellite on Thor-Agena D booster from Vandenberg AFB. (HHN–48)
- Tass reported that a Gorky Univ. team, headed by Vsevolod S. Troitsky, the leading Soviet authority on radio emanations of the moon, had found evidence they believed showed that the upper layer of the moon's surface was saturated with meteoric matter distinguishing it both chemically and in mineral content from deep layers. Results were obtained during the complete lunar eclipse of last Dec. 30 observed from Ussurisk, in far eastern Siberia. Troitsky's group made use of the fact that the lunar surface cooled during eclipses and that the cooling caused a drop in the intensity of radio emanations. The Russians worked out the relationship between the decrease of radio emanations from the obscured part of the moon and the physical properties of the surface. Investigating the properties of the uppermost porous layer of the moon's surface to a depth of about four inches, the Russians established that its electrical conductivity was 50% to 100% greater than that of the underlying rocks and concluded that the two layers differed in chemical composition and mineral content. The conductivity of the top layer was found to correspond to that of meteorites, confirming the hypothesis that bombardment by meteorites and the settling of meteoritic dust played an important part in shaping the uppermost layer of the moon. (*NYT,* 12/21/64)
- The *New York Times* reported an announcement this week by the International Atomic Energy Agency (IAEA) that nuclear energy had gained recognition in 1964 as an economically competitive source of electric power. A spokesman for the IAEA cited a 515,000-kw. nuclear plant to be built by General Electric for the New Jersey Central Power and Light Co. near Toms River, N.J. The bid for a nuclear plant—$66 million—had been lower than those for coal-powered plants despite the abundance of coal in southern New Jersey. (Grimes, *NYT,* 12/20/64, 8)

December 20: MARINER IV, bound for Mars, began passing through a second stream of meteoroids. In its first 23 days of flight, the craft had made 7½ million scientific and engineering measurements in interplanetary space and was 3,748,352 mi. from earth, traveling at 6,970 mph relative to earth. Its velocity relative to the sun was 72,902 mph. The straight-line distance from the spacecraft to Mars was 107,592,980 mi. (NASA Release 64–323)

- NASA announced that it had organized a moonwatch network to assist with development and operation of the "Moon Blink," an instrument designed to verify and locate color on the lunar surface. When alerted by one of the stations, watchers at scattered distant points would train their telescopes on the moon to verify a sighting. Professional and amateur astronomers at some 35 stations linked by long-distance telephone were working with NASA to give volunteer assistance to the Moon Blink program. The instrument was being developed by NASA's Office of Advanced Research and Technology under contract with Trident Engineering Associates, Annapolis, Md., and consisted essentially of rotating color disks which would blink when red color showed on the moon. Although the instrument was being developed for detection of unusual coloration on all types of bodies orbiting in space, the moon's surface was considered of particular interest since the red color could signify a source of gases and possibly of energy. Any source of energy would be significant to the NASA lunar landing program. (NASA Release 64–316)

December 20: Researchers reported the centrifugal experiment conducted for the Air Force by Douglas Aircraft Co.'s Missile and Space Systems Div. indicated that a centrifuge device in a space station could help to recondition astronauts living for prolonged periods in a weightless condition. Tests in which five college students spent 42 days in a horizontal position showed that tendency to faint on standing diminished after periodic centrifuge rides. (*NYT*, 12/20/64, 42)

- Dr. Boris Yegorov, the Russian doctor who orbited the earth in the VOSKHOD I spaceship, suggested that persons suffering from high blood pressure and heart trouble might one day be rocketed into weightless space for therapy. By relieving the gravitational load on the patients and giving them drugs, he said, "it will be possible to heal a person and give his wearied heart a chance to recuperate." Dr. Yegorov said this novel heart treatment would be possible in the "foreseeable future." (Toth, *Boston Sunday Globe*, 12/20/64)

December 21: EXPLORER XXVI, Energetic Particles Explorer D (EPE–D), was launched into an egg-shaped orbit atop a three-stage Delta rocket from Cape Kennedy. The 101-lb. windmill-shaped satellite carried five experiments designed to learn how high energy radiation particles are injected, trapped, and eventually lost in the Van Allen radiation belts. Information obtained from the mission was expected to make important contributions to the Apollo manned lunar landing program, specifically in the design of protective spacecraft shielding and in planning flight trajectories for moon landing. Information on the depth of penetration of the geomagnetic field by high energy solar protons—particles of potential danger to moon-bound astronauts—might also be obtained. Preliminary orbital data: apogee, 16,280 mi.; perigee, 190 mi.; period, 7 hrs. 36 min.; inclination, 20.15°. The satellite was designed for an operational life of one year. (NASA Release 64–302; AP, Benedict, *Wash. Post*, 12/22/64)

- USAF launched an unidentified satellite on a Thor-Agena D booster toward a polar orbit from Vandenberg AFB. Purpose of the shot was not indicated. (UPI, *Wash. Post*, 12/22/64)
- USAF launched Blue Scout, Jr., probe from Vandenberg AFB on a nonorbital space mission. Press sources reported this as the second successful launch of a contact cesium ion engine, developed by Electro-Optical Systems, Inc. (HHN–48; *M&R*, 1/4/65, 11)
- Dr. Winston E. Kock, Director of NASA Electronics Research Center in Cambridge, Mass., announced that Dr. Lester C. Van Atta and Dr. W. Crawford Dunlap would head two of the Center's major operations. Dr. Van Atta would direct research in electromagnetics, including microwave and optical research. Dr. Dunlap would direct research in com-

ponent technology, including solid state and materials and qualifications standards. (NASA Release 64-317)

December 21: NASA Administrator James E. Webb told the Commercial Club of Chicago: "There is an analogy between our national space program and our national history. The exciting adventures of the settlement of North America captured popular fancy. But it was the quiet development of the free spirit of Americans and the growth of our form of self-government that were to weigh heavier in the scales of history than Indian raids and campaigns of the colonial wars.

"Another analogy closer to us has been the change in public attitudes that has led Americans to adopt new scientific ideas and concepts that may have been rejected only a few years ago.

"Today we live in a period of unprecedented scientific and technological explosion. . . . To be visionary is no longer to be labeled as impractical." (Text)

- Sen. Clinton P. Anderson (D.–N. Mex.), Chairman of Senate Committee on Aeronautical and Space Sciences, told *Denver Post* reporters that the manufacture of the MOL spacecraft would be delayed until the Defense Dept. presented an "acceptable definition" of the program. He said that President Johnson was reviewing the 1966 budget which contained $155 million for MOL, but that the money would not be spent until Air Force objectives in space were defined and accepted. (Partner, *Denver Post*, 12/22/64)
- The Air Force version of the controversial F–111 fighter-bomber, formerly called the TFX, made a successful but somewhat abbreviated maiden flight from Carswell AFB, Tex. (the site of the General Dynamics Corp. plant that built the plane). Piloted by Richard L. Johnson and Val E. Prahl, the 1,650-mph aircraft performed all the items on the test program that were possible with the malfunctioning wing flaps. The variable-geometry wings, capable of a 70° sweep, were kept in 26° position during the first flight. In 21 min. of flight, the F–111 achieved an altitude of 17,000 ft. and a speed of 230 mph. On takeoff it required less than half the runway used by conventional jet planes. The Navy version of the F–111 was scheduled to fly by mid-1965. (Witkin, *NYT*, 12/22/64, 12; UPI, *Chic. Trib.* 12/22/64; *Aviation Daily*, 12/23/64)
- A study at Northrop Space Laboratories was reported as indicating that a fluid-filled "bladder" might offer the best form of protection to astronauts during impact acceleration. Using guinea pigs, the investigators found that the fluid-filled garment seemed to offer the same advantages in minimizing internal organ displacement as did water immersion, but without the risk of generating high hydraulic pressure pulses. (*M&R*, 12/21/64, 21)
- Strategic Air Command announced that the Chicago area would be one of two test corridors for an exercise in which a squadron of SAC B–58 Hustler bombers would test for pinpoint accuracy on simulated targets. The three-month exercise would begin Jan. 4. (UPI, *NYT*, 12/22/64, 12)
- NASA announced award of 35 supplemental or new research grants and contracts totaling $1,947,842 to 30 universities, colleges, and private research institutions. (NASA Release 64-313)
- *Missiles and Rockets* reported that studies carried out by Astro Research Corp. for NASA indicated the feasibility of forming large spheres in space

by blowing them as bubbles. The technique promised space structure sizes hitherto unattainable with very low masses. For example, a 100-ft.-dia. ECHO II-type sphere of two-micron-thick film, could be blown from 200 lbs. of viscous material. (*M&R*, 12/21/64, 21)

December 22: X-15 No. 3 was flown by Maj. Robert A. Rushworth (USAF) to 81,000-ft. altitude at maximum speed of 3,545 (mach 5.37) mph to test X-15's skin friction at relatively high speeds and low angles of attack, and to obtain information on ablative materials. The 10-min. flight was one of a series to study heat buildup and air-flow patterns resulting from friction. (NASA X-15 Proj. Off.; AP, *NYT*, 12/24/64)

- United Technology Center successfully test-fired its 120-inch-dia., 1-million-lb.-thrust solid propellant rocket motor for nearly two minutes at Edwards AFB, Calif. The firing was the fourth static test within two months. The 250-ton rockets were being developed for the first stage of the Air Force's Titan III-C standard space launch vehicle, under contract to the AF's Space Systems Div. The Titan III-C was scheduled to orbit a variety of manned and unmanned payloads, including the Manned Orbiting Laboratory (MOL). (*M&R*, 1/4/65, 26)

- USAF's new strategic reconnaissance plane, the SR-71, made its maiden flight at Palmdale, Calif., exceeding an altitude of 45,000 ft. and a speed of 1,000 mph. The one-hour flight was reported to have been very successful; all test objectives were met. Powered by two J-58 jet engines, the SR-71 would eventually fly at more than 2,000 mph and above 80,000 ft. The Lockheed-built plane was flown by civilian test pilot Bob Gilliland. In 1965, the SR-71 would be assigned to Strategic Air Command (SAC) at Beale AFB, Marysville, Calif. (DOD Release 902-64; Sehlstedt, *Balt. Sun*, 12/24/64)

- Secretary of Defense Robert S. McNamara announced that President Johnson had ordered development of the world's largest airplane—a giant military subsonic jet transport able to carry 600 troops or 250 tons of cargo nonstop for 7,000 mi. at 550 mph. The CX (Cargo Experimental) military jet transport was predicted to be ready for the Air Force by later 1968. Development costs would total $750 million. According to McNamara the CX would "maximize the effectiveness" of eight combat-ready divisions stationed in the U.S. and reduce "reaction time in meeting global crisis." It would also "have potential commercial use for both passenger and cargo purposes." In addition the CX would be able to carry any piece of divisional equipment used by an Army division, which present aircraft cannot do. (Mohr, *NYT*, 12/23/64, 1; Glass, *N.Y. Her. Trib.*, 12/23/64)

- It was announced that an advanced experimental radar system, built close to the ground and designed to detect, track, and identify missile warheads and to assign targets to antimissile sites, was undergoing tests at White Sands Missile Range, N. Mex. The radar system had no moving parts and no moving antennas; energy would be transmitted directly from the many elements of the antenna array and would be steered electronically by shifting the electrical phase of the individual antenna elements. (*NYT*, 12/22/64, 45)

- During a background briefing for news representatives on NASA electronics and control research activities, Francis J. Sullivan, Acting Director, Electronics and Control Division, Office of Advanced Research and Technology of NASA Hq., said: "Past analyses of the NASA R&D funds indicate that approximately 40 per cent of our R&D dollars that are utilized in launch vehicle work go into the electronics industry. The same applies to 70 per cent of the spacecraft funds, and 90 per cent of the tracking and data acquisition funds. If you sum up these factors, it will indicate that approximately 50 per cent of NASA's R&D dollars in any one fiscal year find their way into the electronic and control type industries." (Transcript)

December 22: A subcommittee of the House Armed Services Committee recommended that the Department of Defense move ahead with a program to develop fighter planes capable of vertical and short takeoffs and landings (V/Stol). The subcommittee urged that the military services "not over commit themselves to the helicopter to the detriment and interests of V/Stol transport aircraft when they become available." Furthermore, it was the judgment of the subcommittee "that none of these programs presently being supported by the Department of Defense are adequate to determine the operational suitability of a V/Stol fighter aircraft for close tactical support." (Balt. *Sun*, 12/23/64)

- In spite of the successful maiden flight of the F-111, Sen. John L. McClellan (D.-Ark.) said, in a telephone interview, that the Senate investigations subcommittee would not discontinue plans to try to determine whether the $6-billion General Dynamics Corp. contract for production of the F-111 was actually awarded to the high bidder on the basis of an inferior design, and if so, whether influence was involved. No specific date was mentioned for resumption of the investigations. (AP, *NYT*, 12/23/64, 15; AP, Balt. *Sun*, 12/23/64)
- USAF announced it had built the T-27 space flight simulator—a $5,500,000, 15-ton machine which would be used for astronaut training at the Aerospace Research Pilot School at Edwards AFB, Calif. The simulator was built under a contract with the Link Group of General Precision, Inc. (Sehlstedt, Balt. *Sun*, 12/23/64, 24)
- Federal Aviation Agency (FAA) announced that Pakistan International Airlines (PIA) had reserved two delivery positions for the U.S. supersonic transport plane bringing the total of reserved positions for the SST to 93 and the number of airlines holding positions to 21 (14 foreign flag and seven U.S. flag airlines). (FAA Release 64-110)
- USAF successfully fired an Atlas-F ICBM down the WTR from Vandenberg AFB—the third major launching from the base in two days. (UPI, *NYT*, 12/23/64, 28)
- USN launched a Polaris A-3—the 13th consecutive success—from the submerged submarine *Von Steuben* off Cape Kennedy. (*M&R*, 1/4/65, 11)
- Secretary of Defense Robert S. McNamara announced that Gen. John Paul McConnell would be nominated to replace Gen. Curtis E. LeMay, who would retire as Air Force Chief of Staff on Jan. 31, 1965, after 35 years in the armed forces. Gen. McConnell would assume his new position Feb. 1. (*NYT*, 12/23/64, 11; Barrett, HTNS, *Wash. Post*, 12/23/64, 2)

December 23: NASA announced realignment of its manned space flight organization, designed to (1) provide a central organizational element with the overall operations responsibility for manned space flight programs, and (2) make a single organization responsible for the assembly, checkout, and launch of the total Apollo space vehicle in manned space flight programs at Cape Kennedy and the Merritt Island Launch Area (MILA). Key organizational changes would be effective Jan. 1, 1965: (1) appointment of E. E. Christensen to a new position, Mission Operations Director, in NASA Office of Manned Space Flight, supported by two Mission Directors with overall responsibility for the individual mission to which they were assigned; (2) creation of an Operations Support Requirements Office, headed by Porter Brown and staffed by representatives of each NASA program office, appropriate centers, and Office of Tracking and Data Acquisition; (3) transfer of Manned Spacecraft Center's Florida Operations to Kennedy Space Center; and (4) establishment at KSC of the position of Director, Launch Operations. Dr. Kurt H. Debus would also act as Director, Launch Operations; G. Merritt Preston, head of MSC Florida Operations, would become Deputy Director, Launch Operations. Resident offices were provided at KSC for MSC and MSFC program managers. (NASA Ann. 64–301; KSC Release 224)

December 23: NASA and ComSatCorp jointly announced the signing of an agreement covering the launching by NASA of "Early Bird"—the first communications satellite intended for commercial use. The agreement provided that NASA would launch Early Bird from Cape Kennedy, Fla., into synchronous orbit at 22,300-mi. altitude above the Atlantic as soon after Mar. 1 as NASA found feasible. ComSatCorp would pay NASA approximately $3.5 million for the initial launch, whether or not the Thrust Augmented Delta launch vehicle was successful in orbiting the satellite, and for each subsequent launch under the contract. The $3.5 million would reimburse NASA for the cost of the Delta vehicle, other launch services, propellants, guidance and tracking services, and data processing. Should the first launch be unsuccessful, NASA would try again on a second satellite. The contract gave ComSatCorp the option of calling for additional launches, after July 1, of synchronous satellites of the same type. (ComSatCorp Release, 12/23/64)

- Jet Propulsion Laboratory released figures on MARINER IV, valid as of 12:01 pm: total miles to travel—325,000,000; total miles traveled—44,000,000; miles remaining to travel—281,000,000; miles from earth—4,306,180; speed in relation to earth—7,023 mph; speed in relation to sun—72,532 mph; total days required for trip—228; days already in space—26; days remaining to travel—202. (AP, Balt. *Sun,* 12/24/64)

- It was reported that President Johnson would give special attention to science and technology programs in budget-making conferences at his Texas ranch. Dr. Donald S. Hornig, the President's special assistant for science and technology, said in Austin Press Conference that the significance of his presence as a participant in all meetings was that Johnson was looking at the "totality" of the various Government enterprises supporting science and technology in an effort to pursue a balanced program. Hornig said the Federal government was spending about $15 billion—roughly 15% of its budget—on scientific research, technical development, test, and engineering programs with about half this amount spent by DOD. Dr. Hornig also said he believed the hundreds of millions of dollars going into the effort to put an American on the moon by 1970 was well spent because "the moon is a very interesting place scientifically." It will "take a lot of going" to achieve that goal, he said, but present schedules "will allow it." (Wash. *Eve. Star,* 12/23/64)

December 23: Prof. Marcel Florkin of Liège Univ., a member of the Committee on Space Research (COSPAR), said that many American and Soviet scientists shared his belief that men cannot survive weightlessness in space for more than five or six days. He said that Soviet scientists had produced photographs showing abnormalities in cell division attributable to the sustained weightlessness some of their astronauts had experienced. Prof Florkin said that under present conditions, only space robots could visit the moon or Mars and that the only alternative would be space vehicles that could produce artificial gravity to simulate normal human environment. (O'Toole, *NYT*, 12/24/64)

- Dr. Herbert Friedman, chief of astrophysics at the Naval Research Laboratory in Washington, D.C., told a meeting of the American Physical Society at Berkeley that x-ray detectors carried in Aerobee sounding rockets had picked up evidence of radiation above the earth's atmosphere which indicated the possibile existence of a "neutron star." The invisible object, lying in the constellation Scorpius and relatively close to the earth, emitted streams of extremely feeble x-rays. Its size was unknown. The existence of such a star would offer important confirmation of scientific theories about the birth and death of all stars, and the distribution of elements throughout the universe. It would also be the densest collection of matter in the universe: a massive lump of nuclear particles weighing a billion tons per cubic inch. (Perlman, *S.F. Chronicle*, 12/24/64)

- NASA Ames Research Center announced that the Carl N. Swenson Co., San Jose, Calif., had been awarded a $992,000 contract to construct a reinforced-concrete vacuum chamber and Structural Dynamics Laboratory at ARC. The space chamber would be in the form of a ten-story, 30-million-lb., hollow concrete pentagonal tower with walls three feet thick, would accomodate full-scale segments or smaller-scale models of rockets and spacecraft, and could be evacuated to a pressure of ten millimeters of mercury. (ARC Release 64–31)

December 26: A *National Geographic* report, based on findings by NASA and Air Force researchers, estimated that the earth's gravitational field picked up several million tons of dust and small particles every year and that the total accumulation of this material on the earth's surface since its formation two billion years ago was well over ten feet. (*NYT*, 12/27/64)

- It was reported that the Marconi Co., Ltd., in England, was attempting to perfect a safety aid that would make gliders more visible to radar to help avoid air-to-air crashes with aircraft. Carefully calculated patterns of thin foil, detectable by different radar wavelengths, were being mounted in and on a test glider flown by the London Gliding Club. (Sci. Serv., *NYT*, 12/26/64, 9)

- Dr. Hugh L. Dryden, NASA Deputy Administrator, said in an article written for the *Florida Times-Union:* "Expansion's of man's geographical horizons have almost invariably been linked with surges of creativeness in the arts and sciences. The Space Age could well revitalize human affairs even more than did the vigorous world-ranging of Western navigators that ended the stagnation of the Dark Ages in Europe. . . .

"Already some philosophers see the possibility that world-wide interest in the space program and growing international participation in it could be the first steps toward providing our troubled race with a moral and constructive substitute for war.

"It is entirely possible that, along with broadening of perspectives, there will be basic changes for the better in man's aspirations as astronauts of many nations venture deeper and deeper into the vastness of space in which ours is but one small planet circling a sun-star of relatively minor importance. . . ." (Fla. *Times-Union*, 12/27/64)

December 26: Dr. Wernher von Braun, Director of NASA Marshall Space Flight Center, said the *Florida Times-Union:* "Any statement of what space exploration will be like 100 years from now is . . . in the same league as crystal ball gazing. . . . Any specific projects which we may describe . . . will fall far short of activities that will be routine a century from now.

"One hundred years ago the 'iron horse' was the fastest means of travel available. . . . Airplanes were unknown. In this atmosphere Jules Verne sat down and wrote . . . '*From the Earth to the Moon*' published in 1865. . . . His spacecraft carried three men (and two dogs), was launched from the east coast of Florida, circled the Moon, and returned to Earth during a round trip that took 11 days. . . . Jules Verne used a huge cannon buried in the sand to boost his capsule to escape velocity. . . . Today we are using Saturn chemical rockets to propel our Apollo spacecraft outward for its first trip to the Moon. And by the year 2064 altogether new means of propulsion will make our present-day launch vehicles and spacecraft appropriate relics for the Smithsonian Institution. . . ." (Fla. *Times-Union*, 12/27/64)

December 26–30: A plasma torch, designed and built by scientists at the Univ. of California at Los Angeles in a space chemistry program sponsored by the Air Force Office of Scientific Research (AFOSR), was demonstrated publicly for the first time at the annual meeting of the American Association for the Advancement of Science (AAAS), at Montreal. The device provided the first laboratory source of extremely hot, pure plasma which could be precisely modulated, and whose composition could be varied widely, thus making it possible for scientists to simulate the chemistry of the hottest stars. (AFOSR Release 12–16–4S; *M&R*, 1/4/65, 21)

December 27: In the *Journal of the American Physical Society*, Irwin I. Shapiro of MIT's Lincoln Laboratory proposed use of radar as a test for Einstein's general theory of relativity: radar pulses would be sent to one of the solar system's inner planets, either Venus or Mercury, and the time-lapse before the radar echoes' return measured on the earth. According to Einstein's theory, when the radar beam passed close enough to the sun to encounter a very strong gravitational field, the radar waves should slow down. The slowdown should be sufficient to delay return of the radar echoes by as much as 0.0002 seconds, a time interval accurately measurable by electronic techniques. (*Wash. Post*, 12/28/64)

- President of the Delaware Academy of Science, John S. Seney, reported experiments he hoped would lead man to fly under his own power, like a bird. He had captured a turkey buzzard which he hoped to put through a series of tests in a wind-tunnel in the basement of his home. "If we're going to fly, the only way to do it is to find out what this bird's got . . . and then put all this together in relation to what modern materials are available. Then maybe we'll come up with some sort of flying suit." (*NYT*, 12/28/64, 51)

December 28: MARINER IV passed the 50-million-mi. mark on its 325-million-mi. flight to Mars. The spacecraft was operating normally and transmitting items of information at a rate of 33⅓/sec. Nearly 10 million scientific and engineering measurements in space and outside the earth's orbit had been transmitted in its first 29 days of flight. (NASA Release 64-326)

- A spokesman for the Senate's Select Committee on Small Business' Subcommittee, headed by Sen. Russell B. Long (D.-La.), reported it would meet shortly after Congress convened Jan. 4 to consider senatorial requests for an investigation of charges that NASA had given away to business firms valuable patent rights developed at public expense and worth millions of dollars. James E. Webb, NASA Administrator, claimed the agency was following Federal law and guidelines laid down by President John F. Kennedy shortly before his death. (AP, Wash. *Eve. Star*, 12/28/64)

- U.S. Navy was reported to be defining its manned orbiting laboratory (MOL) ocean surveillance experiment despite the proposed merger of the Air Force MOL and NASA's Apollo-X programs. Navy felt its space mission potential should be explored as thoroughly under either military or civilian programs. The Navy space mission was first envisioned to involve five major areas concerning shipping: detection; location; identification; classification; tracking of surface and subsurface craft. (Fink, *Av. Wk.*, 12/28/64, 13)

- Support for theories that extraterrestrial forces influenced rain and snowfall was reported at a meeting of the American Association for the Advancement of Science. Glenn W. Brier of the U.S. Weather Bureau presented evidence that the moon, when new or full, influenced the timing of precipitation by gravitationally nudging unstable atmospheric conditions into rainfall or snowstorms. Dr. E. Keith Bigg of Australia's Commonwealth Scientific and Industrial Research Organization in Sydney reported research which substantiated his theory that meteoric dust particles acted as nuclei for ice crystals to form on and fell to earth as rain or snow, with the whole taking about 30 days. (Osmundsen, *NYT*, 12/29/64, 6)

- U.S. Army announced that the Corps of Engineers awarded a $1,159,000 (NASA funds) fixed-price contract to Carpenter Brothers for construction of a mobile equipment maintenance building at NASA's Mississippi Test Operations. (DOD Release 908-64)

- GAO Comptroller General Joseph Campbell said that GAO would evaluate DOD policies affecting nonprofit organizations performing design, technical direction, and systems engineering functions for the military services. "We have a continuing interest in finding out whether salaries higher than the government scale paid by some of these organizations are really necessary and justified or are only a way of indirectly paying higher [wages] for government work," Campbell said. Another area of prime concern to GAO was relinquishment to nongovernment organizations of technical direction and similar activities it considered Government management functions.

    DOD had already clamped fiscal ceilings on six of these organizations to prevent their growth and was inviting increased industry competition in the areas of technical direction and systems engineering.

Campbell told *Aviation Week and Space Technology* that "if we are faced with any intention to sustain permanently some of these organizations as arms of government, we would have to look into the situation very carefully and thoroughly."

In commenting on NASA's approach to nonprofit organizations, Ernest W. Brackett, Asst. Deputy Associate Administrator, noted that although NASA had "inherited" one nonprofit organization, the California Institute of Technology's Jet Propulsion Laboratory, it had not and would not establish any "captive" nonprofit organizations. Two other nonprofit organizations were also performing work for NASA—RAND Corp., with contracts totaling $3.3 million, and MIT's Lincoln Laboratories, with contracts totaling over $7 million. (*Av. Wk.*, 12/28/64, 31)

December 28: Director General of the International Air Transport Association, Sir William Hildred, said in the association's annual air traffic forecast, "Thanks to increased efficiencies and economies, the balance between revenues and expenditures should continue to improve. In fact, it just seems possible that the industry will succeed in making a positive net profit in 1965 for the first time in its life." (UPI, *NYT*, 12/29/64, 50)

- Among persons listed by *Aviation Week and Space Technology* for significant contributions to the progress of aerospace in the U.S. during 1964 were:

  Harold Finger and Milton Klein, manager and deputy respectively of the joint AEC–NASA Nuclear Space Propulsion Office for "pushing the Rover program through four successful tests of a nuclear rocket reactor using liquid hydrogen and demonstrating the technical feasibility of nuclear propulsion for manned interplanetary flights of the future."

  Harris "Bud" Schurmeier, Ranger program director of the Jet Propulsion Laboratory and his technical team for "their tremendous technical success with the Ranger 7 that produced the first detailed photographs of the lunar surface."

  Paul F. Bikle, Director of NASA Edwards Flight Research Center, and pilots Milton Thompson and Jack Mackay of NASA and Maj. Robert Rushworth and Capt. Joe Engle of USAF for "their continued operation of the North American X–15 research aircraft into a wide variety of unexplored areas both in performance and equipment testing."

  John Mengel of NASA Goddard Space Flight Center for "direction of the global NASA tracking and data acquisition network which produced an average of 50 mi. of taped data per day from payloads in space."

  Col. Harold W. Robbins, USAF Space Systems Div., for "technical direction and support of the large solid rocket program that culminated this year in the successful demonstrations of 156-in.-dia. motors and the initiation of a 260-in.-dia. program." (*Av. Wk.*, 12/28/64, 9)

December 28–30: Dr. John H. Wolfe of NASA Ames Research Center suggested the probable source of Van Allen belt radiation at the meeting of the American Geophysical Union in Seattle. Dr. Wolfe said the Van Allen radiation belts probably had their source in the comet-like tail of the magnetosphere extending from the dark side of earth away from the sun. Spiraling protons and alpha particles emitted by the sun were shielded from solar wind in this area. Affected only by the geomagnetic field, they were picked up by the magnetic lines of force and conveyed to the Van Allen belts.

New data collected by particle energy spectrometers and other instruments aboard EXPLORER XXI IMP-B (Interplanetary Monitoring Platform) and OGO I (Orbiting Geophysical Observatory) led to this conclusion. Data collected from as far into space as 96,000 mi. revealed that interaction of the solar wind and the geomagnetic field around the sunlit side of earth created a boundary region 2½ earth diameters thick at a distance of five earth diameters from the point on earth closest to the sun. This expanded to six diameters thick 90° around earth.

Hydromagnetic effects in this boundary deflected solar radiation around earth and prevented it from penetrating directly to the relatively low altitudes of the Van Allen belts except during the more energetic solar storms.

Satellite data also indicated that the comet-like tail of earth extended directly along the line from the sun through earth, rather than being slightly deflected as had been believed. Wolfe said the tail seemed less tapered and more nearly cylindrical than had been expected, and extended at least as far as the orbit of the moon. (ARC Release 65–2; AP, *Detroit News*, 1/8/65; *M&R*, 1/18/65, 28)

December 29: The first of three giant Pegasus satellites arrived at NASA John F. Kennedy Space Center. A Saturn I space vehicle (SA-9) would launch the 3,200-lb. satellite into an orbit visible from earth and ranging from about 310- to 465-mi. altitude, sometime in the first quarter of 1965. Primary mission of the Pegasus satellites would be to use its 2,000 sq. ft. of capacitator panels to measure, record, and transmit back to earth data on the frequency, size, direction, and speed of meteoroids in near-earth space. This information would be important to development of manned and unmanned spacecraft. (KSC Release 227–64)

- The first Apollo spacecraft adapter was delivered by Army CH-47A helicopter from Tulsa, Okla., to NASA Manned Spacecraft Center-Florida Operations at Merritt Island. The 4,000-lb., 22-ft.-dia., 28-ft. long, cone-shaped unit structurally simulated the adapter section that would house the Apollo lunar excursion module (LEM) on its trip to the moon. It would undergo tests to determine whether the mechanical fit was adequate, if sufficient access was provided to perform service and checkout of the LEM, and to determine its compatibility with test facilities and ground handling. Helicopter was chosen over other forms of transportation because of the awkwardly large diameter of the section. (KSC Release 227–64)

- Soviet radio astronomy expert Iosif Shlovsky said at a symposium on 1964 science conducted by Tass news agency that the biggest success of 1964 in world astronomy was an experiment by Dr. Herbert Friedman of the U.S. Naval Research Laboratory. The experiment showed that a large region of the Crab Nebula produced x-ray radiation. Data obtained by Friedman with an Aerobee rocket launched July 5 from White Sands Missile Range "established how the flux of x-ray radiation coming from the Crab Nebula changed when the moon passed between it and the rocket," Shklovsky said. Some scientists had attributed the radiation to a tiny, superdense "neutron star." But observations from the rocket as the moon shut off direct view of the Nebula showed that a larger area was involved. (AP, Balt. *Sun,* 12/29/64; Sullivan, *NYT,* 12/19/64, 3)

December 29: NASA Flight Research Center requested bids for construction of a high-temperature heat facility at Edwards, Calif. The completed facility would include enough test area to permit heat load testing of actual aircraft of the size of the XB-70 and the proposed supersonic transport. It could produce temperatures up to 3,000° F on small isolated areas of the test aircraft; large areas could be heated up to 600° F. The contract would be administered by the U.S. Army Engineers. (FRC Release 33-64)

- Successful first flight tests at Dallas of XC-142A V/Stol transport aircraft that could take off and land vertically were conducted by Ling-Temco-Vought. Initial hover flights were held to an altitude of five feet.

  Design plans called for the aircraft to attain speeds of 430 mph. Its wings could be tilted upward while the fuselage remained level. Thrust from the engines would lift the plane off the ground to a desired height; the wings could then be tilted back in place for conventional flight. (AP, Wash. *Eve. Star,* 12/30/64)

- Navy radar operators at Patuxent Naval Air Station, Md., reported the sighting of two high-speed unidentified flying objects which approached the base at a speed of approximately 4,800 mph, executed a tight turn, and disappeared. (AP, *Phil. Inq.,* 1/6/65)

December 30: In a New Year's greeting addressed to President Anastas I. Mikoyan and Premier Alexei N. Kosygin of the Soviet Union, President Johnson urged agreement on a series of United States disarmament proposals and arms control: "Arms control remains especially urgent; nothing can contribute more to the hopes of mankind for the future. During the months ahead I hope we can work for practical agreements to this end. We can and should move to limit the spread of nuclear weapons; to achieve a verified worldwide comprehensive test ban; to make a cut-off of fissionable material production for weapons coupled with measures to safeguard the peaceful uses of nuclear power; and to agree on a verified freeze in existing offensive and defensive strategic nuclear delivery systems.

  "By progress in this critical area, our Governments can help to make this a happier and safer world for all people. You may be certain that the American people and their Government will never be second in this effort." (Mohr, *NYT,* 12/31/64, 1-2)

- A policy statement issued by the Committee on Science in the Promotion of Human Welfare of the American Association for the Advancement of Science (AAAS) charged that social and political pressures on the scientific community arising from such major U.S. space programs as Apollo, Starfish, and West Ford constituted a "serious erosion of the integrity of science." The report was issued following a five-day meeting held by the association in Montreal, Canada.

  Commenting on the "spare race" aspects of NASA's program, the committee said the pattern for orderly development of research had been altered by the "essentially political decision to pursue project Apollo." The report concluded that Apollo did not appear to be based on the orderly systematic extension of basic scientific investigations.

  Another NASA program to come under fire at the AAAS meeting was the Voyager Mars exploration project attacked by Dr. Barry Commoner, professor of plant physiology at Washington University. Dr. Commoner said finding life in any form on Mars would have tremendous scientific significance, and that he was not opposed to the basic idea of searching for life on the planet. "I simply feel this country is not yet prepared to make a commitment of this scope," he said. "We could at this time be a lot closer to resolving this question if the scientific community had been given the opportunity to fully discuss it." He said even though the commitment had been made it was not too late to discuss fully the question of life on Mars before the 1971 launch. He said he hoped Congressional space committees would ask for complete justifications on the Voyager project. (*Av. Wk.*, 1/4/65, 18; Abraham, Phil. *Eve. Bull.*, 12/28/64; AP, Wash. *Eve. Star*, 12/28/64)

December 30: Dr. Harold Urey, Nobel prize-winning chemist, told scientists at American Geophysical Union meeting in Seattle that the structure of some of the ridges on the lunar surface photographed by RANGER VII indicated that water might exist on the moon. (AP, *St. Louis Post-Dispatch*, 12/31/64)

- A pea-sized rocket motor, called a cap motor, weighing less than three-hundredths of an ounce, might be given its first trial in space next year as a stabilizing motor for a Tiros weather satellite. Conceived by the Wright Aeronautical Div. of the Curtiss-Wright Corp. in Wood Ridge, N.J., it was made entirely of plastic, contained igniter material which would be touched off by an electrical impulse, a solid propellant, a combustion chamber, and a flared exhaust nozzle. A string of the expendable motors would be mounted on a tape that would feed through a half-pound triggering device and fire them one at a time or in sequence. One cap motor would yield a pound of thrust for about two-hundredths of a second. Two cap motor units, each containing 300 cap motors, would be placed on Tiros for the test in an attempt to lengthen its useful life beyond an expected three months. The satellite must spin at a certain number of revolutions a minute to stabilize its cameras; the slight friction in its trajectory would tend to slow it down. Computers on the ground would calculate the amount of spin to be imparted; aboard Tiros, 100 cap motors would be fired in each burst. (Hudson, *NYT*, 12/30/64, 1, 17)

- Prof. Rene J. Dubos of the Rockefeller Institute charged at the Montreal meeting of the AAAS that present-day biology was almost completely irrelevant to the study of the nature of man. He said that the proper study of man's nature should concern itself with the evolved patterns of physiological responses to a changing environment—what he would call "humanistic biology." Not only would this approach turn up much new knowledge about "humaness," he said, but it would also provide a better understanding of the uniqueness of the individual.

    Prof. Dubos stated he was not discussing psychology, a social science. Man now possessed technical means for manipulating his mental and physiological processes and might soon develop ways to alter his inheritance and so direct the course of his evolutionary fate. "This means that mankind may soon be faced with decisions of immense importance that must be made in the light of the proper kind of biological knowledge," he said. "The glory of the coming age," he continued, "must be conceived within the framework of man's nature—of his biological limitations as well as his potentialities." (Osmundsen, *NYT*, 12/30/64, 11)

December 30: NASA Director of Personnel, Robert J. Lacklen, who had been with NASA since its inception, resigned to accept a position with the Richardson Foundation of Greensboro, N.C. Mr. Lacklen would head a Personnel Research Institute which would conduct research in personnel selection and appraisal. Mr. Grove Webster was named Acting Director of Personnel. (NASA Ann. 64-300)

- USN awarded a $13 million contract to General Dynamics-Pomona for funding, development, and pilot production of both medium and extended range standardized versions of a Tartar/Terrier-type surface-to-air missile. The standardized antiaircraft missile would use the Tartar/Terrier warhead and the same control and launching systems with minor modifications and would offer improved reliability and performance at reduced cost. (DOD Releases 910-64 and 911-64)
- The Paris Municipal Council had voted for creation of a flying fire brigade unit as part of the Paris police budget for 1965. The flying firemen would have three helicopters, would deal with fires in tall new buildings where expandable ladders would be ineffective. (Reuters, *NYT*, 12/30/64, 42)

December 31: NASA announced it had completed its experimental research and development program with SYNCOM II and SYNCOM III and would transfer operation of the synchronous satellites to DOD, starting Jan. 1, 1965. The transfer would be completed Mar. 31, 1965. DOD would operate the satellites for the remainder of their useful lifetimes and would provide NASA with any significant telemetry and ranging data. Communications ground stations used to relay transmissions via the two Syncoms had been furnished by DOD for the past two years. (NASA Release 65-5; DOD Release 1/4/65)

- NASA reported that a magnetometer hammer developed at NASA Marshall Space Flight Center to remove distortions from rocket fuel tank "gore segments" had salvaged rocket hardware valued in the millions of dollars. (NASA Release 64-328)
- Dept. of Commerce announced opening of World Weather Center in Washington, D.C., composed of the Weather Bureau's National Meteorological Center and National Weather Satellite Center with assistance provided by other Weather Bureau components. WWC would gather, process, and distribute weather observations and prepare weather analyses and forecasts for as much of the globe as possible; provide opportunity for the training of meteorologists from the U.S. and abroad; conduct basic and applied research on large-scale weather problems; and archive weather information for use of the scientific community for research purposes. The Center would utilize the new direct Washington-Moscow weather communications link as well as the existing international weather communications systems.

    Plans for the U.N.-sponsored World Weather System would include three World Meteorological Centers: one in Washington, another in Moscow, and a third at a presently undesignated location in the Southern Hemisphere. (Commerce Dept. Release G 64-184)
- Intelligence information reaching Washington revealed that Red China had begun to manufacture its own MiG jet fighters. Significance of the new development was that Red China might have the principal ingredients of a nuclear force—A-bombs and jet fighter bombers. (Fryklund, Wash. *Eve. Star*, 12/31/64)

During December: Boeing Co. was named systems integration contractor for NASA's three-stage Saturn V booster. Work would be performed at the firm's launch systems branch at New Orleans under an $85 million supplement to its contract as prime contractor for the booster's first stage. Systems and engineering integration would be conducted in three major areas: the vehicle, systems documentation, and systems development breadboard facility. On the vehicle itself, Boeing would be responsible for mechanical interface engineering, instrumentation and telemetry systems, aerodynamics and flight mechanics, flight evaluation and operation, atmospheric environmental design criteria, malfunction detection system, structural heating and environmental control system, structural dynamics, and vehicle assembly documentation. (*M&R*, 1/4/65, 10)

- National Weather Satellite Center was seeking sources for services and material for advanced development models of water-borne automatic unmanned weather stations by January 14, 1965. Companies with mooring and static meteorology sensor technology and design capabilities were being sought. (*M&R*, 12/21/64, 8)
- NASA announced that its Office of Technology Utilization had issued Technical Briefs describing two instruments for measuring the rates of human respiration and heartbeat. The devices were a pneumotachometer (Tech Brief 64–10259) which would measure respiration rates per minute on a breath-to-breath basis and a cardiometer (Tech Brief 64–10258) which would compute the heartbeat rate from the wave-form output of an electrocardiograph (EKG) and visually display the data in beats per minute. Both were developed in connection with NASA's manned space flight program. (NASA Release 64–311)

    A NASA technology utilization report described the effects of extremely low temperatures on structural metals. Based on studies at NASA Marshall Space Flight Center, the report outlined some of NASA's experience with various metallic materials and their capabilities at cryogenic temperatures. (NASA Release 64–325)

    NASA Tech Brief 64–10006 described a speed-sensing device to help power crane operators judge payload movements more precisely. Visual and audible signals would be produced in the crane cab to indicate the rate of load movement. The device would eliminate the necessity of a relay man to transmit hand signals to the crane operator. (NASA Release 64–300)

- NASA Flight Research Center completed the first series of flight tests of the Bell Aerosystems Lunar Landing Research Vehicle (LLRV). Program officials observed that the LLRV could simulate many nonaerodynamic space-flight maneuvers in addition to the lunar landing for which it was primarily designed and asked for a full year of experimentation with the craft. (*M&R*, 12/14/64, 10)
- NASA Kennedy Space Center awarded a $1.5-million contract to Monitor Systems, Inc., of Fort Washington, Pa., for two digital telemetry systems to be installed at KSC's Central Instrumentation Facility on Merritt Island. The systems would be used as intermediate data converters and processors during Apollo/Saturn V rocket launches from Merritt Island. (KSC Release 220–64)
- Paul G. Johnson, of the AEC–NASA Space Nuclear Propulsion Office, said in an article in *Astronautics and Aeronautics* that solid-core nuclear rockets

would power the next generation of manned space vehicles. "The exact way in which nuclear rockets will be used in the space program will depend on many factors other than payload comparisons. One of the primary influences will be the relative emphasis placed on manned operations in earth orbit, on the moon, and to the near planets. An equally strong influence will be the experience in the nuclear-rocket development program, which will indicate the possible characteristics and availability dates of propulsion systems." (A&A, 12/64, 22–28)

During December: In the British Interplanetary Society's *Spaceflight*, Brian J. Ford suggested that the first men on the moon might be imperiled by the residue of an electric charge which he said formed its craters. Mr. Ford proposed the theory that the moon's craters were caused by a gigantic electrical discharge when the earth and moon were still hot and close together, and said that he had reproduced the theoretical conditions in miniature in a laboratory. He had found that electrical discharge created craters which showed the characteristic central peak of a lunar crater and formed other corroborating detail. He suggested that the moon might still carry an electrostatic charge from this original discharge which "might seriously imperil" the first men to land there. (UPI, *Chic. Trib.*, 12/28/64)

- In a report to the Atomic Industrial Forum, Frank Deluzio, staff director of the Senate Aeronautical and Space Sciences Committee, called for increased funds for Kiwi and Nerva and also urged that the suspended Rift nuclear rocket program be reactivated. He suggested that studies be begun to substitute a nuclear reactor for the chemical engines in the giant Saturn V rocket. "While it is admitted that nuclear propulsion comes into its own in the large scale manned missions to the planets," Deluzio said, ". . . nuclear stages should be committed to earlier programs, such as lunar landings late in the Apollo program. The integration of a nuclear stage with the Saturn launch vehicle should be done at the earliest possible moment so that proper experience and confidence can be obtained . . . This will increase the utility of the Saturn for a great number of years; it means a high return to the nation for its investment in the Saturn, its ground support, related systems, and industrial capabilities." Deluzio also urged that nuclear rocket development remain the province of AEC. (McHugh, Copley News Service, *San Diego Union*, 12/21/64)

- In an *Astronautics and Aeronautics* editorial, Burton I. Edelson, Chairman of the AIAA Communications Committee, said: "In applying earth satellites to practical tasks, we have long passed the stage of demonstration. Now as we approach the stage of extensive operations, we face another major challenge in what industry terms a 'market survey.' We must be alert to the possibility of entirely new markets which evolve once an initial experiment demonstrates capability. The professional community can and should make major contributions to the prompt exploitation of its brain-childs." (A&A, 12/64, 13)

- In *Fortune* article on "Khrushchev's Paper Bear," Charles J. B. Murphy reviewed the background of U.S. intelligence concerning Soviet missile development as related to the first ICBM's, the first Sputniks, the high-altitude U-2 operations, and the so-called "missile gap." He related these to American decisions and politics. Author submitted that Soviets failed to exploit early lead in missiles and that U.S. miscalculated relative missile strength in 1960. (*Fortune*, 12/64, 114f.)

During December: Karl G. Harr, Jr., Pres. of Aerospace Industries Assn., editorialized: "To those familiar with the technical achievements of the past few years there is no doubt that aviation is on the threshold of its brightest era. . . . Enough is known of the potential of the new aeronautical technology to predict that all types of aviation—military, airline, and general aviation—will be completely transformed in the next two decades. The day of the manned military airplane is far from over. Today's civil jet transports, which enjoy a high reputation, are far from the ultimate in economical flying machines. The usefulness of all types of aircraft will be vastly improved, and they will be much cheaper to operate in the future.

"Perhaps even more important will be the widespread benefits of the new aeronautical technology outside of aviation. In the past, aeronautical technology, which has provided major advances in high performance engines, efficient air flow design, lightweight structures and material, and high reliability, was put to wide use after it was given in aviation. . . . This pattern of transfer of aeronautical technology will not change. In fact, the opportunity should be greater in the future because the jump in efficiency now considered possible for aircraft structure and powerplant technology is greater than anything yet experienced. . . ." (*Aerospace,* Winter 64–8–9)

During 1964: The deepening and broadening of the U.S. space program continued at an accelerated pace in 1964. The U.S. orbited a total of 76 payloads, of which four were deep-space probes, while the U.S.S.R. orbited 35, of which two were deep-space probes. Of the U.S. total NASA orbited 23 payloads. Other major NASA launches (not included in orbital total) were one unsuccessful attempt at orbit (Beacon satellite) and 5 suborbital tests of a burgeoning roster of new launch vehicles, spacecraft, and space equipment.

DOD's launch program continued to be numerically larger than NASA's, with 53 satellites successfully orbited during the year.

NASA's most publicized flight of the year was the RANGER VII flight to the moon with its close-in photography of the moon. Other solid achievements were APT weather photos from TIROS VIII and the much improved clarity and coverage of weather photos from the first advanced research weather satellite, NIMBUS I. The first generation of active communications satellites concluded with SYNCOM III, which was maneuvered into the first truly synchronous orbit of the space age and performed beyond design requirements as a communications relay between Asia and the U.S. A significant scientific contribution was made by EXPLORER XVII, which in 1964 mapped the earth's magnetosphere and found it, instead of the sphere it was previously thought, to be a tremendously elongated comet-like shape. In flight tests Gemini and Apollo spacecraft moved nearer to qualifying as man-rated. The Saturn I completed its seventh (out of 7) successful test flight and orbited a new world weight record of 39,000 lbs. The Centaur continued a successful flight-test program to work the bugs out of the first liquid-hydrogen engines in an operational space vehicle. In ground tests the Saturn IB and V launch vehicles took significant strides in testing of new large engines and of complete stages. In advanced research two achievements of high promise were the full-power runs of solid-core nuclear rocket engines that confirmed the engineering feasibility of nuclear rocket engines with their tremendous increases in specific impulse over chemical-fueled engines, and the successful use of laser in satellite tracking, promising a new order of tracking accuracy for geodetic and other uses.

In DOD's launch program, the most publicized flight was that of Titan III-A on Dec. 10, when the big booster for the proposed MOL successfully flight-tested its crucial transtage. Also DOD continued its policy of launch economy with 8 multiple-satellite launchings, including one (Jan. 11) in which a single Thor-Agena D booster launched 5 satellites.

The most publicized event of the Soviet space program in 1964 was its orbiting of VOSKHOD I, a new 3-man spacecraft, with 2 of the crew not trained pilots.

For both the U.S. and Russia, one major effort in each program remained a question mark at the end of the year. Each country had a deep-space probe (MARINER IV and ZOND II) enroute to Mars, due for flyby in July 1965. Perhaps never in its seven years had the space age offered such a direct confrontation as this promised to be. (HHR-8; HHN-48; NASA Release 64-321; *A&A, 64,* passim.)

During 1964: Soviet scientists P. M. Bayevsky and K. I. Zhukov reported in the U.S.S.R. Academy of Sciences' Journal, *Cosmic Research*, that two of their cosmonauts had suffered a rhythm instability in their heartbeats while in space. The condition, cardiac arrhythmia, was observed in both Valery Bykovsky, who spent five days aloft, and Valentina Tereshkova, who orbited the globe for 71 hrs. Both flights were made in June 1963. Changes were attributed to "normal" reactions to weightlessness, but the doctors warned that the condition should be watched closely on future long-duration flights since it could be a clue to more serious "pathologic changes," including a possibly fatal heart attack.

Dr. W. Randolph Lovelace, NASA Director of Space Medicine, saw the electrocardiograms while in Russia and said they showed only "variations we'd expect." NASA doctors agreed that cardiac arrhythmia should be watched as a potential danger signal but said they had seen "nothing dangerous" so far in U.S. astronauts' heart rhythm changes. (TNS, Bylinsky, *Huntsville Times*, 12/27/64)

- The impact of budgetary restraints upon the Nation's research and development programs was marked in industry, Government agencies, and the university scientific community, John Finney claimed in the *New York Times*. Employment in the aerospace industry dropped from 1,186,000 in August 1963 to 1,112,000 in September 1964, and manufacturers were forced to turn to civilian products to augment sales. Limiting of NASA funds forced postponement of some of the follow-up programs for Apollo, such as a manned orbiting laboratory. Universities whose research programs depended heavily on Federal research support were having difficulty in obtaining funds to complete projects already undertaken and could no longer expect liberal Government support for their research proposals. (Finney, *NYT*, 1/11/65)
- FAA reported record number of applicants for private pilot's licenses— 51,548, an increase of 18% over 1963 total of 44,032. FAA Administrator Najeeb E. Halaby attributed increased public interest in flying to "cooperative efforts on the part of Government and industry in making private flying safer, easier, and cheaper." (FAA Release 65-1)

During 1964: A milestone in ion engine development was achieved when heavy-coated barium oxide filaments operated for up to 5,000 hrs. in bell-shaped vacuum chamber, NASA Lewis Research Center announced on Dec. 31. Until now, filaments for producing ions in electric engines had had a life of only a few hundred hours. (*Lewis News,* 1/8/65, 2)

• Series of six Sandia Corp. Nike-Tomahawk sounding rockets was launched from May to August. The six Nike-Tomahawks, which reached altitudes of about 300 km. (186 mi.), were launched from the Island of Kania, as part of U.S. readiness program for use if space testing of nuclear devices was resumed. Experiment instrumentation for the series was provided by Los Alamos Scientific Laboratory scientists. Objective of the experiment was to obtain precise measurements of radiations in the sun's corona. Another purpose was to demonstrate feasibility of using relatively inexpensive, spin-stabilized, ballistically pointed rockets for solar x-ray spectroscopy. Results of this research were lated reported at the December meeting of the American Geophysical Union (AGU). (OAR *Review,* 12/64, 22)

During the space age: As of the close of 1964, NASA had placed 70 spacecraft into earth orbit, lunar trajectory, or solar orbit since the birth of NASA on Oct. 1, 1958. Of this number 52 were still orbiting as of Dec. 31, 1964, and 21 were still transmitting signals. Some 55 million data points were being received daily at GSFC, recorded, and stored on magnetic tape at the rate of about 100,000 reels per year. Three of the 70 spacecraft impacted on the moon; 15 re-entered the earth's atmosphere and burned up. Of the 31 silent satellites in orbit, 6 were in orbit about the sun, 25 about the earth. Satellites still transmitting meaningful data included one deep-space probe (MARINER IV), 5 communications satellites (plus the beacon on passive communications satellite ECHO II), 3 weather satellites, 10 scientific satellites. (NASA Release 65-1)

Appendix A

SATELLITES, SPACE PROBES, AND MANNED SPACE FLIGHTS

A CHRONICLE FOR 1964

The following tabulation was compiled from open sources by Dr. Frank W. Anderson, Jr., Assistant NASA Historian. Sources included the United Nations Public Registry, the *Satellite Situation Report* issued by the Space Operations Control Center at Goddard Space Flight Center, public information releases of the Department of Defense, NASA, and other agencies, and the *Report to the Congress from the President of the United States: United States Aeronautics and Space Activities, 1964.* Russian data are from the U.N. Public Registry, the *Satellite Situation Report*, translations from Tass News Agency statements in the Soviet press, and international news services' reports.

It might be well to call attention to the terms of reference stated or implied in the title of this tabulation. This is a listing of payloads that have (a) orbited, (b) as probes, ascended to at least the 4,000-mile altitude that traditionally has distinguished probes from sounding rockets, etc., or (c) conveyed one or more human beings in space. Furthermore, only flights that succeeded—or at least are not known to have failed—in doing one of the above are listed. Date of launch is referenced to local time at the launch site. An asterisk by the date marks those dates that are one day earlier in this tabulation than in listings which reference to Greenwich time.

In these categories the world space effort continued to gain momentum, with a total of 111 successfully launched payloads in a total of 90 launches. The U.S. program was up from 65 payloads in 44 launches in 1963 to 76 payloads in 60 launches in 1964 (DOD: 53 payloads in 38 launches; NASA: 23 payloads in 22 launches). The U.S.S.R. doubled its program from a numerical standpoint, with 35 payloads in 30 launches in 1964, as opposed to 17 payloads in 17 launches in 1963. It is interesting that in 1964 the U.S.S.R. blossomed forth with four multiple-payload launches, her first since February 1961.

As we have cautioned in previous years, the "Remarks" column of these appendixes is never complete because of the inescapable lag behind each flight of the analysis and interpretation of scientific results.

Launch Date	Name	International Designation	Vehicle	Payload Data	Apogee (st. mi.)	Perigee (st. mi.)	Period (minutes)	Inclination	Remarks
Jan. 11	DOD Spacecraft (United States)	1964-1A	Thor-Agena D	Total weight: Not available. Objective: Develop spaceflight techniques and technology. Payload: Not available.	582	563	103.5	69.92°	First known successful orbiting of 5 satellites with one launch vehicle; still in orbit.
	and GGSE	1964-1B		Total weight: Not available. Objective: Develop spaceflight techniques and technology. Payload: Not available.	585	560	103.5	69.96°	Still in orbit; still transmitting.
	and EGRS	1964-1C		Total weight: Not available. Objective: Develop spaceflight techniques and technology. Payload: Not available.	582	563	103.5	69.90°	Still in orbit; still transmitting.
	and SOLAR RADIATION	1964-1D		Total weight: Not available. Objective: Develop spaceflight techniques and technology. Payload: Not available.	582	563	103.5	69.89°	Still in orbit; still transmitting.
	and DOD Spacecraft	1964-1E		Total weight: Not available. Objective: Develop spaceflight techniques and technology. Payload: Not available.	591	555	103.5	69.88°	Still in orbit.
Jan. 19	DOD Spacecraft (United States)	1964-2B	Thor-Agena D	Total weight: Not available. Objective: Develop spaceflight techniques and technology. Payload: Not available.	512	501	101.2	99.04°	Both satellites launched with same booster. Still in orbit.
	and DOD Spacecraft	1964-2C		Total weight: Not available. Objective: Develop spaceflight techniques and technology. Payload: Not available.	514	501	101.2	99.06°	Still in orbit.
Jan. 21	RELAY II (United States)	1964-3A	Thor-Delta	Total weight: 184 lbs. Objective: Continue flight research in active repeater communications satellites, radiation damage, and reliability and duration of performance. Payload: 33″×29″ octagonal prism, tapered at one end, containing 2	4,606	1,298	194.7	46.3°	RELAY II was very similar to RELAY I (launched 12/13/62) which was still operating, with improved components and no cut-off timer, since part of objective of RELAY II was to determine durability. Less eccentric orbit provided longer on-station time;

Date	Satellite	Designation	Launch Vehicle	Description	Weight (lbs)	Apogee	Perigee	Inclination	Remarks
Jan. 25	ECHO II (United States)	1964-4A	Thor-Agena B	transponders, duplicated telemetry and command system, attitude control system, tracking beacon; temperature sensors; 8,215 n-on-p solar cells; 3 nickel-cadmium batteries; 5 external antennas. Total weight: 770 lbs., including accessories; 535 lbs. for balloon alone. Objective: Continue communications experiments with passive communications satellite; provide data on orbital environment. Payload: 135' sphere of mylar plastic bonded inside and out with aluminum alloy foil; 2 beacon transmitters and antennas; 8 solar cell modules; 16 nickel-cadmium batteries.	816	642	109	81.5°	excellent transmissions exchanged with Europe and Japan. Still in orbit, still transmitting. Balloon satellite inflated well, seemed to be rigid enough to hold spherical shape well. Most visible of all artifical satellites. Series of international communications experiments between U.S.-U.K.-U.S.S.R. were begun 2/22/64. Still in orbit.
Jan. 29	Saturn (United States)	1964-5A	Saturn I	Total weight: 37,700 lbs. (including S-IV stage and 11,500 lbs. of sand ballast), of which 20,000 lbs. was payload. Objective: Test structure and performance of 2-stage Saturn I; orbit 2d stage. Payload: 84' S-IV stage, with instrument unit, payload adapter, Jupiter nosecone, sand ballast; transmitter; 12 50-lb. batteries; 567 telemetry points and 7 telemetry systems.	467	162	94.8	31.4°	SA-5 was heaviest payload ever launched to that date; 5th straight Saturn success, 1st of the Block II (both stages live) tests; first flight test of the S-IV 2nd stage, which has six liquid hydrogen engines. Confirmed structural integrity of full-scale Saturn I. Still in orbit.
Jan. 30	ELECTRON I (U.S.S.R.)	1964-6A	Not available	Total weight: Not available. Objective: Make simultaneous study of earth's inner and outer radiation belts, cosmic rays, and the upper atmosphere. Payload: Not available.	4,410	252	269	60.89°	Both satellites launched by single booster. Still in orbit.
	and ELECTRON II	1964-6B		Total weight: Not available. Objective: Make simultaneous study of earth's inner and outer radiation belts, cosmic rays, and the upper atmosphere. Payload: Not available.	42,352	285	1,384	59.40°	Still in orbit.

448 ASTRONAUTICS AND AERONAUTICS, 1964

Launch Date	Name	International Designation	Vehicle	Payload Data	Apogee (st. mi.)	Perigee (st. mi.)	Period (minutes)	Inclination	Remarks
Jan. 30	RANGER VI (United States)	1964-7A	Atlas-Agena B	Total weight: 804 lbs. Objective: Take TV photographs of the moon from close range. Payload: 15'-wide and 10'4"-tall (cruise position, with solar panels extended) structure. Hexagonal base contains conical midcourse motor, retrorocket; other elements include command system, radio receiver and 3 transmitters, telemetry system, 4 batteries, 6 TV cameras, 9,792 solar cells, 2 antennas.					Launch was very successful; midcourse correction put spacecraft on impact course to moon. TV cameras failed to perform when turned on in last 15 min. of flight, probably shorted out by arcing. RANGER VI impacted on moon 2/2/64 in Sea of Tranquility, having failed to photograph the moon, as had previous Rangers.
Feb. 15	DOD Spacecraft (United States)	1964-8A	Thor-Agena D	Total weight: Not available. Objective: Develop spaceflight techniques and technology. Payload: Not available.	278	119	90.9	75.06°	Reentered 3/9/64.
Feb. 25	DOD Spacecraft (United States)	1964-9A	Atlas-Agena D	Total weight: Not available. Objective: Develop spaceflight techniques and technology. Payload: Not available.		Not available			Reentered 3/1/64.
Feb. 27	COSMOS XXV (U.S.S.R.)	1964-10A	Not available	Total weight: Not available. Objective: Continuation of Cosmos scientific satellite series. Payload: Not available.	321	169	92	49°	Reentered 11/21/64.
Feb. 28	DOD Spacecraft (United States)	1964-11A	Thor-Agena D	Total weight: Not available. Objective: Develop spaceflight techniques and technology. Payload: Not available.	319	302	95	82.05°	Still in orbit.
Mar. 11	DOD Spacecraft (United States)	1964-12A	Atlas-Agena D	Total weight: Not available. Objective: Develop spaceflight techniques and technology. Payload: Not available.	240	89	89.8	95.75°	Reentered 3/16/64.
Mar. 13	DOD Probe (United States)		Blue Scout, Jr.	Total weight: Not available. Objective: Measure collision of charged particles with earth's magnetic field. Payload: Particle detectors; batteries; telemetry system.	Orbit not intended				4th stage failed to ignite; instead of achieving planned altitude of 23,000 mi., payload impacted 2,000 mi. from launch site.
Mar. 18	COSMOS XXVI (U.S.S.R.)	1964-13A	Not available	Total weight: Not available. Objective: Continuation of Cos-	250	168	91	49°	Reentered 9/28/64.

ASTRONAUTICS AND AERONAUTICS, 1964

Date	Name	Designation	Launch vehicle	Payload/Objective	Weight (lb)	Apogee	Perigee	Period	Inclination	Remarks
Mar. 27	COSMOS XXVII (U.S.S.R.)	1964-14A	Not available	Total weight: Not available. Objective: Continuation of Cosmos scientific satellite series. Payload: Not available.		147	119	88.7	64.80°	Reentered 3/28/64.
Mar. 27	ARIEL II (U.K.–U.S.)	1964-15A	Scout	Total weight: 150 pounds. Objective: Measure ionospheric galactic radio noise, vertical distribution of ozone, and number and size of micrometeoroids. Payload: 35″ × 23″ (dia.) cylinder, plus 2 ferrite rod loop antennas, 130° dipole (55′ per side) antennas, and 4 solar paddles containing 5,400 solar cells; 22 nickel-cadmium batteries; telemetry system; transmitter-receiver; tape recorder; 3 scientific experiments measuring galactic noise, ozone, and micrometeoroids.		843	180	101	52°	Second in series of 3 U.K.–U.S. scientific satellites (ARIEL I was launched 4/26/62, contributed much knowledge on upper atmosphere). U.S. built the satellite and launch vehicle, launched, tracked, and acquired the data. U.K. built the scientific experiments and would interpret the data. Still in orbit; still transmitting.
Apr. 2	ZOND I (U.S.S.R.)	1964-16D	Not available	Total weight: Not available. Objective: Develop a space system for "distant interplanetary flights." Payload: Not available.			In heliocentric orbit.			Still in heliocentric orbit. Identified unofficially by U.S. as Venus probe. Performed successful midcourse maneuver 5/14/64, when 13,000,000 km. from earth; telemetered interplanetary radiation data and engineering data. Should have passed Venus in latter half of July 1964, but Soviet silence at and since that time suggests communications or instrument failure.
Apr. 4	COSMOS XXVIII (U.S.S.R.)	1964-17A	Not available	Total weight: Not available. Objective: Continuation of Cosmos scientific satellite series. Payload: Not available.		245	130	90.38	65°	Reentered 4/12/64.
Apr. 8	Gemini (United States)	1964-18A	Titan II	Total weight: 7,000 lbs. Objective: Test in unmanned flight the structural integrity of Gemini spacecraft and its compatibility with Titan II launch vehicle. Payload: 11′ × 7½′-bell-shaped spacecraft, with telemetry, batteries.		204	99.6	89.27		First flight (G T–1) of flight-rated Gemini spacecraft was highly successful, only surprise being 14 mph excess speed (17,534 mph instead of 17,520 mph) which made orbit 21 mi. higher than planned. Spacecraft and attached Titan II 2nd stage reentered 4/12/64.

Launch Date	Name	International Designation	Vehicle	Payload Data	Apogee (st. mi.)	Perigee (st. mi.)	Period (minutes)	Inclination	Remarks
Apr. 12	POLET II (U.S.S.R.)	1964-19B	Not available	Total weight: Not available. Objective: "Further improve space vehicles capable of maneuvering extensively in all directions, and to work out questions concerning the problem of space rendezvous." Payload: Not available.	311	193	92.4	58.06°	Orbital data are those at completion of maneuvers. Soviets claimed substantial change in orbital inclination was achieved. Reentered 5/30-6/1/64.
Apr. 23	DOD Spacecraft (United States)	1964-20A	Atlas-Agena D	Total weight: Not available. Objective: Develop spaceflight techniques and technology. Payload: Not available.		Not available.			Reentered 4/28-29/64.
Apr. 25	COSMOS XXIX (U.S.S.R.)	1964-21A	Not available	Total weight: Not available. Objective: Continuation of Cosmos scientific satellite series. Payload: Not available.	192	127	89.52	65.07°	Reentered 5/2/64.
Apr. 27	DOD Spacecraft (United States)	1964-22A	Thor-Agena D	Total weight: Not available. Objective: Develop spaceflight techniques and technology. Payload: Not available.	277	109	90.8	79.97°	Reentered 5/26/64.
May 18	COSMOS XXX (U.S.S.R.)	1964-23A	Not available	Total weight: Not available. Objective: Continuation of Cosmos scientific satellite series. Payload: Not available.	238	128	90.24	64.77°	Reentered 5/26/64.
May 19	DOD Spacecraft (United States)	1964-24A	Atlas-Agena D	Total weight: Not available. Objective: Develop spaceflight techniques and technology. Payload: Not available.		Not available.			Reentered 5/22/64.
May 28	Saturn (United States)	1964-25A	Saturn I	Total weight: 37,300 lbs. Objective: Continue qualification of 2-stage Saturn vehicle, orbit first boilerplate Apollo spacecraft. Payload: 80' S-IV stage, with instrument unit and boilerplate Apollo command and service modules; transmitter; 10 telemetry systems; batteries.	149	114	88.6		Flight (SA-6) was excellent; one of booster engines in 1st stage cut off 23 sec. early but guidance system compensated. Orbiting S-IV stage and Apollo spacecraft reentered 5/31/64.

Date	Name	Designation	Launch vehicle	Description					Remarks
Jun. 4	DOD Spacecraft (United States)	1964-26A	Scout	Total weight: Not available. Objective: Develop spaceflight techniques and technology. Payload: Not available.	592	533	103.1	90.49°	Still in orbit.
Jun. 5	DOD Spacecraft (United States)	1964-27A	Thor-Agena D	Total weight: Not available. Objective: Develop spaceflight techniques and technology. Payload: Not available.	239	86	89.1	80.10°	Reentered 6/18/64.
Jun. 6	COSMOS XXXI (U.S.S.R.)	1964-28A	Not available	Total weight: Not available. Objective: Continuation of Cosmos scientific satellite series. Payload: Not available.	309	142	91.6	49°	Reentered 10/20/64.
Jun. 10	COSMOS XXXII (U.S.S.R.)	1964-29A	Not available	Total weight: Not available. Objective: Continuation of Cosmos scientific satellite series. Payload: Not available.	207	131	89.78	50.28°	Reentered 6/18/64.
Jun. 13	DOD Spacecraft (United States)	1964-30A	Thor-Agena D	Total weight: Not available. Objective: Develop spaceflight techniques and technology. Payload: Not available.	225	219	91.7	114.98°	Still in orbit.
Jun. 18	DOD Spacecraft and DOD Spacecraft (United States)	1964-31A 1964-31B	Thor-Agena D	Total weight: Not available. Objective: Develop spaceflight techniques and technology. Payload: Not available.	521 523	516 515	101.6 101.6	99.83° 99.83°	One of 2 spacecraft orbited by single booster. Still in orbit. Still in orbit.
Jun. 19	DOD Spacecraft (United States)	1964-32A	Thor-Agena D	Total weight: Not available. Objective: Develop spaceflight techniques and technology. Payload: Not available.	256	113	90.6	85.02°	Reentered 8/16/64.
Jun. 23	COSMOS XXXIII (U.S.S.R.)	1964-33A	Not available	Total weight: Not available. Objective: Continuation of Cosmos scientific satellite series. Payload: Not available.	182	130	89.38	65°	Reentered 7/1/64.
Jul. 1	COSMOS XXXIV (U.S.S.R.)	1964-34A	Not available	Total weight: Not available. Objective: Continuation of Cosmos scientific satellite series. Payload: Not available.	223	127	90	64.97°	Reentered 7/9/64.
Jul. 2	DOD Spacecraft (United States)	1964-35A	Thor-Agena D	Total weight: Not available. Objective: Develop spaceflight techniques and technology. Payload: Not available.	329	310	94.9	82.08°	Still in orbit.

Launch Date	Name	International Designation	Vehicle	Payload Data	Apogee (st. mi.)	Perigee (st. mi.)	Period (minutes)	Inclination	Remarks
Jul. 6	DOD Spacecraft (United States) and DOD Spacecraft	1964-36A	Atlas-Agena D	Total weight: Not available. Objective: Develop spaceflight techniques and technology. Payload: Not available.		Not available.			One of 2 spacecraft orbited with single booster; reentered 7/8/64.
		1964-36B		Total weight: Not available. Objectives: Develop spaceflight techniques and technology. Payload: Not available.	244	183	91.4	92.99°	Still in orbit.
Jul. 10	DOD Spacecraft (United States)	1964-37A	Thor-Agena D	Total weight: Not available. Objective: Develop spaceflight techniques and technology. Payload: Not available.	282	112	90.9	84.99°	Reentered 8/6/64.
Jul. 10	ELECTRON III (U.S.S.R.) and ELECTRON IV	1964-38A	Not available	Total weight: Not available. Objective: Simultaneous study of the earth's inner and outer radiation belts, cosmic rays, and the upper atmosphere. Payload: Not available.	4,365	251	168	60.87°	Soviet and press sources gave launch date as 7/11/64. One of 2 satellites orbited by single booster. Still in orbit.
		1964-38B		Total weight: Not available. Objective: Simultaneous study of the earth's inner and outer radiation belts, cosmic rays, and the upper atmosphere. Payload: Not available.	41,076	285	1,314	60.87°	Still in orbit.
Jul. 15	COSMOS XXXV (U.S.S.R.)	1964-39A	Not available	Total weight: Not available. Objective: Continuation of Cosmos scientific satellite series. Payload: Not available.	161	130	89.2	51.30°	Reentered 7/23/64.
Jul. 17	DOD Spacecraft (United States) and DOD Spacecraft	1964-40A	Atlas-Agena D	Total weight: Not available. Objective: Develop spaceflight techniques and technology. Payload: Not available.	64,957	63,442	6,023	39.59°	One of 3 spacecraft launched with single booster. Still in orbit.
		1964-40B		Total weight: Not available. Objective: Develop spaceflight techniques and technology. Payload: Not available.	69,445	58,652	6,007	40.88°	Still in orbit.

	TRS II and	1964-40C			65,036	135	2,366	36.73°	Still in orbit; still transmitting.
Jul. 28	RANGER VII (United States)	1964-41A	Atlas-Agena B	Total weight: 4.5 lbs. Objective: Measure radiation levels in Van Allen belts. Payload 9" (per side) tetrahedron; proton and electron detectors; solar cells; telemetry system.					
				Total weight: 806 lbs. Objective: Take TV photographs of the moon from close range, enabling identification of items 10 times smaller than identifiable from earth. Payload: 15'-wide and 104"-tall (cruise position, with solar panels extended) structure. Hexagonal base contains conical midcourse motor, retrorocket; other elements include command system, 1 radio receiver and 3 transmitters, telemetry system, 4 batteries, 6 TV cameras, 9,792 solar cells, 2 antennas.					RANGER VII transmitted to earth 4,316 high-quality photos of the moon's Sea of Clouds area prior to impacting on moon 7/31/64 after excellent flight. Photo time was last 17 min. of flight, from about 1,300 mi. from moon down to 1,600 ft., and improved resolution over telescopic photos from earth by factor of 1,000. Preliminary analysis of photos argues that surface of moon is smoother than expected, covered with no more than thin porous layer (1 ft.) over solid material, all this being suitable for manned landing and exploration.
Jul. 30	COSMOS XXXVI (U.S.S.R.)	1964-42A	Not available	Total weight: Not available. Objective: Continuation of Cosmos scientific satellite series. Payload: Not available.	300	155	91.9	49°	Still in orbit.
Aug. 5	DOD Spacecraft (United States)	1964-43A	Thor-Agena D	Total weight: Not available. Objective: Develop spaceflight techniques and technology. Payload: Not available.	262	113	90.5	79.97°	Reentered 8/31-9/1/64.
Aug. 14	COSMOS XXXVII (U.S.S.R.)	1964-44A	Not available	Total weight: Not available. Objective: Continuation of Cosmos scientific satellite series. Payload: Not available.	186	127	89.45	65°	Reentered 9/22/64.
Aug. 14	DOD Spacecraft (United States)	1964-45A	Atlas-Agena D	Total weight: Not available. Objective: Develop spaceflight techniques and technology. Payload: Not available.	193	97	88.92	95.50°	One of 2 spacecraft orbited by single booster. Reentered 9/23/64.
	and DOD Spacecraft	1964-45B		Total weight: Not available. Objective: Develop spaceflight techniques and technology. Payload: Not available.	2,333	163	127.39	95.60°	Still in orbit.

Launch Date	Name	International Designation	Vehicle	Payload Data	Apogee (st. mi.)	Perigee (st. mi.)	Period (minutes)	Inclination	Remarks
Aug. 18	COSMOS XXXVIII (U.S.S.R.) and COSMOS XXXIX and COSMOS XL	1964-46A 1964-46B 1964-46C	Not available	Total weight: Not available. Objective: Continuation of Cosmos scientific satellite series. Payload: Not available. Total weight: Not available. Objective: Continuation of Cosmos scientific satellite series. Payload: Not available. Total weight: Not available. Objective: Continuation of Cosmos scientific satellite series. Payload: Not available.	544 544 544	130 130 130	95.2 95.2 95.2	56.12° 56.12° 56.12°	One of 3 satellites launched with single booster "of a new type". Reentered 11/8/64. Reentered 11/17/64. Reentered 11/18/64.
Aug. 19	SYNCOM III (United States)	1964-47A	Thrust-Augmented Delta	Total weight: 145 lbs. before firing of apogee-kick motor; after, about half of that. Objective: Place satellite in synchronous equatorial orbit and conduct communications experiments. Payload: 28" (dia.)×15¼" cylinder, plus apogee-kick motor and antennas; 3,840 n-on-p solar cells, 2 transmitter-receivers, telemetry, hydrogen-peroxide tanks and jets for control.	Initial: 23,675 Synchronous: (9/23/64) 22,311 22,164	695 1,436	695 1,436	16° .095°	First satellite to achieve true synchronous orbit, first satellite launch by the uprated Delta booster. After launch came weeks of minor maneuvers until by 9/23/64 SYNCOM III was in almost perfect stationary position above equator and International Date Line. Communications were very successful, including 15 days of transpacific transmission of Olympic Games from Tokyo, beginning 10/7/64. Still in orbit; still transmitting.
Aug. 21	DOD Spacecraft (United States)	1964-48A	Thor-Agena D	Total weight: Not available. Objective: Develop spaceflight techniques and technology. Payload: Not available.	218	206	91.4	114.98°	Still in orbit.
Aug. 22	COSMOS XLI (U.S.S.R.)	1964-49A	Not available	Total weight: Not available. Objective: Continuation of Cosmos scientific satellite series. Payload: Not available.	24,775	245	715	64°	Reentered 9/15/64.

Date	Name	Designation	Launch Vehicle	Payload/Objective	Weight (lbs)	Apogee	Perigee	Inclination	Remarks
Aug. 22	COSMOS XLII (U.S.S.R.) and COSMOS XLIII	1964-50A	Not available	Total weight: Not available. Objective: Continuation of Cosmos scientific satellite series. Payload: Not available.	683	144	97.8	49°	One of 2 satellites launched with single booster. Still in orbit.
		1964-50C		Total weight: Not available. Objective: Continuation of Cosmos scientific satellite series. Payload: Not available.	683	144	97.8	49°	Still in orbit.
Aug. 25	EXPLORER XX (United States)	1964-51A	Scout	Total weight: 97 lbs. Objective: Map irregularities in topside of earth's atmosphere, measure cosmic noise and ion population. Payload: 46¼" long × 26" dia. cylinder with conical ends plus 4"- dia. ball-shaped ion mass-spectrometer on a 10½' boom on one end and 1 set of 122' antennas and 2 sets of 62' antennas; 6 radar sets; transmitter-receiver; tracking beacon; 2,400 solar cells; nickel-cadmium batteries.	625	540	104	80°	Is providing better understanding of F-2 layer of ionosphere and its effect on long-range radio communication. Still in orbit; still transmitting.
Aug. 28	NIMBUS I (United States)	1964-52A	Thor-Agena B	Total weight: 830 lbs. Objective: Demonstrate placement and stabilization of meteorological satellite in a precise, continuous, earth-pointing orbit; evaluate new camera system and radiometer; and provide improved cloud photos. Payload: 10'-high structure with hexagonal control section on top, connected by a truss structure to wheel-like sensory ring and flanked by 2 8' × 2'9" solar paddles containing 10,500 solar cells; 161 nickel-cadmium batteries; active 3-axis control system; 3 vidicon cameras; tape recorder; APT camera system; radiometer; transmitter.	578	252	98	81°	Orbit more elliptical than intended because Agena 2nd stage had short 2nd burn; photos were best cloud cover shots ever received from space; infrared information on night time cloud cover was much more detailed than any previous experiment. On 9/23/64 NIMBUS I ceased transmitting when solar paddles locked, this after 27,000 excellent photos had been transmissions 12/8/64, but satellite tumbling makes photos unusable. Still in orbit.
Aug. 28	COSMOS XLIV (U.S.S.R.)	1964-53A	Not available	Total weight: Not available. Objective: Continuation of Cosmos scientific satellite series. Payload: Not available.	534	384	99.5	65°	Still in orbit.
Aug. 28	DOD Probe (United States)		Blue Scout, Jr.	Not available.		Orbit not intended.			No information available.

Launch Date	Name	International Designation	Vehicle	Payload Data	Apogee (st. ml.)	Perigee (st. ml.)	Period (minutes)	Inclination	Remarks
Sep. 4	OGO I (United States)	1964-54A	Atlas-Agena B	Total weight: 1,073 lbs. Objective: In highly eccentric orbit, make scientific measurements and observations in the earth's atmosphere, magnetosphere, and in interplanetary space beyond the earth's magnetic field. Payload: 59'×50' (cruise position, with antennas and solar panels extended) structure, central portion a rectangular box from which 7 booms extend; 20 scientific experiments; 3 transmitters; 2 receivers; 2 tape recorders; 3 beacons; 32,250 p-on-n solar cells; 2 nickel-cadmium batteries.	92,721	175	3,750	31°	Orbit of geophysical observatory was very close to plan, but 2 booms failed to deploy, interfering with control system's attempt to lock on earth. All 20 experiments functioned at least partially; on 9/10/64 a maneuver succeeded in reorienting solar panels toward sun, providing good power level, but 6 experiments were crippled by non-deployed booms. Still in orbit; still transmitting.
Sep. 13	COSMOS XLV (U.S.S.R.)	1964-55A	Not available	Total weight: Not available. Objective: Continuation of Cosmos scientific satellite series. Payload: Not available.	203	126	89.69	64.87°	Reentered 9/23/64.
Sep. 14	DOD Spacecraft (United States)	1964-56A	Thor-Agena D	Total weight: Not available. Objective: Develop space-flight techniques and technology. Payload: Not available.	286	119	90.81	84.95°	Reentered 10/6/64.
Sep. 18	Saturn (United States)	1964-57A	Saturn I	Total weight: 36,700 lbs. Objective: Continue qualification of 2-stage Saturn I vehicle and of boilerplate Apollo spacecraft. Payload: 80' S-IV stage, with instrument unit and boilerplate Apollo command and service modules; transmitter; 10 telemetry systems; batteries.	145	112	88.6		Flight (SA-7) was excellent, resulted in Saturn I being considered operational after 7 flights instead of the planned 10. Reentered 9/22/64.
Sep. 23	DOD Spacecraft (United States)	1964-58A	Atlas-Agena D	Total weight: Not available. Objective: Develop space-flight techniques and technology. Payload: Not available.		Not available			Reentered 9/28/64.
Sep. 24	COSMOS XLVI (U.S.S.R.)	1964-59A	Not available	Total weight Not available. Objective: Continuation of Cosmos scientific satellite series. Payload: Not available.	168	134	89.2	51.30°	Reentered 10/2/64.

Date	Name	Designation	Vehicle	Description	Weight (lbs)	Perigee (mi)	Apogee (mi)	Period (min)	Inclination	Remarks
pt. 4*	EXPLORER XXI (United States)	1964-60A	Thor-Delta	Total weight: 136 lbs. Objective: In very eccentric orbit, study charged particles from the sun and from sources outside the solar system, study magnetosphere. Payload: 28″ (dia.)×8″ octagonal structure, plus 6′ boom with rubidium-vapor magnetometer on end, 2 flux gate magnetometers deployed on 7′ booms, 4 solar panels; solar cells; 13 silver-cadmium batteries; transmitter.	59,400	120	2,097	33.5°		Partial failure of X-258 third stage of Delta caused insufficient apogee of 59,400 mi., less than half of intended 126,500 mi.; would provide very little data on intended area beyond the earth's atmosphere (40,000 mi.+). Still in orbit; still transmitting.
Oct. 5*	DOD Spacecraft (United States)	1964-61A	Thor-Agena D	Total weight: Not available. Objective: Develop spaceflight techniques and technology. Payload: Not available.	242	109	90.4	79.97°		Reentered 10/26/64.
Oct. 6	COSMOS XLVII (U.S.S.R.)	1964-62A	Not available	Total weight: Not available. Objective: Continuation of Cosmos scientific satellite series. Payload: Not available.	256	110	90	64.77°		Orbital data were nearly identical to those of VOSKHOD I (10/12/64), may have been preparatory flight. Reentered 10/7/64.
Oct. 6	DOD Spacecraft (United States)	1964-63B	Thor-Able-Star	Total weight: Not available. Objective: Develop spaceflight techniques and technology. Payload: Not available.	673	657	106.6	89.93°		One of 3 satellites put into polar orbit by single booster. Still in orbit.
	and DOD Spacecraft	1964-63C		Total weight: Not available. Objective: Develop spaceflight techniques and technology. Payload: Not available.	674	655	106.6	89.93°		Still in orbit.
	and DOD Spacecraft	1964-63E		Total weight: Not available. Objective: Develop spaceflight techniques and technology. Payload: Not available.	673	657	106.6	89.98°		Still in orbit.
Oct. 10	EXPLORER XXII (United States)	1964-64A	Scout	Total weight: 116 lbs. Objective: Study electron behavior and population in the ionosphere, relate ionospheric behavior to solar radiation which causes ionization affecting long-range communications, study geometry and irregularities of the ionosphere, and perform experiments in geodesy. Payload: 18″×12″ octagonal satellite, with 4 solar panels extending like windmill blades; 4 radio transmitters, 4 antennas; magnetometer; nickel-cadmium batteries; 360 1-in. glass-prism reflectors; 2 bar magnets.	669	549	104.7	80°		Satellite was transmitting continuous radio signals down through ionosphere to network of 80 ground stations being operated by 50 scientific groups in 32 countries, largest international effort to date on a space project. On 10/13/64 laser beam aimed at EXPLORER XXII received answering light reflection in geodetic experiment to enable very precise measurement of the satellite's orbit. Still in orbit; still transmitting.

Launch Date	Name	International Designation	Vehicle	Payload Data	Apogee (st. mi.)	Perigee (st. mi.)	Period (minutes)	Inclination	Remarks
Oct. 12	VOSKHOD I (U.S.S.R.)	1964-65A	Not available	Total weight: 11,525 lbs. (5,320 kg.) Objective: Test new multiseat manned spacecraft; investigate under flight conditions the capability and interrelationship of specialist cosmonauts; conduct scientific physiotechnical research under flight conditions; continue studies of spaceflight effects on human organisms; and conduct expanded biomedical investigations under conditions of prolonged spaceflight. Payload: Vostok-type spacecraft with additional life-support equipment; TV; telemetry; new control system; bioscience experiments.	254	110	90.1	65°	First spaceflight of 3-man crew launched from Baikonur, Kazakhstan. Cosmonauts were: pilot-cosmonaut Engr. Col. Vladimir M. Komarov; scientific co-worker cosmonaut Konstantin P. Feoktistov; and physician-cosmonaut Boris B. Yegorov. Spacecraft transmitted TV photos showing crew working in light wool suits instead of spacesuits. VOSKHOD I reentered 10/13/64 and landed near target in U.S.S.R. after 24 hrs, 17 min. of flight (16 orbits). Postflight reports by Russian scientists indicated all 3 cosmonauts had shown effects of weightlessness, especially the scientist and physician.
Oct. 14	COSMOS XLVIII (U.S.S.R.)	1964-66A	Not available	Total weight: Not available. Objective: Continuation of Cosmos scientific satellite series. Payload: Not available.	188	126	89.4	65.67°	Reentered 10/20/64.
Oct. 17	DOD Spacecraft (United States)	1964-67A	Thor-Agena D	Total weight: Not available. Objective: Develop spaceflight techniques and technology. Payload: Not available.	192	109	89.9	74.99°	Reentered 11/4/64.
Oct. 23	DOD Spacecraft (United States)	1964-68A	Atlas-Agena D	Total weight: Not available. Objective: Develop spaceflight techniques and technology. Payload: Not available.		Not available			One of 3 satellites orbited by single booster. Reentered 10/28/64.
	and DOD Spacecraft	1964-68B		Total weight: Not available. Objective: Develop spaceflight techniques and technology. Payload: Not available.	214	193	91.1	95.50°	Still in orbit.
	and DOD Spacecraft	1964-68D		Total weight: Not available. Objective: Develop spaceflight techniques and technology. Payload: Not available.		Not available			Reentered 10/29/64.

Date	Name	Designation	Launch vehicle	Payload/Objective				Remarks
Oct. 24	COSMOS XLIX (U.S.S.R.)	1964-69A	Not available.	Total weight: Not available. Objective: Continuation of Cosmos scientific satellite series. Payload: Not available.	304	162	91.83	49°	Still in orbit.
Oct. 28	COSMOS L (U.S.S.R.)	1964-70A	Not available.	Total weight: Not available. Objective: Continuation of Cosmos scientific satellite series. Payload: Not available.	150	122	88.7	51.30°	Reentered 11/5/64.
Nov. 2	DOD Spacecraft (United States)	1964-71A	Thor-Agena D	Total weight: Not available. Objective: Develop spaceflight techniques and technology. Payload: Not available.	262	114	90.7	79.96°	Reentered 11/28/64.
Nov. 4	DOD Spacecraft (United States)	1964-72A	Thor-Agena D	Total weight: Not available. Objective: Develop spaceflight techniques and technology. Payload: Not available.	327	318	95	82.04°	Still in orbit.
Nov. 5	MARINER III (United States)	1964-73A	Atlas-Agena D	Total weight: 575 lbs. Objective: Investigate interplanetary space between orbits of earth and Mars; perform experiments in vicinity of Mars. Payload: 9'6''-high by 22'7½'' (cruise position, with solar panels and pressure vanes deployed)-structure; basically a 50'' octagon with 46'x3' solar panels extended horizontally and 4 solar pressure vanes extending beyond the panel tips; 28,224 solar cells; 33-lb. silver-zinc battery; 6 experiments to measure interplanetary phenomena; TV camera; transmitters; control system; 2 antennas extend vertically from octagon.		Heliocentric orbit.			Satellite was placed in interplanetary orbit successfully—even though Agena D was 4 sec. short on its second burn—but communications failed permanently when fairing did not jettison, prevented solar panel deployment. This was first NASA use of Agena D 2d stage. Still in orbit.
Nov. 6	EXPLORER XXIII (United States)	1964-74A	Scout	Total weight: 295 lbs. Objective: Investigate meteoroid penetration, solar cell degradation. Payload: 24'' (dia.)×92'' cylinder, containing 4 types of meteoroid-penetration detectors; 2 test groups of solar cells, 1 group protected, one not; 2 transmitters; beacon; 2 nickel-cadmium batteries.	614	286	99	52°	Information expected to be of use in Apollo design. Still in orbit, still transmitting.
Nov. 18	DOD Spacecraft (United States)	1964-75A	Thor-Agena D	Total weight: Not available. Objective: Develop spaceflight techniques and technology. Payload: Not available.	193	111	89.4	70.2°	Reentered 12/6/64.

Launch Date	Name	International Designation	Vehicle	Payload Data	Apogee (st. mi.)	Perigee (st. mi.)	Period (minutes)	Inclination	Remarks
Nov. 21	EXPLORER XXIV (United States)	1964-76A	Scout	Total weight: 19 lbs. Objective: Orbit 2 satellites with single launch vehicle to gather data from same portion of space on (1) air density, (2) radiation, and (3) relationship between the two. Payload: 12' polkadotted sphere, consisting of mylar balloon, with beacon, 260 solar cells, and 10 nickel-cadmium batteries.	1,552	327	116	81°	Data would also be used with those from EXPLORER XIX to provide global air density profile. Still in orbit, still transmitting.
	and EXPLORER XXV	1964-76B		Total weight: 90 lbs. Objective: Orbit 2 satellites with single launch vehicle to gather data from same portion of space on (1) air density, (2) radiation, and (3) relationship between the two. Payload: 2' (dia.) spherical polyhedron, with 3 booms; 5 omnidirectional radiation sensors; 11 directional radiation sensors; magnet; transmitter; telemetry; taperecorder; 2,640 solar cells; 19 nickel-cadmium batteries.	1,548	327	116	81°	Injun satellite would measure flux of corpuscular radiation into the atmosphere and sample concentration and distribution of charged particles in vicinity of its orbit. Still in orbit; still transmitting.
Nov. 28	MARINER IV (United States)	1964-77A	Atlas-Agena D	Total weight: 575 lbs. Objective: Investigate interplanetary space between orbits of earth and Mars; perform experiments in vicinity of Mars. Payload: 9'6"-high by 22'7½" (cruise position, with solar panels and pressure vanes deployed)-structure; basically a 50" octagon with 4 6'×3' solar panels extended horizontally and 4 solar pressure vanes extending beyond the panel tips; 28,224 solar cells; 33-lb. silver-zinc battery; 6 experiments to measure interplanetary phenomena; TV camera; transmitters; control system; 2 antennas extend vertically from octagon.	Heliocentric orbit.				Placed in excellent interplanetary orbit, with all systems functioning well; trajectory would have caused MARINER IV to pass 151,000 mi. ahead of Mars on 7/16/65; on 12/5/64 a midcourse correction maneuver successfully altered course so spacecraft would pass 5,400 mi. behind Mars on 7/14/35 after 325,000,000-mi. flight; on 12/7/34 the solar plasma probe failed but other 5 interplanetary sensors continued to return data. In vicinity of Mars an attempt would be made to take 22 TV photos of Mars and use radio waves in occultation experiment to determine composition of Mars atmosphere. Still in orbit; still transmitting.

ASTRONAUTICS AND AERONAUTICS, 1964

Date	Spacecraft	Designation	Launch vehicle	Description	Weight (lbs)	Apogee	Perigee	Inclination	Remarks
Nov. 30	ZOND II (U.S.S.R.)	1964-78O	Not available	Total weight: Not available. Objective: Test space probe systems in prolonged space flight; carry out scientific investigations in space. Payload: Not available.		In heliocentric orbit.			Launched towards Mars 2 days after MARINER IV launch; on 12/1/64 Soviet reports said communications from satellite were only half of power expected. Still in heliocentric orbit. Still transmitting.
Dec. 4	DOD Spacecraft (United States)	1964-79A	Atlas-Agena D	Total weight: Not available. Objective: Develop space-flight techniques and technology. Payload: Not available.		Not available.			Reentered 12/5/64.
Dec. 10	COSMOS LI (U.S.S.R.)	1964-80A	Not available	Total weight: Not available. Objective: Continuation of Cosmos scientific satellite series. Payload: Not available.	344	164	92.5	48°	Still in orbit.
Dec. 10	DOD Spacecraft (United States)	1964-81A	Titan III-A	Total weight: 9,000 lbs.: 5,250 lb. for transtage; 3,750 lbs. for dummy satellite. Objective: Develop spaceflight techniques and technology. Payload: Transtage plus dummy satellite.	130	111	100		First complete success of Titan III transtage (a 3rd stage with multiple restart capability); midway in 1st orbit transtage performed complete somersault to align guidance system; at end of 1st orbit, expelled dummy satellite into separate orbit. Reentered 12/13/64.
Dec. 11	Centaur (United States)	1964-82A	Atlas-Centaur	Total weight: 6,500 lbs., including 2,100-lb. dummy model of Surveyor spacecraft. Objective: Primary objectives to test structural integrity of total system and test guidance system; secondary objectives to attempt various maneuvers, including restart of Centaur engines. Payload: 46'X10' Centaur stage, including dummy Surveyor spacecraft; all-inertial guidance system; telemetry; beacon signal transponder; batteries.	106	100	87.79	30.67°	Fourth test of liquid-hydrogen Centaur achieved all six primary objectives, orbited Centaur stage and its dummy payload. In orbit, Centaur stage began tumbling, preventing fuel from reaching engines for restart test. Centaur stage reentered after 15 hrs., 12/12/64.
Dec. 13	DOD Spacecraft (United States)	1964-83C	Thor-Able-Star	Total weight: Not available. Objective: Develop spaceflight techniques and technology. Payload: Not available.	673	638	106.3	89.99°	One of 2 satellites orbited with single booster. Still in orbit.
	and DOD Spacecraft	1964-83D		Total weight: Not available. Objective: Develop spaceflight techniques and technology. Payable: Not available.	674	637	106.3	89.97°	Still in orbit.

Launch Date	Name	International Designation	Vehicle	Payload Data	Apogee (st. mi.)	Perigee (st. mi.)	Period (minutes)	Inclination	Remarks
Dec. 15	SAN MARCO I (Italian–U.S.)	1964-84A	Scout	Total weight: 254 lbs. Objective: Train Italian crew in preparation for later sea-platform launches; qualify satellite; measure ionospheric air density and study ionospheric interference with long-range radio transmissions. Payload: 26" spherical satellite consisting of a heavy sphere contained in a much lighter one as a means of measuring air density; instruments for radio-interference experiments; telemetry; batteries.	490	125	94.7	37.77°	First satellite built and instrumented in Western Europe; first satellite launching from U.S. by non-U.S. crew. Still in orbit.
Dec. 19	DOD Spacecraft (United States)	1964-85A	Thor-Agena D	Total weight: Not available. Objective: Develop spaceflight techniques and technology. Payload: Not available.	236	114	90.3	74.99°	Still in orbit.
Dec. 21	EXPLORER XXVI (United States)	1964-86A	Thor-Delta	Total weight: 101 lbs. Objective: Provide data on how high-energy particles are injected, trapped, and eventually lost in the earth's radiation belts. Payload: 17"×27" octagonal satellite, plus external extensions of 4 solar panels, 4 antennas, and 34" tube mounting flux-gate magnetometer; electron-proton angular distribution and energy detectors, directional detectors; ion-electron detector; solar-cell-damage experiment; telemetry; 13 silver-cadmium batteries.	16,290	193	457.7	20.15°	Fourth NASA energetic particles satellite (others were EXPLORERS XII, XIV, and XV). Information gained would be important to Project Apollo in design of spacecraft shielding and planning trajectories to the moon. Still in orbit.
Dec. 21	DOD Spacecraft (United States)	1964-87A	Thor-Agena D	Total weight: Not available. Objective: Develop spaceflight techniques and technology. Payload: Not available.	155	143	89.4	70.09°	Still in orbit.
Dec. 21	DOD Probe (United States)		Blue Scout, Jr.	Total weight: Not available. Objective: Develop spaceflight techniques and technology. Payload: Not available.	Orbit not intended.				Press reports claimed cesium ion engine made 2nd successful flight.

During 1964	DOD Spacecraft (United States)	Thor-Able-Star	Total weight: Not available. Objective: Develop space-flight techniques and technology. Payload: Not available.	Not available.	Pickaback payload, failed to separate from parent payload (parent payload previously reported).

Appendix B

CHRONOLOGY OF MAJOR NASA LAUNCHINGS

JANUARY 1, 1964, THROUGH DECEMBER 31, 1964

This chronology of major NASA launchings in 1964 is intended to provide an accurate and ready historical reference, one compiling and verifying information previously scattered over several sources. It includes launchings of all rocket vehicles larger than sounding rockets launched either by NASA or under "NASA direction" (e.g., NASA provided vehicle, launch facilities, and tracking facilities to the Italian crew who launched SAN MARCO I).

An attempt has been made to classify the performance of both the launch vehicle and the payload and to summarize total results in terms of primary mission. Three categories have been used for vehicle performance and mission results—successful (S), partially successful (P), and unsuccessful (U). A fourth category, unknown (Unk), has been provided for payloads where vehicle malfunctions did not give the payload a chance to exercise its main experiments. These divisions are necessarily arbitrary, since many of the results cannot be neatly categorized. Also they ignore the fact that a great deal was learned from shots that may have been classified as unsuccessful.

A few unique items require separate treatment. Their dates have been kept in sequence but their history has been relegated to footnotes.

Date of launch is referenced to local time at the launch site. A double asterisk by the date marks those dates that are one day earlier in this tabulation than in listings which reference to Greenwich time. Abbreviations of two launch sites change in midyear with the establishment of the USAF's National Range Div. (5/15/64):

Previous
- AMR—Atlantic Missile Range
- PMR—Pacific Missile Range
- WS—Wallops Station
- WSMR—White Sands Missile Range

New
- ETR—Eastern Test Range
- WTR—Western Test Range

Sources used were all open ones, verified where in doubt from the project offices in NASA Hq. and from the NASA Centers. For further information on each item, see Appendix A of this volume and the entries in the main chronology as referenced in the index. Prepared January 1965 by Dr. Frank W. Anderson, Jr., Assistant NASA Historian (APPH).

Date	Name (NASA Code)	General Mission	Launch Vehicle (Site)	Performance			Remarks
				Vehicle	Payload	Mission	
1964 Jan 21	RELAY II (A-16)	Communications earth satellite	Thor-Delta (AMR)	S	S	S	Continued experimentation with medium-altitude active communications satellite; TV test patterns excellent; featured radiation-resistant n-on-p solar cells.
Jan 25	ECHO II (A-12)	Communications earth satellite	Thor-Agena B (PMR)	S	S	S	135-ft. rigidized balloon orbited as passive communications satellite. First U.S.-U.S.S.R. joint space experiments conducted with ECHO II, beginning 2/22/64.
Jan 29	Saturn I (SA-5)	Launch vehicle development test	Saturn (AMR)	S	S	S	First flight test of S-IV 2nd stage; orbited world's heaviest payload (37,000 lbs., of which nearly 20,000 was payload).
Jan 30	RANGER VI (RA-A)* [RA-6]	Lunar photography probe	Atlas-Agena B (AMR)	S	P	U	Crashed onto lunar surface 2/2/64 within 20 mi. of aiming point; outstanding guidance success overshadowed by unexplained failure of TV cameras to take photos of moon during final 15 min. of flight.
Mar 19	Explorer (BE-A) [S-66]	Scientific earth satellite, Ionospheric	Thor-Delta (AMR)	U	Unk	U	3rd stage fired only half normal time, Beacon satellite failed to orbit; first Thor-Delta failure after 22 successes.
Mar 27	ARIEL II (UK-C) [S-52]	Scientific earth satellite, Ionospheric	Scout (WS)	S	S	S	U.S.-U.K. topside sounder satellite to measure micrometeroids, galactic radio noise, and ozone distribution in the ionosphere.
Apr 8	Gemini (GT-1)	Orbital Gemini capsule test	Titan II (AMR)	S	S	S	Confirmed Gemini spacecraft structural integrity and compatibility of spacecraft and launch vehicle. Reentered 4/12/64.
Apr 10 Apr 14	(?) Fire 1	37,000 fps reentry test	Atlas D - X-259 (AMR)	S	S	S	Highest speed reached by a manmade object; Fire capsule driven back into atmosphere by Antares II (X-259) rocket to speed of 37,891 fps (26,000 mph) and exterior heating of 20,000° F.
May 13	Apollo (BP-12)	Suborbital Apollo capsule test	Little Joe II (WSMR)	S	S	S	Boilerplate command and service modules of Apollo spacecraft launched to 17,000 ft.; then booster was exploded and escape tower lifted command module to 24,000 ft.
May 28	Saturn I (SA-6)	Launch vehicle development test	Saturn (ETR)	S	S	S	Orbited Apollo boilerplate spacecraft with S-IV 2nd stage. One of 8 engines in 1st stage shut off 24 sec. early, but other engines and guidance system compensated. Orbiting portion reentered 5/30/64.
Jun 30	Centaur (AC-3)	Launch vehicle development test	Atlas-Centaur (ETR)	P	S	S	Empty rocket casing did not orbit because hydraulic failure in Centaur stage cut engine run 127 sec. short; 5 of 6 flight objectives achieved; 6th—liquid hydrogen engine start and run—partially achieved.

See footnotes at end of table

1964						
Jul 20	Sert I (Sert IA)	Space equipment development test	Scout (WS)	S	P	First flight test (2,500-mi.-altitude) of electric rocket engine met all technical objectives; one of two electric rocket motors operated twice for total of 30 min.; other engine failed to operate.
Jul 28	RANGER VII [RA-B) [RA-7]	Lunar photography probe	Atlas-Agena B (ETR)	S	S	RANGER VII impacted on moon July 31, after relaying 4,316 lunar photos back to earth; closest photos improved resolution of lunar detail by factor of 1,000; complex flight was classic in performance.
Aug 18	Reentry D	Reentry ablation test	Scout (WS)	S	S	Achieved 27,950 fps, telemetered good data; payload impacted 1,265 mi. downrange after 11-min. flight.
Aug 19	SYNCOM III (Syncom C) [A-27]	Communications earth satellite	Thrust-Augmented Delta (ETR)	S	S	First true "hovering" synchronous orbit was achieved 9/11/64; completed a number of communications experiments, relayed TV of Tokyo Olympic Games.
Aug 25	EXPLORER XX (BE-B) [S-48]	Scientific earth satellite, ionospheric	Scout (WTR)	S	S	Ionosphere "Topsi" (topside sounder) began mapping the ionosphere, data to be compared with that from ALOUETTE I.
Aug 28	NIMBUS I (Nimbus A)	Meteorological earth satellite	Thor-Agena B (WTR)	P	S	First launch of Nimbus advanced weather satellite; short burn of Agena stage caused elliptical orbit; photos were of much higher resolution than those from Tiros. Ceased transmitting 9/23/64; on 12/8/64 resumed transmitting but photos were garbled.
Sep 4	OGO I (Ogo-A)	Scientific earth satellite	Atlas-Agena B (ETR)	S	P	First launch of "space bus" geophysical satellite; two antenna booms failed to deploy, preventing OGO from locking on earth, but all 20 experiments returned some data, 6 experiments were crippled by undeployed booms.
Sep 18	Apollo (SA-7)	Orbital Apollo spacecraft test	Saturn I (ETR)	S	S	Seventh straight launch success for Saturn I booster; second orbiting of boilerplate Apollo command and service modules. Orbiting portion reentered 9/22/64.
Oct 4**	EXPLORER XXI (Imp B)	Scientific earth satellite, energetic particles	Thor-Delta (ETR)	U	S	Faulty performance of 4th stage (X-258) caused payload to reach only 59,400 mi. apogee (planned apogee was 161,000 mi.), severely curtailing planned measurements in interplanetary space.
Oct 9** Oct 9	(?)...... EXPLORER XXII (BE-B) [S-66a]	Scientific earth satellite	Scout (WTR)	S	S	Transmissions received by 80 ground stations in 32 countries, largest international effort to date. First of five NASA satellites for geodetic research. Confirmed feasibility of satellite tracking by laser.
Nov 5	MARINER III (Mars 64)	Scientific Mars probe	Atlas-Agena D (ETR)	U	P	Communications failed when fairing did not separate, prevented solar panels from deploying. Agena D burned 4 sec. short. Payload in orbit around the sun, no hope of communication.
Nov 6	EXPLORER XXIII (Micrometeoroid C) [S-55c].	Scientific earth satellite, micrometeoroid	Scout (WS)	S	S	Meteoroid detection satellite, successor to EXPLORER XVI.
Nov 21	EXPLORER XXIV (AD)	Scientific earth satellite, air density	Scout (WTR)	S	S	First NASA launch of two satellites (EXPLORERS XXIV and XXV) with same launch vehicle. EXPLORER XXIV was same 12-ft. sphere used for EXPLORERS IX AND XIX.

See footnotes at end of table.

Date	Name (NASA Code)	General Mission	Launch Vehicle (Site)	Performance			Remarks
				Vehicle	Payload	Mission	
Nov 21	EXPLORER XXV (Injun-B)	Scientific earth satellite	Scout (WTR)	S	S	S	Second of two satellites launched with same Scout vehicle; carried 16 radiation sensors to measure electrons and protons in same part of space as EXPLORER XXIV Air Density satellite.
Nov 28	MARINER IV (Mariner D)	Scientific Mars probe	Atlas-Agena D (ETR)	S	S†	S	Excellent launch, would have brought MARINER IV within 151,000 mi. of Mars; successful midcourse correction cut this to 5,400 mi.; to pass and photograph Mars on 7/14/65.
Dec 8	Apollo (BP-23)	Suborbital Apollo capsule test	Little Joe II (WSMR)	S	S	S	Tested Apollo spacecraft launch escape system at flight altitude of maximum aerodynamic pressure (31,000 ft.).
Dec 9	(¹)	....	....	...	...	...	....
Dec 11	Centaur (AC-4)	Launch vehicle development test	Atlas-Centaur (ETR)	S	S	S	Achieved all six primary objectives in testing structural integrity and guidance system; did not achieve secondary objective of restart of Centaur engines. Centaur stage orbited 15 hrs.
Dec 15	SAN MARCO I (SM-A)	Scientific earth satellite, ionospheric	Scout (WI)	S	S	S	Launched by NASA-trained Italian crew; satellite returned data on air density and communications interference.
Dec 21	EXPLORER XXVI (EPE-D) [S-3c]	Scientific earth satellite, energetic particles	Thor-Delta (ETR)	S	S	S	Put in elliptical orbit to gain data on radiation concentration and behavior in Van Allen and Starfish belts.

*New NASA code name, effective Jan. 1, 1964. Code name in brackets is the same payload under the former code name.
†Conditionally successful as of 12/31/64; success in primary objective would not be certain until July 1965.

¹ Apr 10, 1964. Ram (Radio Attenuation Measurement) 250-lb. payload was launched on suborbital ballistic trajectory from Wallops Island by 3-stage rocket consisting of Castor-Antares-Alcor stages. Flight achieved reentry speed of 18,400 fps in test of communications blackout problems during reentry. Flight was successful; significant data were received; spacecraft impacted 650 mi. east of Bermuda after 10-min. flight.

² Oct 9, 1964. RFD-2 (Reentry Flight Demonstration) was conducted by NASA for AEC. A Scout vehicle from Wallops Station, Va., sent a nonradioactive mockup of nuclear isotopic generator to altitude of 800 mi. Payload reentered 250 mi. SE of Bermuda, providing data on reentry burnup rate of nuclear fuel elements.

³ Dec 9, 1964. Gemini (GT-2) flight from ETR, scheduled for suborbital maximum reentry heating test, was postponed when Titan II launch vehicle experienced nozzle hardover caused by hydraulic-line failure at time of ignition. Shutdown was achieved; spacecraft and vehicle would be launched later.

Appendix C

ABBREVIATIONS OF REFERENCES

Listed here are abbreviations for sources cited in the text. This list does not include all sources provided in the chronology, for some of the references cited are not abbreviated. Only those references which appear in abbreviated form are listed below.

Abbreviations used in the chronology entries themselves are cross-referenced in the Index.

A&A	AIAA's magazine, *Astronautics & Aeronautics*
A&A 64	NASA APPH, *Astronautics and Aeronautics 1964* [this publication]
AEC Release	Atomic Energy Commission News Release
AF Info. Pol. Ltr.	Air Force Information Policy Letter for Commanders
AF Mag.	*Air Force and Space Digest* magazine
AFOSR Release	Air Force Office of Scientific Research News Release
AFSC Release	Air Force Systems Command News Release
AIAA Release	American Institute of Aeronautics and Astronautics News Release
A-N-AF J&R	*Army-Navy-Air Force Journal & Register* (now *Journal of the Armed Forces*)
AP	Associated Press
ARC Release	NASA Ames Research Center News Release
Atlanta Const.	*Atlanta Constitution* newspaper
ATSS-T Trans.	Translation by NASA Scientific and Technical Information Div., Translators
Av. Daily	*Aviation Daily* Newsletter
Av. Wk.	*Aviation Week and Space Technology* magazine
Balt. *Sun*	Baltimore *Sun* newspaper
Chic. Trib.	*Chicago Tribune* newspaper
Commerce Dept. Release WB–	Dept. of Commerce, Weather Bureau News Release
ComSatCorp Release	Communications Satellite Corporation News Release
CR	*Congressional Record*
CSM	*Christian Science Monitor* newspaper
CTNS	Chicago Tribune News Service
DAC Release	Douglas Aircraft Co. News Release
DMSSD *Apogee*	Douglas Missile and Space Systems Div. *Apogee*
DOD Release	Dept. of Defense News Release
FAA Release	Federal Aviation Agency News Release
FonF	*Facts on File*
FRC Release	NASA Flight Research Center News Release
FRC X-Press	NASA Flight Research Center's *FRC X-Press*
GE Forum	*General Electric Forum* magazine
Goddard News	NASA Goddard Space Flight Center's *Goddard News*
GSFC Release	NASA Goddard Space Flight Center News Release
GSFC SSR	NASA Goddard Space Flight Center's *Satellite Situation Report*
HHN-48	NASA APPH, *Satellites, Space Probes, and Manned Space Flights—1964* (HHN-48)

Houston Chron.	Houston Chronicle newspaper
Houston Post	Houston Post newspaper
HTNS	New York Herald Tribune News Service
J/Armed Forces	Journal of the Armed Forces (formerly Army-Navy-Air Force Journal & Register)
JPL Release	Jet Propulsion Laboratory News Release
KSC Release	John F. Kennedy Space Center, NASA, News Release
Langley Researcher	NASA Langley Research Center's Langley Researcher
LaRC Release	NASA Langley Research Center News Release
L.A. Times	Los Angeles Times newspaper
Lewis News	NASA Lewis Research Center's Lewis News
LRC Release	NASA Lewis Research Center News Release
M&R	Missiles and Rockets magazine
Marshall Star	NASA George C. Marshall Space Flight Center's Marshall Star
MSC Release	NASA Manned Spacecraft Center News Release
MSC Roundup	NASA Manned Spacecraft Center's Space News Roundup
MSFC Release	NASA George C. Marshall Space Flight Center News Release
NAA Release	North American Aviation, Inc., News Release
NAA Release	National Aeronautic Association News Release
NAA S&ID Skywriter	North American Aviation, Inc., Space and Information Systems Div., S&ID Skywriter
NANA	North American Newspaper Alliance
NASA Auth. Hearings	NASA Authorization [FY 1965] Hearings
NASA Hq. PB	NASA Headquarters Personnel Bulletin
NASA LAR II/50	NASA Legislative Activities Report, Vol. II, No. 50
NASA Off. Int. Aff.	NASA Office of International Affairs
NASA R&SC Div.	NASA Public Affairs, Reports & Special Communications Div.
NASA Release	NASA (Hq.) News Release
NASA Rpt. SRL	NASA Report of Sounding Rocket Launching
NASA X-15 Proj. Off.	NASA (Hq.) X-15 Project Office
NASC Release	National Aeronautics and Space Council News Release
NAS-NRC Release	National Academy of Sciences-National Research Council News Release
NMI-	NASA Management Instruction-
NSC Release	National Space Club News Release
N.Y. Her. Trib.	New York Herald Tribune newspaper
NYT	New York Times newspaper
NYTNS	New York Times News Service
OAR Release	Office of Aerospace Research (USAF) News Release
Phil. Eve. Bull.	Philadelphia Evening Bulletin newspaper
Phil. Inq.	Philadelphia Inquirer newspaper
Sat. Eve. Post	Saturday Evening Post magazine
SBD	Space Business Daily newsletter
Sci. Amer.	Scientific American magazine
Sci. Serv.	Science Service
SR	Saturday Review magazine
Testimony	Congressional testimony, prepared statements
Text	Prepared report or speech text
Transcript	Official transcript of news conference or Congressional hearing
UPI	United Press International
U.S. Naval Inst. Proc.	U.S. Naval Institute Proceedings magazine
U.S. News	U.S. News and World Report magazine
Wallops Release	NASA Wallops Station News Release
Wash. Daily News	Washington Daily News newspaper
Wash. Eve. Star Wash. Sun. Star	Washington Evening/Sunday Star newspaper
Wash. Post	Washington Post newspaper
WBE Sci. Serv.	World Book Encyclopedia Science Service
WSJ	Wall Street Journal newspaper

Index

A-11 (supersonic aircraft) (see also YF-12A), 85–86, 90, 92, 93, 101, 104, 133, 154, 164, 225, 254, 274, 292, 328, 331
AAAS. See American Association for the Advancement of Science.
AACB. See Aeronautics and Astronautics Coordinating Board, NASA–DOD.
AAS. See American Astronautical Society.
Abella, Dr. Isaac D., 383
Abelson, Dr. Philip, 262
ABL. See Automated Biological Laboratory.
Abrams, Gen. Creighton W., Jr. (USA), 394
Abres. See Advanced ballistic re-entry system (Abres).
AC-3. See Atlas-Centaur 3.
Accelerometer, 18
Accident, 92, 135, 147, 156, 166, 178, 181, 252, 269
Acetaldehyde gas, 180
Acord, James D., 82
AC Spark Plug Div. (General Motors Corp.), 18, 218
"An Accounting of Stewardship, 1961–1964," 318
Activation analysis laboratory (Texas A&M Univ.), 156
Adair, Rep. E. Ross, 55
Adhesive, 360
Advanced ballistic re-entry system (Abres), 403
Advanced Bearing Technology (SP-38), 104
Advanced Manned Precision Strike System (AMPSS) (USAF), 66
Advanced Research Projects Agency (ARPA), 10, 28, 57, 76, 153, 290, 337, 386
Advanced Technological Satellite (ATS), 177
Advanced Vidicon Camera System (AVCS), 298
Advisory Committee on Reactor Safeguards (AEC), 91
Advisory Group for Aeronautical Research and Development (AGARD), 349
AEC. See Atomic Energy Commission.
AEDC. See Arnold Engineering Development Center.
Aero Club of New England, 117
Aerobee (sounding rocket), 82, 95, 254
launch, 210, 363
150, 30, 79, 124, 128, 144, 192, 260, 262, 283, 299, 335, 368, 374, 381, 399, 422
150A, 304, 328, 387
300, 264
300A, 30
350, 415
Aerodynamics, 29, 122, 158, 217, 257, 290
Aeroflot (U.S.S.R. airline), 241
Aerojet-General Corp., 148, 263
contracts, 18, 23, 78, 80, 115, 249, 312, 328, 422
Dade County, Fla. plant, 190, 248
test, 140, 154, 239, 323, 327, 369
Aeronautical Research Institute of Sweden, 194
Aeronautics, 46–47, 78, 154–155, 181, 292, 303
Aeronautics and Astronautics Coordinating Board, NASA–DOD (AACB), 90, 231
Aeronaves de Mexico, 175
Aeronca Manufacturing Corp., 222
Aero Propulsion Laboratory (USAF), 167, 313
Aerospace (subject). See Space.
Aerospace and Defense Service Engineering Dept. (GE), 70
Aerospace Corp., 104, 118, 265
Aerospace Educational Center (USAF Academy), 198
Aerospace Industries Association of America (AIA), 89, 114, 246, 279–280, 442
Aerospace industry, 75, 114, 288, 333, 443
"Aerospace Man of the Year," 315
Aerospace Medical Association (AMA), 174, 175
Aerospace Medical Research Laboratories (AMRL) (AFSC), 70
Aerospace medicine (see also Bioscience, Life science), 45, 48, 52
Aerospace Plane (U.K. space transporter), 270
Aerospace Research Application Center (ARAC), 100
Aerospace Research Center (Univ. of Illinois), 241
Aerospace Research Pilot School (USAF), 197, 206
Aerospace Technical Council (AIA), 89
Aerospace Workshop, 244, 247
AFA. See Air Force Academy.
AFA. See Air Force Association.
AFAA. See American Fighter Aces Association.

AFCEA. See Armed Forces Communications and Electronics Association.
AFCRL. See Air Force Cambridge Research Laboratories.
AFL–CIO (American Federation of Labor–Congress of Industrial Organizations), 141, 323
AFML. See Air Force Materials Laboratory.
AFMTC. See Air Force Missile Test Center.
AFOAR. See USAF Office of Aerospace Research.
Africa, 134, 188, 308
AFSC. See Air Force Systems Command.
AFSSD. See Air Force Space Systems Div.
AFSTC. See Air Force Space Test Center, Provisional.
Agate (rocket), 210
Agena (booster) (see also Atlas-Agena, Thor-Agena), 25, 60, 197, 275, 298, 339
Agreement
   Communications satellite system, global commercial, 260, 293, 431
   France-Portugal, observation station, 132
   France-Spain, satellite tracking station, 204
   labor stabilization, 141
   NAS-Soviet Academy of Sciences, 182
   NASA
      Australia, 103
      Canada, 11
      Commerce, 43, 312
      DOD, 64, 143, 312, 319
      FAA, 312
      India, 70, 97
      Interior, 312
      Malagasy Republic, 306
      Netherlands, 278
      Pakistan, 97
      Sweden, 204
      Treasury, 312
      U.S.S.R., 97, 342
   U.S.-Malagasy Republic, 306
   U.S.-U.S.S.R., 251, 256
Agriculture, U.S. Dept. of (USDA), 102
AGU. See American Geophysical Union.
Ahmedabad, India, 6, 10
AIA. See Aerospace Industries Association.
AIAA. See American Institute of Aeronautics and Astronautics.
AIBS. See American Institute of Biological Sciences.
Air cushion, 8, 289
Air Force. See U.S. Air Force (USAF).
Air Force Academy (AFA), 70, 125, 127, 198
Air Force Aeronautical Systems Div. (AFASD), 181
Air Force Aero Propulsion Laboratory, 167, 313
Air Force and Space Digest, 160, 303
Air Force Association (AFA), 112, 315
Air Force Avionics Laboratory, 178

Air Force Ballistic Systems Div. (AFBSD), 18
Air Force Cambridge Research Laboratories (AFCRL), 74, 132, 147, 159, 178, 230, 241, 265, 288, 354, 409
Air Force Eastern Test Range (ETR). See Eastern Test Range.
"The Air Force in the Years Ahead," 112
Air Force Institute of Technology, 303
Air Force Materials Laboratory (AFML), 153
Air Force Missile Test Center (AFMTC) (now Eastern Test Range), 5, 164
Air Force National Range Div. (AFNRD), 164
Air Force Office of Aerospace Research (AFOAR), 9, 34, 38, 159, 288
Air Force Office of Scientific Research (AFOSR), 379
Air Force Research and Technology Div. (AFRTD), 381, 383, 392
Air Force Space Museum, 261
Air Force Space Systems Div. (AFSSD), 12, 74, 89, 123, 129, 208, 257, 261, 286, 328
Air Force Space Test Center (Provisional) (AFSTC) (see also Western Test Range), 3, 164
Air Force Systems Command (AFSC), 6, 70, 74, 95–96, 135, 146, 153, 210, 217, 327
   contract, 89, 278, 279, 328
   management, 3, 41, 112, 178, 257
   report, 98–99
   research, 303
   test, 167, 313, 337–338, 383, 385
Air Force Western Test Range. See Western Test Range.
Air France, 57
Air-India, 60
"Air pads," 8, 289
Air pollution, 229–230
Air Products & Chemicals, Inc., 14
Air Proving Ground Center (Eglin AFB, Fla.), 70
Air traffic control, 158, 216, 332
Air Transport and Space Meeting and Production Forum (SAE–ASME), 158, 158–159
Air Transport Association of America (ATA), 211, 288, 408
Airbag safety system, 192
Airborne Ranging Orbital Determination (AROD) system, 412
Aircraft, 293, 303
   attack, intercontinental range, 66, 172
   commercial, supersonic. See Aircraft, supersonic transport (Sst).
   counter-insurgency (COIN), 287
   foreign, 209
   interceptor, 72, 79, 92, 98
   reconnaissance, long-range (SR–71), 260–261
   transport, 113–114, 318
Aircraft, supersonic commercial transport (Scat) (see also Aircraft, supersonic transport), 85–86, 108

Aircraft, supersonic transport (Sst), 5, 57, 85–86, 101, 117, 142, 147, 189, 213–214, 220, 223, 246
  appropriations, 115, 119–120
  contract, 201
  cost, 31, 82, 93, 94, 114, 187
  crew training, 200
  development, 4, 13, 28, 43, 78, 115, 116, 135, 148–149, 150, 153, 154, 161, 162, 183, 194–195, 201, 236, 242, 268, 288, 292–293, 318
  Government-industry, 4, 61, 187
  program study report, 76, 85–86, 90, 115, 119–120
  reservations for, 19, 57, 60, 63, 70, 81, 96, 136, 158, 159, 175, 430
  takeoff requirements, 150
AiResearch Manufacturing Co. Div. (Garrett Corp.), 6, 57, 197
Airglow, 11, 15, 99, 243, 290, 398
Airlift, 98–99
Airlines, 2
The Airman, 302, 333
Airports, 133, 332, 404
Airworthiness requirements, 156
Akwa-Downey Construction Co., 328, 424
Al Kahar (U.A.R. rocket) ("Conqueror"), 303
Al Raid (U.A.R. rocket) ("Explorer"), 303
Al Zafar (U.A.R. rocket) ("Victor"), 303
Alabama, Univ. of., 196
Alamagordo, N. Mex., 92, 404, 406
Alaska, 119, 216, 264, 397
Alaska, Univ. of, 133, 137, 145
Albert, Rep. Carl, 274, 318
Aldrin, Maj. Edwin E., Jr. (USAF), 306, 356
Algae experiment, 248
Algeria, 161, 308
Allen, Arthur R., 280
Allott, Sen. Gordon L., 93, 164
Alloy, 291
Alouette (satellite) (program), 11
ALOUETTE I (Canada satellite), 11, 106, 296
Alphonsus (moon crater), 180, 364
Alss. See Apollo Logistic Support System.
Alternate Stability Augmentation System (Asas), 6
Alternating Gradient Synchrotron, 71–72
Altman, David, 166
Alvarez, Luis W., 11
AMA. See American Management Association.
AMA. See American Medical Association.
Amateur rocketry, 255
Ambartsumian, Viktor A., 296–297
AMC. See Army Materiel Command.
Amer, Dr. N. M., 174
American Association for the Advancement of Science (AAAS), 262, 433, 438
American Assembly Student Conference, 125
American Astronautical Society (AAS), 167–168, 170, 170–171
American Bankers Association, 31, 252
American Federation of Labor-Congress of Industrial Organizations (AFL–CIO), 141, 323
American Fighter Aces Association (AFAA), 78
American Geophysical Union (AGU), 146–147, 435–436, 438, 444
American Institute for Biological Sciences (AIBS), 210
American Institute of Aeronautics and Astronautics (AIAA), 21, 139, 140, 166–167, 223, 225–226, 228, 228–229, 230, 234–235, 235–236, 238, 343
  awards, 21, 166, 376–377
  committees, 226
  meetings, 197, 235–236, 350, 375–376, 377
  National Capital Section, 13
American Machine and Foundry Co., 148
American Management Association (AMA), 243
American Medical Association (AMA), 219
American Meteorological Society (AMS), 35, 146–147
American Optical Co., 264–265
American Orthopsychiatric Association (AOA), 110
American Physical Society (APS), 22, 23, 153, 155–156, 432
American Science and Engineering, Inc., 299, 363
American Society of Biological Chemists (ASBC), 110
American Society of Mechanical Engineers (ASME), 403
American Telephone and Telegraph Co. (AT&T), 61, 79, 123, 168, 192, 209, 219, 334, 383–384, 412
American Women in Science symposium, 362
Amery, Julian, 311, 329
Ames, Milton B., Jr., 91
Ames Research Center (ARC), 39, 85, 131, 158, 202, 322
  contract, 65, 264–265, 339, 432
  research, 71, 197, 226, 290
  Structural Dynamics Laboratory, 432
  test, 129
Amine, 181
Amoeba, 72
Amos, Project (ARPA), 57
Amplifier, 295
Ampss. See Advanced Manned Precision Strike System.
AMR. See Atlantic Missile Range.
AMRL. See Aerospace Medical Research Laboratories.
AMS. See American Meteorological Society.
Amu. See Astronaut maneuvering unit.
Anders, Capt. William A. (USAF), 292
Anderson, Dr. Carl C., 423

Anderson, Sen. Clinton P., 33, 176, 350–351
   Apollo-X program, 408, 417, 425
   appropriations, 18
   lunar program, 13, 18, 268–269
   Manned Orbiting Laboratory program, 408, 417, 425
   Ranger failure, 126, 150, 163
   Rift program, 340
Anderson, Dr. Evelyn, 93
Anderson, Jack, 81
Anderson, Dr. Oscar E., Jr., 237
Anderson, William J., 104
Andoeya, Norway, 56, 375
Andover, Me., 32, 146, 219
Andrews AFB, Md., 164, 412
ANL. See Argonne National Laboratory.
Ann Arbor, Mich., 251
Anna (satellite), 222
ANNA IB (satellite), 370
Annapolis, Md., 334
Anniversary, 2, 104, 163, 211, 216, 335, 381, 420
Antares II (motor), 135
Antenna, 33, 38, 222, 264, 297, 321, 424
   parabolic, 31, 216, 396
Antennas and Propagation, Professional Group on, IEEE, 326–327
Antiaircraft missile, 439
Antimissile defense system, 214
Antimissile missile, 381, 388, 394
Antipov, Dr. V. V., 389
Antiradiation compounds, 276
Antisatellite defense system, 197, 330, 370–371
Antonious, Capt. Michael N. (USA), 394
AOA. See American Orthopsychiatric Association.
Aoao. See Orbiting Astronautical Observatory, Advanced.
Aogo. See Orbiting Geophysical Observatory, Advanced.
Aorl. See Apollo Orbital Research Laboratory.
Aoso. See Orbiting Solar Observatory, Advanced.
AP. See Associated Press.
APL. See Applied Physics Laboratory.
Apollo (program), 26, 83, 118, 129, 143, 144, 192–193, 194, 207, 216, 224, 225, 228, 237–238, 297, 326–327, 341
   astronaut, 43, 66, 74–75, 158, 167–168, 173, 206, 209, 224, 239
   criticism, 54, 64, 185–186, 203, 257, 337, 362
   experiment, 258, 313
   facilities, 10, 14, 15, 24, 77, 209, 268
   funds for, 2, 12–13, 18, 45–46, 77, 99–100, 102, 104–105, 221, 224–225, 227–228, 237–238, 242, 247, 274
   lunar landing program, 16, 23, 47, 68–69, 171, 271–272, 275–276, 298–299, 324, 391, 396
   management, 53–54, 120, 364, 375
   military, 258, 302, 316
   plans for, 33, 71, 74–75, 89, 157–158, 185–186, 276, 314, 321
   progress, 3, 13, 52–54, 60
   space hazards, 13, 18, 30–31, 47, 173, 194, 219
   support of, 20–21, 41, 69, 138, 168, 176, 238, 244
   testing, 371
Apollo (spacecraft), 65, 70, 126, 138, 206, 228, 292, 316, 318, 387
   adapter, 436
   Apolo-X, 363, 382
   boilerplate model, 138, 175, 192–193, 196, 314, 321, 325, 466, 467, 468
   contract, 306
   equipment, 6, 68, 81, 134, 218, 239, 299, 300
   escape system, 175, 314
   facilities, 299
   launch vehicle, 89, 229–230, 353
   module, 105
      command, 175
      lunar excursion (Lem), 39, 43, 73, 81, 105, 116, 139, 183
   test, 6, 10, 89, 90, 154, 175, 192–193, 314, 321, 325
   tracking, 216, 305
   use of, 170, 249
Apollo/Apollo Logistic Support System (Alss), 170
Apollo, Extended, (Apollo-X), 145
Apollo Logistic Support System, (Alss), 73, 105, 183
Apollo/Lunar Exploration System for Apollo (Lesa), 105, 170
Apollo Orbital Research Laboratory (Aorl), 145, 258
Apollo, post, study, 388
Apollo/Stay Time Extension Module (Stem), 170
Applied Physics Laboratory (APL) (Johns Hopkins Univ.), 57–58
APS. See American Physical Society.
Apt. See Automatic picture transmission.
ARAC. See Aerospace Research Application Center.
Aral Sea, U.S.S.R., 156
ARC. See Ames Research Center.
Arcas, boosted (sounding rocket), 99, 305, 328
Arcetri, Italy, 180
Archangel, U.S.S.R., 27
Arc-jet engine, 167
Arco, Idaho, 93
ARCS Industries, Inc., 383–384
Arctic Circle, 216
Arents (Advanced Research Environmental Test Satellite) spacecraft, modified. See Orbiting Vehicle 2.
Arfons, Art, 365
Argentina, 403
Argentine National Commission on Space Research (CNIE), 403

Argo D-4 (Javelin) (sounding rocket), 15, 137, 355, 360–361, 380
Argonne National Laboratory (ANL), 4
ARIEL I (satellite), 106, 453
ARIEL II (satellite), 118, 452–453, 482–483
Aristarchus (lunar crater), 78, 106, 180, 203–204, 205, 316
Arizona, 239
Arizona, Univ. of, 148, 301
　Lunar and Planetary Laboratory, 267, 275–276, 331
　Space Sciences Center, 259
Arkansas, 369–370
Arecibo, Puerto Rico, 38, 297, 393
Armed Forces Communications and Electronics Association (AFCEA), 183–184, 185
Arms control, 437
Armstrong, Neil A. (NASA), 102, 244
Army. See U.S. Army.
Army Air Corps, 136
Army Ballistic Missile Agency (ABMA), 37
Army Corps of Engineers, 59, 130, 274
　contract, 8, 13, 15, 28, 31, 38, 61, 62, 63, 74, 102, 121, 127, 164, 203, 216, 268, 291, 328, 329, 400, 410, 434
Army Electronics Command, 241
Army Map Service, 74, 220–221
Army Materiel Command (AMC), 11, 211
Army Missile Command (Redstone Arsenal), 112, 211, 305
Army-Navy-Air Force Journal and Register, 26
Army Signal Corps, 46–47
Arnold Engineering Development Center (AEDC), 215–216
Arnold, H. H., Trophy, 315
Arod. See Airborne Ranging Orbital Determination.
ARPA. See Advanced Research Projects Agency.
Artemia egg experiment, 248
Arthur, George H., 379
Artsimovich, L. A., 198
Asas. See Alternate Stability Augmentation System.
ASBC. See American Society of Biological Chemists.
Ascani, B/G Fred J. (USAF), 172
Ascension Island, 30, 42–43, 62, 135, 137, 277–278, 287
Ashklabad, U.S.S.R., 27
Asimov, Isaac, 163
ASME. See American Society of Mechanical Engineers.
Asset (Aerothermodynamic/elastic Structural Systems Environmental Tests) (see also Start), 19, 114, 208, 258, 286, 307, 392, 411
Associated Press (AP), 77, 143
Association of the United States Army, 388
Asteroid, 28
Astro Electronics Div. (RCA), 193
Astro Research Corp., 428–429

Astrobee 1500 (sounding rocket), 358–359
Astrometric telescope, 75
Astronaut, 81, 120, 173, 183, 190, 236–237, 242, 249, 265, 302, 319
　contract, life story, 119, 287
　death, 370
　performance, 48–49, 112, 208
　politics, 15–16
　quarantine, 215
　rescue, 100, 102, 174, 203
　scientists, 10, 356
　selection, 263
　training, 5, 25, 43, 66, 99–100, 102, 158, 278, 281, 302, 306, 318, 356
　women as, 187, 244, 226
Astronaut Maneuvering Unit (Amu), 89
Astronautics and Aeronautics (magazine), 88, 122, 160, 197–198, 269, 302, 401
Astronautics Award, 375–376
Astronomers, 78, 126, 203, 205, 296, 326
Astronomical Journal (U.S.S.R.), 364
Astronomy, 106, 223, 251, 262, 296, 327, 385, 436
Astrophysical Observatory, Arcetri, Italy, 180
Astrophysics, 386, 426
Astrophysics and Physical Meteorology Laboratory (JHU), 132
Astrorocket, 123
A/SWA. See Aviation/Space Writers Association.
AT&T. See American Telephone and Telegraph Co.
ATA. See Air Transport Association of America.
Atchley, Mrs. Hazel T., 114
Athena (missile) (USAF), 58, 193, 241, 329, 338, 398, 406
Atlantic City, N.J., 56, 297
Atlantic Missile Range (AMR) (see also Eastern Test Range, Kennedy Space Center (NASA), and Cape Kennedy), 5, 9, 19, 37, 38, 72, 113, 116, 144–145
　facilities, 38–39, 63, 77
　launch, 135
　　booster, 13, 82, 113
　　failure, 104
　　Gemini, unmanned, 130
　　missile, 5, 6, 31, 79, 81, 84, 122, 132
　　probe, 34–35
　　satellite, 19
　　Saturn, 32
　　sounding rocket, 122
　management, 127, 192
　prelaunch activities
　　Gemini, 9
　　Saturn, 7, 126, 149
　submarine launch, 11, 19
　test, 114
Atlantic Monthly, 303
Atlantic Ocean, 10, 11, 15, 30, 38, 104, 110, 137, 140, 158, 209, 221, 227, 234, 243, 256, 258, 332–333, 378

Atlas (missile), 111, 157–158, 177, 213, 234, 275, 309
  contract, 308, 319
  launch
    D, 135, 217, 403
    E, 79
    F, 63, 177, 430
    failures, 127, 145–146, 150, 177
    operational, 8, 64, 101, 125, 265
  plans for, 78, 87, 106
  test, 78
Atlas-Agena (booster) (see also Agena), 78, 79, 102, 146, 171, 186, 218, 219, 239, 252–253, 259, 263, 309, 326, 344, 361
  B, 34–35, 60–61, 64, 306, 448, 453, 456, 466, 467
  D, 182, 285, 378, 397–398, 406, 450, 452, 456, 458, 459, 460, 461, 467, 468
Atlas-3 (SLV-3), 95–96, 269
Atlas-Centaur (see also Centaur), 4, 71, 309
  3 (AC-3), 227, 466
  4 (AC-4), 414–415, 461, 468
Atlas 116 D (booster), 221
Atmosphere, 76, 149–150
  artificial, 302
  demarcation line, 14–15
  density, 131–132, 140
  helium-oxygen, 274, 309
  temperature, 22–23, 131–132
  upper, measurement, 14, 47, 50–51, 56, 62, 132, 140, 141, 204–205, 241, 269–270, 273, 277–278, 387
Atmospheric science, institute of (proposed), 100
Atomic bomb, 46–47, 320
Atomic clock, 126, 408–409
Atomic energy. See Nuclear energy.
Atomic Energy Commission (AEC), 40–41, 49, 51–52, 89, 91, 115, 126, 200, 224, 229, 231, 237, 246–247, 307, 369, 388, 400
  contract, 416
  funds, 48, 67, 85, 117, 161, 213, 400
  launch, 173
  report, 37, 72, 91
  test, 3, 63, 93, 104, 125, 175, 188, 256, 338, 360
Atomic Industrial Forum, 400, 405, 441
Atomic Research and Development Authority, New York State, 70
Atomics International, 3, 39, 260
Atoms-for-Peace Conference, Third International, 305
Ats. See Satellite, Advanced Technological.
Attitude control, 82, 257, 264–265, 386, 397–98
Atwood, U.S. Ambassador-to-Kenya, William, 129
Aubiniere, Gen. Robert (France), 161
Auburn, Mass., 104
Auriol, Miss Jacqueline, 200

Aurora, 11, 50–51, 99, 243, 290
Aurora borealis, 107
Australia, 84, 103, 139, 205, 210
  cooperation
    satellite communications system, 189, 218, 292, 330
    space, 96–97, 269–270, 284, 396
  supersonic transport, 19
Australian Commonwealth Scientific and Industrial Organization, 434
Automated Biological Laboratory (ABL), 208, 259, 295
Automatic data processing. See Computer.
Automatic Picture Transmission (Apt) 107
  pictures, 251, 298
  receiving station, 5
Automation, 286
Autopilot, 257
Avcs. See Advanced Vidicon Camera System.
Aviation, 78, 204, 303, 329–330, 332
  civil, 94, 216, 237, 332
Aviation Human Resources Study Board, FAA, 237, 332
Aviation/Space Writers' Association (A/SWA), 190–191, 192, 194, 194–195
Aviation Week and Space Technology, 78, 92, 101, 129, 154, 200, 220, 224, 225, 226, 240, 257, 274, 281, 289, 324, 329, 330, 340, 341, 357, 395, 396, 408, 409, 434, 435
Aviation Writers' Association (AWA), 235
Avionics, 257
AWA. See Aviation Writers' Association.
Awards, 269
  government, U.S., 11, 32, 39, 63, 74, 77, 93, 114, 136, 150, 154, 258, 265, 266, 276, 316, 325, 328
  incentive (see also Inventions), 82, 114, 142
  institutions (univ., etc.), 232
  society, 75, 185, 188, 200, 235–236, 263, 317, 403
    aeronautics, 21
    astronautics, 21, 111, 164, 235–236, 315
    aviation, 317, 334
    foreign, 101, 198, 320
    medicine, 175–176, 266
    space, 21, 111, 235, 334, 392
Azores, 55, 132
B-52 (Stratofortress), 70, 214, 326
B-58 (Hustler), 157–158, 207, 317
B-70 (Valkyrie), 98
  XB-70, 172, 358, 369
  XB-70A, 297, 323, 349
Bacterial compatibility, 396
Bahamas, 386
Baikonur, Kazakhstan, U.S.S.R., 468–469
Bailey, Dr. Victor Albert, 56, 121
Bain, Gordon M., 13, 268
Baker, Burke, Planetarium (Houston), 244
Bakharev, Anatoli, 210
Bal Harbour, Fla., 174
Baldwin, Hanson, 255

Ball Brothers Research Corp., 95, 141
Ball, Ed, 69–70
Balloon
 experiments, 4, 258
 flight, 132–133, 176
 launch, 42, 56, 57, 74, 96, 216, 291, 364, 448–449, 482
Balloon astronomy, 223
Balloon-parachute, 129
Balloon-satellite communications network, 89
Balloon weather station (France), 218–219
Ballute (balloon-parachute), 129
Baltimore, Md., 10, 202
Baltimore *Sun*, 10, 272–273
Bangalore, India, 203
Barabashov, Nikolai, 315
Barbados Island, B.W.I., 10, 29
Barnum, Alan R., 406–407
Barre, Dr. Robert L., 110–111
Barycentric orbit, 159
Baskakov, Edward, 351
Basov, Nikolai G., 367
Bass, Abraham S., 379
Bates, Rep. William H., 108
Bauer, Louis H., Founders Award, 175–176
Baumann, Robert C., 82
Bay of Pigs invasion, 64
Bayevsky, Dr. Roman, 174–175
Beacon, crash, 5, 19
Beacon Explorer (satellite). See Explorer Beacon.
Beadle, George W., 357
Beagle, Project (Mars project), 257
Beale AFB, Calif., 346, 429
Bean, Lt. Alan L. (USN), 336
Bear landing test, 313
Bearing technology, 146, 293
Beattie, Donald A., 170
Beckman Instruments, Inc., 100, 352, 382, 396
Bedwell, M/G T. C., Jr. (USAF), 44
Beet, 204
Beetle experiment, 72
Beischer, Dr. Dietrich E., 174
Belew, Leland F., 377
Belgium, 246, 264, 397
Bell Aerosystems Co., 138, 160, 172, 197, 215, 251, 440
Bell Telephone Laboratories, 154
Bellows, W. S., Construction Corp., 8, 268
Bendix Corp., 148, 183, 251
 Eclipse-Pioneer Div., 22
 Products Aerospace Div., 13
Bendix Field Engineering Corp., 336
Benoist flying boat, 2
Benton, Sen. William H., 195
Berenice (French rocket), 371
Beresford, Spencer, 29
Berge, Glenn, 383
Bergen, William B., 259
Bergmann, Dr. Peter, 155–156
Berkner, Dr. Lloyd V., 155

Berlin, Germany, 41, 59
Bermuda, 39, 132, 176, 239, 252, 487
Bern, Univ. of (Switzerland), 248
Bernardo, James V., 407
Berry, Dr. Charles A., 91, 134, 167–168, 249
Beryllium, 230
Bethesda, Md., 237
Bethpage, N.Y., 299
Bettelheim, Dr. Bruno, 362
Beverly Hills, Calif., 327
Bevins, British Postmaster Gen. Reginald, 136
Bigg, Dr. E. Keith, 434
Bikle, Paul F., 77, 220, 435
Billingham, Dr. John, 230
Bioastronautics, 80, 308–309, 420–421
Biological laboratory, automated (Abl), 208, 259
Biological specimens in space, 72, 178–179, 204
Biology, 168–169, 169, 178–179, 204, 255–256, 438
 gravitational, 309
 space, 208
Biomedicine, 35–36, 245, 262, 305, 310
Bios. See Biosatellite.
Biosatellite (Bios), 72, 168–169, 178–179, 221, 225
Bioscience, 168–169, 208, 210
Bio-Space Technology Training Program, 305
Bird flight, 433
BIS. See British Interplanetary Society.
Biscarosse, France, 132
Bisplinghoff, Dr. Raymond L., 11, 53, 108, 135, 139, 151–152, 160, 202, 275, 293, 322, 333, 405
Bisson, Edmond E., 104
Black & Decker Manufacturing Co., 5
Black, Eugene R., 61, 76, 82, 85–86, 90, 115, 119–120
Black Knight (U.K. rocket), 269–270, 278, 311, 412
Black Prince (U.K. rocket), 311
Black Report (see also Black, Eugene R.), 115
Blackburg, Va., 290
Blagonravov, Anatoli A., 17–18, 30, 54, 171, 175, 204, 208, 342, 349
Blaw-Knox Co., 33
Bleymaier, B/G Joseph S. (USAF), 123, 139, 166, 192, 193, 278, 304, 376, 414
Block II vehicle, 247, 448–449
Blossom Point, Md., 130
Blount Brothers Corp., 224
Blount Construction Co., 28
Blue Scout (rocket), 104
 Jr., 300, 427, 450–451, 476–477
Blue Streak (U.K. booster), 205, 257, 269–270, 358
BNA. See Bureau of National Affairs, Inc.
BNL. See Brookhaven National Laboratory.
Boa constrictors, 56–57

BOAC. See British Overseas Airways Corp.
BOB. See Budget, Bureau of.
Bochum, West Germany, 385
Boeing Co., 10, 12, 73, 105, 124, 127, 171, 269, 352
  Aerospace Div., 183
  contract, 251, 404, 405, 415, 440
  supersonic transport, 13, 28, 183, 201, 373–374, 392, 405
  test, 155
Boeing Scientific Research Laboratories, 424
Boeing 707 (Stratoliner), 150, 158, 392, 408
Boeing 720 (airliner), 233
Boelkow Co. (West Germany), 265
Bogart, L/C Frank A. (USAF), 368
Boggs, Rep. Hale, 324, 361–362
Bomber aircraft, 92, 140–141
  congressional action, 69, 72, 79, 81, 82, 98, 141
  reliability, 28, 136–137
Bomber, rocket-sled-launched orbital, 78
Bone decalcification, 178–179, 309
Bonn Univ. (West Germany), 372
Bonneville Salt Flats, Utah, 353
Bono, Philip, 238, 246, 347
Boosted Arcas. See Arcas, boosted.
Booster, 10, 143, 170–171, 258, 324
Booth, Seth W., 39
Borman, Maj. Frank (USAF), 263
Boron fiber technology, 207, 244, 257
Boscombe Down, U.K., 329
Boston Globe, 43, 109, 141
Boston, Mass., 23–24, 37, 59, 75, 77, 79, 91, 113, 115, 185, 201, 244, 250, 292
Boulder, Colo., 389
Boundary layer stabilization, 122–123
Bounder (Soviet bomber), 196
Bowen, Howard, 387
Boyd, Alan S., 8
Boyle, R/A Francis D., (USN), 358
Boynton, John H., 225
Brackett, Ernest W., 51, 250, 281, 435
Brademas, Rep. John, 145
Brady, Joseph L., 1
Braille translations, 239
Braking techniques, 158, 296
Branch, B/G Irving L. (USAF), 77
Braniff Airways, 136
Bransteller, Robert J., 142
Brazil, 280, 375
Bread mold experiment, 72
Breedlove, Craig, 353
Brennan, Dr. John J., Jr., 337, 363
Brewer, Frank G., National Trophy, 1963 (NAA), 200
Brewster, Kingman, Jr., 191–192
Brezhnev, First Secretary Leonid, 353, 355–356
Brier, Glenn W., 434
Brinich, Paul F., 142
Britain. See United Kingdom (U.K.).
British Aircraft Corp., 170

British Interplanetary Society (BIS), 269–270
British National Committee on Space Research, 118
British Overseas Airways Corp. (BOAC), 57, 158
Broadway Maintenance Corp., 127
Brookhaven National Laboratory (BNL), 71–72, 133
Brooklyn, Polytechnic Inst., 331
Brooks AFB, Tex., 44, 45, 48, 52, 80, 112–113
Brotherhood of Maintenance of Way Employees, 209
Brovtsev, Col. Sergei G., 269
Brown and Root, Inc., 321
Brown, Clinton E., 45
Brown, Dr. Harold, 65
Brown, Dr. Harrison, 314
Brown, Porter, 430–431
Brown Univ., 359
Brown, William F., Jr., 223
Browne, Dudley E., 31
Brun, Pres. Edmond A. (France), 308–309
Brussels, Belgium, 24
Bubble blowing technique, 428–429
Buchanan, L/C Robert S. (USAF), 52
Buchheim, Dr. Robert W., 289
Buckley, Edmond C., 71, 108–109
Budget, Bureau of (BOB), 5, 117
  appropriations, 45–46, 67, 72
Buffalo, N.Y., 199, 251
Bull, Dr. Gerald, 411
Bumper (V–2) (booster), 261
Bunker-Ramo Corp., 259
Burbank, Calif., 85–86, 367
Burch, Dean, 364
Bureau of Aeronautics (USN), 334
Bureau of National Affairs, Inc. (BNA), 250, 252
Bureau of Ships (USN), 305
Burkett, Warren, 258
Burlington, Vt., 207
Bush, Dr. Vannevar, 11
Business Week, 75
Butadiene, 190
Butz, J. S., Jr., 303
Bykovsky, L/C Valery F., 179, 443
C–1 (rocket engine), 399–400
C–5. See CX.
C–97 (Stratocruiser), 214
C–119 (Flying Boxcar), 156, 276, 415
C–130 (Hercules), 158, 264
C–133 (Cargomaster), 176, 245
C–140A (Jetstar), 147, 257
CAB. See Civil Aeronautics Board.
Cable, 1–2, 55, 123
Cairo, U.A.R., 303
Cairo Univ. (U.A.R.), 78
Calcium, 180, 309
California, 40–41, 75, 85–86, 91, 120, 202, 207, 312, 314

California Institute of Technology (Cal Tech), 60, 97–98, 128, 163, 163–164, 179, 291, 314, 326, 383, 435
California, Univ. of (Berkeley), 1, 148, 177–178, 221, 222, 323, 339
California, Univ. of (Los Angeles) (UCLA), 220, 240, 433
California, Univ. of Southern (USC), 243, 245
Calilsto (Jupiter's moon), 179, 326
Cal Tech. See California Institute of Technology.
Cambridge, Mass., 116, 292, 301, 304, 327, 345, 363
Camera, 41, 41–42, 114, 118, 214, 217, 263
Camera station, automatic, 171
Camera System, Advanced Vidicon (Avcs), 298
Cameron, A. G. W., 28, 314
Camp Roberts, Calif., 290–291, 332–333, 408
Campbell, Joseph, 434, 435
Canada, 70, 116, 254, 258–259, 396, 421
  communications, 58, 63, 189, 218
  launch, 10, 29, 30, 411
  satellite program, 11
  sounding rockets, 30, 42–43
Canadian Pacific Air Lines, 70
Canary Islands, 39, 55, 204
Canaveral Council of Technical Studies, 145
Canberra, Australia, 103, 396
Cancer, 144–145, 277
Canning, Thomas M., 290
Canopus (star), 397–398, 398–399, 407–408, 422
Cap motor, 438
Cape Canaveral, Fla., name, 22, 126
Cape Girardeau, Mo., Chamber of Commerce, 63–64
Cape Kennedy Air Force Station, 22
Cape Kennedy, Fla. (see also Cape Canaveral, Atlantic Missile Range, and Kennedy Space Center, NASA), 101, 126, 166, 201, 227, 279, 297, 309, 310, 397, 411, 414
  construction, 15, 39, 51, 59, 61, 62, 63, 65, 77, 124, 130, 131
  earthquake, 119
  labor relations, 209
  lightning damage, 289, 294
  strike, 51, 59, 61, 62, 63, 65, 69–70, 124, 126, 128, 130, 131, 133, 141, 212, 243
"Capitol Cloakroom" (radio program), 2
Capsule, rotating, test, 194
Caracas, Venezuela, 275
Caravelle (jet transport), 411
Carbon 13, 398
Carbon dioxide, 132–133, 382
Carcinogenic research, 277
Cardiac arrhythmia, 443
Cardiometer, 440
Cardiovascular system, 174–175, 177–178, 178–179
Career guidance clinic, 388

Career Service Award (National Civil Service League), 136
Cargomaster. See C-113.
Carlucci, Frank, 129
Carmody, Robert J., 114
Carnarvon, Australia, 39, 222, 284
Carnegie Institution of Washington, 138–139
Carpenter Bros., 203, 434
Carpenter, LCdr. M. Scott (USN), 176, 252, 266, 322, 372, 390
Cartellieri, Wolfgang, 387
Carvalho Silos, Geraldode, 375
Cascade Mountains, Oreg., 296
Case Institute of Technology (CIT), 42, 56, 214–215
Case, Sen. Clifford P., 201
Caspian Sea, 21
Cat. See Turbulence, clear air. 88
Cat-Spy (instrument), 88
CBS. See Columbia Broadcasting System.
CC&S. See Central Computer and Sequence.
Cellular structure, 114
Centaur (booster), 221, 246, 300
  AC-2, 32
  AC-3, 227, 234, 483
  AC-4, 414–415, 476–477, 487
  development, 78, 193, 371–372
  use of, 138, 171, 173
Centaur "E" Stand Project (Plum Brook), 221
Center for European Nuclear Research (CERN) (Switzerland), 71–72
Center for Radio Physics and Space Research (Cornell Univ.), 164–165
Center of Science and Industry, Columbus, O., 119
Central America, 146
Central Computer and Sequence (CC&S), 41
Central Radio Propagation Laboratory (NBS), 363, 373
Centre National d'Etudes Spatiales (CNES). See National Center of Space Research (France).
Centrifuge, human, 226, 243, 245, 249, 427
Ceramics, 261
Ceres Missile Range (France), 233
CERN. See Central for European Nuclear Research.
Cesium, 158–159, 410, 427, 476–479
CH-46C (helicopter), 264
CH-47A (helicopter), 436
Chaff, tracking of, 107
Chamical, Argentina, 403
Chandler, Judge Stephen, 177
Chaney and James Construction Co., Inc., 61, 328
Chanute, Octave, Award, 376–377
Charleston, S.C., 330
Charyk, Dr. Joseph V., 242, 283, 284, 285
Chatenet, Pierre, 25
Chayes, Abram, 63
Checkout equipment, 248
Chemical energy conversion, 151–152

Chemical propulsion, 83, 296, 302
Chemistry, 262, 288
Cheprov, Ivan, 313
Cherry, William R., 276
Chesapeake Bay, 197
Chewing gum, 172
Chicago, Ill., 110–111, 212, 288, 357, 428
Chicago Sun-Times, 211
Chicago, Univ. of, 4, 140, 268, 339, 357, 362
Chick embryo research, 131
Chico, Calif., 78–79
Chile, 153
Chimpanzee, 390
China, Communist, 129, 296–297, 352, 354, 382, 394, 439
Chope, Henry R., 100
Christensen, E. E., 430–431
Christian Science Monitor, 40, 59, 66
Christmas Island, Pacific Ocean, 333
Chrysler Corp., 73, 104, 148, 212, 341–342
 Florida Missile Operations, 201
 Space Div., 84–85
Chubb, Talbot, A., 45
Cincinnati, Univ. of, 139
Circumlunar flight, 173
CIT. See Case Institute of Technology.
Civil Aeronautics Board (CAB), 4, 8
Civil aviation, 94, 216, 237
Clark AFB, Philippines, 290–291
Clark, David B., Co., 197
Clark, Evert, 76
Clark, Sen. Joseph S., 128
Clark Univ., 233, 319
Clark, Maj. William B. (USAF), 48
Clarke, Arthur C., 88
Clay, Cassius, 81
Clear air turbulence (Cat), 88, 291, 383
Clear Lake, Texas, 25, 74, 116, 221
Clemmons, Dewey L., Jr., 419
Cleo, Hurricane, 297
Cleveland, Harlan, 83–84
Cleveland, Rep. James C., 48
Cleveland, Ohio, 91, 163, 166–167, 295, 300
Cloud, 287, 288–289, 464–465
 cosmic-dust, 190
 noctilucent, 3–4, 147, 204–205, 241, 277–278
Cloud cover, 124
Cloudcroft, N.M., 181
CNES (Centre National d'Etudes Spatiales). See National Center of Space Research (France).
CNES–NASA program, 95
CNIE. See Argentine National Commission on Space Research.
Cochran, Jacqueline, 181, 200, 203
CODSIA. See Council of Defense and Space Industry Associations.
COESA. See U.S. Committee on the Extension of the Standard Atmosphere.
COIN. See Aircraft, counter-insurgency.
Cole, Pleasant T., 82

Colleges. See Universities.
Collier, Robert J., Trophy, 328
Collins Radio Co., 176, 216
Collis, Dr. Ronald T. H., 278
Colorado, 58
Colorado Springs, Colo., 125, 198
Columbia Broadcasting System (CBS), 2, 90, 211
Columbia Univ., 188, 238, 277
Columbus, Ohio, 15–16, 30, 33, 100, 119, 142
Combustion, 216
Comet, 210, 398
Commerce, Dept. of, 1–2, 18, 43, 75, 183, 231
Commercial Club, Chicago, 428
Commercial Museum, 322
Commission on Federal Expenditure Policy (proposed), 18
Committee on Potential Contamination and Interference from Space Experiments (SSB), 116–117
Committee on Science and Public Policy (NAS), 110
Committee on Science in the Promotion of Human Welfare (AAAS), 437
Committee on Space Research (COSPAR), 155, 171, 172, 175, 177–178, 178–179, 179, 179–180, 180, 183, 185, 220, 432
Committee to Preserve Cambridge (Mass.) Industries, 327, 363
"Common Action for the Control of Conflict," 59
Commoner, Dr. Barry, 437–438
Communicable disease, 249
Communication (see also Communication satellite systems), 50–51, 113, 142, 189, 258–259, 333
 contracts, 105, 134, 176
 exhibition, USIA, 142, 261
 experiments, 3, 17–18, 22, 55, 60, 75, 105, 132, 143
 extrasensory perception in, 278
 international, 17–18, 22, 30, 60, 75, 79, 111–112, 208
 systems
  balloon-satellite, 89
  microwave, 143
  radio, 140
  satellite, 22, 32, 75, 81, 168, 185, 238
  tracking, 134, 176, 216
  voice, 22, 89, 134, 142, 319
  x-ray, 100
Communications Satellite Act of 1962, 58, 116, 130, 286
Communications Satellite Corp. (ComSat Corp), 255
 agreement, 260, 293, 431
 contract, 61, 105, 209
 criticism, 313
 international applications, 55, 79, 122, 240, 258–259, 293, 347
 management, 219, 260, 286

military applications, 27, 107, 116, 118, 130, 247, 283, 284
operations, 75 113, 168, 192, 202, 209, 279, 285, 291, 310, 317, 321
organization, 242, 265, 318, 321, 323, 331
plans, 38, 58
satellite program (see also Early Bird), 96, 105, 139, 237, 247, 383–384
stock, 75, 279, 291, 405, 407, 421
terminal stations, 365, 383–384
test, 326
Communications satellite system, commercial, 38, 58
Communications satellite system, international, 55, 209
agreement, 205–206, 260, 283, 293
criticism, 313
development, 61, 79, 96, 105, 129, 136, 237, 255, 285
contract, 105, 209
establishment, plans for, 38, 61, 68, 310
impact of, 88
international participation, 26, 58, 61, 63, 122, 129, 136, 189, 205–206, 215, 218, 240
management, 286
military use of, 27, 55, 107, 116, 118, 130, 186, 247, 254, 283, 284
Communications satellite system, military, 27, 55, 107, 116, 118, 130, 186, 247, 254, 279, 283, 284, 289, 332–333, 345, 347, 358
Computer, 142, 269, 383, 412
"satellite alert," 188
systems, 132, 297
use of, 1–2, 161, 237, 252, 284, 411
ComSatCorp. See Communications Satellite Corp.
Concorde (U.K.-France) supersonic transport, 61, 170, 279, 314, 362, 418
Conference, Atoms-for-Peace, Third International, 305
Conference Group of U.S. National Organizations of the U.N., 83–84
Conference, Midwest Governors, 312–313
Conference of Photovoltaic Specialists, 202
Conference of Postal and Telecommunications Administration, European, 26
Conference on aerospace medicine, 45, 48, 52
Conference on Geological Problems in Lunar Research, 180
Conference on Government Contracting for Research and Development, 250, 252
Conference on Human Factors in Electronics, Fifth Annual, 166
Conference on New Technology, 204, 367
Conference on Peaceful Uses of Space, Fourth National, 156–157, 157–158, 162
Conference on Satellite Communications, European, 58, 61, 63, 189, 218, 260

Conference on Space Nutrition and Related Waste Problems, 151–152
Conference on the Role of Simulation in Space Technology, 290, 293
Congress, 2, 12, 23, 23–24, 244, 259, 260, 260–261, 275–276, 279, 300–301, 301, 307, 317, 334, 339
Joint Committee on Atomic Energy, 67, 117, 161
Joint Committee on Economics, 18
NASA budget, 12–13, 23, 181, 221, 227, 300–301
report to, 35, 55, 58, 253, 284
Congress, House of Representatives, 8, 13, 37, 69, 85–86, 115, 117, 189, 201, 221, 237, 259, 274, 284, 291, 292–293, 320, 324, 331–332
bills introduced, 18, 23, 183
bills passed, 72, 101, 108, 185, 259, 285
Committee on Appropriations, 141, 180, 183, 185, 213
Subcommittee on Independent Offices, 185–186
Committee on Armed Services, 28, 44–45, 65, 69, 81, 92, 430
Committee on Education and Labor, Subcommittee on National Research Data Processing and Information Retrieval Center, 144–145
Committee on Government Operations, Subcommittee on Military Operations, 107, 109, 118, 130, 186, 345, 347
Committee on Interstate and Foreign Commerce, 8, 117–118
Subcommittee on Public Health and Safety, 8
Committee on the Judiciary, 43–44
Committee on Rules, 69
Committee on Science and Astronautics, 2–3, 12–13, 15, 64–65, 77, 79, 101, 102, 103, 106, 126, 136, 165, 200, 211, 281, 311, 401–402
hearings, 45–46, 46–47, 49, 50, 50–51, 52–53, 53, 67
Subcommittee on Advanced Research and Tracking, 12, 71, 80, 83, 91–92, 92–93, 95, 101, 103
Subcommittee on Manned Space Flight, 12, 66, 68–69, 73, 80
Subcommittee on NASA Oversight, 107–108, 150, 150–151, 151, 163, 163–164, 215, 220
Subcommittee on Science, Research, and Development, 2–3, 12, 165, 165–166, 186, 191–192, 285, 312, 362
Subcommittee on Space Science and Application, 12, 69, 71, 79, 165
Independent Offices Appropriations Bill, Fiscal Year 1965, 180, 183, 185, 185–186, 242, 266, 276, 284, 285, 286, 300–301
Republican Conference, 203
Select Committee on Government Research, 22, 67

Congress, Senate, 12, 37, 75, 79, 85–86, 93, 187–188, 201, 221, 225, 227, 227–228, 237, 256–257, 277, 284, 310, 329–330, 330–331, 331
  bills introduced, 18, 180, 312
  bills passed, 79, 82, 98, 219, 221, 260, 276, 284
  Commission on Federal Expenditure Policy, 18
  Committee on Aeronautical and Space Sciences, 18, 43–44, 64–65, 94, 96–97, 97–98, 100, 101, 104–105, 106, 108–109, 126, 163, 165, 180, 200, 211, 268–269, 408, 425, 441
  Committee on Appropriations, 237–238, 257, 266
    Subcommittee on Independent Offices, 242
  Committee on Armed Services, 79
    Subcommittee on Preparedness Investigation, 15, 177
  Committee on Commerce, 5
  Committee on Foreign Relations, 67, 68
  Committee on Government Operations, 18
  Committee on the Judiciary, 43–44
  Committee on Labor
    Subcommittee on Employment and Manpower, 128
  Committee on Public Works
    Subcommittee on Air and Water Pollution, 229–230, 230
  Committee on Rules and Administration
    Subcommittee on the Smithsonian Institution, 220
  Committee on Small Business, 434
Congress, International Astronautical Federation (IAF), 15th, 308–309, 313, 314
Congress, 4th, International Council of the Aeronautical Sciences, 296
Congressional Medal of Honor, 271–272
Congressional Record, 22, 24, 35, 84, 145, 164, 187–188, 213–214, 291
  economy, 213–214
  space program, 29, 221, 318, 337
  supersonic transport, 93–94, 213–214
Constellation (Lockheed), 306
Constellation, Super (Lockheed 1049C), 197
Contaminants, 382
Contamination, atmospheric, 76
Contract (see also under agencies, such as NASA, USAF, etc.), 214, 281
  cost-plus-incentive-fee, 164
  fixed-price, 110
  geographical distribution of, 165, 165–166, 261
  incentive, 171, 173, 250, 342
  military, 26, 86, 165–166
  space, 29, 86, 162
Contributions Award, NASA, 276
Control Data Corp., 269, 299
Convair 880 (aircraft), 233
Convention of International Civil Aviation, 216
Convention of Military Electronics, International, 317
Cook Electric Co., 212, 231
Coolant, 125
Cooling techniques, 181, 291
Coons, Dr. D. Owen, 91
Cooper, Maj. L. Gordon (USAF), 4, 33, 48, 98–99, 111, 129, 158, 163, 164, 177–178, 194, 247, 273, 317, 325, 355, 393
Coordinating Research Council (CRC), 81
Copenhagen, Denmark, 96, 173–174
Copernicus (lunar crater), 275–276
Copernicus, Nicolaus, 125, 398
Coriolis effect, 168–169
Cornell Aeronautical Laboratories, Inc., 147, 199
Cornell Guggenheim Aviation Safety Center, 200, 213–214
Cornell Univ., 164–165, 180, 211, 297, 301, 306–307, 393
Corona, 34, 96, 119, 180, 260
Coronie, Surinam, Dutch Guiana, 278
Coronoscope II (project) (ONR-NSF-NASA), 96
Cosmic dust, 190
Cosmic radiation, 174, 216
Cosmic radio communications, 75
Cosmic ray, 4, 42, 50–51, 56, 176, 179–180, 211, 291
Cosmic Research, 443
Cosmonaut, 120, 130, 174–175, 183, 205, 226, 309, 345, 352
Cosmonautics, 353
Cosmonauts' Desk, 131
Cosmos (U.S.S.R. satellite), 151–153, 357, 379
COSMOS V, 399
COSMOS XXIII, 116
COSMOS XXV, 450–451
COSMOS XXVI, 107, 110, 448
COSMOS XXVII, 118, 448
COSMOS XXVIII, 128, 449
COSMOS XXIX, 149, 449
COSMOS XXX, 450
COSMOS XXXI, 205, 451
COSMOS XXXII, 210, 451
COSMOS XXXIII, 220, 451
COSMOS XXXIV, 234, 451
COSMOS XXXV, 249, 452
COSMOS XXXVI, 265, 453
COSMOS XXXVII, 286, 453
COSMOS XXXVIII, 290, 454
COSMOS XXXIX, 290, 454
COSMOS XL, 290, 454
COSMOS XLI, 294, 454
COSMOS XLII, 294, 455
COSMOS XLIII, 294, 455
COSMOS XLIV, 299, 455
COSMOS XLV, 315–316, 456
COSMOS XLVI, 327, 456
COSMOS XLVII, 341, 457
COSMOS XLVIII, 351, 458
COSMOS XLIX, 362, 459
COSMOS L, 365, 459
COSMOS LI, 414, 461

COSPAR. See Committee on Space Research.
Cost-plus-incentive-fee (CPIF) contract, 164
Cotton, Col. Joseph (USAF), 340
Coughlin, William, 58
Council of the Aeronautical Sciences, International, 4th Congress, 296
Council of Defense and Space Industry Associations (CODSIA), 246
Council on World Affairs, 391-392
COURIER (communications satellite), 241
Courtney-Pratt, J. S., 154
Coyote, Calif., 278, 401
Cpif. See Cost-plus-incentive-fee contract.
Crab Nebula, 153, 254, 436-437
Crane, Mayor Edward A., 292
Crash, 178
  aircraft, 8, 41
  beacon, locator, 19
  test, 147, 211, 214, 306
Crater ray, 298-299
CRC. See Coordinating Research Council.
Creede, Colo., 394
Crews, Capt. Albert R. (USAF), 304
Crosby, Harry, 239
Crossfield, A. Scott, 29
Cryogenics, 101, 103
CTA-21 (radio emission source), 364
CTA-102 (radio emission source), 364
Cuba, 81, 195, 200
Cudaback, Dr. David D., 221-222
Cuddeback Dry Lake, Calif., 185
Cunningham, Newton W., 352-353
Cunningham, R. Walter (NASA), 296, 356
Curtis, Rep. Thomas B., 18
Curtiss-Wright Corp., 13, 28, 183, 223, 279, 438
CX (military jet transport C5), 429
Czechoslovakia, 120
Daddario, Rep. Emilio Q., 2-3, 160, 165, 221, 281, 285, 312
Dade County, Fla., 190
Dagle, Sidney J., 141
Dakar, Africa, 240
Dallas, Tex., 30, 106, 187, 203, 437
Dallin, Prof. Alexander, 188
Damping device, 13, 276
Daniel, Mann, Johnson & Mendenhall, 383-384
Darchiya, Sh. P., 203
Darr, Joseph A., Jr., 406
Darwin, George H., 28, 257
Daso. See Demonstration and Shakedown Operations.
Data, 371, 371-372
Data acquisition, 100, 216, 219
Data collection satellite system, 196
Data-filter equipment, 216
Data processing equipment, 378
Data retrieval, 305
Davis, Alonza J., 114
Davis, Ch. M/Sgt. Frank M. (USAF), 315
Davis, Rep. John W., 23

Davis, M/C Leighton I. (USAF), 3, 164, 192
Davis, Dr. Watson, 419
Davis, M/C W. Austin (USAF), 39, 145-146
Dayglow, 378
Dazzle, Project (U.S.-U.K.-Australia reentry research), 269-270, 278
DC-3 (transport), 113-114, 318
DC-7 (transport), 147, 192, 306
DC-8 (transport), 158, 180, 410
De Gaulle, Pres. Charles (France), 161
Debold, LCdr. J. F. (USN), 238
Debrovner, Dr. M., 208
Debus, Dr. Kurt H., 53-54, 109, 352-353, 430-431
Decelerator, 163, 259
Decompression experiment, 274
Decoration for Exceptional Service (FAA), 164
Deep Space Network (NASA), 31
Defence Research Board (Canada), 11
Defender, Project (ARPA), 211
Defense and Space Operations Div. (Westinghouse Corp.), 202-203
Defense Cost Reduction Week, 248
Defense, Dept. of (DOD), 6-7, 9, 23, 35-36, 49, 51, 67, 75, 133, 136-137, 154, 178, 201, 213, 214, 222, 282, 303, 305, 334, 335, 430
  appropriations, 69, 72, 79, 81, 82, 98, 101, 102, 103, 104-105, 106, 108, 115, 141
  budget, 21, 48, 69, 80, 84, 88, 115, 360
  communications satellite system, 27, 107, 109, 116, 130, 247, 254-255, 279, 283, 284, 289, 332-333, 345, 347
  contract, 3, 286, 305, 319, 338, 332, 369, 375
  Contract Audit Agency, 417
  cooperation, 231, 246-247
    NASA, 12, 64-65, 80, 82, 90, 108-109, 143, 145, 166-167, 168, 171, 176, 192, 258-259, 291, 439
    Mol, 1, 8, 10, 68, 417, 428
  NASA-AEC, 231
  cost reduction, 89, 248
  Distinguished Civilian Service Award, 265
  expenditures, 235, 240, 434-435
  information analysis center, 303
  launch, 442-443, 446-479
  management, 26, 49, 51, 96, 115, 164, 183, 281, 311
  R&D, 23, 26-27, 48, 69, 165
Defense Research Laboratories (General Motors), 8, 197
Defense Science Board (DSB), 29
Definition Phase Project (Pdp) (DOD), 96
DeFrance, Dr. Smith J., 136
Degner, Capt. Eugene A. (USAF), 112-113
Dehydration effects, 129
Deimos, Project (Douglas Aircraft Co.), 347
Deissler, Robert G., 403
Delamar Lake, Nev., 29

Delaware Academy of Science, 433
Delta (see also Thor-Delta), 72, 80, 109, 145, 212, 225, 251–252
  x–248, 135, 141, 147, 166
Delta Air Lines, 158
Delta, Thrust-Augmented (Tad) (booster), 80, 212, 225, 290–291, 431, 454, 467
Deluzio, Frank, 441
Demler, m/g Marvin C. (usaf), 174, 392
Democratic National Convention, 246, 297
Democratic Party Platform Space Plank, 294
Demonstration and Shakedown Operations (Daso), 409
Dempsey, J. R., 269, 275
Denmark, 49, 56, 173–174, 397
Denmark-America Foundation, 173–174
Dennis, David H., 226
Dennison, James T., 347
Densitron, 263
Density, atmospheric, 50–51, 131–132, 141, 241
Dentistry, 121, 172
Denver Post, 428
de Orsey, C. Leo, 343
Department of Defense. See Defense, Dept. of.
Derelin, Rep. Lionel Van, 275
Derry, La., 56
Derryberry, Mrs. Mary, 114
Desert survival training, 281
Detection system, malfunction, 13, 121
Detector, electron, 103, 179, 210, 325
Detroit, Mich., 33, 143
Deutsch, Dr. Armin, 180
Dewan, Dr. Edmond M., 409
dh–125 (U.K. jet transport), 209
Diamant (French booster), 161, 210
Dickerson, Joseph T., Jr., 187
Dictionary, astronautical, English-Russian, 128
Dillon, Sec. of Treasury Douglas, 119–120
Dimethylhydrazine, 229–230
Dingeldein, Richard C., 350
Dipole, 116–117
Dirksen, Sen. Everett, 337
Disarmament, 17, 21, 186–187, 437
Disarmament Conference, 17, 21
Discoverer (satellite), 108–109, 150–151
Disease, communicable, 249
Distinguished Civilian Service Award (dod), 265
Distinguished Flying Cross (usaf), 32
Distinguished Flying Cross (usa), 394
Distinguished Service Honor Gold Medal Award and Certificate (afcea), 185
Distinguished Service Medal (usaf), 266
Dixon, Franklin P., 213, 375
Doctors, medical, trained as astronauts, 249
Documentation, Inc., 237
dod. See Defense, Dept. of.
Dolan, Thomas E., 139
Donely, Philip, 156
Donner, Frederick G., 323, 331
Donover Construction Co., 127
Doolittle, l/g James H. (usafr), 206–207
Doppler tracking method, 109, 325
Dornberger, Dr. Walter, 59
Douglas Aircraft Co., 80, 141, 238, 251–252
  contract, 189, 193, 242, 248, 422
  study, 74, 87, 205
  Missile and Space Systems Div., 87, 122, 189, 246, 249, 302, 403, 422, 427
  Saturn, 25–26, 159, 189, 193, 226, 242, 248, 302
  space station, 74, 93, 122, 189, 205, 249
  studies, 81–82
  World Cruiser, 327
Douglas, Donald W., Sr., 327
Douglas, Col. William K. (usaf), 175–176
Douglas-Home, Sir Alec, 349
Dow Chemical Corp., 222
Downey, Calif., 24, 299
Drag measurement, 319
Dragon (French sounding rocket), 273
Drake Univ., 244
Draley, Eugene C., 177
Draper, Prof. Charles Stark, 311, 397
Drinkwater, Fred J., 376
Driscoll, John, 337
Drogue parachute, 131, 156, 175
Drop test, 163, 220, 259
Drummond, Roscoe, 40
Dryden, Dr. Hugh L., 85, 113, 124, 143, 158, 163, 282–283, 300–301, 319, 416, 432–433
  awards by, 77, 82, 276
  awards to, 111, 127, 234, 235–236
  honored, 335
  international cooperation, 96–97
  U.S.-U.S.S.R., 2, 17–18, 175, 203, 204, 208, 342, 349
  space, peaceful use of, 156–157
  space program, 46–47, 74, 149–150, 187, 224, 240, 243, 432–433
dsb. See Defense Science Board.
Duberg, Dr. John E., 177
Dubos, Prof. René J., 438
DuBridge, Dr. Lee A., 176
Duke Univ., 194
Dulles International Airport, Washington, D.C., 411
Duluth, Minn., 116
Dunning, Dr. J. R., 277
Durban, So. Africa, 226–227
Dushanbe, U.S.S.R., 27
Dust, cosmic, 306–307, 331
  extraterrestrial, 263
  lunar, 267, 271, 275–276, 305–307
Dyment, John T., 114
Dynamic Crew Procedures Simulator (msc), 236
Dyna Soar. See x–20.
Dynatronics, Inc., 423
Early Bird (ComSatCorp communications satellite), 105, 139, 237, 247, 411, 431

INDEX 485

Earth, 106–107, 125, 152, 158
 atmosphere, 70, 96, 117
 formation, 28, 170
 gravitational field, 68, 325
 heat, 62
 investigations of, 1
 magnetic equator, 31
 magnetic field, 30, 68, 96, 104, 245, 250, 265
 radiation, solar, effects of, 291
 surface, 21
Earthquake, 119, 146, 354
Eastern Airlines, 88
Eastern Test Range (ETR) (formerly Air Force Missile Test Center (AFMTC) and Atlantic Missile Range (AMR). See also Kennedy Space Center, NASA, and Cape Kennedy), 221–222, 261, 309–310, 336
 launch, 258–259
  booster development, 192
  missile, 266, 328, 330
  plans, 171, 223
  satellite, 252, 264, 290, 305–306
 management, 164, 177, 192
 prelaunch activities
  Ranger, 218
  Saturn, 235
  Titan, 269
 support services, 239
Echo (satellite), 282
ECHO I (communications satellite), 284, 308, 418
ECHO II (communications satellite)
 communications experiment, 54, 60, 75
 launch, 25–26, 447, 449, 466
 performance, 29
 tracking, international cooperation, 30
  U.S.S.R., 17, 27, 96–97
 use of, 85
Eclipse
 lunar, 424, 426
 solar, 180
Eclipse-Pioneer Div. (Bendix Corp.), 22
Economic Club of Detroit, 143
Economy, national, 18, 38, 144, 183, 217, 221, 266
Edelson, Burton I., 441
Edgar, George P., 317
Editor and Publisher, 287
Edmondson, Sen. J. Howard, 57
Education, 56, 197, 217, 247, 263
 funds for, 3, 7, 183, 186
 need for, 263
 programs, 4, 279–280, 407
Edwards, AFB, Calif., 218, 225, 331
 Aerospace Research Pilot School, 197, 206
 engine test stand, 122
 engine tests, 429
 flights, 6, 29, 104, 118, 181, 191, 193, 203, 227, 251, 264–265, 297, 329, 352
 test flight, 85, 278, 323, 349, 425
Eggers, Dr. Alfred J., Jr., 173
Eggleston, John M., 48

Egg experiments, 72, 174, 248
Eglin AFB, Fla., 70
Egypt. See United Arab Republic.
EIA. See Electronic Industries Association.
Eisele, Capt. Donn F. (USAF), 318
Eisenhower, Pres. Dwight D., 33, 64, 133, 188, 213, 246, 274
Eisenhower, Dr. Milton, 194
EJC. See Engineers Joint Council.
Ejection seat (see also Escape system), 129
El Centro, Calif., 129, 317
El Kottamia Observatory, United Arab Republic, 424
ELDO. See European Launcher Development Organization.
Electric propulsion, 228–229, 296, 302
Electric power, nuclear reactor, 229, 305, 333
Electricity, static, 147
Electrojet, 31
Electro Mechanical Research, Inc., 352
Electron (U.S.S.R. satellite), 152
ELECTRON I, 33, 42, 54, 62, 447
ELECTRON II, 33, 42, 54, 77, 447
ELECTRON III, 245, 273, 452
ELECTRON IV, 245, 273, 452
Electron accelerator, 22
Electron beam welder, 184
Electron bombardment, 290
Electron concentration, 4, 22, 50–51
Electron density, 119, 140
Electron detector, 325
Electron miscroscope, 326, 392
Electronic beacon, 325
Electronic Industries Association (EIA), 246
Electronic parts standardization, 334
Electronic stethoscope, 172
Electronics, 166, 183, 300
 industry, 13, 430
 systems, 102
Electronics Instrumentation Div. (Lear Siegler, Inc.), 102
Electronics, Military, International Convention, 317
Electronics Research Center (ERC) (NASA), 43, 59, 388, 393, 427
 appropriations, 115
 congressional action, 77, 100–101, 185
 contract, 410
 location, 13, 37, 79, 91, 201, 292, 327, 339, 345, 337
 organization, 301
 Site Evaluation Committee, 23, 113
Electro-Optical Systems, 341, 392
Electrostatics, 56, 325
Eleuthera Island, Bahamas, 386
Elizabeth II, Queen, 59
Ellington AFB, Tex., 74
Elliott, Rep. Carl, 67
Elms, James C., 16
Ely, L/c William J. (USA), 165
Emme, Dr. Eugene M., 226

Employment, 175–176, 226, 286, 288, 302, 443
Energetic Particles Explorer D (EPE-D). See EXPLORER XXVI.
Energy conversion, 151
Energy transfer, 114
Engine
  arc-jet, 167
  ion, 82, 238, 256, 444
  liquid fuel, 87, 115, 209
  ramjet, 77, 223
  reliability, 235
  rocket, 115, 160
  supersonic transport, 161
  turbofan, 28
  vernier, throttleable, 199
Engineering Experiment Station (USN), 334
Engineering Institute of Canada, 421
Engineers, 84
  and scientists, 83, 74, 94, 302, 308, 345
  space program, 6, 35–36, 184
  training, 255
Engineers Joint Council (EJC), 12
Engineers' Week, 74
Engle, Capt. Joseph H. (USAF), 6, 130, 182, 241, 264, 326, 329, 414
Entry Technology Conference, 350
Environment (see also Space Environment Weightlessness)
  atmospheric pressure, 56
  blood pressure, 129, 177–178
  genetics, 169, 171, 179
  heart, 172, 174
  insect, 72, 117, 174, 179
  nervous system, 169, 178
  oxygen test, 44, 87, 130, 174
  radiation, 48, 169
Environment,
  hazards of, 146
  oxygen, 44, 130, 174
  space cabin, simulation, 87, 130
  space, simulation, 39, 50, 112, 118, 224
  undersea, 176
Environment control, 6, 89
Environmental sensor, 339
Eole, Project (France), 218
EPE-D. See EXPLORER XXVI.
"Epsilon" (meson particle), 352
Equator, 305, 314
ERC. See Electronics Research Center (NASA).
Erevan, U.S.S.R., 27
Erhard, Chancellor Ludwig (W. Ger.), 273
Erpilev, Nicolai P., 121
Escape systems, 89, 129
Esgar, Jack B., 82
ESP Research Associates Foundation, 287
ESRO. See European Space Research Organization.
Esro 1 (satellite), 258, 397
Esro 2 (satellite), 258, 397
Estes, L/G Howell M., Jr. (USAF), 6

ETR. See Eastern Test Range.
Europa 1 (3-stage satellite launcher), 270
Europe, 5, 58, 96, 140, 253
  Western, 297, 303
European Conference of Postal and Telecommunications Administration, 26
European Conference on Satellite Communications, 58, 61, 63, 189, 372
European Economic Community, 25
European Launcher Development Organization (ELDO), 97, 198, 257, 269–270
European Space Research Organization (ESRO), 97, 161, 242, 264
  agreement, 258
  launch
    U.K., 239, 242
  plans, 113, 364, 360
  tracking station, 264, 308, 397
Eurovision (Western Europe TV network), 297
Evans, Albert J., 92
Evans, Thomas C., 170
Evins, Rep. Joe L., 29
Exceptional Civilian Service Medal (USAF), 74
Exceptional Service Decoration (FAA), 164
Exhaust, rocket, 229–230
Exhibit, 145, 214, 230, 247, 266, 300
Exobiology, 50, 366
Explorer (program), 11
EXPLORER I (satellite), 37, 96, 196
EXPLORER VIII, 106
EXPLORER IX, 131–132
EXPLORER X, 172
EXPLORER XII, 479
EXPLORER XIV, 479
EXPLORER XV, 479
EXPLORER XVI, 194, 380
EXPLORER XVIII, 103, 168, 284, 320, 339
EXPLORER XIX, 473
EXPLORER XX, 296, 360–361, 363, 455, 467
EXPLORER XXI, 339, 436, 457, 467
EXPLORER XXII, 346, 351, 386, 407, 457, 467
EXPLORER XXIII, 380, 459, 467
EXPLORER XXIV, 394, 460, 467
EXPLORER XXV, 394, 460, 468
EXPLORER XXVI, 427, 462, 468
Explorer Beacon (satellite), 325, 482, 109
Explorer, Geodetic (satellite), 325
Explorer-Jupiter C (satellite), 234
Explorers' Club, 327
Explosion, 25, 104, 110
Export, 114
Extrasensory perception, 278
Extraterrestrial life, 210
Ezop, James J., 135
F-1 (rocket engine), 114, 120, 122, 197, 242, 346, 377
F-2 (rocket engine), 412
F-4 (Phantom II), 226
F-10F (aircraft), 191
F-12A (A-11 supersonic aircraft), 225, 254

F-104 (Starfighter), 5, 43, 181, 203, 404, 406
F-105 (Thunderchief), 66, 178
F-111 (TFX), 190–191, 428, 430
F-111A (USAF-TFX), 190–191
F-111B (USN-TFX), 191
FAA. See Federal Aviation Agency.
Fafnir Bearing Co., 293
Fairbanks, Alaska, 103, 284
Fairhall, Minister of Supply Allen, 278, 396
Fairhurst, Judge Charles, 345
"Fairy castle" theory, 299
FAITH 7, 367
Falkland Islands, 397
Farman, Henry, 79
Farman, Maurice, 79
Farrell Construction Co., 61
Fassett, John W., 141
Fatigue, metal, 103
FBA. See Federal Bar Association.
FCC. See Federal Communications Commission.
FDA. See Food and Drug Administration.
Fecker, J. W., Co., 264
Federal Aviation Agency (FAA), 60, 80–81, 113, 133, 164, 171, 216, 237, 296, 318, 332, 443
  award, 164
  contract, 56, 201, 259
  cooperation, 231, 233
  organization, 24
  report, 293, 332
  tests, 214
    safety, 147, 211, 306
    sonic boom, 43, 63, 84, 117, 175, 177, 183, 187, 242, 253, 268, 274, 277, 314, 360
  transport, supersonic, 201
    design and development, 5, 28, 57, 61, 148, 183, 318
    management, 115, 154
    reservations, 96, 136, 159, 175
Federal Bar Association (FBA), 250
Federal Communications Commission (FCC), 55, 105, 113, 122, 139, 168, 286, 412
Federal Electric Corp., 201
Federal government. See United States government.
Federal Mediation and Conciliation Service (FMCS), 25
Federal Support of Basic Research in Institutions of Higher Learning, 110
Federal Systems Div. (IBM), 122
Federal Trade Commission (FTC), 333
Federal Women's Award, 93
Federation of Italian-American Democratic Organizations of New York, 101
Fellowships, 308
Fels Planetarium, 205
Fenn College (Cleveland, O.), 214
Fenn, Prof. Wallace Osgood, 266, 315
Feodorov, Yevgeny, 352
Feoktistov, Konstantin Petrovich, 348, 355, 359, 469
Ferguson, E. E., 373
Ferguson, L/G James (USAF), 45
Fermi, Enrico, Award, 395
Fiber metal, 197
Fiberglas, 393
Field, Jennings Pemble, Jr., 342
Field Enterprise Educational Corp., 287
Fighter, tactical. See F-111.
Finger, Harold B., 83, 195, 229, 296, 311, 349, 435
Finney, John, 26–27, 86, 351, 355, 425
Fire, crash, 214
Fire, Project (NASA), 135
Fire 1, (reentry test), 135, 350, 483
Firefly, Project (USAF), 70, 265
Firor, Dr. John W., 389
Fischel, Jack, 122
Fisher County, Tex., 121
Flagstaff, Ariz., 75, 204–205, 218, 223
Flare, solar, 13, 48, 173, 194, 388–389
Flax, Dr. Alexander H., 166, 303
Fleas, 201
Fleming, Arthur S., Award, 63, 173
Flight Achievement Award (AAS), 164
Flight control, 29,
Flight Research Center (FRC) (NASA), 70–71, 77, 138, 145, 147–148, 218, 263
  contract, 201, 319
  flight, 214, 283
Flight Safety Foundation (FSF), 147, 156, 306
Flight simulation, 146, 199
Flight training, 332
Florence, Italy, 171, 175
Florida, 68, 153, 221, 266,
Florida East Coast Railway, 51, 59, 61, 63, 65, 69
Florida Library and Historical Commission, 22
Florida Operations (MSC), 299
Florida Times-Union, 433
Florkin, Prof. Marcel, 432
Flotation collar, 102
Flox (fluorine-liquid oxygen), 77, 106, 219
Fluid mechanics, 235–236
Fluorine, 76, 230
Fluorine-hydrogen, 160, 230
Fly, fruit. See Fruit fly.
Flying Boxcar. See C-119.
Flying saucers, 289, 387
Flying submarine, 397
Flying wing. See MX-324.
FMCS. See Federal Mediation and Conciliation Service.
Food and Drug Administration (FDA), 131
Food, 79, 145
Force measuring instrument, 114
Ford, Brian J., 441
Forecast, Project (USAF), 66, 206, 252, 257
Forgash, Morris, 370
Fort Bliss, Tex., 394

Fort Churchill Research Range, Manitoba, 278, 290
  launch
    balloon, 4
    sounding rocket, 11, 22, 30, 42, 62, 79, 84, 107, 110, 113–114, 133, 140–141, 260, 262, 264, 277, 373, 378, 380
Fort Davis, Tex., 148
Fort Dix, N.J., 146
Ft. Hood, Tex., 182
Ft. Monmouth, N.J., 241
Fort Washington, Pa., 444
Fortune (magazine), 39, 111, 122, 161, 441
Forum on Developments and Techniques for Air Traffic System of the Future (RTCA), 158
Foss, Dr. Ted, 188
Foulois, M/G Benjamin D. (USA, Ret.), 136, 412
Foundation Library Center, 3
Fracture toughness, 223
France, 14, 79, 132, 246, 266, 303, 314, 385
  aircraft, "Concorde" (France-U.K.), 61, 171, 314, 362
  communications, 140
  cooperation, 95
  launch, 210, 371
  rocket engine, 233, 371
  space program, 161
  tracking, 32, 132, 204, 308
  weather balloon, 219
Franchi Construction Co., 38
Frankford Arsenal (USA), 94
FRC. See Flight Research Center.
Freche, John C., 56
Frederick, Okla., 177
Freedman, Dr. Sanford, 9
Freedom Foundation, 75
Freeman, Capt. Theodore C. (USAF), 370, 388
French Guiana, 161
French National Center for Space Studies (CNES), 95
French National Geographic Institute, 308
Frequency control and calibration facility (AMR), 39
Friedl, George, Jr., 224
Friedman, Dr. Herbert, 21, 45, 153, 179, 254, 432, 436
"Friendship Sputnik" (ECHO II), 85
Friloux, Henry, 197
Fritz, John, Medal, 1965, 334
Frog egg experiment, 173
"From the People" (radio program), 354
Fruit fly, 72, 171, 179
Frutkin, Arnold W., 240, 371
Fryd Construction Corp., 63
FSF. See Flight Safety Foundation, 107, 109, 186, 235
FTC. See Federal Trade Commission.
Fubini, Dr. Eugene, 107, 109, 186, 235
Fuel, 32
  high-energy, 115, 167
  safety of, 81
  solid/liquid, 160
  toxic, 230
Fuel assembly, simulated, for nuclear reactor, 25
Fuel cell, 68, 140, 383
Fulbright, Sen. J. William, 84, 173, 221, 225, 227, 237, 257
Fuller, George A., Co., 121, 400
Fullerton, Calif., 100
Fulton, L/C Fitzhugh L. (USAF), 273, 317
Funk, M/G Ben I. (USAF), 12, 29, 162, 269
Furnace, black-body, 142
Furnas, Dr. Clifford C., 29
Fusca, James A., 291
Future Flight Manpower Study Board (FAA), 55
Gable, L. D., 166
Gagarin, Col. Yuri A., 74, 96, 125, 130–131, 340
Galactic radio noise, 118
Galaxy, 297
Galileo, 271
Galveston Bay, Tex., 5, 131, 352
Ganymede (Jupiter's moon), 179
GAO. See General Accounting Office.
Garbacz, Michael L., 113
Garbarini, Robert F., 130
Garner, Howell D., 276
Garrett Corp., 404, 416
  AiResearch Div., 6
Garrick, I. Edward, 352
Gas bearing, 386
Gatland, Kenneth W., 42, 269
Gatlinburg, Tenn., 349
Gazenko, Dr. Oleg, 174, 309, 374, 389–390
GD/A. See General Dynamics/Astronautics.
GE. See General Electric Co.
Geake, Dr. J. E., 403
Geiger counter, 325
Gell-Mann, Dr. Murray, 72
Gem. See Ground effect machine.
Gemini (program), 68, 71, 74, 82, 99, 118, 121, 129, 192, 224, 274, 356
  appropriation, 186
  cooperation, 12, 55, 71, 82
  criticism, 203
  development of, 52, 139, 375–376
  experiment, 258
  plans for, 173, 194, 329
  tracking stations, 222, 306
Gemini (spacecraft), 284, 292, 300, 317
  astronaut training, 25, 81, 209
  equipment, 57, 135, 178, 240
  facilities, 10
  GT-1, 130, 134, 449, 466
  GT-2, 265, 289, 294, 297, 316
  GT-3, 212, 273, 289, 294, 316, 365–366
  GT-4, 263, 265
  launch vehicle (Titan II), 9, 13, 20, 63, 245, 259, 297, 328, 371, 413
  recovery, 5
  landing. See Paraglider.
  rendezvous, 26

test, 131, 156, 167, 174, 181, 193, 197, 264, 276, 278, 318
use of, 249, 307-308
Gemini B/MOL, 145, 261
Geminid stream, 415
Genatt, Saul H., 313
General Accounting Office (GAO), 55, 127, 253, 334
General Atomics Div. (General Dynamics), 248, 380
"General Aviation Aircraft Owners Survey—1962", 293
General Contractor, Inc., 31
General Cultural Exchange Program, 182
General Dynamics Corp., 77, 145, 148, 353
General Dynamics/Astronautics (GD/A), 78, 111, 193, 234, 269, 275, 308, 344, 384
General Dynamics/Convair, 89, 397
General Dynamics/Electric Boat Div., 397
General Dynamics/Electronics Co., 425
General Dynamics/General Atomics Div., 380
General Electric Co. (GE), 68, 73-74, 159
Aerospace and Defense Service Engineering Dept., 70
Apollo Support Dept., 299
Command Systems Div., 2
contract, 59, 201, 219, 221, 248-249, 319, 405
study, 68, 74, 204-205
Re-entry Systems Dept., 337
supersonic transport, 13, 28, 183
Valley Forge Space Technology, 177, 304
General Federation of Women's Clubs, 207
General Motors Corp. (GMC), 323
AC Spark Plug Div., 18
Defense Research Laboratories, 8, 197
General Precision, Inc., 430
General Purpose Airborne Simulator (GPAS), 199
General Telephone Co., 81
Generator, 65, 172
Geneva, Switzerland, 17, 21, 100-101, 116, 122, 187, 205, 215, 255, 305
Genisco, Inc., 65
Geodesy, 109
Geodesy, Intelligence and Mapping Research and Development Agency (GIMRADA) (USA), 74
Geodetic Explorer (satellite), 325
Geodetic research, 146, 222, 325
Geodetic Satellite Policy Board, NASA-DOD-Dept. of Commerce, 325
Geological Problems in Lunar Research, Conference on, 180
Geologist, 66, 100
Geology, 170
Geomagnetic field, 50, 380
Geophysics, 170, 208
Geophysics Corp. of America, 14, 140
George AFB, Calif., 5

George C. Marshall Space Flight Center (NASA). See Marshall Space Flight Center.
George, Col. Clarence J. (USA), 77
George Washington Univ., 178
Georgetown Univ., 148, 405
Georgia Institute of Technology (Georgia Tech), 126, 147
Gerhart, Gen. John K. (USAF), 362
Germanium alloy, 291
German rocket scientists, 254, 262
Germany, East, 15, 41, 129, 344
Germany, West, 265, 273
rocket, 52, 78, 369, 381
scientist, 15, 369, 381
space program, 197
Gerrity, L/G Thomas P. (USAF), 89, 409
Getler, Michael, 317
Giffels and Rossetti, 410
Gibbons Stanley Ltd. of London, 205
Gifford, Frank, 88
Gignoux, Dominique M.P., 66
Gillespie, Ben, 219
Gillespie, Walter F., 114
Gilliland, Bob, 429
Gilruth, Dr. Robert R., 16, 33, 91, 121, 134, 244-245, 322
GIMRADA. See Geodesy, Intelligence and Mapping Research and Development Agency.
Gisell, William G., 215
GISS. See Goddard Institute for Space Studies.
Glantz Contracting Corp., 127
Glass fiber, 393
Glenn, L/C John H., Jr. (USMC), 75, 125, 358, 388, 343, 404
accident, 87, 121, 336
colonelcy, 365
consultant, 237, 245
politics, 16, 87, 121
retirement, 379
Glenn, W. H., 358
Glennan, Dr. T. Keith, 335
Glider, 71, 145, 201
glider, delta-wing (see also Asset), 208, 258
Glow, Project (USA), 211
GMC. See General Motors Corp.
GNP (Gross National Product), 187, 278
Goddard Historical Essay Award, 111
Goddard Institute for Space Studies (GISS) (NASA), 314, 372
Goddard Range and Range Rate System (tracking system), 284
Goddard, Dr. Robert H., 47, 104, 125, 160, 263, 319
Goddard, Robert H., airmail stamp, 160, 341
Goddard, Robert H., Library, 233
Goddard, Robert H., Memorial Dinner, 111
Goddard, Robert H., Memorial Trophy, 111
Goddard Scholarship, 111

Goddard Space Flight Center (GSFC)
   (NASA), 42, 102–103, 181, 188, 200–
   201, 224, 276, 305, 311, 316
 communications, 185, 285
 contract, 202, 239, 330, 336, 396
 cooperation, 10, 38, 118, 289
 experiment, 30, 169, 249, 308, 311, 319,
   375
 Iowa rock, 102
 launch, 19, 109, 118, 290, 297, 306, 339,
   346, 379–380, 394, 427
 management, 356
 test, 137, 140, 278
 tracking, 284
Goett, Dr. Harry J., 240, 282, 319, 344, 352
Goercke, Prof. Paul, 263
Gold alloy, 291
Gold, Harold, 158
Gold, Dr. Louis, 155
Gold, Dr. Thomas, 164, 180, 301, 306
Goldbaum, George C., 238
Goldberg, Delphis C., 24
Goldberg, Dr. Leo, 301
"Golden Age of Political Science", 87
Goldstein, Harold, 149
Goldwater, Sen. Barry M., 9, 69, 136, 253,
   258, 321, 331, 351, 362, 370
Goliath (aircraft), 79
Goodrich, B. F., Sponge Products Div., 52
Goodyear Aerospace Corp., 129, 212, 418
Goonhilly Downs, Cornwall (U.K.), 204,
   270
Gordon, Kermit, 67
Gore, Sen. Albert, 116
Gorki Univ., U.S.S.R., 85, 426
Gorki, U.S.S.R., 54, 75
Government Accounting Office (GAO),
   381, 434
Government Contracting for Research and
   Development, Briefing Conference, 250
Government Employees Incentive Awards
   Act, 142
Government-industry relationship, 165
GPAS. See General Purpose Airborne
   Simulator.
Grain, solid, 248
Grand Canyon, Ariz., 100
Grand Coulee Dam, 152
Grand Rapids, Mich., 343
Granite Construction Co., 402
Grants
 government, 11, 104, 110, 162, 211, 216–
   217, 259, 397
 nongovernment, 3, 111, 219
Graphite, 331
Graphite reactor, 83
Graphite refractory coating, 153
Graveline, Maj. Duane E. (USAF MC), 393
Gravitational biology, 309
Gravitational collapse, 155
Gravity, artificial, 226, 309, 384
Gray, Edward Z., 8, 73, 217
Gray, Robert H., 339
Gray, Prof. Wood, 178
Green Bank, W. Va., 321

Green River, Utah, 58, 193, 241, 329, 338,
   381, 406
Greenacre, J. C., 204
Greenbelt, Md., 185
Greensboro, N.C., 439
Greenstein, Dr. Jesse L., 398
Grissom, Maj. Virgil I. (USAF), 134, 188,
   316, 329, 366, 369
Gronouski, Postmaster Gen. John A., 160
Gross, Rep. H. R., 201
Ground-Based Astronomy-A Ten-Year Program, 385–386
Ground effects machine (Gem), 354
Ground station, transportable, 295
Group Achievement Award (NASA), 77,
   221
Gruene, Hans F., 352–353
Grumman Aircraft Engineering Corp., 100,
   116, 215
GSFC. See Goddard Space Flight Center.
GT–1. See Gemini.
GT–2 See Gemini.
GT–3. See Gemini.
GT–4. See Gemini.
Guam Island, Pacific, 66, 291
Guaymas, Mexico, 39, 369
Gubser, Rep. Charles S., 153
Guggenheim Aeronautical Laboratory, 155
Guggenheim, Daniel and Florence, Award,
   1964 (IAA), 266
Guggenheim, Daniel, Medal, 1964, 263
Guggenheim, Harry F., 387
Guggenheim International Astronautics
   Award, 1964 (IAA), 315
Guidance, 142
 inertial, 121, 193
Gulf Area Development Committee,
   Houston Chamber of Commerce, 277
Gunder, Dr. Dwight F., 360
Gurney, Rep. Edward J., 22
Gus (mouse), 255
Gyros, 18
H–1 (engine), 28, 192, 211
H–10 (atomic clock), 408
Hackes, Peter, 221
Haggerty, James J., Jr., 10, 26, 213, 307,
   339, 347, 443
Hailsham, Lord, 36
Hailstone damage, 227
Halaby, Najeeb E., 4, 55, 84, 119, 148, 183,
   187, 211, 220, 242, 314, 318, 332, 336, 422
Haley, Andrew G., 236
Hall, Dr. Albert C., 51, 68, 74
Hall of Science, 310
Hall, Robert Cargill, 111
Halo, Operation (USA), 32
Hamburg, Germany, 297, 301
Hamburg Univ. (Germany), 296
Hamilton Standard Div. (UAC), 197
Hammack, Jerome B., 63
Hammaguir, Algeria, 161, 210
Hammer, magnetic, 329
Handle, self-latching, 114
Haney, Paul, 248, 280, 294, 384, 395
Hansen, Grant, 234

Hardeman, Paul, Inc., 15
Harmon International Aviator's Trophy, 273, 317
Harp, Project (High-Altitude Research Project) (USA), 29, 411
Harper, Carl Brown, 410
Harper, Charles W., 322
Harper's Dry Lake, Calif., 239
Harrison, Nat, Associates, 52
Harrison, Patrick, Inc., 119
Hart, Mrs. Philip A., 164
Hartman, Edwin P., 139
Hartmann, Dr. Sven R., 383
Harvard Business School Club of Washington, 8
Harvard Univ., 22, 180, 301
Harvey, Gordon, 167
Havana, Cuba, 121
Hawaii, 5, 109, 207, 268, 332
Hawaiian Telephone Co., 412
Hawker-Siddeley DH–125 (U.K. jet), 209
Hawkins, R. S., 230
Haworth, Dr. Leland J., 104
Hawthorne, Calif., 201
Hayes, B/G Thomas J., III (USA), 136, 143
Hazard, outer space, 264–265
Head, Dr. Richard, 4
Health, Education, and Welfare, Dept. of (HEW), 48, 277
Heart, 131, 174, 305
Hearth, D. P., 199
Heat exchanger, 142
Heat-sensitive detector, 179
Heat shield, 9, 146, 181
Heat transfer, 14, 70, 236, 305
Heat Transfer Division Memorial Award (ASME), 403
Heating, 122–123
  aerodynamic, 290
  reentry, 9, 170–171, 226, 258, 289
  solar, 141
Hebrew Univ. (Israel), 125
Height gage, 39
Heinkel Co., 265
Held, Dr. Richard, 9
Helicopter, 216, 392
  record, 394
  speed, 207
  use of, 305, 439
Heliopolis, U.A.R., 303
Helium, 130, 140, 309
Helium-oxygen atmosphere, 309, 373
Helms Soaring Hall of Fame, 244
Henderson, E. P., 133
Herbig, George Howard, 155
Herodotus (lunar crater), 205
Herschel, Sir William, 203
Hershey, Pa., 379
Herskovitz, Sheldon B., 110
Hess, Dr. H. H., 116
Hess, Dr. Victor F., 423
Hetzel, E. P., 264, 278
Heusinger, B. K., 302
HEW. See Health, Education and Welfare, Dept. of.

HF–24 (fighter aircraft), 203
Hibbard, William D., 181
Hibex, Project (Hi-G Booster Experiment) (ARPA), 10
High-altitude Clear-Air Turbulence (Hi-Cat), 383
Hi-Cat. See High-altitude Clear-Air Turbulence.
Hidden Hills, Calif., 70
Hi-Dive (USAF rocket), 317
High, H. J., Construction Co., 164
High Resolution Infrared Radiometer (Hrir), 298
"Highlights of Recent Space Research", 314
Hilburn Board, 151
Hilburn, Earl D., 42, 144, 165, 224
Hill, Louis W., Award, 235
Hindustan Aircraft Co. (India), 203
Hines, Prof. Colin O., 140
Hines, William, 48
History, 255–256
History of Aeronautics and Astronautics, Technical Committee on (AIAA), 226
HITCHHIKER I (satellite), 178
Hjornevik, Wesley L., 353
HL–10 (lifting-body), 71, 201, 220, 308
Hodges, Luther H., 43, 119
Hoffa, James R., 243
Holifield, Rep. Chet. 117, 201
Holloman AFB, N.M., 91, 133, 333, 409
Holmberg, Neal A., 150
Holmes, D. Brainerd, 141
Holzman, B/G Benjamin G. (USAF, Ret.), 366
Honeycomb structure, 13, 197
Honeywell Co., 193, 257, 263
  Aeronautical Div., 65, 224
Hooker, Ray, 139
Hordexum seed experiment, 248
Horizon scanner, 77, 279–280, 241
Horner, H. M., 161
Horner, Richard E., 343
Hornig, Dr. Donald F., 12, 144, 186, 211, 379, 418, 431
"Hot universe" theory, 180
Hotz, Robert B., 225, 240, 274, 324, 357, 374, 409
House, William C., 108
Houston Astros, 409
Houston Chronicle, 258
Houston Museum of Natural Science, 244
Houston Post, 55
Houston, Tex., 8, 35, 65, 81, 121, 206, 209, 235, 244–245, 273, 277, 290, 299, 321, 409
Houston, Univ. of, 199
Hovercraft, 311, 354
Howland Island, 333
Hoyle, Dr. Fred, 23
Huber, William G., 238
Hughes Aircraft Co., 61, 90, 105, 160, 209, 228, 237, 384, 409–410
  Research Laboratory, 159, 256
Human centrifuge, 226, 243

Human factors research, 9, 166
Human Factors in Electronics, Fifth Annual Conference on, 166
Humphrey, Sen. Hubert H., 180, 275, 286, 297, 312, 347
Humphrey, Vice President-elect Hubert H., 398
Humphries, Col. George E., 421
Hunsaker, Dr. Jerome C., 247
Hunter, Maxwell W., Jr., 26
Huntsville, Ala., 29, 35, 70, 104
Huntsville Industrial Expansion Committee, 367
Hurlburt, E. O., Center for Space Research (NRL), 15
Hurricane, 122, 146, 218, 253
Hurricane Cleo, 297, 316
Hurricane Dora, 309–310, 316, 326
Hurricane Ethel, 326
Hurricane Florence, 326
Hurricane Gladys, 321
Hussein I, King (Jordan), 142
Hutchins, Dr. Robert M., 120
Hyatt, Abraham, 247
Hydrazine, 62, 230
Hydrocarbon, 181
Hydrofoil, 370
Hydrogen, 140, 226
  atomic, 15
  liquid, 14, 32
  propellant, 223
Hydrogen bomb, 14, 394
Hydrogen-fire detector, 210
Hydrogen peroxide, 89, 247, 298
Hydrology, 62
Hyman, Hayman, and Harris, 14
Hynek, Dr. J. Allen, 72, 393
Hypersonic flight, 33, 173
Hypersonic Ramjet Experiment Project, 391
Hypersonic vehicle, 257
Hypervelocity impact, 197
Hypervelocity Impact Symposium, Seventh, 392
Hypotension, orthostatic, 129, 177–178
IAA. See International Academy of Astronautics.
IADR. See International Association for Dental Research.
IAF. See International Astronautical Federation.
IATA. See International Air Transport Association.
IAU. See International Astronomical Union.
Iberia Air Lines (Spain), 159
IBM. See International Business Machines Corp.
ICAO. See International Civil Aviation Organization.
Icarus (Intercontinental Aerospacecraft Range Unlimited System) (see also Ithacus), 81–82, 246
ICBM. See Missile, ballistic, intercontinental.

Iceland, 273
Idaho, 306
IEEE. See Institute of Electrical and Electronic Engineers.
IGY. See International Geophysical Year.
IIOE. See International Indian Ocean Expedition.
IISL. See International Institute of Space Law.
IIT Research Institute (Illinois Institute of Technology), 135, 197
Ikeda, Prime Minister Hayato (Japan), 115
Ikeya (comet), 398
Ile du Levant launch center (France), 233, 371
Illinois, Univ. of, 140, 241, 393
Image orthicon tube, 72
Image velocity sensor, 257
IMCC. See Integrated Mission Control Center (MSC).
IMP I (Interplanetary Monitoring Platform). See EXPLORER XVIII.
Impact studies, 197
Imperial College, London, 71
"An Improved Precision Height Gage", 39
India, 60, 70–71
  cooperation, space, 31, 96
  Dept. of Atomic Energy, 31, 70
  launch, 6, 10, 25
  space program, 7
Indian Ocean, 70, 107, 294, 325, 405
Indiana Univ., 100
Indonesia, 89
Industrial applications of space research, 33, 39, 288
Industrial Nucleonics Corp., 100
Industry extension service, 226
Industry Studies on Manned Interplanetary Missions for NASA, Symposium on, 29
Inertial guidance, 121, 193, 263
Information, 132, 312
  distribution of, 243, 255, 305
  exchange of, 1, 310
  scientific and technical, 144, 237, 243, 283, 285, 293
Information and analysis center (DOD), 303
Information retrieval, 58
Infrared, 57
Infrared astronomical observatory (Project Amos), 57
Infrared horizon scanner, 241
Infrared laser rays, 112
Infrared radiation, solar, 76, 230
Injun (satellite), 473
Innovations, 88, 217, 280, 306
Innis, Robert C., 376
Innsbruck, Austria, 32
Insect, 89, 171, 174, 179
Institute for Satellite and Space Research (W. Germany), 385
Institute for Strategic Studies (U.K.), 385
Institute of Astrophysics and Space Research (W. Germany), 372

Institute of atmospheric science (proposed, U. of Miami), 100, 147
Institute of Electrical and Electronic Engineers (IEEE), 51, 300, 306, 366
Institute of Marine Sciences, 100
Institute of Navigation, 231
Institute of Normal and Pathological Physiology (U.S.S.R.), 226
Institute of planetary bioscience (proposed, U. of Miami), 100
Institute of Scientific and Technical Information (U.S.S.R.), 211
Institute of space physics (proposed, U. of Miami), 100
Instituto Nacional de Tecnica Aerospacial (Spain), 31
Insulation, 103
Integrated Mission Control Center (IMCC) (MSC), 206
Interceptor aircraft, 72, 154
Intercontinental Aerospacecraft, Range Unlimited System. See Icarus.
Interferometer, 269
Interim Communications Satellite Committee, 337
Interior, Dept. of, 231
  Office of Solar Energy, 180, 275
International Academy of Astronautics (IAA), 266, 311, 389
International agreements
  communications satellite system, 260, 293
  India-NASA, 70–71
  NAS-U.S.S.R., 182
  NBS-U.S.S.R., 41
  U.S.-ESRO, 264
  U.S.-Malagasy Republic, 306
  U.S.-U.S.S.R., 251, 256
International Air Transport Association (IATA), 114
International Association for Dental Research (IADR), 121
International Association of Man in Space, 372
International Astronautical Federation (IAF), 14
  15th Congress, 308–309–310, 313–314
International Astronautics Award, Guggenheim, 1964, 315
International Astronomical Union (IAU), 296–297–298–301
International Atomic Energy Agency (IAEA), 426
International Atoms-for-Peace Conference, Third, 305
International Business Machines Corp. (IBM), 73, 142, 284
  Advanced Systems Development Div., 239, 255
  Federal Systems Div., 122
International Civil Aviation Conference, 216
International Civil Aviation Organization (ICAO), 406

International communications satellite system. See Communications satellite system, international.
International controls, space experiments, 76
International Convention of Military Electronics, 317
International cooperation, 16, 84, 96–97, 198, 251, 287, 295, 308–309, 324
  civil aviation, 216
  communications, 25, 32, 49, 55, 58, 61, 75, 129, 185, 205, 215, 218, 247, 260, 282–283, 285, 293, 313, 315, 330, 449
  IQSY, 45
  meteorology, 1–2
  nuclear, 72
  space (see also International space programs), 6–7, 16, 27, 29, 35, 38, 75, 97, 112–113, 132, 194, 195, 203, 235, 239–240, 250, 269–270, 271–272, 287–288, 311, 345
  tracking stations, 264, 306
International Council of the Aeronautical Sciences, 4th Congress, 296
International Date Line, 105, 294, 305, 310, 314
International Geophysical Year (IGY), 45, 70
International Indian Ocean Expedition (IIOE), 71, 107
International Institute of Space Law (IISL), 14
International Satellites for Ionospheric Studies. See Isis
International Scientific Radio Union, 140
International space law, symposium, 313
International space programs (see also International cooperation, space), 36–37, 269–270, 371
  France-Spain, 204
  U.K.-Australia-ELDO, 269–270
  U.K.-Italy, 311
  U.S.-Argentina, 403
  U.S.-Australia, 84, 139, 222, 284, 292
  U.S.-Canada, 10–11, 30, 42, 47, 63, 189, 218
  U.S.-Denmark, 56, 173
  U.S.-ESRO, 264
  U.S.-France, 311
  U.S.-India, 31, 70–71
  U.S.-Italy, 311, 476,
  U.S.-Japan, 115, 140, 294
  U.S.-Netherlands, 278
  U.S.-Norway, 56, 216
  U.S.-Spain, 31
  U.S.-Sweden, 99, 147, 205, 241
  U.S.-U.K., 118, 311, 453, 482
  U.S.-U.S.S.R., 2, 17, 30, 59–60, 97, 172, 175, 204–205, 208, 240
International Symposium on Bioastronautics and the Exploration of Space, Third, 388, 393
International Telecommunications Union (ITU), 37, 260, 313

International Telephone and Telegraph Co. (ITT), 61, 74, 79, 379, 412
International Union of the Physiological Sciences (IUPS), 266, 315
International Year of the Quiet Sun (IQSY), 34, 42, 140, 356, 395
Interplanetary flight, 176, 301
Interplanetary Monitoring Platform (Imp). See EXPLORER XVIII, EXPLORER XXI.
Inventions (see also Patents), 12, 25, 82, 114, 142
  policy, national, 5, 293
  property rights, 5, 43
  waiver, 294, 329
Inventors, 5, 12, 25, 66, 82
Inventors' Congress and Space Symposium, 369
Ion engine, electrostatic, 142, 159, 238
Ion measurement, 15, 30, 50–51, 140
Ion propulsion, 82, 324, 410
Ion thrustor, 159
Ionization gage, 77
Ionized calcium, 180
Ionized plasmas, 132
Ionosphere, 76, 140
  effects of, 251
  measurements, 11, 210, 249, 296–297, 393
  research, 30, 50, 110, 125, 241, 468
Ionospheric D region, 51, 56, 140, 422
Ionospheric E region, 51, 140
Iinospheric Research Laboratory (Royal Technical Univ., Denmark), 56
Iowa, 102, 128
Iowa, Univ. of, 387
Iowa Wesleyan College, 232
IQSY. See International Year of the Quiet Sun.
Irish International Airlines, 81
Iron Workers Union (IWU) (see also United Iron Workers), 126, 131
Isakov, P. K., 309
Isis (International Satellites for Ionospheric Studies), 11
Israel, 55
Italian-American Democratic Organizations of New York, Federation of, 101
Italy, 227, 246
  cooperation, 311
  satellite launch, 420, 476, 487
  tracking station, 264, 381
Ithacus (rocket transport) (formerly Icarus), 238, 246
ITT. See International Telephone and Telegraph Co.
ITU. See International Telecommunications Union.
IU data adapter. See Computer.
IUPS. See International Union of the Physiological Sciences.
Iwan, Lawrence S., 236
IWU. See Iron Workers Union.
Izvestia, 1, 104, 204, 272, 282, 348, 396
J–2 (rocket engine), 159, 248, 294, 403, 416

J–4 (engine), 215
J–58 (engine), 86, 261, 429
Jack, John R., 142
Jackass Flats, Nev., 175, 300
Jackson, Nelson P., Aerospace Award, 111
Jacobson, Col. R. K. (USAF), 41
Jaffe, Leonard, 130, 352
Jane's All the World's Aircraft, 421
Japan, 19, 210, 258–259, 347
  communications, 185, 189, 219
  launch, 245, 258–259
  meteorologists, 112
  TV transmission, comsat, 115, 140, 294, 314
Japan Broadcasting Corp., 259
Japanese Ministry of Posts and Telecommunications, 315
Japanese Research Council, 112
Jasperson, Capt. Robert E. (USN, Ret.), 188
Jastrow, Dr. Robert, 314
Javelin (sounding rocket). See Argo D–4.
Javits, Sen. Jacob K., 18, 35, 201
Jefferson, Howard B., 319
Jeffries, John, Award, 376
Jenkins, Dr. Dale, 178
Jet airborne simulator system, 147
Jet control, 248
Jet Propulsion Laboratory (JPL) (Cal Tech), 37, 135, 60, 137
  awards, 246
  facilities, 177
  launches, 34, 64, 263, 378, 397
  Mariner project, 300–301, 338
  NASA management, 58, 60–61, 97–98, 128, 137–138, 163–164, 422–423, 435
  organization, 137, 220, 224
  Ranger project, 42, 58, 64, 128, 137–138, 151, 163–164, 215, 218, 266–267, 275, 309
  research, 62, 144, 167, 297, 399–400
Jetstar. See C–140A.
Jet stream, 140
JHU. See Johns Hopkins Univ.
Jodrell Bank Experimental Station (U.K.), 54, 75–76, 85, 97, 242, 246
Jodrell Bank Radio Astronomy Station, (U.K.), 239, 253, 270
Job classification handbook (Dept. Labor), 218
Johannesburg, So. Africa, 174, 226
John F. Kennedy Space Center, NASA. See Kennedy Space Center, NASA.
Johns Hopkins Univ. (JHU), 57, 79, 131–132, 408
Johnson, Clarence L., 315–316, 328
Johnson, David S., 101
Johnson, Gerald W., 19
Johnson, John A., 265, 337
Johnson, President Lyndon B., 6, 55, 117, 129, 143, 164, 188, 218, 234–235, 245, 247, 257, 294, 297, 307, 323, 330, 397–398, 408, 437

aircraft, 287
  A-11 (YF-12A), 85-86, 90, 93, 133, 255
  bomber, 140
  SR-71, 260-261, 274
appointments, 12-13
awards by, 11, 184, 316, 328
budget, 12, 20-21, 244, 300-301
communications satellite system, 58, 278, 405
defense, 320-321
disarmament, 17, 21
economy, 18, 38, 40, 257, 265-266
education, 2
international cooperation, 16, 265, 287
peace, 17, 21, 50, 143, 201, 272, 333
RANGER VII, 272, 287, 300
space program, 32, 40, 85-86, 223, 267, 272, 278-279, 317-318, 335
supersonic transport, 85, 149, 153, 183, 405
technological change, 387
Johnson, Mrs. Lyndon B., 109, 114, 294
Johnson, Paul G., 235, 440
Johnson, Richard L., 428
Johnson, Vincent L., 210, 371
Johnson, Dr. Woodrow E., 108
Johnston Island, Pacific, 161, 394
Johnston, Maj. John A. (USA), 394
Joint Chiefs of Staff, 260
Joint Navigation Satellite Committee (JNSC), 312, 347
Jones, B/G David M. (USAF), 375
Jones, Dr. Walton L., Jr., 361
Jordan, 142
Josias, Conrad, 82
Journal of the Armed Forces, 246, 254, 262, 307-308, 339, 347
Journal of the National Education Association (NEA), 2
JP-4 (fuel), 81
JPL. See Jet Propulsion Laboratory (Cal Tech).
Judi-Dart (NASA-Pakistan sounding rocket) (see also Dart), 107
Juergensen, Klaus, 114
Jupiter (missile), 32
Jupiter (planet), 153, 399
  atmosphere, 140, 179
  radiation belt, 387
  temperature, 326
Jupiter C (booster), 37
Justice, Dept. of, 220
Kabanov Effect, 330
Kabanov, Prof. Nikolai, 330
Kagoshima Space Center, Japan, 245
Kalimin, Prof. Yu. D., 374
Kaman Aircraft Corp., 207
KAMANIN, L/G N., 365
Kane, Col. Francis X. (USAF), 161
Kania, Island of, 444
Kantrowitz, Arthur, 230
Kaplan, Lewis D., 35, 48
Kappel, Frederick R., 334
Karachi, Pakistan, 107
Kardashev, Nikolai S., 364

Karlsruhe, West Ger., 26
Karrer, Dr. Sebastian, 148
Karth, Rep. Joseph E., 69, 71, 114, 150, 163, 313
Karume, Pres. Abeid (Zanzibar), 129
Kashima, Japan, 185, 315, 326
Katz, Lester, 114
Katz, Dr. Ludwig, 178, 325
Kaufman, Harold R., 82, 142
KC-130F (Hercules), 371
KC-135 (Stratolifter), 182, 208
Keating, Sen. Kenneth B., 187, 201
Keats, E. S., 158
Keldysh, Mstislav V., 198, 359
Kelley Board. See NASA Board of Inquiry.
Kelley, Dr. Albert J., 43, 111, 184, 275, 368, 388, 393
Kemp, Richard H., 82
Kennedy Airport International Hotel, 326
Kennedy, David M., 242
Kennedy, President John F., 22, 47, 58, 60-61, 101, 189, 206, 208, 223, 246, 272, 329, 377
Kennedy Space Center, NASA (KSC) (see also Merritt Island Launch Area), 81, 101, 120, 263, 317
  accident, 135, 141, 147
  contract, 88, 328-329, 336
  facilities, 39, 77, 276
  organization, 53-54
  support services, 85, 331
Kepler (lunar crater), 275
Kerbel, Lev, 132
Kerosene, 81, 229-230
Kerr, Breene M., 195, 395
Kerr, Clark, 323, 331
Kerr, Sen. Robert S., 201
Kharkov Univ. Observatory (U.S.S.R.), 315
Khrushchev, Premier Nikita (U.S.S.R.), 17, 143, 195, 253, 300, 323, 351, 353
Kidd, Dr. Charles V., 308
Kidde, Walter, and Co., 247
Kiess, Dr. Carl C., 48, 148
Kiewit, Peter, and Sons, Inc., 8, 13, 62
Kiker, John, 361
Killian, Dr. James R., Jr., 154, 355
Killion, George L., 242
Kimball, Daniel A., 55, 332
Kimball, Neb., 268
Kindelberger, James Howard, Memorial Laboratories, NAA Science Center, 298
Kinetic energy, 151
Kinzel, Dr. Augustus B., 154, 416
Kirchner, Dr. Werner R., 369
Kiruna, Sweden, 360
Kistiakowsky, Dr. George B., 35, 281
Kiwi (nuclear reactor), 385
Kiwi B-4A (nuclear reactor), 195, 320
Kiwi B-4D (nuclear reactor), 195, 229
Kiwi B-4D-CF (nuclear reactor), 63
Kiwi B-4E (nuclear reactor), 229, 311
Kiwi B-E (nuclear reactor), 299
Klein, Milton, 435
Kleinknecht, Kenneth S., 265

Kleinsaechter, Dr. Hans, 263
Knauf, Dr. George M., 80, 423
Knepton, James C., Jr., 174
Knudsen, Morrison, 102
Kock, Dr. Winston Edward, 244, 304, 427
Koelle, Dietrich E., 197
Koelle, H. H., 401
Kollmorgen Corp., 269
Kollsman Instrument Co., 218
Komarov, Vladimir Mikhailovich, 348–349, 355, 359, 469
Komsomolskaya Pravda, 30
Konecci, Dr. Eugene B., 124, 166, 309–310, 361, 376, 419, 420–421
Koppers Co., Inc., Malan Construction Dept., 247, 406
Korovin, Y. A., 396
Kosygin, Premier Alexi (U.S.S.R.), 353, 413, 437
Kotelnikov, Vladimir, 321
Kozyrev, Dr. Nikolai A., 180, 316
Kraft, Christopher C., Jr., 63, 265
Kraft, Dr. Robert, 401
Kramer, James J., 328
Kramer, S. B., 200
Kranish, Arthur, 88
Krasnaya Zvezda (see also *Red Star*), 74, 118, 131, 253–254, 273
Kreplin, Robert W., 45
Kronogård Range, Sweden, 241, 277, 283, 287–288
Krylov, Marshal Nikolai I., 394
Krzeminski, Dr. W., 401
KSC. See Kennedy Space Center, NASA.
Kubesch, Maj. Sidney J. (USAF), 315
Kuhrt, Welsey, 26
Kuiper, Dr. Gerard P., 148, 267, 271, 275, 299, 301, 331, 336, 347
Kurashika Astronomical Observatory (Japan), 210
Kurnit, Norman A., 383
Kuzin, Alexander, 202
Kuznetsov, Prof. A. G., 309
Kuznetsov, Dr. Oleg, 355
Kwajalein Island, Pacific, 239
L-29 (U.S.S.R. aircraft), 212
Labor, 24, 141
Labor, Dept. of, 24, 218, 302
Lacklen, Robert J., 341, 439
Lagos, Nigeria, 158
Laika (dog, U.S.S.R.), 152
Laird, Rep. Melvin, 115
Lambda III (Japanese sounding rocket), 245
Laminar flow, 123
Landing, computer, 411
Landing impact test, 313
Landing instrumentation, 259
Landing requirements, jet transport, 150
Landing pad, instant, 232
Landing site, 303
Land's End (U.K.), 110
Lang, Dave W., 277
Langley AFB, Va., 176

Langley Medal (Smithsonian Institution), 166
Langley Research Center (LaRC) (NASA), 42, 71, 135, 142, 181–182, 188, 216, 230, 260, 276
   contract, 52, 93
   experiment, 132, 135, 290
   facilities, 159, 182
   inspection week, 181, 188
   management, 177
   organization, 177
   research, 189–190, 223, 319, 327
   test, 150
Langmuir probe, 373
Lankeneau Hospital, 129
Lansche, Jane Elizabeth, 237
Lanzo, Chester D., 25
LaRC. See Langley Research Center.
Large Orbital Research Laboratory (Lorl), 122, 145
Lariat (Laser Radar Intelligence Acquisition Technology), 181
Las Cruces, N. Mex., 353
Laser, 155, 257, 346
   deep space acquisition, 5
   tracking techniques, 5, 109–110, 386, 407
   infrared rays, 112, 325
   use of, 93, 121–122, 168, 181, 258, 383
Laser plasma diagnostic probe, 159
Laser Radar Intelligence Acquisition Technology (Lariat), 181
Laser radar system, 113, 278
Laser rifle, 94
LASL. See Los Alamos Scientific Laboratory.
Lasv. See Low Altitude Supersonic Vehicle.
Launch Complex 19 (KSC-MILA), 20, 38
Launch Complex 20 (KSC-MILA), 269
Launch Complex 39 (KSC-MILA), 291
Launch vehicle
   development, 309,
      contract, 253, 260, 319
   expenditure, 99
   military, 213
   operations and capability, 176, 234, 237
Laurence, William L., 19
Lausche, Sen. Frank, 201
Lava fields, lunar surface, 276, 299, 301
Lavrentev, Mikhail A., 21
Lawrence Radiation Laboratory (Univ. of Calif.), 1, 19
Lawyer, Capt. Richard E. (USAF), 304
Lazar, James, 229
Lear, John, 307, 347
Lear Siegler, Inc., Electronics Instrumentation Div., 102
Leavell, C. H., and Co., 13, 62, 102
Lebedev, Dr. Vladimir, 355
Legislative Reference Service, Library of Congress, 301, 310
Leighty, George W., 69

Lem. See Lunar Excursion Module.
Lem Truck (Lunar Excursion Module) (see also Apollo), 105
LeMay, Gen. Curtis E. (USAF), 44, 66, 69, 72, 92, 140, 315, 401, 430
Lemmon, R. M., 215
Leningrad, U.S.S.R., 261
Lerner, Max, 352
Lesa (Lunar Exploration System for Apollo), 74, 170
Leslie Miller, Inc., 127
Lessman, Richard C., 236
Leutkirch, West Germany, 283
Levin, Boris, 272
Levitt, Dr. I. M., 205
Levy, Lillian, 86
Lewis Research Center (LRC) (NASA), 25, 62, 78, 115, 142, 158, 214, 219, 256, 276, 403
  contract, 119, 367
  experiment, 124
  management, 311-312
  organization, 328
  Plum Brook Station, 221, 224, 317
  programs, 230, 236, 444
  research, 293
  test, 163, 197, 216, 444
  zero-gravity facility, 62
Lexington, Va., 188
Ley, Dr. Willy, 64
Libonati, Rep. Roland V., 70
Library of Congress, 239, 301, 310
Lick Observatory, 155
Liege Univ. (Belgium), 432
Lievens, Edward J., Jr., 150
Life detector instruments, 169
Life (magazine), 405
Life preserver, 354
Life science (see also Bioscience), 194, 208, 309-310, 350
  atmospheric pressure, 56
  blood pressure, 178, 243
  disease, communicable, 249
  genetics, 169, 171, 179
  heart, 172, 174, 241, 243
  helium environmental effects, 309
  nervous system, 169, 178
  nutrition, 151
  optics, 48
  oxygen environmental test, 44, 87, 130, 174
  radiation, 48, 72, 169, 248
  weightlessness, 56, 89, 129, 171, 174, 179, 248, 309
Life support systems, 122
Lift-fan aircraft (XV-5A), 251
Lifting-body glider, 71
Lifting body space vehicle, 224
Light beam communications system, 142
Light intensity measurement, 13, 243
Lighthill, Michael V., 279
Lightning damage, 289, 294
Ligida, Dr. Myron G. H., 278
Lil. See Lunar International Laboratory.

Lincoln, Neb., 247
Lincoln Laboratory, 269, 433, 435
Lindbergh, Charles A., 326-327
Linde Div., Union Carbide Corp., 401
Ling-Temco-Vought, Inc., Range Systems Div., 88, 106, 216, 217, 236, 437
Lingle, Walter L., Jr., 144, 173
Link Foundation of New York, 200
Link Group, General Precision, Inc., 430
Link, Marilyn, 200
Lipscomb, Rep. Glen P., 59
Liquid Missile Propellants Symposium, 379
Liston, Sonny, 81
Little Joe II (booster), 89, 90, 175, 410, 483, 487
Little Rock, Ark., 278, 369-370
Littlewood, William, 236
Litton Industries, 159, 224
Livermore Radiation Laboratory, 246-247
Lloyd-Davies, David, 253
Llrv. See Lunar Landing Research Vehicle.
Loads, aircraft, 156
  helicopter, 216
Lockheed Aircraft Corp., 31, 216
  A-11 (YF-12A), 85-86, 92, 154, 254, 328
  contracts, 85-86, 405
  SR-71, 429
  supersonic transport, 13, 28, 183, 201, 373-374, 405
  U-2, 383
  XH-51A, 420
  XH-51N, 216, 422
Lockheed-California Co., 80, 135, 291-292
Lockheed-Georgia Co., 207, 379
Lockheed Missiles & Space Co., 9, 103, 247, 281, 343, 393, 406
Lockheed Propulsion Co., 193, 331, 402
Lof, George, 275
Loginov, Yevgeny F., 241
Loki-Dart (sounding rocket), 70-71
London, U.K., 5, 46-47, 59, 129, 158, 161, 189, 205, 218, 314, 330
London Gliding Club, 432
London Univ. (U.K.), 424
Long, Dr. S. Burman, 207-208
Long, Sen. Russell B., 79, 434
Longmont, Colo., 166
Loory, Stuart H., 145
Lorl. See Large Orbital Research Laboratory.
Los Alamos Scientific Laboratory (LASL), 63, 175, 229, 299-300, 320, 444
Los Angeles, Calif., 5, 8, 66, 74-75, 121, 165-166, 168, 244
Los Angeles Times, 262
Losey, Robert M., Award, 376-377
Louden, Frederic A., 90
Louisiana, 324
Love, Gov. John A., 58
Lovelace Foundation for Medical Education and Research, 175-176

Lovelace, Dr. W. Randolph, II, 143, 168–169, 175–176, 187, 443
Lovell, Sir Bernard, 76, 109, 242, 246, 266, 273, 340, 357
Lovell, Cdr. James A., Jr. (USN), 102, 167, 263, 336
Low Altitude Supersonic Vehicle (Lasv), 303
Low, George M., 17, 30–31, 65, 68–69, 206, 209, 347, 356
Lowell Observatory, 78, 106, 203–204, 205, 316, 401
Lowery Electric, Inc., 39
Lowman, Dr. Paul, 170
Lowry, Dr. Romney H., 24
LRBM. See Missile, ballistic, long-range.
LRC. See Lewis Research Center (NASA).
Lucas, William R., 352–353
Luedecke, M/G Alvin R. (USAF Ret.), 224
Lufthansa-German Airlines, 96
Lunar (see also Moon)
 base, manned, 73, 95, 105, 143, 389
 crater, 154, 180, 267, 271, 275–276, 298–299, 364
 environment, 13
 exploration, 20–21, 170, 180, 253, 262, 276, 366–367
 gravity, 70
 photographs, 271, 272, 274, 275–276, 287–288, 301, 306–307, 307, 357
 probe, 152
 surface, 164–165, 180, 197, 203, 263, 267, 271, 272, 275–276, 287–288, 298–299, 301, 306–307, 331
Lunar and Planetary Laboratory, 223
Lunar clock, 205
Lunar eclipse, 424, 426
Lunar Excursion Module (Lem), (see also Apollo spacecraft), 39, 73, 100, 139, 183, 267–268, 298–299, 306, 344, 357, 358
Lunar Excursion Module (Lem) Test Facilities (NASA), 31
Lunar Exploration Symposium, 215
Lunar Exploration System for Apollo. See Lesa.
Lunar hopper (helicopter device), 251
Lunar International Laboratory (Lil), 308–309, 314, 389
Lunar Landing Research Facility (LaRC), 181–182
Lunar Landing Research Vehicle (Llrv), 138, 197, 218, 368, 377, 396, 440
Lunar map, 136, 220–221, 223
Lunar Orbiter (probe), 47, 69, 171, 276
Lunar program (U.S.) (see also Moon and Lunar), 6–7, 35–36, 131, 152, 188–189, 189, 254, 314
 appropriations, 12–13, 20–21, 40, 45–46, 244
 cooperation, 12, 35–36
 criticism, 131, 194, 245–246, 258, 362
 importance of, 63–64, 271, 271–272
 manned flight and landing, 13, 20–21, 71, 77, 141, 158, 199, 206, 215, 228, 262, 267–268, 268–269, 275–276, 276, 316–317, 324, 354–355, 355, 379
 simulation, 10, 181–182
Lunar research, 362
Lunar Research, Conference on Geological Problems in, 180
Lunar science, 271–272
Lunar surface locomotion, 197
Lunar telescope, 72
Lunar transport ferry, reusable, 281
Lundberg, B. K. O., 194–195
Lunde, Barbara, 181
Lundin, Bruce T., 163
LUNIK III (U.S.S.R. lunar probe), 150–151, 152
LUNIK IV (U.S.S.R. lunar probe), 159
Luster, Project, 387
Luyten, Dr. Willem J., 154
Lyman-alpha radiation experiment, 15
Lynch, James J., 229
Lyster, Theodore C., Award, 175–176
M–1 (rocket engine), 78, 80
M–2 (lifting-body), 71, 201, 220, 308
MA–4 (Project Mercury), 225
MA–5, 225
MA–6, 225
MA–7, 225
MA–8, 225
MA–9, 225, 247, 273
Macalester College, 33
McClellan, Sen. John L., 430
McCone, John, 119
McConnell, Gen. John P. (USAF), 333, 430
McCormick, B. B., and Sons, Inc., 329
McCusker, Donald F., 425
MacDill AFB, Fla., 404
McDivitt, Capt. James A. (USAF), 263
McDonald, Frank B., 352
McDonald Observatory, 148
McDonnell Aircraft Corp., 111, 182, 226, 240, 261
McDonnell, James S., 240
McElroy, Neil H., 358
McGill Univ. (Canada), 10, 29, 411
McGovern, Sen. George, 75, 81–82
McGrath, Thomas L., 59
McKay, John B., 103, 174, 227, 297, 352, 380, 399
Mackay Radio and Telegraph Co., 55
McKee, Col. Daniel (USAF), 55
McKee, Robert E., 31
McKee, Gen. William F. (USAF, Ret.), 266, 304
MacKenzie, John P., 61
McKenzie Pass, Cascade Mountains, Oreg., 296
MacKenzie, Vernon G., 230
McLafferty, George H., 26, 108
McMillan, Brockway, 354
McMorrow, M/G Francis J. (USA), 103
McNamara, Robert S., 9, 28, 66, 92, 98, 119, 135, 206, 240, 248, 257, 287, 322–323, 345, 353, 388, 404, 414, 417, 429–430

Macomber, John W., 25
Madagascar, Malagasy Republic, 306
Madrid, Spain, 31
Magic Mountain Laboratory, 368
Maglic, Dr. Bogdan C., 352
Magnet coil, 380
Magnetic equator, 31, 50
Magnetic field, 2, 50, 169, 174, 245, 250, 379
Magnetic fluid, 142
Magnetic hammer, 329
Magnetic storm, 50
Magnetic tape, 82, 196
Magnetometer, 31, 103, 325
  rubidium vapor, 210, 395
Magnetometer hammer, 439
Magnetosphere noise, 125
Magnuson, Sen. Warren G., 38
Maguire, Charles A., Associates, 410
Majunga, Madagascar, 306
Malagasy Republic, 306
Malan Construction Co., 247, 406
Malaysia, 263
Malfunction detection system, 13, 121
Malina, Dr. Frank J., 389
Malta Test Station (GE), 70
Manpowered flight, 433
Manchester, U.K., 85
Manchester College of Science and Technology (U.K.), 403
Mandelkorn, Joseph, 202, 276
Mandiovsky, Dr. Boris, 390
Manila, Philippines, 332
Manned Interplanetary Missions Studies Performed by Industry for NASA in 1963/64, Symposium on, 29
Manned Orbital Research Laboratory (MORL), 93, 189
Manned Orbiting Laboratory (MOL), 74, 146, 206, 307, 429
  booster, 245, 429
  contract, 192, 197, 257
  Gemini B, 261
  hearings, 44, 65, 92, 108-109
  management, 11, 41, 123
  plans for, 26, 45, 90, 96, 139-140, 209
  proposed merger with Apollo-X, 382, 418
  purpose, 2, 4-5, 8-9, 44, 68, 91-92, 99, 112, 211
  test, 174, 195
Manned Orbiting Telescope (MOT), 327
Manned Spacecraft Center (MSC) (NASA), 16, 33, 66, 116, 121, 134, 207, 223, 263, 265, 302
  astronaut training, 5, 25, 43, 66, 100, 159, 263, 336
  contracts, 277
    construction, 8, 10, 102, 268, 328
    service, 320-321
    study, 81, 292, 404
  facilities, 3, 25, 74, 77, 115, 237
    Center Support Facilities, 8, 268
    Florida Operations, 54, 299
    Mission Control Center, 10, 206
    Mission Simulation and Training Facility, 8
  support services, 81-82
  manned space missions
    Apollo, 24, 43, 77, 116, 237, 239, 244, 292, 297, 299, 314
    Gemini, 5, 25, 43, 57, 131, 168, 182, 193, 212, 239, 259, 264-265, 276, 292, 294, 329
    Mars, 9
    space stations, 80, 122, 287
    tracking, 5
  organization, 17, 54, 91, 242
  research, 81, 103, 158, 172, 188, 230
Manned space flight, 26, 96, 118, 152, 197, 294, 301, 326, 340
  appropriations, 20, 46, 185
  capability, 184
  circumlunar, 173
  cooperation, 2
  cost of, 20-21, 23, 99, 104
  hazards, 48
  lunar bases, 95
  lunar exploration, 67, 105
  lunar landing (see also Lunar program), 69, 274, 301
    appropriations, 20, 285
    criticism, 194, 269
    goal, 23, 46, 52, 300
    training, 296
  objectives, 217
  program, 52-53, 156
  R&D, 65
Manned Space Flight Meeting, Third, 375, 377
Manned space shuttle service, 24, 135
Manned space station. See Space station, manned.
Manov, Dr. George G., 102
Manpower
  civil aviation, 55
  scientific and engineering, 7, 246, 262-263
Mansfield, Sen. Mike, 221
Manton, William I., 174
Map
  lunar, 220, 223
  weather, 240
Maples, Hoyt, 415
Mar. See Radar, Multi-function Array.
Marconi Co., 424
Mare Cognitum (RANGER VII lunar impact site), 304
Marine Sciences, Institute of, 100
Mariner (spacecraft), 82, 199, 212, 276, 300-301
MARINER II (Venus probe), 152, 172, 184, 318
MARINER III (Mars probe), 378, 459, 467
MARINER IV (Mars probe), 393, 443, 444
  course, 398-399, 407, 422
  Geminid stream, 415, 426
  launch, 397-398, 460, 468
  performance, 431, 434
Markey, Dr. Winston R., 289

500 INDEX

Marks, Leonard H., 242
Marquardt Corp., 368
Mars (planet), 1, 105, 180, 187, 393
  atmosphere, 8, 35, 48, 88–89, 106, 148, 168–169, 180, 336
  exploration, 106, 359, 391
  flight
    manned, 9, 158, 185, 235
    unmanned, 171, 199, 217, 257, 276, 295, 301, 324, 470–471, 474–475
  life, 8, 48, 143–144, 148, 168–169, 180, 208, 210, 259, 437–438
  surface, 88–89, 106, 148
MARS I (U.S.S.R. probe), 172
Mars spacecraft, 9, 106, 143–144, 168–169, 309
Marshak, Prof. Robert E., 348
Marshall, Gen. George C., 77
Marshall, George C., Research Library (VMI), 188
Marshall Space Flight Center (MSFC) (NASA), 29, 114, 122, 291–292, 302, 320, 329, 397
  C–1 rocket engine, 399–400
  contract, 73, 120, 424
    construction, 102, 402
    study, 183, 248–249, 253, 281, 302, 352
    support services, 14, 59, 84–85, 238–239
  facilities, 77
  management, 73, 120, 367
  organization, 53–54, 109
  Saturn (booster), 8, 14, 49, 73, 84, 104, 109, 114, 120, 149, 159, 189, 212, 213, 235, 247, 248, 294, 302, 330
Saturn Flight Evaluation Group, 49
  studies, 105, 224
Marshall Star, 8
Martin Co., 5, 10, 39, 155–156, 192, 197, 205, 217, 240, 245, 259
Martin Marietta Corp., 123, 171
  contract, 3, 74, 308
  Space Systems Div., 286
Martlet (Canada-U.S. rocket), 10, 29
Maryland, Univ. of, 162
Maser, 295
Maser Optics, Inc., 94
Mason-Rust Co., 238–239, 414
Massachusetts, 312
Massachusetts Institute of Technology (MIT), 9, 22, 56, 154, 237, 245–246, 247, 289, 331, 336, 339, 362, 379, 397
  Lincoln Laboratory, 269, 433
Massie, Sir Harrie, 118
Matador, 241
Mateos, Pres. Adolfo Lopez (Mexico), 273
Materials, 156, 181, 244, 303
  heat-resistant, 181–182, 264, 289
  structural, 77, 147, 207, 257
Mathematics, 255–256
Mathews, Charles W., 20, 212, 329, 365–366

Matthews Corp., 224
Matthews, Prof. Paul T., 71–72
Matthews, Dr. Thomas A., 120
Maui Island, Hawaii, 57
Mauna Loa, Hawaii, 275–276
Maxwell, B/G Jewell C. (USAF), 3
Mayevsky, Victor, 352
Mayo, Ivan W., 358
Mazdiyasni, Khodabakhsh S., 153
Mazur, Daniel G., 352
MBA. See Missouri Bankers Association.
MBS. See Mutual Broadcasting System, Inc.
Mds. See Satellite, Micrometeoroid Detection.
Mead, Merrill H., 202
Meany, George, 323, 331
Medal For Outstanding Leadership, NASA, 258
Medal of Recognition (USA), 136
Medicine (see also Life sciences), 204, 265, 397
Medium Orbital Research Laboratory (Morl). See Orbital Research Laboratory, Medium.
Meehan, Dr. John P., 243
Meeker, Leonard C., 116
Meinel, Dr. Aden B., 3–4
Meinel, Carolyn P., 3–4
Melbourne, Australia, 139
Memorandum of Understanding
  international, 70–71, 95, 258, 362, 375
  national, 57, 237, 319
Mengel, John, 435
Mercury (planet), 1, 105
Mercury (project), 33, 46–47, 99–100, 111, 129, 157, 194, 207, 225, 228, 247, 265, 384
  award, 317, 325
  flight, 4, 190, 273, 317, 318, 325
Mercury propellant, 158–159
Mercury-Atlas exhibit, 145
Merritt Island Launch Area (MILA) (KSC), 109, 120, 354
  facilities
    construction, 28, 38, 39, 131, 164, 181, 268, 328, 329, 369
    support services, 54–55, 88, 148, 201, 222, 299
  launch complex, 38, 268, 291
  strike, 59, 65, 124, 126, 128, 131, 133, 141, 424
Message Multiplexer, 330
Messerschmitt Co. (W. Germany), 265
Metal fatigue, 103
Metal fiber, 197
Metal fracture toughness, 223
Meta-planning, 161
Meteor, 14–15, 52, 171, 174, 241, 414, 415
Meteorite, 121, 128, 132, 133, 171, 190, 280
Meteoroid, 194, 197, 286
  hazards, 210, 389
Meteoroid detection satellite, 380

Meteorological data
  acquisition and processing, 43, 49, 196, 223
  exchange, 1-2
Meteorological Institute (Univ. of Stockholm), 288-289
Meteorological satellite. See Nimbus, Tiros, Tiros Operational System.
Meteorology (see also Weather), 190, 204, 208
Metric system, 5, 339
Metrology, 41
Metzler, Allen J., 142
Mexico, 46
Mexico City, Mexico, 39, 241
Mexico, Gulf of, 49, 102, 167
Mexico-U.S. Commission for Space Observations, 12th Session, 39
Miami, Fla., 112, 175, 190, 194, 323
Miami, Univ. of, 100
Michigan, 13
Michigan, Univ. of, 30, 135, 137, 210, 264, 391, 393
Michoud Operations (NASA), 84, 127, 149, 173, 211, 212, 230, 238-239, 304, 316-317, 324, 402, 405, 414, 424
Micrometeoroid, 118, 263, 283
Micrometeoroid Detection Satellite (Mds) (see also Pegasus), 138
Micrometeoroid dust impact, 131
"Micromouse burps" calibration, 324
Microscope, electron, 326
Microwave communications system, 143
Midwest, 75, 211, 261, 330-331
Midwest Governors' Conference, 312-313, 313
Midwest Office (NASA), proposal for, 102-103
MiG (jet fighter), 439
Mikhailov, A. A., 203
Mikoyan, President Anastas I. (U.S.S.R.), 437
Mikoyan, Artem I., 363
MILA. See Merritt Island Launch Area.
Milan, Univ. of (Italy), Polytechnic Institute, 127
Military bases, satellite photographs, 195, 210
Military capability, 89, 333
Military Electronics Convention (IEEE), 51-52
Military procurement and R&D authorization bill, 69, 72, 79, 81, 82, 98, 101
Military Requirements Review Group, USAF/NASA, 319
Military Sea Transportation Service (MSTS), 64-65
Milky Way Galaxy, 23, 314
Miller, Bernie P., 353
Miller, Mrs. Betty, 273, 317
Miller, Rep. George P., 2-3, 12-13, 15, 67, 126, 136, 150, 163, 176, 292-293, 298, 311
Miller, Col. Robert C. (USAF), 376-377
Miller, Rep. William, 313
Millikan, Dr. Clark Blanchard, 155

Milovidov, I. V., 374
Milwaukee, Wis., 47
Minnaert, Marcel Gilles Josef, 155
Minneapolis Honeywell Regulator Co. See Honeywell Co.
Minneapolis, Minn., 312-313, 313
Minnesota, Univ. of, 154
Minot AFB, N. Dak., 51
Minton, Dr. John Peter, 112
Minuteman (missile), 8, 9, 66, 89, 166-167, 314-315
  bases, 255
  launch, 15, 32, 62, 63, 78-79, 79, 84, 89, 122, 264, 330, 378
  failure, 116
  I, 330
  II, 328, 367, 424
Mira (red star), 223
Mishkol'tse, U.S.S.R., 253
Miskovsky, George, 175, 244
Missile, 303
  air-to-air, 86, 331
  antisatellite, 330, 370-371
  ballistic, 46-47, 98-99, 129, 136-137, 182-183
    intercontinental (Icbm), reliability, 9, 69, 112, 136-137, 170-171, 238, 409
    long-range (Lrbm), 15
  explosion, 181
  guided, 52, 280, 328
  military application, foreign, 52
  nuclear, 213, 246-247
  reliability, 6, 9, 28, 33, 69, 136-137, 141
  transport system, 238
Missile, Ballistic, and Space Technology, Ninth Symposium, 284-285
Missile-bearing submarine, 136-137
"Missile gap", 441
Missile installations, 393
Missile Sites Labor Commission, President's (MSLC), 24-25, 65, 124, 131
Missiles and Rockets, 269, 281, 317
  missile, 52, 282
  space
    cooperation, 311
    facilities, 58, 67
    plans, 19, 143-144, 144, 199, 220, 247, 295
    programs, 29, 164, 165, 181, 419
    research, 78, 129, 274, 383, 409, 428-429, 429
Mission Control Center (MSC), 10
Mission Simulation and Training Facility (MSC), 8
Mississippi Test Facility (MTF) (NASA), 51, 77
  construction, 13, 61, 121, 127, 203, 216, 247, 434
  support services, 14
Missouri Bankers Association (MBA), 173
Missouri Cotton Producers Association, 195

Missouri Valley College (Marshall, Mo.), 196
MIT. See Massachusetts Institute of Technology.
Mmu. See Modular Maneuvering Unit.
Moberly, Mo., 291
Mock, Mrs. Geraldine, 142, 164
Modular Maneuvering Unit (Mmu), 313
Moffatt, Dr. John W., 155
Moffett Field, Calif., 231
Mohole, Project (NSF), 21
Mojave, Calif., 19
Mol. See Manned Orbiting Laboratory.
Molab (mobile laboratory), 251
Mold, 73
Molecular densities, 241
Molybdenum, 146
Moms. See Satellite, Multiple Orbit-Multiple.
Mongolia, 354
Monitor Systems, Inc., 440
Monomethylhydrazine, 400
Monroney, Sen. A. S. (Mike), 78, 82, 181, 310
Montoya, Rep. Joseph M., 320
Monument, 132, 384
Moon (see also Lunar program), 28, 70, 102, 125, 188
  and cloud relationship, 288
  atmosphere, 106–107
  craters, 154, 180, 267, 271, 299, 364, 441
  exploration of, 2, 35, 69, 72, 81, 106, 156, 170, 230, 240, 251, 272, 275, 297
  landing
    manned, 2, 6, 16, 20, 45, 52, 54, 95, 100, 104–105, 116, 132, 152, 173, 189, 204, 206–207, 258, 261, 268, 278, 321
    unmanned, 156–157, 176–177, 258
  map of, 220, 223
  photographs, 271–272, 273, 307, 460
  surface of, 69, 78, 103, 106, 128, 156, 164–165, 180, 203, 220–221, 223, 237, 267, 272–273, 275–276, 299, 307, 331, 426–427
  volcanic activity, 180
  water on, 223, 438
"Moon Blink", 426
Moon Harvest, Project (NASA), 128
Moon spots, red, 78, 203–204, 205
Moore, George S., 19
Morgan, William C., 82
Morl. See Manned Orbital Research Laboratory.
Morl. See Orbital Research Laboratory, Medium.
Moron Air Base, Spain, 5
Morris, David, 383
Morris, Thomas D., 291
Morrison-Knudsen Co., Inc., 15
Morrison, Dr. Richard B., 210
Morse, Rep. F. Bradford, 339
Morse, Sen. Wayne L., 51
Moscow (U.S.S.R.), 10, 21, 34, 54, 59, 130, 131, 132, 143, 174, 204–205, 208, 210, 226, 289

Moses, Robert, 310
Moss, William W., 349
Mosser Construction Co., 367
Mot. See Manned Orbiting Telescope.
Motion picture, 382
Motorola, Inc., Western Military Electronics Div., 412
Mt. Haleakala, Maui Island, Hawaii, 57
Mt. Kobau, B.C., 396
Mt. Palomar Observatory, 120, 179–180, 326, 398
Mt. Wilson Observatory, 148, 180, 398
Mrazek, Dr. William A., 109
MSC. See Manned Spacecraft Center.
MSFC. See Marshall Space Flight Center.
MSTS. See Military Sea Transportation Service.
MTO. See Mississippi Test Operations, NASA.
Mud Lake, Nev., 66
Muden, So. Africa, 226
Mueller, Dr. George E., 120, 187, 218, 234
  Apollo, 54, 74, 77, 104–105, 157–158, 173, 194, 237, 299–300, 364, 387
  Gemini, 68–69, 74, 316, 356
  hearings, 69, 104
  lunar program, 52–53, 369
  Saturn, 138, 173, 199
  space flight, manned, 184, 206
  space program, 53, 326, 368–369
Muhleman, Dr. Duane O., 165
Murphy, Charles J. B., 441
Murphy, Dr. Thomas P., 4, 395
Murray, Dr. Bruce C., 179
Mururoa, Tuamotu Islands, 14, 266
Museum of History and Technology (Smithsonian Institution), 241, 282
Museum of Modern Art, 230
Museum of National Science (Houston), 244
Mutual Broadcasting System, Inc. (MBS), 149
Muzzaffarpur, India, 133
MX-324 (Northrop flying wing), 239
Mylar, 141
Myrtle Beach, S.C., 309
N-on-p (negative-on-positive). See Solar cell, radiation-resistant.
NAA. See National Aeronautic Association.
NACA. See National Advisory Committee for Aeronautics.
NAE. See National Academy of Engineering.
NAFEC. See National Aviation Facilities Experimental Center. FAA.
Nakamura, Seitaro, 394
NAPCA. See National Association of Professional Contract Administrators.
NAR. See National Association of Rocketry.
NAS. See National Academy of Sciences.
NAS-NRC. See National Academy of Sciences-National Research Council.
NASA. See National Aeronautics and Space Administration.

NASA-AEC Space Nuclear Propulsion Office (SNPO), 63, 140, 248, 311, 327, 385
NASA Authorization Act (see also NASA, appropriations), 247
NASA Board of Inquiry (Kelley Board), 151
NASA Central Instrumentation Facility, MILA, 28
NASA Contributions Award, 276
NASA Deep Space Network, 31
NASA-DOD Aeronautics and Astronautics Coordinating Board (AACB), 90
NASA-DOD agreement on operational tracking ships, 64–65
NASA-DOD—Dept. of Commerce Geodetic Satellite Policy Board, 325
NASA Electronics Research Center Site Evaluation Committee (see also Electronics Research Center, NASA), 23, 113
NASA Electronics Research Task Group, 304
NASA Group Achievement Award, 77, 221
NASA Historical Advisory Committee, 178, 322
NASA-JPL Tracking Station, Goldstone, Calif., 37, 143
NASA Manned Space Flight Experiments Board, 137
NASA Medal for Outstanding Leadership, 258
NASA Midwest Office, proposal for, 102, 145
NASA New Orleans Coordinating Committee, 316, 324
NASA *Ninth Semiannual Report to the Congress*, 235
NASA North Eastern Office, 304
NASA Policy Planning Board, 166
NASA Science and Technology Advisory Committee for Manned Space Flight, 120, 237
NASA Summer Faculty Fellowship, 214
NASA-SUPARCO Project (see also Judi-Dart), 107
NASA Test Complex Beta, 403
NASC. See National Aeronautics and Space Council.
Natal, Brazil, 375
National Academy of Engineering (NAE), 12, 154, 416
National Academy of Sciences (NAS), 12, 31, 154–155, 245, 281, 416
  agreements, 2, 182
  Panel on Astronomical Facilities, 385
  report, 109
  sonic boom, 183, 227
  Space Science Board, 116, 151, 158, 210, 356, 366, 391
National Academy of Sciences-National Research Council (NAS-NRC), 308
National Advisory Committee for Aeronautics (NACA), 136, 147, 154, 171, 182, 326

National Aeronautics and Space Act of 1958, 5, 44, 217, 282, 294, 311, 331, 338, 377
National Aeronautics and Space Administration (NASA), 25–26, 35, 51–52, 54, 71–72, 102, 117, 136, 154, 156, 171, 206–207, 214, 216, 235, 261, 294–295, 300–301, 308, 310–311
  agreement, 43, 64, 237, 258
  international, 11, 59, 67, 71
  appropriations, 12, 20, 46, 115, 180, 183, 185–186, 228, 237, 242, 266, 276, 285–286
  astronaut, 10, 209, 263, 306, 318
  award, 77, 82, 150, 180, 276, 352
  budget, 7, 49, 66, 181, 223
  contract, 86, 162, 189, 214, 237, 261, 281
    booster, 27, 84, 160, 211, 212, 213
    engine, 115, 116, 120, 242, 294
    facilities, 39, 61, 63, 143, 224
    fuel, 14, 77
    ground systems, 103, 132, 176, 216, 325
    incentive, 171, 173, 342
    spacecraft, 65, 90, 116, 134, 171, 193, 306
    space equipment, 142, 143, 177, 218
    study, 33, 105, 208, 248, 259, 292
    support services, 58, 85, 148, 168, 236, 237
  cooperation, 176, 231
    DOD, 51, 64, 167, 176, 325
    application, 33
    Gemini, 12, 82, 192
    Mol, 10, 68, 90
    DOD-AEC, 231
    USAF, 55, 80, 96, 117, 171, 206, 237, 257, 292, 302, 319, 322, 354
    USN, 71, 292–93
    ONR, 34, 96, 406
  cooperation, international, 96–97, 240, 324
    Argentina, 403
    Australia, 84, 103
    Canada, 11, 23, 30, 42, 47, 62, 141, 296
    Denmark, 56
    ELDO, 269–270
    ESRO, 258
    France, 95
    India, 6, 31, 70
    Italy, 115
    Japan, 115
    Netherlands, 278
    Norway, 56
    Pakistan, 107, 403
    Spain, 31
    Sweden, 99, 205, 299
    U.S.S.R., 17, 27, 30, 54, 59, 75
  criticism of, 56, 201, 203, 207
  economy, 38
  education, 136
  expenditures, 202
  experiment, 71, 77, 137, 196

National Aeronautics and Space Administration—Continued
facilities, 103, 109
   construction, 15, 38, 61, 63, 77, 102, 136, 164
   ground system, 132
   location, 91, 327
   support services, 54
funds, 48
grants, 11, 156, 162, 199, 211, 216, 220, 239, 241, 245, 259, 280, 331
information dissemination, 255, 282
launch, 8, 258
   balloon, 96, 291
   development, launch vehicle, 32, 192, 227, 321, 414–415
   failure, 9, 41, 109, 215, 227
   probe, interplanetary, 378, 397
   probe, lunar, 34, 263
   satellite, 8, 118, 290, 296, 298, 305–306, 339, 346, 380, 394–395, 420, 427
   sounding rocket, 11, 13, 14, 22–23, 30, 30, 42–43, 47, 62, 62, 79, 99, 114, 115, 124, 128, 137, 140, 141, 145, 192, 210, 243, 249, 251, 260, 262, 264, 277–278, 283, 299, 304, 305, 328, 335, 343, 355, 360–361, 363, 368, 374, 374–375, 379, 380, 384, 384, 387, 387, 391, 395, 415, 422, 422
   test, 192–193, 256
      Apollo, 175, 192–193, 321
      equipment, 256
      Gemini, 130
   reentry, 132, 135, 289–290
management, 11, 24–25, 25, 41, 49, 51, 60, 73, 137–138, 173, 187, 195, 220, 281, 304, 311–312
organization, 4, 11, 12, 16, 17, 22, 24–25, 43, 49, 51, 53–54, 77, 102, 113, 120, 143, 144, 154, 173, 187, 195, 202, 210, 213, 224, 265, 266, 268, 274, 287, 301, 304, 306
patent, 56, 95
   legislation, 5, 43–44
   policy, 201, 293, 434
   waiver, 214, 293, 329
personnel, 6, 35, 52–53, 230, 237, 244
policy, 165, 173, 250
procurement, 49, 51, 80, 143, 189, 257, 293, 329
programs, 181–182, 196, 219, 243, 257
   Apollo, 18, 185–186, 224, 225, 228, 237, 316
   astronomy, 327
   bioscience, 72, 80, 151–152, 178–179, 210
   Gemini, 18, 55, 68–69, 71, 273, 316, 325
   lunar orbiter, 69
   Oao, 212, 215
   Ogo, 249–250, 308
   Ranger, 34–35, 41, 41–42, 42, 60–61, 64, 150, 150–151, 151, 215, 263, 267, 267–268, 298–299

   Saturn, 238, 330
   space, 8, 20–21, 26, 35–37, 49, 50–51, 52–53, 69, 71, 77, 94–95, 98, 106, 117, 141, 199, 206, 228, 258, 324
   space station, 8, 80, 90
   Syncom, 286, 290–291, 326
   Tos, 193
   Tracking and data acquisition, 71, 202–203
report, 5, 35, 39, 43–44, 67, 95, 101, 104, 142, 217, 235, 257, 263, 306
research, 47, 53, 78, 277, 293, 326–327
   aeronautical, 78, 108, 117–118, 187–188, 292–293
   fuel, 246
   nuclear, 26–27, 83, 103, 229, 248
   propulsion, 23, 223–224
scientist and engineer, 94–95, 110–111, 196
standards, reliability and quality, 143, 281
studies, 45, 95, 98, 241
supersonic transport, 4, 43, 57, 78, 115, 117–118, 119–120
test, 71, 137, 147, 182, 192, 296
   communications, 3, 17–18, 22, 32, 146
tracking station, 10, 27, 55, 103, 129, 159, 306
universities, 4, 100, 406, 428
   training, 4, 419
X–15, 29, 33, 77, 263
National Aeronautics and Space Council (NASC), 136, 186–187, 269, 279–280, 350, 354–355, 361
National Air and Space Museum (Smithsonian Institution) (formerly National Air Museum), 260
National Air Museum (Smithsonian Institution), 35, 260
National Association of Professional Contract Administrators (NAPCA), 281
National Association of Rocketry (NAR), 292
National Automation Conference (American Bankers Association), 252
National Aviation Facilities Experimental Center, FAA (NAFEC), 56, 214
National Aviation System Symposium, 332, 336
National Broadcasting Co. (NBC), 146, 221, 294, 347
National Bureau of Standards (NBS), 5, 41, 339, 363, 373
National Business Aircraft Association (NBAA) Award, 334
National Cancer Institute, PHS, 112, 277
National Center for Atmospheric Research (NCAR), 42, 96, 176, 399
National Center of Space Research (CNES-Centre National d'Etudes Spatiales, France), 95, 273
National Civil Service League Career Service Award, 1964, 136

National Commission on Technology, Automation, and Economic Progress, 259, 286
National Conference on the Peaceful Uses of Space, Fourth, 156–157
National Defense Transportation Association, 370
National Editorial Writers' Conference, 387
National Education Association (NEA) *Journal*, 2
National Electronics Conference, 357
National Geographic Society (NGS), 16–17, 325, 432
National Humanities Foundation (NHF) (proposed), 240
National Institutes of Health (NIH), 165–166, 308
National Labor Relations Board (NLRB), 24–25, 61, 126, 128, 133, 209, 212
National Magnet Laboratory, 379
National Management Association, 342
National Medal of Science, 11, 397
National Merit Scholarship Corp., 255–256
National Meteorological Center, 439
National Multipurpose Space Station, 388
National Operational Meteorological Satellite System (Nomss) (USWB), 43
National Orbiting Space Station (Noss) (DOD), 68
National Radio Astronomy Observatory, NSF, 321
National Range Division (NRD) (AFSC), 3, 164
  Eastern Test Range (ETR), 178, 192, 239
  Western Test Range (WTR), 178, 192, 239
National Reactor Testing Station, AEC, 93
National Science Foundation (NSF), 21, 48, 96, 104, 112, 161, 165–166, 220, 223, 240, 362, 401–402
National Security Industrial Association (NSIA), 246, 344
National Space Club, 34, 68, 111, 144, 182–183, 217, 259, 290, 319, 358, 397, 420–421
National Space Development Center (Japan), 259
National Space Research Commission (Spain), 31
National Space Station Planning Subpanel, AACB, 90
National University Extension Association (NUEA), 149–150
National Weather Satellite Center USWB, 101, 439, 440
NATO (North Atlantic Treaty Organization), 25, 349
Nature (British journal), 121, 278
Naugle, Dr. John E., 98
Naval Research Laboratory (NRL), 15, 45, 179–180, 254, 271, 432, 436–437
Naval School of Aviation Medicine, 174, 219
Navigation, 111–112, 125, 146–147, 305, 333
Navigation satellite. See Satellite, Navigational.
Navy. See U.S. Navy.
Navy Oceanographic Meteorological Automatic Device. See Nomad.
NBAA. See National Business Aircraft Association.
NBC. See National Broadcasting Co.
NBS. See National Bureau of Standards.
NCAR. See National Center for Atmospheric Research.
NEA Journal (National Education Association), 2
Nebraska, 247
Neeper, Mr. & Mrs. Bernard W., 121
Negative-on-positive (n-on-p). See Solar cell, radiation resistant.
Negro, space-age training, 263
Nelson, CWO Emery E. (USA), 394
Nelson, Sen. Gaylord, 81, 214, 261, 330–331
Nelson, Richard H., 235
NEO. See North Eastern Office, NASA.
Nerva (Nuclear Engine for Rocket Vehicle Application), 83, 140, 151–152, 236, 327, 353
Ness, Dr. Norman F., 420
Netherlands, 249–250, 278
Netherlands Organization for the Advancement of Pure Research, 278
Neutron flux mapping device, 25
Neutron star, 153, 254, 432, 436–437
Nevada, 281
Nevada Test Site, AEC, 104
New Hampshire, 64
New Hampshire, Univ. of, 25, 31, 408
New Jersey, 165
New Jersey Central Power and Light Co., 426
New Mexico, 34, 254
New Mexico State Univ., 206
New Orleans, La., 212, 238, 304, 316–317, 324
New Orleans *Times-Picayune*, 281–282
New Product of the Year Award (*Missiles and Rockets*), 269
New York, 75, 231
  Atomic Research and Development Authority, 70
New York Herald Tribune, 82, 122, 145
New York Medical College, 226
New York, New York, 1, 5, 19, 81, 129, 142, 165, 200, 230, 240, 247, 288, 310
  meetings, 9, 21, 22, 23, 112, 114, 143, 149, 154, 158, 158, 164, 167, 180, 231, 243
New York Post, 352
New York Society of Newspaper Editors, 215
New York Society of Security Analysts, 161

New York Times, 10, 14, 16, 19, 26, 27, 66, 76, 86, 92, 156, 200, 220, 227–228, 255, 256, 272, 288, 382, 411, 436
New York Univ., 331
New York World's Fair, 101, 145, 214, 310
New Zealand, 66, 414
Newell, Dr. Homer E., 176, 176–177
  moon, 106
  post-Apollo, 366–367
  Ranger, 150–151, 267, 271
  space programs, scientific, 50–51, 69, 71, 79
Newkirk, Dr. Gordon, 96
News, dissemination of, 248
Newsom, Dr. Bernard D., 384
Newsweek, 28
Newton, Sir Isaac, 125
NGS. See National Geographic Society.
NHF. See National Humanities Foundation.
Nichols, Jack J., 114
Nickel alloy, 56
Nickerson, Col. John C. (USA), 92
Nicks, Oran W., 151, 352
Nicolaides, John D., 163
Nicolet, Prof. Marcel, 266
NIH. See National Institutes of Health.
Nii Jima Island, Japan, 259
Nike-Apache (sounding rocket)
  launch
    foreign, 56
      Canada, 72, 84, 110, 113, 114, 115, 133, 137, 145, 373, 378, 380
      India, 6, 10, 382, 384
      Pakistan, 107, 399, 403
      Sweden, 277–278, 283, 287, 288–289
    U.S., 11, 13, 14, 42–43, 62, 140, 192, 202, 210, 222, 243, 249, 251, 287, 297, 343, 344, 379, 384, 391, 392, 395, 422
  use of, 31, 241, 278, 290
Nike-Cajun (sounding rocket), 147
  launch
    foreign, 277–278, 283, 287, 403, 406
    U.S., 22–23, 30, 42–43, 47, 62, 99, 141, 277–278, 283, 287, 374–375, 391, 392
  use of, 241
Nike-Javelin (sounding rocket), 122
Nike-Tomahawk (sounding rocket), 444
Nike-X (anti-Icbm missile system), 250, 328, 332, 388
Nike-Zeus (antimissile missile system), 211, 250, 322
Nikolayev, L/c Andrian G., 10, 74, 177–178, 209, 248, 286
Nikolayeva, Yelena Andrianovena, 248
Nikolayeva-Tereshkova, Valentina, 10, 59, 209, 226, 248, 443
Nimbus (meteorological satellite), 283
NIMBUS I (meteorological satellite), 298, 305, 326, 442, 464–465, 485
  photographs, use of, 309, 310
Nitric acid oxidizer, 233

Nitrogen, 217
Nitrogen-oxygen atmosphere, 309
Nitrogen tetroxide, 48, 229–230, 399–400
Nitze, Paul H., 353, 379
NLRB. See National Labor Relations Board.
Nobel Prize, 367
Noctilucent clouds, 3–4, 147, 241
Noise (see also Sonic boom), 236
  aircraft, reduction, 187–188
  boundary layer, 174, 305
  galactic, radio, 118
  radio frequency, 125
  turbulent air, 103
Nomad (Navy Oceanographic Meteorological Automatic Device), 49
Nomss. See National Operational Meteorological Satellite System.
Noordwijk, Netherlands, 364
NORAD. See North American Defense Command.
Norair Div., Northrop Corp., 145, 201
Nord 262 (French transport), 318
Norman, Lee, 156
North American Aviation, Inc., 6, 66, 172, 198, 214, 264, 278, 299, 402
  contract, 24, 29
  research, 194, 197, 280
  Rocketdyne Div., 45, 120, 159, 211, 219, 236, 242, 248, 249, 294, 346
  Science Center, James Howard Kindelberger Memorial Laboratories, 298
  Space and Information Systems Div., 56–57, 89, 158, 160, 186, 216, 239, 306
  supersonic transport design, 13, 28, 183
North American Defense Command (NORAD), 61, 198, 362
  Space Detection and Tracking Center (SPADATS), 254
North Atlantic Treaty Organization. See NATO.
North Carolina Science and Technology Research Center, 194
North Carolina State College, 194
North Carolina, Univ. of, 194
North Eastern Office, NASA (NEO), 102
Northeastern Univ., 163
Northrop Aircraft Corp., 321
  Norair Div., 145
  Ventura Div., 129, 144
Northwestern Univ., 393
Norway, 56, 216, 246, 397
Norwegian Space Research Committee, 56, 375
Nose cone, missile, 57, 113, 171, 403
Noss. See National Orbiting Space Station.
Notre Dame, Univ. of, 342
Nova (booster), 158
Novae Sagittae, 401
Novins, Stuart, 211
Nozzle, gimbaled, 89, 236, 331
NRD. See National Range Division.
Nrds (Nuclear Radiation Detection Satellite). See Vela, Project.

NRL. See Naval Research Laboratory.
NRX-A (experimental nuclear reactor), 327, 385
  A1, 140
  A2, 327, 353
NSF. See National Science Foundation.
NSIA. See National Security Industrial Association.
Nuclear energy, 17, 19, 21, 37, 104, 126, 161, 266, 288, 295, 337, 360, 426
  peaceful use, 1, 25, 40–41, 143, 251
  power, 25, 40–41, 49, 145, 151–152, 296, 324, 333
  propulsion system, 83, 89, 93, 228–229, 296, 302, 305, 320, 333
  reactor, 25, 40–41, 83, 195, 228–229, 248, 296, 305, 311, 320
  rocket, gas-core, 26–27, 83
Nuclear energy conversion, 151–152
Nuclear-powered Low Altitude Supersonic Vehicle (Lasv), 303
Nuclear radiation detection. See Vela, Project, and Sentry.
Nuclear Rocket Systems Analysis Program, Advanced, 236
Nuclear submarine, 221, 266, 302, 406
Nuclear test ban treaty, 266
NUEA. See National University Extension Association.
Nutley, N.J., 22, 130
Nutrition, 151
Nylon parachute, 144
Oakland, Calif., 317
Oakwood, Tex., 176
Oao. See Orbiting Astronomical Observatory.
OAR. See USAF Office of Aerospace Research.
Oberth, Hermann, Award, 392
Oberth, Hermann, Medal (West German Association for Rocket Research Technique and Space Flight), 320
Oceanography, 49, 146–147, 196, 262
Odessa, Tex., 268
O'Donnell, Kenneth, 363
Offenbach, West Germany, 1
Office National d'Etudes et de Recherches Aerospatiales (ONERA) (formerly Office National d'Etudes et de Recherches Aeronautiques), 155, 233, 371
Office of Naval Research (ONR), 34, 96, 140, 222, 330
Office of Science and Technology, President's, 12, 186
Office of Solar Energy (Dept. of Interior), 180, 275
Ogo. See Orbiting Geophysical Observatory.
Ogo-E. See Orbiting Geophysical Observatory.
Ohio, 15, 87, 121, 289
Ohio State Univ., 33, 217
Ohnuki, Prof. Y., 71
O'Keefe, Dr. John A., 102, 128

Oklahoma City, Okla., 43, 63, 84, 90, 175, 177, 183, 187, 242, 244, 253, 268, 274, 279, 314
Oklahoma Frontiers of Science Foundation, 340
Olling, Edward H., 140
Olson, Walter T., 293
Olympic Games, 1964, 32, 146, 258–259. 261, 347
Olympus (space station), 247
Omaha *World-Herald*, 120
Omega-minus particle, 71
ONERA. See Office National d'Etudes et de Recherches Aerospatiales.
ONR. See Office of Naval Research.
Optical degradation, 118
Optical Re-entry Instrumentation System, 211
Optical reflector, 325–326
Orange County, Calif., 140
Orbital operations, earth, 253
Orbital Research Laboratory, Large (Lorl), 122, 145
Orbital Research Laboratory, Medium (Morl), 145
Orbital shuttle transport, 249
Orbital systems, 145
Orbital tanker, 281
Orbiting Astronomical Observatory (Oao), 78, 106, 196, 201, 212, 215, 231
Orbiting Astronomical Observatory, Advanced (Aoao), 98
Orbiting Geophysical Observatory (Ogo), 196, 200, 436
  OGO-I, 306, 308, 311, 319, 456, 467
  Ogo E, 249
Orbiting Geophysical Observatory, Advanced (Aogo), 98
Orbiting observatory program, 258, 318
Orbiting Solar Observatory (Oso), 55, 95, 135, 196
  OSO I, 82
Orbiting Solar Observatory, Advanced (Aoso), 98
Orbiting Vehicle 1 (OV-1), 383
Orbiting Vehicle 2 (OV-2), 269
Ordahl, Douglas D., 190
Orion (constellation), 374
Orion, Project, 419
Orlando, Fla., 281
Orroral Valley, Canberra, Australia, 103
Orthicon, image, 72
Orthostatic hypotension, 129, 177
Osborne, Dr. Louis S., 22
Osborne, Stanley de J., 76, 90
Oscar satellite series, 254
Oso. See Orbiting Solar Observatory.
OSSA. See NASA Office of Space Science and Applications.
Ostrander, M/G Don R. (USAF), 34
O'Sullivan, William J., 418
Ousley, Gilbert W., 336
Outstanding Achievement Award (AFSC), 39

Outstanding Leadership Medal (NASA), 258
OV-1. See Orbiting Vehicle 1.
OV-2. See Orbiting Vehicle 2.
Oxford, N.C., 99
Oxidizer, 77, 79, 125, 190
Oxygen, 15, 230, 290
　environment test, 44, 87, 130
　liquid, 190, 246
　measurement, 56–57
Oxygen difluoride propellant, 78
Oxygen-helium atmosphere, 309
Oxygen-hydrogen fuel, 230
Ozma, Project, 321
Ozone, 76, 118
P-on-n. See Positive-on-negative.
PAA. See Pan American Airways.
Pace, Prof. Nello, 177–178
Pacific Missile Range (PMR) (see also Western Test Range), 3–4, 95, 161
　budget, 21
　launch
　　missile
　　　Atlas F, 63, 127, 145–146
　　　Minuteman, 15, 62, 78–79, 79, 116
　　　Titan II, 24, 67, 104
　　satellite mission, 10, 17, 25, 66, 78–79, 85, 102
　management, USAF, 3–4, 178, 192
Pacific Northwest, 75
Pacific Ocean, 37, 105, 158, 178, 185, 196, 266, 268, 273–274, 290–291, 305, 305, 314, 326, 333
Pack, back, jet-propelled, 313
Pads. See Precision Aerial Delivery System.
Page Communications Engineers, Inc., 383–384
Pageant, 268–269
Pageos (Passive Geodetic Satellite), 419
Pakistan, 96–97, 107, 399, 403
Pakistan International Airlines (PIA), 430
Palestine, Tex., 42, 56, 57, 96, 140, 176, 397
Palewski, Gaston, 14
Palm Springs, Calif., 197
Palmdale, Calif., 172, 297, 323, 429
Pan American Airways (PAA), 233, 398
Pan American-Grace Airways, Inc. (Panagra), 175
Panama Canal Co., 19
Panama Canal Zone, 221
Panel on Astronomical Facilities (NAS), 385–386
Pangborn, W. V., and Co., Inc., 39
Papeete, Tahiti, 14
Papell, S. Stephen, 142
Parabola, zero-gravity, 70
Parachute, 129, 144, 193, 317, 367
　drogue, 131, 156, 175
　jumps, 32, 213, 283
Parade of Progress. Cleveland, O., 300
Parade magazine (*Washington Post*), 207
Paraglider, 68–69, 280
　test, 158, 193, 210, 239, 264, 278, 425

Parasail, 131, 156, 182, 352, 361
Parin, Prof. Vasily V., 177–178, 226
Paris, France, 14, 25, 79, 113, 264, 273, 326–327, 439
Parker, Alvin H., 268
Particles, energetic, 103, 249–250, 306–307
Pasadena, Calif., 139, 139, 176
PASSIM. See President's Advisory Staff on Scientific Information Management.
Pastore, Sen. John O., 19, 40–41, 117
Patents (see also Inventions), 56, 95, 100, 102, 153, 184
　award, 66, 172, 188, 287
　legislation, 5, 43–44
　policy, 5, 43–44, 201, 293, 329, 434
　waiver, 214, 293, 329
Patrick AFB, Fla., 3, 164, 192, 424
Patrick AFB Contractors' Association, 126, 141
Patuxent Naval Air Station, Md., 437
Paul VI, Pope, 273
Pdp. See Project Definition Phase (DOD).
Pearson, Drew, 81, 262–263
Pegasus (micrometeoroid detection satellite), 138, 286, 436
Pell, Sen. Claiborne, 56
Pelly, Rep. Thomas M., 23
Pendleton, Oreg., 116
Pendray, G. Edward, Award, 236
Pennsylvania, 75
Pennsylvania Military College Engineering Centennial Medal, 206–207
Pennsylvania, Univ. of, 395
Pepper plant experiment, 72
Perini Corp., 15
Perkin-Elmer Co., 211
Perkinson, William J., 10
Pershing (missile), 394
Personnel exchange agreement (NASA-USAF), 319
Peters, L/C John D. (USAF), 358
Pettingill, Dr. Gordon H., 297
Phantom II. See F-4 aircraft.
Philadelphia, Pa., 129, 143, 205, 322
Philco Corp., 10, 61, 176, 259
Philippines, 109
Phillips, Franklyn W., 301
Phillips, M/G Samuel C. (USAF), 315, 364, 371
Phoebus (graphite reactor), 320
Phoenix, Ariz., 147, 306
Phoenix (missile), 190–191
Photoelectric sensor, 339
Photography, 55, 271, 315
Photon echo effect, 383
Photovoltaic cell, 202
Physical Review Letters, 4, 383
Physical sciences, 151–152, 255–256, 344–345
Physical stress measurement, 241
Physics, 198, 241, 262, 265
Physiological research, 29
PIA. See Pakistan International Airlines.
Piccard, Dr. Jean Felix, 132
Piccard, Dr. Jeannette, 132

Pickering, Dr. William H., 41, 128, 176, 224, 267–268, 271, 308–309, 315, 416
Pierce, Dr. J. A., 263
Pierce, John R., 11
Piezo-electric transducer, 131
Pioneer (space probe), 225
Pilot, 110, 233, 288, 443
Piper Apache (aircraft), 317
Pittsburgh, Univ. of, 328
Pitz, Prof. Wolfgang, 262–263
Planaria, 89
Planetarium, 244
Planetary spaceflight, 176, 297, 300–301
Plans for Progress Seminar, 212
Plant experiment, 179
Plasma, 82, 132, 159
Plasma accelerator, 424
Plasma-jet generator, 238
Plasma physics, 241
Plasma probe, 325
Plasma torch, 433
Plastic, 303
Plastic honeycomb matrix, 181–182
Platform, stabilized, 22
Pleumeur-Bodou, France, 32, 140
Plum Brook Station, LRC, 221, 224, 317, 367
Plummer, William E., 340
Pluto (nuclear missile), 213, 246–247, 317
Plutonium, 143
PMR. See Pacific Missile Range.
Pneumotachometer, 440
Point Arguello, Calif., 102, 106, 114, 146, 182, 203, 220, 239
Point Loma, Calif., 78
Pt. Mugu, Calif., 178, 315, 326
Poland, 311
Polaris (missile), 150, 268, 307, 314–315, 330, 349
 launch
  A–2, 11, 68, 221
  A–3, 6, 19, 238, 266, 279, 406
 submarine firing, 209, 430
POLET II (U.S.S.R. spacecraft), 134, 450
Politics and science, 86–87
"The Politics of the Space Age," 86–87
Polonium, 200
Polyamide prepolymer, 228
Polynesia, 14, 266
Polyuretanes, 228
Pompidou, Premier Georges (France), 266
Ponnamperuma, Dr. Cyril, 390
Poodle, Project (USAF-AEC) (see also Rover), 200, 209
Popovich, Marina, 212, 282
Popovich, L/C Pavel R., 177, 212, 282, 286, 348
Port Columbus Airport, O., 142
Porter, Richard W., 172
Portsmouth, N.H., 9
Port Tobacco, Md., 364
Portugal, 55, 132
Positive-on-negative (p-on-n), 19–20
Positron, 4

Potassium, 62
Potrero, Calif., 193
Power, Gen. Thomas S. (USAF), 400
Power Engineering Co., 127
Prahl, Val E., 428
"Prairie Network" (meteorite tracking stations), 171
Prather, Victor A., Award, 164
Pratt & Whitney Div. (United Aircraft Corp.), 77, 248–49
 construction contract, 85–86, 193
 supersonic transport contract, 13, 28, 161, 183, 201, 373–74, 405
Pravda, 105, 142, 272, 308, 333, 381
Precision Aerial Delivery System (Pads) (See also Paraglider), 239
Preflight Operations Div., MSC, 54
Presidential Medal of Freedom, 316
President's Advisory Committee on Supersonic Transport, 135, 142, 149, 162, 183
President's Advisory Staff on Scientific Information Management (PASSIM), 2, 71
President's Board of Inquiry, 51
President's Missile Sites Labor Commission, 24–25
President's Office of Science and Technology, 12, 186, 362
President's Safety Award, 184
President's Science Advisory Committee (PSAC), 35
Press conference, 12, 40, 130, 140, 171, 200, 319, 323, 327
 A–11, 85, 133
 Apollo, 43
 international cooperation, 74–75, 129
 JPL management, 137
 missile reliability, 9, 33
 Mol project, 417
 radiation hazards, 45, 131
 RANGER VII, 42
 Saturn, 32, 40
 supersonic transport, 119, 148–149, 318
Pressly, Eleanor, 109
Pressure measurements, 379
Pressure suit, 146, 159, 197, 239–240
Pressure valve, helium, 304
Pressure vessel, thin-walled, 82
Preston, G. Merritt, 120, 431
Price, Rep. Melvin, 317
Priester, Dr. Wolfgang, 372
Primus Club, 412
Princeton Univ., 154, 223, 374
Pritchard, Wilbur L., 118
Prittie, Terence, 303
Probe, 20, 263
 atmospheric, 42–43, 103, 159, 325
 lunar (see also Ranger), 47–48, 143, 152, 308
 planetary (see also Mariner), 198, 208, 295
Products Aerospace Div. (Bendix Corp.), 13

Professional Group on Antennas and Propagation, IEEE, 326
Prokhorov, Aleksandr M., 367
Prom (Program Management, Resources Management, and Objectives Management), 285
Propellant grain, 248
Propellant, 78, 229, 235
Propellant tanks, 82
Propellant valve, 9, 181
Propulsion, 190, 241, 302–303
  chemical, 229
  electric, 229, 302, 324
  hydrogen peroxide, 89
  ion, 82, 324, 410, 429
  jet, 47
  nuclear, 229–230, 296, 320, 324, 333
  ramjet, 77, 224
  rocket, 334
Propulsion Award (AIAA), 166
Propulsion system, 22, 66, 107, 228, 296
Proton, 22, 34, 50, 178, 219
Proton detector, 325
Proxmire, Sen. William, 81, 93, 127, 276, 329
PSAC. See President's Science Advisory Committee.
Pucinski, Rep. Roman C., 117, 144, 211
Puckett, Dr. Allen E., 237
Pulkovo Observatory (U.S.S.R.), 180, 203
Pulse code modulation, 423
Purdue Univ., 216, 274
Pygmy (tetrahedral satellite), 252
Qantas Empire Airways (Australia), 19
"Quality Assurance Provisions for Government Agencies" (NPC 200–1A), 231, 291
Quality control, 6, 143
Quarantine, astronaut, 215
Quasar (quasi-stellar object), 155, 340, 419
Quasar 3C–273, 155
Quasi-Stellar Force (Qsf), 426
Quasi-stellar radio source, 155
Queen Mary College, London, 424
Quesada, Elwood R., 213
Rabatin, Michael J., 114
Raborn, V/A William F., Jr. (USN, Ret.), 126
Radar, 30, 168, 239, 429
  laser, 112, 278
  Multi-function Array (Mar), 250
  over-the-horizon, 320, 330
Radar astronomy, 105, 297
Radar interference, 340
Radar reflector, 325
Radford, Adm. Arthur (USN, Ret.), 370
Radiation, 55, 138–139, 194, 202, 291
  cosmic, 174, 216
  damage, 8, 131
  effects, 72, 169, 171, 179, 219, 248, 265
  energetic particle, 250
  ionizing, 174, 248
  measurement, 125, 169, 245, 265

solar
  effects of, 13, 18, 34, 45, 48, 75, 173, 250
  measurement, 45, 140, 444
  shielding from, 13, 31
  study of, 18, 132
  space, 146, 169, 349
  measurement, 28, 32, 245
  ultraviolet, 76
Radiation, Inc., 299
Radiation belt, 35, 37, 42, 50, 152, 168, 245
  Van Allen, 37, 50, 168, 169, 173, 178, 273, 325, 435
Radiation detector, 8
Radio and Television News Directors Association, 384
Radioastronomy, 106, 116, 321
Radio Attenuation Measurement (Ram), 132, 319, 487
Radio blackout prevention, 319
Radio communication treaty, 68
Radio Corp. of America (RCA), 3, 61, 79, 168, 193, 209, 252, 412
  computer, 326, 412
  contracts, 3, 39, 134, 297
Radio frequency, 37, 68, 125
Radio noise, galactic, 118
Radio propagation, 2
Radio pulses, high-frequency, 239
Radio Research Laboratory Station (Japan), 185, 355
Radio Technical Commission for Aeronautics (RTCA), 158
Radio thermal measurements, 164
Radio waves, 34, 103, 319, 330
Radioactive fallout, 146, 202
Radiobiologist, 169
Radioisotope, 100
Radiometer, 125, 411
Radiometer, Infrared, High Resolution (Hrir), 298
Radiometric observations, 82
Radiotelephony, 76
Radiotelescope, 34, 38, 224, 239, 424
Radkowsky, Dr. Alvin, 369
Radnofsky, Mathew I., 354
RAE. See Royal Aircraft Establishment.
Raffensperger, M. J., 213, 397
Rahman, Prime Minister Tunku Abdul (Malaysia), 263
Railroad Telegraphers Union (RTU), 61, 62, 65
Raines, Col. Martin L. (USA, Ret.), 339
Rain-making rocket, 227
Raisting, West Germany, 22
Ram (Radio Attenuation Measurement), Project, 132, 319, 487
Ramjet engine, 77, 223
Ramey, James T., 400
RAND Corp., 89, 435
Ranger (program), 128, 163, 197, 300
  failure (see also RANGER VI), 215
  management, 138, 163–164
  objective, 47
  program review, 34–35, 42, 58, 150, 150–151

Ranger B (see also RANGER VII), 218
RANGER VI (lunar probe), 37, 82
    failure, 54, 56, 58, 60–61, 64, 65, 138, 451, 482
        hearings, 150, 150–151, 151, 163, 215
    landing, 41, 41–42
    launch, 34–35, 448, 466
    Review Board, 42, 126
RANGER VII (lunar probe), 42, 64, 128, 176, 218, 272, 275, 275–276, 276, 279, 287–288, 300, 304, 309
    briefing, 275–276, 298–299
    comments, 267–268, 271–272, 272, 272–273, 273, 300, 313
    data, 267, 271, 301, 331
    launch, 263, 453, 467
    performance, 264, 267
    photographs, 306–307, 307, 318, 419, 421, 435, 438
Rao, Sweden, 397
RAS. See Royal Aeronautical Society of Great Britain.
Rat experiment, 305, 328
Ravenholt, Albert, 281–282
Ray, cosmic, 4, 42, 50, 56, 176, 179–180
Raytheon Co., 216, 218
RCA. See Radio Corp. of America.
Reaction Motors Div. (Thiokol Chemical Corp.), 29, 78
Reactor, 3, 25, 40–41, 83, 93, 143, 176, 248, 260, 296, 299–300, 305, 311, 317
    graphite, 83, 151–152, 320
Reactor-in-Flight Test. See Rift, Project (NASA-AEC).
Reader, Paul D., 142, 238
Records, 66, 268
    helicopter, 394
    soaring, 268
    speed, 321
    women, 181, 200, 203
Recovery technique, spacecraft, 158
Red China. See China, Communist.
Red Star (see also *Krasnaya Zvezda*), 41, 253–254
Redding, Calif., 78
Redeye (anti-aircraft missile), 328
Redstone (missile), 368
Redstone Arsenal, Ala., 12, 103, 332
Reed, Sylvanus Albert, Award, 21, 173
Reed, Wilmer H., III, 276
Reentary, 99, 132, 159
    heating, 135, 173
    research, 8, 103, 278, 307–308
    vehicle, 3, 9, 11, 135, 161, 241, 307–308, 309
Re-entry Flight Demonstration, 289–300, 468
    RFD–1, 3
    RFD–2, 346, 468
Reeves Instrument Co., 336
Reeves, John, 406
Reflector, 325
Regeneration, 89
Reid, Edwin, 316
Reid, Henry J. E., Jr., 276

Reiffel, Dr. Leonard, 135
Reischauer, Edwin O., 142
Relativity theory, 433
Relay (communications satellite), 130, 282–283
RELAY I (communications satellite), 3, 19, 38, 286, 353, 373, 397, 418
RELAY II (communications satellite), 22
    launch, 19, 446
    transmissions, 32, 81, 115, 185, 246, 397
    use of, 270, 353, 373
Reliability, 45, 143, 234–235, 281
    missile system, 6, 129
    space system, 6
    symposium, 6, 6–7
Religion and space, 207–208
Remote Maneuvering Unit (Rmu), 217
Rendezvous, 224, 228, 241, 259, 301–302, 357, 419
"Reporters' Roundup" (radio program), 149
Republic Aviation Corp., 98, 196, 289
Republican Citizens Committee, Republican Critical Issues Council, 194
Republican National Convention, 246, 248, 250–251
Republican study group on space and aeronautics, 153
Rescue receiver/transmitter system, 146
Rescue ships, 203
Research Airplane Committee, FRC, 77
Research and development, 33, 78, 181–182, 279–280, 295, 300–301, 303, 326–327
    aeronautical, 292–293
    Federal support, 18, 108, 165, 165–166, 185–186, 211, 312, 330–331, 374
    funds for, 20–21, 21, 33, 48, 69, 72, 160, 300–301, 401–402, 429–430
    industry, 7, 154
    information, distribution of, 144–145, 312, 386
    management, 160, 344
    planning, long-range, 67, 160, 166–167, 344
Research Award (AIAA), 21
Research, basic, 8, 24, 33, 53, 91–92, 160
Research Institute for Advanced Studies (Martin Co.), 155
Research Triangle Institute, 194
Retrometer, 142
Retrorocket, 106, 121
Reusable spacecraft, 123, 249, 391–392
Reuss, Rep. Henry S., 115
Reynolds, Maj. Herbert H. (USAF), 390
Reynolds, James J., 69–70
Reynolds, Dr. Orr E., 63, 168–169, 295
RFD–1. See Re-entry flight demonstration.
Rhodes, Gov. James A., 261
Rice Univ., 11, 107, 110, 113, 162, 243, 290
Richardson Foundation, 439
Rickover, V/A Hyman G. (USN), 40, 395
Rieke, William B., 347

Rift, Project (Reactor-in-Flight Test) (NASA-AEC), 103, 340, 441
Righini, Dr. Guglielmo, 180
Ripley, Dr. S. Dillon, II, 56
Ritland, M/G O. J. (USAF), 170–171
Riverdale, the Bronx, 205
RL-10 (liquid hydrogen rocket engine), 32, 67, 193, 414
A3, 227
Rmu. See Remote Maneuvering Unit.
RN-6 nozzles, 236
Roanoke, Ala., 56
Robbins, Col. Harold W. (USAF), 435
Rochester, Univ. of, School of Medicine, 266, 315
Rock, Vincent P., 59
Rockefeller Institute, 438
Rocket, 59, 159, 190, 215, 229–230, 248, 258
  chemical, 83, 125
  foreign, 268, 303
  nuclear, 83, 107–108, 115, 195, 229, 234–235, 235, 236, 296, 311
Rocket case, solid propellant, 279
Rocket Day (U.S.S.R.), 394
Rocket engine (see also Rocket motor), 25–26, 171, 219, 239
  electric, 256
  hybrid, 233
  ion, 82, 256
  liquid propellant, 47, 67, 248–249
  nuclear, 115, 200, 234
  reliability, 45, 234
  solid propellant, 23, 39, 80, 167
Rocket launcher installation, 253
Rocket, model, 292
Rocket motor, solid propellant (see also Rocket engine), 193, 235, 311–312, 323, 328
"Rocket Propulsion of Long Range Bombers," 59
Rocket sled, 78, 224
Rocketdyne Div. See North American Aviation, Inc., Rocketdyne Div.
Rodino, Rep. Peter W., 291
Roediger Construction, Inc., 367
Rogallo, Vernon L., 131
Rogallo wing, 198
Roganville, Tex., 57
Rogers Dry Lake, Calif., 6, 285
Rogers, Thomas F., 49
Rom, Frank E., 337
Roman, Dr. Nancy G., 163, 327
"Romashka" ("Daisy") (U.S.S.R. nuclear reactor), 333
Romatowski, Ray, 154
Rome, Italy, 58, 61, 63
Rome, N.Y., 30, 33
Rome Air Development Center (RADC), 381, 406
Romo, L/C Peter (USAF), 5
Rose, V/A Rufus E. (USN, Ret.), 124
Rosen, Dr. Harold A., 111
Rosenberg, Dr. Norman W., 265
Rosenblum, Dr. Louis, 184

Rosman, N.C., 103, 202–203, 308, 311
Roswell, N.M., 64, 101
Rotary Club, Los Angeles, 244
ROTC (Reserve Officers Training Corps), 319
Rotor downwash, 214
Round-the-world flight, 327
Roush, Rep. J. Edward, 38, 102–103, 145, 331–332
Rover (nuclear rocket engine) (see also Poodle, Project), 209, 435
Rowan, Carl T., 282–283
Roy, Maurice, 155
Royal Aeronautical Society of Great Britain (RAS), 198
Royal Aircraft Establishment (RAE), (U.K.), 230
Royal Crown Cola Co., 358
Royal Technical Univ., Ionospheric Research Laboratory (Denmark), 56, 375
RS-70 (aircraft), 260–261
RS-71 (aircraft), 292
RTCA. See Radio Technical Commission for Aeronautics.
RTU. See Railroad Telegraphers Union.
Rubidium vapor magnetometer, 210
Rubis (French rocket), 210
Rumsfeld, Rep. Donald, 79, 115
Runway traction equipment, 296
Rushworth, Maj. Robert (USAF), 29, 57, 118, 156, 214, 222, 285, 321, 331, 429
Rusk, Secretary of State Dean, 343
Russell, Sen. Richard B., 90
Russell, Terrence E., 259
Russia. See U.S.S.R.
Russian Institute (Columbia Univ.), 188
Rutter, Richard, 288
Ryan Aeronautical Corp., 39
Ryan, Rep. Harold M., 13
Ryan, L/C John D. (USAF), 400
Ryan, Tom, 147
S-48 (ionosphere explorer satellite), 106, 296
S-51 (U.K. satellite), 270
S-52 (U.K. satellite), 270
S-53 (U.K. satellite), 270
S-66. See Explorer, Beacon.
S-I. See Saturn I.
S-IC. See Saturn V.
S-I-8. See Saturn I (SA-8).
S-IV. See Saturn I, Saturn IB.
SA-5, SA-6, etc. See Saturn I.
SAA. See South African Airways.
SAC. See Strategic Air Command.
Sacramento, Calif., 25, 154, 193, 320, 403
Sacramento Peak Observatory, 34
"Saddlebags" for Lem, 139
SAE. See Society of Automotive Engineers.
Saenger, Dr. Eugen, 59, 78
Safety, air, 156, 171, 192, 211, 214, 291, 306, 332
Safety Award, President's, 184
Sagan, Dr. Carl, 180
Saigon, S. Vietnam, 332
Sailplane, record, 268

St. Louis, Mo., 173, 223
St. Petersburg-Tampa Airboat Line, 2
Salisbury, Australia, 84, 292
Salmon, Project (AEC), 360
Salt water conversion, 55
Salto di Quirra, Sardinia, 239, 242
SAM. See USAF School of Aerospace Medicine.
Sammy (sounding rocket experiment), 290
Sammy I (sounding rocket), 11
Sammy II (sounding rocket), 107
Samos (Satellite Military Observation System), 207, 281–282
San Antonio, Tex., 388
San Bernardino, Calif., 78
Sandia Corp., 444
San Diego, Calif., 166, 253, 284
San Francisco, Calif., 36, 78, 165
SAN MARCO I (SM–1) (Italian satellite), 420, 462, 468
San Miguel Observatory, Buenos Aires, 387
San Nicholas Island, Calif., 178
San Salvador, 386
Sandage, Allan R., 426
Sandar, Dr. Herman J., 74
Sanders Associates, Inc., 330
Sandia Corp., 3
Sands, B/G Harry J., Jr. (USAF), 3
Santa Ana, Calif., 224
Santa Clara, Calif., 360
Santa Clara, Cuba, 121
Santa Rita Technology, 339
Santa Rosa Island, Fla., 70
Santa Susana Field Laboratory, 416
Santiago, Chile, 168
SARC. See Systems Analysis and Research Corp.
Sarnoff, David, 252
Sas. See Stability Augmentation System.
Satellite, 2, 254, 255, 281, 284, 302
   armed, interception and destruction, 322, 323
   cooperation, international, 97, 264, 282–283
   foreign, 203, 264
   surveillance system, 102, 181
   use of, 2, 30, 118, 146–147, 324
Satellite, Advanced Technological (Ats), 50, 90, 203, 223
"Satellite alert" computer, 188
Satellite, biological (see also Biosatellite), 257
Satellite, communications, 20, 140, 204, 209, 238, 240, 241, 269
   cooperation, international, 49, 84, 96–97, 189, 282–283, 313
   U.S.-U.S.S.R., 2, 17–18
   medium altitude, 44, 92, 186
   military use of, 27, 118, 130, 186, 247, 254–255, 279, 289
   synchronous-orbit (see also Syncom C and SYNCOM III), 84, 139, 209, 290–291
Satellite Control Facility (SCF), 164
Satellite, data collection system, 196
Satellite Div. (Union Carbide Corp.), 56
Satellite, ESRO, 113, 161
Satellite, geodetic program, 325
Satellite, ionosphere explorer (S–48). See Topsi.
Satellite, Ionospheric Studies, International. See Isis.
Satellite launcher (see also Europa 1), 269–270
Satellite launching center (French), 161
Satellite, meteorological, 20, 193, 224
   cooperation, 43
   international, 1–2, 30, 204
   use of, 50, 95, 157
Satellite, micrometeoroid detection. See Pegasus.
Satellite, navigational, 68, 95, 157, 158, 188
   civilian system, 68, 158, 231, 312, 381
Satellite, nuclear powered, 8
Satellite, nuclear radiation detection. See Vela, Project; Sentry.
Satellite observation station (U.S.S.R.), 81
Satellite, orbiting observatory (see also Oao, Ogo, Oso, Aoao, Aoso), 103
Satellite, polar orbit, 231
Satellite, solar-powered, 8
Satellite, solar x-ray monitor, 45
Satellite stabilization, 68
Satellite Techniques Laboratory (Rice Univ.), 107
Satellite Telemetry Automatic Reduction Systems (Stars), 352, 396
Satellite, tetrahedral (see also Pygmy), 252
Satellite Tracking and Data Acquisition Network. See STADAN.
Satellite, twin (ARPA). See Vela, Project.
Satellite, unidentified, 275
   launch vehicle
      Atlas-Agena, 79, 102, 146, 182, 239, 285, 326, 406
      Scout, 203
      Thor-Able-Star, 173, 342, 418
      Thor-Agena, 10, 17, 66, 85, 114, 151, 205, 213, 216, 218, 238, 244, 277, 294, 316, 340, 354, 373, 378, 392, 426, 427
Satellite, weather (see also Satellite, meteorological), 188, 204, 240, 258, 309
Saturday Evening Post, 76, 86, 133, 164
Saturday Review, 307
Saturn (program), 4, 13, 124, 170, 226, 238, 302
   achievements, 38, 40, 281, 317, 318, 379
   contracts, 14, 39, 62
   facilities, construction of, 62, 211, 347
   management, 73, 169

Saturn I, 114, 318
  booster, 80, 104, 149, 286, 341
  engine, 25–26, 211
  propulsion, 229–230, 246, 302
  stage, 84, 246
    s-iv, 25, 192, 196, 242, 321, 325, 454
Saturn I (sa-5), 32, 40, 54
  launch, 27, 32, 447, 449, 466
  performance, 32, 49
  preflight preparation, 26
  stage, 27, 32
Saturn I (sa-6), 196, 199, 238, 314, 321
  launch, 192–193, 450, 466
  preflight preparation, 70, 126
Saturn I (sa-7), 309, 325, 386
  launch, 321, 456, 467
  preflight preparation, 238, 317
Saturn I (sa-8), 138, 149, 212
Saturn I (sa-9), 104, 138, 436
Saturn I (sa-10), 138
Saturn IB, 213
  contract, 22, 84, 142, 189, 193, 213, 248, 302
  development, 157, 192–193, 193, 318
  engine, 302, 402
    h-1, 211
    j-2, 159, 294
  equipment, 22, 132, 142, 297
  plans for, 53, 140, 145, 189
  stage, 84
    s-ivb, 159, 189, 193, 242, 248, 302
Saturn IB/Centaur (booster), 173
Saturn V, 38, 145, 189, 195, 234, 235, 248, 329
  booster, 87, 105, 213, 223, 229–230, 247, 377
  contract, 15, 22, 39, 120, 127, 142, 160, 189, 213, 242, 248, 248–249, 297, 330, 440
  development, 20–21, 157, 193, 235, 318
  engine, 229–230, 402
    f-1, 120, 242, 346, 377
    j-2, 248, 294
  equipment, 22, 132, 142
  facility, 15, 39
  ground support system, 297, 330, 400
  plans for, 53, 80, 140, 145, 157, 223, 242, 247, 295, 350
  stage, 8, 127, 160, 248
Saturn V/Apollo, 105, 124, 132, 139, 183, 196
Saturn V, post-, booster, 247
Saturn Flight Evaluation Group (msfc), 49
Saturn Vertical Assembly Building (vab) (mila), 131
Saudi Arabia, 52
Saunders, Benjamin M., 114
Savino, Joseph M., 25
Scama (communications system), 222.
Scandinavia, 99, 375, 397
Scanner, horizon, 77, 224–225, 241
Scanner, Project (nasa), 65, 224–225
Scat. See Aircraft, Supersonic Commercial Air Transport.
Scat wing design, 189
scf. See Satellite Control Facility.
Schaefer, Dr. Hermann J., 219
Scheer, Julian, 395
Schell, Dr. Allan, 38
Schenectady, N.Y., 255
Schilling, David C., Trophy, 315
Schiro, Mayor Victor H., 324
Schirra, Cdr. Walter M., Jr. (usn), 99, 129, 134, 178, 188, 317, 393
Schjeldahl, G. T., Co., 284, 397
Schleicher, Richard L., 198
Schmidt, Dr. Maertin, 120
Schmitz, Bruce W., 276
Schneider, William C., 371
Schnitzer, Emanuel, 287
School of Aerospace Medicine (sam). See usaf School of Aerospace Medicine.
School of Environmental and Planetary Sciences (proposed, Univ. of Miami), 100, 101
Schriever, Gen. Bernard A. (usaf), 98, 112–113, 164, 182–183, 190–191, 207, 211–212, 244, 252, 284–285
Schulte, William J., 334
Schultz, Col. Kenneth (usaf), 90
Schurmeier, Harris ("Bud"), 275–276, 309, 353, 435
Schwab, William B., 25
Schwarzschild, Dr. Martin, 223
Schweikart, Russell L., 356, 363
Schwinghamer, Robert J., 329
Science, 2, 3, 16, 18, 24, 110–111, 149, 199, 221, 240, 277, 280, 318, 344, 348, 357, 413, 437
Science (magazine), 3, 62, 262, 306, 314, 414
Science Advisory Committee, President's (psac), 35
Science and Mechanics, 331
Science and politics, 86–87
Science and technology. See Science; Technology.
Science and Technology Advisory Committee for Manned Space Flight (nasa), 120, 237
Science Development Program (nsf), 104
Science Policy Research Div., Library of Congress, 301, 310
Scientific and Technical Information Facility, nasa, 237
Scientific Association for Air and Space Flight, West Berlin (Germany), 320
Scientific Committee on the Effects of Atomic Radiation (U.N.), 202
Scientist-astronaut, 395, 424
Scientists, 71, 110–111, 253–254
  and engineers, 74, 94, 128, 149, 156–157, 161, 196, 291, 302, 308
  as astronauts, 10, 237, 245–246
  emigration of, 15, 308
  exchange program (U.S.-U.S.S.R.), 182
  space program, 7, 36, 139, 184, 196
  training, lunar mission, 158

Scientists and Engineers for Johnson and Humphrey, 374
SCORE (communications satellite), 241
Scorpio (constellation), 153
Scott, Capt. David R. (USAF), 306
Scott, Walter C., 150
Scout (booster), 71, 106, 176
  launch, 4, 118, 203, 289, 394, 451, 455, 457, 459, 460, 462, 466, 467, 468
  use of, 4, 95, 113, 132, 161, 258, 270, 296
"Scramjet" (supersonic combustion ramjet), 385
SDI. See Selective Dissemination of Information.
Sea Lab I, Project (USN), 176, 252, 266
Sea of Clouds (moon), 263, 267
Sea of Tranquility (moon), 37, 41, 451
Sea urchin egg experiment, 72, 174
Sea water desalinization, 251, 256
Seaborg, Dr. Glenn T., 40–41, 67
Seacat (U.K. guided missile), 280
Seagoing tracking system, 84
Seal Beach, Calif., 375
Sealant, 228
Seamans, Dr. Robert C., Jr., 12–13, 32, 42, 50, 97–98, 149, 162, 163, 176, 184, 228, 254, 295, 297, 300–301, 335, 344
Seamster, Dr. Aaron P., 407
Seattle, Wash., 124, 251, 279
SEC. See Securities and Exchange Commission.
Secor (Sequential Collation of Range) (geodetic satellite) (USA), 68, 74, 222, 365
Securities and Exchange Commission (SEC), 168, 333
Sedov, Dr. Leonid I., 315, 348
See, Elliott M., 306
Seed experiment, 248
Segner, Donald, 420
Seismograph, 106
Seitz, Dr. Frederick, 29, 154, 416
Selective Dissemination of Information (Sdi), 255
Selenodetic measurements (lunar surface survey), 81
Semipalatinsk region, U.S.S.R., 388
Sen, Dr. Hari K., 119
Seney, John S., 433
Sensor, 146, 196, 305
  image velocity, 257
  meteoroid penetration, 210
  radiation receiving, 251
Sentry (nuclear radiation detection satellite) (see also Vela, Project), 253, 256
Sert (Space Electric Rocket Test), 82, 300
Sert I, 82, 158, 197, 256, 354, 467
Seychelles Islands (West Indian Ocean), 159, 405
SFOF. See Space Flight Operations Facility (JPL).

Shafrir, Dr. Uri, 220
Shapiro, Irwin I., 433
Shea, Joseph F., 47, 206, 314
Sheldon, Dr. Charles S., 152, 376
Shell Oil Co., 62
Shepard, Cdr. Alan B., Jr. (USN), 16, 134, 166, 188, 242
Sherman, Harold, 278
Shetland Island, Scotland, 264
Shewmake, Glenn A., 354
Shielding, 13
Shiina, Foreign Minister Etsusaburo, 343
Ships, Bureau of, (USN), 305
Ships, tracking, 64–65, 168
Shlovsky, Dr. Iosif S., 23, 436
Shock absorber, 13, 136
Shock wave, 103, 156
Shoemaker, Dr. Eugene M., 223, 271, 299
Short takeoff capability, 191
Shuey, Henry M., 21
Shuttle transport, orbital, 135, 249, 307
Siberia launch site, 333
Sikeston, Mo., 195
Sikorsky, Igor, 317
Silicon cells, 124
Silicon elastomer, 181–182
Silverstein, Dr. Abe, 21, 204, 214, 245
Simkin, William, 124
Simons, Howard, 1, 21, 57, 342, 351, 423
Simplon Tunnel, 248
Simpson, Dr. George L., 100, 140, 195, 395
Simulation in Space Technology, Conference on Role of, 290, 293
Simulator, jet airborne, 147
Simulator, satellite attitude control, 264–265
Singer, Dr. S. Fred, 62, 100, 101
Sissakian, N. M., 309, 348
Sissenwine, Norman, 14–15
Sisu 1A (sailplane), 268
Skidmore, Owings & Merrill, 56
Skrivanek, R. A., 147
Skua (U.K. meteorological rocket), 270
Skybolt (missile), 150
Skyhook (ONR balloon flight program), 291
Skylark (U.K. sounding rocket), 239, 242, 270
Slayton, Donald K., 66, 188, 242, 345
Sled, monorail, 333
Sleighride, Project (USA), 11
Slemon, Air Marshal C. R. (Canada), 198
Sloan, Alfred P., Fellowships, 1964–65, 150
Slone, Henry O., 404
Sloop, John L., 11, 229–230, 360
SLV–3. See Atlas–3.
Small business and Government, 29, 261
Smith and Sapp Construction Co., 369
Smith, Prof. Alex G., 153
Smith, Francis B., 177
Smith, Sen. Margaret Chase, 57
Smith, Wendell S., 141
Smith, William W., 276

Smithsonian Astrophysical Observatory, 116, 171
Smithsonian Institution, 35, 56, 133, 155, 166, 220, 241, 260, 280, 282
Snakes, experiments, 56–57
Snap (System for Nuclear Auxiliary Power), 67, 72
Snap, 7D, 49
Snap 8, 260, 404
Snap 9A, 173, 188
Snap 10A, 72, 333
  funds for, 67, 72, 85, 117, 161
  test, 39, 93, 117, 125
Snap 50, 67, 416
Snow, 434
Snow goose, 388
SNPO. See Space Nuclear Propulsion Office (NASA-AEC).
Soaring record, 268
Soaring Society of America, 244
Soberman, R. K., 147
Social sciences, 255–256
Society of Automotive Engineers (SAE), 158, 344
Society of British Aerospace Companies, 311
Society of Photographic Scientists and Engineers, 154
Society of Security Analysts, New York (SSA), 161
Sodium-lithium experiment, 99
Sodium potassium, 125
Sodium vapor experiment, 6, 10, 249, 384
Sognnaes, Dr. Reidar F., 121
Sohier, Walter D., 166
Sohio Tracking Station, 289
Solar cell
  radiation-resistant (n-on-p), 20, 202, 276, 482
  thin-film, flexible, 124
Solar collector (mirror), 212
Solar corona, 34, 96, 119, 180, 260
Solar cycle, 13, 48, 95
Solar eclipse, 180
Solar energy conversion, 151
Solar Energy Laboratory (Univ. of Wis.), 275
Solar Energy, Office of (Dept. of Interior), 180, 275
Solar flare, 13, 48, 173, 194, 388–389
Solar gases, 250
Solar heating, 141
Solar paddles, 326
Solar panel, 311
Solar particles, energetic, 307
Solar plasma, 50
Solar radiation. See Radiation, solar.
Solar system, 125, 157, 170, 366
Solar wind, 119
Solar x-ray emission, 45
Solid propellants, 39, 247, 331, 369, 377, 416–417, 429, 435

Sonic boom (see also Noise), 277
  damage, 90, 117, 175, 242, 253
  results, 274, 279
  studies, 43, 63, 84, 90, 117–118, 175, 177, 183, 187, 236, 242, 244, 253, 268, 314, 360, 392, 404, 406
Sonmiani Range (Karachi, Pakistan), 107, 399, 403
Sound, speed of, 15
Sounding rocket, 153
  experiments, 106, 258, 355
  international programs, 97
    NASA-India, 31, 70, 380, 384
    NASA-Japan, 355
    NASA-Netherlands, 278
    NASA-Norway-Denmark, 56
    NASA-Pakistan, 107, 399
    NASA-Sweden, 99, 204–205, 241, 277–278
    U.S.-ESRO, 264
  launch
    France, 273
    Japan, 245, 259
    NASA, 11, 13, 14, 15, 22–23, 30, 42–43, 62, 79, 99, 114, 115, 124, 128, 137, 140, 141, 145, 210, 243, 249, 251, 260, 262, 264, 278, 283, 299, 304, 305, 328, 335, 343, 344, 360–361, 363, 368, 374, 379, 384, 387, 391, 392, 395, 399, 415, 422
    Sweden, 99, 277
    U.K., 239, 242, 278
    USAF, 122, 444
South Africa, 25
South African Airways (SAA), 226
South America, 134, 146
South Pacific, 14
South Pole, 211
Southern Interstate Nuclear Board, 126
Southwest Research Institute (SRI), 388
Southwestern Bell Telephone Co., 81
Soviet Academy of Sciences, 17, 27, 97, 177, 198, 204, 321, 443
  agreement, U.S. (NAS), scientist exchange, 182
  Astronomical Council, 121
Soviet Committee on Standards, Measures, and Measuring Instruments, 41
Soviet Institute of Normal and Pathological Physiology, 177
Soviet Russia, 351
Space (medium), 179
  environment, 50, 117
  exploration of, 2, 6, 67, 87, 95, 125, 126, 133–134, 139, 149, 194, 277, 298, 321
Space Act of 1958. See National Aeronautics and Space Act of 1958.
Space Aeronautics, 291
Space age, 86, 240, 243, 257, 320, 432
  economics, 144, 157, 221
Space and Upper Atmosphere Research Committee, Pakistan (SUPARCO), 107, 399
Space biology, 208

Space bus (see also Orbiting Geophysical Observatory, OGO I), 306, 308
Space Business Daily, 4, 8, 15, 18, 30, 45, 77, 138, 140
Space cabin environment simulation, 44, 87, 130, 304
Space capability, 20, 32, 38, 92, 151–152, 156–157, 184, 199, 207, 223, 302, 324, 326, 333
Space, challenge of, 6, 87, 92, 127, 155, 298
Space Congress, 143
Spacecraft, 90, 100, 102, 289
　international agreement, 100, 102, 116
　systems, 224, 228–229, 237, 361
Spacecraft Technology and Advanced Reentry Test (Start) (see also Asset), 161, 286, 307
"Space czar," 388
Space Detection and Tracking Center, NORAD (SPADATS), 254
Space Electric Rocket Test. See Sert.
Space-General Corp., 317, 404
Space environment, effects of, 112–113, 168–169, 174–175, 194, 204
Space ferry, 224, 307
Space flight, 9, 20, 125, 173, 217, 258, 298, 304, 309
　effects of, 171, 183, 194, 309, 328
Space Flight Operations Facility (JPL), 176
Space: Its Impact on Man and Society, 86
Space law, 313
Space Law and Government, 236
Space lifeboats, 174
Space maneuvering systems, 89
Space medicine. See Aerospace medicine; Bioscience; Life science.
Space Medicine Liaison Office, NASA, 80
Space, military use of (see also Space race, military; Space station, manned), 7, 34, 194, 197, 206, 258, 333
　manned space flight, 302
　objectives, 12, 51
Spacemobile, NASA, 247
Space Nuclear Auxiliary Power. See Snap.
Space Nuclear Propulsion Office (SNPO) (NASA-AEC), 63, 140, 248, 311, 353, 385
Space Park (N.Y. World's Fair), 101, 145, 214
Space, peaceful use of, 12, 64, 67, 157, 190, 268, 277, 288, 311
Space physics, institute of, proposed (Univ. of Miami), 100
Space, politics and, 86–87
Space port, 157
Space program, national, 112, 172, 207, 215–216, 217, 258, 279–280, 292–293, 295, 312–313, 324
　accomplishments, 4, 35, 35–36, 40, 42, 169, 111–112, 122, 151–153, 156–157, 182–183, 274, 279, 282–283, 294, 313, 318, 338

　cost of, 23, 33, 38, 82, 128, 181, 234–235, 278–279
　criticism of, 64, 133–134, 185–186, 225, 250–251, 258, 337, 351
　military, 4, 12, 44, 51–52, 64, 181, 331–332, 351
　need, 36–37, 44, 94–95, 127, 136, 256, 277, 322, 324
　objectives, 27, 36–37, 86–87, 133–134, 144, 195–196, 236, 297, 368, 371
　policy, 2, 6–7, 12, 36–37, 118
　prospects for, 109, 221, 355
　requirements, 232–233, 302
　support of, 67, 177, 186–187, 238, 333
Space Propulsion Facility (Plum Brook Station), 224
Space race, 7, 215–216, 294, 303
　U.S.-U.S.S.R., 35, 111–112, 122, 136, 215–216, 259, 294, 295, 318, 324, 348, 379
　aircraft, 133, 241
　booster, 32, 35, 38, 111, 122, 190, 195, 223, 318, 351, 355, 376
　manned space flight, 111–112, 118, 152, 190, 354–355, 357, 374
　　to moon, 40, 64, 120, 125, 152, 196, 199, 221, 268–269, 322, 350, 352, 357, 437
　military, 51–52, 118, 136, 181, 190, 250–251, 258, 268, 322, 323, 351, 385, 421, 442
　payloads, 61, 111, 152–153, 254, 321, 376
　probes, 150–151, 152
　surveillance, satellites, 207, 357
　weather, satellites, 35, 112
Space Radiation Effects Laboratory (LRC), 159
Space Research Committee (Norway), 56
Space research results, 79, 172, 306
Space science, 50–51, 102, 156–157, 177, 301, 318
Space Science Award (AIAA), 21
Space Science Board (SSB) (NAS), 116, 151, 158, 194, 210
Space Science Dept. (Rice University), 11
Space Science Technology Center (Georgia Tech), 147
Space shuttle vehicle (space tug), 24, 135, 249, 307
Space station, manned (see also Large Orbital Research Laboratory; Manned Orbiting Laboratory; Manned Orbital Research Laboratory; and Olympus), 52, 80, 90, 95, 169, 211–212, 249, 258, 287, 307, 324, 375
　foreign, 199
Spacesuit (see also Pressure suit), 230, 239, 296, 355, 363, 389, 390, 415–416
Space surveillance system (USN), 102
Space technology, 20, 25, 33, 40, 112, 157, 181, 204, 211, 240, 243, 255, 284–285, 295, 302, 312, 318, 324
Space Technology Center, Valley Forge (GE), 177

Space Technology Laboratories, Inc., 160, 209, 292
Space tools, 5, 344
Space Tracking and Data Acquisition Network. See STADAN.
Space Transporter (U.K. Aerospace Plane), 270
Space tug, 135, 249
Space vehicle research, 91, 224, 281
Spaceyards, 157
SPADATS. See Space Detection and Tracking Center, NORAD.
Spain, 31, 55, 159
  tracking stations, 31, 204
Spar, Dr. Jerome, 258
Sparks, Brian, 224
Spectrogram, 180
Spectrograph, 78, 398
Spectrographic analysis, 48, 138
Spectrometer, 15, 231, 263
Spectroscopy, 444
Speed record, women's, 181, 200, 203
Sperry, Lawrence B., Award, 236
Sperry Rand Corp., 305
  Electro-Optics Laboratory, 113
  Univac Div., 68, 330, 423
Sphere, drop test, 220
Spin adjusting mechanism, 82
Spin Test Facility (STF) (KSC), 147
Spinrad, Dr. Hyron, 148
Spitzer, Dr. Lyman, Jr., 45
Spivak, Jonathan, 85
Sprint (antimissile missile), 388
Sproull, Dr. Robert L., 153, 290, 386
SPSE. See Society of Photographic Scientists and Engineers.
SPUTNIK I, 86, 111, 155, 268, 270, 340, 441
Sputtering apparatus, 124
SR-71 (long-range reconnaissance aircraft), 260–261, 274, 346, 429
SRI. See Southwest Research Institute.
SRI. See Stanford Research Institute.
SSA. See Society of Security Analysts.
SSB. See Space Science Board (NAS).
SSD. See Air Force Space Systems Div.
Sst. See Aircraft, supersonic transport.
ST-124-M platform system, 22, 142
STAB. See Supersonic Transport Advisory Board.
Stability Augmentation System (Sas), 6
Stabilization system, gravity gradient, 68, 177, 223
Stabilized platform, 22
Stack, John, 5, 196
STADAN (Satellite/Space Tracking and Data Acquisition Network), 130, 202–203, 284
Stadimeter, 178
Stafford, H. N., 286
Stafford, Maj. Thomas P. (USAF), 102,
Standards, 41
Stamps, postage, 120, 160, 205
Stanford, Dr. Henry King, 100

Stanford Research Institute (SRI), 278, 332
Stanford Univ., 231
Stanford Worldwide Acquisition of Meteorological Information (Swami), 231
Stapp, Col. John P. (USAF), 313
Star
  dwarf, 218
  intensity, 381, 387
  neutron, 153, 254, 432, 436
"The Star Spangled Banner," 290
Star tracker, 77
Starfighter. See F-104.
Start. See Spacecraft Technology and Advanced Re-entry Tests.
State, Dept. of, 264
State of the Union message, 6
Static electricity, 147
Staisical Handbook, FAA, 204
Stawikowski, Dr. Antoni, 398
Stay Time Excursion Module (Stem), 138
Stead AFB, Nev., 281
Steel, 279
STEG. See Supersonic Transport Evaluation Group.
Steinberg, Dr. Menahem, 125
Seinberg, Robert, 25
Steller curve of growth, 155
Stellar fusion, 155
Stem. See Stay Time Excursion Module.
Stennis, Sen. John, 15, 277
Stern, Dr. Ralph H., 121
Stethoscope, 172
STF. See Spin Test Facility (KSC).
Stockholm, Sweden, 169
Stockholm, Univ. of (Sweden), 147, 288–289
Stone, Edward Durell, 410
Stonewall, Tex., 279
Strategic Air Command (SAC), 86, 260, 274, 428, 429
  failures, Atlas, 145
  launch
    Minuteman, 15, 62, 378
    Titan II, 409
Stratocruiser. See C-97.
Stratoscope II (balloon), 223, 397
Strawley, John E., 223
Strikes
  Cape Kennedy, 51, 59, 61, 62, 65, 69–70, 124, 126, 128, 131, 133, 209, 212, 243
  White Sands Missile Range, 14
Strong, Dr. John, 364
Structural materials, 77, 429
Structures, 29, 77, 147
Structures and Materials Conference (AIAA), 197
Strughold, Dr. Hubertus O., 164
Student Space Conference, 163
Stuhlinger, Dr. Ernst, 302, 320, 352
Stump Neck, Md., 30
Sturman, John C., 25
Stuttering cure, 79

Styles, Paul L., 24
Styrofoam, 259
Subbotin, M., 105
Submarine
  flying, 397
  missile-bearing, 136
  nuclear, 221, 266, 302
Submarine launching, 68, 266
Submarine tracking, 84
"Subscale 1" (solid rocket grain), 248
Sud-Aviation (France), 170
Sudets, Soviet Air Marshal Vladimir Aleksandrovich, 104
Sullivan, Francis J., 368, 393, 429
Sumatra, Indonesia, 292
"Summary of Lessons Learned from Air Force Management Surveys," 198
Summerfield, Dr. Martin, 382
Sun (see also Solar cell, etc.), 1, 219, 271
  cosmic rays, 211
  electric charge, 121
  corona, 260, 444
  radiation, 76, 444
Sundlan, Bruce G., 242
Sunnyvale, Calif., 164
"Sunrise." See VOSKHOD I (U.S.S.R. spacecraft).
Suomi, Prof. Verner E., 290
SUPARCO. See Space and Upper Atmosphere Research Committee, Pakistan.
Supernovae theory, 4
Supersonic Advisory Committee (Transport Association), 114
Supersonic Transport Advisory Board (STAB), 119
Supersonic transport aircraft. See Aircraft, supersonic transport (Sst).
Supersonic Transport Evaluation Group (STEG), 4, 148
Support Facility (MSC), 8
Surveyor (program), 10, 41, 106, 171, 200
  lunar lander, 47, 275–276, 276
  lunar probe, 71, 160, 197
Survival and Special Training Squadron, USAF, 3637th, 281
Survival training, 5, 25, 102, 221, 281, 315
Suspension system, 276
Sustained Superior Performance award, 114
SV-5 (ablative-type reentry body), 308
Swami (Stanford Worldwide Acquisition of Meterological Information), 231
Sweden, 246, 397
  experiments, 99, 147, 277–278, 288–289
Swedish Space Committee, 99, 204–205, 241
Sweet, L/C Floyd J. (USAFR), 244
Swenson, Carl N., Co., 432
Switzerland, 246
Sybaris, Greece, 395
Sydney, Australia, 317
Sylvania Co., 89
Sylvania Electric Products, Inc., 196
Sylvania Electronics Systems-East, 295
Sylvester, Arthur, 284

Symington, Sen. Stuart, 63–64, 256–257, 278–279
Symposium on Ballistic Missile and Space Technology, Ninth, 284
Symposium on Lunar Exploration, 215
Symposium on Manned Interplanetary Missions Studies Performed by Industry for NASA in 1963/64, 29
Symposium on Reliability and Quality Control, 10th National, 6,
Symposium on Space Stations and Their Logistic Support (AIAA), 139
Symposium on the Establishment of a European Network for Artificial Satellites, 419
Synchrotron, 72
Syncom (program), 4, 90, 109, 228, 237, 282
SYNCOM II (communications satellite), 82, 105, 146, 238, 247, 286, 305, 332, 373, 397, 439
SYNCOM III (communications satellite), 84, 96, 139, 259, 286, 290–291, 294, 298, 305, 310, 311, 314, 315, 326, 332, 347, 398, 408, 439, 442, 454, 467
Syrtis Major (on Mars), 144
Systems analysis, 144
Systems Analysis and Research Corp. (SARC), 113
Syvertson, C. A., 226
T-27 (space flight simulator), 430
T-33 (jet trainer), 5, 306
T-38 (jet trainer), 302
T-39 (jet trainer), 41
TAC. See Tactical Air Command.
Tactical Air Command (TAC), 70, 191
Tad. See Delta, Thrust Augmented.
Takeoff requirements, jet transport, 150
Tampa, Fla., 2, 151
Tananarive, Madagascar, 306
Tank, Dr. Kurt, 203
Tanker, orbital, 281
Taper, Project (Turbulent Air Pilot Environment Research) (NASA-FAA), 233
Tarter/Terrier (missile program), 439
Tashkent, U.S.S.R., 27
Tass, 30, 34, 35, 42, 62, 75, 82, 85, 103, 107, 118, 128, 130, 132, 138, 149, 159, 165, 174, 176, 190, 200, 210, 220, 234, 239, 245, 286, 290, 315, 321, 424, 426, 436
Tat. See Cable, transatlantic communications.
Tat. See Thor, Thrust-Augmented.
Tat-4 (cable), 55
Tat-5 (cable), 55
Tat-Delta. See Thor-Delta, Thrust-Augmented.
Tatum Salt Dome, Miss., 360.
Taylor, George, Gold Medal, 1963 (RAS), 198
Taylor, Hugh M., 392
Taylor, Gen. Maxwell D. (USA), 214
Taylor, Dr. Theodore B., 108
Teague, Rep. Olin E., 66, 107–108, 185, 316

Technology (see also Space technology), 18, 20, 25, 53, 75, 83, 160, 167, 173, 181, 194, 199, 206–207, 234–235, 243, 257, 277, 284–285, 286, 303, 312, 318, 324, 336, 344
Technology, Automation, and Economic Progress, National Commission on, 259, 286
Technology commission, national, 313
Technology utilization, 39, 140, 142, 217, 243, 306, 329
Tel Aviv, Israel, 205
Telecommunication system (see also Communications), 26, 49, 270
Telemetry, 3, 11, 30, 143, 150, 176, 264
 data evaluation, 168
 system, 64, 134, 142, 440
Telephone communications via satellite, 185
Telephone contract (MSC), 81
Telescope, 12, 45, 72, 198, 326, 393
 astrometric, 75, 218,
 balloon-borne, 74, 132
 manned orbiting (Mot), 327,
 orbiting, 301, 327
 radio, 224
  U.S.S.R. 34, 198
 stellar photometer, 231
 x-ray rocket-borne, 179–180
Television camera, 217, 384
Television transmission, 3, 19, 115, 146, 168, 305, 315
Tellep, Daniel M., 236
Teller, Dr. Edward, 22, 303
TELSTAR I (communications satellite), 168, 261, 270
TELSTAR II (communications satellite), 140, 168, 297
Temperature measurements, 305
 atmosphere, 23, 30, 42, 47, 99, 131, 141, 241, 379
Tennessee Valley Post of AOA, 12
Terent'yev, Col. G., 118
Tereshkova, Valentina. See Nikolayeva-Tereshkova, Valentina.
Terselic, Richard A., 184
Test facility, solid-motor (Edwards AFB), 162
Test stand, F–1 rocket engine, 122
Texas A&M, College of, 57, 64, 66, 156
Texas Southern Univ., 263
Texas Symposium on Relativistic Astrophysics, 426
Texas Technological College, 154
Texas, Univ. of, 326
Tfx (Tactical Fighter Experimental). See F–111.
Thermal conductivity, 15
Thermodynamics, 257
Thermonuclear weapons, 14, 37
Thiokol Chemical Corp., 23, 80, 312, 400
 Reaction Motors Div., 29, 78
 Wasatch Div., 416
Thomas, Rep. Albert, 81
Thomas, Numa E., 142

Thompson, Floyd L., 166, 282, 319, 343
Thompson, J. I., Co., 296
Thompson, Milton O., 14, 70, 185, 283, 305, 367, 368, 412, 435
Thompson, R. A., 289
Thompson-Ramo-Wooldridge, Inc., 61, 160, 209
Thor (booster), 87, 129, 197, 212, 234, 322, 411
Thor-Able-Star (booster), 173, 342, 418, 457, 461, 463
Thor-Agena (booster), 124, 234
 launch, 10, 17, 66, 114, 151, 213, 216, 218, 283, 294, 482
  B, 25, 277, 298, 447, 464, 465, 466, 467
  D, 238, 316, 340, 354, 373, 378, 392, 446, 448, 450, 451, 452, 453, 454, 456, 457, 458, 459, 460, 462
Thor-Delta (booster) (see also Delta), 114, 309, 434
 launch, 19, 446, 457, 462, 466, 467, 468
Thor-Delta, Thrust Augmented (Tat-Delta) (booster), 96
Thor-Jupiter controversy (USA–USAF), 92
Thorium, 40–41, 62
Thousand Oaks, Calif., 298
Thrust-Augmented Delta (Tad). See Delta, Thrust Augmented.
Thrust-Augmented Thor-Delta (Tat). See Thor-Delta, Thrust Augmented.
Thrustor, contact-ionization, 159
Thumba Equatorial, Rocket Launching Station (India), 6, 10, 25, 31, 70–71, 375, 380, 382, 384
Thunderchief. See F–105.
Thurman, Harry L., 168
Time, Inc., 287
Times-Picayune (New Orleans), 281
Tipton, Stuart G., 211, 288
Tiros (meteorological satellite) 43, 62, 240, 286, 298, 305, 438
 accomplishments, 84, 124, 179, 318
TIROS I (meteorological satellite), 124
TIROS II (meteorological satellite), 230
TIROS VII (meteorological satellite), 218, 253
TIROS VIII (meteorological satellite), 5, 100, 251, 298
Tiros Operational Satellite (Tos), 193
Tischler, Adelbert O., 11, 80, 181, 228, 377
Titan missile squadrons, 8, 314–315
Titan II (missile and booster), 18, 66, 74, 87, 289, 297, 309
 launch
  operational, 13, 24, 63, 67, 81, 101, 283, 337
  R&D, 113, 130, 132
 launch vehicle, Gemini, 9, 11, 121, 259, 294, 361, 371, 413–414, 449, 466
 propulsion system, 18, 328
 reliability, 9, 409
 tests, 20, 318, 409
Titan III (booster), 92, 112, 124, 171, 173, 186, 240
 contract for, 3, **164**

facility, 39, 131, 209, 212
payload, 269, 376
plans for, 78, 129, 205, 255
progress, 29, 44, 92
test, 162, 245, 278, 386
Titan III-A (booster), 269, 304, 414, 443, 474
Titan III-C (booster), 197, 206, 213, 269, 279, 289, 358, 386, 401, 429
Titanium, 85
Titusville, Fla., 63
Tobias, Dr. Cornelius, 174
Tokuyasu, Communications Minister Zitsuzo, 343
Tokyo Astronomical Observatory, 210
Tokyo, Japan, 146, 258, 291
Tokyo Univ. Institute for Nuclear Physics, 394
Toll, Thomas A., 122
Tomato, 204
Tonkin, Gulf of, 279
Tools, space, 5, 344
Topeka, Kans., 351
Topsi (Topside sounder satellite), 106, 296, 484
Tory II C (reactor), 89
Tos. See Tiros Operational Satellite.
Toth, Robert C., 262
Tow Test Vehicle (Ttv), 264
Tower, Sen. John, 321
Townes, Dr. Charles H., 237, 367
Townsend, Dr. John W., Jr., 63
Toxicity, 230, 382
Tracking, 121, 170, 176, 200, 222, 270
  Doppler method, 109
  laser, 5
  optical, 3
  radar, 38
  S-band, 216
  seagoing system, 84
  stations
    Alaska (ESRO), 264, 397
    Australia, 39, 103
    Belgium, 397
    Falkland Islands, 397
    France-ESRO, 308
    France-Spain, 204
    Malagasy Republic, 306
    Mexico, 39
    Spain, 31
    U.S., 10, 39, 55, 159, 332
    Zanzibar, 129
Tracking and Data Acquisition (see also STADAN), 71, 176
Traction, runway, equipment, 296
Tradescantia (plant), 179
Trajectory control, 146
Trajectory-Correction Propulsion System, 276
Trans-American Aeronautical Corp. of Washington, 63, 136
Trans World Airlines (TWA), 58
Transatlantic flight capability, 327
Transducer, 82, 131, 146
Transit (satellite), 222

Transistors, radiation damage, 8
Transport, reusable shuttle, orbital, 249, 291–292
Transport, troop, rocket (see also Ithacus), 238, 246
Transportation system, air, 332
Transporter, air-cushion, 289
Treaty
  astronaut rescue and aid, 100–101, 101–102, 116
  radio communications, 68
  spaceship damage protection, 101, 116
Trew, Francis, 189
Trident Engineer Associates, 364,
Trinity Bay, Tex., 156, 361, 415
Troitsky, Vsevolod S., 426
Tropic Survival School, USAF, 221
TRS II (tetrahedral satellite), 252, 453
Truax, Capt. R. C. (USN, Ret.), 334
Trud, 340, 353
Trudeau, Gen. Arthur (USA, Ret.), 370
TRW. See Thompson-Ramo-Wooldridge.
Tsiranana, President Philibert (Malagasy Republic), 265
TSR-2 (U.K. supersonic aircraft), 329
Ttv. See Tow Test Vehicle.
Tuckerman, Arnold J., 228
Tuscon, Ariz., 3, 223, 259
Tufts Univ., 9
Tulane Univ., 111
Tungsten, 19, 115, 248
Turbofan engine, 28
Turbojet engine, 28
Turbulence, 8, 14, 103, 233, 278
  atmospheric, 146
  clear air, 88, 291, 383
Turner, Dr. B. A., 263
TWA. See Trans World Airlines.
"20th Century Engineering" (exhibit), 230
Twining, Gen. Nathan (USAF, Ret.), 370
Tycho (lunar crater), 275, 299
Tyler, Tex., 361
Tyndall AFB, Fla., 181
Typhon (missile), 26
Typhoon, 112, 218
Typhoon Ruby, 326
Typhoon Sally, 326
Tyura Tam range, U.S.S.R., 375–376
U-2 (photographic airplane), 282, 316, 383, 441
UAC. See United Aircraft Corp.
U.A.R. See United Arab Republic.
UCLA. See California, Univ. of (Los Angeles).
Ufo. See Unidentified Flying Objects.
UIW. See United Iron Workers.
U.K. See United Kingdom.
Ultraviolet astronomy, 106
Ultraviolet observations, 212
Ultraviolet photography, 271, 315
Ultraviolet radiation, 76, 210, 374
U.N. See United Nations.
Underground nuclear test, 388

UNESCO (United Nations Educational Scientific and Cultural Organization), 308, 387
Unidentified Flying Objects (Ufo), 110, 289, 387, 437
Union Carbide, 56, 401
Union of Soviet Socialist Republics. See U.S.S.R.
Unitary transformations, 72
United Air Lines, 411
United Aircraft Corp. (UAC), 13, 26, 197, 405
United Arab Republic (U.A.R.), 203, 303
  Egyptian rocket development, 78, 262–263, 303
United Arab Republic Astronomical Service, 424
United Brotherhood of Carpenters and Joiners of America, 424
United Engineering Trustees, Inc., 334
United Iron Workers (UIW) (see also Iron Workers Union), 130, 131
United Kingdom (U.K.), 52, 59, 61, 79, 94, 158, 246, 253–254, 254, 280, 303, 311, 314
  aircraft, 209, 329
  communications, 26, 75, 85, 136, 204, 218, 405
  "Concorde" (U.K.-France supersonic airliner), 170, 279, 314
  House of Commons, 136
  launch, 118, 205, 239, 242, 257, 278
  Office of the Minister for Science, 118
  Post Office, 26, 270
  space cooperation, 97, 196, 249, 311, 412, 453
  space program, 269–270, 278, 412
United Nations (U.N.), 219, 266, 269, 288, 313
  Committee on Peaceful Uses of Outer Space, 57, 187, 205, 362, 374, 375, 381
  Legal Committee, 348
  General Assembly, 21, 37
  Scientific Committee on the Effects of Atomic Radiation, 202
  Space library, 269
  World Weather Center, 439
United Press International (UPI), 273, 383
United Plant Guards Workers Union of America, 124, 126, 128, 133
United States (U.S.), 40, 46, 47, 58, 61, 85–86, 90, 93–94, 128, 143, 154–155, 156–157, 218, 254, 282–283, 285, 292–293, 308, 332
  budget, 16, 18, 20–21, 45–46, 48, 108, 244
  defense, 126, 214, 320, 322, 323, 331–332, 333
  economy, 18, 38, 40, 240
  goal, 6, 160
  Government, 28
  contracts, R&D, 165, 165–166, 261, 313
    criticism of, 22, 93–94, 114, 122, 245–246, 253–254
  R&D, 18, 22, 24, 48, 108, 165, 186, 211, 262, 281, 312, 313, 330–331
  science and technology, 110–111, 165
  missile reliability, 9, 10, 15, 28, 33, 69
  policy, 188–189, 324, 329
  security, national, 7, 9, 12, 36, 99, 217, 250–251, 258, 280, 284–285, 295, 298, 322
  space
    accomplishments, 4, 33, 40, 61, 111–112, 151–152, 172, 271–272, 272, 282, 300, 313, 318
    activities, 27, 184, 268–269, 302, 377
    capability, 32, 38, 94, 158, 271–272
    power, 8, 12, 268–269
    race, 133–134, 295
United States Freight Co., 370
United Technology Center (UTC), 39, 162, 166, 190, 278, 386, 401, 429
Univac Div. (Sperry Rand Corp.), 68, 330, 423
"Universal" (upper stage vehicle), 87
Universe, 180, 239
Universities, 4, 18, 24, 186, 196, 308, 312, 397, 406, 419, 428
  federal support, 110, 443
University Club, 9
University Explorer program, 357
University Park, N. Mex., 206
UPI. See United Press International.
Uppsala Ionospheric Observatory (Sweden), 99
Uranium, 62, 143, 248
Uranium-235, 143
Urey, Dr. Harold, 299, 438
Ursini, G. A., 347
U.S. See United States.
USA. See U.S. Army.
USACDA. See U.S. Arms Control and Disarmament Agency.
U.S. Aeronautics and Space Activities, 27
USAF. See U.S. Air Force.
USAF Aeronautical Chart and Information Center, 223
USAF Aerospace Medical Division (Brooks AFB, Tex.), 44, 48, 80
USAF Aerospace Research Pilot School (Edwards AFB), 197, 206, 395
USAF Air Defense Command, 420
USAF Cambridge Research Laboratories. See Air Force Cambridge Research Laboratories.
USAF Electronics Systems Command, 404
USAF Missile Development Center, 333
USAF/NASA Military Requirements Review Group, 319
USAF OAR. See Air Force Office of Aerospace Research.
USAF School of Aerospace Medicine (Brooks AFB, Tex.), 87, 174
USAF Space Physics Laboratory, 325
USAF 3637th Survival and Special Training Squadron, 281
USAF Tropic Survival School, 221

U.S. Air Force (USAF), 6, 29, 41, 66, 70, 78, 135, 178, 190–191, 203–204, 207, 211–212, 261, 278, 288, 289, 292, 302, 315, 322, 325
  aircraft, 69, 90, 101, 331
  antisatellite system, 197
  award, 32
  contract, 39, 74, 197, 233, 308
  cooperation
    AEC, 200
    FAA, 214, 277
    NASA, 80, 117, 171, 181, 206, 237, 277, 302, 319, 354
      Gemini, 55, 121, 182, 206, 209
    NASA-USN, 33
  facilities, 39
  launch, 258, 304
    failure, 104, 144, 145, 193, 222
    missile
      operational, 15, 32, 62, 63, 79, 81, 84, 116, 122, 125, 132, 264, 265
      R&D, 13, 24, 58, 67, 79, 81, 104, 113, 217, 241, 283, 330
    plans, 78, 254, 279
    probe, 104
    rocket, 300, 329
    satellite, 17, 27–28, 66, 79, 85, 102, 114, 145, 146, 151, 178, 182, 203, 205, 213, 239, 244, 252–253, 277, 294, 316, 326, 342, 373, 392
    sounding rocket, 122, 125
  management, 8, 11, 23, 92, 98–99, 127, 198, 289, 311–312
  missile program, 112, 252
  MOL, 8, 11, 44, 74, 92, 96, 99, 112, 139, 192, 197, 205, 209, 257, 261
  report, 159, 198
  rocket, 317
  rocket motor, solid propellant, 23, 39, 80, 311–312, 323
  space museum, 261
  space program, 18, 33, 170–171, 181, 286, 333
  supersonic transport, 4, 43, 82
  tests, 10, 19, 114, 117, 129, 323, 328
  Thor, 129
  Titan II, 66, 74
  Titan III, 92, 129, 164, 245, 278
U.S. Arms Control and Disarmament Agency (USACDA), 59
U.S. Army (USA), 80, 92, 207, 239, 241, 275, 322, 392
  contracts, 11, 328
  cooperation, 328
    international, Canada, 10, 29
    NASA, 182
  Corps of Engineers, 31, 116, 188, 274, 328
    contracts, 8, 13, 15, 247, 268, 406
    launch, 68
  radar (Mar), 250
  XV-5A (Vtol), 251, 391
U.S. Army Yuma Proving Ground, Ariz., 239
U.S. Board of Geographic Names, 22, 126
U.S. Bureau of Labor Statistics, 149
U.S. Bureau of Sport Fisheries and Wildlife, 109
U.S. Committee for International Quiet Sun Year (IQSY), 45
U.S. Committee on the Extension of the Standard Atmosphere (COESA), 14
U.S. Congress. See Congress.
USDA. See Agriculture, U.S. Dept. of.
U.S. District Court (Fla.), 61
U.S. District Judge
  Florida, 212
  Oklahoma City, Okla., 187, 244
U.S. Embassy (Moscow), 122
U.S. Embassy (Venezuela), 275
U.S. Fighter Aces Association, 188
U.S. Geological Survey, 100, 299
U.S. Geological Survey Observatory, 223
USIA. See U.S. Information Agency.
U.S. Information Agency (USIA), 142, 282–283
U.S. Marine Corps (USMC), 16, 87, 328, 330
USMC. See U.S. Marine Corps.
USN. See U.S. Navy.
U.S. National Museum (Smithsonian Institution), 121
U.S. Naval Air Equipment Center, 404
U.S. Naval Institute Proceedings, 334
U.S. Naval Observatory (Flagstaff, Ariz.), 75, 218
U.S. Naval Research and Development Satellite Communications Group, 238
U.S. Naval School of Aviation Medicine, 174, 219
U.S. Naval Training Center, 284
U.S. Navy (USN), 10, 29, 33, 45, 64, 71, 84, 178, 190–191, 203–204, 214, 218, 253, 287, 305, 334
  Polaris A-2, 11, 68
  Polaris A-3, 6, 19, 279
  satellite, 8,
  submarine, 330, 397
  space surveillance system, 102
  supersonic transport, 4
U.S. Navy Engineering Experiment Station, 334
U.S. News and World Report, 199, 341
USNS *Croatan*, 384, 391, 393, 395
USNS *Kingsport*, 238
USNS *Rose Knot*, 197
U.S. Public Health Service, 80, 230, 277
U.S.S. *Daniel Boone*, 266
U.S.S. *Daniel Webster*, 209, 307, 330
U.S.S. *Enterprise*, 246, 307
U.S.S. *John Adams*, 221
U.S.S. *Nathan Hale*, 11
U.S.S. *Norton Sound*, 218
U.S.S.R. (Union of Soviet Socialist Republics) (see also Soviet Academy of Sciences, etc.), 21, 188, 189, 190, 210, 275, 283, 340, 348, 353, 367, 394, 396, 413, 427

U.S.S.R.—Continued
  achievements, 86, 111–112, 132, 133, 142, 152, 172, 190, 254, 283, 322, 350, 375–376, 442–443
  aircraft, 41, 104, 133, 142, 289
  astronomy, 198, 203–204, 315, 364, 436–437
  bioastronautics, 130, 171, 172, 174–175, 177–178, 179, 204, 209, 226, 248, 309, 345, 350, 352, 355, 357, 359, 389, 393, 396, 418, 419, 421, 443
  booster capability, 32, 35, 38, 111, 122, 189, 195, 242, 302, 318, 351, 353, 355, 359, 379
  cooperation, international, 34, 42, 101–102, 269, 308–309
    comsat network, 122, 205–206, 255, 313
    Lil (Lunar Interplanetary Laboratory), 314
  cooperation, U.S.-U.S.S.R., 2, 41, 59, 116, 128, 142, 195–196, 203, 251, 295, 342, 379
    bioastronautics, 204, 208, 310, 342
    communications, 208, 215
    ECHO II, 17, 17–18, 27, 54, 75, 97
    disarmament, 17, 21, 266, 437
    lunar exploration, 2, 59, 60
    meteorology, 204, 208, 349, 362, 366
  competition, U.S.-U.S.S.R. See Space race.
  cosmonaut, 10, 74, 96, 130, 131, 134, 174–175, 177–178, 190, 209, 213, 226, 248, 282, 286, 309, 340, 345, 348, 349, 352, 355, 359, 381, 389, 393, 396, 418, 427, 443
  launch (see also COSMOS V–LI; ELECTRON I–IV; POLET II; VOSKHOD I; ZOND I and II), 156, 171, 336, 385
  lunar surface, 103, 180, 203, 426, 436–437
  manned space flight, 118, 190, 220, 282, 286, 302, 309, 340, 348, 349, 350, 357, 359, 385, 393, 396
    to moon, 40, 64, 111, 118, 125, 154, 156, 196, 199, 205, 221, 242, 268–269, 315, 353, 354–355, 357, 362, 365, 374, 379
  planetary exploration, 105, 112, 120, 126, 149, 152, 172, 176, 199, 200, 246, 253, 262, 315, 348, 359, 365, 399, 403
  radiation, 169, 171, 245
  scientists, 253–254, 379
  space program predictions, 2, 171, 205, 220, 379
  weapons, 250, 318, 323, 349, 381, 385, 388, 441
  Icbm's, 136–137, 170, 255
U.S.S. *Stonewall Jackson*, 406
Ussurisk, Siberia, 426
U.S.S. *Von Steuben*, 430
U.S. Weather Bureau, 1, 43, 193, 240, 251, 253, 258, 290, 366
UTC. See United Technology Center.
Utrecht, Univ. of, 155

v–2 (missile), 47, 70
v–8 (helicopter, U.S.S.R.), 142
Vab. See Vertical Assembly Building (KSC).
Vacuum chamber, 3
Valkyrie. See B–70.
Valley Forge, Pa., 75, 304
Valley Forge Space Technology Center (GE), 177, 304
Van Allen, Dr. James A., 37, 266, 388
Van Allen radiation belt, 50, 103, 173, 178, 252, 273, 325, 435–436
Van Atta, Dr. Lester C., 427
van Neil, Cornelius B., 11
Vandenberg AFB, Calif., 51, 254, 330,
  launch. See Pacific Missile Range; Western Test Range.
  management, 164
Vanguard (program), 377
VANGUARD I (satellite), 107
Van Nuys, Calif., 368
VARC. See Virginia Associated Research Center.
Varian Associates, 409
Variable-geometry wing, 428
Vatican, 260, 293
VC–10 (U.K. transport), 158
Vecchietti, George J., 51
Vehicle Antenna Test Facility (LaRC), 52
Vela Hotel. See Vela, Project.
Vela, Project (ARPA),
  launch, 153
  performance, 27–28, 252–253, 255, 262, 290
Venezuela, 275
Ventilation, 57
Venus (planet), 74, 105, 176, 200, 297, 318, 364, 408
  flights to, 120, 149, 184, 235, 246, 253, 262, 453
  surface, 132–133, 180, 239, 314, 408
Vermont, Univ. of, 207
Vernier, 121
Vernier engine, 160
Vernon, Sergei, 131
Verrett, Dr. Jacqueline, 131
Vershinin, Air Marshal Konstantin A., 41, 289
Vertical Assembly Building (Vab) (KSC), 209
Very low frequency. See Vlf.
Vespucci, Amerigo, 101
Veterans of Foreign Wars (VFW), 295
VFW. See Veterans of Foreign Wars.
Vibration, 63
Vibration test laboratory (MSC), 102
Vienop, Edna, 1
Vinson, Rep. Carl, 69
Virginia Associated Research Center (VARC), 159
Virginia Military Institute (VMI), 188
Virginia Polytechnic Institute (VPI), 159, 290
Virginia, Univ. of, 159
Visual observations, 82

INDEX 525

Vivian, Howard C., 82
VJ 101 (West Germany Vtol), 265
Vlf (Very low frequency), 50, 95
VMI. See Virginia Military Institute.
Vogel, Col. Lawrence W. (USA), 268, 274
Voice of America, 373
Volcanoes, 180, 411
Vologda, U.S.S.R., 27
von Braun, Dr. Wernher, 43, 66, 71, 73, 114, 173, 199, 232, 346, 349, 352, 433
von Kármán, Theodore, Lecture, 230
von Kármán, Theodore, Trophy, 315
von Neumann, John, Committee, 377
Voris, Dr. Frank B., 168, 169
Vorkuta, U.S.S.R., 27
Vorontsov, Yuri M., 348
VOSKHOD I (U.S.S.R. spacecraft), 348, 349, 350, 351, 352, 355, 357, 363, 365, 374, 381, 389, 396, 419, 421, 427, 443, 458
Vostok (U.S.S.R. spacecraft), 111, 152, 190, 226, 309, 365, 421
VOSTOK III, 177
VOSTOK IV, 177
VOSTOK V, 179
Voyager (program), 199, 208, 259, 295, 437-438
VPI. See Virginia Polytechnic Institute.
Vredefort, South Africa, 174
V/STOL (Vertical/Short-TakeOff-and-Landing) aircraft, 216, 303, 430, 437
Vtol (Vertical TakeOff-and-Landing) aircraft, 66, 223, 251, 391
Wac Corporal (Army missile), 261
Wackenhut Corp., 124
Wagner, Dr. Bernard M., 226
Walker, Dr. Eric A., 12
Walker, Joseph A., 321, 368, 377, 396
Walker, Robert P., 111
Wall Street Journal, 84, 85, 93
Wall, self-sealing, 228
Wallops Station (NASA), 35, 65, 109, 132, 176, 292, 323
  contract, 219, 222, 225
  launch
    re-entry experiment, 289-290
      Ram, 132, 487
      Reentry Flight Demonstration, 3, 346, 487
  satellite, 118, 379, 420
  sounding rocket, 355
    Aerobee 150, 374, 387
    Aerobee 150A, 304
    Aerobee 300A, 30
    Arcas, Boosted, 99, 305, 328
    Argo D-4 (Javelin), 15, 137, 343, 378
    Nike-Apache, 11, 13, 14, 42-43, 62, 140, 202, 210, 222, 243, 249, 251, 297, 343, 344, 384, 395, 408
    Nike-Cajun, 22-23, 30, 42-43, 47, 62, 99, 141, 277-278, 283, 287, 392
    Scout, 256
Warhead, 37, 125, 255
Warren AFB, Wyo., 51
Warrior Constructors, Inc., 121
Warsaw, Poland, 308, 309, 311, 313, 314
Washington Daily News, 144-145, 272,
Washington, D.C., 16, 21, 22, 40, 96, 129, 140, 153, 165, 208, 262, 270, 293, 325
  meetings, 4, 34, 63, 68, 78, 94, 111, 116, 140, 144, 146, 148, 149, 154, 155, 158, 183, 223, 226, 234, 237, 238, 242, 259, 260, 290, 316, 319, 332
  press conferences, 13, 41, 74-75, 131, 148-149, 260, 317, 320
Washington *Evening Star*, 8, 48, 325, 351
Washington, George, Award, 75
"The Washington Merry-Go-Round" (newspaper column), 81
Washington Post, 21, 40, 54, 57, 61, 81, 342, 351, 423
Washington *Sunday Star*, 188, 323
Washington Univ., 438
Washington World (magazine), 48, 181, 224
Water, 217, 383
  desalinization, 251, 256
Water pollution, 229-230
Water Resources Research Center, 201
Waterman, Dr. Alan T., 322
Watson, James Craig, Medal (NAS), 154
Watts, CWO Joseph C., 394
Wayne State Univ., 33, 100, 184
Weapon system development, 186-187, 250-251, 320, 322, 323
Weapons and Air Equipment Co., W. Ger., 52
Weapons platform, 392
Weapons Research Establishment, Woomera, Australia, 270
Weather (see also Meteorology), 146-147, 156-157
  forecast, 2, 5, 62, 112, 333
Weather Bureau. See U.S. Weather Bureau.
Weather map, Tiros, 240
Weather observation network, U.S., 150th anniversary, 163
Weather prediction system, global satellite, 231
Weather satellite system, global, 36-37, 240
  operational. See Tiros Operational Satellite.
Weather station
  balloon, 218-219
  nuclear powered, 49
Weaver, Rep. James D., 50
Webb, James E., 43, 67, 81, 101, 154-155, 182, 189, 222, 237, 245, 256, 259, 302, 304, 310, 312-313, 322, 335, 340
  appropriations, 45-46, 94-95, 95, 242, 300-301
  award by, 136, 258
  award to, 185
  economy, 38, 256
  JPL management, 137-138, 163, 220

Webb, James E.—Continued
  lunar landing, manned, 99–100, 316–317, 343, 367
  NASA Electronics Research Center, 23–24, 37, 67, 292
  RANGER VI, 126, 137–138, 150, 163, 272
  space achievements, 9, 183–184, 223, 344, 361
  space cooperation, communications (U.S.-U.S.S.R.), 60
  space program, 36–37, 86, 94–95, 95, 125, 195–196, 196, 207, 212, 338, 342, 391–392
  supersonic transport, 119
  U.S.S.R. space launching, 357
Weber, Dr. Thomas B., 382
Weber Aircraft Co., 367
Webster, Grove, 439
Weightlessness, effects of, 72, 208
  humans, 89, 129, 172, 174–175, 179, 243, 248, 427, 432, 443, 469
  insects, 72, 171
  plants, 72
  reptiles, 56–57
Weinstein, Dr. Roy, 22
Weizmann Institute of Science, 55
Welch, Dr. Billy, 174
Welch, Leo D., 242, 255
Welder, 184, 269
Welsh, Dr. Edward C., 4, 6, 35, 40, 111–112, 136, 186–187, 188–189, 190, 215–216, 217, 254, 269, 279–280, 298, 315, 322, 350, 353–354, 375–376, 384
Wenk, Dr. Edward, Jr., 310
West Coast, U.S., 5, 146, 225, 313
West Ford, Project (USAF), 116
Welter, Capt. William A., Jr. (USA), 394
West German Association for Rocket Research Technique and Space Flight, 320
West Germany. See Germany, West.
Western Aerospace, 88
Western Devices, Inc., 291
Western Electric Co., 11, 328
Western Operations Office (WOO), NASA, 102
Western Materials Co., 216
Western Test Range (WTR), formerly Air Force Space Test Center (see also Pacific Missile Range and Vandenberg AFB), 164, 178, 239
  launch, 283
    missile
      Atlas D, 217, 403
      Atlas F, 265, 430
      Minuteman, 264, 378
      Titan II, 337, 409
    satellite missions, 182, 203, 205, 216, 218, 238, 239, 244, 277, 285, 294, 296, 298, 316, 326, 340, 342, 354, 361, 373, 378, 392, 406, 418, 426, 427
Western Union International, 412
Western Union Telegraph Co., 79, 143

Westinghouse Electric Corp., 68, 118, 140, 148, 158, 168, 202, 327, 384
Wharton, Dr. George, 217
Wheat, 72
Whipple, Dr. Fred L., 389
Whitaker, E. A., 223
White, Alvin S., 349
White, Capt. Edward H., II (USAF), 263
White House, 11, 76, 85, 90, 133, 140, 143, 164, 183, 251, 258, 271, 273, 286, 287, 316, 323, 388
White, Dr. M. Samuel, 24
White, Dr. Robert M., 1, 101, 146
White Sands Missile Range (WSMR), N. Mex., 14, 31, 241, 250, 339
  launch, 30, 124, 128, 144, 153, 192, 210, 283, 299, 335, 363, 368, 374, 381, 387, 422
  test, 58
    Apollo, 89, 90, 100, 175, 314
    Athena, 338, 406
    F–2, 412
    sonic boom, 360
White, Gen. Thomas D. (USAF, Ret.), 177
White, Gen. Thomas D., Space Trophy, 325
Whiteman AFB, Mo., 51
Wideband transmission, 3
Wiener, Dr. Norbert, 11, 109
Wiesner, Dr. Jerome B., 11, 12, 22
Wildey, Dr. Robert L., 179
Wilkes, Mayor Jack S., 90
William and Mary, College of, 159
Williams, Capt. Clifton C., Jr. (USMC), 237
Williams, Donald D., 111
Williams, Walter C., 18, 104, 130, 265, 376
Williamsburg, Va., 350
Williamsport, N.C., 42
Wilson, Rep. Bob, 95
Wilson, Prime Minister Harold, 349, 412
Wilson, Dr. Olin, 138
Wilson, Dr. Richard, 22
Wilson, Woodrow, School of Public and International Affairs, 154
Wind, 15, 241
  upper atmosphere, 14, 23, 30, 42, 47, 99, 141, 249, 379, 384
Wind tunnel, 171, 230
Wing, design, 28, 189, 191, 198, 239
Wing-shaped craft, 290
Winkler, Leopold, 82
Wirtz, W. Willard, 70
Wisconsin, Univ. of, 212, 231, 275, 290, 304
Wise, Donald U., 28
Witkin, Richard, 66
Witwatersrand, Univ. of the (Johannesburg, S. Africa), 174
WMO. See World Meteorological Organization.
Wolfe, Allen E., 353
Wolfe, Dr. John H., 435

Women, speed record for, 181
Women's Advisory Committee on Aviation (FAA), 164
Women's Space Symposium, Fifth, 63
Wong, Thomas Jefferson, 85
WOO. See Western Operations Office, NASA.
Woodworth, J. L., 347
Woomera Rocket Range, Australia, 205, 257, 270, 278
Worcester Polytechnic Institute, 344
World Affairs Council of Northern California, 36
World Book Encyclopedia, 119
World Book Encyclopedia Science Service, 316, 345
World flight, first woman, 142
World Meteorological Organization (WMO), 37, 84, 251
World War I, 46, 79, 154
World War II, 47, 59, 78, 112, 154, 203, 254
World Weather Center, 439
World's Fair, N.Y., 145, 310
WOWO radio station, 145
Wright Aeronautical Div. (Curtiss-Wright Corp.), 438
Wright brothers, 46, 155
Wright Brothers Memorial Trophy, 387
Wright-Patterson AFB, Ohio, 167, 303, 318
WSMR. See White Sands Missile Range.
WTR. See Western Test Range.
Wydler, Rep. John W., 185
Wyld, James H., Memorial Award, 166
Wyman, Rep. Louis C., 185, 274
X-5 (research aircraft), 191
X-15 (rocket research aircraft), 6, 122–123, 145, 207, 258, 292–293, 315, 320–321
  award, 77, 198, 435
  flight
    canceled, 326, 383
    successful
      No. 1, 6, 29, 33, 118, 130, 156, 182, 227, 352, 414
      No. 2, 214, 222, 285, 321, 329, 331, 380, 388, 399
      No. 3, 14, 70, 103, 174, 185, 241, 264, 283–284, 297, 305, 329, 368, 412, 429
  Memo of Understanding (NASA–USAF–USN), 29
  modification, 66, 263
  test, 6, 57, 77, 103, 182, 241, 319
  heat-resisting materials, 264, 283–284, 329
  heat-transfer, 14, 70, 174, 297, 305–306
  noise, boundary layer, 174, 297, 305–306
XV-5A (Vtol), 251, 391
X-19 (Vtol transport), 223
X-20 (Dyna Soar), 8, 78, 206, 208
  cancellation, 24, 26
  replacement of, 26, 44, 99, 109, 146, 263
X-248 (rocket motor), 135, 141, 147, 166
X-258 (rocket motor), 137
XB-70 (supersonic aircraft), 172, 340
XB-70A (supersonic aircraft), 297, 323
XC-142A (V/stol aircraft), 216, 437
XH-51A (compound helicopter), 420
XH-51N (helicopter), 216, 422
Xerox Corp., 386
X-ray, 153, 254, 296–297
  use of, 100, 179–180
X-ray telescope, rocket borne, 179–180
X-ray unit, miniature, 135
Yale Univ., 191
Yarborough, Sen. Ralph, 116
Yarymovych, Dr. Michael I., 90, 145
Yeager, Col. Charles E. (USAF), 164, 390
Yegorov, Dr. Boris Borisovich, 348, 355, 359, 427, 469
Yerpylev, Nikolai, 81
Yeshiva Univ. (N.Y.), 155
YF-12A (A-11 aircraft), 225, 254
York, Dr. Herbert F., 35
Yost, Charles F., 357
Young, Judge George C., 128, 133
Young, LCDR. John (USN), 134, 182, 316, 366, 369
Young President's Organization Eastern Area Conference, 338
Young, Col. Robert P. (USA), 113, 258, 268, 274
Young, Sen. Stephen M., 91
Zanzibar, 10, 129, 159
Zero-gravity, 62, 70, 119
Zimenki Observatory (U.S.S.R.), 75, 85
Zimney Corp., 392
Zirconium, 19, 153
ZOND I (U.S.S.R. space probe), 126, 127, 128, 131, 138, 149, 165, 176, 200, 246, 253, 262, 449
ZOND II (U.S.S.R. space probe), 399, 403, 412, 424, 443, 461
Zuckert, Eugene M., 8, 44, 66, 67, 92, 127, 205, 319, 325, 353